# Achieving sustainable cultivation of tropical fruits

It is widely recognised that agriculture is a significant contributor to global warming and climate change. Agriculture needs to reduce its environmental impact and adapt to current climate change whilst still feeding a growing population, i.e. become more 'climate-smart'. Burleigh Dodds Science Publishing is playing its part in achieving this by bringing together key research on making the production of the world's most important crops and livestock products more sustainable.

Based on extensive research, our publications specifically target the challenge of climate-smart agriculture. In this way we are using 'smart publishing' to help achieve climate-smart agriculture.

Burleigh Dodds Science Publishing is an independent and innovative publisher delivering high quality customer-focused agricultural science content in both print and online formats for the academic and research communities. Our aim is to build a foundation of knowledge on which researchers can build to meet the challenge of climate-smart agriculture.

For more information about Burleigh Dodds Science Publishing simply call us on +44 (0) 1223 839365, email info@bdspublishing.com or alternatively please visit our website at www.bdspublishing.com.

**Related titles:**

Achieving sustainable cultivation of bananas Volume 1: Cultivation techniques
Print (ISBN 978-1-78676-156-9); Online (ISBN 978-1-78676-158-3, 978-1-78676-159-0)

Achieving sustainable cultivation of mangoes
Print (ISBN 978-1-78676-132-3); Online (ISBN 978-1-78676-134-7, 978-1-78676-135-4)

Chapters are available individually from our online bookshop:
https://shop.bdspublishing.com

climate•SMART•publishing
IN AGRICULTURAL SCIENCE

BURLEIGH DODDS SERIES IN AGRICULTURAL SCIENCE
NUMBER 65

# Achieving sustainable cultivation of tropical fruits

Edited by Professor Elhadi M. Yahia, Universidad Autónoma de Querétaro, Mexico

burleigh dodds
SCIENCE PUBLISHING

Published by Burleigh Dodds Science Publishing Limited
82 High Street, Sawston, Cambridge CB22 3HJ, UK
www.bdspublishing.com

Burleigh Dodds Science Publishing, 1518 Walnut Street, Suite 900, Philadelphia, PA 19102-3406, USA

First published 2020 by Burleigh Dodds Science Publishing Limited


Library of Congress Control Number: 2019951621

British Library Cataloguing in Publication Data
A catalogue record for this book is available from the British Library

ISBN 978-1-78676-284-9 (Print)
ISBN 978-1-78676-287-0 (PDF)
ISBN 978-1-78676-286-3 (ePub)
ISSN 2059-6936 (print)
ISSN 2059-6944 (online)

DOI 10.19103/AS.2019.0054

Typeset by Deanta Global Publishing Services, Dublin, Ireland
Printed and bound by CPI Group (UK) Ltd, Croydon, CR0 4YY

# Contents

## Part 2 Other tropical fruits

# Series list

| Title | Series number |
|---|---|
| **Achieving sustainable cultivation of maize - Vol 1**<br>From improved varieties to local applications<br>Edited by: Dr Dave Watson, CGIAR Maize Research Program Manager, CIMMYT, Mexico | 001 |
| **Achieving sustainable cultivation of maize - Vol 2**<br>Cultivation techniques, pest and disease control<br>Edited by: Dr Dave Watson, CGIAR Maize Research Program Manager, CIMMYT, Mexico | 002 |
| **Achieving sustainable cultivation of rice - Vol 1**<br>Breeding for higher yield and quality<br>Edited by: Prof. Takuji Sasaki, Tokyo University of Agriculture, Japan | 003 |
| **Achieving sustainable cultivation of rice - Vol 2**<br>Cultivation, pest and disease management<br>Edited by: Prof. Takuji Sasaki, Tokyo University of Agriculture, Japan | 004 |
| **Achieving sustainable cultivation of wheat - Vol 1**<br>Breeding, quality traits, pests and diseases<br>Edited by: Prof. Peter Langridge, The University of Adelaide, Australia | 005 |
| **Achieving sustainable cultivation of wheat - Vol 2**<br>Cultivation techniques<br>Edited by: Prof. Peter Langridge, The University of Adelaide, Australia | 006 |
| **Achieving sustainable cultivation of tomatoes**<br>Edited by: Dr Autar Mattoo, USDA-ARS, USA & Prof. Avtar Handa, Purdue University, USA | 007 |
| **Achieving sustainable production of milk - Vol 1**<br>Milk composition, genetics and breeding<br>Edited by: Dr Nico van Belzen, International Dairy Federation (IDF), Belgium | 008 |
| **Achieving sustainable production of milk - Vol 2**<br>Safety, quality and sustainability<br>Edited by: Dr Nico van Belzen, International Dairy Federation (IDF), Belgium | 009 |
| **Achieving sustainable production of milk - Vol 3**<br>Dairy herd management and welfare<br>Edited by: Prof. John Webster, University of Bristol, UK | 010 |
| **Ensuring safety and quality in the production of beef - Vol 1**<br>Safety<br>Edited by: Prof. Gary Acuff, Texas A&M University, USA & Prof. James Dickson, Iowa State University, USA | 011 |
| **Ensuring safety and quality in the production of beef - Vol 2**<br>Quality<br>Edited by: Prof. Michael Dikeman, Kansas State University, USA | 012 |
| **Achieving sustainable production of poultry meat - Vol 1**<br>Safety, quality and sustainability<br>Edited by: Prof. Steven C. Ricke, University of Arkansas, USA | 013 |
| **Achieving sustainable production of poultry meat - Vol 2**<br>Breeding and nutrition<br>Edited by: Prof. Todd Applegate, University of Georgia, USA | 014 |

# Foreword

Practically all tropical crops are grown in developing countries The exquisite taste, nutritional value and, in many cases, the nutraceutical properties of these fruits make them highly attractive to consumers. In recent years, these fruit crops have gained importance due to growing consumption all over the world, particularly in developed countries with high purchasing power. According to FAOSTAT data, tropical fruit crop production has grown by 35% in the last 10 years. A striking example is avocado which has been called *'green gold'* because of its high economic and nutritional value, with production growing by a spectacular 52% over the last 10 years.

In parallel to the increase in the planted surface area, there is an increasing interest in more sustainable production of these fruit crops, including a growing demand for technical information on cropping, production and postharvest management. Edited by Prof. Elhadi Yahia, a well-known expert in this area, this book provides an up-to-date review of the state of the art in tropical and subtropical fruit production systems. It focuses on the most successful strategies to ensure long-term sustainability in cultivation of tropical woody species, including careful soil management and optimal soil nutrition both to minimize environmental impact and ensure the long-term economic viability of farming operations.

As well as a wide-ranging coverage of key aspects of citrus production, the book reviews advances in papaya growing and breeding and discusses the ecological issues in intensive banana production systems. It provides valuable information on avocado, its propagation and growing systems, including the use of rootstocks to improve resistance to *Phytophthora cinnamomi*, the devastating fungus causing root rot. It analyzes coconut growing systems and presents novel information on sustainable production of important crops which are barely mentioned in scientific literature, including litchi, pomegranate, guava and jackfruit. In these chapters, the reader will learn about crucial aspects of production, including irrigation management, the use of effective rootstocks, canopy management and planting density, and the use of biological control agents. Combining these strategies provides a sustainable approach to production which fully respects the environment.

This magnificent book will become a valuable reference for technicians, consultants, professors, students and growers who want to use the latest advances to optimize yield and quality of these fruit species, while ensuring long-term sustainability; an essential requirement in every modern agricultural system.

*Dr. Javier Z. Castellanos*
*President of the Institute for Agricultural Innovation (INTAGRI), Mexico.*
*September 2019*

# Acknowledgements

We wish to acknowledge the following for their help in reviewing particular chapters:

- Chapter 1: Dr Maria Islas-Osuna, CIAD, Mexico
- Chapter 4: Dr Lisa Neven, USDA-ARS, USA
- Chapter 7: Dr Tokurou Shimizu, Institute of Fruit Tree and Tea Science – NARO, Japan
- Chapter 11: Dr Carlos Oropeza, CICY, Mexico
- Chapter 13: Dr Manoj Raj, Indira Ghandi National Tribal University, India
- Chapter 16: Dr Prakash Chandra Tripathi, ICAR-Indian Institute of Horticultural Research, India
- Chapter 19: Dr Diego Fernando Marmolejo Cortes, Universidade Nacional da Colômbia (UNAL), Colombia; and Dr Janna Ong Binti Abdullah, University Putra Malaysia

# Introduction

Tropical fruits have been cultivated for centuries. The Chinese have grown lychee/litchi for thousands of years while mango, for example, has been cultivated in India for more than 4000 years (Yahia et al., 2006). Carambolas have been popular in Malaysia for centuries and papaya was consumed throughout Latin America long before Columbus's arrival in 1492. While customers in many developed countries may just be discovering some tropical fruits, they have been staple foods for many ethnic cultures for generations.

It is not easy to clearly distinguish between most tropical and subtropical tree fruits but, with some exceptions, the growth of these trees is practically impossible below 10°C. Tropical trees require a climate with average mean temperatures higher than 10°C during the coldest month (Watson and Moncur, 1985). In addition, most tropical trees require humid environmental conditions. The tropics extend between the Tropic of Cancer and Tropic of Capricorn, at 23° north and south of the equator. However, in terms of actual cultivation, these boundaries are too rigid. Not only do they contain areas, especially at higher altitudes, that do not conform to the climatic characteristics generally assigned to the tropics, but also regions outside this belt have coastal areas or inland climates with similar climatic conditions to areas in the tropics. As a result, the region for tropical fruit cultivation has been extended to the 30th parallels (Nakasone and Paull, 1998). The most important feature characterizing the tropics is less a specific minimum or maximum temperature than steady warm temperatures throughout the year. Samson (1986) defined the tropical climate as having temperatures averaging 27°C. He emphasized the importance of stable temperatures, with the warmest month being only a few degrees higher than the coldest month and temperature differences between night and day being greater than those between winter and summer. He also highlighted a little variation in day length.

Subtropical fruit crops are commonly grown in climatic conditions between temperate and tropical. The subtropics have hotter summers and cooler winters than the tropics and the humidity is generally lower. The limit for the subtropics is those areas that are able to maintain an average temperature of 10°C for the coldest month (Nakasone and Paull, 1998). Subtropical fruits may be either deciduous or evergreen and are usually able to withstand a low temperature but not frost. Since differences in day length become greater with increased latitude, subtropical fruit plants are able to adapt to fluctuations in light during the day. Some subtropical fruit plants require chilling for flower bud differentiation, as in the case of grape, citrus, durian and jackfruit.

Tropical and, in particular, subtropical fruits, grow in all continents, but most of the best-known fruits are commercially produced in the tropical and subtropical regions of the Americas (such as avocado, banana, citrus, guava, mango, papaya and pineapple) and Asia (such as banana, lychee/litchi, mango, jackfruit, mangosteen and durian). Almost no tropical fruits and limited quantities of subtropical fruits are grown in Europe.

Key factors in the spread of many tropical fruits have been migration and trade. As an example, mango spread this way throughout Southeast Asia from its origin in the Indo-Burman region and then spread to Africa possibly via Arab traders (Galán Saúco, 1999; Yahia et al., 2006). Malaysian travelers introduced banana to Madagascar. Arab conquest brought the orange to the Mediterranean and southern Europe. In the Americas, evidence suggests exchanges between the cultures of Mexico and Peru dating as far back as 1000 B.C.E. (Purseglove, 1968), allowing tropical and subtropical fruits to disseminate in the region.

The arrival of Europeans in the Americas accelerated the exchange of plant species between the Old and New Worlds. The sixteenth-century monk Bartolomé de las Casas recorded that orange seeds were carried from the island of La Gomera (Canary Islands, Spain) to Haiti on Christopher Columbus's second voyage in 1493 (Amador de los Ríos, 1851–1855). Banana was carried to the Caribbean island of 'Santo Domingo' from the Canary Islands in 1516. In the sixteenth and seventeenth centuries, Portuguese trading routes linked Europe, Latin America, Africa and the Far East, including Brazil, Goa, Malacca, the Moluccas, Canton, Macao Japan and Formosa. Similarly, Spanish traders connected countries such as Mexico and the Philippines. Dutch, British and French traders were also important in spreading tropical and subtropical fruits around the world.

Several factors have influenced the distribution of tropical and subtropical fruits, such as crop adaptability to different climatic conditions, ease of propagation, sensory qualities and fruit postharvest life. For example, the taste of pineapple, the long-term viability of the plant's suckers as planting material and the rapidity with which it produces fruit have been factors in its distribution (Nagy and Shaw, 1980; Galán Saúco, 2001). Similar factors also helped in the distribution of other fruits such as banana, papaya, mango and guava. However, some true tropical trees, such as mangosteen, rambutan and durian, have strict climatic requirements and are difficult to propagate, meaning that they have been confined almost exclusively to their area of origin in Southeast Asia. The mangosteen is notable among tropical fruits and in that it has proven particularly intractable to most attempts to establish it outside of its area of origin via selection of cultivars capable of adapting to different environments. The species consists of a single genotype, which, in essence, means no genetic variation exists with which to breed or improve stock (Yaacob and Tindall, 1985).

## Classifying tropical and subtropical fruits

Tropical and subtropical fruits not only include woody plants, such as mango or oranges, but also herbaceous crops such as banana and vines such as passion fruit. Most botanical families can lay claim to at least one species of tropical or subtropical fruit. Martin et al. (1987) listed some 137 families, with the best-known shown in Table 1. Tropical and subtropical fruits are botanically and morphologically diverse, including single fruits such as avocado or orange, drupes such as mango, pomes such as loquat, capsules such as durian, nutlets such as litchi and longan, to compound fruits such as the typical syncarpium of the pineapple, or a bunch of individual berries such as in the banana.

**Table 1** Well-known tropical and subtropical fruits and their botanical families

| Family | Common names of species |
|---|---|
| Anacardiaceae | Mango, Cashew |
| Annonaceae | Cherimoya, Guanábana, Custard apple |
| Bombacaceae | Durian |
| Bromeliaceae | Pineapple |
| Cactaceae | Pitaya |
| Caricaceae | Papaya |
| Ebenaceae | Caki |
| Guttiferae | Mangosteen |
| Lauraceae | Avocado |
| Malphigiaceae | Acerola |
| Meliaceae | Langsat or Lanson |
| Moraceae | Breadfruit, Jackfruit |
| Musaceae | Banana, Plantain |
| Myrtaceae | Guava |
| Oxalidaceae | Carambola |
| Palmaceae | Coconut, Date |
| Passifloraceae | Passion fruit, Granadilla |
| Proteaceae | Macadamia |
| Rosaceae | Loquat |
| Rutaceae | Orange, Grapefruit, Mandarin |
| Sapindaceae | Litchi/Lychee, Longan, Rambutan |
| Sapotaceae | Chicosapote, Lucuma |
| Solanaceae | Sweet pepino, Lulo, Tamarillo |

Tropical and subtropical fruits are also very diverse in size, shape and other morphological and anatomical characteristics. Some of these fruits have large seeds, inedible skin or tough rind, which can contribute to a relatively high waste index (compared with temperate-zone fruits). For example, waste indexes in papaya, banana and pineapple are 25, 30 and 41%, respectively. Fruit weight can range from 5–7 gm in acerola to up to 20 kg in jackfruit. Tropical fruits can also be classified on the basis of their variation in texture as follows (G = green, R = ripe): 'hard' such as guava (G), mango (G), passion fruit (G) and pineapple, (G), 'semi-hard', such as banana (G), cherimoya (G), papaya (G) and 'soft' such as date, banana (R), guava (R) and lychee (R).

## Uses and nutritional value

Most tropical and subtropical fruits are consumed fresh. Some fruits need further preparation such as breadfruit, which is cooked and plantain, which is commonly fried. Flour is also made from some tropical and subtropical fruits such as durian and banana. Other fruits are made into pickles, chutneys, jams, dips or spreads (Yahia et al., 2006). Commercial processed products include juices, sauces, ice cream and other desserts as well as confectionery and baby foods

Tropical and subtropical tree crops also have a range of non-food uses. On a commercial scale, tropical and subtropical fruits are used to make soaps, bath gels and shampoos. The fruit and flowers of some species, particularly oranges, are used to manufacture essential oils. There is an increasing interest in the potential use of leaves or flower extracts as biological products for the control of pests and diseases. Parts of plants have traditionally been used as animal feed such as banana leaves, dried dates and their pits, breadfruit leaves and mango seeds kernels. Pineapple and banana leaves are used for papermaking and cloth, while avocado wood is used for decorative carvings. Banana and date palm leaves have traditionally been used as roofing material, while the wood of breadfruit, citrus, guava, longan, mango, mangosteen and persimmons has been used for interior paneling.

While most tropical and subtropical fruits are characterized by relatively low caloric values, they are rich in carbohydrates, vitamins, minerals and phytochemicals, ensuring they play a very important role in human diet and nutrition (Yahia et al., 2011a). Bananas, plantains and breadfruit are widely used as a source of starch. Acerola fruit contains the highest known ascorbic acid (vitamin C) content among all fruits (1000–3300 mg/100 gm fresh weight). Other fruits, such as citrus, guava, lychee, papaya and passion fruit are good sources of ascorbic acid. Mango and papaya are good sources of vitamin A. Breadfruit and cherimoya contain relatively high amounts of niacin and thiamin. Guava contains fair amounts of niacin and iron. Most tropical fruits are good sources of minerals, especially potassium and iron. Banana is a good source of vitamin A, B and C, riboflavin and potassium. Papaya fruit has high quantities of vitamin C and A as well as potassium and calcium. Papaya also contains papain, an enzyme that promotes digestion. Pineapple is rich in vitamin C and carbohydrates and is a good source of calcium, phosphorus, iron, potassium and thiamine. Litchi/lychee, longan and durian are good sources of carbohydrates and vitamin C and durian has fair amounts of iron and niacin. Carambola is low in calories, rich in vitamin C and an adequate source of vitamin A. Macadamia nut is rich in protein, oil, iron, calcium, thiamine, riboflavin and niacin. Dates have a high nutritive value, being rich in carbohydrates and are a good source of vitamin A, potassium and iron. Coconut is high in phosphorus and iron. Tropical and subtropical fruits are also rich in pectin and fiber, which promote intestinal health. In common with other fruits, tropical and subtropical fruits are good sources of antioxidants (Martin et al., 1987; Yahia et al., 2011a). Many tropical and subtropical fruits, notably mango and papaya, are good sources of carotenoids. Traditionally, this has meant use of the fruit and other parts of the plant to treat a wide range of ailments, from cold, asthma and other lung conditions to intestinal disorders. Modern science has now linked the phytochemicals in many tropical and subtropical fruits to the prevention of a range of chronic diseases, resulting in their being marketed as 'superfruits' to western consumers (Yahia et al., 2018).

## Tropical and subtropical fruit production

Production and trade figures assume a division of tropical and subtropical fruits into the following three main categories:

- Major fruits, such as banana and plantain, avocado, citrus, kiwi, mango, papaya and pineapple. These are cultivated in most tropical and subtropical countries and are well-known in local and export-import markets.
- Minor fruits, many such as abiu, atemoya, breadfruit, cactus pears, caimito (star apple), carambola (star fruit), cashew apple, cherimoya, custard apples, durian, guanabana (soursop), guava, jaboticaba, jackfruit, langsat, litchi, longan, macadamia, mabolo (velvet apple), mamey sapote, mangosteen, monstera (ceriman), papaya, passion fruit, pejibaye (peach palm), pitaya, pulusan, rambutan, sapodilla, soursop, sugar apple (annona), white sapote. and wax apple (wax jambu). Minor fruits are not so extensively cultivated and consumption and trade tend to be limited, both geographically and quantitatively. Most are commonly known only in very

specific regions. Many are of considerable economic importance in their respective regional markets such as durian and mangosteen.

- Wild fruits belonging to diverse botanical families such as jamun (Syzygium cumini L), among several others, which are not cultivated commercially.

As noted above, examples of truly tropical fruit tree crops include mangosteen, durian and rambutan. Some fruit crops, such as banana, avocado and citrus can be cultivated equally well in either the tropics or the subtropics. Tropical fruit plants are generally evergreen and are extremely sensitive to cold. They do well with few fluctuations in diurnal temperature or periods of daylight. They require a moist warm climate but are capable of withstanding dry weather in some cases for example, mango, banana, papaya and sapota.

Global aggregate production of major tropical fruits was estimated at 93.7 million tons in 2017 (Table 2). The sector saw 3% growth in 2015 and 2016, a slight decrease on the 4% average annual growth rate seen over the previous decade, which has been related to poorer weather conditions in 2015 and 2016 (Altendorf, 2019). About 98% of tropical fruit production is in developing countries. Around 50% of tropical fruit production is consumed fresh and the rest is consumed in different processed forms. There is a considerable gap between world per capita consumption (less than 60 kg per year) and estimated potential consumption (about 100–120 kg per year) (Jansen and Subramaniam, 2000) which suggests significant potential for growth.

The leading producer of major tropical fruits is India, accounting for an estimated 30% of total global production of major tropical fruits in 2017, due to its strong position in the production of fruits such as mango and papaya. India accounts for approximately 40% of total global mango and papaya production, which is predominantly destined for domestic market. Other significant producers of tropical fruits include China and Brazil, with most production also destined for the domestic market and Costa Rica and Mexico, which rank as the two biggest exporters of tropical fruits.

About 50 million tons of mango, mangosteen and guava were produced globally in 2017 (Table 2), an increase of 4% over 2016. It is estimated that mango accounts for approximately 75% of total production volume, guava for 15% and mangosteen for the remaining 10%. In 2017 approximately 75% of mango, mangosteen and guava production was in Asia, 15% in Africa and 10% in Latin America and the Caribbean. Production in India, the largest producer of these fruits, increased by an estimated 1 million tons in 2017. With an estimated output of 19.7 million tons in 2017, India accounted for approximately 40% of total global production.

Global pineapple production was approximately 27 million tons in 2017 (Table 2). Output in 2017 only increased fractionally because of weather-related crop damage in Costa Rica, the main producer of pineapple (accounting for

**Table 2** Global cultivated area (1000 Ha) and production (1000 tons) of some tropical and subtropical fruits in 2017 (FAOSTAT, 2019)

| Fruit | Area (1000 Ha) | Production (1000 Tons) | Major producing countries |
|---|---|---|---|
| Avocados | 587.278 | 5,924.398 | Brazil, Chile, Dominican Republic, Colombia, Kenya, Mexico, Peru, USA, Venezuela |
| Bananas | 5,637.508 | 113,918.763 | Angola, Brazil, China, Colombia, Ecuador, Guatemala, India, Indonesia, Mexico, Tanzania, The Philippines |
| Coconuts | 12,303.924 | 60,773.435 | Brazil, India, Indonesia, Papua New guinea, Mexico, Sri Lanka, Thailand, Tanzania, The Philippines, Vietnam |
| Figs | 36.535 | 307.685 | Algeria, Brazil, China, Egypt, Greece, India, Iran, Libya, Morocco, Spain, USA, Turkey |
| Grapes | 6,931.353 | 74,276.583 | Argentina, Brazil, Chile, China, France, India, Italy, South Africa, Spain, Turkey, USA |
| Grapefruits | 348.212 | 9,063.143 | China, India, Mexico, South Africa, Thailand, Turkey, USA, Vietnam |
| Kiwi fruit | 247.793 | 4,038.872 | China, Italy, New Zealand, Chile, USA |
| Lemons and limes | 1,084.505 | 17,218.173 | Argentina, Brazil, China, Colombia, Egypt, India, Iran, Italy, Mexico, South Africa, Spain, Sudan, Turkey, USA |
| Mangos, mangosteens and guava | 5,681.310 | 50,649.147 | Brazil, China, Egypt, India, Indonesia, Mexico, Pakistan, Thailand |
| Olives | 10,804.517 | 20,872.788 | Argentina, Algeria, Egypt, Greece, Italy, Libya, Morocco, Portugal, Spain, Turkey, USA |
| Oranges | 3,862.449 | 73,313.089 | Brazil, China, Egypt, India, Indonesia, Mexico, Pakistan, Spain, Turkey, USA |
| Papayas | 440.629 | 13,016.281 | Brazil, Colombia, Cuba, India, Indonesia, Mexico, Venezuela |
| Pineapples | 1,098.705 | 27,402.956 | Brazil, China, Colombia, Costa Rica, India, Indonesia, Mexico, The Philippines |
| Plantains | 5,522.745 | 39,241.376 | Cameron, Colombia, Cote d` Ivoire, Ghana, Myanmar, Nigeria, Peru, The Philippines, Uganda, |

over 11% of global production). This marked a significant curtailment of global production growth compared to the average annual rate of 4% seen over the previous decade. The country's prolonged and intense 2016 rainy season, which lasted until December and the drought that followed in January and February of 2017, caused significant damage to production in the first half of 2017, resulting in an estimated 6% decline in output. Other major producers of pineapple include Brazil and the Philippines, with an estimated 11% and 10% share of production in 2017 respectively, followed by China, India and Thailand. Thailand is the world's leading producer and exporter of processed pineapple.

In global production terms, avocado remains one of the less important of the major tropical and subtropical fruits. Global production of avocado reached an estimated almost 6 million tons in 2017, representing a 2% increase from 2016. Avocado has seen very fast production growth over the last decade, primarily due to increases in harvested area among the major producers. Nearly half of all production takes place in Central America and the Caribbean, notably in Mexico, the Dominican Republic, Colombia, Peru and Chile. Production in Mexico alone accounted for more than one-third of the global output in 2017. Although adverse weather conditions in the country affected both the quantity and quality of the harvest in the first 9 months of the year, overall production increased by 6% from 2016, reaching a new peak of nearly 2 million tons. Production in the Dominican Republic, hitherto the world's second leading producer of avocado, experienced significant hurricane damage in October 2017 with an estimated 55% decline in output in 2017 compared with 2016. As a result, the share of the Dominican Republic in global production dropped to 5% in 2017. The peak harvest period in the Dominican Republic is between October and March, when approximately 80% of production is harvested, making the crop particularly vulnerable to hurricane damage. Production in Peru, meanwhile, registered a 22% increase over 2016, following a near equivalent increase in harvested area. This was stimulated by strong import demand from the European Union, the largest export destination. Consequently, Peru ranked as the second leading producer of avocado, with an estimated 10% volume share in 2017.

Global production of papaya reached an estimated 13.7 million tons in 2017, a 5% increase from 2016. Due to the crop's greater resilience to adverse weather conditions, papaya production in 2017 experienced significantly less disruption from adverse weather events compared to other tropical fruits, exceeding the average annual growth rate of 3% achieved over the past 10 years. In terms of regional distribution, an estimated 56% of global papaya production originated from Asia in 2017, 33% from Latin America and 11% from Africa. India, the largest papaya producer in the world, with an estimated 45% output share in 2017, achieved 7% production growth from 2016, largely due to increases in harvested area. Production in Brazil, the second largest producer in the world, increased by an estimated 6%, after recovering from a 4% drought-induced decline in production in 2016.

## The trade in tropical and subtropical fruit

Many tropical and subtropical fruits are of great importance in the trade in fresh fruit, such as citrus, bananas, pineapples, mango, avocado and papaya. Several tropical and subtropical fruits are also very important in the trade in processed fruits and vegetables, such as citrus and pineapples. The main producers of subtropical and, in particular, tropical fruits are developing countries, while the main importers are developed countries in Europe as well as the United States, Canada, Japan and China.

While a small number of tropical and subtropical fruits, such as banana, pineapple and citrus, have a long history of international trade, other fruits such as mango and avocado have only started to develop significant export markets in the last few decades. Most tropical fruits were considered 'exotic' in developed countries until a few years ago, but many are now familiar products in supermarkets around the world such as guava, lychee/litchi, carambola, mangosteen, passion fruits and rambutan.

There are a number of reasons for the growth in export markets for tropical and subtropical fruits, including:

- Overall growth in world trade;
- Improved breeding, production, postharvest handling and processing systems in developed countries, resulting in a sufficient volume and quality of produce for export;
- Better international distribution and storage technologies and systems (such as improved cold chain technologies, or the use of modified ad controlled atmosphere technologies);
- The growing ethnic diversity of populations in developed countries, increasing interest in new foods;
- Growing consumer interest in the perceived nutraceutical properties and health benefits of tropical and subtropical fruits.

As with other products, the tropical and subtropical fruit trade benefitted from the World Trade Organization (WTO) agreement in Marrakesh on 15 April 1994 following the conclusion of the Uruguayan round of talks on the General Agreement on Tariffs and Trade (GATT). This agreement helped to reduce trade barriers and promote free trade.

Globally, banana, pineapple, mango and avocado are the four most significantly traded tropical fruits. Other fruits such as papaya are traded in much lower volumes but have shown significant export growth. Global trade in major tropical fruits was an estimated 6.6 million tons in 2017, following a small 1% contraction on the previous year (Altendorf, 2019). Adverse weather conditions in two leading exporting countries, Mexico (for avocado, mango and papaya) and Costa Rica (for pineapples) has been cited as the main cause of low production.

Global exports of mango, guava and mangosteen were an estimated 1.7 million tons in 2017 with mango accounting for 90% of these exports. Exports in 2017 represented a 9% increase from 2016, significantly higher than the 5% average annual growth registered between 2007 and 2016 (Altendorf, 2019). The expansion follows strong growth in demand in the main export destinations, namely the United States, which expanded its import share to 34% of global imports and the European Union, which held a 20% global share in 2017. The latest available figures indicate that per capita availability of mangoes reached 1.8 kg in the United States and 0.7 kg in the European Union in 2017, up from 1 kg and 0.4 kg in 2007, respectively. The leading exporter of these fruits is Mexico, presenting 24% of overall exports by volume in 2017, followed by Thailand (14%), Brazil (10%) and Peru (9%). Mexico primarily exports mangoes to the United States (approximately 80% of the country's entire shipments in 2017) and benefited strongly from a combination of favorable weather in the main production areas and higher import demand for the fruit in 2017, expanding overall shipments by 19%. Mango exports from Brazil are mainly destined for the European Union market, where the country maintains a strong position through its ability to produce mangoes perennially, including a number of varieties that are popular in European markets such as Tommy Atkins, Keitt and Kent. In 2017, mango exports from Brazil additionally benefited from growing demand in the United States. Thailand, which also ranks as one of the leading exporters of mangosteen, ships the bulk of its mango production to foreign markets, primarily to destinations within the confines of the Southeast Asian region. Peru exports mainly fresh mango and mangosteen, with approximately 40% of shipments going to the Netherlands, 30% to the United States and 10% to the United Kingdom in 2017.

Global exports of pineapple fell to an estimated 2.8 million tons in 2017, representing a 9% decline from 2016 (Altendorf, 2019). Costa Rica, which accounted for approximately 65% of global exports by volume, experienced an estimated 10% fall in exports from 2016 due to adverse weather conditions. The Philippines, the second largest exporter with an approximately 17% share in 2017, also experienced an estimated 9% decline in exports. Despite the supply shortages in Costa Rica, imports by the United States, the largest importer of pineapple globally, increased by an estimated 7% in 2017 from the previous year on the back of strong demand. Imports reached approximately 1.2 million tons in 2017, representing 40% of total global import volumes.

Global exports of avocado reached 1.6 million tons in 2017, representing a 3% increase from 2016, significantly slower than the 11% average annual growth rate seen over the previous decade. Mexico accounted for an estimated 58% of global avocado exports in 2017, with around 77% of shipments going to the United States of America. Exports in 2017 represented a 4% decline from 2016 due to poor weather, in contrast to the 15% annual average growth rate achieved between 2007 and 2016. Other significant exporters are Peru, Chile, South Africa, Israel and Kenya, which export primarily to the European Union and have benefited from significantly increased demand from customers in the European Union.

Global exports of papaya reached an estimated 350 000 tons in 2017, an annual increase of 5% from 2016 (Altendorf, 2019). This represented a healthy recovery from the estimated 6% decline in export volumes experienced in 2016, caused by weather-related supply shortages in Brazil and Guatemala. The largest importers of papaya in 2017 were the United States, with an estimated 57% volume share and the European Union, with an approximate 13% volume share. Mexico accounts for approximately half of all the global papaya exports, with shipments almost exclusively destined for the United States. Rapid growth in demand from the United States gave Mexican producers a strong incentive to improve production practices, resulting in a 6% improvement in average yields between 2010 and 2015, though overall papaya exports from Mexico declined slightly in 2017 (Altendorf, 2019). While the largest share of papaya production in Brazil is destined for the domestic market, the country ranks as the second largest exporter behind Mexico and has rapidly raised its export volumes in response to growing demand from the United States and the European Union. In terms of market potential, the latest available data indicate a per capita availability of 0.6 kg in the United States in 2017, up from 0.4 kg in 2005. Despite the European Union ranking as the second largest importer, consumer awareness of the fruit remains low, with per capita consumption of papaya remaining at a negligible 0.1 kg in 2017. This indicates the potential for future growth as well as the need for more promotion of the fruit and its nutritional benefits to customers, particularly in the European Union.

## Key challenges facing tropical and sub-tropical fruit production and trade

As the discussion of production and trade in tropical and subtropical fruits shows, there is significant potential to develop export markets but production can be very vulnerable to weather fluctuations, a problem that is likely to be exacerbated by climate change. Other challenges include the availability of good-quality planting materials, the need to achieve more consistent yields and quality in cultivation, improving postharvest practices to maintain quality as well as threats from pests and diseases (also linked to the effects of climate change, which is allowing pests and diseases to spread into new areas).

Tropical and subtropical fruit species have been neglected areas of research, for example in the quality of germplasm collections and the application of modern biotechnological techniques (such as tissue culture techniques) in conservation and breeding. They therefore represent a wealth of as yet untapped potential for development.

Since the 1990s, there have been increased efforts to improve this situation. As an example, Bioversity International (formerly IPGRI) initiated research on conservation and use of tropical fruit tree species in Asia in 1993 in collaboration with national partners, subsequently followed by large regional projects funded by donor agencies to strengthen Bioversity's work in genetic resources management. The International Tropical Fruits Network (TFNet) was established in 2000 under the auspices of the Food and Agriculture Organization of the United Nations (FAO) and now plays a major role in promoting the health benefits of tropical fruits and facilitates cooperation among countries, the private sector and other stakeholders. Two species in particular, papaya and citrus, have recently been the focus of extensive research and have become the models for the application of biotechnology to other tropical fruit species. Most citrus research has been conducted on subtropical citrus species but is highly relevant to commercially important tropical citrus species.

A major focus of research is improving the postharvest handling of tropical and subtropical fruits. The tropics are characterized by warm, humid climatic conditions that promote spoilage of fruit (Yahia et al., 2011b). Most tropical fruits have a limited postharvest life or shelf life due to their high perishability, sensitivity to low (chilling) temperatures and vulnerability to attack by spoilage organisms and pests (Kader and Yahia, 2011). The main causes of postharvest losses for tropical and subtropical crops include:

- Mechanical damage due to mishandling along the supply chain;
- Loss of moisture which causes shriveling of fruit;
- Early senescence when stored in higher or lower than optimum temperatures;
- Sensitivity to chilling (lower than optimum) temperature;
- Spoilage due to attack by pathogens and pests.

While some fruits (such as grapefruit, mandarin, orange and pomegranate) have a relatively long shelf life others are highly perishable, such as avocado, guava, lychee/litchi, longan, rambutan, loquat and kiwifruit. This makes it critical to implement the correct postharvest practices for individual fruit species (Yahia, 2011; Yahia et al., 2011b).

While some are non-climacteric (such as citrus fruits and pineapples), most tropical fruits and several subtropical fruits are climacteric, that is, characterized by a marked and sudden rise in the respiratory rate just prior to full fruit ripening (Yahia, 2011). These characteristics are critical in establishing the best time to harvest fruit as well as appropriate postharvest treatments. For example, bananas (climacteric fruit) can be harvested when mature green and ripened later, while pineapple (non-climacteric fruit) should be harvested in a partially ripe state to attain good eating quality when ripe. As examples, maturity indices commonly used for banana, mangoes, papaya and pineapples are shown in Table 3.

Chilling injury (CI) is one of the main sources of deterioration and losses in all tropical and several subtropical fruits (Kader and Yahia, 2011). Sensitive fruits can be injured at temperatures from 0 to 18°C, depending on the type of fruit and the duration of exposure to the incorrect temperature. Some fruits are highly sensitive to CI, such as banana, breadfruit, cherimoya, jackfruit, mamey, mango, mangosteen, papaya, pineapple, rambutan and soursop. Others are moderately sensitive such as carambola, durian, guava, sugar apples and tamarillo. Subtropical fruits that are less sensitive to chilling include grapes and kiwifruit. Table 4 shows CI symptoms in banana, mango, papaya and pineapple.

Tropical and subtropical are also sensitive to several other physiological disorders (Kader and Yahia, 2011). For example, endogenous brown spot (EBS) of pineapples can be induced by holding the fruit at 8°C for one week followed

**Table 3** Maturity indices of banana, mango, papaya and pineapple fruits

| Fruit | Maturity indices |
|---|---|
| Banana | Fullness of finger (disappearance of angularity), days after flowering or fruit set |
| Mango | Fruit shape, internal color, formation of shoulder, external color change in some cultivars |
| Papaya | Surface color (6–33% yellow), soluble solids content (11.5% or higher) |
| Pineapple | Shell color: 'color break' to 25% yellow stage |

**Table 4** Chilling injury symptoms in banana, mango, papaya and pineapple fruits

| Fruit | Minimum safe temperature (°C) | Chilling injury symptoms |
|---|---|---|
| Banana | 13–15 | Surface discoloration, dull color, failure to ripen, browning of flesh. |
| Mango | 10–12 | Grayish scald-like discoloration of skin, uneven ripening, poor flavor, increased decay |
| Papaya | 10–12 | Pitting, failure to ripen, off-flavor development, accelerated decay. |
| Pineapple | 8–10 | Uneven ripening, dull color, water-soaked flesh, off-flavor, increased acidity, loss in ascorbic acid content, wilting of crown leaves, endogenous brown spot. |

by holding for another week at 20°C. EBS may also occur before harvest during cool and overcast periods. Exposure of pineapple fruit to high temperatures (35–38°C) for one day before or after transit reduces the development of EBS symptoms. Waxing pineapples also reduces the severity of EBS because of increased endogenous concentration of $CO_2$.

Diseases are the major cause of postharvest losses in tropical and subtropical fruits. The most important decay-causing organisms are *Colletotrichum gloeosprioides* (causing anthracnosis), *Diplodia natalensis* (causing stem end rot), *Ceratocystis paradoxa* (causing black rot in banana and pineapples) and *Penicillium* and *Fusarium* (causing brown rot on pineapple). Anthracnose is the major postharvest problem in tropical fruits and latent infection often occurs in green fruit before harvest. Preharvest treatments and methods of disease control are important. These include the use of healthy seedlings as well as proper field sanitation practices. Some pesticides are still allowed for applications before and after harvest to control decay, however other non-chemical treatments, such as postharvest heat treatments are effective and safer for consumers. For example, hot water treatments (48-55°C for 5-15 minutes) are commonly used for mango and papaya.

Many insects infect tropical and subtropical fruits (Yahia et al., 2011c). Some of these are restricted to certain regions, such as the Mediterranean fruit fly (*Ceratitis capitata*) which has spread from Africa to Australasia and the Americas, several *Anastrepha* species which are restricted to South, Central and parts of North America and the West Indies and the genus *Dacus* in Africa and Asia. This has led to strict quarantine measures to prevent such pests being accidentally introduced into new areas. Traditionally control of such pests has been via the use of fumigants such as methyl bromide but methyl bromide is now being restricted because of its environmental impact. This has led to the search for alternative methods of control.

Heat treatments are an effective, non-chemical method of treatments, widely used to treat mangoes, papaya and some other tropical fruits grown for export. Heat treatments maintains produce at a certain (high) temperature for a fixed period of time, using a hot water or steam vapor system. For example, a hot water treatment at 46.1°C for 65–90 minutes (depending on fruit weight) is commonly used for mango in several countries (Yahia et al., 2006, 2011c). For papaya a 2-stage heating process has been used to eradicate the Mediterranean fruit fly consisting of treatment at 42°C for 30 min, followed by 49°C for 20 min. Vapor heat treatments have also been developed. Irradiation at low doses (about 300 Grays) are also been used in some countries to disinfect fruits such as guava and mango (Yahia et al., 2011c).

Tropical and subtropical fruits should be pre-cooled as soon as possible after harvest and should be transported at ideal temperature. Several tropical and subtropical fruits, especially banana and mango, are transported by sea in modified (MA) and controlled (CA) atmospheres (Yahia, 1998; (Yahia et al., 2011b). Some tropical fruits, such as bananas, are artificially ripened after arrival to destination market and ethylene gas is commonly used for this process (Kader and Yahia, 2011). Ideal temperature for ripening is 20–24°C in high relative humidity (RH = 90 to 95%). Temperatures above 27°C accelerate softening and may cause tissue discoloration, excessive decay and off-flavors. Temperatures above 35°C inhibit ethylene production and action and consequently inhibit the process of ripening. Ripening is also impaired when fruits are exposed to chilling temperatures. Ethylene concentrations commonly used for fruit ripening are 10 to 100 ppm. Removal of ethylene, using different agents such as potassium permanganate, is used to delay ripening in different tropical and subtropical fruits (Yahia et al., 2011b). The application of waxes is a also very common practice in postharvest handling of some fruits such as citrus. Waxes modify the internal atmosphere of the fruit, reduce weight loss and improve surface appearance. However, the selection of a wax or coating is very crucial, because improper waxes or coatings may lead to development of off-flavors and surface pitting.

## Conclusions

Cultivation, production and trade of tropical and subtropical fruits are gaining increased importance and interest globally. Important tropical and subtropical fruits in world trade include banana/plantain, citrus, grapes, pineapples, mango, papaya and avocado. However, many others such as jackfruit, lychee/litchi, durian, mangosteen, longan, cherimoya, among others, are growing in importance but are still not widely known among consumers in developed countries who constitute the key export markets, suggesting strong potential for the future.

This book reviews recent research on key tropical and subtropical fruits such as avocado, banana, citrus, coconut, guava, jackfruit, lychee, mango, papaya and pomegranate. Chapters cover topics such as advances in production systems, breeding, propagation, soil nutrition and management, irrigation, use of growth regulators, understanding insects pests and their control, harvesting and postharvest handling practices. The book should be of great interest and benefit to producers and traders, as well as researchers and students.

## References

Altendorf, S. 2019. *Major Tropical Fruits Market Review 2017*. FAO, Rome, Italy.

Amador de Los Ríos, J. 1851–1855. *Historia General y Natural de las Indias, Islas y Tierra-Firme del Mar Océano, por el Capitán Gonzalo Fernández de Oviedo y Valdés (General and natural history of the Indies, Islands and Terra Firme of the Ocean Sea).* Critical edition. 4 vols. Impr. de Real Academia de la Historia, Madrid.

FAOSTAT. 2019. Food and Agriculture Organization of the United Nations, Rome, Italy. http://www.fao.org/faostat/en/#home (accessed on 19 September 2019).

Galán Saúco, V. 2001. Greenhouse cultivation of tropical fruits. International Symposium on Tropical and Subtropical Fruits. *Acta Horticulturae* 575, 727–35.

Galán Saúco, V. 1999. *El Cultivo Del Mango (The Cultivation of Mango)*. Ediciones Mundi-Prensa, Madrid.

Jansen, M. J. J. and Subramaniam, B. 2000. Long-term perspectives of fruit and other tree crops in the new century. *Acta Horticulturae* 531, 23–7.

Kader, A. A. and Yahia, E. M. 2011. Postharvest biology of tropical and subtropical fruits. In: Yahia, E. M. (Ed.), *Postharvest Biology and Technology of Tropical and Subtropical Fruits. Volume 1. Fundamental Issues*. Woodhead Publishing, England, pp. 79–111. ISBN: 978 1 84569 733 4.

Martin, F. W., Campbell, C. W. and Ruberté, R. M. 1987. *Perennial Edible Fruits of the Tropics: an Inventory*. United States Department of Agriculture, Agricultural Research Service, Washington DC.

Nagy, S. and Shaw, P. E. 1980. *Tropical and Subtropical Fruits: Composition, Properties and Uses*. AVI Publishing Co., Inc., Westport, CT.

Nakasone, H. Y. and Paull, R. E. 1998. *Tropical Fruits*. CAB International, New York.

Purseglove, J. W. 1968. *Tropical Crops: Dicotyledons* (2 vols). Wiley, New York.

Samson, J. A. 1986. *Tropical Fruits* (2nd edn.). Longman, New York.

Watson, B. J. and Moncur, M. 1985. *Criteria for Determining Survival: Commercial and Best Minimum July Temperatures for Various Tropical Fruits in Australia (S. Hemisphere)*. Wet Tropical Regional Publication, Queensland, Australia.

Yaacob, O. and Tindall, H. D. 1985. *Mangosteen Cultivation. FAO Plant Production and Protection*. Paper 129. Food and Agriculture Organization of the United Nations, Rome, Italy.

Yahia, E. M. 1998. Modified and controlled atmospheres for tropical fruits. *Horticultural Reviews* 22, 123–83.

Yahia, E. M. (Ed.). 2011. *Postharvest Biology and Technology of Tropical and Subtropical Fruits* (4 vols). Woodhead Publishing, Cambridge, UK.

Yahia, E. M., Ornelas Paz, J. J. and Ariza, R. 2006. *Mango: Production, Handling and Processing* (in Spanish). Trillas, Mexico.

Yahia, E. M., De Jesús Ornelas-Paz, J. and Gonzalez-Aguilar, G. A. 2011a. Nutritional and health-promoting properties of tropical and subtropical fruits. In: Yahia, E. M. (Ed.), *Postharvest Biology and Technology of Tropical and Subtropical Fruits. Volume 1. Fundamental Issues*. Woodhead Publishing, England, pp. 21–78. ISBN: 978 1 84569 733 4.

Yahia, E. M., de Jesús Ornelas-Paz, J. and Elansari, A. 2011b. Postharvest technologies to maintain the quality of tropical and subtropical fruits. In: Yahia, E. M. (Ed.), *Postharvest Biology and Technology of Tropical and Subtropical Fruits. Volume 1. Fundamental Issues*. Woodhead Publishing, England, pp. 142–93. ISBN: 978 1 84569 733 4.

Yahia, E. M., Jones, R. and Thomas, D. B. 2011c. Quarantine pests of tropical and subtropical fruits and their control. In: Yahia, E. M. (Ed.), *Postharvest Biology and Technology of Tropical and Subtropical Fruits. Volume 1. Fundamental Issues*. Woodhead Publishing, England, pp. 224–87. ISBN: 978 1 84569 733 4.

Yahia, E. M., Maldonado, M. E. and Svendsen, M. 2018. The contribution of fruit and vegetable consumption to human health. In: *Fruit and Vegetable Phytochemicals: Chemistry and Human Health*. Wiley-Blackwell, Oxford, UK, pp. 1–52. ISBN: 978-1-119-15794-6.

# Part 1

## Citrus fruits

# Chapter 1

## Mapping and exploiting the citrus genome

*Patrick Ollitrault, CIRAD, France*



## 1 Introduction

Citrus belongs to the Aurantioideae subfamily in the Rutaceae family. Most of the cultivated citrus species are part of the *Citrus* genus. According to traditional taxonomy, the *Citrus* genus included between 16 (Swingle and Reece, 1967) and 156 species (Tanaka, 1961). The phenotypical variability of citrus for morphological and agronomic traits is very large. Several sources of tolerance have been identified for abiotic stresses (Krueger and Navarro, 2007). Tolerance to salinity is present in Rangpur lime (*C. limonia* Osb.), Cleopatra mandarin and *C. macrophylla* Wester. The *Microcitrus* and *Eremocitrus* species and at lower level the Rangpur lime display tolerance to water deficit. Adaptation to iron chlorosis is found in Rough lemon (*C. jambhiri* Lush), Volkamer lemon (*Citrus limonia* Osbeck), *C. amblycarpa* (Hassk.) Ochse and *C. macrophylla*. The Satsuma mandarins, the Kumquats (*Fortunella* sp.) and *Poncirus trifoliata* (L.) Raf. are cold tolerant. Important variability is also observed for resistance/susceptibility to important pests and diseases. Sour orange, Volkamer lemon, *C. amblycarpa,* trifoliate oranges and certain pummelos and mandarins are tolerant to *Phytophthora* sp., while tolerances to the African cercosporiosis (fungal disease due to *Phaemularia angolensis* De Carvalho and O. Mendes

http://dx.doi.org/10.19103/AS.2019.0054.02

P.M. Kirk) are found in pummelos, lemons and Satsuma and Beauty mandarins. The susceptibility to another fungal disease, *Alternaria alternata* of tangerine, is only found in a limited set of mandarins, principally Dancy and its hybrids including Orlando tangelo and Fortune mandarin. Resistance to the citrus tristeza virus (CTV) is highly variable. Immunity is found in trifoliate orange, partial resistance in some pummelos and kumquats and tolerance in different germplasm used as rootstocks such as *C. amblycarpa,* Cleopatra mandarin, Rangpur lime, Rough lemon and Volkamer lemons. Different levels of tolerance to citrus canker (bacteriosis due to *Phoma tracheiphila* (Petri) L. A. Kantsch. & Gikaschvili) are found in Kumquats, Yuzu, Satsuma and Dancy mandarins. *P. trifoliata* is a source of resistance to the nematode *Tylenchulus semipenetrans* Cobb. Recently true resistance to Huanglongbing (the most devastating citrus disease due to the bacteria Candidatus Liberibacter sps.) was described in *Microcitrus* and *Eremocitrus*. The variability for fruit morphology and quality traits is also huge and is closely linked with the evolution of cultivated citrus as we will discuss below.

The important variability for biotic and abiotic tolerance traits opens very broad prospects for the exploitation of citrus genetic resources for adaptation breeding, particularly at rootstock level (Cimen and Yesiloglu, 2016). The identifications of molecular markers linked to adaptation traits for abiotic stresses and to disease and pest resistances are now essential to drive efficiently multi-traits breeding projects. In the same way, it is essential to develop molecular markers for early selection of fruit traits. Indeed, due to the juvenile phase and size of evaluation plots, conventional breeding for quality is very costly and the numbers of hybrids to work with are very limited compared with annual crops.

The recent development of genomics in citrus and particularly the availability of reference genome sequences for sweet orange (Xu et al., 2013), clementine (Wu et al., 2014) and pummelo (Wang et al., 2017) and ongoing re-sequencing projects (Wu et al., 2018; Wang et al., 2017) pave the way for a more efficient exploitation of citrus genetic diversity in conventional breeding programmes. The present challenge is to associate genome polymorphisms with the diversity of useful traits. The most common genetic approaches are to perform quantitative trait loci (QTLs) analysis or genome-wide association studies (GWAS), both based on genetic linkage between genes implied in the phenotype diversity and genetic markers (i.e. close genomic position of the implied genes and genetic markers). QTL analysis is conducted on sexual recombining populations, while GWAS explore associations in more diverse germplasms. GWAS has the advantage of allowing identification of key genes that should be monomorphic in a single hybrid population and to bypass the necessity to create designed segregating progenies. It should therefore offer advantages in citrus-breeding programmes where breeding materials are even derived from several parental combinations. The uses of GWAS in fruit crops has increased in recent years (Farneti et al., 2017; Minamikawa et al., 2017; Iwata et al., 2016; Moriya et al., 2017). However, the genetic structure of the germplasm under investigation, and particularly the decay of linkage disequilibrium (LD) with genetic distance, are key elements in applying GWAS. These elements are strongly affected by the historical evolution of the gene pools being used.

The application of GWAS is also strongly dependent on the advance in genomics and genetic marker resources development. In recent years, with the development of next-generation sequencing (NGS) methods, new high-throughput genotyping methods have been developed combining NGS and reduced genome representation such as allelic capture sequencing (Holtz et al., 2016) and methods based on restriction site-associated DNA sequencing (RADseq (Miller et al., 2007); GBS (Elshire et al., 2011)). These make it possible to perform high-throughput phenotyping at the same time as SNP mining, while microarray genotyping can identify useful polymorphisms. High-throughput genotyping at affordable costs open the way to the analysis of large segregating progenies and marker-trait association studies (Baxter et al., 2011; Ipek et al., 2016; Montero-Pau et al., 2017) and further to genomic selection (GS).

This chapter reviews the recent developments in genetics and genomics to decipher the citrus genomes and optimize their exploitation. It starts with a reminder about important and distinctive characteristics that determine the citrus gene pool which define priorities for the international citrus research community and the challenge of marker-traits association in citrus. The recent developments of genetics and genomics resources and knowledge are then reviewed discussing studies of the links between molecular and phenotypic diversity and resulting concrete applications for marker-assisted selection (MAS) and perspectives for GS.

## 2 Characteristics of the citrus gene pool

### *2.1 The taxonomic position*

While previously it was included in the Geraniales order (Swingle and Reece, 1967), taxonomists now agree that *Citrus* L. genus is part of the Sapindales Berchtold & J. Presl. order in the Rutaceae Jussieu family (Stevens, 2017; NCBI, 2017). The genus *Citrus* belongs to the subfamily Aurantioideae, which is divided into two sub-tribes (Swingle and Reece, 1967): the Clauseneae (five genera) and the Citreae (28 genera). The Clauseneae tribe is considered more primitive than the Citreae tribe and there is no sexual compatibility between the two tribes. Citreae are divided in three sub-tribes: Triphasiinae, Balsamocitrinae and Citrinae. Swingle and Reece (1967) recognized three groups in the Citrinae. One of them, the 'true citrus' group includes the *Citrus* genus and five other genera: *Clymenia*, *Eremocitrus*, *Fortunella*, *Microcitrus* and *Poncirus*. Maternal phylogenetic studies (Bayer et al., 2009) confirmed the monophyly of the 'true citrus group' and suggested including the neo-caledonian *Oxanthera* genus in the same group.

## 2.2 A large interspecific and intergeneric sexual compatibility

The different species of the other 'true citrus' genera display sexual compatibility with the *Citrus* species (Iwamasa et al., 1988). Particularly, many fertile hybrids have been produced between different *Citrus* species *and P. trifoliata* which is a key germplasm for citrus rootstock improvement. Hybrids between sweet orange, grapefruit, lemon, sour orange and mandarin and *P. trifoliata* are, respectively, called citrange, citrumelo, citremon, citradia and citrandarin. *Fortunella* species were also used as breeding parents during the twentieth century (mandarinquat, limequat) and the natural origin of Calamondin from intergeneric hybridization between *Fortunella* and *Citrus* was recently demonstrated from WGS data (Wu et al., 2018). Several hybrids have been created between *Citrus* and *Microcitrus* (e.g. 'Australian blood' lime). Hybrids between *Citrus* and *Eremocitrus glauca* were also obtained (eremorange and eremolemon, respectively, with sweet orange and lemon). The different 'true Citrus' genera present interesting traits for citrus breeding such as cold tolerance, dwarfing, disease and pest tolerances (including phytophthora, nematodes) and overall CTV immunity in *Poncirus*; cold tolerance and citrus canker (bacteriosis due to *Xanthomonas axonopodis*) resistance in *Fortunella*; high tolerance and resilience to water deficit and recently found (Ramadugu et al., 2016) resistance to citrus greening in *Microcitrus* and *Eremocitrus*. Therefore, the 'true citrus' group can be considered as the fundamental germplasm to improve citrus by sexual hybridization and needs to be integrated in the international efforts to map and exploit the citrus genome.

## 2.3 The reticulate evolution of the cultivated citrus and facultative apomixes resulted in a highly structured gene pool

Within the *Citrus* genus, the taxonomy remained confused and conflicted for a long time with two principal classifications that are still used today: the Tanaka system (Tanaka, 1954) which recognizes 156 species of *Citrus* while the Swingle and Reece classification (Swingle and Reece, 1967) is limited to 16 species. The reticulate evolution of the cultivated citrus combined with partial apomixis and vegetative propagation were the main sources of difficulties to establish *Citrus* classifications. Indeed many taxonomists considered interspecific hybrids, propagated true to ness by apomictic seeds, as new species (Lawrence, 1951). Recently, phylogenetic and phylogenomic studies revealed the origins and admixtures of modern cultivars and wild types and clarified the organization of the citrus gene pool.

Most recent molecular studies (Nicolosi et al., 2000; Garcia-Lor et al., 2013b; Ollitrault et al., 2012b; Barkley et al., 2006; Curk et al., 2015, 2016; Wu et al., 2018; Carbonell-Caballero et al., 2015) agree that the cultivated *Citrus* results from a reticulate evolution involving four ancestral species: *C. maxima* (the pummelos), *C. medica* (the citrons), *C. micrantha* (a wild papeda) and *C. reticulata* (very closely related with the mandarin horticultural group). The differentiation between these four sexually compatible ancestral taxa resulted from a founder effect in four geographic zones and initial allopatric evolution (Swingle and Reece, 1967; Scora, 1975; Webber, 1943; Wu et al., 2018). *C. reticulata* originated and initially diversified in Vietnam, Southern China and Japan; *C. maxima* evolved in Indonesia and the Malay Archipelago; *C. medica* originated in Northeastern India and the nearby region of Burma and China; and *C. micrantha* is native to the southern Philippines. Interestingly, while most actual pummelos, citrons and 'Small flowered papedas' appear, respectively, as pure representative of *C. maxima, C. medica* and *C. micrantha,* the recent phylogenomic studies (Wu et al., 2014, 2018; Oueslati et al., 2017; Curk et al., 2015) revealed that most of the cultivated mandarin where introgressed by *C. maxima*.

All other edible *Citrus* secondary species such as *C. sinensis* (sweet oranges), *C. aurantium* (sour oranges), *C. paradisi* (grapefruits), *C. aurantiifolia* ('Mexican lime' type) and *C. limon* (lemons), for the most important horticultural groups, resulted from admixture of these ancestral taxa (Curk et al., 2015, 2016; Oueslati et al., 2017; Wu et al., 2018; Ahmed et al., 2019). The *C. reticulata/C. maxima* admixture gene pool is particularly important from a horticultural point of view as it includes mandarins, pummelos, sweet and sour oranges, grapefruits and recent hybrids such as tangelos (mandarin × grapefruit), tangors (mandarin × sweet oranges) and orangelo (orange × grapefruits). In many cases, the interspecific hybrids resulting from the reticulate evolution were polyembryonic and the facultative apomixes greatly limited further sexual recombinations resulting in mosaics of large genome fragments from different species including frequent interspecific heterozygosity (Wu et al., 2014, 2018; Oueslati et al., 2017).

Another consequence of this reticulate evolution coupled with apomixes, and resulting limited interspecific recombination, is the very high structuration of the genetic diversity of the gene pool when considering the *Citrus* genus globally. An important part of the actual phenotypic diversity of edible citrus should be related to the interspecific differentiation during the allopatric evolution of the ancestral taxa. As a result, the organization of phenotypic polymorphisms is closely associated with the pattern of genetic diversity. Such links with genetic organization were observed for morphological and pomological characters (Ollitrault et al., 2003; Barrett and Rhodes, 1976), flavone constitution (Mizuno et al., 1991), peel oil volatile compounds (Liu et al., 2013a), carotenoid contents (Fanciullino et al., 2006), coumarin and furanocoumarin constitution (Dugrand-Judek et al., 2015) and fingerprinting of secondary metabolites (Matsukawa and Nito, 2017). It results from a generalized LD, when considering the global *Citrus* genus gene pool, also revealed from molecular marker analysis (Garcia-Lor et al., 2012). As mentioned before, this generalized LD should preclude GWAS considering only the natural germplasm. However, the importance of the interspecific component (between the ancestral taxa) in overall phenotypic polymorphism, opens the way for marker-traits association studies based on ancestral haplotype using composite populations resulting from diverse hybridizations.

In horticultural groups such as sweet and sour oranges, grapefruit, lemons and limes, vegetatively propagated by apomictic seeds or horticultural methods (grafting, cutting) the intra-group variability results from mutations, transposable element activation or epigenetic features. As we will discuss below, catching the molecular bases of this phenotypic variability generally needs specific approaches in relation with the considered mechanisms of diversification.

## 2.4 Ploidy manipulation is an important component of citrus breeding

Most of the citrus germplasm is diploid with a basic chromosome number of 9 (Krug, 1943). However, spontaneous reproductive features can lead to ploidy variation. Chromosome doubling of nucellar cell is relatively frequent in apomictic lines (Aleza et al., 2011), producing doubled-diploid genotypes. Unreduced (2n) gamete are produced both in pollen and ovules. Second division restitution (SDR) was found to be the main mechanisms of 2n ovules formation in mandarin (Cuenca et al., 2011, 2015), in tangors including clementine (Xie et al., 2014; Aleza et al., 2015) and lemon (Rouiss et al., 2017b) while first division restitution (FDR) was predominant in the pollen of a Clementine × Sweet orange hybrid (Rouiss et al., 2017a). In relation with the objectives of citrus genetic improvement and taking advantage of the natural mechanisms of polyploidization, ploidy manipulation became an important component of citrus genetics and breeding. For varieties, the main objective is to develop triploid seedless cultivars. For rootstock, the aim is to develop interesting traits in tetraploid hybrids and to improve adaptation to biotic and abiotic stresses. Thus, beside diploid hybrids, citrus breeders manage recombining triploid and tetraploid progenies requiring, as discussed below, specific tools and approaches for marker-traits association studies, MAS and GS.

## 3 Forty years of molecular markers development to study and exploit the citrus genome

Citrus has a relatively small genome size varying among ancestral taxa from 398 to 360 Mb/haploid genome for *C. medica* and *C. reticulata*, respectively (Ollitrault et al., 1994). *C. maxima* had an intermediate genome size of 383 Mb and the one of *C. sinensis* was estimated to be 372 Mb (Gmitter et al., 2012).

### 3.1 The dominant polymerase chain reaction-based markers

Citrus genome study started during the 1980s with isozymes markers (Torres et al., 1982). Several kinds of dominant nuclear markers based on polymerase chain reaction (PCR) were developed such as random amplified polymorphic DNA (RAPDs; Xiao et al., 1995; Luro et al., 1992), inter-simple sequence repeat (ISSR; Fang and Roose, 1997; Fang et al., 1997b, 1998b) and amplified fragment length polymorphisms (Fang et al., 2009; Liang et al., 2007; de Oliveira et al., 2007). These methods had the advantage of being easy to perform and do not require previous knowledge on citrus genome. They were used for germplasm studies (Krueger and Roose, 2003; Pang et al., 2007; Kumar et al., 2010; Biswas et al., 2010a,b; Yang et al., 2010; Lombardoa et al., 2012; Dorji and Yapwattanaphun, 2015; Munankarmi et al., 2018), nucellar/zygotic differentiation (Rao et al., 2007) and somatic hybrid genome characterization (Scarano et al., 2002; Guo and Deng, 2001; Fu et al., 2004), genetic mapping and marker-traits association (de Simone et al., 1998; Sankar and Moore, 2001; Choi et al., 2007; de Oliveira et al., 2007; Gulsen et al., 2010; Roose et al., 2000). However, today, their dominant character and the lack of knowledge of their sequence and position on the citrus genome strongly limit their usefulness compared with the new generation of co-dominant markers.

### 3.2 The mono-locus co-dominant markers

Restriction fragment length polymorphisms (RFLPs) were the first single locus co-dominant markers developed in citrus (Durham et al., 1992; Luro et al., 1996; Federici et al., 1998, 2000), but, due to technical difficulties and the development of co-dominant PCR markers, their application remained limited. Simple sequence repeat (SSRs) markers have been derived from genomic data (Kijas et al., 1995; Barkley et al., 2006; Froelicher et al., 2008; Ollitrault et al., 2010; Liu et al., 2013b) and from transcriptomic data (Chen et al., 2006; Luro et al., 2008; Liu et al., 2013b; Liang et al., 2015; Shimizu et al., 2016). They have been included in citrus genetic maps (Ruiz and Asins, 2003; Chen et al., 2008b; Lyon et al., 2007; Luro et al., 2008; Bernet et al., 2010; Raga et al., 2012; Ollitrault et al., 2012a; Cuenca et al., 2016). They were also used for the analysis of genetic diversity (Shimizu et al., 2016; Luro et al., 2001, 2012; Corazza-Nunes et al., 2002; Barkley et al., 2006; Garcia-Lor et al., 2012, 2015; Curk et al., 2016), the discrimination of zygotic and nucellar seedlings (Ruiz et al., 2000; Rao et al., 2008; de Oliveira et al., 2002), the control of the origin of plants obtained by induced gynogenesis and androgenesis (Froelicher et al., 2007; Germana et al., 2013; Aleza et al., 2009) and somatic hybridization (Chen et al., 2008a; Bassene et al., 2009), the molecular characterization of triploid cultivars (Aleza et al., 2010) and the analysis of the origin of 2n gametes (Luro et al., 2004; Chen et al., 2008b; Ferrante et al., 2010; Cuenca et al., 2011, 2015; Rouiss et al., 2017a,b).

With the development of NGS methods SNPs have become the most important resources for molecular marker development. They were mined from trancriptomic (Chen and Gmitter, 2013) or genomic data (Ollitrault et al., 2012b; Wu et al., 2014, 2018). Efficient SNP genotyping methods have been developed for scalable experiments using competitive allele amplification (KASPar© technology; Garcia-Lor et al., 2013b; Cuenca et al., 2013a). Curk et al. (2015, 2016) developed a set of KASPar SNP markers specific to the four ancestral taxa and studied the phylogenetic origin of the modern varieties. The ancestor diagnostic SNPs were also very useful to analyse the meiotic behaviour of allotetraploids and to identify zygotic plants in the seedlings of interspecific or intergeneric hybrids (Ollitrault et al., 2016; Bruyere et al., 2016). In Japan, special emphasis was given to the development of cleaved amplified polymorphic sequences targeting previously mined SNPs (Shimada et al., 2014; Omura and Shimada, 2016).

### 3.3 High-throughput genotyping

For high-throughput studies, several SNP arrays have been developed. Ollitrault et al. (2012b) focussed on markers useful to establish the reference genetic map of clementine and therefore heterozygous for this cultivar. They were mined in Clementine BAC-end sequences (Terol et al., 2008) to implement an Illumina GoldenGate assay including 1494 SNPs. Fujii et al. (2013) genotyped Japanese germplasm and recombining populations with an Illumina array of 384 SNPs while Yu et al. (2016) used a 1536-SNP Illumina GoldenGate assay to establish the genetic maps of Fortune and Murcott mandarins. Recently, two Affymetrix Axion SNP arrays with about 1.5 million and 56 000 SNPs were developed from the polymorphisms mined in 30 re-sequenced citrus accessions including the relative genera *Poncirus, Fortunella, Microcitrus and Eremocitrus* (Eck et al., 2016). GBS, RAD-sequencing and DARTSeq were also successfully developed in citrus and applied for germplasm analysis (Penjor et al., 2014; Oueslati et al., 2017; Ahmed et al., 2019), genetic mapping (Guo et al., 2015; Curtolo et al., 2017a,b; Huang et al., 2018), QTL studies (Curtolo et al., 2017a) and GWAS (Imai et al., 2018).

### 3.4 Markers for polyploid citrus studies

The genetics of polyploid species remains much more challenging than diploid genetics. Indeed, accurate polyploidy population genetic studies require the estimation of molecular marker allele copy number of heterozygous genotypes. Moreover, allelic dosage can affect gene expression and phenotype and its determination is therefore particularly important for marker-trait association studies (De Jong et al., 2003). Several methods have been developed to estimate allele dosage in polyploid citrus genotypes. For SSRs markers, the MAC-PR method (Esselink et al., 2004) was proposed to deal with differential amplification intensities among alleles in polyploid plants and was successfully transferred in citrus (Ferrante et al., 2010; Cuenca et al., 2011). For SNP markers, Cuenca et al. (2013a) demonstrated the efficiency of the KASPar SNP genotyping technique, to assign heterozygous allelic configurations within polyploid citrus populations. MAC-PR approach for SSRs markers and KASPar technology for SNPs were routinely applied for the study of 2n gamete origin and their genetic structures (Cuenca et al., 2015; Aleza et al., 2015; Rouiss et al., 2017b; Rouiss et al., 2017a), tetraploid meiosis analysis (Rouiss et al., 2018; Aleza et al., 2016; Ollitrault et al., 2016), centromere mapping (Aleza et al., 2015) and marker-trait association in a triploid progeny (Cuenca et al., 2013b). To achieve high throughput, Ahmed et al. (2019) developed a pipeline, based on a maximum likelihood analysis from the number of GBS reads, to infer ancestor doses along the genome of triploid and tetraploid individual. Beside molecular markers, the estimation of ploidy level by flow cytometry (Ollitrault and Michaux-Ferriere, 1992; Ollitrault et al., 1994) constituted a decisive step to manipulate efficiently the ploidy of the citrus genome. Indeed, it allowed the screening of large populations to identify spontaneous doubled diploids in polyembryonic seedlings (Aleza et al., 2011), to select triploids issued from 2n gametes in 2x × 2x hybridizations (Aleza et al., 2010) and to identify triploid and tetraploid plants in interploid progenies (2x × 4x and 4x × 2; Aleza et al., 2012a,b).

### 3.5 Markers of asexual diversification processes

Within vegetatively propagated horticultural groups such as sweet orange, where phenotypical diversity is not related with sexual recombination, targeted studies of transposable elements make it possible to reveal polymorphisms among varieties (Breto et al., 2001; Tao et al., 2005; Bernet and Asins, 2003; Bernet et al., 2004b) and transposition events directly linked with phenotypic variation such as the production of anthocyanin in blood oranges (Butelli et al., 2012) and apomixes in mandarins (Wang et al., 2017). IRAP and REMAP markers have also been used for studies of genetic similarity based on retro-transposon within the genus Citrus and its relatives (Biswas et al., 2010a,b).

# 4 Citrus genetic mapping

## 4.1 Early citrus genetic mapping work

Due to the important heterozygosity of citrus germplasm, most of the citrus genetic maps were developed from F1 crosses and segregation analyses allowed developing genetic maps for each of the parents and eventually consensus genetic maps (Table 1). Early genetic maps (1990, 2005) were reviewed by Ruiz and Asins (2003), Chen et al. (2008b) and Roose (2007). Many of them were based on intergeneric hybrids between *Citrus* and *Poncirus*, due to the importance of *Poncirus trifoliata* for rootstock breeding and particularly its immunity to CTV. Most of these maps suffered from relatively low numbers of analysed hybrids and from the multilocus and dominant nature of the markers (RAPD, ISSR, AFLP), even if a few isozymes and RFLPs were included. In most cases they resulted in more or less than nine linkage groups (LGs). Cai et al. (1994) using 189 markers (RAPD, RFLP, isozymes) were the first to propose a map with nine LGs from a small population of 65 hybrids resulting from an intergeneric backcross (*C. grandis* cv. Thong Dee × USDA 17-40 (*C. grandis* cv. Thong Dee × *P. trifoliata* cv Pomeroy)). It was extended to 310 markers using ISSR (Sankar and Moore, 2001). The first map of co-dominant markers grouped in nine LGs was based on 120 CAPS markers genotyped in a progeny of 125 BC1 hybrids of Kiyomi (*C. unshiu* cv. Miyagawa × *C. sinensis* cv. Trovita) × *C. unshiu* cv. Miyagawa (Omura et al., 2000). Subsequently, several maps were generated using co-dominant markers, particularly SSRs (Bernet et al., 2010; Chen et al., 2008b; Lyon et al., 2007) but failed to result in nine LGs due to insufficient marker numbers. During the same period, de Oliveira et al. (2007) and Gulsen et al. (2010) published new genetic maps of nine LGs mostly based in dominant markers, respectively, for Murcott tangor (227 AFLPs) and clementine and Orlando tangelos (with, respectively, 215 and 189 SRAP, RAPD, SSR, ISSR and POPG markers). However, none of these maps encompassed enough markers with published sequences to establish a reference citrus map which could be combined with whole-genome sequence data.

## 4.2 Saturated and reference genetic maps of citrus species

Given its aim to establish the citrus genome reference sequence from a haploid clementine, the International Citrus Genome Consortium (ICGC) identified the construction of a saturated genetic map of clementine as essential to improve the whole-genome sequence assembly in pseudomolecules (chromosomes).

**Table 1** Major citrus genetic maps and their application for phenotypic traits association

| Reference | Map | Cross | Female parent | Male parent | N | Mk | LG | Mk type | Phenotypical traits |
|---|---|---|---|---|---|---|---|---|---|
| Durham et al. (1992) | Intergeneric consensus | BC1 | ThongDee Pummelo | T.D. Pummelo × Poncirus | 65 | 33 | 8 | RFLP | NA |
| Cai et al. (1994) | Intergeneric consensus | BC1 | ThongDee Pummelo | T.D. Pummelo × Poncirus | 60 | 161 | 9 | IRAP, RFLP, SRAP, RAPD, SSR | Morphological characters |
| Gmitter et al. (1996) | Citrus consensus | 6 intergeneric F1 and BC/BSA | Multiple | Multiple | 56 | 9 | 1 | RAPD | CTV resistance |
| Luro et al. (1996) | Pummelo | F1 | Seedless pummelo | Cleopatra Md × Poncirus | 52 | 34 | 7 | RFLP, RAPD, Isozymes | NA |
| Luro et al. (1996) | Cleopatra X Poncirus | F1 | Seedless pummelo | Cleopatra Md × Poncirus | 52 | 95 | 12 | RFLP, RAPD, Isozymes | NA |
| Mestre et al. (1997) | Poncirus | AF/BSA | Flying dragon Poncirus | Flying dragon Poncirus | 24 | 5 | 1 | RAPD | CTV resistance |
| Fang et al. (1998b) | Poncirus consensus | 10 populations/BSA | Citrus and Poncirus | Poncirus and citrange | 554 | 9 | 1 | RAPD, SSR, ISSR | CTV resistance |
| de Simone et al. (1998) | Sour orange | F1 | Sour orange | Khasi papeda | 120 | 247 | 20 | RAPD, RFLP, AFLP | NA |
| de Simone et al. (1998) | Kashi papeda | F1 | Sour orange | Khasi papeda | 120 | 92 | 12 | RAPD, RFLP, AFLP | NA |
| Cristofani et al. (1999) | Sunki mandarin | F1 | Citrus Sunki | Poncirus Rubidoux | 80 | 63 | 9 | RAPD | Phytophthora resistance |
| Cristofani et al. (1999) | Rubidoux Poncirus | F1 | Citrus Sunki | Poncirus Rubidoux | 80 | 62 | 8 | RAPD | Phytophthora, CTV res. |
| García et al. (1999) | Rubidoux Poncirus | F1 | Volkamer lemon | Poncirus Rubidoux | 50 | 17 | 5 | SSR, IRAP | Apomixis |

(*Continued*)

**Table 1** (*Continued*)

| Reference | Map | Cross | Female parent | Male parent | N | Mk | LG | Mk type | Phenotypical traits |
|---|---|---|---|---|---|---|---|---|---|
| García et al. (1999) | Volkamer lemon | F1 | Volkamer lemon | Poncirus Rubidoux | 50 | 32 | 9 | SSR, IRAP | Apomixis |
| Ling et al. (2000) | Citrus consensus | BCI | ThongDee Pummelo | T.D. Pummelo × Poncirus | 65 | 337 | 11 | AFLP, RFLP, isozyme | Nematodes |
| Omura et al. (2000) | Citrus consensus | BC1 | Kiyomi tangor | *C. unshiu* cv. Miyagawa | 125 | 120 | 9 | CAPS | NA |
| Roose et al. (2000) | Intergeneric consensus | F1 | Sacaton citrumelo | Troyer citrange | 60 | 156 | 16 | RAPD, RFLP, ISSR | NA |
| Sankar and Moore (2001) | Intergeneric consensus | BCI | ThongDee Pummelo | T.D. Pummelo × Poncirus | 65 | 310 | 9 | ISSR, RAPD, RFLP, isozymes | NA |
| Bernet and Asins (2003) | Sour Orange | F1 | Afin Verna sour orange | Flying dragon Poncirus | 66 | 102 | 12 | IRAP, SSR, RAPD | NA |
| Bernet and Asins (2003) | Flying dragon Poncirus | F1 | Afin Verna sour orange | Flying dragon Poncirus | 66 | 15 | 2 | IRAP, SSR, RAPD | NA |
| Ruiz and Asins (2003) | Volkamer lemon | F1 | Volkamer | Rubidoux | 80 | 124 | 22 | IRAP, SSR, RAPD | NA |
| Ruiz and Asins (2003) | Sour orange | F1 | Afin Verna sour orange | Flying dragon Poncirus | 66 | 151 | 24 | IRAP, SSR, RAPD | NA |
| Ruiz and Asins (2003) | Flying dragon Poncirus | F1 | Poncirus FD | FD | 57 | 41 | 11 | IRAP, SSR, RAPD | NA |
| Weber et al. (2003) | Pummelo DP16 | F1 | Pummelo DP16-4 | Rubidoux | 99 | 117 | 9 | CAPS, RAPD, SSR, RBC, SCAR, STS | Cold |

| | | | | | | | | | |
|---|---|---|---|---|---|---|---|---|---|
| Weber et al. (2003) | Rubidoux Poncirus | F1 | Pummelo DP16-5 | Rubidoux | 99 | 100 | 9 | CAPS,RAPD, SSR, RBC, SCAR, STS | Cold |
| Bernet et al. (2004) | Flying dragon Poncirus | F1 | Afin Verna sour orange | Flying dragon Poncirus | 63 | 27 | 2 | IRAP, SSR, RAPD | CTV resistance |
| de Oliveira et al. (2007) | Murcott | F1 | Pera sweet orange | Murcott | 87 | 65 | 9 | AFLP | NA |
| de Oliveira et al. (2007) | Pera sweet orange | F1 | Pera sweet orange | Murcott | 87 | 55 | 5 | AFLP | NA |
| de Oliveira et al. (2007) | Murcott | F1 | Pera sweet orange | Murcott | 87 | 227 | 9 | AFLP | NA |
| Chen et al. (2007) | Sweet Orange | F1 | Sweet orange | Poncirus Argentina | 97 | 113 | 11 | Est-SSRs | NA |
| Chen et al. (2007) | Poncirus | F1 | Sweet orange | Poncirus Argentina | 97 | 45 | 8 | Est-SSRs | NA |
| Lyon (2008) | Sweet orange | 13 F1 sweet orange × Poncirus | Sweet orange | Poncirus | 208 | 988 | 9 | SSR, SNP | NA |
| Lyon (2008) | Poncirus | 13 F1 sweet orange × Poncirus | Sweet orange | Poncirus | 208 | 220 | 9 | SSR, SNP | NA |
| Gulsen et al. (2010) | Clementine | F1 | Tangelo Orlando | Clementine | 164 | 215 | 9 | SRAP, RAPD, SSR, ISSR, POPG | NA |
| Gulsen et al. (2010) | Tangelo Orlando | F1 | Tangelo Orlando | Clementine | 164 | 189 | 9 | SRAP, RAPD, SSR, ISSR, POPG | Alternaria resistance |
| Bernet et al. (2010) | Fortune mandarin | 2 F1 reciprocal progenies | Fortune mandarin | Chandler pummelo | 174 | 95 | 13 | SSR, IRAP, SCAR | NA |
| Bernet et al. (2010) | Chandler | 3 F1 reciprocal progenies | Chandler pummelo | Fortune | 174 | 77 | 11 | SSR, IRAP, SCAR | NA |
| Ollitrault et al. (2012a) | Chandler | F1 | Chandler pummelo | Clementine | 156 | 151 | 9 | SSRs, Indels, SNP (array) | NA |

(*Continued*)

**Table 1** (*Continued*)

| Reference | Map | Cross | Female parent | Male parent | N | Mk | LG | Mk type | Phenotypical traits |
|---|---|---|---|---|---|---|---|---|---|
| Ollitrault et al. (2012a) | Sweet orange | F1 | Sweet orange | Poncirus | 147 | 569 | 9 | SSRs, Indels, SNP (array) | NA |
| Ollitrault et al. (2012a) | Pink pummelo | F1 | Clementine | Pink pummelo | 140 | 81 | 9 | SSRs, Indels, SNP (array) | NA |
| Ollitrault et al. (2012a) | Clementine | F1 | Clementine/Chandler | Clementine/Pink | 296 | 959 | 9 | SSRs, Indels, SNP (array) | NA |
| Ollitrault et al. (2012a) | Citrus consensus | 3 F1 populations | NA | NA | 436 | 1158 | 9 | SSRs, Indels, SNP (array) | NA |
| Raga et al. (2012) | Cleopatra mandarin | F1 | Cleopatra | Flying dragon Poncirus | | 78 | 9 | SSR, IRAP, SCAR, ISSR, RAPD | Seed reproductive traits |
| Raga et al. (2012) | Flying dragon Poncirus | F1 | Rich Poncirus | Flying dragon Poncirus | | 68 | 10 | SSR, IRAP, SCAR, RAPD | Seed reproductive traits |
| Çevik and Moore (2012) | Citrus consensus | F1 | Pummelo × Citrumelo | Citrumelo × mandarin | 98 | 69 | 9 | RAPD | Morphological traits |
| Shimada et al. (2014) | Citrus consensus | F1 | Sweet Spring tangor | Lee tangelo | 87 | 706 | 9 | STS, CAPS | NA |
| Guo et al. (2015) | Guanzi pummelo | F1 | Guanzi pummelo | Pingshan pummelo | 124 | 1549 | 9 | SSR, SNP (RADSeq) | NA |
| Guo et al. (2015) | Pinshang pummelo | F1 | Guanzi pummelo | Pingshan pummelo | 124 | 500 | 9 | SSR, SNP (RADSeq) | NA |
| Guo et al. (2015) | Pummelo consensus | F1 | Guanzi pummelo | Pingshan pummelo | 124 | 1563 | 9 | SSR, SNP (RADSeq) | NA |

| | | | | | | | | | |
|---|---|---|---|---|---|---|---|---|---|
| Hong et al. (2015) | Citrus-Li2/Wan2- | F1 | Citrus-Li2 | Wan | 68 | 452 | 9 | SSRs | Freeze |
| Ohta et al. (2014) | Citrus × Poncirus Nou-8 | F1 | Nou-8 | Siamese acidless | 93 | 189 | 10 | SNPs (caps) | CTV resistance |
| Raga et al. (2016) | Cleopatra mandarin | F1 | Cleopatra | *P. trifoliata* | 68 | 81 | 9 | IRAP, SSR, SCAR, ISSR, RAPD | Salt tolerance |
| Raga et al. (2016) | Cleopatra mandarin | F1 | Cleopatra | *P. trifoliata* | 83 | 73 | 12 | IRAP, SSR, SCAR, ISSR, RAPD | Salt tolerance |
| Yu et al. (2016) | Fortune mandarin | F1 | Fortune | Murcott | 116 | 189 | 22 | SNPs (array) | Fruit characters |
| Yu et al. (2016) | Fortune mandarin | F1 | Fortune mandarin | Murcott | 116 | 380 | 24 | SNPs (array) | Fruit characters |
| Yu et al. (2016) | Murcott | F1 | Fortune mandarin | Murcott | 116 | 104 | 15 | SNPs (array) | Fruit characters |
| Curtolo et al. (2017a) | Citrus consensus | F1 | Murcott | Pera sweet orange | 278 | 661 | 13 | DARTSeq | Fruit characters |
| Curtolo et al. (2018). | Sunki mandarin | F1 | Sunki mandarin | Rubidoux Poncirus | 276 | 2778 | 10 | DARTSeq | NA |
| Curtolo et al. (2018). | Rubidoux Poncirus | F1 | Sunki mandarin | Rubidoux Poncirus | 276 | 3084 | 10 | DARTSeq | NA |
| Huang et al. (2018) | Sweet Orange | F1 | Sweet orange | Poncirus | 170 | 754 | 9 | RADSeq | HLB tolerance |
| Huang et al. (2018) | Poncirus | F1 | Sweet orange | Poncirus | 170 | 647 | 9 | RADSeq | HLB tolerance |

N: number of individuals; Mk: number of markers; LG: number of linkage groups.

An international collaboration involving France, Morocco, Spain, Turkey and the United States resulted in a reference Clementine genetic map, including 961 co-dominant markers (677 SNPs, 258 SSRs and 26 Indels) spread among nine LGs (Ollitrault et al., 2012a). This map spans 1084.1 cM, with an average marker spacing of 1.13 cM. A sweet orange genetic map (569 markers) and two pummelos ('Chandler' and 'Pink' cultivars with, respectively, 151 and 81 markers) maps of nine LGs were also published in the same study. Saturated maps of sweet orange with 943 markers (Xu et al., 2013, based on the PhD by Lyon, 2008) and mandarin with 706 markers (Shimada et al., 2014) were also published at the same time. More recently, NGS applied with complexity reduced genomes was used to produce medium- to high-density genetic maps. Guo et al. (2015) produced two pummelos maps from an F1 cross using RADSeq (around 1500 markers each). DARTSeq was also successfully applied to produce a synthetic map of Murcott and sweet orange (661 markers; Curtolo et al., 2017a) as well as a 'Sunki' mandarin and a 'Rubidoux' trifoliate orange maps from an F1 intergeneric cross (2778 and 3084 markers, respectively; Curtolo et al., 2017b).

## 4.3 Synteny is high among citrus species and related genera

A few comparative genetic mapping studies analysed the synteny and collinearity in the species of the citrus genus and in Poncirus. The conservation of synteny was complete between sweet orange, pummelo and clementine and the linear order of markers also appeared to be highly conserved between these species (Ollitrault et al., 2012a). These observations extended to the whole-genome conclusions of Bernet et al. (2010) based on partial maps of sour orange, pummelo and Fortune mandarin (a Clementine hybrid). The same authors also reported high collinearity between Fortune 'mandarin' and *Poncirus trifoliata* as previously revealed by Chen et al. (2008b) from a comparative study of *C. sinensis* and *P. trifoliata* genetic maps. However, the latter also observed a few inversions between shared loci that might reveal chromosomal rearrangement events, such as translocations or inversions. More recently, the availability of a reference whole-genome sequence allowed the comparison of genetic and physical maps and globally confirmed the good conservation of marker order over the different species of *Citrus* and *Poncirus trifoliata* (Curtolo et al., 2017b; Shimada et al., 2014; Huang et al., 2018; Yu et al., 2017).

## 4.4 Non-Mendelian segregations are frequent and can mask the genetic determinism of phenotypic traits

Important proportions of distorted loci (significant deviation to Mendelian segregations) have been observed in many citrus genetic mapping studies (Luro et al., 1996; Oliveira et al., 2004; Bernet et al., 2010; Raga et al., 2012;

Ollitrault et al., 2012a). Ollitrault et al. (2012a) reported higher percentage of skewed loci in clementine when used as the male parent than when used as the female parent. Similar observation was made previously by Bernet et al. (2010) for Fortune mandarin. In these two studies, most segregation distortions affected the allele frequencies without disturbing the genotypic frequency equilibrium. Therefore, the authors concluded that the deviations resulted from gametic selection before fertilization rather than zygotic selection. Considering that upon cross-pollination with compatible parents the proportion of fertilized ovules is much greater than the proportion of successful male gametes, it is quite logical that gametic selection would be much more pronounced in male gametes than in female ones. This selection can result from several mechanisms such as gamete abortion, pollen competition or the citrus gametophytic incompatibility system. Similarly, Raga et al. (2012) reported a pre-zygotic selection in an F1 cross between 'Cleopatra' mandarin and Poncirus, while for the self-fertilization of the intergeneric hybrid (F2 population) they observed a post-zygotic selection favouring heterozygous genotypes. Considering the importance of non-Mendelian segregation in large parts of the genome, it appears difficult to infer the genetic control of phenotypic traits correctly from their segregations without concomitant marker segregation analysis.

## *4.5 Inheritance by diploid gametes*

The occurrence of unreduced (2n) gametes in citrus opens the way for alternative methods of genetic mapping based on half tetrad analysis (HTA; Mendiburu and Peloquin, 1979; Bastiaanssen et al., 1996). In particular they allow genetic mapping of centromeres (Park et al., 2007; Nie et al., 2012). Centromeres constitute a very important component of sexual reproduction and centromere mapping allows the development of improved linkage maps, the identification of chromosome arms and the analysis of crossover interference. Unreduced gametes mainly result from two mechanisms, FDR and SDR (Peloquin et al., 1989; Lim et al., 2004; Dewitte et al., 2012). A FDR 2n gamete contains non-sister chromatids, while a SDR 2n gamete contains two sister chromatids (Bretagnolle and Thompson, 1995; Tavoletti et al., 1996). FDR and SDR mechanisms result in opposite patterns of parental heterozygosity restitution (PHR) along the genome. However, for both mechanisms, PHR is a direct function of the genetic distance to the centromere (Tavoletti et al., 1996; Zhao and Speed, 1998). Therefore, molecular marker analysis of 2n gamete population is a powerful means to locate centromeres genetically (Kauffman et al., 1995; Park et al., 2007; Nie et al., 2012). Cuenca et al. (2011) proposed a new method to locate the positions of the centromeres in LGs from PHR patterns and located the chromosome II centromere of 'Fortune' mandarin. With the same method, Aleza et al. (2015) located the centromeres of the nine citrus chromosomes in the Clementine reference genetic map by SSR and SNP genotyping of a triploid progenies derived from SDR 2n ovules of clementine. The inferred physical locations of centromeres revealed one acrocentric, four metacentric and four sub-metacentric chromosomes. The identification of centromere positions was a useful way to confirm, using centromeric markers, in a large range of genotypes that SDR was the main mechanism of unreduced ovule formation (Cuenca et al., 2015; Rouiss et al., 2017b) while FDR was found predominant in the 2n pollen of a tangor (Rouiss et al., 2017a). Molecular marker inheritance studies were also performed to analyse the meiotic behaviour of tetraploid parents. They revealed segregation patterns related to the divergence of the genomes constituting the tetraploids under review. Preferential tetrasomic inheritance was found for a doubled-diploid clementine (Aleza et al., 2016) whose genome is principally derived from *C. Reticulata's* ancestor while a doubled-diploid 'Mexican lime' (interspecific *C. micrantha* × *C. medica* hybrid) displayed a preferentially disomic inheritance (Rouiss et al., 2018). The meiotic inheritance patterns were also evaluated in citrus allotetraploid somatic hybrids. Inheritance was found to be intermediary with preferential tetrasomic inheritance for 'Nova' tangelo + 'HB' pummelo (Xie et al., 2015) and 'Willow leaf' mandarin + Eureka lemon hybrid (Kamiri et al., 2011) while preferential disomic inheritance was described for an intergeneric 'Willow leaf' mandarin + 'Pomeroy' trifoliate orange hybrid (Kamiri et al., submitted). Such knowledge is essential to optimize triploid and tetraploid citrus breeding programmes because the meiotic mechanisms involved in the production of the diploid gamete strongly determine the inheritance of the genome fragment and therefore of the phenotypical traits. HTA analysis from triploid progenies also makes it possible to locate the resistance gene for *Alternaria alternata* (Cuenca et al., 2013b).

## 5 Recent development of genomic resources

### *5.1 Reference genome sequences of citrus species*

Very important progress has been achieved during the last 5 years for structural genomics with the release of reference whole-genome sequences. Interestingly, haploid and dihaploid citrus have been key resources for *de novo* whole-genome sequencing (WGS) projects. Indeed, heterozygosity in a whole-genome sequence complicates the assembly process, affecting quality, assembly contguity and completenesss (Vinson et al., 2005; Huang et al., 2017). Since commercial citrus varieties are characterized by high heterozygosity, the ICGC selected a homozygous clementine derived from induced gynogenesis to establish the citrus reference genome (Aleza et al., 2009; Germana et al., 2013). At the same time, a Chinese consortium produced a draft genome of sweet orange (Xu et al., 2013) from a doubled haploid line derived from the anther culture of Valencia sweet orange (Cao et al., 2011). More recently, a *de novo* assembly of *C. maxima* was performed with a pummelo haploid line (Wang et al., 2017).

Xu et al. (2013) were the first to release a whole-genome reference sequence for citrus from a dihaploid sweet orange. It was established from 785 million high-quality paired-end-tag sequencing reads from various DNA fragment sizes (~300 bp, 2, 10 and 20 kb) using the Illumina GAII sequencer. After assembly, the total contig sequence length (320.5 Mb) covered 87.3% of the estimated sweet orange genome. 20% of the contig sequence was composed of repetitive elements. Pseudomolecule assembly, with the help of the sweet orange genetic map, encompassed 239 Mb. A total of 29 445 protein-coding genes were predicted.

The ICGC produced a high-quality reference genome assembled in nine pseudomolecules from 4.6 million Sanger reads (including 469 000 fosmid-end and 73 000 BAC-end reads) of a haploid clementine (Wu et al., 2014). Ninety-six per cent of the assembly (301.4 Mb) were located in the nine chromosomes. Despite the compactness of the citrus genome, 45% was found repetitive. A total of 24 533 protein-coding gene loci were predicted with 33 929 transcripts.

Recently, a high-quality genome of a haploid pummelo was generated from a 56.8× coverage of single-molecule sequences on the PacBio RS II platform via a shotgun approach (Wang et al., 2017). Additionally, 370.3× Illumina sequencing data were used to correct sequencing errors and to fill in the gaps. A total of 117 scaffolds, accounting for 87% of the assembled genome (300.03 Mb), were anchored by a genetic map of 1563 molecular markers. A total of 30 123 protein-coding genes with 42 886 transcripts were identified. In addition, the same authors released the draft genomes of the Citrinae species *Atalantia buxifolia* and two citrus species *C. ichangensis* a wild citrus and *C. medica,* sequenced and assembled with Illumina shotgun sequencing reads.

## 5.2 Re-sequencing data and their exploitation for phylogenomy

Beside *de novo* sequencing, a large number of *Citrus* species and varieties were re-sequenced principally from shotgun sequence pair-ends reads in Illumina platforms. The principal re-sequencing release and application results are discussed below (Table 2). In addition to the sweet orange genome, Xu et al. (2013) re-sequenced three pummelos and three mandarins. Wu et al. (2014) re-sequenced the genomes of four mandarins (including Clementine), two pummelos, one sour orange and one sweet orange. Mikeal Roose and Yoko Eck released in 2016 the raw sequences of 30 *Citrus* (including seven mandarins, six pummelos, four citrons, two papedas, two grapefruits, two lemons and one lime) and related genera accessions (*Poncirus, Fortunella, Eremocitrus* and *Microcitrus*) used to establish two Affymetrix Axion SNP arrays

**Table 2** Re-sequencing data in public NCBI databases (https://www.ncbi.nlm.nih.gov/)

| | Xu et al. (2013) | Wu et al. (2014) | Wang et al. (2017) | Eck et al. (2016) | Wu et al. (2018) |
|---|---|---|---|---|---|
| *Total* | *7* | *8* | *100* | *30* | *30* |
| Pummelo | 3 | 2 | 15 | 6 | |
| Mandarin | 3 | 4 | 22 | 7 | 9 |
| Citron | | | 9 | 4 | 4 |
| Papeda | | | 12 | 2 | 2 |
| Sweet orange | 1 | 1 | 6 | | |
| Sour orange | | 1 | 10 | | |
| Grapefruit | | | 3 | 2 | 1 |
| Lemon | | | 8 | 2 | 3 |
| Lime | | | | 1 | 1 |
| Other citrus | | | | | 1 |
| Poncirus | | | | 2 | 1 |
| Fortunella | | | | 1 | 1 |
| Microcitrus | | | | 2 | 4 |
| Eremocitrus | | | | 1 | 2 |
| Other Citreae | | | 15 | | 1 |
| BioProjects | PRJNA86123 | PRJNA225969; PRJNA225967; PRJNA225965; PRJNA225968; PRJNA225835; PRJNA225964; PRJNA225966; PRJNA225963 | PRJNA320985; PRJNA318855; PRJNA321100; PRJNA320023; PRJNA321657; PRJNA327148 | PRJNA357637 | PRJNA414519 |
| SRA | NA | SRP032500; SRP032496; SRP032491; SRP032497; SRP032460; SRP032490; SRP032492; SRP032489 | SRP073648; SRP077805; SRP077806; SRP077845; SRP078015; SRP078015; SRP078016; SRP078382; SRP080384 | SRP095606 | SRP120362 |

BioProject: reference of the project at NCBI; SRA: Sequence Read Archive at NCBI.

with about 1.5 million and 56 000 SNPs (Eck et al., 2016). Part of these data were also used by Wu et al. (2018), combined with previous data and new sequences for 30 accessions (nine mandarins, four citrons, two papedas, one grapefruit, three lemons, one lime, one calamondin, one *Poncirus*, one *Fortunella,* two *Eremocitrus*, four *Microcitrus* and one *Atalantia buxifolia*) to study the evolution, domestication and admixture structures of citrus. They proposed two main periods of radiation for the 'true citrus'. A first one during the late Miocene epoch with a rapid Southeast Asian migration and a second radiation in the early Pliocene epoch across the Wallace line giving rise to the Australian citrus. The analysis of the admixed genome structures based on diagnostic SNPs (DSNPs) of the ancestral taxa provided fundamental knowledge on the complex phylogenomic mosaic structures of modern horticultural groups and cultivars. They confirmed that some horticultural types as sour orange, rough lemon, Rangpur lime, calamondin and Mexican lime derived from direct hybridization between two ancestral taxa. Other horticultural groups, such as sweet oranges and grapefruit that display homozygosity for one species in addition to interspecific heterozygosity or lemons with the contribution of three ancestral taxa, have more complex history, following the previous findings from molecular marker analysis (Nicolosi et al., 2000; Garcia-Lor et al., 2013a; Ollitrault et al., 2012b; Curk et al., 2016). Cultivated mandarins and sweet orange displayed important networks of relatedness. Moreover, the generalized introgression of *C. maxima* in cultivated mandarins confirmed the observation from 454 haplotyping and markers studies (Curk et al., 2015) and GBS analysis (Oueslati et al., 2017), suggesting that these introgression contributed to the domestication of palatable mandarins. Oueslati et al. (2017) and Ahmed et al. (2019) adapted the DSNPs approach to decipher complex interspecific genomic structure from GBS data, respectively, in the *C. reticulta/C. maxima* admixture gene pool and the whole cultivated citrus germplasms taking into account the four ancestral taxa. These developments from GBS open the way to the analysis of interspecific recombination in large populations at affordable costs.

Wang et al. (2017) released a set of re-sequencing data including 85 from *Citrus* genus (22 mandarins, 6 sweet oranges, 10 sour oranges, 15 pummelos, 3 grapefruit, 9 citrons, 8 lemons and 12 papedas) and 15 from *Atalantia* genus. Comparative population analysis revealed genomic regions with energy- and reproduction-associated genes probably under selection in cultivated citrus. They also identified a genomic area of 80-kb region containing 11 candidate genes associated with apomixes in mandarin. Apomixis could result from the insertion of a miniature inverted-repeat transposable element (MITE) in the promoter region of one of these candidate genes; the CitRWP that encodes a RWP-RK domain-containing protein similar to the Arabidopsis RKD family of proteins, implied in the regulation of egg cells (Koszegi et al., 2011).

## 6 Linking phenotypic and molecular variability and MAS

### *6.1 Identification of zygotic/nucellar seedlings*

The first concrete application of molecular markers in citrus breeding was the identification of zygotic and nucellar seedlings in preferential apomictic female line progenies. Isozymes (Torres et al., 1982), SSRs (Ruiz et al., 2000; Rao et al., 2008; de Oliveira et al., 2002) and SNPs (Ollitrault et al., 2016; Bruyère et al., 2016) were and are still routinely used to control the conformity of rootstock seedling used for research trials and rootstock agronomic evaluation and to select hybrids in breeding programmes using preferential apomictic line as female parents. For zygotic eliminations in rootstock trials, the selection is generally based on heterozygous markers in the maternal line and considering the most defavourable situation for detection (self-fertilization) the use of six independent loci allow identifying more than 96% ($1-0.5^6$) of the zygotic plantlets. For the selection of controlled hybrids in preferentially apomictic female lines, one co-dominant marker totally differentiating the two parents should be enough.

### *6.2 The quest of the CTV immunity from* **Poncirus trifoliata**

Many of the mapping studies focussed on the resistance to CTV including fine mapping of the targeted genomic area (Gmitter et al., 1996; Deng et al., 1997; Mestre et al., 1997; Fang et al., 1998a; Cristofani et al., 1999; Bernet et al., 2004a; Asins et al., 2004, 2012; Fang and Roose, 1999). A set of SSRs markers identified in a BAC clone containing the immunity gene (Yang et al., 2003) from *Poncirus trifoliata* was successfully developed and used (M. Roose, pers. comm.) for the molecular selection of the resistant alleles, in intergeneric progenies. More recently, Otha et al. (2014) developed and applied a graphical genotyping approach in a pseudo-backcrossed F2 (BC′ 2) progeny to introduce CTV resistance from *Poncirus trifoliata* (L.) Raf. into *Citrus* by introgression breeding.

### *6.3 Concrete MAS application for monogenic-controlled traits and major genes*

Besides the markers for the CTV immunity, very efficient SNP markers for the recessive monogenic resistance to *Alternaria alternata* (fungal disease) of mandarins were also developed following the identification of a 366 Kb genomic area from bulk segregant and HTA analysis from triploid progenies (Cuenca et al., 2013b) and fine mapping in diploid populations (Cuenca et al., 2016). They are routinely used in the framework of the Spanish (Pablo Aleza, pers. comm.) and French (Yann Froelicher, pers. comm.) triploid mandarin breeding programmes. Efficient SAM has also been developed and routinely applied for two other traits with mono-locus determinant resulting from transposable elements insertion. Butelli et al. (2012) revealed that the high anthocyanin content of blood orange was associated with the insertion of a Copia-like retrotransposon adjacent to a gene encoding Ruby, a MYB transcriptional activator of anthocyanin production. Sequence diversity analysis in the vicinity of the Ruby gene indicated that most of the natural variation of anthocyanins in *Citrus* species can be explained by differences in the activity of the *Ruby* gene, caused by point mutations and deletions and insertions of transposable elements (Butelli et al., 2017) and favoured the development of selection markers. The markers developed by Wang et al. (2017) to analyse the role of the MITE insertion associated with apomixes in mandarins can also be usefully applied to select monoembryonic female parents (non-apomictic) in the framework of pre-breeding activities. SAM was also developed for a major gene locus 'Tyr1' controlling citrus nematode resistance from trifoliate orange (Xiang et al., 2010). One marker (4L17R/CfoI) displayed full match with resistance to citrus nematode when the USDA hybrid rootstocks were screened.

## 6.4 QTL studies

For multilocus-controlled traits, despite consequent genomic and genetic developments during the last years, the identification of QTLs related with useful traits and mostly the concrete application of MAS for phenotypic traits remains limited in citrus breeding. Early QTL analysis concerned cold tolerance (Cai et al., 1994; Weber et al., 2003), salt tolerance, (Tozlu et al., 2002) phytophthora resistance (Siviero et al., 2006), fruit acidity (Fang et al., 1997a), apomixes (Garcia et al., 1999), leaf miner susceptibility (Bernet et al., 2005), nematode resistance (Ling et al., 2000) and morphological traits (Cevik and Moore, 2012). However, most of these studies suffered from insufficient numbers of progenies and markers as well as the frequent use of anonymous dominant markers and did not result in concrete tools for MAS. Improved marker molecular studies were more recently performed for morphological and fruit quality traits (Asins et al., 2015; Yu et al., 2016), aroma volatile contents (Yu et al., 2017), carotenoid contents and gene of the carotenoid pathway expression (Sugiyama et al., 2011, 2014), freeze tolerance (Hong et al., 2015) and salt tolerance (Raga et al., 2016). A quantitative break in the density of marker was made with the adoption of genotyping methods derived from NGS such as RADSeq, GBS and DARTSeq. Curtolo et al. (2017a) performed a QTL study of fruit quality traits based on DARTSeq genotyping. QTL and eQTL mapping related to citrandarins' resistance to citrus gummosis disease (Lima et al., 2018) was based on a saturated genetic map generated with DARTSeq (Curtolo et al., 2017b). Saturated genetic maps based on GBS genotyping were also recently used to perform QTL analysis of the HLB tolerance identified in *Poncirus trifoliata* (Huang et al., 2018). Today the citrus genome database (https://www.citrusgenomedb.org/) has an inventory of 579 published QTLs.

### 6.5 GWAS and GS promises

Due to the high structuration of the citrus gene pool, GWAS remain very limited in citrus. From WGS data and in relation with *C. maxima* introgression, Wu et al. (2018) identified several candidate genes, and particularly a gene coding for mitochondrial $NAD^+$ isocitrate dehydrogenase ($NAD^+$ IDH) that regulates citric acid synthesis, potentially linked with palatability in mandarin. However, their varietal set was limited. A very large GWAS for fruit quality traits based on germplasm (111 varieties) and hybrids from different crosses (676 individuals from 35 full-sib families) involving the *C. reticulata/C. maxima* gene pool, genotyped with two SNP arrays (768 or 1536 SNPs) was performed by Minamikawa et al. (2017). It resulted for the first time for *Citrus* in GS models. Imai et al. (2018) successfully analysed 110 citrus accessions, including landraces, modern cultivars and breeding lines, for eight fruit quality traits with GBS data and revealed seven QTLs, including four novel ones, opening the way for concrete application of GS with affordable genotyping method.

## 7 Future trends and conclusion

During the last 10 years, the rapid evolution of genomic tools and their application to citrus has led to the production of crucial genomic and genetic resources and particularly de novo genome reference sequences and increasing number of re-sequencing data. It has resulted in a new understanding of citrus evolution and contributed to elucidating the molecular determinants of agronomically important traits. The progress regarding molecular marker developments, particularly SNP markers, MAS made it possible to establish saturated genetic maps and opened the way to carry out QTLs and association genetics studies. New approaches based on competitive allele-specific PCR or sequencing of reduced genome representation (GBS, RADSeq) were developed to optimize polyploid citrus genetics. In breeding programmes, MAS is already applied to traits such as CTV and *Alternaria alternata* resistance or fruit anthocyanin synthesis. Development of selection models has already been initiated and should assume increasing importance for multi-trait breeding projects. In the future, the implementation of high-throughput phenotyping methods will be key for GS development. In addition to genotypic polymorphisms, phenotype and breeding value models should usefully include biological traits such as metabolomics data. Systematic and integrative approaches of systems biology should help to incorporate multiple genomic data and assemble the information towards cellular or physiological behaviours related to a particular phenotype. So-called systems genomic or genomic systems biology will pave the way to decipher the complexity of quantitative traits by assuming models that are biologically more explicit.

## 8 Where to look for further information

At the international level, the main research groups involved in citrus genetics and genomic include:

In Europe:

- Instituto Valenciano de Investigacion Agrarias (IVIA, Spain): 'Centre de genomica': http://www.ivia.gva.es/va/centro-de-genomica and 'Centre de Citricultura i Producció Vegetal': http://www.ivia.gva.es/va/centro-de-citricultura-y-produccion-vegetal
- Agricultural Research for Development (CIRAD, AGAP research unit, France): SEAPAG team: https://umr-agap.cirad.fr/en/equipes-scientifiques/evolution-et-amelioration-des-agrumes/contexte-et-enjeux
- Consiglio per la ricerca in agricoltura e l'analisi dell'economia agraria (CREA-Italy): Research Centre for Olive, Citrus and Tree Fruit (https://www.crea.gov.it/en).

In the Americas:

- University of Florida (USA): UF/IFAS Citrus Research and Education Center: https://crec.ifas.ufl.edu/
- University of California, Riverside (USA): department of botany and plant sciences: https://plantbiology.ucr.edu/research/ggb.html
- Instituto Agronomico de Campinas (IAC; Brazil): 'Centro de citricultura Sylvio Moreira': http://www.iac.sp.gov.br/areasdepesquisa/citros/.

In Asia:

- Huazhong Agricultural University (HZAU: China), Institute of Citrus Science (http://ics.hzau.edu.cn/ics/en/)
- National agricultural and food research organization (NARO, Japan), Institute of Fruit Tree and Tea Science (http://www.naro.affrc.go.jp/english/laboratory/nifts/index.html).

In addition to NCBI (https://www.ncbi.nlm.nih.gov/), important databases for citrus genomics data can be found at:

- https://www.citrusgenomedb.org/
- http://citrus.hzau.edu.cn/orange/
- https://phytozome.jgi.doe.gov/pz/portal.html

Two books including the most current information on citrus genomic will be released in October 2019:

- Gentile, A., La Malfa, S. and Deng, Z. (Eds). 2019. *The Citrus Genome* (1st edn.). Compendium of Plant Genomes series. Springer Publisher, 320p. ISBN-10: 303010799X.
- Talon, M., Caruso, M. and Gmitter, F. G. 2019. *The Genus Citrus* (1st edn.). Woodhead Publishing and Elsevier, 720p. ISBN-10: 0128121637.

## 9 References

Ahmed, D., Comte, A., Curk, F., Costantino, G., Vignes, H., Luro, F., Froelicher, Y., Dereeper, A., Mournet, P. and Ollitrault, P. 2019. A pipe-line for phylogenomic inference from genotyping by sequencing data in large diploid and polyploid population; an application to *Citrus* germplasm. *Annals of Botany* mcz029, https://doi.org/10.1093/aob/mcz029.

Aleza, P., Juarez, J., Hernandez, M., Pina, J. A., Ollitrault, P. and Navarro, L. 2009. Recovery and characterization of a *Citrus clementina* Hort. ex Tan. 'Clemenules' haploid plant selected to establish the reference whole *Citrus* genome sequence. *BMC Plant Biology* 9, 110. doi:10.1186/1471-2229-9-110.

Aleza, P., Juarez, J., Cuenca, J., Ollitrault, P. and Navarro, L. 2010. Recovery of citrus triploid hybrids by embryo rescue and flow cytometry from 2x × 2x sexual hybridisation and its application to extensive breeding programs. *Plant Cell Reports* 29(9), 1023–34. doi:10.1007/s00299-010-0888-7.

Aleza, P., Froelicher, Y., Schwarz, S., Agusti, M., Hernandez, M., Juarez, J., Luro, F., Morillon, R., Navarro, L. and Ollitrault, P. 2011. Tetraploidization events by chromosome doubling of nucellar cells are frequent in apomictic citrus and are dependent on genotype and environment. *Annals of Botany* 108(1), 37–50. doi:10.1093/aob/mcr099.

Aleza, P., Juarez, J., Hernandez, M., Ollitrault, P. and Navarro, L. 2012a. Implementation of extensive citrus triploid breeding programs based on 4x × 2x sexual hybridisations. *Tree Genetics and Genomes* 8(6), 1293–306. doi:10.1007/s11295-012-0515-6.

Aleza, P., Juarez, J., Cuenca, J., Ollitrault, P. and Navarro, L. 2012b. Extensive citrus triploid hybrid production by 2x × 4x sexual hybridizations and parent-effect on the length of the juvenile phase. *Plant Cell Reports* 31(9), 1723–35. doi:10.1007/s00299-012-1286-0.

Aleza, P., Cuenca, J., Hernandez, M., Juarez, J., Navarro, L. and Ollitrault, P. 2015. Genetic mapping of centromeres in the nine *Citrus clementina* chromosomes using half-tetrad analysis and recombination patterns in unreduced and haploid gametes. *BMC Plant Biology* 15, 80. doi:10.1186/s12870-015-0464-y.

Aleza, P., Cuenca, J., Juarez, J., Navarro, L. and Ollitrault, P. 2016. Inheritance in doubled-diploid clementine and comparative study with SDR unreduced gametes of diploid clementine. *Plant Cell Reports* 35(8), 1573–86. doi:10.1007/s00299-016-1972-4.

Asins, M. J., Bernet, G. P., Ruiz, C., Cambra, M., Guerri, J. and Carbonell, E. A. 2004. QTL analysis of citrus tristeza virus-citradia interaction. *TAG. Theoretical and Applied Genetics. Theoretische und Angewandte Genetik* 108(4), 603–11. doi:10.1007/s00122-003-1486-7.

Asins, M. J., Fernández-Ribacoba, J., Bernet, G. P., Gadea, J., Cambra, M., Gorris, M. T. and Carbonell, E. A. 2012. The position of the major QTL for *Citrus tristeza virus* resistance is conserved among *Citrus grandis*, *C. aurantium* and *Poncirus trifoliate*. *Molecular Breeding* 29(3), 575–87. doi:10.1007/s11032-011-9574-x.

Asins, M. J., Raga, V., Bernet, G. P. and Carbonell, E. A. 2015. Genetic analysis of reproductive, vegetative and fruit quality traits to improve *Citrus* varieties. *Tree Genetics and Genomes* 11(6), 117. doi:10.1007/s11295-015-0949-8.

Barkley, N. A., Roose, M. L., Krueger, R. R. and Federici, C. T. 2006. Assessing genetic diversity and population structure in a citrus germplasm collection utilizing simple sequence repeat markers (SSRs). *TAG. Theoretical and Applied Genetics. Theoretische und Angewandte Genetik* 112(8), 1519–31. doi:10.1007/s00122-006-0255-9.

Barrett, H. C. and Rhodes, A. M. 1976. A numerical taxonomic study of affinity relationships in cultivated *Citrus* and its close relatives. *Systematic Botany* 1(2), 105–36. doi:10.2307/2418763.

Bassene, J. B., Froelicher, Y., Dhuique-Mayer, C., Mouhaya, W., Ferrer, R. M., Ancillo, G., Morillon, R., Navarro, L. and Ollitrault, P. 2009. Non-additive phenotypic and transcriptomic inheritance in a citrus allotetraploid somatic hybrid between *C. reticulata* and *C. limon*: the case of pulp carotenoid biosynthesis pathway. *Plant Cell Reports* 28(11), 1689–97. doi:10.1007/s00299-009-0768-1.

Bastiaanssen, H. J. M., Ramanna, M. S., Sawor, Z., Mincione, A., Steen, A. V. and Jacobsen, E. 1996. Pollen markers for gene-centromere mapping in diploid potato. *TAG. Theoretical and Applied Genetics. Theoretische und Angewandte Genetik* 93(7), 1040–7. doi:10.1007/BF00230122.

Baxter, S. W., Davey, J. W., Johnston, J. S., Shelton, A. M., Heckel, D. G., Jiggins, C. D. and Blaxter, M. L. 2011. Linkage mapping and comparative genomics using next-generation RAD sequencing of a non-model organism. *PLOS ONE* 6(4), e19315. doi:10.1371/journal.pone.0019315.

Bayer, R. J., Mabberley, D. J., Morton, C., Miller, C. H., Sharma, I. K., Pfeil, B. E., Rich, S., Hitchcock, R. and Sykes, S. 2009. A molecular phylogeny of the orange subfamily (Rutaceae: Aurantioideae) using nine cpDNA sequences. *American Journal of Botany* 96(3), 668–85. doi:10.3732/ajb.0800341.

Bernet, G. P. and Asins, M. J. 2003. Identification and genomic distribution of *gypsy* like retrotransposons in *Citrus* and *Poncirus*. *TAG. Theoretical and Applied Genetics. Theoretische und Angewandte Genetik* 108(1), 121–30. doi:10.1007/s00122-003-1382-1.

Bernet, G. P., Breto, M. P. and Asins, M. J. 2004a. Expressed sequence enrichment for candidate gene analysis of *Citrus tristeza virus* resistance. *TAG. Theoretical and Applied Genetics. Theoretische und Angewandte Genetik* 108(4), 592–602. doi:10.1007/s00122-003-1479-6.

Bernet, G. P., Mestre, P. F., Pina, J. A. and Asins, M. J. 2004b. Molecular discrimination of lemon cultivars. *HortScience* 39(1), 165–9. doi:10.21273/HORTSCI.39.1.165.

Bernet, G. P., Margaix, C., Jacas, J., Carbonell, E. A. and Asins, M. J. 2005. Genetic analysis of citrus leafminer susceptibility. *TAG. Theoretical and Applied Genetics. Theoretische und Angewandte Genetik* 110(8), 1393–400. doi:10.1007/s00122-005-1943-6.

Bernet, G. P., Fernandez-Ribacoba, J., Carbonell, E. A. and Asins, M. J. 2010. Comparative genome-wide segregation analysis and map construction using a reciprocal cross design to facilitate citrus germplasm utilization. *Molecular Breeding* 25(4), 659–73. doi:10.1007/s11032-009-9363-y.

Biswas, M. K., Xu, Q. and Deng, X. 2010a. Utility of RAPD, ISSR, IRAP and REMAP markers for the genetic analysis of *Citrus* spp. *Scientia Horticulturae* 124(2), 254–61. doi:10.1016/j.scienta.2009.12.013.

Biswas, M. K., Baig, M. N. R., Cheng, Y. and Deng, X. 2010b. Retro-transposon based genetic similarity within the genus *Citrus* and its relatives. *Genetic Resources and Crop Evolution* 57(7), 963–72. doi:10.1007/s10722-010-9533-0.

Bretagnolle, F. and Thompson, J. D. 1995. Gametes with the somatic chromosome number: mechanisms of their formation and role in the evolution of autopolyploid plants. *New Phytologist* 129(1), 1–22. doi:10.1111/j.1469-8137.1995.tb03005.x.

Breto, M. P., Ruiz, C., Pina, J. A. and Asins, M. J. 2001. The diversification of *Citrus clementina* Hort. ex Tan., a vegetatively propagated crop species. *Molecular Phylogenetics and Evolution* 21(2), 285–93. doi:10.1006/mpev.2001.1008.

Bruyere, S., Luro, F., Froelicher, Y., Morillon, R. and Ollitrault, P. 2016. Poncirus phylogenetic diagnostic SNPs markers are useful to analyse zygotic rates in diploid and tetraploid *Citrus* × *Poncirus* rootstock seedlings. In: *Abstract Book Sustainable Citriculture: the Role of Applied Knowledge. International Citrus Congress*. IAPAR, Londrina, p. 125.

Butelli, E., Licciardello, C., Zhang, Y., Liu, J., Mackay, S., Bailey, P., Reforgiato-Recupero, G. and Martin, C. 2012. Retrotransposons control fruit-specific, cold-dependent accumulation of anthocyanins in blood oranges. *The Plant Cell* 24(3), 1242–55. doi:10.1105/tpc.111.095232.

Butelli, E., Garcia-Lor, A., Licciardello, C., Las Casas, G., Hill, L., Reforgiato Recupero, G. R., Keremane, M. L., Ramadugu, C., Krueger, R., Xu, Q., et al. 2017. Changes in anthocyanin production during domestication of *Citrus*. *Plant Physiology* 173(4), 2225–42. doi:10.1104/pp.16.01701.

Cai, Q., Guy, C. L. and Moore, G. 1994. Extension of the genetic linkage map in *Citrus* using random amplified polymorphic DNA (RAPD) markers and RFLP mapping of cold-acclimation responsive loci. *Theoretical and Applied Genetics* 89, 606–14.

Cao, H., Biswas, M. K., Lu, Y., Amar, M. H., Tong, Z., Xu, Q., Xu, J., Guo, W. and Deng, X. 2011. Doubled haploid callus lines of Valencia sweet orange recovered from another culture. *Plant Cell, Tissue and Organ Culture* 104(3), 415–23. doi:10.1007/s11240-010-9860-z.

Carbonell-Caballero, J., Alonso, R., Ibanez, V., Terol, J., Talon, M. and Dopazo, J. 2015. A phylogenetic analysis of 34 chloroplast genomes elucidates the relationships between wild and domestic species within the genus *Citrus*. *Molecular Biology and Evolution* 32(8), 2015–35. doi:10.1093/molbev/msv082.

Cevik, M. and Moore, G. A. 2012. Quantitative trait loci analysis of morphological traits in *Citrus*. *Plant Biotechnology Reports* 6(1), 47–57. doi:10.1007/s11816-011-0194-z.

Chen, C. and Gmitter Jr., F. G. 2013. Mining of haplotype-based expressed sequence tag single nucleotide polymorphisms in citrus. *BMC Genomics* 14, 746. doi:10.1186/1471-2164-14-746.

Chen, C. X., Zhou, P., Choi, Y. A., Huang, S. and Gmitter Jr., F. G. 2006. Mining and characterizing microsatellites from citrus ESTs. *TAG. Theoretical and Applied Genetics. Theoretische und Angewandte Genetik* 112(7), 1248–57. doi:10.1007/s00122-006-0226-1.

Chen, C. X., Bowman, K. D., Choi, Y. A., Dang, P. M., Rao, M. N., Huang, S., Soneji, J. R., McCollum, T. G. and Gmitter, F. G., Jr. 2007. EST-SSR genetic maps for *Citrus sinensis* and *Poncirus trifoliata*. *Tree Genetics and Genomes* 4(1), 1–10. doi:10.1007/s11295-007-0083-3.

Chen, C. X., Grosser, J. W., Calovic, M., Serrano, P., Pasquali, G., Gmitter, J. and Gmitter, F. G.,Jr. 2008a. Verification of mandarin and pummelo somatic hybrids by expressed sequence tag-simple sequence repeat marker analysis. *Journal of the American Society for Horticultural Science* 133(6), 794–800. doi:10.21273/JASHS.133.6.794.

Chen, C. X., Lyon, M. T., O'Malley, D., Federici, C. T., Gmitter, J., Grosser, J. W., Chaparro, J. X., Roose, M. L. and Gmitter Jr., F. G. 2008b. Origin and frequency of *2n* gametes in *Citrus sinensis* × *Poncirus trifoliata* and their reciprocal crosses. *Plant Science* 174(1), 1–8. doi:10.1016/j.plantsci.2007.08.005.

Choi, Y. A., Chen, C. X., Huang, S. and Gmitter, F. G.,Jr. 2007. Identification of AFLP markers associated with *Citrus* Canker resistance within *Citrus*. *Acta Horticulturae* 738(738), 201–6. doi:10.17660/ActaHortic.2007.738.18.

Cimen, B. and Yesiloglu, T. 2016. Rootstock breeding for abiotic stress tolerance in *Citrus*. In: Shanker, A. K., Shanker, C. (Eds), *Abiotic and Biotic Stress in Plants – Recent Advances and Future Perspectives*. InTech, Rijeka. Chapter 23.

Corazza-Nunes, M. J., Machado, M. A., Nunes, W. M. C., Cristofani, M. and Targon, M. L. P. N. 2002. Assessment of genetic variability in grapefruits (*Citrus paradisi* Macf.) and pummelos (*C. maxima* (Burm.) Merr.) using RAPD and SSR markers. *Euphytica* 126(2), 169–76. doi:10.1023/A:1016332030738.

Cristofani, M., Machado, M. A. and Grattapaglia, D. 1999. Genetic linkage maps of *Citrus sunki* Hort. ex. and *Poncirus trifoliata* (L.) Raf. and mapping of *citrus tristeza virus* resistance gene. *Euphytica* 109(1), 25–32. doi:10.1023/A:1003637116745.

Cuenca, J., Froelicher, Y., Aleza, P., Juarez, J., Navarro, L. and Ollitrault, P. 2011. Multilocus half-tetrad analysis and centromere mapping in citrus: evidence of SDR mechanism for 2n megagametophyte production and partial chiasma interference in mandarin cv 'Fortune'. *Heredity* 107(5), 462–70. doi:10.1038/hdy.2011.33.

Cuenca, J., Aleza, P., Navarro, L. and Ollitrault, P. 2013a. Assignment of SNP allelic configuration in polyploids using competitive allele-specific PCR: application to citrus triploid progeny. *Annals of Botany* 111(4), 731–42. doi:10.1093/aob/mct032.

Cuenca, J., Aleza, P., Vicent, A., Brunel, D., Ollitrault, P. and Navarro, L. 2013b. Genetically based location from triploid populations and gene ontology of a 3.3-mb genome region linked to alternaria brown spot resistance in citrus reveal clusters of resistance genes. *PLOS ONE* 8(10), e76755. doi:10.1371/journal.pone.0076755.

Cuenca, J., Aleza, P., Juarez, J., Garcia-Lor, A., Froelicher, Y., Navarro, L. and Ollitrault, P. 2015. Maximum-likelihood method identifies meiotic restitution mechanism from heterozygosity transmission of centromeric loci: application in citrus. *Scientific Reports* 5, 9897. doi:10.1038/srep09897.

Cuenca, J., Aleza, P., Garcia-Lor, A., Ollitrault, P. and Navarro, L. 2016. Fine mapping for identification of citrus alternaria brown spot candidate resistance genes and development of new SNP markers for marker-assisted selection. *Frontiers in Plant Science* 7, 1948. doi:10.3389/fpls.2016.01948.

Curk, F., Ancillo, G., Ollitrault, F., Perrier, X., Jacquemoud-Collet, J. P., Garcia-Lor, A., Navarro, L. and Ollitrault, P. 2015. Nuclear species-diagnostic SNP markers mined from 454 amplicon sequencing reveal admixture genomic structure of modern citrus varieties. *PLOS ONE* 10(5). doi:10.1371/journal.pone.0125628.

Curk, F., Ollitrault, F., Garcia-Lor, A., Luro, F., Navarro, L. and Ollitrault, P. 2016. Phylogenetic origin of limes and lemons revealed by cytoplasmic and nuclear markers. *Annals of Botany* 117(4), 565–83. doi:10.1093/aob/mcw005.

Curtolo, M., Cristofani-Yaly, M., Gazaffi, R., Takita, M. A., Figueira, A. and Machado, M. A. 2017a. QTL mapping for fruit quality in *Citrus* using DArTseq markers. *BMC Genomics* 18(1), 289. doi:10.1186/s12864-017-3629-2.

Curtolo, M., Soratto, T. A. T., Gazaffi, R., Takita, M. A., Machado, M. A. and Cristofani-Yaly, M. 2017b. High-density linkage maps for *Citrus sunki* and *Poncirus trifoliata* using DArTseq markers. *Tree Genetics and Genomes* 14(1), 5. doi:10.1007/s11295-017-1218-9.

De Jong, W. S., De Jong, D. M. and Bodis, M. 2003. A fluorogenic 5′ nuclease (TaqMan) assay to assess dosage of a marker tightly linked to red skin color in autotetraploid potato. *TAG. Theoretical and Applied Genetics. Theoretische und Angewandte Genetik* 107(8), 1384–90. doi:10.1007/s00122-003-1420-z.

de Oliveira, C. A., Novac Garcia, A., Cristofani, M. and Machado, M. A. 2002. Identification of citrus hybrids through the combination of leaf apex morphology and SSR markers. *Euphytica* 128(3), 397–403. doi:10.1023/A:1021223309212.

de Oliveira, C. A., Bastianel, M., Cristofani-Yaly, M., Morais do Amaral, A. and Machado, M. A. 2007. Development of genetic maps of the citrus varieties 'Murcott' tangor and 'Pera' sweet orange by using fluorescent AFLP markers. *Journal of Applied Genetics* 48(3), 219–31. doi:10.1007/BF03195216.

de Simone, M., Russo, M. P., Puelo, G., Marsan, P. A., Lorenzoni, C., Marocco, A. and Recupero, G. R. 1998. Construction of genetic maps for *Citrus aurantium* and *C. latipes* based on AFLP, RAPD and RFLP markers. *Fruits* 53, 383–90.

Deng, Z., Xiao, S., Huang, S. and Gmitter, F. G. 1997. Development and characterization of SCAR markers linked to the *citrus tristeza virus* resistance gene from *Poncirus trifoliata*. *Genome* 40(5), 697–704. doi:10.1139/g97-792.

Dewitte, A., Huylenbroeck, J. V. and Laere, K. V. 2012. Use of 2n gametes in plant breeding. In: Abdurakhmonov, I. (Ed.), *Plant Breeding*. InTech, Rijeka.

Dorji, K. and Yapwattanaphun, C. 2015. Assessment of the genetic variability among mandarin (*Citrus reticulata* Blanco) accessions in Bhutan using AFLP markers. *BMC Genetics* 16, 39. doi:10.1186/s12863-015-0198-8.

Dugrand-Judek, A., Olry, A., Hehn, A., Costantino, G., Ollitrault, P., Froelicher, Y. and Bourgaud, F. 2015. The distribution of coumarins and Furanocoumarins in *Citrus* species closely matches citrus phylogeny and reflects the organization of biosynthetic pathways. *PLoS ONE* 10(11), e0142757. doi:10.1371/journal.pone.0142757.

Durham, R. E., Liou, P. C., Gmitter, F. G., Jr. and Moore, G. A. 1992. Linkage of restriction fragment length polymorphisms and isozymes in *Citrus*. *TAG. Theoretical and Applied Genetics. Theoretische und Angewandte Genetik* 84(1-2), 39–48. doi:10.1007/BF00223979.

Eck, Y., Moragues, M., Ferrante, S. and Roose, M. 2016. Development and application of affymetrix SNP arrays for citrus. In: *Book of Abstract; Sustainable Citriculture, The Role of Applied Knowledge. 13 th International Citrus Congress*, Foz do Iguaçu, Brazil.

Elshire, R. J., Glaubitz, J. C., Sun, Q., Poland, J. A., Kawamoto, K., Buckler, E. S. and Mitchell, S. E. 2011. A robust, simple genotyping-by-sequencing (GBS) approach for high diversity species. *PLoS ONE* 6((5)), e19379. doi:10.1371/journal.pone.0019379.

Esselink, G. D., Nybom, H. and Vosman, B. 2004. Assignment of allelic configuration in polyploids using the MAC-PR (microsatellite DNA allele counting–peak ratios) method. *TAG. Theoretical and Applied Genetics. Theoretische und Angewandte Genetik* 109(2), 402–8. doi:10.1007/s00122-004-1645-5.

Fanciullino, A. L., Dhuique-Mayer, C., Luro, F., Casanova, J., Morillon, R. and Ollitrault, P. 2006. Carotenoid diversity in cultivated citrus is highly influenced by genetic factors. *Journal of Agricultural and Food Chemistry* 54(12), 4397–406. doi:10.1021/jf0526644.

Fang, D. Q. and Roose, M. L. 1997. Identification of closely related citrus cultivars with inter-simple sequence repeat markers. *Theoretical and Applied Genetics* 95(3), 408–17. doi:10.1007/s001220050577.

Fang, D. Q. and Roose, M. L. 1999. A novel gene conferring *citrus tristeza virus* resistance in *Citrus maxima* (Burm.) Merrill. *HortScience* 34(2), 334–5. doi:10.21273/HORTSCI.34.2.334.

Fang, D. Q., Federici, C. T. and Roose, M. L. 1997a. Development of molecular markers linked to a gene controlling fruit acidity in citrus. *Genome* 40(6), 841-9. doi:10.1139/g97-809.

Fang, D. Q., Roose, M. L., Krueger, R. R. and Federici, C. T. 1997b. Fingerprinting trifoliate orange germ plasm accessions with isozymes, RFLPs, and inter-simple sequence repeat markers. *Theoretical and Applied Genetics* 95(1-2), 211-9. doi:10.1007/s001220050550.

Fang, D. Q., Federici, C. T. and Roose, M. L. 1998a. A high-resolution linkage map of the *citrus tristeza virus* resistance gene region in *Poncirus trifoliata* (L.) Raf. *Genetics* 150(2), 883–90.

Fang, D. Q., Krueger, R. R. and Roose, M. L. 1998b. Phylogenetic relationships among selected *Citrus* germplasm accessions revealed by inter-simple sequence repeat (ISSR) markers. *Journal of the American Society for Horticultural Science* 123(4), 612–7. doi:10.21273/JASHS.123.4.612.

Fang, J., Wang, C., Yu, H., Li, X., Song, C. and Chen, J. 2009. Identification of 57 sweet orange cultivars using AFLP markers. *The Journal of Horticultural Science and Biotechnology* 84(6), 585–90. doi:10.1080/14620316.2009.11512569.

Farneti, B., Di Guardo, M., Khomenko, I., Cappellin, L., Biasioli, F., Velasco, R. and Costa, F. 2017. Genome-wide association study unravels the genetic control of the apple volatilome and its interplay with fruit texture. *Journal of Experimental Botany* 68(7), 1467–78. doi:10.1093/jxb/erx018.

Federici, C. T., Fang, D. Q., Scora, R. W. and Roose, M. L. 1998. Phylogenetic relationships within the genus *Citrus* (Rutaceae) and related genera as revealed by RFLP and RAPD analysis. *Theoretical and Applied Genetics* 96(6–7), 812–22. doi:10.1007/s001220050807.

Federici, C. T., Roose, M. L. and Scora, R. W. 2000. RFLP analysis of the origin of *Citrus bergamia*, *Citrus jambhiri*, and *Citrus limonia*. *Acta Horticulturae* 535(535), 55–64. doi:10.17660/ActaHortic.2000.535.6.

Ferrante, S. P., Lucretti, S., Reale, S., Patrizio, A. d., Abbate, L., Tusa, N. and Scarano, M. T. 2010. Assessment of the origin of new citrus tetraploid hybrids ($2n = 4x$) by means of SSR markers and PCR based dosage effects. *Euphytica* 173(2), 223–33. doi:10.1007/s10681-009-0093-3.

Froelicher, Y., Bassene, J. B., Jedidi-Neji, E., Dambier, D., Morillon, R., Bernardini, G., Costantino, G. and Ollitrault, P. 2007. Induced parthenogenesis in mandarin for haploid production: induction procedures and genetic analysis of plantlets. *Plant Cell Reports* 26(7), 937–44. doi:10.1007/s00299-007-0314-y.

Froelicher, Y., Dambier, D., Bassene, J. B., Costantino, G., Lotfy, S., Didout, C., Beaumont, V., Brottier, P., Risterucci, A. M., Luro, F., et al. 2008. Characterization of microsatellite markers in mandarin orange (*Citrus reticulata* Blanco). *Molecular Ecology Resources* 8(1), 119–22. doi:10.1111/j.1471-8286.2007.01893.x.

Fu, C. H., Chen, C. L., Guo, W. W. and Deng, X. X. 2004. GISH, AFLP and PCR-RFLP analysis of an intergeneric somatic hybrid combining Goutou sour orange and Poncirus trifoliata. *Plant Cell Reports* 23(6), 391–6. doi:10.1007/s00299-004-0828-5.

Fujii, H., Shimada, T., Nonaka, K., Kita, M., Kuniga, T., Endo, T., Ikoma, Y. and Omura, M. 2013. High-throughput genotyping in citrus accessions using an SNP genotyping array. *Tree Genetics and Genomes* 9(1), 145–53. doi:10.1007/s11295-012-0542-3.

Garcia, R., Asins, M. J., Forner, J. and Carbonell, E. A. 1999. Genetic analysis of apomixis in *Citrus* and *Poncirus* by molecular markers. *TAG. Theoretical and Applied Genetics. Theoretische und Angewandte Genetik* 99(3–4), 511–8. doi:10.1007/s001220051264.

Garcia-Lor, A., Luro, F., Navarro, L. and Ollitrault, P. 2012. Comparative use of InDel and SSR markers in deciphering the interspecific structure of cultivated citrus genetic diversity: a perspective for genetic association studies. *Molecular Genetics and Genomics: MGG* 287(1), 77–94. doi:10.1007/s00438-011-0658-4.

Garcia-Lor, A., Ancillo, G., Navarro, L. and Ollitrault, P. 2013a. *Citrus* (Rutaceae) SNP markers based on Competitive Allele-Specific PCR; transferability across the *Aurantioideae* subfamily. *Applications in Plant Sciences* 1(4). doi:10.3732/apps.1200406.

Garcia-Lor, A., Curk, F., Snoussi-Trifa, H., Morillon, R., Ancillo, G., Luro, F., Navarro, L. and Ollitrault, P. 2013b. A nuclear phylogenetic analysis: SNPs, indels and SSRs deliver new insights into the relationships in the 'true citrus fruit trees' group (Citrinae, Rutaceae) and the origin of cultivated species. *Annals of Botany* 111(1), 1–19. doi:10.1093/aob/mcs227.

Garcia-Lor, A., Luro, F., Ollitrault, P. and Navarro, L. 2015. Genetic diversity and population structure analysis of mandarin germplasm by nuclear, chloroplastic and mitochondrial markers. *Tree Genetics and Genomes* 11(6), 123. doi:10.1007/s11295-015-0951-1.

Germana, M. A., Aleza, P., Carrera, E., Chen, C., Chiancone, B., Costantino, G., Dambier, D., Deng, X., Federici, C. T., Froelicher, Y., et al. 2013. Cytological and molecular characterization of three gametoclones of *Citrus clementina*. *BMC Plant Biology* 13, 129. doi:10.1186/1471-2229-13-129.

Gmitter, F. G.,Jr., Xiao, S. Y., Huang, S., Hu, X. L., Garnsey, S. M. and Deng, Z. 1996. A localized linkage map of the *citrus tristeza virus* resistance gene region. *TAG. Theoretical and Applied Genetics. Theoretische und Angewandte Genetik* 92(6), 688–95. doi:10.1007/BF00226090.

Gmitter, F. G., Chen, C., Machado, M. A., de Souza, A. A., Ollitrault, P., Froehlicher, Y. and Shimizu, T. 2012. Citrus genomics. *Tree Genetics and Genomes* 8(3), 611–26. doi:10.1007/s11295-012-0499-2.

Gulsen, O., Uzun, A., Canan, I., Seday, U. and Canihos, E. 2010. A new citrus linkage map based on SRAP, SSR, ISSR, POGP, RGA and RAPD markers. *Euphytica* 173(2), 265–77. doi:10.1007/s10681-010-0146-7.

Guo, W. W. and Deng, X. X. 2001. Wide somatic hybrids of *Citrus* with its related genera and their potential in genetic improvement. *Euphytica* 118(2), 175–83. doi:10.1023/A:1004147208099.

Guo, F., Yu, H., Tang, Z., Jiang, X., Wang, L., Wang, X., Xu, Q. and Deng, X. 2015. Construction of a SNP-based high-density genetic map for pummelo using RAD sequencing. *Tree Genetics and Genomes* 11(1), 2. doi:10.1007/s11295-014-0831-0.

Holtz, Y., Ardisson, M., Ranwez, V., Besnard, A., Leroy, P., Poux, G., Roumet, P., Viader, V., Santoni, S. and David, J. 2016. Genotyping by sequencing using specific allelic capture to build a high-density genetic map of durum wheat. *PLOS ONE* 11(5). doi:10.1371/journal.pone.0154609.

Hong, Q., Ma, X., Gong, G., Peng, Z. and He, Y. 2015. QTL mapping of citrus freeze tolerance. *Acta Horticulturae* 1065, 467–74.

Huang, S., Kang, M. and Xu, A. 2017. HaploMerger2: rebuilding both haploid sub-assemblies from high-heterozygosity diploid genome assembly. *Bioinformatics* 33(16), 2577–9. doi:10.1093/bioinformatics/btx220.

Huang, M., Roose, M. L., Yu, Q., Du, D., Yu, Y., Zhang, Y., Deng, Z., Stover, E. and Gmitter, F. G. 2018. Construction of high-Density genetic maps and detection of QTLs associated with Huanglongbing tolerance in citrus. *Frontiers in Plant Science* 9, 1694. doi:10.3389/fpls.2018.01694.

Imai, A., Nonaka, K., Kuniga, T., Yoshioka, T. and Hayashi, T. 2018. Genome-wide association mapping of fruit-quality traits using genotyping-by-sequencing approach in citrus landraces, modern cultivars, and breeding lines in Japan. *Tree Genetics and Genomes* 14(2), 24. doi:10.1007/s11295-018-1238-0.

Ipek, A., Yilmaz, K., Sikici, P., Tangu, N. A., Oz, A. T., Bayraktar, M., Ipek, M. and Gulen, H. 2016. SNP discovery by GBS in olive and the construction of a high-density genetic linkage map. *Biochemical Genetics* 54(3), 313–25. doi:10.1007/s10528-016-9721-5.

Iwamasa, M., Nito, N. and Ling, J. 1988. Intra and intergeneric hybridization in the orange subfamily. In: *Proceedings of 6th International Citrus Congress*, pp. 123–30.

Iwata, H., Minamikawa, M. F., Kajiya-Kanegae, H., Ishimori, M. and Hayashi, T. 2016. Genomics-assisted breeding in fruit trees. *Breeding Science* 66(1), 100–15. doi:10.1270/jsbbs.66.100.

Kauffman, E. J., Gestl, E. E., Kim, D. J., Walker, C., Hite, J. M., Yan, G., Rogan, P. K., Johnson, S. L. and Cheng, K. C. 1995. Microsatellite centromere mapping in the zebrafish (*Danio rerio*). *Genomics* 30(2), 337–41. doi:10.1006/geno.1995.9869.

Kamiri, M., Stift, M., Srairi, I., Costantino, G., El Moussadik, A., Hmyene, A., Bakry, F., Ollitrault, P. and Froelicher, Y. 2011. Evidence for non-disomic inheritance in a *Citrus* interspecific tetraploid somatic hybrid between *C. reticulata* and *C. limon* using SSR markers and cytogenetic analysis. *Plant Cell Reports* 30(8), 1415–25. doi:10.1007/s00299-011-1050-x.

Kamiri, M., Stift, M., Costantino, G., Dambier, D., Kabbage, T., Ollitrault, P. and Froelicher, Y. 2018. Preferential homologous chromosome pairing in a tetraploid intergeneric somatic hybrid (*Citrus reticulata* + *Poncirus trifoliata*) revealed by SSR marker inheritance. *Frontiers in Plant Science* 9, 1557. doi:10.3389/fpls.2018.01557.

Kijas, J. M., Fowler, J. C. and Thomas, M. R. 1995. An evaluation of sequence tagged microsatellite site markers for genetic analysis within *Citrus* and related species. *Genome* 38(2), 349–55. doi:10.1139/g95-045.

Koszegi, D., Johnston, A. J., Rutten, T., Czihal, A., Altschmied, L., Kumlehn, J., Wust, S. E. J., Kirioukhova, O., Gheyselinck, J., Grossniklaus, U., et al. 2011. Members of the RKD transcription factor family induce an egg cell-like gene expression program. *The Plant Journal: for Cell and Molecular Biology* 67(2), 280–91. doi:10.1111/j.1365-313X.2011.04592.x.

Krueger, R. R. and Navarro, L. 2007. *Citrus* germplasm resources. In: *Citrus Genetics, Breeding and Biotechnology*. CAB International, Wallingford, UK, pp. 45–140.

Krueger, R. R. and Roose, M. L. 2003. Use of molecular markers in the management of *Citrus* germplasm resources. *Journal of the American Society for Horticultural Science* 128(6), 827–37. doi:10.21273/JASHS.128.6.0827.

Krug, C. A. 1943. Chromosome numbers in the subfamily *Arantioideae*, with special reference in the genus *Citrus*. *Botanical Gazette* 104(4), 602–11. doi:10.1086/335173.

Kumar, S., Jena, S. N. and Nair, N. K. 2010. ISSR polymorphism in Indian wild orange (*Citrus indica* Tanaka, Rutaceae) and related wild species in North-East India. *Scientia Horticulturae* 123(3), 350–9. doi:10.1016/j.scienta.2009.10.008.

Lawrence, G. 1951. *Taxonomy of Vascular Plants*. The Macmillan Company, New York.

Liang, G., Xiong, G., Guo, Q., He, Q. and Li, X. 2007. AFLP analysis and the taxonomy of *Citrus*. *Acta Horticulturae* 760(760), 137–42. doi:10.17660/ActaHortic.2007.760.17.

Liang, M., Yang, X., Li, H., Su, S., Yi, H., Chai, L. and Deng, X. 2015. *De novo* transcriptome assembly of pummelo and molecular marker development. *PLOS ONE* 10(3), e0120615. doi:10.1371/journal.pone.0120615.

Lim, K., Shen, T., Barba-Gonzalez, R., Ramanna, M. S. and Van Tuyl, J. M. 2004. Occurrence of SDR 2N-gametes in *Lilium* hybrids. *Breeding Science* 54(1), 13–8. doi:10.1270/jsbbs.54.13.

Lima, R. P. M., Curtolo, M., Merfa, M. V., Cristofani-Yaly, M. and Machado, M. A. 2018. QTLs and eQTLs mapping related to citrandarins resistance to citrus gummosis disease. *BMC Genomics* 19(1), 516. doi:10.1186/s12864-018-4888-2.

Ling, P., Duncan, L. W., Deng, Z., Dunn, D., Hu, X., Huang, S. and Gmitter, F. G.,Jr. 2000. Inheritance of citrus nematode resistance and its linkage with molecular markers. *Theoretical and Applied Genetics* 100(7), 1010–7. doi:10.1007/s001220051382.

Liu, C., Jiang, D., Cheng, Y., Deng, X., Chen, F., Fang, L., Ma, Z. and Xu, J. 2013a. Chemotaxonomic study of *Citrus*, *Poncirus* and *Fortunella* genotypes based on peel oil volatile compounds – deciphering the genetic origin of Mangshanyegan (*Citrus nobilis* Lauriro). *PLOS ONE* 8(3), e58411. doi:10.1371/journal.pone.0058411.

Liu, S. R., Li, W. Y., Long, D., Hu, C. G. and Zhang, J. Z. 2013b. Development and characterization of genomic and expressed SSRs in citrus by genome-wide analysis. *PLOS ONE* 8(10), e75149. doi:10.1371/journal.pone.0075149.

Lombardoa, G., Schicchia, R., Marinoa, P. and Pallaa, F. 2012. Genetic analysis of *Citrus aurantium* L. (Rutaceae) cultivars by ISSR molecular markers. *Plant Biosystems – an International Journal Dealing with All Aspects of Plant Biology* 146(sup1), 19–26. doi :10.1080/11263504.2011.557101.

Luro, F., Laigret, F., Bovè, J. M. and Ollitrault, P. 1992. Application of random amplified polymorphic DNA (R.A.P.D.) to *Citrus* genetics and taxonomy. In: *Proceedings of the 7th International Citrus Congress. Taxonomy, Breeding and Varieties, Rootstocks and Propagation, Plant Physiology and Ecology*. Tribulato, E., Gentile, A., Reforgiato, G. and ISC. ISC, Riverside, pp. 225–8. International Citrus Congress. 7, Acireale, Italie, 8 March 1992/13 March 1992.

Luro, F., Laigret, F., Lorieux, M. and Ollitrault, P. 1996. *Citrus* genome mapping with molecular markers: two maps obtained by segregation analysis of progeny of one intergeneric cross. In: *Proceedings of the International Society of Citriculture* (vol. 2), Nelspruit, South Africa, pp. 862–6.

Luro, F., Rist, D. and Ollitrault, P. 2001. Evaluation of genetic relationships in *Citrus* genus by means of sequence tagged microsatellites. *Acta Horticulturae* 546(546), 237–42. doi:10.17660/ActaHortic.2001.546.27.

Luro, F., Maddy, F., Jacquemond, C., Froelicher, Y., Morillon, R., Rist, D. and Ollitrault, P. 2004. Identification and evaluation of diplogyny in clementine (*Citrus clementina*) for use in breeding. *Acta Horticulturae* 663(663), 841–8. doi:10.17660/ActaHortic.2004.663.152.

Luro, F. L., Costantino, G., Terol, J., Argout, X., Allario, T., Wincker, P., Talon, M., Ollitrault, P. and Morillon, R. 2008. Transferability of the EST-SSRs developed on Nules clementine (*Citrus clementina* Hort ex Tan) to other Citrus species and their effectiveness for genetic mapping. *BMC Genomics* 9, 287. doi:10.1186/1471-2164-9-287.

Luro, F., Venturini, N., Costantino, G., Paolini, J., Ollitrault, P. and Costa, J. 2012. Genetic and chemical diversity of citron (*Citrus medica* L.) based on nuclear and cytoplasmic markers and leaf essential oil composition. *Phytochemistry* 77, 186–96. doi:10.1016/j.phytochem.2011.12.013.

Lyon, M. P. 2008. A genomic genetic map of the common sweet orange and *Poncirus trifoliate*. PhD dissertation. University of California, Riverside.

Lyon, M. T., Federici, C. T., Kacar, Y., Chen, C., O'Malley, D., Chaparro, J. X., Gmitter, F. G. and Roose, M. L. 2007. SSR-based linkage maps for sweet orange and trifoliate orange, In: *Plant & Animal Genome XVth Conference*, 13–17 January 2007. Town & Country Convention Center, San Diego, CA.

Matsukawa, T. and Nito, N. 2017. Metabolic fingerprinting of citrus cultivars and related genera using HPLC and multivariate analysis. *Journal of Plant Studies* 7(1), 1–9. doi:10.5539/jps.v7n1p1.

Mendiburu, A. O. and Peloquin, S. J. 1979. Gene-centromere mapping by 4x-2x matings in potatoes. *TAG. Theoretical and Applied Genetics. Theoretische und Angewandte Genetik* 54(4), 177–80. doi:10.1007/BF00263048.

Mestre, P. F., Asins, M. J., Carbonell, E. A. and Navarro, L. 1997. New gene(s) involved in the resistance of *Poncirus trifoliata* (L.) Raf. to *citrus tristeza virus*. *Theoretical and Applied Genetics* 95(4), 691–5. doi:10.1007/s001220050613.

Miller, M. R., Dunham, J. P., Amores, A., Cresko, W. A. and Johnson, E. A. 2007. Rapid and cost-effective polymorphism identification and genotyping using restriction site associated DNA (RAD) markers. *Genome Research* 17(2), 240–8. doi:10.1101/gr.5681207.

Minamikawa, M. F., Nonaka, K., Kaminuma, E., Kajiya-Kanegae, H., Onogi, A., Goto, S., Yoshioka, T., Imai, A., Hamada, H., Hayashi, T., et al. 2017. Genome-wide association study and genomic prediction in citrus: potential of genomics-assisted breeding for fruit quality traits. *Scientific Reports* 7(1), 4721. doi:10.1038/s41598-017-05100-x.

Mizuno, M., Iinuma, M., Ohara, M., Tanaka, T. and Iwamasa, M. 1991. Chemotaxonomy of the genus *Citrus* based on polymethoxyflavones. *Chemical and Pharmaceutical Bulletin* 39(4), 945–9. doi:10.1248/cpb.39.945.

Montero-Pau, J., Blanca, J., Esteras, C., MartÃnez, P. Ã., Gómez, P., Monforte, A. J., Cañizares, J. and Picó, B. 2017. An SNP-based saturated genetic map and QTL analysis of fruit-related traits in *Zucchini* using genotyping-by-sequencing. *BMC Genomics* 18(1), 94. doi:10.1186/s12864-016-3439-y.

Moriya, S., Kunihisa, M., Okada, K., Iwanami, H., Iwata, H., Minamikawa, M., Katayose, Y., Matsumoto, T., Mori, S., Sasaki, H., et al. 2017. Identification of QTLs for flesh mealiness in apple (*Malus domestica* Borkh.). *The Horticulture Journal* 86(2), 159–70. doi:10.2503/hortj.MI-156.

Munankarmi, N. N., Rana, N., Bhattarai, T., Shrestha, R. L., Joshi, B. K., Baral, B. and Shrestha, S. 2018. Characterization of the genetic diversity of acid lime (*Citrus aurantifolia* (Christm.) Swingle) cultivars of Eastern Nepal using inter-simple sequence repeat markers. *Plants* 7(2), 46. doi:10.3390/plants7020046.

NCBI. 2017. Taxonomy/browser. Available at: http://www.ncbi.nlm.nih.gov/Taxonomy/Browser/wwwtax.cgi.

Nicolosi, E., Deng, Z. N., Gentile, A., La Malfa, S., Continella, G. and Tribulato, E. 2000. *Citrus* phylogeny and genetic origin of important species as investigated by molecular markers. *Theoretical and Applied Genetics* 100(8), 1155–66. doi:10.1007/s001220051419.

Nie, H., Li, Q. and Kong, L. 2012. Centromere mapping in the Pacific abalone (*Haliotis discus hannai*) through half-tetrad analysis in gynogenetic diploid families. *Animal Genetics* 43(3), 290–7. doi:10.1111/j.1365-2052.2011.02254.x.

Ohta, S., Endo, T., Shimada, T., Fujii, H., Shimizu, T., Kita, M., Kuniga, T., Yoshioka, T., Nesumi, H., Yoshida, T., et al. 2014. Construction of genetic linkage map and graphical genotyping of pseudo-backcrossed F2 (BC2) progeny to introduce a CTV resistance from *Poncirus trifoliata* (L.) Raf. into *Citrus* by introgression breeding. *Tree Genetics and Genomes* 11(1), 797.

Oliveira, R. Pd, Aguilar-Vildoso, C. I., Cristofani, M. and Machado, M. A. 2004. Skewed RAPD markers in linkage maps of *Citrus*. *Genetics and Molecular Biology* 27(3), 437–41. doi:10.1590/S1415-47572004000300021.

Ollitrault, P. and Michaux-Ferriere, N. 1992. Application of flow cytometry for citrus genetic and breeding. In: *Proceeding of the International Citrus Congress (7th)* (vol. 1), pp. 193–8.

Ollitrault, P., Dambier, D., Luro, F. and Duperray, C. 1994. Nuclear genome size variations in *Citrus*. *Fruits (Paris)* 49(5/6), 390–3.

Ollitrault, P., Jacquemond, C., Dubois, C. and Luro, F. 2003. *Citrus*. In: *Genetic Diversity of Cultivated Tropical Plants*. CIRAD, Montpellier, pp. 193–217.

Ollitrault, F., Terol, J., Pina, J. A., Navarro, L., Talon, M. and Ollitrault, P. 2010. Development of SSR markers from *Citrus clementina* (Rutaceae) BAC end sequences and interspecific transferability in *Citrus*. *American Journal of Botany* 97(11), e124–9. doi:10.3732/ajb.1000280.

Ollitrault, P., Terol, J., Chen, C., Federici, C. T., Lotfy, S., Hippolyte, I., Ollitrault, F., Berard, A., Chauveau, A., Cuenca, J., et al. 2012a. A reference genetic map of *C. clementina* Hort. ex Tan.; citrus evolution inferences from comparative mapping. *BMC Genomics* 13, 593. doi:10.1186/1471-2164-13-593.

Ollitrault, P., Terol, J., Garcia-Lor, A., Berard, A., Chauveau, A., Froelicher, Y., Belzile, C., Morillon, R., Navarro, L., Brunel, D., et al. 2012b. SNP mining in *C. clementina* BAC end sequences; transferability in the *Citrus* genus (Rutaceae), phylogenetic inferences and perspectives for genetic mapping. *BMC Genomics* 13(1). doi:10.1186/1471-2164-13-13.

Ollitrault, P., Curk, F., Ollitrault, F., Garcia-Lor, A., Navarro, L. and Morillon, R. 2016. Usefulness of phylogenetic diagnostic SNP markers of citrus ancestral taxa for genetics and breeding. In: *Book of Abstract; Sustainable Citriculture, the Role of Applied Knowledge*. International Citrus Congress, pp. 126–7.

Omura, M. and Shimada, T. 2016. Citrus breeding, genetics and genomics in Japan. *Breeding Science* 66(1), 3–17. doi:10.1270/jsbbs.66.3.

Omura, M., Ueda, T., Kita, M., Komatsu, A., Takanokura, Y., Shimada, T., Endo-Inagaki, T., Nesumi, H. and Yoshida, T. 2000. EST mapping of *Citrus*. In: *Proceedings of the International Society of Citriculture, vol. IX congress*. Orlando, USA.

Oueslati, A., Salhi-Hannachi, A., Luro, F., Vignes, H., Mournet, P. and Ollitrault, P. 2017. Genotyping by sequencing reveals the interspecific *C. maxima/C. reticulata* admixture along the genomes of modern citrus varieties of mandarins, tangors, tangelos, orangelos and grapefruits. *PLOS ONE* 12(10), e0185618. doi:10.1371/journal.pone.0185618.

Pang, X., Hu, C. G. and Deng, X. X. 2007. Phylogenetic relationships within *Citrus* and its related genera as inferred from AFLP markers. *Genetic Resources and Crop Evolution* 54(2), 429–36. doi:10.1007/s10722-006-0005-5.

Park, T. H., Kim, J. B., Hutten, R. C., van Eck, H. J., Jacobsen, E. and Visser, R. G. 2007. Genetic positioning of centromeres using half-tetrad analysis in a 4x × 2x cross population of potato. *Genetics* 176(1), 85–94. doi:10.1534/genetics.107.070870.

Peloquin, S. J., Yerk, G. L., Werner, J. E. and Darmo, E. 1989. Potato breeding with haploids and 2n gametes. *Genome* 31(2), 1000–4. doi:10.1139/g89-174.

Penjor, T., Mimura, T., Matsumoto, R., Yamamoto, M. and Nagano, Y. 2014. Characterization of limes (*Citrus aurantifolia*) grown in Bhutan and Indonesia using high-throughput sequencing. *Scientific Reports* 4, 4853. doi:10.1038/srep04853.

Raga, V., Bernet, G. P., Carbonell, E. A. and Asins, M. J. 2012. Segregation and linkage analyses in two complex populations derived from the citrus rootstock Cleopatra mandarin. Inheritance of seed reproductive traits. *Tree Genetics and Genomes* 8(5), 1061–71. doi:10.1007/s11295-012-0486-7.

Raga, V., Intrigliolo, D. S., Bernet, G. P., Carbonell, E. A. and Asins, M. J. 2016. Genetic analysis of salt tolerance in a progeny derived from the citrus rootstocks Cleopatra mandarin and trifoliate orange. *Tree Genetics and Genomes* 12(3), 34. doi:10.1007/s11295-016-0991-1.

Ramadugu, C., Keremane, M. L., Halbert, S. E., Duan, Y. P., Roose, M. L., Stover, E. and Lee, R. F. 2016. Long-term field evaluation reveals huanglongbing resistance in citrus relatives. *Plant Disease* 100(9), 1858–69. doi:10.1094/PDIS-03-16-0271-RE.

Rao, M. N., Soneji, J. R., Chen, C. X., Choi, Y. A., Huang, S. and Gmitter Jr., F. G. 2007. Identification of zygotic and nucellar seedlings from citrus rootstock candidates using RAPD. *Acta Horticulturae* 738, 255–60.

Rao, M. N., Soneji, J. R., Chen, C. X., Huang, S. and Gmitter, F. G.,Jr. 2008. Characterization of zygotic and nucellar seedlings from sour orange-like citrus rootstock candidates using RAPD and EST-SSR markers. *Tree Genetics and Genomes* 4(1), 113–24.

Roose, M. L. 2007. Mapping and marker assisted selection in *Citrus*. In: Khan, I. A. (Ed.), *Citrus Genetics, Breeding and Biotechnology*. CAB International, Wallingford, UK, pp. 275–86.

Roose, M. L., Feng, D., Cheng, F. S., Tayyar, R. I., Federici, C. T. and Kupper, R. S. 2000. Mapping the *Citrus* genome. *Acta Horticulturae* 535(535), 25–32. doi:10.17660/ActaHortic.2000.535.1.

Rouiss, H., Cuenca, J., Navarro, L., Ollitrault, P. and Aleza, P. 2017a. Tetraploid citrus progenies arising from FDR and SDR unreduced pollen in 4x × 2x hybridizations. *Tree Genetics and Genomes* 13(1), 10. doi:10.1007/s11295-016-1094-8.

Rouiss, H., Cuenca, J., Navarro, L., Ollitrault, P. and Aleza, P. 2017b. Unreduced megagametophyte production in lemon occurs via three meiotic mechanisms, predominantly second-division restitution. *Frontiers in Plant Science* 8, 1211. doi:10.3389/fpls.2017.01211.

Rouiss, H., Bakry, F., Froelicher, Y., Navarro, L., Aleza, P. and Ollitrault, P. 2018. Origin of *C. latifolia* and *C. aurantiifolia* triploid limes; the preferential disomic inheritance of doubled-diploid 'Mexican' lime is consistent with an interploid hybridization hypothesis. *Annals of Botany* 121(3), 571–85. doi:10.1093/aob/mcx179.

Ruiz, C. and Asins, M. J. 2003. Comparison between *Poncirus* and *Citrus* genetic linkage maps. *TAG. Theoretical and Applied Genetics. Theoretische und Angewandte Genetik* 106(5), 826–36. doi:10.1007/s00122-002-1095-x.

Ruiz, C., Paz Breto, M. and Asins, M. J. 2000. A quick methodology to identify sexual seedlings in citrus breeding programs using SSR markers. *Euphytica* 112(1), 89–94. doi:10.1023/A:1003992719598.

Sankar, A. A. and Moore, G. A. 2001. Evaluation of inter-simple sequence repeat analysis for mapping in *Citrus* and extension of the genetic linkage map. *Theoretical and Applied Genetics* 102(2–3), 206–14. doi:10.1007/s001220051637.

Scarano, M.-T., Abbate, L., Ferrante, S., Lucretti, S. and Tusa, N. 2002. ISSR-PCR technique: a useful method for characterizing new allotetraploid somatic hybrids of mandarin. *Plant Cell Reports* 20(12), 1162–6. doi:10.1007/s00299-002-0450-3.

Scora, R. W. 1975. On the history and origin of *Citrus*. *Bulletin of the Torrey Botanical Club* 102(6), 369-369-375. doi:10.2307/2484763.

Shimada, T., Fujii, H., Endo, T., Ueda, T., Sugiyama, A., Nakano, M., Kita, M., Yoshioka, T., Shimizu, T., Nesumi, H., et al. 2014. Construction of a citrus framework genetic map anchored by 708 gene-based markers. *Tree Genetics and Genomes* 10(4), 1001–13. doi:10.1007/s11295-014-0738-9.

Shimizu, T., Kitajima, A., Nonaka, K., Yoshioka, T., Ohta, S., Goto, S., Toyoda, A., Fujiyama, A., Mochizuki, T., Nagasaki, H., et al. 2016. Hybrid origins of citrus varieties inferred from DNA marker analysis of nuclear and organelle genomes. *PLOS ONE* 11(11), e0166969. doi:10.1371/journal.pone.0166969.

Siviero, A., Cristofani, M., Furtado, E. L., Garcia, A. A., Coelho, A. S. and Machado, M. A. 2006. Identification of QTLs associated with citrus resistance to *Phytophthora* gummosis. *Journal of Applied Genetics* 47(1), 23–8. doi:10.1007/BF03194595.

Stevens, P. F. 2017. Angiosperm phylogeny website. last update 14 July 2017. Available at: http://www.mobot.org/MOBOT/research/APweb.

Sugiyama, A., Omura, M., Matsumoto, H., Shimada, T., Fujii, H., Endo, T., Shimizu, T., Nesumi, H. and Ikoma, Y. 2011. Quantitative trait loci (QTL) analysis of carotenoid content in *Citrus* fruit. *Journal of the Japanese Society for Horticultural Science* 80(2), 136–44. doi:10.2503/jjshs1.80.136.

Sugiyama, A., Omura, M., Shimada, T., Fujii, H., Endo, T., Shimizu, T., Nesumi, H., Nonaka, K. and Ikoma, Y. 2014. Expression quantitative trait loci analysis of carotenoid metabolism-related genes in citrus. *Journal of the Japanese Society for Horticultural Science* 83(1), 32–43. doi:10.2503/jjshs1.CH-054.

Swingle, W. T. and Reece, P. C. 1967. The botany of *Citrus* and its wild relatives. In: Reuther, W., Webber, H. J. and Batchelor, D. L. (Eds), *The Citrus Industry*. University of California Press, Berkeley, CA, 190–430.

Tanaka, T. 1954. *Species Problem in Citrus (Revisio Aurantiacearum IX)*. Japanese Society for Promotion of Science, Tokyo, Japan.

Tanaka, T. 1961. *Citologia: Semi-Centennial Commemoration Papers on Citrus Studies*. Citologia Supporting Foundation, Osaka, Japan.

Tao, N., Xu, J., Cheng, Y., Hong, L., Guo, W., Yi, H. and Deng, X. 2005. Isolation and characterization of *copia*-like retrotransposons from 12 sweet orange (*Citrus sinensis* Osbeck) cultivars. *Journal of Integrative Plant Biology* 47(12), 1507–15. doi:10.1111/j.1744-7909.2005.00188.x.

Tavoletti, S., Bingham, E. T., Yandell, B. S., Veronesi, F. and Osborn, T. C. 1996. Half tetrad analysis in alfalfa using multiple restriction fragment length polymorphism markers. *Proceedings of the National Academy of Sciences of the United States of America* 93(20), 10918–22. doi:10.1073/pnas.93.20.10918.

Terol, J., Naranjo, M. A., Ollitrault, P. and Talon, M. 2008. Development of genomic resources for *Citrus clementina*: characterization of three deep-coverage BAC libraries and analysis of 46,000 BAC end sequences. *BMC Genomics* 9, 423. doi:10.1186/1471-2164-9-423.

Torres, A. M., Soost, R. K. and Mau-Lastovicka, T. 1982. Citrus isozymes: genetics and distinguishing nucellar from zygotic seedlings. *Journal of Heredity* 73(5), 335–8. doi:10.1093/oxfordjournals.jhered.a109664.

Tozlu, I., Guy, C. L. and Moore, G. A. 2002. QTL analysis of salt stress related traits in citrus and related genus *Poncirus*. *Acta Horticulturae* 573, 237–46. doi:10.17660/ActaHortic.2002.573.27.

Vinson, J. P., Jaffe, D. B., O'Neill, K., Karlsson, E. K., Stange-Thomann, N., Anderson, S., Mesirov, J. P., Satoh, N., Satou, Y., Nusbaum, C., et al. 2005. Assembly of polymorphic genomes: algorithms and application to *Ciona savignyi*. *Genome Research* 15(8), 1127–35. doi:10.1101/gr.3722605.

Wang, X., Xu, Y., Zhang, S., Cao, L., Huang, Y., Cheng, J., Wu, G., Tian, S., Chen, C., Liu, Y., et al. 2017. Genomic analyses of primitive, wild and cultivated citrus provide insights into asexual reproduction. *Nature Genetics* 49(5), 765–72. doi:10.1038/ng.3839.

Webber, H. J. 1943. Cultivated varieties of citrus. In: Webber, H. J. and Batchelor, L. D. (Eds), *The Citrus Industry. History, World Distribution, Botany and Varieties*. University of California Press, California, pp. 475–668.

Weber, C. A., Moore, G. A., Deng, Z. N. and Gmitter Jr., F. G. 2003. Mapping freeze tolerance quantitative trait loci in a *Citrus grandis* × *Poncirus trifoliata* $F_1$ pseudo-testcross using molecular markers. *Journal of the American Society for Horticultural Science* 128(4), 508–14. doi:10.21273/JASHS.128.4.0508.

Wu, G. A., Prochnik, S., Jenkins, J., Salse, J., Hellsten, U., Murat, F., Perrier, X., Ruiz, M., Scalabrin, S., Terol, J., et al. 2014. Sequencing of diverse mandarin, pummelo and orange genomes reveals complex history of admixture during citrus domestication. *Nature Biotechnology* 32(7), 656–62. doi:10.1038/nbt.2906.

Wu, G. A., Terol, J., Ibanez, V., López-García, A., Pérez-Román, E., Borredá, C., Domingo, C., Tadeo, F. R., Carbonell-Caballero, J., Alonso, R., et al. 2018. Genomics of the origin and evolution of *Citrus*. *Nature* 554, 311.

Xiang, X., Deng, Z. A., Chen, C. X., Gmitter, F. G.,Jr. and Bowman, K. 2010. Marker assisted selection in citrus rootstock breeding based on a major gene locus *'Tyr1'* controlling citrus nematode resistance. *Agricultural Sciences in China* 9(4), 557–67. doi:10.1016/S1671-2927(09)60129-2.

Xiao, S., Gmitter, F. G., Jr., Grosser, J. W. and Shu, H. 1995. RAPD analysis: a rapid method to identify citrus somatic hybrids. *Hereditas (Beijing)* 17(4), 40–2.

Xie, K. D., Wang, X. P., Biswas, M. K., Liang, W. J., Xu, Q., Grosser, J. W. and Guo, W. W. 2014. 2n megagametophyte formed via SDR contributes to tetraploidization in polyembryonic ˜Nadorcott™ tangor crossed by citrus allotetraploids. *Plant Cell Reports* 33(10), 1641–50. doi:10.1007/s00299-014-1643-2.
Xie, K., Xia, Q., Wang, X., Liang, W., Wu, X., Grosser, J. W. and Guo, W. 2015. Cytogenetic and SSR-marker evidence of mixed disomic, tetrasomic, and intermediate inheritance in a citrus allotetraploid somatic hybrid between 'Nova' tangelo and 'HB' pummelo. *Tree Genetics and Genomes* 11(6), 112. doi:10.1007/s11295-015-0940-4.
Xu, Q., Chen, L. L., Ruan, X., Chen, D., Zhu, A., Chen, C., Bertrand, D., Jiao, W. B., Hao, B. H., Lyon, M. P., et al. 2013. The draft genome of sweet orange (*Citrus sinensis*). *Nature Genetics* 45(1), 59–66. doi:10.1038/ng.2472.
Yang, Z. N., Ye, X. R., Molina, J., Roose, M. L. and Mirkov, T. E. 2003. Sequence analysis of a 282-kilobase region surrounding the *citrus tristeza virus* resistance gene (*Ctv*) locus in *Poncirus trifoliata* L. Raf. *Plant Physiology* 131(2), 482–92. doi:10.1104/pp.011262.
Yang, Y., YueZhi, P., Xun, G. and MouTian, F. 2010. Genetic variation in the endangered Rutaceae species *Citrus* hongheensis based on ISSR fingerprint. *Genetic Resources and Crop Evolution* 57, 1239–48.
Yu, Y., Chen, C. and Gmitter, F. G. 2016. QTL mapping of mandarin (*Citrus reticulata*) fruit characters using high-throughput SNP markers. *Tree Genetics and Genomes* 12(4), 77. doi:10.1007/s11295-016-1034-7.
Yu, Y., Bai, J., Chen, C., Plotto, A., Yu, Q., Baldwin, E. A. and Gmitter, F. G. 2017. Identification of QTLs controlling aroma volatiles using a Fortune × Murcott (*Citrus reticulata*) population. *BMC Genomics* 18(1), 646. doi:10.1186/s12864-017-4043-5.
Zhao, H. and Speed, T. P. 1998. Statistical analysis of half-tetrads. *Genetics* 150(1), 473–85.

# Chapter 2

# Harvesting citrus: challenges and perspectives in an automated world

*Marcos David Ferreira, Embrapa Instrumentação, Brazil*



## 1 Introduction

Oranges (*Citrus sinensis*) are an important global fruit crop, with an estimated worldwide production of 47 754 metric tons (MT). The top 11 main producers are, in order: Brazil, China, the European Union (mainly Spain), Mexico, the United States, Egypt, Turkey, South Africa, Morocco, Vietnam and Australia. Annual production (2017/2018) varies from 16 034 MT in Brazil and 7300 MT in China to 480 MT in Australia (USDA, 2018; AGRIANUAL, 2017). Fruit goes to both the fresh market and to industry for processing (e.g. into fruit juice).

Even allowing for national differences in production practices, harvesting oranges has been a challenge for all these countries. Labour shortages, high costs and a short season have been problems all over the world, with each country trying to find a solution for efficiently harvesting citrus at low cost. *Citrus sinensis* is the most grown species of the *Citrus* genus, though other species are important, depending on the region and country. Most countries have a similar harvesting system, which is mainly manual harvesting using canvas bags and ladders. This chapter discusses orange harvesting for both the fresh market and for industry.

http://dx.doi.org/10.19103/AS.2019.0054.04

A major issue when harvesting is ensuring that the fruit is easily detached from the tree without damage, and is rapidly transported to the consumer or industry. Harvesting is not an easy task, even for small growers, as orange trees can be very tall and require ladders and other harvest-aid equipment. The cost of citrus harvesting ranges from 30–47% of the overall production budget and has been steadily increasing. In the case of Brazil, AGRIANUAL (2017) reported that, depending on the crop production system, harvesting costs could be as high as 47% of production, Neves et al. (2013) reported costs increasing in a 10-year period (2000/01 to 2009/10) from 30 to 44%, while Sanders (2005) estimated 35–45%. In Florida, USA, harvesting costs have increased due to the advance of Huanglongbing (HLB), also known as citrus greening, which decreases yield and fruit size and, during 2015–2016, increased harvest costs by approximately 23% from the previous season (Trejo-Pech et al., 2018). As a result, harvest costs in Florida are highly variable and depend upon the incidence of HLB in the orchard. Singerman (2017) reported typical harvesting costs for the 2015–2016 harvest could reach up to 34%.

The main methods employed for harvesting are manual picking, use of harvesting aids, and the use of mechanical and robotic devices, each of which has advantages and disadvantages. Although manual harvesting is the most widely used, it has become uneconomical in the main citrus-growing countries due to factors such as increased labour costs, seasonality and availability. Replacing manual harvesting is a difficult task, as a well-trained picker can select and distinguish colour and size, ensuring good fruit selection. It is important to emphasize that the market type, that is, the fresh market (for consumers) or juice (for the industrial market), can impact harvest speed and yield. Mobile platforms are an option for many growers and give better ergonomic conditions for pickers than traditional harvesting, which may improve productivity. However, certain issues need to be overcome in order to improve yield efficiency, for example, appropriate crew training. Mechanical harvesting is an alternative, but implementation costs, orchard design, yield efficiency and tree damage have limited its use in many countries. Harvesting with robotics, using machine vision, is still at the research stage for oranges and is very expensive specially for large-scale outdoor harvesting. In this chapter, all of these harvesting methods are discussed, highlighting their advantages and disadvantages.

## 2 Manual harvesting

Manual harvesting remains the main harvesting method used worldwide to pick oranges (Castro-Garcia et al., 2018; Ferreira et al., 2018; Liu et al., 2018). However, as Whitney and Harrell (1989) highlighted, attracting sufficient labour has long been a challenge, even during times of economic hardship. Manual harvesting is the most complete harvesting system, as only the picker has all the necessary abilities to select the fruit by size, colour and defects. An important point to highlight is that trained pickers can handle fruits with better care, more accurate selection and less fruit injury, compared to a picker with no training (Mitchell et al., 1964). Ampatzidis et al. (2011) suggested monitoring pickers via a wearable system, consisting of a pedometer and bar code, to track picker activity in order to improve harvest efficiency and training, which could be an alternative to a global positioning system (GPS). However, cost must be considered when the system is employed on a large scale. Whitney et al. (2001) also reported a system using button identification and bar codes integrated into a differential global positioning system (DGPS) to monitor pickers, with buttons showing efficiency and tracking picker progress.

Canvas bags and ladders comprise the main harvesting equipment used by pickers in many countries including Brazil. Canvas bags are made of polypropylene with an upper opening for fruit to enter the bag and a false bottom to exit, which is opened via buckles (Miranda et al., 2015; Narimoto and Belussi, 2018) (Fig. 1). The bag is carried by a transversal strap and, in general, supports between 23–27 kg of fruit (Miranda et al., 2015; Narimoto and Belussi, 2018). Although several varying ladder types exist, the 'A' model is the main ladder model in use, which has a wider base than the top (Fig. 1) and is mainly made of hollow metal (Narimoto and Belussi, 2018). Modifications can be made to improve the main equipment used for orange harvesting. Fathallah (2010) reported that musculoskeletal disorders (MSDs) resulting from repetitive

**Figure 1** Manual orange harvesting using canvas bags and 'A' type ladders. Accredited to: Pedro Campaner Hernandez.

activities of manual harvesting operations, such as bending (stooping) and carrying heavy loads, are very common (Fig. 1); hence, alternatives have to be pursued to mitigate this situation.

Narimoto and Belussi (2018) reported 15 modifications to canvas bags, mainly by pickers. Shoulder straps were modified in length with some extended up to 10 cm, and an additional coat was added to protect the neck and shoulders. The main bag structure was also altered to strengthen seam resistance, as well as bag opening and false bottom size. These changes made the bag more robust, alleviated pressure on the shoulders and provided more stability to the worker, as well as increasing fruit picking efficiency. As ladders are the main cause of accidents, changes and improvements are typical, and one of the most common is to add a support pin to avoid slipping.

The main harvesting model used in Brazil is canvas bags and ladders, both for fresh market and industry uses of fruit. Harvesting is usually carried out by groups of 45 people and payment is based on individual yield, which is measured multiple times using plastic boxes (Fernandes et al., 2017). Fruit is transferred from bags to 'big bags' (Fig. 2), which have a maximum capacity of 600 kg, and then to trucks for internal transport to a temporary bulk storage facility on a farm, and finally loaded for transport (Miranda et al., 2015; Fernandes et al., 2017). Big plastic bowl boxes are an alternative to big canvas bags (Fig. 3).

Other devices can be used to avoid reliance on ladders, platforms and manual detachment methods, and improve citrus harvesting, especially for small farm growers. Hamam et al. (2011) described the use of two prototypes of rotary cone harvesting and harvesting electrical scissor heads that can be used to facilitate orange harvesting, especially for fruits at the top of the

**Figure 2** Picker opening a canvas bag to transfer fruit to a field box. 'Big bags' are usually used to load trucks. Accredited to: Pedro Campaner Hernandez.

**Figure 3** Trucks being loaded with fruit destined for the juice industry. Accredited to: Marcos David Ferreira.

canopy. Both methods increased harvest efficiency and decreased harvest costs.

## 3 Harvesting aids

Harvesting aid equipment, also referred to as semi-mechanized harvesting, has the potential to improve orange harvesting and provide better conditions for pickers, diminishing MSDs (Fathallah, 2010) and increasing productivity. Costa and Camarotto (2012) evaluated mobile platforms as a harvesting aid and observed a 60% increase in productivity compared with manual labour.

Mobile platforms for citrus harvesting have been studied for a number of years (Jutras and Coppock, 1958; Seamount and Opitz, 1973). Whitney and Harrell (1989) described a harvesting platform for one man, which allowed canopy access at different levels and flexibility of use on the same or different crops. These devices can also be used for inspection purposes, such as, for field disease detection (Ferreira et al., 2018).

### *3.1 Ergonomics of mobile platforms*

Mobile platforms provide better ergonomic conditions for pickers and prevent the need to carry heavy loads compared with traditional manual harvesting (Costa and Camarotto, 2012); an additional benefit is avoiding the use of ladders which could reduce accidents. Vidal et al. (2015) highlighted that it takes longer for pickers to harvest the middle third of the canopy compared with the bottom or top, further promoting the use of harvesting platforms as it enables pickers to reach fruit at high canopy levels.

Another important benefit of using harvesting aids, such as platforms, is that it enables pickers with less strength to carry out harvesting operations (Elkins, 2012; Sarig, 2012; Fathallah, 2010). Vidal et al. (2015) reported that male pickers were faster than female pickers when manually harvesting on the high part of the canopy. Costa and Camarotto (2012) observed that manual harvesting resulted in 29% more postural harvesting movements when compared to an aid platform, especially relating to static posture and grip-related muscle use.

Changing from individual to group harvesting using platforms can cause problems (Sanders, 2005; Costa and Camarotto, 2012), for example the issue of stop breaks, as it is necessary to work as a group, as well as issues relating to payment of individuals in the group. More productive workers can be delayed by pickers that are not well trained or physically weaker. In this case, training and dividing into working groups, based on work ability, could be the solution. Costa and Camarotto (2012) have described the use of a bonus system to fit in with group working.

### 3.2 Harvesting aid use for oranges

Harvesting aid machines have been employed for different crops and multiple uses in some countries, for example Italy, Spain and France (Sarig, 2012; MSA, 2015). Cutini et al. (2017) reported that there are numerous mobile platforms in Italy, with about 20 factories and 110 different units, each with a distinctive use. Countries with labour shortages, such as Australia and Israel, have been harvesting fruit and vegetables using this device, whereas in the United States, mobile platforms are mainly used for vegetable crops (Sarig, 2012).

In Brazil, some effort has been made to search for alternatives to traditional harvesting methods for oranges. Since orchards are not suitable for mechanized or even semi-mechanized harvesting, machine design has to be based on orchard configuration and tree format. Over the last few years, harvesting aid equipment has been available for growers in Brazil. It typically comprises a mobile platform with hydraulic arms that move vertically to reach different levels at the top of the tree canopy, small boxes for temporary orange storage and conveyor belts to move the fruit to temporary storage, and from there to a big bag that can then be moved to trucks. A group of pickers follow the mobile unit and harvest fruits close to the ground. Figure 4 shows a general overview of a mobile platform with different levels for placing pickers, accessed by ladders.

**Figure 4** Mobile platform showing the main features that allow harvesting at different levels. The picker can be placed on different levels, which are accessed by ladders. Accredited to: Poliana Cristina Spricigo.

Costa and Camaroto (2012) ergonomically evaluated one type of mobile platform and observed improved working conditions by decreased walking, weight carried, discarding ladder use and diminishing physical stress. Another type of mobile platform that has multiple uses for orange harvesting was described by Ferreira et al. (2018). This mobile platform (Fig. 5) is a hybrid vehicle (fuel and electricity) with independent wheels (Fig. 6) and a retractable chassis, to which other devices can be added. The main goal of this machine was to facilitate orange harvesting, with the top fruits harvested by pickers in a crib and pickers following the machine to harvest fruits on the ground. In this study, yield productivity was similar to manual harvesting, which could be due to using only one crib.

Even though there has been an increase in the number of mobile platforms available for growers, adoption is still low; however, many farmers are testing this equipment and making modifications and adaptations. Besides feasibility,

**Figure 5** Mobile platform for harvesting oranges. The pickers are located on the ground and on a crib to pick fruits from the high canopy. Accredited to: Pedro Campaner Hernandez.

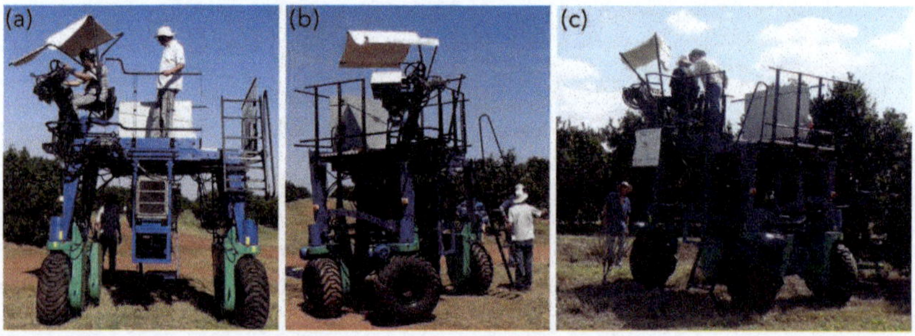


**Figure 6** Mobile platform. Clockwise. Machine movement positions. (a) Harvest: wheels are placed at a 180° angle to the chassis. (b) Spinning on its own axis: the machine turns around its own axis. The wheels are placed at angles from 45° to 155°. (c) Transport: the wheels are placed at a 90° angle to the chassis. Accredited to: Pedro Campaner Hernandez.

some other issues to be considered include the following. Pickers work as a group and not as individuals, training is fundamental, and the machine needs to be designed to enable pickers to catch the fruit and reduce fruit impact. Mobile platform use is a new harvesting method for many growers and pickers, with numerous advantages and appears to be very reliable in Brazilian orchards. However, when considering a new system, both growers and pickers must embrace the change.

### 3.3 Sorting and selection on mobile platforms

A considerable amount of poor-quality fruit comes to the packing houses during fruit harvesting which could be more efficiently separated in the field. Selection at the point of harvesting decreases losses and improves system efficiency. Some initiatives for mobile platforms to improve this situation have been reported, with the use of an electronic (Vidal et al., 2013; Cubero et al., 2014) or mechanical platform unit (Carmelo et al., 2015).

One approach that can improve mobile platform use is related to the employment of computer vision for fruit selection under field conditions. Based on computer vision, fruit can be efficiently sorted and separated by size and colour. Electronic detection on a mobile unit is very different to a fixed structure, such as a packing line. This approach faces the challenge of working effectively under field conditions, as detailed by Cubero et al. (2014) and Vidal et al. (2013). As previously mentioned, Ferreira et al. (2018) described a mobile platform with multiple uses for orange harvesting. Initially, this platform was designed for fresh market tomatoes, and cleaning and sorting was achieved mechanically in the same machine, in a similar manner to a traditional packing line. However, this system encountered many problems, primarily related to physical damage associated with the length and complexity of the machine. This platform was later modified and was tested using oranges. Compact fruit sorter equipment was initially planned to be added to this unit, composed of a three-way compact brush system (Fig. 7), with the purpose of cleaning without water and sorting mechanically (Carmelo et al., 2015), but this was tested separately. Sorting in the field can considerably decrease fruit volume in packing houses, diminishing costs and resulting in better efficiency of renewable resources.

### 3.4 Preventing accidents on mobile platforms

Accidents on mobile platforms can occur, as with any machine. However, as more pickers may be involved in harvesting on the machine, accidents can harm more people than an individual mechanical harvester. MSA (2015) described a study in France involving accidents and ways of preventing them that could be applied to platforms used for activities such as harvesting and pruning. Approximately 325 accidents were notified from 2002 to 2009, and it was found that many of the accidents could be prevented by modifying the machine design, with special attention to picker protection. Cutini et al. (2017) reported that one problem of the mobile platform, for self-propelled fruit harvesters (SPFHs), is related to machine stability. Since these machines often have lateral extension arms and other devices that can affect stability, accident prevention could be achieved by thoroughly evaluating machine stability.

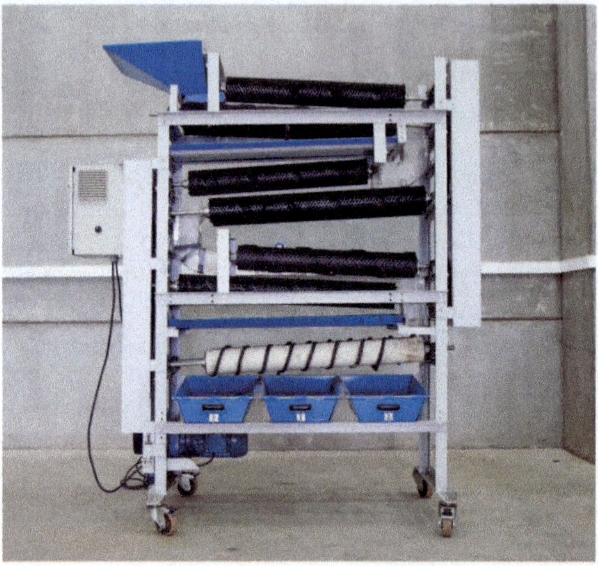

**Figure 7** Compact vertical sorting of fruits and vegetables. Accredited to: Luiza H. de M. Stalder.

### *3.5 Deciding on the use of a harvesting aid*

One important point to be considered when deciding on the use of a harvesting aid unit is related to costs and benefits. Zhang and Heinemann (2017) studied an apple harvest assisting unit and noted that it is important to analyse the profits and costs of using such equipment. For small farmers, even though this equipment can be used for other farm activities and replace ladder use, the savings often do not outweigh the costs. Therefore, an economic evaluation has to be undertaken before purchase. An alternative, in the case of oranges and small farmers, is to form an association or cooperative to share equipment costs. Sarig (2012) reported that harvesting performed by hand can restrict harvest yield improvement, leading to an impact on costs, and is dependent upon the picker's ability. Costa and Camarotto (2012) reported that in larger Brazilian orange orchards, mobile platform use can achieve a return on investment in one to three years.

## 4 Mechanical harvesting

Labour shortage and a search for citrus harvesting alternatives have been ongoing for many years, with most studies involving fruits destined for the juice industry. Whitney and Harrell (1989) reported a number of possibilities for mechanical harvesting, based mainly on the shake-catch principle, for example limb shaker-catch system, the self-propelled limb shaker and a conical scan air shaker. The harvesting efficiency of these systems was considered low at that time and highly dependent upon abscission chemicals to increase fruit dropping. Some work was undertaken to detach fruits via rotating augers inside the tree to pull the fruit out, but without much success among growers at that time; however, it was the basis for studies involving a system for mass fruit removal. Since then, the machines have undergone some changes but the main principles remain the same. Sanders (2005) reported that the main mechanical harvesters are based on shaking the trunk or limb (Figs. 8 and 9), and air shaking (Fig. 10).

The most readily available mechanical harvesting system is based on canopy shake, either by limb shake and catch (Fig. 8) or using a tractor (Fig. 9) (Sanders, 2005; Ebel et al., 2010; Gupta et al., 2016). The necessary force applied for fruit removal is very important, as well as time-shaking frequency and depth of penetration of fruit removal, which will influence fruit

**Figure 8** Continuous canopy shake and catch system. Clockwise. Top. Overview of the machine. Orange harvesting and catch system showing fruits and branches. Accredited to: Marcos David Ferreira.

**Figure 9** Tractor drawn canopy shake. Clockwise. Machine overview showing branches caught by the machine. Fruit detachment. Machine harvest. Shaking detachment. Accredited to: Marcos David Ferreira.

detachment using low power energy, affecting costs and efficiency (Liu et al., 2018). Gupta et al. (2016) reported that shaking enables efficient fruit removal but that tree damage was still high; changes to limb frequency, amplitude and material use allowed adequate fruit detachment and incurred less injury to tree trunks and branches. Consequently, a proposed model was designed using a Pareto-optimal solution search technique. One important point, which was stressed by the authors, was that the tree canopy is not of uniform design and structure; therefore, shaking forces have to be specific for the citrus canopy. For most of these techniques, abscission chemicals can be an important method to help increase harvest efficiency (Ebel et al., 2010; Moreno et al., 2015).

As previously mentioned, the prevailing method for the fresh market is manual harvesting, since fruit quality needs to be higher. Torregrosa et al. (2009) reported that for Spain, manual harvesting is the most widely used

**Figure 10** Air-shaking harvester. Accredited to: Marcos David Ferreira.

harvest method as most fruit production is for fresh market. Detachment, using trunk or hand-held shakers, may be an alternative. Fruit detachment varies among cultivars and canopy fruit position, which is related to fruit maturity. A tractor shaker was found to be more effective than a hand-held shaker, which could be related to the type of equipment used by the pickers. Sanders (2005) reported that most mechanical methods detach the fruit from the tree without any selection process, which can be harmful to the crop due to the removal of small immature fruits. Therefore, fruit uniformity on the day of harvest is an important feature to be considered.

Mechanical harvesting was widely used in the orange groves of Florida, USA, but the use of machines for fruit harvesting has decreased due to greening (HLB) and has been in decline since the harvest of 2008–2009 (Roka and Hyman, 2012). Orange production decreased from 242 million boxes (90 pound boxes), before greening, to 104.6 million boxes in 2014 (Singerman and Useche, 2016), and is still decreasing, being one-third of the production prior to greening (2003–2004) (Trejo-Pech et al., 2018).

### *4.1 Fruit detachment and quality*

Fruit detachment, mechanically or manually, can improve harvest speed and has been a common practice for a number of years (Whitney and Harrell, 1989). However, it can have some disadvantages, such as fruit impact damage

(Miranda et al., 2015) and contamination (Goodrich et al., 2006). Miranda et al. (2015) reported that, when comparing manual harvesting and detachment, fruits harvested by detachment can be subjected to 4-fold increased acceleration forces compared with manual harvesting. Using a laboratory simulation, from harvest to truck loading to industry, fruits subjected to impact drop forces showed a decrease in titratable acidity, soluble solids, ascorbic acid and weight by 8.7%, 5.5%, 4.6% and 0.5%, respectively, when compared to fruits not subjected to impact. Ortiz et al. (2011) described different drop surfaces showing specific capacities to absorb impact during citrus detachment. To decrease fruit impact, the use of elevated canvases inserted onto the tree canopy is recommended. Weeds and mulch showed an intermediate capacity of absorbing impact, which was better than fruit falling on bare soil.

### 4.2 Automation of mechanical harvesters

Sola-Guirado et al. (2017) reported the potential use of automated canopy shakers tested on olives that may increase field yield, compared with the manual method, with potential use for other crops. Two distinctive automatic systems composed of electronic devices with shaker heads were added to a machine harvester for olives. An increase of 5.9% in detachment efficiency was observed, with no increased tree damage. An LED laser was used to measure canopy distance, but it was not as efficient as human vision. There is the potential to use this system for other crops, but orchard design modifications have to be considered as well as improvements to increase the harvesting efficiency of the automated system.

## 5 Robotic harvesting

Robotics and automation systems for harvesting fruits destined for the fresh market are not currently ready for commercialization. Zujevs et al. (2015) reported that robotics has been applied to greenhouse harvesting, but not for field commercial harvesting. However, there is a considerable amount of research using different sensors, based on machine vision, applied mainly to robotics and even harvesters that indicates some alternatives and possibilities. The main feature of robotic harvesting is to harvest fruits from the tree with a mechanical arm using machine vision. Consequently, there are a considerable number of studies on robotic machine vision applied to citrus harvesting, considering different methods and sensors (Harrell, 1987; Whitney and Harrell, 1989; Harrell et al., 1990; Pla et al., 1993; Jiménez et al., 1999; Hannan and Burks, 2004; Mehta and Burks, 2014, 2016; Cubero et al., 2016; Mehta et al., 2017).

Robotic harvesting has to be very efficient to compete with other harvest systems. Cost is still a barrier, but a number of possibilities have been investigated. Harrell (1987) reported that some obstacles to robotic adoption were related to the cost of machine implementation and low harvest efficiency, as well as changes to harvesting operations and orchard design. Li et al. (2011) has reported the issues, history, challenges and benefits of using automatic robotic harvesting. Machine vision systems are the most used, and identification rate can be between 70–90%, with variation related to environmental conditions; however, harvest speed is one of the main challenges to be solved in these systems. Furthermore, Changhui et al. (2017) reported that overlapping fruit is a major problem when using vision by robotics for automatic harvesting, though improvements in programming can reduce errors significantly. Another problem reported from using robotic machine vision was the effect caused by wind, canopy movement and fruit detachment. Mehta and Burks (2016) proposed an adaptive visual servo control in order to alleviate this problem.

### *5.1 Sensors*

Li et al. (2011) highlighted that the main points for automatic harvesting are the sensors used to capture the image, as well as processing computer capacity to distinguish fruit. Congal et al. (2015) have categorized sensors and systems for fruit detection and localization. There are a number of camera types for fruit detection, from black and white (B/W), colour, and more recently spectral (Okamoto and Lee, 2009; Congal et al., 2015) and thermal types (Bulanon et al., 2008).

Image classification methods for fruit detection can be defined as: (1) unsupervised (K-means clustering), (2) supervised (Bayesian classifier and KMN clustering) and (3) soft computing methods (Artificial Neural Network (ANN) and Support Vector Machine (SVM)). For fruit localization, an important factor is velocity of catching fruit from the tree, hence single camera, laser range finder, stereovision and time-of-flight (TOF) light-based 3D cameras have been used (Li et al., 2011; Congal et al., 2015). Lee and Ehsani (2015) reported that sensing systems can be important in such areas as yield mapping, disease incidence, monitoring nutritional stress and so on. There are some differences on sensors type applied to grain and for perennial crops, such as citrus, which need to be considered.

## 6 Precision agriculture to estimate citrus yield

Initial studies in this area suggested that it was possible to design yield mapping for manual harvesting using GIS/GPS with high accuracy (Whitney et al., 1999). There has been considerable research since then using different methods to estimate citrus yield, as a method to help harvest planning. Colaço et al. (2015) reported that it is possible to generate yield maps employing various different methods to determine bag location according to optimum likely yield, which could be an important contribution to harvesting, especially manual. There are reports related to fruit detection by image processing (Mehta et al., 2017; Sengupta and Lee, 2014) which can also be used to estimate yield (Dorj et al., 2017).

## 7 Future trends

Sanders (2005) reported that dwarf trees could be an alternative to improve mechanization. There have been some studies (Wheaton et al., 1991, 1995; Donadio and Stuchi, 2001; Stuchi and Girardi, 2010; Stuchi, 2012; Soares Filho et al., 2014; Martínez-Cuenca et al., 2016) on the use of dwarf orange trees, however, little has been applied to commercial orchards. A dwarf tree provides potential opportunities for mechanical harvesting, such as using tunnel harvesters, and can also facilitate manual harvesting. However, it is not an easy situation to change. The main varieties in production in Brazil are Pera Rio, Valência, Hamlin and Natal, which all have an elevated canopy. However, there is a significant move to enhance tree density, with a current average of 668 trees/hectare, which is approximately double the density of 30 years ago and has been increasing year by year (FUNDECITRUS, 2018). Advanced planning is essential in order to change the harvesting of a perennial crop, such as oranges, and growers are very concerned about harvesting, which takes at least 35% of production costs and is increasing year by year. A plan for change is complex and involves many sectors and multidisciplinary work. For olives changing from elderly trees to rows in high density was a challenge (Sola-Guirado et al., 2017). Ferguson et al. (2010) described the mechanical harvesting of olives mainly by shaking the canopy and trunk, and modifications of this system have made it more efficient. However, some issues have been reported such as specific orchard management requirements and a possible increase in pest and disease incidence (Rallo et al., 2013).

## 8 Conclusion

Currently, there is no single best method for harvesting oranges. It is important to integrate the work of all those involved in harvesting research and development, that is, growers, industry, the research community and the public sector. For example, in Brazil, harvesting is mainly manual. All attempts to harvest mechanically have failed, due to technical and economic issues, such as machine costs, orchards not designed for mechanization, tree damage, low yield efficiency and so on. Brazilian orange orchards historically have focussed disease control and any damage to trees could potentially increase the chance of disease and pests. Semi-mechanized harvesting using mobile platforms could be an option for Brazilian growers. However, many Florida growers invested in mechanical harvesting, but have had to discontinue this method of harvesting due to greening (HLB) disease. The orange system is dynamic and has overcome many challenges, and with cooperation can find technologically based solutions. It seems that with the advance of sensors and genetics, new possibilities can emerge, for example mechanical systems, such as tunnel harvesters, or even automated, robotic systems. Advances in tree format (e.g. use of dwarf varieties) may facilitate these developments. A short and narrow canopy could enable tunnel harvester use, as for olives and even for grapes, and research is ongoing. In the meantime, possibilities include improving the available methods and equipment, such as the use of mobile platforms, or even making machine harvesting more efficient. Sensor use connected to machine vision on an automated system could complement and improve this situation. Mass harvesting has to be considered as oranges are harvested in many countries for juice. The pressure for alternatives is very high, with commercial automation harvesting showing potential, but integrated planning with all involved will be crucial. In a world in which automation is adopted on a daily basis, automated harvesting has potential, but barriers such as cost and efficiency have to be overcome.

## 9 Where to look for further information

Citrus is an important crop for a number of countries, and each country search for alternatives for harvesting and crop production. There are number of centers for citrus research around world, studying from production, harvesting and consumption, for fresh market and fruit juice.

Information about crop production data is available on:

- Citrus world markets and trade information from United States and other producer countries – Biannual report published by United States Department of Agriculture https://www.fas.usda.gov/data/citrus-world-markets-and-trade
- Fundo de Defesa da Citricultura – Fundecitrus https://www.fundecitrus.com.br/. Fundecitrus annually publish reports on citrus Brazilian harvest estimation, in Portuguese and English.
- Brazilian Association of Citrus Exporters – CitrusBR http://www.citrusbr.com/ is another important source of data information related to production, yield, and so on.

Citrus research studies have been developed in Centers and Universities around world. Further information also including harvesting can be found on the following:

- In Brazil: Embrapa Mandioca e Fruticultura - https://www.embrapa.br/mandioca-e-fruticultura; Fundo de Defesa de Citricultura - Fundecitrus https://www.fundecitrus.com.br/. Centro de Citricultura Sylvio Moreira - http://ccsm.br/. On United States: University of Florida - https://irrec.ifas.ufl.edu/citrus-horticulture/ and https://irrec.ifas.ufl.edu/postharvest/ and on USDA - Agricultural Research Service - (https://www.ars.usda.gov/southeast-area/fort-pierce-fl/us-horticultural-research-laboratory/).

There are important research centers with intensive citrus studies on Australia, Israel and Spain.

A number of recent publications can be found on the following research centers:

- One recent publication that refers to this chain is: "Orange Juice Chain: Past, Present and Future". Authors: Trombin, V. G.; Neto, L. C. M.; Kalaki, R. B. 175p. Available for free download on https://www.fruit-processing.com/wp-content/uploads/2019/04/OJC_Neves_et_al_2019.pdf
- Another recent contribution is: Citrus. Chapter 9. Chen, Y.; Barzee, T. J.; Zhang, R.; Pan, Z. p.217-241. On: Integrated Processing Technologies for Food and Agricultural By-Products. Editors Zhongli Pan, Ruihong Zhang, Steven Zicari
- Postharvest and harvest issues in Brazil: Embrapa Instrumentação - http://poscolheita.cnpdia.embrapa.br/livros.

Conferences:

- International Citrus Conference occurs every 4 years and it is an important forum for this crop. The last one was in Brazil - Foz do Iguaçu - www.icc2016.com and the next one will be held on Mersin, Turkey - www.icc2020.org.

## 10 References

Agrianual - ANUÁRIO DA AGRICULTURA BRASILEIRA. 2017. FNP, São Paulo.

Ampatzidis, Y. G., Vougioukas, S. G. and Whiting, M. D. 2011. A wearable module for recording worker position in orchards. *Computers and Electronics in Agriculture* 78(2), 222–30.

Bulanon, D. M., Burks, T. F. and Alchanatis, V. 2008. Study on temporal variation in citrus canopy using thermal imaging for citrus fruit detection. *Biosystems Engineering* 101(2), 161–71.

Carmelo, L. G., Miranda, M., Spricigo, P. C., Pilon, L. and Ferreira, M. D. 2015. Evaluation of potential use of a compact sorter equipment for small vegetables growers. *Chemical Engineering* 44, 91–6.

Castro-Garcia, S., Sola-Guirado, R. R. and Gil-Ribes, J. A. 2018. Vibration analysis of the fruit detachment process in late-season 'Valencia' orange with canopy shaker technology. *Biosystems Engineering* 170, 130–7.

Changhui, Y., Youcheng, H., Lin, H., Sa, L. and Yanping, L. 2017. Overlapped fruit recognition for citrus harvesting robot in natural scenes. In: *2017 2nd International Conference on Robotics and Automation Engineering (ICRAE)*. IEEE, pp. 398–402.

Colaço, A. F., Trevisan, R. G., Karp, F. H. S. and Molin, J. P. 2015. Yield mapping methods for manually harvested crops. In: *Precision Agriculture'15*. Wageningen Academic Publishers, pp. 39–44.

Costa, S. E. A. and Camarotto, J. A. 2012. An ergonomics approach to citrus harvest mechanization. *Work* 41(Suppl. 1), 5027–32. doi:10.3233/WOR-2012-0794-5027.

Cubero, S., Aleixos, N., Albert, F., Torregrosa, A., Ortiz, C., García-Navarrete, O. and Blasco, J. 2014. Optimised computer vision system for automatic pre-grading of citrus fruit in the field using a mobile platform. *Precision Agriculture* 15(1), 80–94.

Cubero, S., Lee, W. S., Aleixos, N., Albert, F. and Blasco, J. 2016. Automated systems based on machine vision for inspecting citrus fruits from the field to postharvest–a review. *Food and Bioprocess Technology* 9(10), 1623–39.

Cutini, M., Brambilla, M., Bisaglia, C., Melzi, S., Sabbioni, E., Vignati, M., Cavallo, E. and Laurendi, V. 2017. A study of the lateral stability of self-propelled fruit harvesters. *Agriculture* 7(11), 92.

Donadio, L. C. and Stuchi, E. S. 2001. *Adensamento de plantio e ananicamento de citros*. FUNEP, Jaboticabal.

Dorj, U. O., Lee, M. and Yun, S. S. 2017. An yield estimation in citrus orchards via fruit detection and counting using image processing. *Computers and Electronics in Agriculture* 140, 103–12.

Ebel, R. C., Burns, J. K., Morgan, K. T. and Roka, F. 2010. Abscission agent application and canopy shaker frequency effects on mechanical harvest efficiency of sweet orange. *HortScience* 45(7), 1079–83.

Elkins, R. B. 2012. A Historical timeline of research efforts to implement mechanized pear harvest in California. In: *I International Symposium on Mechanical Harvesting and Handling Systems of Fruits and Nuts 965*, pp. 225–30.

Fathallah, F. A. 2010. Musculoskeletal disorders in labor-intensive agriculture. *Applied Ergonomics* 41(6), 738–43. doi:10.1016/j.apergo.2010.03.003.

Ferguson, L., Rosa, U. A., Castro-Garcia, S., Lee, S. M., Guinard, J. X., Burns, J., Vossen, P. M. and Glozer, K. 2010. Mechanical harvesting of California table and oil olives. *Advances in Horticultural Science* 24(1), 53–63.

Fernandes, W. R., Silva, A. L. V. and Martins, M. E. A. 2017. Colheita de laranja, o negócio. In: Ferreira, M. D. (Ed.), *Instrumentação pós-colheita em frutas e hortaliças*. Embrapa Instrumentação, São Carlos, pp. 89–101.

Ferreira, M. D., Sanchez, A. C., Braunbeck, O. A. and Santos, E. A. 2018. Harvesting fruits using a mobile platfom: a case study applied to citrus. *Engenharia Agrícola* 38(2), 293–9.

FUNDECITRUS. 2018. Tree inventory of São Paulo and west-southwest of Minas Gerais citrus belt: snapshot of groves in March 2018/ Fundo de Defesa da Citricultura... [et al.]. Fundecitrus, Araraquara, SP, 111p.

Gongal, A., Amatya, S., Karkee, M., Zhang, Q. and Lewis, K. 2015. Sensors and systems for fruit detection and localization: a review. *Computers and Electronics in Agriculture* 116, 8–19.

Goodrich, R. M., Ehsani, R. and Friedrich, L. M. 2006. Effect of fruit pick up systems on the microbial quality of citrus fruit surfaces. *Proceedings of the Florida State Horticultural Society* 119, 195–7.

Gupta, S. K., Ehsani, R. and Kim, N. H. 2016. Optimization of a citrus canopy shaker harvesting system: mechanistic tree damage and fruit detachment models. *Transactions of the ASABE* 59(4), 761–76.

Hamam, A. S., El-Iraqi, M. E., Sharibim, Y. and Awais, T. R. 2011. Design and evaluation of mechanical picking heads for citrus fruit harvesting. *Egypt Journal of Agricultural Research* 89(4), 1545–62.

Hannan, M. W. and Burks, T. F. 2004. Current developments in automated citrus harvesting. In: *2004 ASAE Annual Meeting*. American Society of Agricultural and Biological Engineers, p. 1.

Harrell, R. 1987. Economic analysis of robotic citrus harvesting in Florida. *Transactions of the ASAE* 30(2), 298–0304.

Harrell, R. C., Adsit, P. D., Munilla, R. D. and Slaughter, D. C. 1990. Robotic picking of citrus. *Robotica* 8(4), 269–78. doi:10.1017/S0263574700000308.

Jiménez, A. R., Jain, A. K., Ceres, R. and Pons, J. L. 1999. Automatic fruit recognition: a survey and new results using range/attenuation images. *Pattern Recognition* 32(10), 1719–36.

Jutras, P. J. and Coppock, G. E. 1958. Mechanization of citrus fruit picking. *Florida State Horticultural Society* 71, 201–4.

Lee, W. S. and Ehsani, R. 2015. Sensing systems for precision agriculture in Florida. *Computers and Electronics in Agriculture* 112, 2–9.

Li, P., Lee, S. H. and Hsu, H. Y. 2011. Review on fruit harvesting method for potential use of automatic fruit harvesting systems. *Procedia Engineering* 23, 351–66.

Liu, T. H., Luo, G., Ehsani, R., Toudeshki, A., Zou, X. J. and Wang, H. J. 2018. Simulation study on the effects of tine-shaking frequency and penetrating depth on fruit detachment for citrus canopy-shaker harvesting. *Computers and Electronics in Agriculture* 148, 54–62.

Martínez-Cuenca, M. R., Primo-Capella, A. and Forner-Giner, M. A. 2016. Influence of rootstock on citrus tree growth: effects on photosynthesis and carbohydrate distribution, plant size, yield, fruit quality, and dwarfing genotypes. In: Rigobelo, E. C. (Ed.), *Plant Growth*. InTech, pp. 107–24.

Mehta, S. S. and Burks, T. F. 2014. Vision-based control of robotic manipulator for citrus harvesting. *Computers and Electronics in Agriculture* 102, 146–58.

Mehta, S. S. and Burks, T. F. 2016. Adaptive visual servo control of robotic harvesting systems. *IFAC-PapersOnLine* 49(16), 287–92.

Mehta, S. S., Ton, C., Asundi, S. and Burks, T. F. 2017. Multiple camera fruit localization using a particle filter. *Computers and Electronics in Agriculture* 142, 139–54.

Miranda, M., Spricigo, P. C. and Ferreira, M. D. 2015. Mechanical damage during harvest and loading affect orange postharvest quality. *Engenharia Agrícola* 35(1), 154–62.

Mitchell, F. G., Maxie, E. C. and Greathe, A. S. 1964. *Handling Strawberries for Fresh Market*. University of California, Davis, 16pp.

Moreno, R., Torregrosa, A., Moltó, E. and Chueca, P. 2015. Effect of harvesting with a trunk shaker and an abscission chemical on fruit detachment and defoliation of citrus grown under Mediterranean conditions. *Spanish Journal of Agricultural Research* 13(1), 1-12. Instituto Nacional de Investigación y Tecnología Agraria y Alimentaria (INIA).

MSA. 2015. La sécurité des plates-formes de récolte et de taille en arboriculture. Caisse Centrale MSA, 60p. Available at: https://ssa.msa.fr/wp-content/uploads/2018/12/La-securite-des-plates-formes-de-recolte-et-de-taille-en-arboriculture.pdf.

Narimoto, L. R. and Belussi, S. E. A. C. 2018. The role of design in use in agriculture: the case of Brazilian crops. In: *Congress of the International Ergonomics Association*. Springer, Cham, pp. 1316-27.

Neves, M. F., Trombin, V. G., Kalaki, R. B. and Lopes, F. F. 2013. Competitiveness of the orange juice chain in Brazil. *International Food and Agribusiness Management Review* 16(4), 141-58.

Okamoto, H. and Lee, W. S. 2009. Green citrus detection using hyperspectral imaging. *Computers and Electronics in Agriculture* 66(2), 201-8.

Ortiz, C., Blasco, J., Balasch, S. and Torregrosa, A. 2011. Shock absorbing surfaces for collecting fruit during the mechanical harvesting of citrus. *Biosystems Engineering* 110(1), 2-9.

Pla, F., Juste, F. and Ferri, F. 1993. Feature extraction of spherical objects in image analysis: an application to robotic citrus harvesting. *Computers and Electronics in Agriculture* 8(1), 57-72.

Rallo, L., Barranco, D., Castro-García, S., Connor, D. J., Gómez del Campo, M. and Rallo, P. 2013. High-density olive plantations. *Horticultural Reviews* 41, 303-84.

Roka, F. M. and Hyman, B. R. 2012. Mechanical harvesting of sweet oranges for juice processing. In: *I International Symposium on Mechanical Harvesting and Handling Systems of Fruits and Nuts 965*, pp. 241-3.

Sanders, K. F. 2005. Orange harvesting systems review. *Biosystems Engineering* 90(2), 115-25.

Sarig, Y. 2012. Mechanical harvesting of fruit-Past achievements, current status and future prospects. In: *I International Symposium on Mechanical Harvesting and Handling Systems of Fruits and Nuts 965*, pp. 163-9.

Seamount, D. T. and Opitz, K. W. 1973. Sunkist 2C platform harvester. *Citrograph* 59, 55-8.

Sengupta, S. and Lee, W. S. 2014. Identification and determination of the number of immature green citrus fruit in a canopy under different ambient light conditions. *Biosystems Engineering* 117, 51-61.

Singerman, A. 2017. Cost of production for processed oranges grown in Central Florida (Ridge), 2015/2016. UF/IFAS Extension. FE1006, 4p.

Singerman, A. and Useche, P. 2016. Impact of citrus greening on citrus operations in Florida. University of Florida/Institute of Food and Agricultural Sciences Extension. FE983.

Soares Filho, W. D. S., Ramos, Y. C., da Silva, A. L. V., Fadel, A. L., Stuchi, E. S., Girardi, E. A., de Leão, H. C., Gesteira, A. da S. and Passos, O. S. 2014. Dwarfing rootstocks for 'Valencia' sweet orange. In: *XXIX International Horticultural Congress on Horticulture: Sustaining Lives, Livelihoods and Landscapes (IHC2014): 1178*, pp. 85-8.

Sola-Guirado, R. R., Ceular-Ortiz, D. and Gil-Ribes, J. A. 2017. Automated system for real time tree canopy contact with canopy shakers. *Computers and Electronics in Agriculture* 143, 139–48.

Stuchi, E. 2012. Controle do tamanho de plantas cítricas. *Cordeirópolis* 33(2), 91–112.

Stuchi, E. and Girardi, E. 2010. Utilização de práticas culturais na citricultura frente ao huanglongbing. Embrapa Mandioca e Fruticultura-Documentos (INFOTECA-E).

Torregrosa, A., Orti, E., Martín, B., Gil, J. and Ortiz, C. 2009. Mechanical harvesting of oranges and mandarins in Spain. *Biosystem Engineering* 104(1), 18–24.

Trejo-Pech, C. J., Spreen, T. H. and Zansler, M. L. 2018. Is growing oranges in Florida a good investment? *American Journal of Agricultural Economics* 100(2), 625–39.

USDA. 2018. Citrus: world markets and trade. United States Deptartment of Agriculture, Foreign Agricultural Service, Government Printing Office, Washington DC.

Vidal, A., Talens, P., Prats-Montalbán, J. M., Cubero, S., Albert, F. and Blasco, J. 2013. In-line estimation of the standard colour index of citrus fruits using a computer vision system developed for a mobile platform. *Food and Bioprocess Technology* 6(12), 3412–9.

Vidal, D. O., Compagnon, A. M., da Silva, R. P., Raveli, M. B. and Cavichioli, F. A. 2015. Medidas antropométricas e eficiência na colheita manual de citros. *Multi-Science Journal* 1(2), 111–6.

Wheaton, T. A., Castle, W. S., Whitney, J. D. and Tucker, D. P. H. 1991. Performance of citrus scion cultivars and rootstock in a high-density planting. *HortScience* 26(7), 837–40.

Wheaton, T. A., Whitney, J. D., Castle, W. S., Muraro, R. P., Browning, H. W. and Tucker, D. P. H. 1995. Citrus scion and rootstock, topping height, and tree spacing affect tree size, yield, fruit quality, and economic return. *Journal of the American Society for Horticultural Science* 120(5), 861–70.

Whitney, J. D. and Harrell, R. C. 1989. Status of citrus harvesting in Florida. *Journal of Agricultural Engineering Research* 42(4), 285–99.

Whitney, J. D., Miller, W. M., Wheaton, T. A., Salyani, M. and Schueller, J. K. 1999. Precision farming applications in Florida citrus. *Applied Engineering in Agriculture* 15(5), 399.

Whitney, J. D., Ling, Q., Wheaton, T. A. and Miller, W. M. 2001. A citrus harvesting labor tracking and yield mapping system. *Applied Engineering in Agriculture* 17(2), 121.

Zhang, Z. and Heinemann, P. H. 2017. Economic analysis of a low-cost apple harvest-assist unit. *HortTechnology* 27(2), 240–7.

Zujevs, A., Osadcuks, V. and Ahrendt, P. 2015. Trends in robotic sensor Technologies for Fruit Harvesting: 2010–2015. *Procedia Computer Science* 77, 227–33.

# Chapter 3

# Advances in postharvest handling of citrus fruit

*John Golding and John Archer, New South Wales Department of Primary Industries and University of Newcastle, Australia*



## 1 Introduction

Citrus is a major world horticultural crop with over 92 million tons of fruit produced annually. Oranges have traditionally been the primary citrus

http://dx.doi.org/10.19103/AS.2019.0054.08

grown for fresh fruit consumption (30 million tons) with significant volumes of oranges specifically grown for processing, that is juice production (18 million tons). However an increasing trend is the production of other citrus types such as mandarins and tangerines which are increasing in popularity, where world production in these 'soft' citrus types is now 29 million tons per year. The increased production of these citrus types is being driven by consumer demand which generally wants smaller, easy-to-peel and sweeter fruit. The transition in production from oranges to mandarin-type citrus has been a global trend for many years, where the production of mandarins and tangerines is now nearly equal to the production of fresh oranges. While lower in production than oranges and mandarins, the production of grapefruit (6 million tons) and lemons and limes (5 million tons) also significantly contributes to the economic importance of global citrus production. Each of these different citrus types has its own special handling, storage and physiological needs.

The drive and demand of consumers are not only changing citrus types but also orchard and postharvest management such as reduced chemical use and organics. These changes in postharvest management will be discussed in more detail in this chapter.

Citrus fruit have a characteristic maturation and ripening physiology, where some citrus types (such as early season mandarins and oranges) undergo unique postharvest treatments such as degreening. These techniques are described and compared in more detail in this chapter.

Citrus is a very important global horticultural commodity which is widely traded. Indeed citrus is the highest value horticulture crop but the global trade of fresh citrus must also meet phytosanitary (quarantine) requirements. Indeed, overcoming phytosanitary barriers for the export of citrus around the world is critical to many export-focussed countries. Global citrus exports account for 15% of world citrus production, where 10.2 million tons of fruit (primarily oranges) are traded each year. While much of world citrus trade is conducted without any quarantine barriers, many important export markets have phytosanitary restrictions to prevent the entry and spread of quarantine pests and diseases. For example, the presence of fruit flies in a citrus-growing region can restrict trade to markets where these fruit flies are not present. While there can be several market access treatments available to allow trade, these can often have negative effects on fruit quality or can be too costly. These market access treatments and their effects on fruit quality will be discussed.

Food safety is non-negotiable and is fundamental to the safe consumption of fruit and vegetables, including citrus. While citrus has often been regarded as a low-risk horticultural crop, microbial and chemical contamination risks can occur and need to be managed. Other trends in global citrus trade include chemical use, sustainability and the increasing role of consumers. The role of postharvest meeting consumer expectations, reducing chemical use and improving the sustainability of citrus production is described.

## 2 Background

The most commonly grown citrus species around the world are sweet orange (*Citrus sinensis*), mandarin (*C. reticulate*), grapefruit (*C. paradisi*), lemon (*C. limon*), calamondin (*C. mitis*), citron (*C. medica*), Kaffir lime (*C. hystrix*), sour orange (*C. aurantium*), Mexican lime (*C. aurantifolia),* pomelo (*C. grandis*) and Persian (Tahitian) limes (*C. latifolia*). It was generally considered that citrus originated in the area between India, South China and Indonesia, but more recent genomic, phylogenetic and biogeographic analyses indicate that the centre of origin of citrus species was the southeast foothills of the Himalayas (in a region that includes the eastern area of Assam, northern Myanmar and western Yunnan) (Wu et al., 2018). Indeed Wu et al. (2018) contend there may be a new evolutionary framework for citrus, a scenario that challenges current taxonomic and phylogenetic thoughts and points towards a reformulation of the genus *Citrus*.

Citrus fruit are botanically classified as modified berry, a specialised form of berry (hesperidium) resulting from a single ovary and have a thick peel/skin (pericarp), which surrounds the fleshy edible portion of the fruit. The outermost layer of the pericarp is the flavedo or exocarp. This is the coloured portion of the peel and contains the distinctive oil glands which can either be sunken or bulge from the surface contributing to the characteristic 'dimples' on the fruit rind (Fig. 1). The size of oil glands ranges from 10 to 100 μm or more. Terpenes (mainly D-limonene) and sesquiterpenes of oils in these glands give the characteristic aroma of different citrus species. The orange colour of the fruit is due to the presence of carotenoids such as β-carotene and cryptoxanthin. Interestingly, the colour 'orange' was named after the fruit. The outer layer of the peel is protected by a cuticle which itself is protected with an outer layer of natural waxes (formed of alcohols, paraffin, aldehydes, ketones etc.). It is these outer layers which give protection against water loss. The underside of the peel (mesocarp) is the albedo or pith which is characterised by soft white tissue comprising a web of thin-walled cells with large air spaces. The albedo in mature citrus can range from 1 to 2 mm thick in limes and tangerines, 2 to 5 mm thick in sweet oranges and up to 20 mm thick in pomelos. However, there is generally no albedo in kumquats and very little in mandarins. The edible portion of citrus fruit contains segments which are enclosed with a continuous endocarp membrane (Fig. 1). The interior of the segment contains many small thin-walled sacs called juice vesicles (pulp) which are attached to a thin wall called the carpellary septum which surround the segments. The juice sacs can contain varying numbers of seeds (ovules).

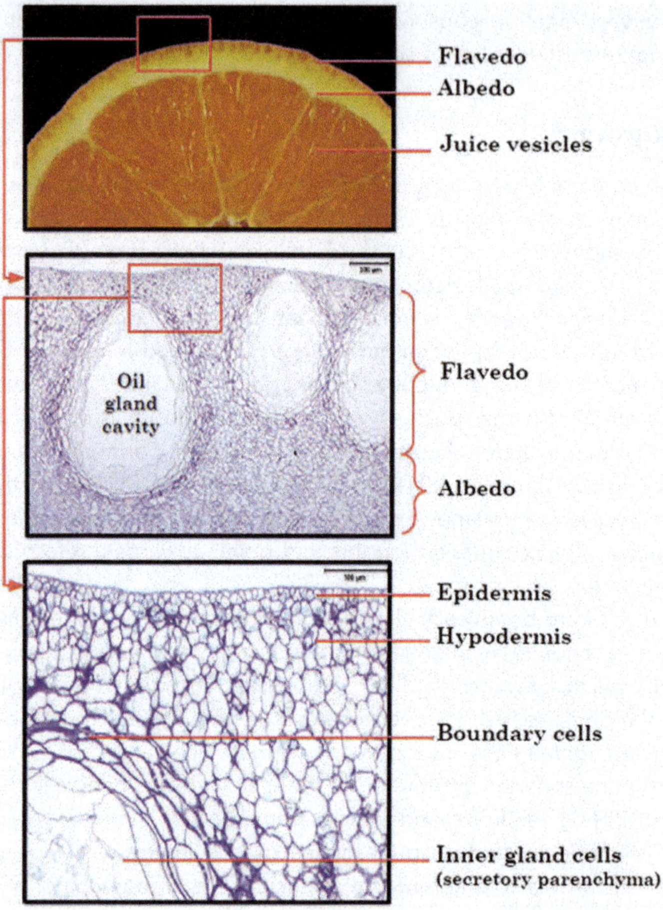


**Figure 1** Anatomy of navel orange rind. Source: used with permission from Lindhout (2007).

## 3 Nutritional content

Citrus fruit are considered nutritious healthy fruit which contain a range of key nutrients such as vitamin C, vitamin A, carotenoids of various kinds (such as β carotene, lutein, zeaxanthin), folate and fibre. In addition, citrus fruit also contain many non-nutrient phytochemicals including dietary fibre and phytochemical classes such as flavonoids, glucarates, coumarins, monoterpenes, triterpenes and phenolic acids and include individual components such as hesperidin, naringin, tangeretin, limonene, nomilin, perillyl alcohol, myricetin, quercetin, sinsensetin, tangeretin and nobiletin. Indeed it is these phytochemical constituents that give the different citrus fruit and types their flavours and unique colour. For example oranges are orange in colour due to carotenoids while blood oranges are red due to the presence of anthocyanins.

Oranges have reported over 170 different phytochemicals and more than 60 flavonoids. Many of these phytochemicals have not only been shown to have antioxidant effects but also, in cellular and animal experiments, to have anti-inflammatory and anti-tumour activity, as well as blood clot-inhibiting activity. However in this latter area, human data are sparse. The antioxidant (free radical scavenging) ability of citrus fruit is generally high, where protection against free radical damage to proteins, lipids and DNA is an integral part of the body's defence mechanism. It is possible that at least some of the antioxidants in fruit such as citrus may contribute to enhancing these endogenous antioxidant defences. However of increasing interest is the possibility that these substances interact with dietary fibre, and gut microbiota contribute to optimal health and may protect against some of the common chronic diseases and general damage caused by ageing and degenerative diseases. For example a 15-year long-term study showed that the consumption of oranges (containing flavonoids) reduced the prevalence and incidence of late age-related macular degeneration in older consumers (Gopinath et al., 2018).

## 4 Fruit physiology

Citrus are classified as 'non-climacteric' fruit, where mature citrus fruit show no respiratory peak and little ethylene production during storage. Therefore once the fruit is harvested, it will not further ripen. This is in contrast to climacteric fruit such as banana which go through rapid ripening and quality transformations after harvest and treatment with ethylene to initiate ripening (Wills and Golding, 2016). Therefore it is important that citrus fruit are harvested at their optimum maturity, as the internal fruit quality will not substantially alter during storage, there may be some changes but these will be relatively small. While citrus are classified as non-climacteric, it is important to recognise that ethylene is used in degreening mature citrus fruit (see Section 5) and ethylene can also have negative effects on fruit quality during long storage.

## 5 Degreening

One of the unique postharvest treatments that are applied to citrus fruit is degreening. In some growing situations, citrus fruit can be internally mature and ready-to-eat but due to the warm growing conditions, the peel remains green. Degreening is a process of using ethylene to initiate the removal of the green colour (chlorophyll) from the skin of fruit after harvest and it also hastens the development of carotenoid synthesis. Thus degreening enables ready-to-eat fruit to meet market requirements for a full orange skin colour. However in many tropical countries/markets, the green-coloured skin of citrus fruit is considered normal and citrus fruit are not normally degreened in this way. But in markets which expect orange-coloured skin, degreening is common place, particularly in the early parts of the season where the night orchard temperatures are high (>15°C) and chlorophyll content in the peel naturally remains in the skin.

The degreening treatment itself involves treating citrus fruit at elevated storage temperatures and relative humidity in the presence of ethylene for a short period before packing and marketing. The optimum temperature for degreening depends on initial fruit colour and local climatic conditions (especially temperature) at the time of harvest. In general, 20–25°C treatment temperature is recommended for fruit grown in the Mediterranean-type climates and up to 27°C for fruit grown in the humid subtropical regions. Treating the fruit with a concentration of ethylene between 1 and 5 parts per million is normally required for degreening, and it is also essential to ensure the levels of carbon dioxide are kept low. The success of degreening can vary with variety, the initial colour of the fruit, concentration of ethylene used and the duration of exposure to ethylene.

It is essential that only mature fruit are degreened. It is important that immature fruit are not harvested and degreened as this immature fruit will have very high acid levels and low sugar levels and will not be suitable for eating. But equally important is that immature dark green fruit will not successfully degreen. Ethylene treatment only helps with the breakdown of the green colour in the skin. As citrus are a non-climacteric fruit, ethylene does not ripen the fruit, that is the levels of sugars and acids are not affected by degreening. Therefore when degreening, it is critical to selectively pick mature fruit based on colour.

## 6 Physiological storage disorders

Physiological disorders involve the breakdown of plant tissue that is not directly caused either by pests and diseases or by mechanical damage. There are many different (and related) described disorders and injuries in citrus. The names and classification of these disorders are very descriptive and reflect their often indeterminate and seasonal nature. The biochemical and biophysical mechanisms that give rise to physiological disorders are extremely complex and not well understood (Wills and Golding, 2016).

There are numerous physiological disorders described in the literature which are sporadic in nature but can cause substantial postharvest losses. Different storage disorders develop in response to various preharvest growing and postharvest storage conditions such as nutrient deficiency that occurs in the field and low temperature stress during storage. The effect of the preharvest growing environment on the development of many disorders cannot be overstated where often unknown preharvest growing conditions affect postharvest storage life and quality. There are numerous postharvest disorders described in citrus namely creasing, granulation, oleocellosis, peteca, puffiness, rind staining, soggy bottom, stem end breakdown, zebra skin and so on. These are described in the literature (Ladaniya, 2008; Korsten and Taverner, 2012; Magwaza et al., 2014) and are not discussed here. However one of the most important physiological disorders is chilling injury which can affect a significant volume of citrus exports due to the demands of cold treatment for quarantine purposes.

## 7 Chilling injury

Chilling injury is an important physiological disorder that can affect most citrus fruit and is a good example of the expression and complexity of physiological disorders. Chilling injury occurs in many crops of tropical and subtropical origin and occurs where fruit are exposed to chilling-sensitive temperatures, less than 10°C and above freezing. Chilling injury is distinct from freezing injury since no water/ice phase transition occurs in chilling. Some citrus types are susceptible to chilling injury where the typical symptoms include pitting of the peel and superficial scald-like symptoms of the peel, browning of the albedo and watery breakdown. The severity of chilling injury is related to both the storage temperature and the duration of treatment, whereby symptoms are increased with lower storage temperatures for longer treatment times. The symptoms of chilling injury can occur while the fruit is at low temperature but will generally increase when the fruit is removed from the chilling temperature to ambient temperatures.

The major challenge with chilling injury is that the biochemical and physiological basis for its development is not known. While the biochemical and molecular mechanisms involved in chilling tolerance in different citrus types have been extensively studied there seems to be a complex interplay of different metabolic pathways which operate in the induction of cold tolerance (lipid metabolism, oxidative stress, dehydrins, osmoprotectants, metallothioneins, defence responses etc.).

Predicting the onset and severity of chilling injury is also difficult, due to the unpredictable nature of the time × temperature combinations required to produce the onset of symptoms. A short cold treatment time may not develop any symptoms, but longer cold storage times may express the disorder. The easiest way to avoid chilling injury is to avoid storing citrus at <5°C; however, many of the phytosanitary cold treatments against fruit flies require cold treatment (1°C) for up to 3 weeks (see Section 15).

Although the mechanisms of chilling injury are not fully understood, there are a number of pre- and postharvest factors which interact to affect the development of chilling injury symptoms.

## 8 Preharvest factors affecting chilling injury

Preharvest factors contributing to the development of chilling injury include citrus type and cultivar, preharvest orchard temperatures, harvest times, growing locations and so on. The susceptibility to chilling injury differs among species and different citrus types. For example limes and lemons are generally more susceptible to chilling injury than oranges and mandarins. But even within a citrus type, there are differences in susceptibility to chilling injury. For example comparing the susceptibility of different navel oranges, it was observed that 'Navelina' fruit have stronger tolerance to chilling temperatures while 'Thomson' are moderately sensitive and 'Navelate' and 'Roberts' fruit are highly sensitive to chilling temperatures (Lindhout, 2007). Grapefruit are one of the most chilling-susceptible citrus types but there are significant differences in susceptibility between red and regular grapefruit types. Lado et al. (2015) showed there is a positive relationship between carotene content, mainly lycopene, and chilling tolerance during storage of grapefruit at low temperature (Fig. 2). The potential protective role of lycopene in chilling injury has been shown to be related to the antioxidant properties of lycopene (Lado et al., 2015).

The incidence of chilling injury also depends on preharvest factors, including maturation stage and environmental conditions during fruit growth such as field temperatures prior to cold storage. For example in some situations, fruit harvested earlier and later in the season are more susceptible to chilling injury than mid-season fruit, but this does not always hold in different seasons and growing conditions.

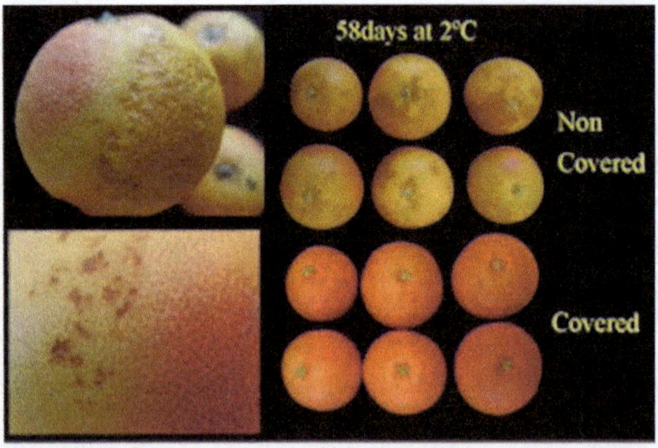


**Figure 2** Chilling injury symptoms present in yellow chilling injury-sensitive rind zones/ fruit and completely absent in lycopene-containing chilling injury-tolerant rind zones/ fruit. Source: adapted from Lado et al. (2015).

# 9 Postharvest management to reduce chilling injury symptoms

## 9.1 Temperature treatments

Short-term high temperature treatments in the form of hot water dips, hot water brushing, curing or conditioning with hot humid air have been shown to reduce the incidence of chilling injury in citrus. It is considered that these heat treatments are a form of stress which helps protect the fruit from another stress, that is cold stress and chilling injury. The protective nature of heat treatments have been shown to occur through various physiological mechanisms such as the development of heat shock proteins and induction of other defence mechanisms in the citrus albedo. Several curing and heat treatment conditions have been tested and show a range of results depending on the fruit type and treatment. Conditioning citrus fruits with hot air for three days at 37°C has been consistently found to be effective in increasing chilling tolerance without inducing heat damage. This has been demonstrated for different citrus seasons and in fruits harvested at all maturity stages, despite the variable susceptibility of fruit throughout the season. In addition to hot water dip treatments, intermittent warming, rinsing and brushing at temperatures of 60°C for a few seconds have been also reported to be effective at reducing the symptoms of chilling injury.

## 9.2 Plant growth regulators

A range of different plant growth regulators such as jasmonic acid, salicylic acid have been implicated in the development of chilling injury. Jasmonic acid and its various metabolites have been shown to regulate plant responses to abiotic and biotic stresses including chilling injury in a range of crops, including citrus. Jasmonic acid and salicylic acid have been shown to reduce the incidence of chilling injury in citrus possibly through enhanced antioxidant activities, increased production of total phenolics and/or suppressed membrane lipid peroxidation and reactive oxygen species production.

## 9.3 Fungicides

Common postharvest fungicides such as thiabendazole are applied to control postharvest decay such as green mould (see Section 10), but thiabendazole has also been shown to reduce the incidence of chilling injury. Furthermore the application of thiabendazole as a warm dip (40°C) has been shown to further decrease the expression of chilling symptoms. It is thought that the increased temperature results in higher deposition of the chemical into the fruit, but it is not known how thiabendazole confers increased chilling tolerance.

### 9.4 Waxes

The application of waxes is commonly used to replace the natural waxes/cuticles which are removed with postharvest brushing, handling and packing. Waxes reduce water loss from the fruit and have also been shown to reduce the incidence of chilling injury but with different outcomes with different wax and fruit types.

### 9.5 Modified atmospheres

The use of modified atmosphere packaging is not widely commercially used, but it has been shown to reduce chilling injury. It is thought that the high relative humidity within the storage carton reduces water loss and chilling injury.

## 10 Postharvest decay

Fungal pathogens are the main cause of citrus diseases. Preharvest infections include brown rot (*Phytophthora* spp.), Alternaria rot (*Alternaria* spp.), stem end rot (*Diplodia natalensis* Pole-Evan, *Phomopsis citri* Fawcett), grey mould (*Botrytis cinerea* Pers.) and anthracnose (*Colletotrichum gloeosporioides* Penz.). However it is the postharvest infection pathogens, the green and blue mould, caused by *Penicillium digitatum* (Pers.: Fr.) Sacc. and *P. italicum* Wehmer, respectively, that are the most economically important postharvest diseases of citrus, particularly in production areas with a Mediterranean-type climate with low summer rainfall. These postharvest pathogens are wound pathogens that only infect the fruit through peel injuries produced in the field during harvest and during packing and handling. When not adequately controlled, the postharvest losses caused by *Penicillium* can be enormous. But these losses are variable and depend upon climate and orchard factors, citrus cultivar, the extent of physical injury to the fruit during harvest and subsequent handling, the effectiveness of postharvest fungicides and treatments and the storage environment. Green mould typically causes larger losses throughout the supply chain as it tends to dominate at ambient temperatures, but blue mould becomes more important when citrus fruit are cold-stored for long periods, as *P. italicum* grows faster than *P. digitatum* below 10°C.

Sour rot (*Geotrichum citri-aurantii* (Ferraris)) is another major postharvest decay in many areas especially during fruit degreening and in wet and high rainfall seasons. However the control of this disease is more difficult as there are relatively few postharvest fungicides available.

## 11 Fungicides

Postharvest fungicides are the common control measure to effectively control postharvest decay. Some common active ingredients in citrus postharvest fungicides include fludioxonil, guazatine, imazalil, propiconazole, pyrimethanil, sodium *ortho*-phenylphenate and thiabendazole. These synthetic postharvest fungicides are efficacious, typically 'relatively' inexpensive, easy to apply and have curative action against pre-existing or established infections and persistent preventative action against potential new infections that can occur after their application in the packinghouse. Many of these fungicides also inhibit the sporulation from lesions on decaying fruit that reduces airborne inoculum production to break infection cycles. However it is often the case that not all these fungicides are available or used in all markets due to regulatory restrictions, cost and availability. Indeed, increasing consumer and regulatory resistance to the use of synthetic chemicals and the development of resistant strains of pathogens to these fungicides is leading to increasing interest in the development of alternative treatments. Some alternatives to traditional synthetic fungicides such as the use of sodium bicarbonate and sodium carbonate are already commercially used to control postharvest decay. While these treatments have some practical issues with their application and efficacy, they are widely used in an integrated decay control programme. Other registered biological control treatments such as 'Bio-Save® 10 LP' (JET Harvest Solutions, USA) which contain *Pseudomonas syringae* (Strain ESC-10 2) are available in some markets but are not widely applied. These treatments can be effective but have not been shown to completely replace the application of synthetic fungicides which provide the high level of decay control required.

## 12 Alternative decay control treatments

A range of different approaches to the control of postharvest pathogens have been developed and assessed. Palou et al. (2008) compiled an excellent practical review on alternatives to conventional synthetic fungicides for the control of green and blue moulds. Alternative decay control methods can either be physical, chemical or biological, or combinations of each.

Physical control measures such as heat treatments (curing and hot water) and irradiation treatments (UV-C and ionising irradiation) have the advantages of leaving no chemical residues, but have limited efficacy and may lack persistence or preventative activity. The development of alternative chemical control treatments have focussed on food additives and natural compounds with minimal effects on human health and the environment. There has been a focus on the application of natural organic acid salts such as sodium benzoate and potassium sorbate applied in aqueous solutions. While these treatments can be effective, they have limited persistence and lack preventative activity. Other natural products which have been evaluated as postharvest decay control treatments include essential oils (such as terpenes) and other plant extracts, chitosan oligosaccharides and antimicrobial peptides and so on.

Other approaches have examined enhancing natural resistance with both chemical elicitors (such as salicylic acid) and non-chemical treatments (such as light-emitting diode blue light) to increase the natural defences against postharvest infection. These chemical treatments have shown some promise in the laboratory, but their commercial application is currently limited.

Biological control of citrus postharvest decay has been extensively studied and reviewed (Lima et al., 2015; Droby et al., 2016). Indeed biological control agents have been the focus of considerable research for the last 30 years by many scientists and commercial companies around the world. Several antagonists to blue and green mould have been identified and tested but their full commercial potential has not been realised. The low success rate of biological control products has been attributed to several problems, including difficulties in mass production and formulation of the antagonist, the physiological status of the harvested commodity and its susceptibility to specific pathogens. All these factors play a major role in the reduced and inconsistent performance of the biocontrol product when used under commercial conditions (Droby et al., 2016).

The use of combination treatments offer some hope of an integrated approach to successfully control postharvest pathogens.

## 13 Sanitisers

Sanitation is a critical step in the control of postharvest pathogens but they do not provide any ongoing protection against these pathogens. Sanitisers are very reactive and not specific where they control fungi, bacteria and yeasts. They kill pathogens in water on contact but once they are inactivated there is no residual protection. Therefore the additional use of fungicides, which deposit a residue of the active fungicidal chemical onto the fruit is required to provide ongoing postharvest protection.

Harvested fruit coming in from the orchard is a continuous source of postharvest pathogens. The surface of virtually every citrus fruit that arrives at the packinghouse is contaminated with conidia and the inoculum can build up to high levels if sanitation is not properly conducted. The highest level of inoculum generally occurs in the fruit receival and bin tipping/emptying area, but all areas including the storage and cool rooms are sources of postharvest fungi. Sanitation occurs primarily by the sanitation of wash water, produce surfaces, equipment and storage rooms rather than direct control of infections of the decay pathogens within the produce. The aim of sanitation is to reduce the levels of postharvest pathogen inoculum in the packinghouse and fruit.

In addition to postharvest fruit pathogens, sanitisers can also assist in the control of human microbial food safety risks such as *Salmonella*. This aspect of sanitation is discussed in more detail in the next section.

A number of sanitisers with different chemistries are currently available in the market and each sanitiser has its own merits and limitations. The selection of sanitiser is influenced by a number of factors such as price, water source and quality, method of application and monitoring, packing line machinery and so on.

Chlorine has been a very popular sanitiser for years. Chlorine-based sanitisers include sodium hypochlorite, calcium hypochlorite, bromo-chloro-dimethylhydantoins and chlorine dioxide. The use of chlorine in wash water requires more effort and management in monitoring the pH, chlorine concentration and sanitation potential of water. In addition to the pH, organic load also impacts the efficacy of chlorine because active chlorine gets neutralised due to its reaction with debris and dirt present in water and effective monitoring of sanitiser concentration and efficacy is critical. Chlorine dioxide is another form of chlorine sanitiser which is currently available in different formulations with stabilised active ingredient. Because of its very high reactivity, a low concentration and less contact time is required for wash water treatment and is effective against bacteria, moulds and yeast over a pH range up to 8.5. Bromo-chloro-dimethylhydantoins are another sanitising option and provide broad-spectrum control of microbial pathogens. They represent a unique blend of chlorine and bromine chemistries which provide sanitising activity in a broad pH range.

The use of other sanitisers such as peracetic acid (PAA)-based sanitisers are growing in use due to the environmental impacts of chlorine by-products and management issues associated with chlorine. PAA is very reactive and quickly decomposes to oxygen, water and acetic acid and is effective in reducing microbial load and achieving product sanitation. Unlike chlorine, it is effective over a wide range of pH and requires less management. However, the effective monitoring and recording of PAA concentrations in wash water is critical as for any other sanitiser. The corrosiveness of PAA should be considered for machinery and appropriate protection of workers.

Ozone is a gaseous sanitiser which can be dissolved in wash water by generating it on-site using an ozone generator. Ozone is very effective in killing microbial pathogens over a wide range of pH, but it also reacts with organic matter. However it is highly corrosive with potential risks to workers' health and safety, but it has successfully been used in many packinghouses.

## 14 Food safety

Food safety is a critical component of the postharvest handling and storage of citrus. However the production of safe citrus requires a whole-of-chain approach with good agricultural practice. Food safety begins in the orchard where there is significant potential for microbial and chemical contamination. Human pathogens such as *Salmonella* and *Listeria* are inherent in the environment. Reducing the risk of these human pathogens coming into the packinghouse is critical to reduce food safety risks. These microbial risks can only be minimised by washing and sanitisation processes in the packinghouse (as described earlier) but cannot guarantee the complete elimination of microbes. Therefore a preventative approach is the most economical and effective way in minimising food safety risks. While citrus fruit are generally not considered a good carrier of food-borne pathogens, they can still pose food safety risks. For example, the peel of many citrus fruit is generally discarded; however, recent research has shown that human pathogens present on the citrus rind may get transferred to hands and mouth of consumers during peeling and eating. In addition, the untreated peel of some citrus types such as lemons and limes are routinely used as garnishes, zests or in drinks.

## 15 Market access

Exports are a growing sector of many citrus-growing countries. Global citrus exports account for 15% of world citrus production, where over 10.2 million tons of fruit (primarily oranges) are traded each year (USDA-FAS, 2018). While much of this fruit is traded between countries without any quarantine/phytosanitary barriers, many countries have quarantine barriers embedded within the trading practice to prevent the introduction and spread of exotic quarantine pests and diseases, such as Mediterranean fruit fly (*Ceratitis capitata* Wiedermann), red scale (*Aonidiella aurantii*) and lightbrown apple moth (*Epiphyas postvittana*) and citrus canker (*Xanthomonas axonopodis*) which are present in some citrus-producing countries. The presence of these quarantine pests in citrus-growing areas restricts market access to many export markets where these pests are not found. These quarantine barriers are critical for importing countries to protect their natural environment and stop the potential spread of quarantine pests into their countries and environment.

While the presence of these quarantine pests can totally restrict trade, in some cases, these quarantine barriers can be overcome with the use of effective in-field and/or end point quarantine treatments. These postharvest end point treatments are applied to the fruit after harvest and kill or ensure no viable quarantine insects can emerge from the fruit. There is a range of end point treatments available, but their use and application depends on the importing country, costs, applicability and their effects on fruit quality. This section will outline three market access treatments for citrus: cold treatment, fumigation and low-dose irradiation.

*Cold treatment.* Cold treatment is the most widely used treatment to allow market access of citrus. Major exporting countries such as South Africa routinely use cold disinfestation to kill quarantine pests in citrus fruit. For example the export of oranges from South Africa to phytosanitary markets requires the fruit to be treated for 12 days at a fruit pulp temperature of −0.6 ± 0.6°C. Cold treatment has several advantages in that the treatment does not use chemicals or leave any chemical residues, it can be applied in-transit on the ship for long-distance shipping and it fits into the current export supply chain. However, while this treatment is effective in killing any potential fruit fly larvae inside the fruit, this extended cold treatment can result in chilling injury to the fruit (see Section 7). This is a significant limitation of the export of many citrus types such as lemons, which are susceptible to chilling injury.

*Fumigation.* Methyl bromide still remains one of the most widely used fumigants for quarantine uses. While methyl bromide is a very effective fast-acting fumigant, it is also an ozone-depleting substance. Under the Montreal Protocol on Substances that Deplete the Ozone Layer (which is an international treaty designed to protect the ozone layer), many countries are phasing out the production of numerous substances that are responsible for ozone depletion. However due to the lack of alternative treatments, the use of methyl bromide is still currently permitted for quarantine and pre-shipment uses. While it is still widely used for quarantine treatments, there are growing environmental and health concerns with the continued use of methyl bromide and many countries have restrictions requiring complete recapture or destruction. In addition to the growing regulatory issues, the loss of the in-field uses of methyl bromide under the Montreal Protocol has resulted in reduced economic interest by manufacturers in continued methyl bromide production and put pressure on its cost. Therefore the continued use of methyl bromide is limited. Thus the development of alternative treatments to methyl bromide has been an active area of development.

*Other fumigants.* Potential methyl bromide alternatives include ethyl formate, methyl iodide, sulphuryl fluoride, phosphine and others. This has been an active area of research with work done on both the effects of toxicity of the fumigant on quarantine insect mortality and the tolerance of the citrus fruit to these fumigants. Ethyl formate is registered as a postharvest fumigant in some countries for the control of various surface insects such as light brown apple moth (*Epiphyas postvittana*), red scale (*Aonidiella aurantii*) and bean thrips (*Caliothrips fasciatus*).

*Irradiation.* Irradiation is a technologically proven, viable and scientifically sound disinfestation treatment. Moreover, irradiation is increasingly becoming an approved and agreed treatment in world trade of food and horticultural products. Generic irradiation treatments have been approved by the U.S. Department of Agriculture – Animal and Plant Health Inspection Service (USDA-APHIS) at doses of 150 Gy (Gray) for tephritid fruit flies and 400 Gy for all insects except pupal and adult Lepidoptera (USDA-APHIS, 2012). Further APHIS rulings and new rulings by the International Plant Protection

Convention (IPPC) have approved new minimum doses for 6 fruit fly pests and 14 other plant insect pests regardless of the host product, at doses between 60 and 300 Gy (USDA-APHIS, 2012). Irradiation does not necessarily kill the quarantine insect, but it breaks chemical bonds in DNA and other molecules, thereby sterilising the pest or preventing it from achieving sexual maturity. The major benefit of the use of irradiation as a market access treatment is that it overcomes any issues of chilling injury with cold treatment for export. However there are a few inconsistent negative effects of irradiation on final citrus fruit quality following treatment and storage. Citrus is relatively sensitive to irradiation and the response to treatment is highly variable and dependent on species, hybrid and cultivar. Physical injury can occur at low doses (<1000 Gy) and usually occurs in the peel as peel pitting and fruit softening in some types of citrus fruit but the contributing factors to this damage are not known and could potentially be managed. However the major issue for the use of low-dose irradiation is its acceptance on export markets. While some countries allow limited treatment with irradiation for quarantine treatment, its acceptance is currently limited.

To allow trade and market access of citrus, numerous alternative treatments to kill quarantine pests have been studied. The main alternatives are the use of alternative fumigants, temperature (high or low), forced air (hot air blown over the commodity), vapour heat treatments (hot air saturated with water blown over the fruit) and controlled atmosphere treatments which manipulate the levels of oxygen and carbon dioxide and treatment temperatures. These treatments have had limited success with limited commercial applications.

## 16 Consumer and retail trends driving changes in citrus production and postharvest

Consumers are driving changes in world horticultural and citrus production. In the last 20 years, fresh fruit consumption has significantly increased but the increase in the demand for citrus has not been as strong as compared to other fresh fruit, such as mangoes. And even within the citrus category, consumers are switching from oranges which have traditionally been the dominant citrus fruit, to more easy peel, 'soft' citrus types such as mandarins and tangerines. In addition, consumer preference is also for fresh citrus with few or no seeds. These consumer changes and demands are leading to significant changes in production areas and cultivars. Indeed consumers and therefore retailers are the drivers for changes in production and postharvest technologies. However despite these trends in citrus preference and production, it has been shown that heterogeneous preferences exist among consumers for citrus fruit and both demographic and behavioural variables which have significant impacts on preferences (Gao et al., 2011). Thus there is potential to develop specific marketing strategies based on demographics; however, more research is required to focus on these strategies.

While research has shown that there are no dominant best predictors of consumer preferences for fresh citrus fruit, consumer acceptability is strongly linked to fruit taste. The role of citrus consumption and human health has been well studied and documented, but consumer attitudes and changing consumption behaviours need to be more carefully examined.

Other major consumer and regulatory trends driving changes in citrus production and postharvest are focused on lower chemical use, organics and sustainability. Many consumers are now demanding lower chemical use in agriculture including in citrus production and processing. While reduced chemical use in the orchard can be managed with integrated pest and disease management, there are fewer options in the packinghouse to control decay and maintain quality. Some of these options have already been described in previous sections, but a general summary of these treatments is briefly outlined below (see Section 16.3).

Finally, the sustainability of citrus production and the supply chain is becoming an increasing focus of many consumers and retailers. Lowering environmental impacts of citrus production, processing and distribution of citrus to consumers is becoming a priority for many citrus supply chains and postharvest handling, processing and distribution are key factors in managing this environmental responsibility. Therefore this final section will outline the major consumer and regulatory trends in (1) consumer focus on taste, (2) nutrition, (3) low chemicals/organics and (4) sustainability.

## 16.1 Consumer focus on taste

The main consumer preferences for fresh citrus products are freshness, flavour and appearance. Freshness and flavour have been traditionally quantitatively assessed by measuring the sugar and acid content of the fruit. The sugar and organic acid concentration of the fruit affect taste characteristics and organoleptic quality. Sugars, consisting of sucrose, glucose and fructose, are the major components of citrus juice-soluble solids content (SSC) (70–80%) and are responsible for sweetness. Fruit acidity as measured by titratable acidity (TA) is composed of the organic acids, citric acid (70–90%) and malic acid. The ratio of SSC to TA (SSC/TA) has traditionally been the basis of harvest maturity and harvest times. However the SSC/TA ratio does not always correlate well with the perception of sweetness or tartness in the fruit (Fig. 3). One difficulty is that the same SSC/TA ratio may be derived from widely differing levels of SSC and TA, leading to different flavour perceptions for the same SSC/TA ratio. Jordan et al. (2001) recognising that sugar and acid have the opposite effect on flavour and that the tongue is more sensitive

and acidity, proposed subtracting TA from SSC after multiplying TA by a constant that differs by fruit type. This measurement index, given the name of 'BrimA', has been found to be more closely related to flavour than SSC/TA (Obenland et al., 2009), and is used as a maturity index in many markets such as California and Australia.

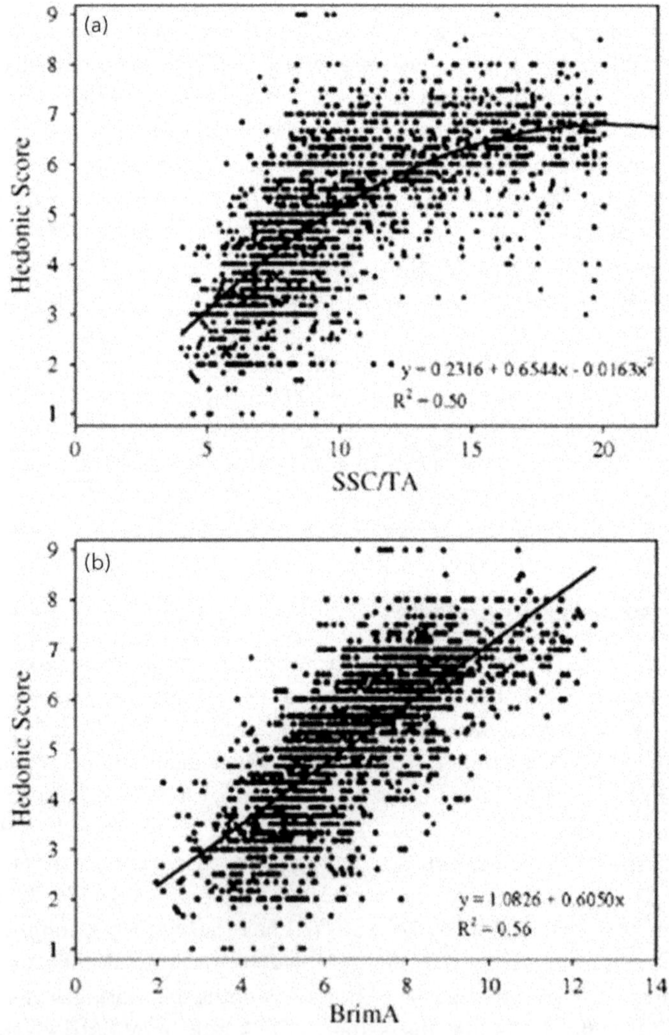


**Figure 3** Relationship between hedonic score and SSC/TA (a, top) or BrimA (b, lower). BrimA = SSC − 3(TA). Points indicate individual fruit (n = 2124) that were tasted and measured for SSC and TA over three seasons. Listed equations were those that best fit the data. Source: adapted from Obenland et al. (2009).

## 16.2 Consumer focus on nutrition and health

There have been many reviews and papers describing the phytochemical composition of citrus fruit and further epidemiological and experimental studies linking the consumption of citrus and juices to health outcomes. Indeed Section 3 of this chapter outlined the major nutrients and phytochemicals in citrus fruit; however, there have been few conclusive studies linking consumer attitudes to health outcomes with the consumption of citrus. Many consumers are largely ill-informed about the attributes of the citrus fruit that they purchase. But even among people who know that citrus is nutritious and good for their health, the consumption of citrus is often very low. The reasons for this are varied, but it indicates that knowledge of a nutritional benefit is just one of the many factors that influence food choices. Economos and Clay (1999) outlined some other factors that influence citrus fruit preference and consumption. These include an individual's food preferences and previous experience with citrus; cultural values, perceptions, attitudes and societal influences including the media and advertising; and, most directly, the availability, taste and price of the fruit. For these reasons, it is difficult to bring about widespread behavioural change to increase citrus consumption. Some of the barriers to purchasing and consuming citrus are high cost, fear of harmful pesticides and quick spoilage. Postharvest horticulture has a role in overcoming many of these barriers (chemical residues, postharvest storage and shelf life) and further work is essential to continue to reduce postharvest chemical use, improve fruit quality and market access, which will help drive increased consumption of citrus.

## 16.3 Reduced chemical use and organics

There has been a growing trend in world agriculture, including citrus production, back to 'organic' farming strategies. While citrus growers have generally developed and implemented integrated pest and disease management with minimal/no use of synthetic pesticides, the development and application of postharvest organic treatments have been limited.

The organic standards for postharvest (and all agriculture) are specific for each country, and if exporting, care must be taken to match the needs of the importing country's regulations. For example, ethylene is a natural ripening hormone that is produced naturally by all horticultural produce. However its application for artificial degreening (see Section 5) is often restricted. Operators must contact their own organic certification office for advice before using ethylene gas for degreening, as some countries have restrictions on its use.

A similar situation also occurs with the application of postharvest waxes. The postharvest application of waxes is a useful way of maintaining the quality and shelf life of fruit by reducing water loss. However organic fruit waxes must not contain prohibited substances. The use of carnauba wax is acceptable in Australia but fruits coated with these waxes must be labelled on the shipping container or packaging and comply with importing country requirements. However many fruit waxes are not acceptable in the European Union market. The onus is therefore on operators and exporters to ensure conformance with specific regulations for the importing country.

The basic principles of good postharvest handling and storage of citrus such as maintaining hygiene and avoiding damage and wounding of the fruit are especially critical for organic citrus fruit handling. However the postharvest storage and handling of organic produce presents some specific challenges, particularly in decay control. Postharvest synthetic fungicides are very effective and allow the transport and marketing of citrus fruit around the world. While organic treatments to control postharvest decay are available, their efficacy and application is limited. There are several organic physical, chemical and biological postharvest treatments to minimise postharvest decay losses which have been discussed in Sections 10 and 13.

## 16.4 Sustainability

### 16.4.1 Role of postharvest technology in sustainability

The environmental impacts of food production, including citrus production, handling and distribution, are becoming an increasingly important aspect of sustainability. Studies by the Food and Agriculture Organization of the United Nations consistently show that roughly one-third of all food produced for human consumption is lost or wasted globally, which amounts to about 1.3 billion tons per year (Gustavsson et al., 2011). These losses mean that huge amounts of the resources used in food production, including citrus production (e.g. water use, fertilisers, labour etc.), are wasted and that the greenhouse gas emissions caused by production of food that gets lost or wasted are also emissions in vain. Therefore preventing food loss and waste is critical to combating global hunger and improve food security, as reducing food losses will increase food availability, security and promote environmental sustainability (Porat et al., 2018).

Food losses can occur during the entire supply chain including at production, postharvest storage, transportation and processing, whereas food waste takes place towards the end of the food supply chain (including retail and consumption). While there are differences in food waste between developing and industrialised economies, there are some similarities which are common to both supply chains. In industrialised economies, food losses are mainly in the agricultural production domain, mostly due to postharvest fruit and vegetable grading caused by quality standards set by retailers. But waste after harvest is still considerable, contributing to over 50% of all food waste.

In industrialised economies approximately 15–30% of purchases by mass are discarded by consumers. This food waste is largely related to improper behaviour and decision making of suppliers, retailers and consumers, which results in discarding of food that still contains adequate nutritional value which could have been consumed (Porat et al., 2018).

In developing regions, losses in agricultural production dominate total losses throughout the supply chains but postharvest losses during distribution and marketing can be severe (Gustavsson et al., 2011). This can be due to poor postharvest practices where rapid deterioration of perishable crops in the warm and humid climates can occur and the lack of storage facilities limits marketing options of seasonal produce and leads to unsaleable gluts. Given that many smallholder farmers in developing countries live on the margins of food insecurity, a reduction in food losses could have an immediate and significant impact on their livelihoods.

Postharvest technology is fundamental to reduce waste and maintain fruit quality during storage, handling and transport of citrus fruit to the market. This chapter has already outlined some of the postharvest technologies, such as the use of fungicides to minimise decay during storage. While the current postharvest losses for citrus at the retail level have been estimated at around 10% (Porat et al., 2018), there is still significant potential to reduce this waste and its effect on the environment.

### 16.4.2 Sustainability and life cycle analysis

Life cycle impact assessment can be used as a measure of the total environmental impacts associated with citrus production (Beccali et al., 2010). Lo Giudice et al. (2013) estimated that only 35% of the total environmental impacts of citrus production and distribution were attributed to the actual growth and production of the oranges (Fig. 4). They estimated that the remaining 65% of environmental impacts were associated with postharvest operations with transport of the fruit from the orchard to the packinghouse (11%), processing (20%) and transport of fruit from packinghouse to distribution warehouses (32%). While this single life cycle impact assessment of citrus example is from the production and marketing of Sicilian blood oranges (cv. Tarocco) in Sicily to markets in the north of Italy, similar findings may be expected in other production areas and markets. Indeed the environmental impacts of postharvest operations would be even greater with longer storage and distribution networks, such as the addition of market access treatments, exports and distribution to distant markets (e.g. Spain to the United States). Thus there are significant opportunities to increase the sustainability of citrus production and distribution through increased processing and transport efficiencies and reduction in packaging. For example the use of 100% recycled polypropylene in the production of re-useable plastic

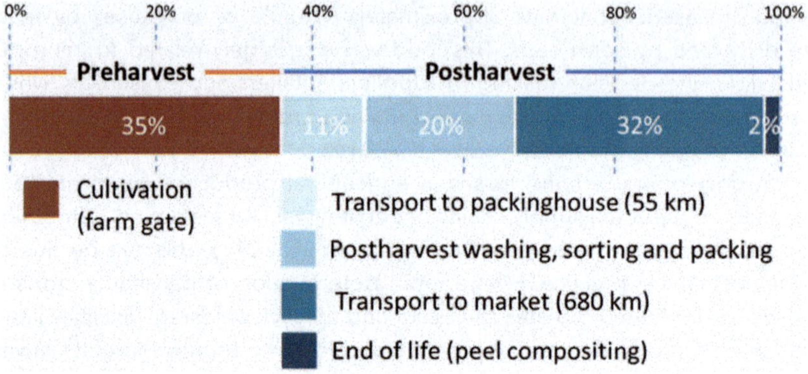


**Figure 4** Contribution of the different life cycle stages (growing through to the consumer) to carbon footprint of 'Tarocco' blood oranges grown in Sicily and sold in markets in the north of Italy. Source: adapted from Lo Giudice et al. (2013).

containers can reduce total environmental impacts by up to 50% (Lo Giudice et al., 2013). Thus postharvest technology has a major role in improving the sustainability of citrus production and distribution not only by reducing waste with improved decay control measures, but also with improved postharvest processing and handling efficiencies.

## 17 Conclusions

Citrus fruit are one of the major world horticultural crops with over 30 million tons of fruit consumed as a fresh produce each year. Consumers are driving major trends in citrus production and postharvest handling, storage and distribution. World production of 'soft' citrus types such as mandarins and tangerines is now nearly equal to those of oranges for fresh consumption. The increased production of these citrus types is being driven by consumer demand which generally wants smaller, easy-to-peel and sweeter fruit. The drive and demand of consumers are not only changing citrus production but also orchard and postharvest management such as reduced chemical use and organics. Postharvest decay primarily caused by green and blue mould and sour rot is the main storage issue for the storage and marketing of citrus. These pathogens are currently controlled with the judicious use of synthetic postharvest fungicides; however, consumers and the citrus industry are looking for 'green' alternatives to these chemicals. This has been an active area of research and development using both chemical and physical treatment combinations with varying degrees of success.

Citrus is the highest value horticulture crop but the global trade of citrus must ensure that all phytosanitary (quarantine) requirements are met. Cold treatment is the major end point quarantine treatment used around the world to ensure no living insects are accidently moved around the world. This treatment is a very effective end point treatment which uses no chemicals and can be conducted in-transit to distant markets. While other market access treatments are often available, cold treatment is well established and widely used. However long storage periods of cold treatment required for quarantine disinfestation can sometimes induce chilling injury in some citrus types. This can significantly downgrade fruit quality and grower returns. Understanding and managing this physiological disorder has been a major research area for many years, but as yet few commercial options are available to reliably prevent this physiological disorder.

Major consumer and regulatory trends are driving changes in supply chains and trade. The increased awareness of chemical residues and the role of production and supply chain sustainability are becoming greater issues for consumers and the citrus industry. There have been considerable research efforts in applying reliable alternatives to synthetic postharvest fungicides. This work has been successful in implementing a range of 'reduced risk' fungicides for commercial application; however, more work is required to find and apply alternative treatments and combinations.

Future trends in applied postharvest citrus research and development will be driven by consumers, regulators, industries and funding agencies. Current consumer trends of convenience, healthy eating, low chemical use, market access, sustainability and so on will continue to drive industry changes and demand new solutions to current issues. Indeed improved understanding of consumer needs, and demands are necessary to increase the demand for citrus. It is critical that industries meet consumers' expectations of citrus and change buying behaviours to increase demand. This will involve multidisciplinary teams in psychology, consumer and behavioural science, nutrition and postharvest.

In addition fundamental issues such as food safety which are non-negotiable and essential to the safe consumption of citrus will require ongoing research and development. While citrus has often been regarded as a low-risk horticultural crop, microbial and chemical contamination risks can occur and need to be managed with science-based solutions. In addition, improved market access treatments and protocols are also necessary to increase local and world trade of citrus, but fruit quality issues and costs associated with these treatments still need to be managed.

These issues will continue to evolve and new aspects of postharvest science and technology will need to be developed and applied. A growing trend in the way this research is done and applied is through multidisciplinary teams with collaboration between academia and industry. For example the development and application of an improved edible coating for citrus or a new postharvest treatment to control decay requires a team of chemists, food scientists and technologists, pathologists, development engineers, consumer and sensory scientists, postharvest scientists and many more. It is essential that these different disciplines work closely with commercial partners and industry to ensure the application of this research is not only commercially efficacious which can fit into the supply chain, but that it is also cost-effective, safe and has regulatory approvals. This is a complex task but is necessary to ensure commercial outcomes for industry.

While the results of the majority of postharvest research studies are not immediately commercially applied, it is important to ensure applicability of research is always considered. It may be that the results of a postharvest study do not currently have immediate direct outcomes, but these research results may then lay the basis for future studies and developments for later commercial applications. The evolution and development of science will result in improvement in postharvest handling, treatment and storage practices.

While emphasising the necessity for applied collaborative research, it is also necessary to maintain and encourage theoretical research which will enhance our deeper understanding of citrus physiology and pathology. For example, developments in advanced plant and molecular sciences will be assessed and applied for their later applicability to the storage and handling of citrus. It is essential to continually develop all levels of science to improve our understanding of citrus and provide solutions to practical industry problems.

## 18 Future trends

Future trends in applied postharvest citrus research and development will be driven by consumers, regulators, industries and funding agencies. Current consumer trends of convenience, healthy eating, low chemical use, market access, sustainability and so on will continue to drive industry changes and demand new solutions to current issues. Indeed improved understanding of consumer needs and demands are necessary to increase the demand for citrus. It is critical that industries meet consumers' expectations of citrus and change buying behaviours to increase demand. This will involve multidisciplinary teams in psychology, consumer and behavioural science, nutrition and postharvest.

Citrus is the highest value horticultural crop in the world which is supported by government- and industry-funded research and development around the world. Leading citrus producing countries, such as South Africa, Spain, China, Brazil, Argentina, Egypt, the United States and Australia have active postharvest R&D programmes to support their local citrus industries. A growing trend in the way this research is done and applied is through multidisciplinary teams with collaboration between academia and industry. It is essential that these different disciplines work closely with commercial partners and industry to ensure the application of this research is not only commercially efficacious which can fit into the supply chain, but it is also cost-effective, safe and have regulatory approvals. This is a complex task but is necessary to ensure commercial outcomes for industry.

## 19 Acknowledgements

This chapter is a contribution from the Australian Citrus Postharvest Science Program (CT15010) funded by Horticulture Innovation and NSW Department of Primary Industries.

## 20 References

Beccali, M., Cellura, M., Iudicello, M. and Mistretta, M. 2010. Life cycle assessment of Italian citrus-based products. Sensitivity analysis and improvement scenarios. *Journal of Environmental Management* 91(7), 1415–28. doi:10.1016/j.jenvman.2010.02.028.

Droby, S., Wisniewski, M., Teixidó, N., Spadaro, D. and Jijakli, M. H. 2016. The science, development, and commercialization of postharvest biocontrol products. *Postharvest Biology and Technology* 122, 22–9. doi:10.1016/j.postharvbio.2016.04.006.

Economos, C. and Clay, W. D. 1999. Nutritional and health benefits of citrus fruits. In: Albert, J. L. (Ed.), *Food Nutrition and Agriculture 24*. Food and Agriculture Organization of the United Nations, Food and Nutritional Division, Rome, Italy, pp. 13–26.

Gao, Z., House, L. O., Gmitter, F. G., Valim, M. F., Plotto, A. and Baldwin, E. A. 2011. Consumer preferences for fresh citrus: impacts of demographic and behavioral characteristics. *International Food and Agribusiness Management Review* 14, 23–40.

Gopinath, B., Liew, G., Kifley, A., Flood, V. M., Joachim, N., Lewis, J. R., Hodgson, J. M. and Mitchell, P. 2018. Dietary flavonoids and the prevalence and 15-y incidence of age-related macular degeneration. *The American Journal of Clinical Nutrition* 108(2), 381–7. doi:10.1093/ajcn/nqy114.

Gustavsson, J., Cederberg, C., Sonesson, U., van Otterdijk, R. and Meybeck, A. 2011. Global food losses and food waste: extent, causes and prevention. Interpack 2011. Food and Agriculture Organization of the United Nations, Rome, Italy.

Jordan, R., Seelye, R. and McGlone, A. 2001. A sensory-based alternative to brix/acid ratio. *Food Technology* 55, 36–44.

Korsten, L. and Taverner, P. 2012. Citrus. In: Rees, D., Farrell, G. and Orchard, J. (Eds), *Crop Post-Harvest: Science and Technology. Volume 3: Perishables*. Blackwell Publishing, West Sussex, UK, pp. 43–87.

Ladaniya, M. S. 2008. *Citrus Fruit: Biology, Technology and Evaluation*. Elsevier Academic Press, New York, NY.

Lado, J., Rodrigo, M. J., Cronje, P. and Zacarías, L. 2015. Involvement of lycopene in the induction of tolerance to chilling injury in grapefruit. *Postharvest Biology and Technology* 100, 176–86. doi:10.1016/j.postharvbio.2014.10.002.

Lima, G., Sanzani, S. M., De Curtis, F. and Ippolito, A. 2015. Biological control of postharvest diseases. In: Wills, R. B. H. and Golding, J. B. (Eds), *Advances in Postharvest Fruit and Vegetable Technology*. CRC Press, USA, pp. 65–88.

Lindhout, K. 2007. Physiology of chilling-related postharvest rind breakdown of navel oranges (*Citrus sinensis* (L.) Osbeck). PhD thesis. La Trobe University, Australia.

Lo Giudice, A. L., Mbohwa, C., Clasadonte, M. and Incrao, C. 2013. Environmental assessment of the citrus fruit production in Sicily using LCA. *Italian Journal of Food Science* 25, 202–12.

Magwaza, L. S., Opara, U. L., Cronjé, P. J., Landahl, S., Terry, L. A. and Nicolaï, B. M. 2014. Nonchilling physiological rind disorders in citrus fruit. In: Janick, J. (Ed.), *Horticultural Reviews* (vol. 41). John Wiley & Sons Inc., West Sussex, UK, pp. 131–76. Chapter 3.

Obenland, D., Collin, S., Mackey, B., Sievert, J., Fjeld, K. and Arpaia, M. L. 2009. Determinants of flavor acceptability during the maturation of navel oranges. *Postharvest Biology and Technology* 52(2), 156–63. doi:10.1016/j.postharvbio.2009.01.005.

Palou, L., Smilanick, J. L. and Droby, S. 2008. Alternatives to conventional fungicides for the control of citrus postharvest green and blue molds. *Stewart Postharvest Review* 2, 2.

Porat, R., Lichter, A., Terry, L. A., Harker, R. and Buzby, J. 2018. Postharvest losses of fruit and vegetables during retail and in consumers' homes: quantifications, causes, and means of prevention. *Postharvest Biology and Technology* 139, 135–49. doi:10.1016/j.postharvbio.2017.11.019.

USDA-APHIS. 2012. U.S. Animal and Plant Health Inspection Service. Plant Protection and Quarantine (USDA APHIS PPQ), 2012. Treatment Manual. Section 3-8. Nonchemical treatments. Irradiation. United States Department of Agriculture, Washington DC.

USDA-FAS. 2018. Citrus: World Markets and Trade 2017/18. United States Department of Agriculture Foreign Agricultural Service. July 2018. http://usda.mannlib.cornell.edu/usda/current/citruswm/citruswm-07-26-2018.pdf (accessed on September 2018).

Wills, R. B. H. and Golding, J. B. 2016. *Postharvest: an Introduction to the Physiology and Handling of Fruit and Vegetables*. NewSouth Publishing, Sydney Australia, 293pp. ISBN: 9781742247854.

Wu, G. A., Terol, J., Ibanez, V., López-García, A., Pérez-Román, E., Borredá, C., Domingo, C., Tadeo, F. R., Carbonell-Caballero, J., Alonso, R., et al. 2018. Genomics of the origin and evolution of Citrus. *Nature* 554(7692), 311–6. doi:10.1038/nature25447.

# Chapter 4

# Advances in understanding the ecology of citrus insect pests

*Robert W. Jones, Universidad Autónoma de Querétaro, Mexico; Pablo Liedo, Departamento de Entomología, El Colegio de la Frontera Sur, Mexico; and Santiago Vergara-Pineda, Universidad Autónoma de Querétaro, Mexico*



## 1 Introduction

The genus *Citrus* L. is one of the earliest known domesticated perennial fruit crops. Having apparently originated in the southeastern slopes of the Himalayas, approximately, eight million years ago, the genus subsequently expanded and diversified into the much of Southern Asia including China, Malaysia, the Philippines, Indonesia and into Japan and northern Australia (Wu et al., 2018a). Following domestication in these regions, citrus species accompanied early human migrations into Micronesia and Polynesia, and later into the Mediterranean beginning in the fifth and fourth century BC (Langgut, 2017). Citrus species presently are grown in all tropical and subtropical regions and together represent one of the most important fruit crops of the world.

Because of its long history and enormous importance in many cultures, citrus has been a seminal crop in the development of technology to manage insect pests. The first known pest management strategy is reported from China in 304 AD, describing the placement of ant nests (*Oecophylla smaragdina* Fabr.) in citrus trees to control insect herbivores (Huang and Yang, 1987). Likewise, the first successful classical biological control project was carried out in 1889

http://dx.doi.org/10.19103/AS.2019.0054.06

against the cottony-cushion scale (*Icerya purchase* Mask.) which had invaded California from Australia. Virtual total success was accomplished by collecting a natural enemy of the scale in its native Australia, a beetle predator, *Rodolia cardinalis* (Mulsant), and releasing it in citrus groves of California (Caltagirone and Doutt, 1989). Biological control in citrus continues to be in the forefront of this management tactic and in understanding the basic ecology of the complex interactions of parasitoids (Murdoch et al., 2006). Other pest management techniques which were initiated in citrus culture include early fumigation practices, the establishment of plant quarantine regulations, biological control through natural enemy conservation, region-wide eradication programs, and early insecticide trials (Jeppson, 1967).

Citrus cultivation was also at the forefront in the formulation of what Paul DeBach (1951) called the 'Ecological Approach to Pest Control'. This approach was borne out of the realization of the consequences of dependence and overuse of chemical pesticides with the resultant development of pest resistance and loss of natural enemies (DeBach, 1951; DeBach and Bartlett, 1951; Stern el al., 1959; Smith, 1962). Although the basic concepts of the 'ecological approach' were practiced before the discovery of synthetic insecticides (Kogan, 1998), the need to reduce the dominance of pesticide use as a sole pest management strategy led to the formulation of the concepts of integrated pest management (IPM). IPM attempts to incorporate all available tactics in a unified and coherent strategy of pest management which is cognizant of the myriad interactions of the agroecosystem. IPM continues to be the moniker of pest management programs worldwide, despite criticisms and calls for new conceptual frameworks (Zalucki et al., 2009; Menalled et al., 2016; Peterson et al., 2018), and citrus IPM continues to be at the vanguard of innovative and biological based strategies.

## 2 Important pests of citrus

The basic characteristics of citrus culture, an evergreen tree cultivated worldwide in monocultures in subtropical and tropical climates, make it especially attractive to arthropods and favorable to their population growth (Jeppson, 1967). In India alone, more than 250 species of arthropods are listed as pests, (Deka et al., 2015) and even regions outside the center of origin support large communities of citrus herbivores. Some of the most important arthropod pests of the world are given in Table 1. The high richness of herbivore species and the benign tropical environment of citrus cultivation also result in generally high diversity of natural enemies of citrus pests. This is favorable for the development of management tactics using biological control.

There are two basic types of associations of arthropod pests and their cultivated host plants. The first are those species that have a long evolutionary

**Table 1** Selected insect and mites pests of cultivated citrus of the world

| Common name(s) | Scientific name | Origin | Present distribution | Comments |
|---|---|---|---|---|
| *Insects* | | | | |
| *Hemiptera (Leafhoppers, Aphids, Psyllids and Scale Insects)* | | | | |
| Leafhopper | *Hishimonus phycitis* (Distant) | Unknown | SE Asia, India, Pakistan, Middle East (Oman, Iran, United Arab Emirates)[a] | Vector of witches' broom disease |
| Citrus Aphid=Brown Citrus Aphid= Tropical Citrus Aphid | *Toxoptera citricida* (Kirkaldy) | Southeast Asia | Global, Tropics and Subtropics, see map of distribution EPPO (2019)[a] | Vector of citrus tristeza virus (CTV) |
| Green citrus aphid, Spira aphid | *Aphis citricola* (=*spiraecola*) van der Groot | Eastern Asia | Worldwide[b] | Vector of CTV |
| Citrus blackfly | *Aleurocanthus woglumi* (Ashby) | India | Asia, Africa, Central and South America, Mexico, West Indies, USA (Florida, Texas, Hawaii)[a] | A serious citrus pest of Asian origin (Dietz and Zetek, 1920), it is usually under effective biological control in Florida |
| Asian citrus psyllid | *Diaphorina citri* Kuwayama | Southern Asia | Asia, Middle East, South and Central America, Mexico, West Indies, southern United States, see map of distribution EPPO (2019)[a] | Vector of citrus greening disease ('Huanglongbing'), various forms of the bacteria, *Candidatus* L. |
| Citrus mealybug | *Planococcus citri* (Risso) | Asia | Worldwide. Asia, North and South America, Europe, and Oceania[b] | Usually under control by natural enemies |
| Green Scale | *Coccus viridis* Green | Possibly Brazil | Worldwide in tropics, see map of distribution EPPO (2019)[a] | Especially damaging to young trees |

(*Continued*)

**Table 1** (Continued)

| Common name(s) | Scientific name | Origin | Present distribution | Comments |
|---|---|---|---|---|
| Cotton-cushion(y) scale | *Icerya purchasi* (Maskell) | Australia | Worldwide, on citrus, see map of distribution EPPO (2019)[a] | Usually under biological control |
| California red scale | *Aonidiella aurantii* Maskell | Southeast Asia | USA (California), Australia, New Zealand, Europe, Mexico, Central America, Chile, Argentina, Brazil, Israel, and South Africa[b] | Key pest of citrus in California, Australia, South Africa; important biological control programs with *Aphytis* spp.; see Murdoch et al. (2006) for review |
| Florida red scale | *Chrysomphalus aonidum* (L.) | Southern Asia | Worldwide[a] | The parasitoids, *Aphytis holoxanthus* and *Pteroptrix* (= *Casca) smithi* are important biological control agents |
| Purple scale = mussel scale | *Lepidosaphes beckii* (Newman) | Southern Asia (presumed) | Worldwide; Central and South America, Mediterranean region, South Africa, Australia, USA (Hawaii, Florida, Texas, California) | Highly specific to citrus, controlled in most regions were introduced by the parasitoid, *Aphytis lepidosaphe. s* (Kennett et al., 1999) |
| *Thysanoptera (Thrips)* | | | | |
| Citrus thrips | *Scirtothrips dorsalis* Hood | Southeast Asia or India | Asia, Africa, Australia, USA (Texas, Florida) , Caribbean[a] | Generalist feeder on many crops |
| *Coleoptera (Beetles)* | | | | |
| Citrus longhorn beetle, Citrus-root cerambycid | *Anoplophora chinensis* (Forster) | Southeast Asia | Southeast Asia, Hong Kong, Korea Republic, Malaysia, Myanmar<br>China, Korea, and Japan, Taiwan, the Philippines, Indonesia, Malaysia, Myanmar, Vietnam[b] | Feed on many hardwood tree species, larvae tunnel under bark and in wood, reported in the United States (Washington State) in 1999, declared eradicated in 2002[b] |

| | | | | |
|---|---|---|---|---|
| Fuller Rose Weevil | *Naupactus cervinus* Boheman | Southern South America | Worldwide[b] | Parthenogenic outside region of origin (Rodriguero et al., 2013). Of quarantine concern in parts of Asia |
| Diaprepes root weevil (Fig. 1, lower right) | *Diaprepes abbreviatus* (L.) | Caribbean | USA: (Florida, Texas, Louisiana, California)<br>Caribbean[b] | Highly polyphagous root feeder in larval stage |
| *Lepidoptera (Moths and Butterflies)* | | | | |
| Citrus leafminer | *Phyllocnistis citrella* Stainton | Asia | USA: (Alabama, California, Florida, Louisiana, and Texas).[b] Worldwide, see map of distribution EPPO (2019)[a] | Damage usually related to reduced growth of nursery and young trees, and not fruit production (Heppner and Fasulo, 2010) |
| False codling moth= Citrus codling moth | *Thaumatotibia* (=*Cryptophlebia*) *leucotreta* (Meyrick) | Sub-Saharan Africa | Sub-Saharan Africa, Madagascar, see map of distribution EPPO (2019)[a] | Feeds on wide variety of trees |
| *Diptera (Flies)* | | | | |
| Mediterranean fruit fly ('Medfly') | *Ceratitis capitata* Wiedemann | Sub-Saharan Africa | Africa, Middle East, South and Central America, southern Europe, Western Asia (Iran), Western Australia, see map of distribution EPPO (2019)[a] | Feeds on a wide variety of fruit. Introduced and eradicated from Florida and California several times. Focus of sterile male control programs (see text) |
| Oriental fruit fly | Bactrocera dorsalis (Hendel)<br>(= *B. invadens*; San Jose et al., 2013) | Southeast Asia and Pacific Islands, Hawaii | Southern Asia, Africa, Pacific islands including Hawaii, see map of distribution EPPO (2019)[a] | Recorded from more than 440 types of fruit; of high quarantine importance for the Americas |
| Queensland fruit fly | *Bactrocera tryoni* (Froggatt) | Australia | Australia, New Zealand, French Polynesia, New Caledonia, see map of distribution EPPO (2019)[a] | Very destructive pest of citrus, male sterile method has been used for local eradication of the species in Western Australia[b] |

(*Continued*)

**Table 1** (Continued)

| Common name(s) | Scientific name | Origin | Present distribution | Comments |
|---|---|---|---|---|
| Mexican fruit fly ('Mexfly') | *Anastrepha ludens* (Loew) | Mexico, Central America | Mexico, Central America[a] | Eradicated from Texas, periodically captured in Florida, endemic in Mexico and under control by sterile male method |
| South American fruit fly | *Anastrepha fraterculus* (Wiedemann) | South America | South America, Central America) and Mexico, USA (Texas)[a] | Apparently a complex of eight cryptic species (Hernández-Ortiz et al., 2012) |
| Caribbean fruit fly | *Anastrepha suspensa* (Leow) | Originally described from Cuba | Cuba, Jamaica, Hispaniola, Puerto Rico, French Guiana and Florida[a] | Almost 100 host plants reported |
| *ARACNIDA* *Acari (Mites)* | | | | |
| Citrus rust mite | *Phyllocoptruta oleivora* (Prostigmata: Eriophyoidea) | | Worldwide, USA (Texas, Florida), Mexico, Central America, South America, Africa, Asia, Southern Europe, see map of distribution EPPO (2019)[a] | Considered important in humid, tropical areas. Phytoseiid mites and fungal infections check infestations. (Kennett et al., 1999) |
| Citrus red mite | *Panonychus citri* (MacGregor) | Asia | Worldwide | Causes a bronzing or silvering effect. Severe infestations may cause a partial defoliation of tree. Phytoseiid mites are common predators. (Kennett et al., 1999) |

[a] EPPO (2019).
[b] Weeks et al. (2019).

history with the wild cultivars and/or their relatives and were present and adapted to their host plants during the process of domestication. An example of this type is the citrus leaf miner, *Phyllocnistis citrella,* which occurs naturally on citrus and other plants of the citrus family, Rutaceae, in Southeast Asia (Latif and Muhammad Yunus, 1951; Argov and Rossler, 1996) and its subsequent expansion around the globe has been closely documented since the beginning of the last century (Clausen, 1931; Schauff et al., 1998). As is common for indigenous herbivores, it is a specialist feeder and is attacked by a diverse guild of indigenous natural enemies (Schauff et al., 1998; Costa et al., 1999) (Fig. 1).

The second type of pests are those that have adapted more recently to cultivated species and varieties after domestication. This can occur in regions far from the crop origins, where the latent pest is found on native host plants or other cultivated species and adapts to and feeds on the introduced cultivar.

**Figure 1** Upper left, Mediterranean fruit fly (Medfly), *Ceratitis capitata* (Wiedemann), photo by Scott Bauer. Image Number K8898-2, USDA-Agricultural Research Service (public domain); Upper right, A female oriental fruit fly, *Bactrocera dorsalis*, photo by Scott Bauer. Image Number K9588-6, USDA-Agricultural Research Service (public domain). Bottom. Lower left, citrus leaf miner larva, *Phyllocnistis citrella,* Photo by Dan Papacek, with permission Bugs for Bugs P/L (Australia). Lower right, Diaprepes root weevil, *Diaprepes abbreviatus* (L.).

These newly adapted pests are often generalists, though not necessarily, and feed on a wide variety of host plants. An example of this type of pest is the Mediterranean fruit fly (Medfly), *Ceratitis capitata* (Wiedemann), which originated in Sub-Saharan Africa and is now found in much of Africa, western Asia, western Australia, Central America, the Southern Europe, and parts of South America (Szyniszewska and Tatem, 2014; Thomas et al., 2001). As for many recently adapted pest species, Medfly is polyphagous and it is not only a threat to citrus but can infest more than 250 varieties of fruits and vegetables including many ornamental fruits (Thomas et al., 2001).

Both types of pests can pose threats as invasive exotic species in regions where they do not occur. Besides direct damage to agricultural crops, invasions of nonindigenous pest species can threaten native biodiversity, ecosystem functioning, animal and plant health, and human economies (Myers et al., 2000). One of the pests of most concern is the Medfly, which is closely monitored in many countries. Generally successful eradications were implemented, following separate incidents of detection of the pest, in the United States, Mexico, Chile, Western Australia, and New Zealand (De Meyer et al., 2008; Szyniszewska and Tatem, 2014). These eradication programs relied heavily on sterile male control programs, discussed in detail below. Another important fruit fly, is *Bactrocera dorsalis* (=*invadens*) (Hendel) from Southern Asia which has been introduced in the Pacific Islands, including Hawaii, but yet to invade North or South America, the Mideast or Europe (De Meyer et al., 2008). However in 2003, *B. dorsalis* was reported in Kenya (Lux et al., 2003) and has subsequently invaded 40 countries in Africa (De Villiers et al., 2016).

Several pests are under strict surveillance programs not because of the direct damage they cause but for their role as citrus disease vectors. This is the case for the brown citrus aphid, *Toxoptera citricida* (Kirkaldy) and the green citrus aphid, *Aphis citricola* (=*spiraecola*) van der Groot, which are vectors of citrus tristeza virus (CTV). Of recent concern, the Asian citrus psyllid, *Diaphorina citri* Kuwayama vector of citrus greening disease ('huanglongbing'), caused by various forms of the bacteria, *Candidatus* L. has been expanding its distribution in various parts of the world (Halbert and Manjunath, 2004). Insect vectors such as *D. citri* are especially hard to manage because economic thresholds of populations are very low given that few individuals are required to transmit the pathogen. This pest and huanglongbing is discussed more in detail below and in the subsequent chapter.

## 3 New tools in citrus pest research

### *3.1 Modeling species distribution (SDM)*

Modeling of species distribution (SDM) has become an important tool in predicting the distribution and invasion dynamics of exotic citrus pests. These models estimate present or future distributional areas of a species based on correlations of known occurrences with abiotic and biotic variables to estimate probabilities of the presence and establishment of organisms (Elith et al., 2006; Peterson and Soberón, 2012). Whether analyses are 'species distribution models' or 'ecological niche models' is debated (Peterson and Soberón, 2012), but all utilize computer algorithms to process and associate distributional data with environmental variables. One of the most commonly used software for SDM is 'Maxent', maximum entropy modelling software (Phillips and Dudík, 2008). Climate data for analyses can be obtained from country data bases or WorldClim, a free dataset of global climate layers (WorldClim, 2019).

Various studies of important citrus pests demonstrate the utility of SDM analyses. This approach can give a worldwide perspective of present and potential distributions of pests under different climate scenarios. For example, Szyniszewska and Tatem (2014) modeled the world distribution and seasonal occurrence of the Medfly, *Ceratitis capitata*, and were able to produce probabilities of occurrence of the pest on a global scale, as well as on a seasonal basis. De Meyer et al. (2008) also characterized the ecological niches and potential geographic distributions of Medfly, and compared the predictions with a similar lesser known sibling species from Africa, *Ceratitis rosa*. The two species were found to have broadly similar potential distributions but data suggested that *C. capitata* was apparently more tolerant of dryer and warmer conditions than *C. rosa*, which may explain the greater distribution and important world pest status of *C. capitata*. In a similar study, De Meyer et al. (2010) modeled the relatively recent (2003) invasion into Africa by the fruit fly, *Bactrocera invadens* (Hendel), now considered conspecific with *B. dorsalis* (San Jose et al., 2013). They found that much of equatorial Africa has high environmental suitability for the insect, and the data also accurately predicted seasonal occurrences.

Given the significant global threat posed by the citrus greening disease (huanglongbing) vectored by the Asian citrus psyllid, *Diaphorina citri* (Fig. 2), several modeling studies have demonstrated the wide-ranging predictive power of SDM analyses. Aurambout et al. (2009) developed predictive models to assess the risk of introduction of the disease and vector into Australia under changing climate regimes. The models predicted that the seasonal flush of new growth in citrus, a key requirement for psyllid development, would be of shorter duration and reduce the risk of *D. citri* establishment. However, more southern areas of citrus production would become more suitable to colonization of the pest. In a global analysis, Narouei-Khandan et al. (2016) also combined SDM models of both the psyllid vector and huanglongbing to indicate regions of suitable areas for the two organisms and the climate factors involved. The models also indicated that rainfall was more important than temperature in determining

**Figure 2** Adult Asian citrus psyllid, *Diaphorina citri* Kuwayama vector of citrus greening disease ('huanglongbing'), photo by David Hall, Image Number D595-1, SDA-Agricultural Research Service (public domain).

climate suitability. Geographically, high suitability of climates for the disease and its vector were predicted for large areas of Africa, Latin America, Northern Australia and New Zealand, and more in-depth analysis differentiated the suitability for regions within the United States.

### *3.2 Stable isotope analysis (SIA)*

Another tool in ecological research is stable isotope analysis (SIA). This is a relatively new technique that can be utilized to study trophic structure in agroecosystems. As organic material moves up through the food chain, the ratio of naturally occurring stable (non-radioactive) isotopes changes, with heavier isotopes being incorporated more in the body of the consumer at each level (Ehleringer and Rundel, 1988). The ratios of isotopes can be used to determine the energy source of a trophic web and also determine the tropic position of a particular organism in the trophic sequence (Vander Zanden and Rasmussen, 2001; McNabb et al., 2001; Hood-Nowotny and Knols, 2007). One of the applications of these methods is determining the prey and importance of predators within a specific ecosystem, which is often difficult from observational field studies (McNabb et al., 2001; Hood-Nowotny and Knols, 2007).

Several examples of the use of SIA come from citrus agroecosystems. Platner et al. (2012) were able to place ant species within trophic levels in citrus orchards in Spain. They found a large range of isotope proportions among the 18 ant species analyzed. Ants found at the lowest levels were considered plant feeding and non-predatory, thus not to be considered natural enemies. A large number of the ant species were found at an intermediate levels suggesting both predatory and nectar feeding, whereas ants at the highest levels were considered specialized predators. This study also found marked seasonal differences in isotope signatures, indicating changes in resource use over time. Another study found similar results for spiders in citrus (Mestre et al., 2013), with a wide range of isotope proportions in tropic levels of the spiders in citrus orchards, and there was also similar seasonal changes in isotope signatures suggesting changes in feeding behaviors.

Identifying the base of a trophic chain with SIA has several applications. Santi-Júnior et al. (2018) were able to infer the source of plant feeding for two species of beetles that were considered as pests of citrus in Brazil, *Macrodactylus pumilio* and *Naupactus cervinus*. They found that *M. pumilio* had an isotope signature that suggested a mixed diet of grasses and citrus, whereas results suggested *N. cervinus* fed more exclusively on citrus. These results also gave insight into feeding preferences of their soil inhabiting larval stages, a notoriously difficult behavior to study with standard observational methods. Identifying the food source with SIA can also be used to differentiate laboratory reared individuals from those of wild populations, which is especially important in the sterile male management programs. For example, Botteon et al. (2018) were able to clearly discriminate between diet -reared and wild, fruit -fed adults of the South American fruit fly, *Anastrepha fraterculus* based on $\delta^{13}C$ signatures. This application of SIA, as well as other uses in citrus entomology will certainly increase in the future as a result of the reductions in the cost of analysis combined with improved analytical methods and technology (Hood-Nowotny and Knols, 2007).

## 4 Case study: understanding the ecology of *Diaphorina citri*, vector of citrus greening disease (huanglongbing)

The Asian citrus psyllid, *Diaphorina citri* Kuwayama (Hemiptera: Liviidae) is a widely distributed pest of citrus (Fig. 2). Apparently indigenous to the Indian subcontinent (Yang et al., 2006), the psyllid has expanded its distribution and is presently found in 26 Asiatic countries, four African countries, 17 countries of the American continent, many of the Caribbean islands, two European countries (probably only through interception) and several islands of the Oceania (CABI, 2019). Most of this distribution is within the tropics but it is also found above the tropic of cancer in such areas as Florida, USA and even above 30° N into Japan and below 30° S latitude into South Africa. The insect is also limited in distribution by altitude. Aubert (1987) reported that this insect was not found above 1300–1500 m in elevation in various regions of Asia. This latitudinal and altitudinal distribution is apparently limited by temperature, as the psyllid is generally freezing intolerant (Halbert and Manjunath, 2004). The host plants of this psyllid are confined to the cultivated and wild species of the citrus family, Rutaceae. These include cultivated citrus such as, especially, lemon (*Citrus limon*), rough lemon (*C. jambhuri*), sour orange (*C. aurantium*), grapefruit (*C. paradise*), limes (*C. aurantifolia*), orange jasmine (*Murraya paniculata*), and wild rutaceous hosts from at least six different genera besides *Citrus* (Yang et al., 2006). This psyllid normally spreads locally by natural flight dispersal. However, budwood, grafted trees, and rootstock seedlings from infested areas can carry eggs and/or nymphs over longer distances (CABI, 2019). Adults and nymphs damage plants by sucking sap from the phloem of tender shoots and buds, and adults can also feed on the phloem of mature leaves. Populations on citrus leaves can be high, and severe infestations in orchards and gardens lead to withering, distortion, and loss of immature leaves and irregular shaped canopies (Yang et al., 2006).

The Asian citrus psyllid would be considered a minor pest if it were not the principal vector of citrus greening disease (huanglongbing) which is probably the most important vectored disease of citrus in the world. The pathogen of huanglongbing is considered to be a phloem-inhabiting, unculturable bacteria of the genus *Candidatus* Liberibacter, presently consisting of four isolates (Texeira et al., 2005; Manjunath et al., 2008; Keremane et al. 2015; Suaste-Dzul et al., 2017). The disease causes mottling, chlorosis, yellow shoots, twig dieback, reduced fruit size, and bitter fruit taste (Halbert and Manjunath, 2004). The disease is very destructive and can render citrus orchards unproductive after several years. The various species that produce huanglongbing are presently found in southern Asia, Japan, New Zealand, central and southern Africa, Europe, Brazil, Central America and the southern United States (Haapalainen, 2014).

*Diaphorina citri* was described from Taiwan in 1907. Of the six other species of *Diaphorina* associated with *Citrus*, *D. citri* can be distinguished by a distinct pattern of markings on the forewings and differences in genitalia (Halbert and Manjunath, 2004). Based on whole mitochondrial genome sequences from 23 locations, Wu et al. (2018b) found two major mitochondrial groups in China: the southwest group MG1 and the southeast group MG2 suggesting that development and spread in China were likely through two different routes, the western associated with high-altitude regions and the eastern route with low-altitude regions point out since this insect was first described and studied in southeastern China. Previously Wu et al. (2017) found the similitude of *D. citri* populations from California and Florida, forming a distinct MG3 and clustering with populations from Pakistan (Wu et al., 2018a).

The effects of light intensity and duration significantly affect the preoviposition period and realized fecundity of female *D. citri* and when light duration was below 18 h a day the number of eggs laid increases (Yang, 1989). This factor cannot be used for this management because citrus produce areas usually are located at tropical and subtropical areas with eight to twelve day light h/day. Yang (1989) also found no effects of humidity and temperature on egg hatching, but nymphal mortality was notorious at higher temperatures with high humidity and low at moderate temperatures (20–30 °C) with lower humidity (43–75%), but these conditions are relatively normal for citrus production areas. *Diaphorina citri* can overwinter when the mean temperature exceeds 6.3°C and and leads to infestations on spring growth flush which is critical for transmission and spread of HLB (Yang et al., 2006) However, Halbert and Manjunath (2004) observed that populations have overwintered in northern Florida, where temperatures dropped to at least −5°C on several nights. It seems to be a key condition for IPM to reduce populations of the psyllid when citrus production areas are in cold conditions due to winter but of course the tree phenology should be taken in to account. Liu and Tsai (2000) found that temperature has a significant effect on developmental rate, survival, reproduction, and longevity of *D. citri*. For example, there were failures in development of nymphs at extreme temperatures (10°C and 33°C), also high temperature (30°C) caused a decrease in development rate for immature stages with 26.3% of mortality as well as lower production of 316 eggs per female. The best temperature range for its development is between 25° C and 28° C and the maximum egg production was 748 at 28° C.

The initial search for biological control agents of *D. citri* were carried out in the region of origin. One of the most important parasitoids is the idiobiont *Tamarixia radiata* (Hymenoptera: Eulophidae). This species was first described by Waterston in 1922, in the genus *Tetrastichus*, from specimens collected in northwestern India in the province of Punjab, currently part of Pakistan (Postali Parra et al., 2016). This parasitoid has a natural distribution in various citrus-growing regions in Asia and the Arabian Peninsula. Lama et al. (1988) report more than 90% parasitism from this eulophid on mandarin and sweet orange commodities in Nepal. However, indigenous predators in regions where this parasitoid has been released may have more impact on *D. citri*, and may have counterproductive interactions. This was reported by Michaud (2004) who found that the coccinellid beetles (*Harmonia axyridis*, *Olla v-nigrum*, *Cycloneda sanguinea* and *Exochomus children*) were the most important biological control agents of high-density *D. citri* populations in central Florida and that intraguild predation caused >95% mortality of immature stages of the introduced parasitoid, *T. radiata*, within the nymphs of *D. citri*. In exclusion experiments, the absence of these coccinellid predators improved the survival of *D. citri* by a factor of 120, despite improving the survival of the *T. radiata* by a factor of 40 (Michaud, 2004). Another important parasitoid is the endoparasitic *Diaphorencyrtus aligarhensis* (Hymenoptera: Encyrtidae) recorded in Guangdong and Taiwan (Yang et al., 2006). Rohrig et al. (2012) tried to establish *D. aligarhensis* in Florida during 2000–2002 with Taiwan populations of the parasitoid which did not become established and this apparent failure was attributed, in part, to the predation of parasitized hosts by generalist predators.

Biological control programs based on the introduction of the natural enemies of *D. citri* from its place of origin requires an understanding of the role of indigenous natural enemies in order to establish management efforts aimed to reduce the dispersion of 'huanglongbing'.

## 5 Case study: fruit fly management and the sterile insect technique (SIT)

### *5.1 Fruit flies and their management*

Fruit flies are among the most important pests of fruits worldwide and citrus are of particular concern. They belong to the family Tephritidae, order Diptera. The most important genera are *Anastrepha* Schiner, *Bactrocera* Macquart, and *Ceratitis* Macleay in tropical and suptropical regions, and *Rhagoletis* Loew in temperate regions (Bateman, 1972; White and Elson-Harris, 1992). Not all species in these genera are considered as pests of economic importance. For example, in Mexico there are approximately 30 species of *Anastrepha* (Norrbom and Kim, 1988; Hernández-Ortiz, 2007). However, extensive surveys have demonstrated that only four species are of economic importance. Of these, only one, *Anastrepha ludens*, is a pest of citrus fruits (Aluja, 1994; Reyes et al., 2000; Hernandez-Ortiz et al., 2010).

In a typical life cycle, mated females forage for mature fruits. When they find a suitable host, they insert their eggs. After a few days, the eggs hatch and the larvae feed in the fruit. The larvae in the fruit represent the main damage of these pests. When the larvae are mature, fruits fall to the ground, and the larvae come out of the fruit, burrow in the soil, and pupate. The pupae develop in the soil, and afterwards, adults emerge and forage for food and mates, repeating the cycle. Developmental time depends on the fruit fly species, the host species and availability of adult food and the environmental conditions. In addition to the direct damage caused by the worms in the fruit, the threat of invasion or establishment of this dangerous pest can result in the implementation of barriers to trade, difficult quarantine and regulation measures, as well as costly surveys (Christenson and Foote, 1960).

Because of the economic importance of fruit flies, many efforts have been invested in developing control methods and management strategies (Robinson and Hooper, 1989). As with most insect pests, IPM has been accepted as the most viable strategy (Kogan, 1998). The components or tools for fruit flies IPM are divided into two main groups: (1) detection or survey methods, and (2) control methods (Aluja, 1993, 1994). Fruit sampling is a common method used to detect immature stages, mainly larvae. This consists of the collection of fruit from the trees or soil and their subsequent dissection to see if they are infested. Another detection method is trapping. There are different trap designs and attractants available (Shelly et al., 2014). The use of them depends on the target species and the aim of the trapping system. There are food attractants that can be used for almost all species of economic importance. Parapheromones are very powerful attractants that are available for some species. These are species, or species-group specific (Diaz-Fleischer and Castrejón-Gómez, 2012).

The methods used for control of fruit flies are: (a) sanitation, (b) biological control, (c) mass trapping, (d) insecticide bait sprays, (e) legal or regulatory measures, and (f) the SIT. Sanitation is the collection and destruction of fallen fruit that could be infested, to avoid the emergence of flies in the orchard. Biological control of fruit flies could be by parasitoids or entomopathogens (Montoya and Toledo, 2010). Mass trapping using hydrolyzed enzymatic proteins as Ceratrap® has been found recently to be an effective suppression method (Villalobos et al., 2017). Application of insecticide bait sprays, by ground or by air, has been the most common control method. In the past, Malathion was the insecticide mixed with the protein bait. More recently it has been replaced by spinosad (Prokopy et al., 2003). Legal or regulatory measures include quarantines, postharvest treatments, pest risk assessment, systems approach and the development and application of international standards to minimize the risk of invasion and at the same time, facilitate trade (FAO, 2005; Dannug, 2018).

Typical IPM is a localized, farm-by-farm approach. Since fruit flies are highly mobile insects that can easily move from one farm to another or to backyard gardens or areas of wild vegetation, a different strategic approach to deal with this type of pests is required. The area-wide integrated pest management (AW-IPM), which considers the total population, currently represents the most feasible and convenient strategy. For biological control methods, including the SIT, this is a much more suitable approach (Vreysen et al., 2007).

## 5.2 Management of fruit flies with the sterile insect technique (SIT)

The SIT is a pest control method that has been used for the suppression or eradication of fruit fly infestations. In this case, control is effected through competition for copulation between sterile males, artificially reared, and females of the natural population (Knipling, 1979; Dyck et al., 2005). This implies that a large number of insects have to be reared, sterilized, and subsequently released in the indicated area.

Since the effect of STI is through reproduction, reducing the birth rate of the population, it is highly specific. This significantly minimizes the impact on other nontarget organisms and unwanted side effects. For the same reason, it is a control method that requires a deep knowledge of the biology of the insect pest, especially on factors that influence their behavior, reproduction, and population dynamics (Liedo et al., 2010).

Since its initial development in the 1950s, and given its success in the eradication of the New-world screwworm (*Cochiomyia homnivorax* Coquerel) (Knipling, 1955), SIT application has extended gradually to other pests (Dyck et al., 2005). The first large scale application of the SIT against fruit flies was made in the southern border of Mexico to avoid the introduction of the Mediterranean fly (*Ceratitis capitata* Wied.). In the late 1970s, a mass-rearing and sterilization facility was built in Metapa, Chiapas, close to border with Guatemala. The goal of this facility was to produce 500 million sterile flies per week. Wild *C. capitata* populations were initially suppressed using a combination of bait sprays and releases of sterile flies. As the number of detections decreased, the barrier of sterile flies was moved South-East, from the Mexican side toward the Guatemalan side of the border. By 1982, approximately half of the production of sterile flies was released in Guatemala and by 1985 it increased to two -thirds of total production (Schwarz et al., 1989). Budgetary constraints led to maintain the sterile flies' releases as a containment barrier on both sides of the Mexico-Guatemala border (Baker, 1984). In 1995 a new mass-rearing facility began operations at El Pino, Guatemala, with the capacity to produce 1000 million sterile *C. capitata* males per week. The greater availability of sterile flies, together with technological innovations, such as the production of sterile males only, a better knowledge of *C. capitata* population ecology in the region, and accumulated experience in the field, have all contributed to the increased success of the program and Mexico remains a country free of this pest (Enkerlin et al., 2015, 2017).

Another use of the SIT in Mexico has been to control native *Anastrepha* species. The National Fruit Fly Program began in the early 1990s with a North to South strategy based on the natural environmental conditions that determine fruit fly population densities. The dry regions of the North have lower host plant diversity and are characterized by lower fruit fly population densities (Gutiérrez, 2010). Here, eradication of fruit fly pest was considered feasible using SIT within the area-wide IPM approach. A new mass-rearing facility, the MOSCAFRUT facility, was built in Metapa, next to the original *C. capitata* rearing facility. In 1993 this facility began operations, producing 100 million sterile *A. ludens* flies per week. Later, methods were developed to produce *A. obliqua* flies and the parasitoid *Diachasmimorpha longicaudata* (Ashmead). This resulted in the application of the SIT and augmentative biological control under the AW-IPM approach. A new *A. ludens* genetic sexing strain was developed to release only males (Zepeda-Cisneros et al., 2014).

This strategy has been successful. From 1992 to date, over half of the Mexican territory has been recognized as fruit -fly -free areas, and about 10% as low -prevalence areas (Gutiérrez, 2010). Fruit can be exported from fruit-fly-free areas without the need for quarantine treatments. Orchards in low-prevalence areas can still export their fruits if they meet phytosanitary requirements such as maintaining monitoring trapping and fruit sampling, orchard sanitation, and postharvest treatments (hot water, irradiation, fumigation). There is also a certification available for orchards that are temporarily free of fruit flies. These orchards can export their products if they follow IPM protocols such as monitoring fruit fly populations, orchard sanitation, bait spraying in case of detection, and postharvest treatment.

Climate change will impact all economic, social and biological aspects of citrus cultivation. The effects of climate change on citrus arthropods is in speculative and based on studies of arthropod response in general, and on a limited number of specific studies of crop pests. Some of the factors considered to be of greatest influence on the challenges and responses of insect pests to climate change are: increased atmospheric levels of CO2 and host plant responses; increased and greater seasonal variation in ambient temperatures; drought and greater variations in seasonal precipitation; changes in plant phenology; shifts in tri-trophic interactions of plant, pest, and natural enemies; and changes in the interactions of insect vectors and plant diseases (Robinet and Roques 2010; Bentz et al., 2010; Robinson et al., 2012; Bebber et al., 2013; Donkersley et al., 2018; Trebicki and Finlay, 2019; Ponti et al., 2016). How all these the climate induced effects will play out for citrus agriculture will require enormous resilience and innovative IPM solutions, especially given that the impacts in topics and subtropics will be highly variable depending on the region (Cerri et al. 2007) and complicated by the prevalence of singular extreme weather events.

## 6 Future trends and conclusion

Three issues will play critical roles in the transformation of ecological understanding and management of citrus insect pests in the future: globalization together with the resultant emergence of new exotic pests, generalized loss of biodiversity, and climate change. All three will be incrementally important and synergistic. An example of the first phenomenon is the present crisis in world citrus production imposed by the Asian citrus psyllid and its associated pathogen causing citrus greening disease. Originally restricted on native citrus and its relatives in Asia (Batool et al., 2007), the insect and bacteria have spread around the world, as reported herein and in subsequent chapters. The impact of the disease is currently at such a critical level that it threatens virtually all aspects of citrus cultivation of the globe. The success of the research and management programs to manage the vector and the pathogen will determine when, where and how future research programs will be modified. Such programs will require regional strategies and international coordination of in management tactics and the rapid incorporation of technological advances (Santivañez et al., 2014). New exotic pests will require this regional approach combined with innovative technologies.

The general loss of biodiversity has been evidenced through the well-publicized decline of pollinators and the more recent reports of the general loss of arthropod abundance as evidenced from varied regions of the globe (Hallman et al., 2017; Lister and Garcia, 2018). What this general trend portends for citrus pest management is a concurrent simplification of agroecosystems with reduced diversity of predators and parasitoids (Janzen and Hallwachs, 2019), greater difficulty in the search for new biological control agents and probable increased extremes in outbreaks of primary and secondary insect pests. These consequences indicate a need for continued effort to reduce the use of synthetic insecticides with new approaches and organisms for biological control programs. This requires special emphasis on conservation of native and established natural enemies through habitat diversification of the citrus orchard and its surrounding environment.

Climate change will impact all economic, social and biological aspects of citrus cultivation. As discussed previously, how the effects of climate change will play out for citrus is still speculative, especially given that the impacts in the tropics and subtropics will be highly variable depending on the region (Cerri et al., 2007) and complicated by the prevalence of singular extreme weather events.

## 7 Where to look for further information

Many institutions around the globe have important research programs and regional management initiatives in citrus entomology. The Food and Agricultural Organization of the United Nations has several international programs. In particular, the Regional Office for Latin America and the Caribbean has initiated a strong regional management program for Huanglongbing in Latin America and the Caribbean. Also, the Centre for Agriculture and Bioscience International (CABI) has centers in 12 countries and actively promotes crop health world-wide through field programs, the publication of scientific literature and importantly, the maintenance of an online, detailed data base of invasive species of the world.

In North America, the United States Department of Agriculture, Animal and Plant Health Inspection Service (USDA APHIS) has many research projects, and a Citrus Health Response Program (CHRP) directed at implementing standards for survey and mitigation of diseases and their vectors. The North American Plant Protection Organization (NAPPO) which includes Canada, USA and Mexico plant health organizations, has pest alert system for several crops including citrus. The states of Florida and California have solid and diverse research and management programs. In Florida, the Citrus Research & Education Center of the University of Florida and Institute of Food and Agricultural Sciences (UF/IFAS) is an important research institution of basic and applied studies of insect pest management. The California Department of

Food and Agriculture -California Citrus Program and the Division of Agriculture and Natural Resources of the University of California also have active research programs and are heavily investing in the preventing further spread of HLB in the state. Research funding in California is also aided by the grower-funded Citrus Research Program (CRB). Institutions involved with pest management and quarantine in Mexico include the Servicio Nacional de Sanidad, Inocuidad y Calidad Agroalimentaria (SENASICA) within the Secretaría de Agricultura y Desarrollo Rural (SADER) which has been monitoring and working to contain the spread of HLB in the country.

In South America, besides national programs in individual countries, a regional organization for plant protection, COSAVE (Comité de Sanidad Vegetal) was created in 1989 and participating countries now include Argentina, Brasil, Chile, Paraguay, Peru and Uruguay (COSAVE). In 2011, this organization approved a regional plan for the containment (HLB) in the region.

In Africa, two of the notable research institutions are Citrus Research International which is based in South Africa and focusses primarily on improving citrus cultivation, including studies of citrus plant protection. The other program is the Centre for Insect Physiology and Ecology (ICIPE) which was established in 2010 in Nairobi, Kenya and has programs with impact on the citrus entomology of the region.

Asian research institutes include the Citrus Research Institute of Chinese Academy of Agricultural Sciences in Chongqing, China. This institute is especially important as it is located within the region of origin of citrus and center of genetic diversity of citrus cultivars. In India, the National Research Centre for Citrus is an institute for research in citrus fruits located at Nagpur in the state of Maharashtra which also within a region of citrus diversity. Another important Asian institute is the Institute of Fruit Tree and Tea Science, NARO (NIFTS) of Japan.

## 8 References


Aluja, M. (1993). *Manejo integrado de la mosca de la fruta*. Trillas, México D.F., 251p.

Aluja, M. (1994). Bionomics and management of *Anastrepha*. *Annual Review of Entomology* 39: 155–78.

Argov, Y. and Rossler, Y. (1996). Introduction, release and recovery of several exotic natural enemies for biological control of the citrus leaf miner, *Phyllocnistis citrella*, in Israel. *Phytoparasitica* 24(1): 33–8.

Aubert, B. (1987). *Trioza erytreae* Del Guercio and *Diaphorina citri* Kuwayama (Homoptera: Psylloidea), the two vectors of citrus greening disease: Biological aspects and possible control strategies. *Fruits* 42(3): 149–62.

Aurambout, J. P., Finlay, K. J., Luck, J. and Beattie, G. A. C. (2009). A concept model to estimate the potential distribution of the Asiatic citrus psyllid (*Diaphorina citri* Kuwayama) in Australia under climate change–a means for assessing biosecurity risk. *Ecological Modelling* 220(19): 2512–24.

Baker, P. (1984). Insect machismo in Central America. *New Scientist* 16: 41.

Bateman, M. A. (1972). The ecology of fruit flies. *Annual Review of Entomology* 17: 439-518.

Batool, A., Iftikhar, Y., Mughal, S. M., Khan, M. M., Jaskani, M. J., Abbas, M. and Khan, I. A. 2007. Citrus greening disease-a major cause of citrus decline in the world-a review. *Horticultural Science (Prague)* 34(4), 159–66.

Bebber, D. P., Ramotowski, M. A. and Gurr, S. J. 2013. Crop pests and pathogens move polewards in a warming world. *Nature Climate Change* 3(11): 985.

Bentz, B. J., Régnière, J., Fettig, C. J., Hansen, E. M., Hayes, J. L., Hicke, J. A., Kelsey, R. G., Negrón, J. F. and Seybold, S. J. (2010). Climate change and bark beetles of the western United States and Canada: direct and indirect effects. *BioScience* 60(8): 602–13.

Botteon, V., Costa, M. D. L. Z., Kovaleski, A., Martinelli, L. A. and Mastrangelo, T. (2018). Can stable isotope markers be used to distinguish wild and mass-reared *Anastrepha fraterculus* flies? *PLoS one* 13(12): e0209921.

CABI. (2019). *Invasive Species Compendium*. Centre for Agriculture and Biological Information. Available at: https://www.cabi.org/isc/datasheet/18615#82C3C952-BBD3-429B-BCD8-C4D9B102D116 (Accessed on January 2019).

Caltagirone, L. E. and Doutt, R. L. (1989). The history of the vedalia beetle importation to California and its impact on the development of biological control. *Annual Review of Entomology* 34(1): 1–16.

Cerri, C. E. P., Sparovek, G., Bernoux, M., Easterling, W. E., Melillo, J. M. and Cerri, C. C. (2007). Tropical agriculture and global warming: impacts and mitigation options. *Scientia Agricola* 64(1): 83–99.

Christenson, L. D. and Foote, R. H. (1960). Biology of fruit flies. *Annual Review of Entomology* 5: 171–92.

Clausen, C. P. (1931). *Two Citrus Leaf Miners of the Far East* (No. 252). US Department of Agriculture.

Costa, V. A., De Sá, L. A. N., LaSalle, J., De Nardo, E. A. B., Arellano, F. and Fuini, L. C., (1999). Indigenous parasitoids (Hym., Chalcidoidea) of *Phyllocnistis citrella* Stainton (Lep., Gracillariidae) in Jaguariúna, São Paulo State, Brazil: Preliminary results. *Journal of Applied Entomology* 123(4): 237–40.

Dannug, D. T. (2018). *Government Regulations and Interventions for the Production of High Quality Tropical Fruits for Export*. In: *Proceedings, International Symposium on Tropical Fruits (ISTF2017), Food Security amidst a Changing Climate: Towards a Sustainable and Resilient Tropical Fruit Industry*, 23–25 October 2017, Nadi, Fiji (pp. 118–23). International Tropical Fruits Network (TFNet).

DeBach, P. (1951). The necessity for an ecological approach to pest control on citrus in California. *Journal of Economic Entomology* 44(4).

DeBach, P. and Bartlett, B. (1951). Effects of insecticides on biological control of insect pests of citrus. *Journal of Economic Entomology* 44(3): 372–83.

Deka, S., Kakoti, R. K. and Sabir, N. (2015). Survey and surveillance of insect pests of *Citrus* and their natural enemies in Assam. *Journal of Insect Science*. 29(1): 158–61.

De Meyer, M., Robertson, M. P., Peterson, A. T. and Mansell, M.W. (2008). Ecological niches and potential geographical distributions of Mediterranean fruit fly (*Ceratitis capitata*) and Natal fruit fly (*Ceratitis rosa*). *Journal of Biogeography* 35: 270–81.

De Meyer, M., Robertson, M. P., Mansell, M. W., Ekesi, S., Tsuruta, K., Mwaiko, W., Vayssières, J. F. and Peterson, A. T. (2010). Ecological niche and potential geographic

distribution of the invasive fruit fly *Bactrocera invadens* (Diptera, Tephritidae). *Bulletin of Entomological Research* 100(1): 35–48.
De Villiers, M., Hattingh, V., Kriticos, D. J., Brunel, S., Vayssières, J. F., Sinzogan, A., Billah, M. K., Mohamed, S. A., Mwatawala, M., Abdelgader, H. and Salah, F. E. E. (2016). The potential distribution of *Bactrocera dorsalis*: considering phenology and irrigation patterns. *Bulletin of Entomological Research* 106(1): 19–33.
Díaz-Fleischer, F. and Castrejón-Gómez, V.R. (2012). El papel de los semioquimicos en el manejo de las moscas de la fruta (Diptera: Tephritidae). In: Rojas-León J., Malo-Rivera E. A. (Eds), *Temas Selectos en Ecología Química de Insectos*. El Colegio de la Frontera Sur, Chiapas, Mexico, pp. 401–26.
Dietz, H. F. and Zetek J. (1920). The blackfly of citrus and other subtropical plants. *USDA Bulletin* 885: 1–55.
Donkersley, P., Silva, F. W., Carvalho, C. M., Al-Sadi, A. M. and Elliot, S. L. (2018). Biological, environmental and socioeconomic threats to citrus lime production. *Journal of Plant Diseases and Protection*. 125(4): 339–56.
Dyck, V. A., Hendrichs, J. and Robinson, A. S. (Eds). (2005). *Sterile Insect Technique: Principles and Practice in Area-Wide Integrated Pest Management*. Springer, Dordrecht, The Netherlands, 787p.
Ehleringer, J. R. and Rundel, P. W. (1988). Stable isotopes: history, units and instrumentation. In: Rundel, P. W., Ehleringer, J. R. and Nagy, K. A. (Eds), *Stable Isotopes in Ecological Research*. Springer-Verlag, New York, pp. 2–15.
Elith, J., Graham, C. H., Anderson, R., Dudík, M., Ferrier, S., Guisan, A., Hijmans, R. J., Huettmann, F., Leathwick, J. R., Lehmann, A. and Li, J. (2006). Novel methods improve prediction of species' distributions from occurrence data. *Ecography*. 29(2): 129–51.
Enkerlin, W. R. (2005). Impact of fruit fly control programmes using the sterile insect technique. In: Dyck, V. A., Hendrichs, J. and Robinson, A. S. (Eds), *Sterile Insect Technique: Principles and Practice in Area-Wide Integrated Pest Management*. Springer, Dordrecht, The Netherlands, pp. 651–76.
Enkerlin, W., Gutierrez-Ruelas, J. M., Villaseñor-Cortes, A., Cotoc-Roldan, E., Midgarden, D., Lira, E., Zavala-López, J. L., Hendrichs, J., Liedo, P. and Trujillo-Arriaga, F. J. (2015). Area freedom in Mexico from Mediterranean fruit fly (Diptera: Tephritidae): A review of over 30 years of a successful containment program using an integrated area-wide SIT approach. *Florida Entomologist*. 98(2): 665–81. DOI: 10.1653/024.098.0242.
Enkerlin, W. R., Gutiérrez-Ruelas, J. M., Pantaleón, R., Soto-Litera, C., Villaseñor-Cortes, A., Cotoc-Roldán, E., Hernández-López, F., Arenas-Castillo, A., Castellanos-Dominguez, D., Valle-Mora, A., Rendón-Arana, P., Cáceres-Barrios, C., Midgarden, D., Villatoro, C., Lira-Prera, E., Zelaya-Estrade, O., Castañeda-Aldana, R., López-Culajay, J., Ramírez y Ramírez, F., Liedo-Fernández, P., Ortiz-Moreno, G., Reyes-Flores, J. and Hendrichs, J. (2017). The Moscamed Regional Programme: review of a success story of area-wide sterile insect technique application. *Entomología Experimentalis et Applicatta. SI Sterile Insect Technique* 164(3): 188–203. DOI: 10.1111/eea.12611.
EPPO (2019). EPPO Global Database (available online). Available at: https://gd.eppo.int.
FAO international standards for phytosanitiary measures. (2005). Rome. http://www.fao.org/3/a0785e/a0785e00.htm (accessed 29 May 2019).

Gutiérrez, J. M. (2010). El Programa Moscas de la Fruta en México. In: Montoya, P., Toledo, J. and Hernández, E. (Eds), *Moscas de la Fruta: Fundamentos y Procedimientos para su Manejo*. S y G Editores, México D. F., pp. 3–10.

Haapalainen, M. (2014). Biology and epidemics of *Candidatus* Liberibacter species, psyllid-transmitted plant-pathogenic bacteria. *Annals of Applied Biology*. 165(2), 172–98.

Halbert S. E. and K. L. Manjunath. (2004). Asian citrus psyllids (Sternorrhyncha: Psyllidae) and greening disease of citrus: a literature review and assessment of risk in Florida. *Florida Entomologist* 87(3): 330–53.

Hallmann, C. A., Sorg, M., Jongejans, E., Siepel, H., Hofland, N., Schwan, H., Stenmans, W., Müller, A., Sumser, H., Hörren, T. and Goulson, D. 2017. More than 75 percent decline over 27 years in total flying insect biomass in protected areas. *PLoS One* 12(10), e0185809.

Hendrichs, J., Franz, G. and Rendon, P. (1995). Increased effectiveness and applicability of the sterile insect technique through male-only releases for control of Mediterranean fruit flies during fruiting seasons. *Journal of Applied Entomology*. 119: 371–77.

Heppner, J.B. and Fasulo, T.R. 2010. Citrus leafminer, *Phyllocnistis citrella* Stainton (Insecta: Lepidoptera: Phyllocnistinae). Online. Univ. Florida Publ.# EENY038. http://edis. ifas. ufl. edu/in165.

Hernández-Ortiz, V. (2007). Diversidad y biogeografía del género *Anastrepha* en México. In: Hernández-Ortiz, V., Morales, P., Gonzalez, E., Hernández-Ortiz, V., Ortiz, V., Zucchi, R. A., Velasquez, V. and Gonzalez, L. (Eds), *Moscas de la fruta en Latinoamérica (Diptera: Tephritidae): Diversidad, biología y manejo*. S y G Editores, Mexico D.F., pp. 53–76.

Hernandez-Ortiz, V., Guillén-Aguilar, J. and López, L. (2010). Taxonomía e identificación de moscas de la fruta de importancia económica en América. In: Montoya, P., Toledo, J. and Hernandez, E. (Eds). *Moscas de la Fruta: Fundamentos y Procedimientos para su Manejo*. S y G Editores, México, D.F., pp. 49–80.

Hernández-Ortiz, V., Bartolucci, A.F., Morales-Valles, P., Frías, D. and Selivon, D. (2012). Cryptic species of the *Anastrepha fraterculus* complex (Diptera: Tephritidae): a multivariate approach for the recognition of South American morphotypes. *Annals of the Entomological Society of America*.105(2): 305–18.

Hood-Nowotny, R. and Knols B. G. J. (2007). Stable isotope methods in biological and ecological studies of arthropods. *Entomologia Experimentalis et Applicata*. 124(1): 3–16.

Huang, H. T. and Yang, P. (1987). Ancient cultured citrus ant used as biological control agent. *BioScience*. 37 (9): 665–71.

Janzen, D. H. and Hallwachs, W. (2019). Perspective: Where might be many tropical insects? *Biological Conservation*. 233:102–8.

Jeppson, L. R. (1967). Chapter 1, Biology of citrus insects, mites and mollusks. In: Reuther, W. (Ed.), *The Citrus Industry: Crop Protection, Postharvest Technology, and Early History of Citrus Research in California* (vol. V). UCANR Publications, pp. 1–10.

Kennett. C. E., McMurtry, A. and Beardsley, W. 1999. Biological control in subtropical and tropical crops. In: Bellows, S. and Fisher, T. W. (Eds), *Handbook of Biological Control*. Academic Press, San Diego, CA, pp. 713–42

Keremane, M. L., Ramadugu, C., Castaneda, A., Díaz, J. E., Peñaranda, E. A., Chen, J., Duan, Y. P., Halbert, S. E. and Lee, R. F. (2015). Report of '*Candidatus* Liberibacter caribbeanus', a new citrus- and psyllid associated Liberibacter from Colombia, South America. (Abstr.) *Phytopathology*. 105: S4.71.

Knipling, E. F. (1955). Possibilities of insect control or eradication through the use of sexually sterile males. *Journal of Economic Entomology*. 48: 459–62.

Knipling, E. F. (1979). *The Basic Principles of Insect Suppression and Management*. U.S. Dept. Agric. Handbook 512. U.S. Government Printing Office, Washington DC.

Kogan, M. (1998). Integrated pest management: historical perspectives and contemporary developments. *Annual Review of Entomology*. 43: 243–70.

Lama T. K., Regmi C. and Aubert B. (1988). Distribution of the citrus greening disease vector (*Diaphorina citri* Kuw.) in Nepal and attempts to establish biological control. In: *Proceedings, 10th Conference of the International Organization of Citrus Virologists (IOCV)*, Valencia, University of California, pp. 255–7.

Langgut, D. (2017). The citrus route revealed: from Southeast Asia into the Mediterranean. *HortScience*. 52(6): 814–22.

Latif, A. and C. Muhammad Yunus. (1951). Food-plants of Citrus Leaf-miner (*Phyllocnistis citrella* Stn.) in the Punjab. *Bulletin of Entomological Research*. 42(2): 311–16.

Liedo, P., Enkerlin, W. and Hendrichs, J. (2010). Fundamentos de la Técnica del Insecto Estéril (TIE). In: Montoya, P., Toledo, J. and Hernández, E. (Eds), *Moscas de la Fruta: Fundamentos y Procedimientos para su Manejo*. S y G Editores, México D.F., pp. 243–55.

Lister, B. C. and Garcia, A. 2018. Climate-driven declines in arthropod abundance restructure a rainforest food web. *Proceedings of the National Academy of Sciences* 115(44), E10397–406.

Liu, Y. H. and Tsai, J. H. (2000). Effects of temperature on biology and life table parameters of the Asian citrus psyllid, *Diaphorina citri* Kuwayama (Homoptera: Psyllidae). *Annals of Applied Biology*. 137: 201–6.

Lux, S. A., Copeland, R. S., White, I. M., Manrakhan, A. and Billah, M. K. (2003). A new invasive fruit fly species from the *Bactrocera dorsalis* (Hendel) group detected in East Africa. *Insect Science and its Applications*. 23: 355–61.

Manjunath, K. L., Halbert, S. E., Ramadugu, C., Webb, S. and Lee, R. F. (2008). Detection of 'Candidatus Liberibacter asiaticus' in *Diaphorina citri* and its importance in the management of citrus huanglongbing in Florida. *Phytopathology*. 98(4), 387–96.

McNabb, D. M., Halaj, J. and Wise, D. H. (2001). Inferring trophic positions of generalist predators and their linkage to the detrital food web in agroecosystems: a stable isotope analysis. *Pedobiologia*. 45(4): 289–97.

Menalled, F., Peterson, R., Smith, R., Curran, W., Páez, D. and Maxwell, B. (2016). The eco-evolutionary imperative: revisiting weed management in the midst of an herbicide resistance crisis. *Sustainability*. 8(12): 1297.

Mestre, L., Piñol, J., Barrientos, J. A., Espadaler, X., Brewitt, K., Werner, C. and Platner, C. (2013). Trophic structure of the spider community of a Mediterranean citrus grove: a stable isotope analysis. *Basic and Applied Ecology*. 14(5): 413–22.

Michaud, J. P. (2004). Natural mortality of Asian citrus psyllid (Homoptera: Psyllidae) in central Florida. *Biological Control*. 29: 260–9.

Montoya, P. and Toledo J. (2010). Estrategias de control biológico. In: Montoya, P., Toledo, J. and Hernández, E. (Eds), *Moscas de la Fruta: Fundamentos y Procedimientos para su Manejo*. S y G Editores, México D.F., pp. 169- 82.

Murdoch, W. W., Swarbrick, S. L. and Briggs, C. J. (2006). Biological control: lessons from a study of California red scale. *Population Ecology*. 48(4): 297–305.

Myers, J. H., Simberloff, D., Kuris, A. M. and Carey, J. R. (2000). Eradication revisited: dealing with exotic species. *Trends in Ecology and Evolution*. 15(8): 316–20.

Narouei-Khandan, H. A., Halbert, S. E., Worner, S. P. and van Bruggen, A. H. (2016). Global climate suitability of citrus huanglongbing and its vector, the Asian citrus psyllid, using two correlative species distribution modeling approaches, with emphasis on the USA. *European Journal of Plant Pathology*. 144(3): 655–70.

Norrbom, A. L. and Kim, K. C. (1988). *A List of the Reported Host Plants of the Species of Anastrepha (Diptera: Tephritidae)*. APHIS 81 (USA).

Peterson, A. T. and Soberón, J. (2012). Species distribution modeling and ecological niche modeling: getting the concepts right. *Natureza and Conservação*. 10(2): 102–7.

Peterson, R. K., Higley, L. G. and Pedigo, L. P. (2018). Whatever happened to IPM? *American Entomologist*. 64(3): 146–50.

Phillips, S.J. and Dudík, M. (2008). Modeling of species distributions with Maxent: new extensions and a comprehensive evaluation. *Ecography*. 31(2): 161–75.

Platner, C., Piñol, J., Sanders, D. and Espadaler, X. (2012). Trophic diversity in a Mediterranean food web–stable isotope analysis of an ant community of an organic citrus grove. *Basic and Applied Ecology*. 13(7): 587–96.

Ponti, L., Gutierrez, A.P. and Iannetta, M. (2016). *Climate Change and Crop-Pest Dynamics in the Mediterranean Basin*. Italian National Agency for New Technologies, Energy and Sustainable Economic Development, RT/2016/27/ENEA.

Postali Parra, J. R., Rodrigues Alves, G., Ferreira Diniz, A. J. and Mendes Vieira, J. (2016). *Tamarixia radiata* (Hymenoptera: Eulophidae ) × *Diaphorina citri* (Hemiptera: Liviidae): mass rearing and potential use of the parasitoid in Brazil. *Journal of Integrated Pest Management*. 7 (1): 1–11.

Prokopy, R. J., Miller, N. W., Piñero, J. C., Barry, J. D., Tran, L. C., Oride, L. and Vargas, R. I. (2003). Effectiveness of GF-120 fruit fly bait spray applied to border area plants for control of melon flies (Diptera: Tephritidae). *Journal of Economic Entomology*. 96: 1485–93.

Reyes J., Santiago G. and Hernandez P. (2000). The Mexican fruit fly eradication programme. In: Tan, K. H. (Ed.), *Area-Wide Control of Fruit Flies and Other Pests*. Penerbit Universiti Sains Malaysia, Penang, pp . 377–80.

Robinet, C. and Roques, A. (2010). Direct impacts of recent climate warming on insect populations. *Integrative Zoology*. 5(2): 132–42.

Robinson, A. S. and Hooper, G. (Eds). (1989). *Fruit Flies, Their Biology, Natural Enemies and Control. World Crop Pests*, volume 3. Elsevier Science Publishers B.V., Amsterdam, The Netherlands.

Robinson, E. A., Ryan, G. D. and Newman, J. A. (2012). A meta-analytical review of the effects of elevated $CO_2$ on plant-arthropod interactions highlights the importance of interacting environmental and biological variables. *New Phytologist*. 194(2): 321–36.

Rodriguero, M. S., Lanteri, A. A. and Confalonieri, V. A. (2013). Speciation in the asexual realm: is the parthenogenetic weevil *Naupactus cervinus* a complex of species in statu nascendi? *Molecular Phylogenetics and Evolution*. 68(3): 644–56.

Rohrig, E. A., Hall, D. G., Qureshi, J. A. and Stansly, P. A. (2012). Field release in Florida of *Diaphorencyrtus aligarhensis* (Hymenoptera: Encyrtidae), an endoparasitoid of *Diaphorina citri* (Homoptera: Psyllidae) from Mainland China. *Florida Entomologist*. 95(2): 479–81.

San Jose, M., Leblanc, L., Geib, S. M. and Rubinoff, D. (2013). An evaluation of the species status of *Bactrocera invadens* and the systematics of the *Bactrocera dorsalis* (Diptera: Tephritidae) complex. *Annals of the Entomological Society of America*. 106(6): 684–94.

Santi-Júnior, A. D., Botteon, V. W., Mastrangelo, T. and Moreira, M. Z. (2018). Trophic ecology of citrus pests based on stable isotope analysis. *Scientia Agricola*. 75(6): 504–8.

Santivañez, C. T., Vernal, H. P., Mora-Aguilera, G., Diaz, P. G. and Lopez-Arroyo, J. I., (2014). Marco Estrategico para la Gestion Regional del Huanglongbing en America Latina y el Caribe. FAO, Rome, Italy. Online publication: http://www. fao. org/3/a-i3319s. pdf.

Schauff, M. E., LaSalle, J. and Wijesekara, G. A. (1998). The genera of chalcid parasitoids (Hymenoptera: Chalcidoidea) of citrus leafminer *Phyllocnistis citrella* Stainton (Lepidoptera: Gracillariidae). *Journal of Natural History*. 32(7): 1001–56.

Schwarz, A., Liedo, P. and Hendrichs, J. (1989). Sterile Insect Technique. 9.5.3 Current Programme in México. In: Robinson, A. S . and Hooper, G. (Eds). *Fruit Flies, Their Biology, Natural Enemies and Control*. World Crop Pests, Volume 3B. Elsevier-Science Publishers B.V., Amsterdam, pp. 375–86.

Shelly, T., Epsky, N., Jang, E. B., Reyes-Flores, J. and Vargas, R. (Eds). (2014). *Trapping and the Detection, Control, and Regulation of Tephritid Fruit Flies: Lures, Area-Wide Programs, and Trade Implications*. Springer, Dordrecht, The Netherlands, 643p.

Smith, R. F. (1962). Integration of biological and chemical control: introduction and principles. *Bulletin of the Entomological Society of America*. 8(4): 188–9.

Stern, V. M. R. F., Smith, R., Van den Bosch, R. and Hagen, K. 1959. The integration of chemical and biological control of the spotted alfalfa aphid: the integrated control concept. *Hilgardia* 29(2), 81–101.

Suaste-Dzul, A., Gallou, A., Félix-Portillo, M., Moreno-Carrillo, G., Sánchez-González, J., Palomares-Pérez, M. and Arredondo-Bernal, H. (2017). Seasonal incidence of '*Candidatus* Liberibacter asiaticus' (Rhizobiales: Rhizobiaceae) in *Diaphorina citri* (Hemiptera: Liviidae) in Colima, Mexico. *Tropical Plant Pathology*. 42(5): 410–15.

Szyniszewska, A. M. and Tatem, A. J. (2014). Global assessment of seasonal potential distribution of Mediterranean fruit fly, *Ceratitis capitata* (Diptera: Tephritidae). *PLoS One*. 9(11), e111582.

Texeira, D. D. C., Ayres, J., Kitajima, E. W., Danet, L., Jagoueix-Eveillard, S., Saillard, C. and Bové, J. M. (2005). First report of a huanglongbing-like disease of citrus in São Paulo State, Brazil and association of a new Liberibacter species, "*Candidatus* Liberibacter americanus", with the disease. *Plant Disease*. 89(1): 107.

Thomas, M. C., Heppner, J. B., Woodruff, R. E., Weems, H. V., Steck, G. J. and Fasulo, T. R. (2001). *Mediterranean Fruit Fly*, Ceratitis capitata (*Wiedemann*) (*Insecta: Diptera: Tephritidae*). Florida Department of Agriculture and Consumer Services, Division of Plant Industry. Entomology Circulars 4, 230 and 263, http://entnemdept.ufl.edu/creatures/fruit/mediterranean_fruit_fly.htm (accessed 2 May 2019).

Trebicki, P. and Finlay, K. (2019). Chapter 11. Pests and diseases under climate change. In: Shyam, S., Yadav, R. J., Redden, J. L., Hatfield, A. W. and Hunter, E. D. (Eds), *Food Security and Climate Change*. Wiley Press, NY.

Vander Zanden, M. J. and Rasmussen, J. B. (2001). Variation in δ15N and δ13C trophic fractionation: implications for aquatic food web studies. *Limnology and Oceanography*. 46(8): 2061–6.

Villalobos, J., Flores, S., Liedo, P. and Malo, E. A. (2017). Mass trapping is as effective as ground bait sprays for the control of *Anastrepha* (Diptera: Tephritidae) fruit flies in mango orchards. *Pest Management Science*. 73(10): 2105–10.

Vreysen, M. J., Robinson, A. S. and Hendrichs, J. (Eds). (2007). *Area-wide Control of Insect Pests: From Research to Field Implementation*. Springer Science & Business Media.

Waterston, J. (1922). On the chalcidoid parasites of psyllids (Hemiptera, Homoptera). *Bulletin of Entomological Research*. 13(1): 41–58.

Weeks, J. A., Martin, K. W., Hodges, A. C. and Leppla, N. C. 2019. Fact sheets; Citrus Pests. Available at: https://idtools.org/id/citrus/pests/factsheet.php?name=California %20red%20scale (Accessed on May 2, 2019).

White, I. M. and Elson-Harris, M. M. (1992). *Fruit Flies of Economic Significance: Their Identification and Bionomics*. CAB International.

WorldClim. (2019). https://www.worldclim.org/ (accessed on 10 May 2019).

Wu, F., Kumagai, L., Cen, Y., Chen, J., Wallis, C. M., Polek, M., Jiang, H., Zheng, Z., Liang, G. and Deng, X. (2017). Analyses of mitogenome sequences revealed that Asian citrus psyllids (*Diaphorina citri*) from California were related to those from Florida. *Science Report*. 7: 10154.

Wu, F. H., Jiang, G. A., Beattie, C., Holford, P., Chen, J., Wallis, C. M., Zheng, Z., Denga, X. and Cena, Y. (2018a). Population diversity of *Diaphorina citri* (Hemiptera: Liviidae) in China based on whole mitochondrial genome sequences. *Pest Managament Science*. 74: 2569–77.

Wu, G. A., Terol, J., Ibanez, V., López-García, A., Pérez-Román, E., Borredá, C., Domingo, C., Tadeo, F. R., Carbonell-Caballero, J., Alonso, R. and Curk, F. (2018b). Genomics of the origin and evolution of *Citrus*. *Nature*. 554 (7692): 311.

Yang, Y. (1989). Effects of light, temperature and humidity on the development, reproduction and survival of citrus psylla. *Acta Ecological Sinica*. 9(4): 348–54.

Yang, Y., Huang, M., Beattie, G. A. C., Xia, Y., Ouyang, G. and Xiong, J. (2006). Distribution, biology, ecology and control of the psyllid *Diaphorina citri*, Kuwayama, a major pest of citrus: a status report for China. *International Journal of Pest Management*. 52(4): 343-52.

Zalucki, M. P., Adamson, D. and Furlong, M. J. (2009). The future of IPM: Whither or wither? *Australian Journal of Entomology*. 48(2): 85–96.

Zepeda-Cisneros, C. S., Meza-Hernández, J. S., García-Martínez, V., Ibañez-Palacios, J., Zacharopoulou, A. and Franz, G. (2014). Development, genetic and cytogenetic analyses of genetic sexing strains of the Mexican fruit fly, *Anastrepha ludens* Loew (Diptera: Tephritidae). *BMC Genetics*. 15 (Suppl 2): S1.

# Chapter 5

# Advances in integrated pest management (IPM) in citrus cultivation

*Lukasz Stelinski, University of Florida, USA*



## 1 Introduction

Citrus is grown worldwide with the highest production per country (million tons) occurring in Brazil (20.7), China (19.6), the United States (10.0), Mexico (6.9) and India (6.3) (WorldAtlas, 2019). Other large citrus-producing countries include Spain, Iran, Italy, Nigeria, Turkey and South Africa. Citrus is a low input crop and thus integrated pest management (IPM) traditionally flourished in regions where it was implemented. While citrus production occurs globally, the defined practice of IPM in citrus historically occurred in only a small portion of worldwide citrus-producing regions, with Florida serving as a prime example. However, pest management in citrus has been significantly transformed where citrus greening disease or Huanglongbing (HLB) became invasive (Wang et al., 2017). Florida is currently perhaps the best example of a citrus industry where IPM had been implemented at a groundbreaking level in the absence of HLB and where it has now diminished to calendar-based sprays or marginal input management under HLB. The insect vector of greening, the Asian citrus psyllid (*Diaphorina citri* Kuwayama, ACP), was first detected in Florida in 1998, and it rapidly established throughout the state (Grafton-Cardwell et al., 2013). Currently, the Florida citrus industry struggles for its very existence while management practices are being adapted for living with the endemic HLB.

http://dx.doi.org/10.19103/AS.2019.0054.07

Today, as HLB has spread globally, IPM in citrus faces challenges associated with managing this disease. The goal of this chapter is to discuss IPM in citrus production and thus it has been prepared using Florida as a case study. For historical and statistical overviews of global citrus production, please see Reuther et al. (1967), Davies and Albrigo (1994) and FAO (2017).

There is no current cure for citrus greening and the impact of the disease on Florida citrus production has been substantial since confirmation of the disease in 2005 (Spreen et al., 2014). The disease limits citrus production by reducing yield and decreasing fruit quality, and can eventually kill trees (Bovè, 2006). Citrus canker (*Xanthomonas axonopodis* pv. *citri*) is another severe disease currently affecting citrus industries worldwide. Like greening, canker is closely associated with an insect pest, the citrus leafminer (*Phyllocnistis citrella* Stainton, CLM), present in Florida since 1993 (Gottwald et al., 2007). While not a true vector of citrus canker, CLM larvae damage young citrus leaves when feeding, making them highly susceptible to canker infection (Hall et al., 2010). The threat of these two diseases has required much more intense chemical pest management in Florida citrus than was required in the entire industry's history (Qureshi et al., 2014). In areas where these two pest/disease complexes affect citrus production, all other insect pests are considered secondary and are often acceptably controlled by treatments targeting *D. citri* and *P. citrella*.

## 2 Integrated pest management (IPM) in Florida citrus: historical perspective

Although citrus trees were introduced to Florida more than 400 years ago, interest in commercial citrus production began to expand significantly in the 1920s (Simanton, 1996). The industry experienced significant expansion after adopting a new process for making frozen juice concentrate in 1946. By 1950, there were approximately 28 million trees grown on over 177 000 ha (Simanton, 1996).

Initially, growers had limited knowledge of optimal citrus-growing practices and often planted trees in areas poorly suited for commercial production (Simanton, 1996), which highlighted the need for better research. Both the U.S. Department of Agriculture (USDA) and the University of Florida Agricultural Experiment Station established facilities where entomologists, plant pathologists and horticulturalists could develop best practices for citrus production. The University of Florida Citrus Research and Education Center was established in 1917. This resulted in significant progress by 1950 in such areas as dealing with nutritional deficiencies. However, insect pests and diseases have remained a challenge because the Florida peninsula is highly susceptible to invasion by exotic pests and pathogens.

In 1950, the director of the Florida Citrus Experiment Station, Dr A. F. Camp, proposed the 'Ecological Survey of Citrus Pests and Disorders' to provide a comprehensive survey of the ecology of citrus groves throughout Florida (Fisher, 1950; Simanton, 1996). The project included a detailed survey of both harmful and beneficial organisms, their interactions and the impacts of pesticides. The survey led to the identification of previously unknown pest species, including the spider mite, *Eutetranychus banksi* (McGregor) (Muma et al., 1953). Important discoveries were made concerning the existence and impact of biological control agents regulating pest populations. This included the accidental introduction of a parasitoid, *Aphytis lepidosaphes* Compere, of the purple scale, *Lepidosaphes beckii* (Newman), which was the most important citrus pest prior to 1959 (Simanton, 1960). It was realized that excessive sulphur use had a negative impact on natural enemies of the purple scale. Advising growers to use less sulphur allowed populations of this parasitoid to rebound, which decreased the purple scale populations below economic thresholds.

This success led to the implementation of classical biological control programmes, including the importation of *Aphytis holoxanthus* Debach against Florida red scale, *Chrysomphalus ficus* (Ashmead), essentially eliminating it as a pest (Simanton, 1974). The project lasted 16 years and led to the development of one of the most comprehensive databases on the seasonal cycles of organisms for a single crop in the United States (Simanton, 1996). The database helped to highlight the importance of biological control in Florida citriculture with numerous successful classical biological control programmes that are still in use. This research-based programme has significantly improved citrus production in Florida, making this American state a world leader of citriculture.

Formally defined IPM programmes for citrus production in Florida began in the 1970s and were based on their initial inception in the 1950s (Knapp et al., 1996). These programmes led to a substantial rise in citrus production, reaching over 330 000 ha of crop in the early 1970s, and included the following: the modern application of host-plant resistance, classical and conservation biological control programmes, development of effective horticultural/cultural control tactics and emphasis on the selective application of pesticides (Knapp et al., 1996). From the 1970s to the 2000s, the pesticides applied to Florida citrus were restricted to a few fungicide, herbicide and horticultural oil sprays with notable reduction of contact poison insecticide or miticide use. Resistance rootstocks, weed control via cultural control practices and significant reliance on parasitoids, predators and pathogens for herbivore regulation were the primary components of the pest management in an ecosystem that was continually faced with the potential introduction of new pests and diseases. Growers have employed the extensive knowledge gathered by scientists investigating various aspects of the Florida citrus ecosystem, which has led to publications, statewide training programmes, on-farm educational visits and the web-based Electronic Data Information Source (edis.ifas.ufl.edu).

## 3 The arrival of HLB

The management of certain pests and pathogens, for example, citrus root weevil species (e.g. *Diaprepes abbreviatus* (L.)), citrus leafminer (*Phyllocnistis citrella* Stainton) and citrus bacterial canker (*Xanthomonas axonopodis* (Hasse)) in Florida citrus has proven to be problematic, requiring extensive measures that include chemical input or tree removal. However, in the 1990s, the arrival of the ACP, *Diaphorina citri* Kuwayama, in Florida and the identification of the tree-killing disease, HLB, in the same state in 2005 (Grafton-Cardwell et al., 2013), eliminated years of effort to develop effective IPM practices. The disease is caused by a bacterial pathogen that is transmitted by the psyllid vector and, due to the severity of the disease, results in tree decline and possible tree death, which has transformed Florida citriculture. Maintaining productive trees, via intensive vector management, has nearly eliminated previously effective biological control (Grafton-Cardwell et al., 2013).

Maintaining low HLB incidence in Florida, and remaining profitable, has become increasingly difficult for citrus growers. Inoculum removal by removing infected trees, multiple insecticide application strategies and enhanced nutritional programmes all aim to reduce the impact of disease symptoms. However, many growers are abandoning the strategy of inoculum removal and have chosen the more costly approach of prolonging the life and productivity of diseased trees via intense applications of supplemental micronutrients. Insecticide resistance is a current threat to an industry that relies on insecticides for ACP suppression and the management of other significant pests. Long-term solutions for HLB control, such as resistant citrus cultivars and other methods of preventing HLB transmission by the vector, may succeed in the future but will take time to develop.

The ACP transmits three species of bacteria belonging to the genus *Candidatus* Liberibacter. In the United States, *Ca.* Liberibacter asiaticus (CLas) is the most probable cause of the citrus disease HLB, although Koch's postulates have not been satisfied (Tylor et al., 2009; Pelz-Stelinski et al., 2010). ACP is currently found in Southern United States but was first discovered in Florida in 1998 (Grafton-Cardwell et al., 2013). In 2005, HLB-infected plants were detected in Florida and within 4 years all 32 citrus-growing counties in the state had HLB-infected citrus (Anon., 2008). HLB has also been reported in both California and Texas, with positive trees being detected in about 30% of the groves. On a global scale, HLB is considered to be the most destructive disease of citrus crops as all known citrus cultivars are susceptible to it (Folimonova et al., 2009) and prevention of disease transmission has proven problematic (Bové, 2006). Without intervention, infected mature trees drop foliage, weaken and may die within 5–10 years (Roistacher, 1996). Direct effects of HLB may be exacerbated by secondary affliction(s), including infestation by Diaprepes root weevil (*Diaprepes abbreviates*) and indigenous root weevils (e.g., *Compsus* sp.,

*Epicaerus* sp.), which may lead to root girdling and infection by *Phytophthora* spp. Infected mature trees yield significantly less fruit, with some fruits being small, misshapen or with uneven or zero colour development. Furthermore, the juice from affected fruit tastes bitter and unbalanced due to low soluble solids and high acid content.

HLB can spread rapidly within and between groves via short- and long-range dispersal capabilities as a result of the efficient pathogen acquisition and transmission capabilities of ACP (Pelz-Stelinski et al., 2010; Tiwari et al., 2010). Nymphs are primarily responsible for pathogen acquisition, which may be introduced to the shoot by an infected adult that, in the majority of cases, has emerged from an infected nymph (Lee et al., 2015). Thus, tree infection and the next generation of ACP can occur within a single vector generation (Lee et al., 2015). Once the ACP acquires the bacterium, the ACP is able to transmit bacteria into another host plant within 7–25 days (Pelz-Stelinski et al., 2010). However, an infected ACP can transmit the bacteria into another host plant by continuous feeding for only 5–7 h (Xu et al., 1998). The time frames involved in acquisition, latency and transmission provide a window of opportunity to control infected ACP and prevent disease spread. Gravid ACP females only lay eggs on new flush, and hatched early instar nymphs feed on new flush, proceeding through five instars before becoming adults (Halbert and Manjunath, 2004; Grafton-Cardwell et al., 2013). This results in a rapid and very high build-up of ACP populations during flushing periods that are capable of acquiring and transmitting HLB-causing bacteria. Furthermore, the multigenerational and season-long occurrence of arthropod pests, such as ACP, is common within perennially growing citrus in subtropical climates.

The development of insecticide resistance in citrus is a recurrent worldwide problem and has been well documented. The use of organophosphate for >40 years in California to control the citrus thrips, *Scirtothrips citri*, and the California red scale, *Aonidiella aurantii*, resulted in insecticide failures in the 1980s and 1990s, respectively (Morse and Brawner, 1986; Immaraju et al., 1989; Grafton-Cardwell and Vehrs, 1995). In addition, organophosphate resistance in California has been reported for yellow scale (*Aonidiella citrina*) and citricola scale (*Coccus psuedomagnoliarum*) (Grafton-Cardwell and Vehrs, 1995). Furthermore, in Florida, the citrus rust mite (*Phyllocoptruta oleivora*) has shown resistance to dicofol, which is managed by restricting its use to one application per year (Omoto et al., 1995).

Soil and foliar applied insecticides play a vital role in controlling ACP populations that are infected or capable of acquiring and transmitting HLB bacteria (Qureshi et al., 2014). Currently, only two modes of action (neonicotinoids and cyantraniliprole) are available, as a soil-applied, systemic treatment for ACP control. The neonicotinoids have been a particularly critical tool for the protection of young trees and are the most widely used insecticides on newly planted groves or young tree resets (Langdon et al., 2018a,b). The aim of vector control is to keep ACP population densities as low as possible (Qureshi et al., 2014). The threshold for a disease vector that causes significant fruit drop, kills trees and/or renders fruit unsuitable for consumption is extremely low or non-existent, although economic injury levels for different conditions are demonstrable (Monzo and Stansly, 2017). Consequently, intense insecticide use has been deemed necessary with up to 12 applications per year for ACP in Florida. However, this use has already caused widespread indications of insecticide resistance in ACP populations throughout this state (Tiwari et al., 2011) and requires refinement through the use of realistic thresholds to eliminate unnecessary or poorly timed sprays on mature trees.

Despite efforts to control the disease, HLB has rapidly spread in Florida due to the mobility of *D. citri* (Boina et al., 2009a,b; Tiwari et al., 2010). *D. citri* adults are able to disperse over long distances during periods of peak activity, which results in the widespread movement of the causal pathogen of HLB, particularly from unmanaged, abandoned areas into well-managed groves (Tiwari et al., 2010). Neighbouring citrus growers regularly made independent decisions regarding production and pest management; however, through outreach and demonstrating the benefits of cooperative actions, this mind-set is being overcome. It has been shown that a spray programme effectively applied to a single block of citrus and not to neighbouring blocks is ineffectual, as untreated blocks serve as sources of infected ACP and disease (Tiwari et al., 2010). Thus, it became clear that area-wide, cooperative management of the vector was necessary. In 2010, the U.S. National Academy of Science published a strategic plan for effective HLB management and identified the development and implementation of area-wide, cooperative 'Citrus Health Management Areas (CHMAs)' as the most important organizational priority (National Research Council, 2010), and significant progress has been made to develop successful CHMAs (www.flchma.org). Similarly, meetings were held in other states, such as California, to organize CHMAs for citrus-producing areas.

In Florida, baseline susceptibility data has been gathered for the majority of currently registered insecticides used for the control of ACP, and the potential resistance development has been monitored for 12 consecutive years. Resistance has already been documented and has been shown to develop rapidly. In 2009, field populations of ACP in Florida were monitored for insecticide susceptibility, with decreased susceptibility to several important insecticides being observed, including up to a 34-fold decrease in susceptibility to imidacloprid, which is a critical tool for the protection of young trees (Tiwari et al., 2011a). In 2010, the problem became more widespread throughout the state of Florida, with decreased susceptibility to the majority of insecticides used for ACP control (Tiwari et al., 2011b, 2012, 2013a,b). Artificial selection for imidacloprid resistance in greenhouse trials and experimental fields has shown that an ACP population can become 100-fold resistant to this insecticide after only five generations of constant selection pressure (Chen et al., 2018). Furthermore, six continuous generations of relaxed selection (no insecticide exposure) are required for ACP populations to return to normal susceptibility levels (Chen and Stelinski, 2017).

## 4 Root weevil complex

With the exception of ACP and CLM, the root weevil, *Diaprepes abbreviatus* (Linnaeus) (Coleoptera: Curculionidae), is the most destructive arthropod to citrus trees in Florida (Campos-Herrera et al., 2015). *D. abbreviatus* was first introduced to Florida in 1964 (Beavers and Selhime, 1975). Over the past 40+ years, it has significantly contributed to the spread of disease and damage to citrus, ornamental plants and other crops. *D. abbreviatus* is a native economic pest of the Caribbean where at least 19 additional species are known within the genus (Weissling et al., 2002). *D. abbreviatus* has spread over a large area of central and southern Florida where it causes approximately US$70 million in damage annually (Lapointe, 2000). The initial area of infestation was an estimated 6500 acres in Apopka, Florida, and has now increased to an estimated 164000 acres over 20 counties in central and southern Florida (Weissling et al., 2002).

*D. abbreviatus* has a wide host range, attacking approximately 293 different plant species including citrus, sugarcane, potatoes, strawberries, woody field-grown ornamentals, sweet potatoes, papaya, guava, mahogany, containerized ornamentals and non-cultivated wild plants (Simpson et al., 1996, 2000). *D. abbreviatus* damage to the vegetative portion of plants is most often seen as notching of young leaves (Fennah, 1940). This is a key trait characterizing *D. abbreviatus* infestation. Adults continue to feed on foliage and lay eggs between older leaves (Schroeder, 1992; Fennah, 1940). However, the greatest damage is caused by larvae feeding below ground. Upon hatching, the larvae fall to the soil and make their way to the roots of plants where later instars feed and develop (Schroeder, 1992). This feeding can girdle the taproot and rhizosphere causing damage that disables the plant from taking up water and nutrients resulting in plant death (Schroeder, 1992). This type of damage also facilitates secondary infections by *Phytophthora* fungus species (Graham et al., 1996). Young hosts can be killed by a single larva while several larvae can result in serious decline of older, established hosts (Weissling et al., 2002).

## 5 Response to HLB

In quick response to the recommendations made by the National Academy of Sciences, CHMAs were established in the fall of 2010 in Florida (Rogers et al., 2012). From the programme's inception in 2010, the number of active CHMAs in Florida grew from approximately a dozen to 38 by the summer of 2012, encompassing a total of slightly over 200 000 ha (Rogers et al., 2012). The basic purpose of CHMAs was to encourage neighbouring citrus growers to work collaboratively and on an area-wide basis to manage the *D. citri* vector of HLB through coordinated sprays of insecticides. Given the movement capability of *D. citri* and propensity for resistance development, area-wide control is considered an effective method for combating both problems. In the summer of 2011, the 'CHMA ACP (Asian citrus psyllid) Monitoring' programme was officially initiated by scouts from the USDA's APHIS-PPQ Citrus Health Response Program and the Florida Department of Agriculture and Consumer Services' Division of Plant Industry (Rogers et al., 2012). This programme employed staff to monitor approximately 6000 blocks of citrus within all of the CHMAs throughout the state. All blocks were monitored at 3-week intervals. The purpose of the monitoring was both to validate that the coordinated area-wide vector management programme was working and perhaps more importantly to provide growers with real-time updates regarding psyllid populations to aid their pest management decision-making regarding the need for pesticide application. From 2011 to 2012, 30–70% reductions in psyllid populations were recorded in citrus groves where growers were participating in CHMAs as compared with non-CHMA areas (Rogers et al., 2012). While the CHMA programme was not a traditional IPM-based outreach programme, it was analogous to such IPM programmes and designed for management of a plant pathogen disease vector and a resultant necessity for intense vector suppression in the HLB era of citriculture in Florida.

Although the CHMA programme no longer functions, citrus growers continue to receive the latest pest management information from the University of Florida extension service through the more traditional communications such as formally organized meetings, pamphlets and handbooks. However, web-based approaches are heavily emphasized and new tools have been developed in addition to the above-mentioned, online EDIS database. The CHMA programme and associated literature on the latest citrus research was available on the CHMA website mentioned above. The website provided detailed information for how to join a local CHMA, it had a directory of all currently functioning CHMAs, including maps and up-to-date scouting reports of psyllid populations, as well as, the most recent information regarding statewide CHMA meetings and planned coordinated sprays for the psyllid. It was also a comprehensive, web-based collection on modern extension literature regarding citrus production in Florida from diverse fields, including entomology/nematology, plant pathology, horticulture, economics, and food science and nutrition.

## 6 Current management practices in the HLB era

Several management options are available for ACP and CLM in Florida and continue to evolve for the protection of young, non-bearing, as well as, mature citrus. IPM involves employing multiple methods of pest control in a coordinated action plan. The goal is sustainable pest control that also returns an economic gain. Individually, these methods may not be enough to achieve needed pest reductions and not all methods are applicable in all environments. However, when used in concert, the benefits of an integrated approach are realized. For example, conventional insecticides play an integral role in integrated systems available for commercial production systems, but may not be suitable for urban areas or organic production. Biological control, on the other hand, may play a larger role in these areas and, in general, will contribute more to overall pest suppression in groves employing an integrated approach rather than sole reliance on insecticides for pest management.

IPM also focuses on pest identification and monitoring, prevention and decision-making based on economic injury levels so that insecticide use is prescribed when it is necessary. In the case of pathogen vectors, such as ACP, application thresholds are exceedingly low since very few individuals can cause disease outbreak. Therefore, many current research projects involve citrus production either with complete vector exclusion or by reducing vector populations sufficiently to maintain economically viable yield. For example, growing citrus under screen protection and therefore excluding the vector from the production system completely is an idea that is gaining momentum based on ongoing investigation. Furthermore, yield improvements have been demonstrated by reducing ACP populations in standard mature groves with HLB as compared directly with counterparts where vector populations are not well managed and allowed to re-inoculate trees (Stansly et al., 2014). These results suggest that vector control is still important even when nearly 100% of a block is already expressing symptoms of HLB.

### *6.1 Pest monitoring*

Regular scouting for pests is important for making need-based decisions regarding application of insecticides for pests in order to minimize unnecessary investment in control measures and potential collateral damage to beneficial insects and mites. At present, most spray applications are made for ACP. The adults spread the pathogens causing HLB from tree to tree and grove to grove. Therefore, ACP should be monitored. Sticky traps, sweep nets, suction devices and stem tap sampling are all options for tools in estimating ACP populations. Stem tap sampling is easy to conduct compared to other methods and provides instant data for making management decisions. A tap sample is made by using a white clipboard (60 × 100 cm) or laminated white paper sheet held horizontally under randomly chosen branches. One should strike the branch three times with a length of PVC pipe (https://edis.ifas.ufl.edu/pdffiles/IN/IN111600.pdf). Psyllids can be counted as they fall onto the clipboard. It is also possible to monitor other insects with this method, such as CLM, weevils and beneficial insects such as ladybeetles, spiders and lacewings, if desired. Complete elimination of ACP is not possible without exclusion cages, and a threshold is needed to use such monitoring data for making spray decisions. Investigations in Florida of HLB-infected blocks have revealed that even in blocks where 0.1-0.2 adults are counted per infected tree yields can be maintained with proper nutrition. These numbers translate to 10 or 20 ACP adults per 100 tap samples. Eggs and nymphs of ACP and similarly larvae of CLM are also important to monitor; however, they are hard to target but are eaten by small predators and parasitoids which are able to reach in locations where they develop.

## 6.2 Cultural control

Preventing ACP from colonizing the citrus, particularly non-bearing trees, is paramount to minimize HLB and maximize chances of these trees to enter production. Growing citrus under screens offers a possibility of complete protection from ACP and HLB. This technique is currently under investigation and shows promise. However, there may be some risk from other pests such as thrips, mites and CLM, which are manageable. Growers have already begun establishing these protected structures for growing citrus by physically excluding vector and pathogen. Significant progress has also been made towards reducing psyllid access to young citrus by planting seedlings on UV-reflective mulches. Success has been demonstrated with reduction in pest and disease incidence and increased profits (Croxton and Stansly, 2014).

## 6.3 Chemical control

Insecticides are critical components of management programmes for ACP and CLM. However, it is important to understand the class of the product being used and proper timing of application. During the period between November and February, most mature trees go through dormancy and do not produce new growth that is required by psyllids for reproduction. Therefore, psyllid populations are mainly comprised of adults during this period and, generally, in comparatively lower densities because reproduction is not taking place widely. CLM is also not generally reproducing at this time for the same reason and overwintering as adults. Beneficial insects are also less common due to lack of prey. Therefore, broad-spectrum organophosphate (Chlorpyrifos, Imidan, Dimethoate) and pyrethroid (Mustang, Danitol, Baythroid) insecticides are good candidates for use during this time. These products are comparatively less expensive than other chemistries and are effective adulticides for ACP. If populations of ACP adults are effectively killed off during this time with these broad-spectrum chemistries, it can take many weeks for populations to re-establish (Qureshi and Stansly, 2010). One application during November–December and a second one before the spring flush in January or early February can be sufficient to significantly reduce populations of ACP (Qureshi and Stansly, 2010). If these sprays are applied in a coordinated fashion over a large contiguous area, it may take psyllid populations 4–6 weeks (or more) to effectively recolonize those areas (Qureshi and Stansly, 2010). A third spray may be needed if temperatures are high and if trees should flush during November–December (Qureshi et al., 2014). The products in the two classes of insecticides are hard on beneficial insects and mites and therefore restricting their use to this time of the year is best for integrating chemical and biological control because natural enemy populations begin to increase during spring. These dormant sprays are very important and are known for providing significant reductions in psyllid populations into the growing season demonstrated through individual and coordinated programmes such as CHMAs (Qureshi et al., 2014). If adults are controlled effectively during dormant season, then most flush can be maintained with much lower infestation during spring. Heading off a major populations increase directly prior to bloom is important because of the restrictions imposed on insecticide use during bloom when pollinators are present.

Selective insecticides, such as horticultural crop oils and soaps are expected to be less hazardous to beneficial insects and mites and therefore better choices for use during the growing season when those organisms are common and contributing to control of multiple pests (Qureshi and Stansly, 2007). Selection of products should be made according to the labels and IFAS guidelines which provide details on product use rates, timing and expected efficacy against multiple pests in order to minimize unnecessary applications (http://www.crec.ifas.ufl.edu/extension/pest/PDF/2017/ACP%20and%20Leafminer.pdf). Psyllids generally colonize the borders of blocks in greatest abundance and become more abundant directly after unexpected flush events as seen after hurricanes. Therefore, supplemental border sprays throughout the growing season or additional cover sprays following an unexpected flush event with an organophosphate insecticide may be warranted.

Soil-applied systemic neonicotinoids belong to the same insecticide chemical class and thus share the same mode of action (MOA 4A). These have been used for well over a decade to protect non-bearing citrus from ACP and CLM. Pest populations can develop resistance to insecticides following repeated exposure to the same MOA. This has been documented with certain populations of ACP in Florida for several products that have been extensively used in spray programmes, particularly the neonicotinoids commonly used as soil drenches. Diversifying and strictly rotating MOAs can allow populations of resistant pests to reverse back to susceptibility and this practice is very useful for ACP (Chen and Stelinski, 2017). It is extremely important to integrate sprays with soil drenches of a different MOA to conserve effectiveness of neonicotinoids. Programmes are being expanded for additional area-wide monitoring of ACP resistance in order to maintain efficacy of available tools.

For CLM, insecticides that kill the larvae are currently the most important tools for the control of this pest. The early larval stage is the most susceptible and best target to avoid injury as larvae emerge from the egg and begin to feed on flush. Young trees are often the most susceptible to CLM damage. Soil applications of neonicotinoid insecticides can provide multi-week control of CLM in non-bearing trees and should be applied up to 2 weeks prior to a leaf flush in order to allow lethal concentrations of insecticide to accumulate in the foliage. However, it is important to remember that all neonicotinoid insecticides share the same MOA, so back-to-back applications of these insecticides may hasten development of insecticide resistance in both pests, which must be avoided.

Insecticides recommended for leafminer control can be found in the IFAS Florida Citrus Pest Management Guide. Proper timing of foliar sprays to coincide with flushing cycles is critical to optimize CLM management. The goal is to kill larvae as soon as they begin mining. Although broad-spectrum insecticides targeting ACP adults may kill adult CLM, typically populations rebound in groves soon after foliar sprays of organophosphate or pyrethroid insecticides and may even increase under intense ACP management due to suppression of natural enemies. Applications of foliar insecticides should be made during a window when CLM larvae hatch and begin feeding to maximize larval kill. In general, the earliest applications should occur between 13 and 30 days after budbreak. The duration of control may be shorter if a heavy flush occurs soon after the foliar application. It may be advantageous to target spring flush, even though CLM damage is not evident, to prevent build-up of populations that will cause damage later in the year.

There are opportunities for combining control of CLM with that of other pests. Most recommended products for CLM control are also effective against ACP, which also develops in young flush. Well-timed summer applications of Agri-Mek or Micromite can be useful to manage citrus rust mite. Products with the active ingredient, cyazypyr, are highly effective against CLM (as well as ACP). Several (8+) weeks of protection against CLM have been documented following soil-applied treatments of this active ingredient. It is also highly effective as a foliar treatment against CLM.

Some newly registered insecticides for use in Florida citrus are mixtures of two modes of action that allow for simultaneous CLM and ACP control. Agri-Flex combines the active ingredient of Agri-Mek with that of Actara (a neonicotinoid) in a single formulation to target CLM larvae and ACP nymphs and adults. However, any consecutive application of an insecticide employing the same MOA, whether it is a single or dual MOA, increases the probability of insecticide resistance developing in the target pests. Intrepid 2F (Methoxyfenozide) + 2% v/v horticultural spray oil (435) is effective against CLM larvae by interfering with the molting hormone. Intrepid can be used alone or in a tank mixed with another insecticide without selecting for resistance to ACP upon which it has no effect.

## 6.4 Biological control

Biological control has been practised for many years in citrus (Browning et al., 1996). In many cases, researchers discovered that biological control was already occurring and learned how to manage it more efficiently. Many of these natural enemies were probably imported with citrus plants just as the pests were. Purposeful importations of predators and parasites have occurred also. Significant investments were made to import and establish species of predators, which attack not only ACP and CLM, but also several other pests of citrus. Parasitoids have also been introduced and established which specifically attack ACP and CLM. For example, the parasitoids *Ageniaspis citricola* and *Tamarixia radiata* are highly specific to CLM and ACP, respectively (Qureshi et al., 2009). Historically, high levels of parasitism (60–80%) by these species were observed at different times of the year, but this has been significantly impacted by increased use of insecticides for ACP management over the past several years (Qureshi and Stansly, 2009). Following consistent releases, *T. radiata* has been established in greater abundance in organic citrus production systems with parasitism averaging up to 31% (Qureshi et al., 2009). Populations of this parasitoid and associated higher attack rates of ACP are fewer when conventional insecticide sprays are applied (Qureshi and Stansly, 2009). Also, fungi such as *Hirsutella*, *Beauvaria* and *Paecilomyces* contribute to pest control. These beneficial organisms can be conserved by limiting the use of broad-spectrum insecticides to the dormant season, as supplemental border sprays, or with targeted use following unexpected flush events. Furthermore, use of selective insecticides, oils, soaps and release of parasitoids during the growing season will contribute to development of functioning citrus IPM.

## 6.5 Area-wide management in the era of HLB

ACPs are able to disperse several kilometres over fallow fields, lakes and roads (Tiwari et al., 2010; Martini et al., 2013; Lewis-Rosenblum et al., 2015). Measures for HLB management have consisted the following: (1) reduction of ACP populations by intensive insecticide treatments (Qureshi et al., 2014), (2) removal of infected trees to reduce HLB inoculum (Bassanezi et al., 2013) and (3) re-planting healthy citrus plants produced under insect-proof nurseries. Soil- and foliar-applied insecticides play a vital role in suppressing ACP populations that are infected or capable of acquiring and transmitting HLB bacteria. Among these recommended measures for HLB management, the removal of symptomatic trees is the most controversial and difficult to accomplish in the United States. First, it increases per-acre grove maintenance costs by approximately US$400 annually, and second, newly planted young trees designated to replace removed trees are more susceptible to HLB infection than mature trees (Martini et al., 2015). While vector control is still a key component of HLB management in Texas, the disease incidence continues to exponentially progress in citrus groves and backyard trees, and some growers are voicing their concerns about the usefulness of vector control. Some growers in Florida are also on the brink of abandoning ACP vector management in Florida. HLB infection rates are hovering around 100% at all times and newly established plantings are quickly becoming infected prior to reaching the bearing stage. Growers are ceasing citrus production and citrus processing plants are closing at unprecedented rates. The future industry in Florida may require a clean-slate approach by re-planting with the first HLB-tolerant or -resistant varieties, when those are available, appropriately vetted by regulatory agencies and met with public acceptance. However, even then, partially resistant trees will likely require some protection from pathogen inoculation by ACP.

Although area-wide ACP control programmes were initially adopted in Florida in coordinated scale and incorporated resistance management strategies, their adoption was short-lived (2–3 years). While growers continue to implement area-wide dormant sprays in Texas, no spray coordination is done during the active growing season. Therefore, a comprehensive, long-term management approach beyond only insecticides requires further investigation. By effectively optimizing spray schedules, improving the effectiveness of biological control, and affecting bacterial acquisition and transmission, the prospect of producing profitable citrus will increase.

# 7 Conclusion and future trends

## 7.1 Management and quarantine

Introduction of ACP and HLB has challenged the previous norm of managing exotic pest introductions in U.S. citrus production through rapid spread, difficulty in detection and lack of preventive or therapeutic treatments. Industries in all three states responded to the appearance of ACP and HLB by coordinating committed funding within and between citrus states to initiate new research. They also enlisted help in coordination, response, funding from USDA-ARS and

USDA-APHIS. Emergent from this request for federal help was the evolution in 2006–07 of APHIS CHRP Citrus Health Response Program, addressing all citrus states. There also was a large-scale redirection of funding from marketing and other self-funded programmes to fuel and accelerate needed research efforts. Florida has spearheaded research efforts to understand the HLB problem and initially sought assistance (California and Texas involved) from the National Research Council of NAS and USDA-ARS National Program Leaders to (1) assist in prioritizing within-year research projects for immediate funding in 2008, (2) develop short-, medium- and long-term research priorities to address HLB/ *D. citri* and (3) develop a management plan for industries to organize, manage and communicate research projects and results. The strategic plan was published by NAS in 2010 (www.nap.edu/catalog/12880.html, ISBN: 0-309-15208-9). The number one research priority is to 'Improve insecticide-based management of Asian citrus psyllid (ACP)'. The number one organizational recommendation is to 'Create "Citrus Health Management Areas" in Florida'.

Area-wide management practices to control HLB and its psyllid vector in Florida may help to mitigate the impact of the immigration of infected ACP into commercial citrus. Citrus groves that are within 2 km of any other citrus plantings are at risk for ACP infestation and HLB disease introduction from those areas. Implementing practical use of any single or multiple integrated technologies will require cooperation of many growers on a large scale to overcome the natural and human-assisted dispersal capabilities of this vector. Quarantine protocols must also account for the possibility of hundreds of km of movement by the vector, even when these long-range movements are rare, they occur through human- or wind-assisted movement events.

## 7.2 Abandoned citrus and management

An important factor in the spread of HLB is the role of abandoned citrus as a reservoir for ACP and HLB. The number of abandoned groves in Florida has increased in the past decade because of socio-economic pressures, freezing events and the loss of production due to canker and HLB diseases (Cumming and George, 2009). Abandoned groves are not treated with insecticides or managed in other ways to limit the populations of ACP. It is clear that abandoned groves can act as reservoirs of ACP (Tiwari et al., 2010). The extent of the impact of these abandoned groves on the commercial citrus industry in Florida is difficult to quantify. Knowing how devastating HLB can be to citrus production, the presence of these areas of refuge should be of immediate concern, especially in locations such as California where quarantine and eradication methods are still being pursued. This issue should also be considered with respect to the citrus grown in residential areas by homeowners. While intensity and coordination of control measures for ACP are an important step in mitigating the damage caused by HLB, the presence of a large number of abandoned groves in Florida may undermine these efforts (US Department of Agriculture, 2013).

Consequently, there are recommendations to remove and destroy abandoned citrus (US Department of Agriculture, 2013). As a result, the acreage of abandoned groves in Florida has only recently decreased from 56 055 to 51 386 ha over the past several years (US Department of Agriculture, 2013, 2015a,b). Management of an adjacent abandoned citrus acreage should be practiced by growers in the future as if that acreage were part of their production land. Geographical barriers may have little impact on dispersal of ACP given that this psyllid has been captured in a dense forest 2.3 km away from the nearest citrus grove, which was abandoned (Martini et al., 2013).

Recently, in reaction to the record low 2014 citrus production in Florida (US Department of Agriculture, 2015a), the 'USDA's Tree Assistance Program' (US Department of Agriculture, 2014) has been launched to help growers to plant new citrus trees. Private companies have also undertaken similar initiatives. These planting programmes designated to recoup Florida citrus production impacted by HLB will probably include the implementation of large blocks of solid sets. Therefore, particular attention should be taken to intensively manage the ACP vector given that these blocks will be particularly susceptible to psyllid infestation. These re-planting initiatives should also favour reset plantings, as well as, the establishment of windbreaks on the edges of the new solid sets. This cultural control method may reduce the need for the intense insecticide input that is currently required to manage populations of this phytopathogen vector (Qureshi et al., 2014).

In the context of IPM for ACP and associated HLB, Martini et al. (2015) advocated conservation of windbreaks and hedgerows and support the implementation of new ones. Although there are some constraints to implementing windbreaks, including cost, shading, competition for water and nutrients, obstruction for agricultural equipment and loss of exploitable surface, windbreak protection against citrus canker and possibly ACP, in addition to other ecological benefits that include increase of habitats for natural enemies, as well as, protection from wind and soil erosion (Baudry et al., 2000), likely outweigh the disadvantages. Further investigations to evaluate potential changes in efficacy of windbreaks to decrease ACP populations depending on windbreak structure (living versus artificial), species, size, cardinal orientation and distance from the orchard are needed. Implementation of solid sets as a method of re-planting citrus in highly HLB-infected regions will likely require much more intense management of the ACP vector than that for resets within mature citrus orchards.

A potential bridge to allow for growth of natural windbreaks would be erecting temporary artificial windbreaks surrounding solid sets to protect young citrus at planting. These artificial windbreaks could be removed as the natural windbreaks develop into mature trees that subsequently protect citrus at the fruit-bearing stage, when more protection from sun exposure will likely be needed to reduce populations of ACP.

## 8 Where to look for further information

For a historical perspective on worldwide variation in production practices, please refer to Reuther et al. (1967) and Davies and Albrigo (1994). A comprehensive overview of global citrus harvest and production can be found in FAO (2017).

## 9 References

Anon. 2008. USDA-APHIS. (2008). *Citrus Greening National Quarantine Map*. Available at: http://www.aphis.usda.gov/plant_health/plant_pest_info/citrus_greening/downloads/pdf_files/nationalquarantinemap.pdf (accessed on 16 January 2019).

Bassanezi, R. B., Belasque, J. and Montesino, L. H. 2013. Frequency of symptomatic trees removal in small citrus blocks on citrus huanglongbing epidemics. *Crop Prot.* 52, 72-7.

Baudry, J., Bunce, R. G. H. and Burel, F. 2000. Hedgerows: an international perspective on their origin, function and management. *J. Environ. Manage.* 60(1), 7-22. doi:10.1006/jema.2000.0358.

Beavers, J. B. and Selhime, A. G. 1975. Development of *Diaprepes abbreviatus* on potted citrus seedlings. *Fla Entomol.* 58(4), 271–3. doi:10.2307/3493688.

Boina, D. R., Meyer, W. L., Onagbola, E. O. and Stelinski, L. L. 2009a. Quantifying dispersal of *Diaphorina citri* (Hemiptera: Psyllidae) by immunomarking and potential impact of unmanaged groves on commercial citrus management. *Environ. Entomol.* 38(4), 1250–8. doi:10.1603/022.038.0436.

Boina, D. R., Onagbola, E. O., Salyani, M. and Stelinski, L. L. 2009b. Antifeedant and sublethal effects of Imidacloprid on Asian citrus psyllid, *Diaphorina citri*. *Pest Manag. Sci.* 65(8), 870–7. doi:10.1002/ps.1767.

Bovè, J. M. 2006. Huanglongbing: a destructive, newly-emerging century-old disease of citrus. *J. Plant Path.* 88, 7–37.

Browning, H., Peña, J. E. and Stansly, P. 1996. Evaluating impact of indigenous parasitoids on populations of citrus leafminer. In: Hoy, M. (Ed.), *Proceedings of International Conference on Managing Citrus Leafminer*. University of Florida, pp. 14–15.

Campos-Herrera, R., El-Borai, F. E. and Duncan, L. W. 2015. It takes a village: entomopathogenic nematode community structure and conservation biological control in Florida (US) orchards. In: Campos Herrera, R. (Ed.), *Nematode Pathogenesis of Insects and Other Pests*. Springer, Switzerland, pp. 329–52.

Chen, X. D. and Stelinski, L. L. 2017. Resistance management for Asian citrus psyllid, *Diaphorina citri* Kuwayama, in Florida. *Insects* 8(3), 103. doi:10.3390/insects8030103.

Chen, X. D., Gill, T. A., Ashfaq, M., Pelz-Stelinski, K. S. and Stelinski, L. L. 2018. Resistance to commonly used insecticides in Asian citrus psyllid: stability and relationship to gene expression. *J. Appl. Entomol.* 142(10), 967–77. doi:10.1111/jen.12561.

Croxton, S. D. and Stansly, P. A. 2014. Metalized polyethylene mulch to repel Asian citrus psyllid, slow spread of Huanglongbing and improve growth of new citrus plantings. *Pest Manag. Sci.* 70(2), 318–23. doi:10.1002/ps.3566.

Cumming, G. S. and George, A. 2009. Historical influences dominate the composition of regenerating plant communities in abandoned citrus groves in north-central Florida. *Landsc. Ecol.* 24(7), 957–70. doi:10.1007/s10980-009-9368-2.

Davies, F. S. and Albrigo, L. G. (Eds) 1994. *Citrus*. CAB International, Wallingford, UK, 254pp.

Fennah, R. G. 1940. Observations on behaviour of citrus root-stocks in St. Lucia, Dominica and Montserrat. *Trop. Ag.* 17, 72–6.

Fisher, F. E. 1950. Entomogenous fungi attacking scale insects and rust mites on citrus in Florida. *J. Econ. Entomol.* 43(3), 305–9. doi:10.1093/jee/43.3.305.

Folimonova, S. Y., Robertson, C. J., Garnsey, S. M., Gowda, S. and Dawson, W. O. 2009. Examination of the responses of different genotypes of citrus to Huanglongbing (Citrus Greening) under different conditions. *Phytopathology* 99(12), 1346–54. doi:10.1094/PHYTO-99-12-1346.

Food and Agriculture Organization of the United Nations. 2017. *Citrus Fruit-Fresh and Processed Statistical Bulletin*. Food and Agriculture Organization of the United Nations, Rome, p. 63.

Gottwald, T. R., Bassanezi, R. B., Amorim, L. and Bergamin-Filho, A. 2007. Spatial pattern analysis of citrus canker-infected plantings in São Paulo, Brazil, and augmentation of infection elicited by the Asian leafminer. *Phytopathology* 97(6), 674–83. doi:10.1094/PHYTO-97-6-0674.

Grafton-Cardwell, B. E. and Vehrs, S. L. C. 1995. Monitoring for organophosphate- and carbamate- resistance in San Joaquin Valley citrus. *J. Econ. Entomol.* 88, 495–504.

Grafton-Cardwell, E. E., Stelinski, L. L. and Stansly, P. A. 2013. Biology and management of Asian citrus psyllid, vector of the huanglongbing pathogens. *Annu. Rev. Entomol.* 58, 413–32. doi:10.1146/annurev-ento-120811-153542.

Graham, J. H., McCoy, C. W. and Rogers, J. S. 1996. Insect-plant pathogen interactions: preliminary studies of *Diaprepes* root weevils injuries and *Phytophthora* infections. *Proc. Fla. State Hort. Soc.* 109, 57–62.

Halbert, S. E. and Manjunath, K. L. 2004. Asian citrus psyllids (Sternorrhyncha: Psyllidae) and greening disease of citrus: a literature review and assessment of risk in Florida. *Fla. Entomol.* 87, 330–53.

Hall, D. G., Gottwald, T. R. and Bock, C. H. 2010. Exacerbation of citrus canker by citrus leafminer *Phyllocnistis citrella* in Florida. *Fla Entomol.* 93(4), 558–66. doi:10.1653/024.093.0413.

Immaraju, J. A., Morse, J. G. and Kersten, D. J. 1989. Citrus thrips (Tysanoptera: Thripidae) pesticide resistance in the Coachella and San Joaquin valleys of California. *J. Econ. Entomol.* 82(2), 374–80. doi:10.1093/jee/82.2.374.

Knapp, J. L., Noling, J. W., Timmer, L. W. and Tucker, D. P. H. 1996. Florida citrus IPM. In: Rosen, D., Bennett, F. F. and Capinera, J. L. (Eds), *Pest Management in the Subtropics: Integrated Pest Management–A Florida Perspective*. Intercept Press, Andover, UK, pp. 317–47.

Langdon, K. W., Schumann, R., Stelinski, L. L. and Rogers, M. E. 2018a. Spatial and temporal distribution of soil-applied neonicotinoids in citrus tree foliage. *J. Econ. Entomol.* 111, 1788–98. doi:10.1093/jee/toy114.

Langdon, K. W., Schumann, R., Stelinski, L. L. and Rogers, M. E. 2018b. Influence of tree size and application rate on expression of thiamethoxam in citrus and its efficacy against *Diaphorina citri* (Hemiptera: Leviidae). *J. Econ. Entomol.* 111(2), 770–9. doi:10.1093/jee/toy001.

Lapointe, S. L. 2000. History and importance of *Diaprepes* to agriculture in the Caribbean region. Diaprepes short course. Citrus Research and Education Center, Lake Alfred, FL, pp. 8–12.

Lee, J. A., Halbert, S. E., Dawson, W. O., Robertson, C. J., Keesling, J. E. and Singer, B. H. 2015. Asymptomatic spread of huanglongbing and implications for disease control. *Proc. Natl. Acad. Sci. U.S.A.* 112(24), 7605–10. doi:10.1073/pnas.1508253112.

Lewis-Rosenblum, H., Martini, X., Tiwari, S. and Stelinski, L. L. 2015. Seasonal movement patterns and long-range dispersal of Asian citrus psyllid in Florida citrus. *J. Econ. Entomol.* 108(1), 3–10. doi:10.1093/jee/tou008.

Martini, X., Addison, T., Fleming, B., Jackson, I., Pelz-Stelinski, K. and Stelinski, L. L. 2013. Occurrence of *Diaphorina citri* (Hemiptera: Liviidae) in an unexpected ecosystem: the Lake Kissimmee state park forest, Florida. *Fla Entomol.* 96(2), 658–60. doi:10.1653/024.096.0240.

Martini, X., Pelz-Stelinski, K. S. and Stelinski, L. L. 2015. Absence of windbreaks and replanting citrus in solid sets increase density of Asian citrus psyllid populations. *Agric. Ecosyst. Environ.* 212, 168–74. doi:10.1016/j.agee.2015.06.027.

Monzo, C. and Stansly, P. A. 2017. Economic injury levels for Asian citrus psyllid control in process oranges from mature trees with high incidence of huanglongbing. *PLOS ONE* 12(4), e0175333. doi:10.1371/journal.pone.0175333.

Morse, J. G. and Brawner, O. L. 1986. Toxicity of pesticides to *Scirtothrips citri* (Thysanptera: Thripidae) and implications to resistance management. *J. Econ. Entomol.* 79(3), 565–70. doi:10.1093/jee/79.3.565.

Muma, M. H., Holtsberg, H. and Pratt, R. M. 1953. *Eutetranychus banksi* (McG) recently found on citrus in Florida (Acarina: Tetranychidae). *Fla Entomol.* 36(4), 141–4. doi:10.2307/3492151.

National Research Council. 2010. *Strategic Planning for the Florida Citrus Industry: Addressing Citrus Greening Disease (Huanglongbing).* National Academies Press, pp. 101–2. Available at: http://www.nap.edu/catalog.php?record_id=12880 (accessed on 16 January 2019).

Omoto, C., Dennehy, T. J., McCoy, C. W., Crane, S. E. and Long, J. W. 1995. Management of citrus rust mite (Acari: Eriophyidae) resistance to dicofol in Florida citrus. *J. Econ. Entomol.* 88(5), 1120–8. doi:10.1093/jee/88.5.1120.

Pelz-Stelinski, K. S., Brlansky, R. H., Ebert, T. A. and Rogers, M. E. 2010. Transmission parameters for Candidatus Liberibacter asiaticus by Asian citrus psyllid (Hemiptera: Psyllidae). *J. Econ. Entomol.* 103, 1531–41.

Qureshi, J. A. and Stansly, P. A. 2007. Integrated approaches for managing the Asian citrus psyllid *Diaphorina citri* (Homoptera: Psyllidae) in Florida. *Proc. Fla. State Hort. Soc.* 120, 110–5.

Qureshi, J. A. and Stansly, P. A. 2009. Exclusion techniques reveal significant biotic mortality suffered by Asian citrus psyllid *Diaphorina citri* (Hemiptera: Psyllidae) populations in Florida citrus. *Biol. Contr.* 50(2), 129–36. doi:10.1016/j.biocontrol.2009.04.001.

Qureshi, J. A. and Stansly, P. A. 2010. Dormant season foliar sprays of broad spectrum insecticides: an effective component of integrated management for *Diaphorina citri* (Hemiptera: Psyllidae) in citrus orchards. *Crop Prot.* 29(8), 860–6. doi:10.1016/j.cropro.2010.04.013.

Qureshi, J. A., Rogers, M. E., Hall, D. G. and Stansly, P. A. 2009. Incidence of invasive *Diaphorina citri* (Hemiptera: Psyllidae) and its introduced parasitoid *Tamarixia radiata* (Hymenoptera: Eulophidae) in Florida citrus. *J. Econ. Entomol.* 102(1), 247–56.

Qureshi, J. A., Kostyk, B. C. and Stansly, P. A. 2014. Insecticidal suppression of Asian citrus Psyllid *Diaphorina citri* (Hemiptera: Liviidae) vector of huanglongbing pathogens. *PLOS ONE* 9(12), e112331. doi:10.1371/journal.pone.0112331.

Reuther, W., Weber, H. I. and Batchelor, L. D. (Ed.). 1967. *The Citrus Industry, Volume I*. University of California Board of Regents, University of California, Riverside, 1028pp.

Rogers, M. E., Carlton, G. and Riley, T. D. 2012. Results from the 'CHMA ACP Monitoring' program: statewide psyllid populations are considerably lower in 2012 compared to one year ago. *Citrus Ind.* 93, 12–6.

Roistacher, C. N. 1996. The economics of living with citrus disease: huanglongbing (greening) in Thailand. In: da Graça, J. V., Moreno, P. and Yokomi, R. K. (Eds), *Proceedings of the 13th Conference of the International Organization of Citrus Virologists (IOVC)*. Riverside, CA, pp. 279–85.

Schroeder, W. J. 1992. Entomopathogenic nematodes for control of root weevils of citrus. *Fla Entomol.* 73, 563–7.

Simanton, W. A. 1960. The reduced status of purple scale as a citrus pest. *Proc. Fla. State Hort. Soc.* 73, 64–9.

Simanton, W. A. 1974. Occurrence of insect an mite pests of citrus, their predators and parasitism in relation to spraying operations. *Proceedings of the Tall Timbers Conference on Ecology and Animal Control by Habitat Management*. Tall Timbers Research Station, Tallahassee, Florida, pp. 135–63.

Simanton, W. A. 1996. Foundations of IPM: ecological survey of citrus groves. In: Rosen, D., Bennett, F. F. and Capinera, J. L. (Eds), *Pest Management in the Subtropics: Integrated Pest Management–A Florida Perspective*. Intercept, Andover, UK.

Simpson, S. E., Nigg, H. N., Coile, N. C. and Adair, R. A. 1996. *Diaprepes abbreviatus* (Coleoptera: Curculionidae). *Environ. Entomol.* 25(2), 333–49. doi:10.1093/ee/25.2.333.

Simpson, S. E., Nigg, H. N. and Knapp, J. L. 2000. Host plants of Diaprepes root weevil and their implications to the regulatory process. Diaprepes short course. Citrus Research and Education Center, Lake Alfred, FL, pp. 19–37.

Spreen, T. H., Baldwin, J.-P. and Futch, S. H. 2014. An economic assessment of the impact of Huanglongbing on citrus tree plantings in Florida. *HortScience* 49(8), 1052–5. doi:10.21273/HORTSCI.49.8.1052.

Stansly, P. A., Arevalo, H. A., Qureshi, J. A., Jones, M. M., Hendricks, K., Roberts, P. D. and Roka, F. M. 2014. Vector control and foliar nutrition to maintain economic sustainability of bearing citrus in Florida groves affected by huanglongbing. *Pest Manag. Sci.* 70(3), 415–26. doi:10.1002/ps.3577.

Tiwari, S., Lewis-Rosenblum, H., Pelz-Stelinski, K. and Stelinski, L. L. 2010. Incidence of *Candidatus* Liberibacter asiaticus infection in abandoned citrus occurring in proximity to commercially managed groves. *J. Econ. Entomol.* 103(6), 1972–8. doi:10.1603/EC10149.

Tiwari, S., Mann, R. S., Rogers, M. E. and Stelinski, L. L. 2011a. Insecticide resistance in field populations of Asian citrus psyllid in Florida. *Pest Manag. Sci.* 67(10), 1258–68. doi:10.1002/ps.2181.

Tiwari, S., Gondhalekar, A. D., Mann, R. S., Scharf, M. E. and Stelinski, L. L. 2011b. Characterization of five CYP4 genes from Asian citrus psyllid and their expression levels in *Candidatus* Liberibacter asiaticus-infected and uninfected psyllids. *Insect Mol. Biol.* 20(6), 733–44. doi:10.1111/j.1365-2583.2011.01103.x.

Tiwari, S., Stelinski, L. L. and Rogers, M. E. 2012. Biochemical basis of organophosphate and carbamate resistance in Asian citrus psyllid. *J. Econ. Entomol.* 105(2), 540–8. doi:10.1603/EC11228.

Tiwari, S., Killiny, N., Mann, R. S., Wenninger, E. J. and Stelinski, L. L. 2013a. Abdominal color of the Asian citrus psyllid, *Diaphorina citri*, is associated with susceptibility to various insecticides. *Pest Manag. Sci.* 69(4), 535–41. doi:10.1002/ps.3407.

Tiwari, S., Killiny, N. and Stelinski, L. L. 2013b. Dynamic insecticide susceptibility changes in Florida populations of *Diaphorina citri* (Hemiptera: Psyllidae). *J. Econ. Entomol.* 106(1), 393–9.

Tyler, H. L., Roesch, L. F. W., Gowda, S., Dawson, W. O. and Triplett, E. W. 2009. Confirmation of the sequence of '*Candidatus* Liberibacter asiaticus' and assessment of microbial diversity in Huanglongbing-infected citrus phloem using a metagenomic approach. *Mol. Plant-Microbe Int.* 22, 1624–34.

US Department of Agriculture. 2013. Citrus abandoned acres (WWW Document). Available at: http://www.nass.usda.gov/Statistics_by_State/Florida/Publications/Citrus/aban/CitAA13.pdf (accessed on 19 February 2015).

US Department of Agriculture. 2014. Tree assistance program (WWW Document). Available at: http://www.fsa.usda.gov/Internet/FSA_File/tap_2014.pdf (accessed on 24 February 2015).

US Department of Agriculture. 2015a. Citrus February forecast maturity test and results and fruit size (WWW Document). Available at: http://www.nass.usda.gov/Statistics_by_State/Florida/Publications/Citrus/cit/2014-15/cit0215.pdf (accessed on 23 February 2015).

US Department of Agriculture. 2015b. Citrus abandoned Acres. Available at: http://www.nass.usda.gov/Statistics_by_State/Florida/Publications/Citrus/aban/CitAA15.pdf). (Accessed 1.16.2019).

Wang, N., Stelinski, L. L., Pelz-Stelinski, K. S., Graham, J. H. and Zhang, Y. 2017. Tale of the huanglongbing disease pyramid in the context of the citrus microbiome. *Phytopathology* 107(4), 380–7. doi:10.1094/PHYTO-12-16-0426-RVW.

Weissling, T. J., Peña, J. E., Giblin-Davis, R. M. J. R. and Knapp, J. L. 2002. *Sugarcane Rootstock Borer Weevil, Diaprepes abbreviatus (L.). Featured Creatures*. University of Florida, Gainesville, FL.

WorldAtlas. 2019. The world's top citrus producing countries. Available at: https://www.worldatlas.com/articles/the-world-s-top-citrus-producing-countries.html (accessed on 16 January 2019).

Xu, C. F., Xia, Y. H., Li, K. B. and Ke, C. 1998. Further study of the transmission of citrus huanglongbing by a psyllid, *Diaphorina citri* Kuwayama. In: Timmer, L. W., Garnsey, S. M. and Navarro, L. (Eds), *Proceedings of the 10th Conference of the International Organization of Citrus Virologists*. Riverside, CA, pp. 243–8.

# Chapter 6

# Lime breeding: a way forward

*Muhammad Usman, University of Agriculture, Pakistan; Muhammad Mumtaz Khan and Rashid Al-Yahyai, Sultan Qaboos University, Oman; and Bilquees Fatima, University of Agriculture, Pakistan*



## 1 Introduction

### *1.1 Origin, botanical classification, domestication and reproductive biology of limes*

Lime is an important group in the genetically-diverse citrus family and comprises both acidic limes and sweet limes (SL). Limes have been cultivated for over 2000 years (Curk et al., 2016; Kahn et al., 2001). Lime originated in the tropical area of the Malay Archipelago and is the most sensitive of citrus species to freezing temperatures (Yelenosky, 1985). Hence, lime distribution is limited to tropical, warm and humid subtropical areas where temperatures don't fall below 4°C. Acidic lime types are commercially more widely cultivated compared with non-acidic ones. Acid limes include Mexican (Key, West Indian) lime (ML) [*Citrus aurantifolia* (Christm.) Swing.] group and Persian (Tahiti, Bears) lime (PL) (*Citrus latifolia* Tanaka).

Swingle and Reece (1967) classified lime [*C. aurantifolia* (Christm.) Swing.] and lemon [*C. limon* (L.) Burm. F.] as two separate species while Tanaka classified them as 37 different species. Hodgson (1955) categorized citron, lime and lemon as a single group. This complexity is mainly due to interspecific

http://dx.doi.org/10.19103/AS.2019.0054.09

sexual compatibility, presence of apomixis, polyembryony and many clonally propagated interspecific hybrids that were classified as separate species (Barrett and Rhodes, 1976).

Recent studies have revealed that all available Citrus species are the outcome of interspecific hybridization events between mandarin, pummelo, citron and papeda (Garcia-Lor et al., 2013; Ollitrault et al., 2012; Curk et al., 2015). The lime group of citrus is genetically complex due to the involvement of four ancestral species and having a diverse ploidy status compared with other horticultural groups of citrus which have originated from a diploid genome (Ollitrault et al., 2012; Garcia-Lor et al., 2013). It was believed that limes were related to lemon and citron (Barkely et al., 2006; Bayer et al., 2009). ML is a direct hybrid of citron (*C. medica*) x papeda (*C. micrantha*) confirmed using SNP analysis (Nicolosi et al., 2000; Garcia-Lor et al., 2013; Curk et al., 2015; Wang et al., 2017). However, there is still controversy in establishing the origin of other lime varieties.

Lime trees grow vigorously, upright, are spreading in nature and are mostly thorny. Trees show an average plant growth pattern and have low tree height (2–3 m tall) compared with other edible citrus species. However, when grown in fertile loamy soils, it may attain a height up to 7–8 m. Lime trees are evergreen, and leaves remain functional throughout the year. Plants have all three types of buds including vegetative, flowering and mixed buds. Productive branches that can bear flowers are normally 5–6 months old. Flowers of *C. aurantifolia* and other lime species are perfect and staminate types of flowers. The pistil development can be restrained at different stages of flower development and pistil may be absent in many flowers. Percentage of functional pollen grains varies in different citrus species. Many meiotic abnormalities have been reported in limes and lemons which may increase with temperatures below 20°C leading to increased sterility. Self- and cross-incompatibility is found in citrus in varying degrees. Self-incompatibility and pollen sterility were reported in *C. limettioides* (Frost and Soost, 1968). Reproductive behaviour of different lime cultivars and species varies under different environmental conditions.

## 2 Lime varieties

### 2.1 Acid limes

#### 2.1.1 Mexican lime

Mexican lime also originated in the Indo-Malayian region and is referred to as the true lime. It was presumed to be a hybrid developed from pummelo, citron and finger lime (Barrett and Rhodes, 1976). Molecular marker analysis revealed equal contribution of citron and papeda (Ramadugu et al., 2013, 2017) while sequencing revealed papeda (*C. micrantha*) as the maternal parent (Carbonell-Caballero et al., 2015). Admixture analysis using markers and sequence information have confirmed ML as hybrid of papeda and citron (Curk et al., 2016; Wang et al., 2017). West Indian and Kaghzi lime are clonal selections of ML. The Peruvian lime cultivated in Peru is also presumed to be genetically close to ML.

Trees are medium in size, grow vigorously like a bush and have many small thorns. Leaves are small, light green to pale green in colour and pointed at the apex. Young foliage is light purple in colour. Trees are more sensitive to cold than PL. Leaf lamina is round to ovate in shape. In tropical areas, ML keeps blooming all year round depending upon the intensity and frequency of rainfall while in the subtropics it blooms once in January–February and the crop matures in the late summer or early winter season. During July–August there is no flowering due to the presence of the developing fruit, rainy season, high humidity and high temperature which collectively hinders flower differentiation. In coastal areas of central Mexico, abundant flowering has been observed during November–April (Martinez, 2002). Flowering intensity depends upon the number of mature branches available at the time of flowering. In humid areas, maximum reproductive buds are available in the spring season. In tropical areas, water stress plays an important role in flower induction and regulating flowering intensity (Davenport, 1990; Alamguer-Vargas et al., 2011). Pruning of branches (20–30%) and foliar application of 10% urea can also be used to regulate flowering and has enhanced fruit yield in both ML and PL (Alamguer-Vargas et al., 2011). Citrus breeders have been trying to induce early flowering in different citrus species including ML. Seedlings have been raised under controlled environmental conditions in chambers, at day/night temperatures 30°C/25°C and 16 h photoperiod, up to 2 m plant height. Seedling growth was arrested using paclobutrazol leading to enhanced lateral branching and flowering at the age of less than 18 months (Snowball et al., 1988; Davies and Albrigo, 1994). Alternatively, long stems of the seedlings raised under greenhouse conditions can be bent down to force flowering.

During the summer season, fruit development and maturity is faster (100–110 days) compared with the winter season (130–140 days; Martinez, 2002). Fruit holding capacity is low and over mature fruit normally drops. Lime has small (2–5 cm in diameter), round-shaped fruit with a slight neck at the fruit apex. Rind is thin and has pungent oil that differentiates it from Palestinian SL. Usually fruit is harvested when green in colour, very juicy and highly acidic compared with PL. It can be seed propagated due to higher nucellar embryony and may have 15 or more seeds. Many land races of ML having wide diversity are available in the Indo-Pakistan region (Munankarmi et al., 2014) that need to be collected, characterized and conserved. Kaghzi lime is also a selection of ML and is a commercial variety in the region. Its fruit has many seeds and higher polyembryony. Hence, it may also be raised by seeds and seedling population will be morphologically identical to the mother plants (Davies and Albrigo, 1994). Other popular ML selections include Everglad, Yung and Palmetto.

### 2.1.2 Persian lime

Persian lime originated in South East Asia including north east India, the Malay Archipelago and south west China (Moore, 2001). PL was introduced to the United States from Tahiti during 1850–80 and due to its higher association with the Tahitian citrus industry, it was also named as the Tahiti lime (TL). A selection of PL seedlings is also known as Bears lime in California. It was presumed to be a hybrid of small acid lime and citron (Reece and Childs, 1962). DNA sequencing data revealed sour orange as the maternal parent (sharing same mitotype) with other species including *C. medica*, *C. maxima*, *C. reticulata* and *C. micrantha* as ancestral taxa. It is also believed that PL might have developed by fertilization of unreduced gamete of ML with haploid ovule of lemon (Curk et al., 2015, 2016; Wang et al., 2017).

Its trees are larger than ML, the branches are thornless, leaf size is larger with a dark green colour, elliptical in shape and has serrations at the leaf margins near the apex. Petioles in both ML and PL are small and not very prominent. The plant is cold sensitive. It bears flowers and fruit throughout the year. Flower petals are purple on the outside like lemon. There is no need for cross-pollination for fruit setting and seeds produced are highly monoembryonic. Fruit is larger (up to 7 cm in diameter) in size, resembles ML and small immature lemon. Mature fruit peel skin colour is greenish yellow to pale yellow and rind is thick. Its seedless form was first noticed in 1975 in California, which was triploid, produces no viable pollens and has ovule sterility (Crane and Osborne, 1991). Tetraploid of PL is available having fruit size larger than ML. Fruit is juicy and highly acidic. It's easy to ship and has a longer storage life hence it is more popular for transportation to distant markets. There is no more cultivation of PL in Florida due to climate change and susceptibility to citrus canker; however, overall its demand and production have increased over time in the United States. It is mainly used for production of lime juice concentrate. Other seedless cultivars developed from PL include Idemor (limb sport) and Sakhesli lime (clonal selection). It requires less heat and is more cold tolerant than ML.

## 2.2 Sweet limes

Sweet limes (*C. limettioides*) is commonly referred as Palestine or Indian SL or Mitha nimboo in the local language. It is mainly cultivated in Pakistan, India, the Middle East and parts of central and southern America. This SL is different from Mediterranean or Tunisian SL (*C. limetta*). It is native to north east India and has many variants. It was presumed to be a hybrid of ML and sweet lemon; however, molecular marker-based admixture analysis has revealed it as a hybrid of citron and papeda (Barkley et al., 2006) which was further confirmed by sequence analysis as a citron-mandarin hybrid (Ramadugu et al., 2013). Chloroplast sequence of the SL are similar to pummelo, hence sour lime may not be the maternal parent of SL (Curk et al., 2016).

Several SL strains have low sugar content compared with other acid citrus varieties and are the most insipid variety due to very low acidity (<0.1%). Trees have vigorous growth habit, irregular branching, are thorny with leaves that are rolled or cup shaped. Flowers are pure white in colour and are self-compatible. Cross-pollination using grapefruit and orange have shown metaxenial effects on fruit size, hence interplantation may be useful (Morton, 1987). Fruits are medium sized (5–7 cm in diameter) and round shaped with smooth skin. Fruits have a small nipple on the fruit apex and a few polyembryonic seeds. Rind oil has a distinct aroma. Fruits are very juicy; however, its taste is like acid-less orange and does not appeal to everyone. SL is popular in Pakistan, India and Middle East. It has also been widely used as rootstock in India, Israel and lowers the fruit acidity in the scion variety (Morton, 1987).

### 2.3 Australian lime types

Australian citrus species are diverse in nature and are present in both rainforests and areas with dry climatic conditions. These include round lime (*Microcitrus australasica* Swingle), finger lime (*Microcitrus australis* (Planch.) Swingle) and desert lime (*Eremocitrus gluaca* (Lindl.) Swingle). The taxonomic differences in these genera have little significance and have been reclassified under citrus as *C. australasica* F. Mueller, *C. glauca* (Lindley) Burkill and *C. australis* (Mudie) Planchon (Saunt, 1990; Mabberley, 1998). Desert lime is the only xerophytic plant in the Aurantioideae sub-family. These are evergreen plants and the fruit is used to prepare sauces and marmalades. These citrus types can hybridize with other species and help in development of biotic and abiotic stress-tolerant varieties. Sydney hybrid is highly drought tolerant. Desert lime can tolerate very high (45°C) and low (−25°C) temperatures, hot, dry and saline conditions.

### 2.4 Small-fruited acid citrus

Small-fruited acid citrus fruits are not true limes and are important in Far Eastern countries. These include Yuzu (*C. junos*), mainly grown in China and Japan, hybrids of Kabosu (*C. spaerocarpa*) and Sudachi (*C. sudachi*), which are Japanese acid citrus fruit. Yuzu (*C. junos*) is grown in Japan for both fruit and as rootstock (Saunt, 1990). It originated in China and is cold-hardier compared with Satsuma mandarin.

Kabosu fruit is larger in size than Yuzu. It is round shaped and harvested in August-September. It has better shelf life and is sold from winter to April. Its flesh is juicy and acidic with 10–12 seeds per fruit. Sudachi (*C. sudachi*) has similar features to Kabosu; however, the fruit is much smaller. It is harvested in late summer. Yuzu fruit is small, round, quite juicy and seedy. Its rind is thick and used in many dishes. Its harvesting span is like Sudachi and Kabosu. Makrut lime (*C. hystrix*) is used for flavouring food. Its tree and fruit are small. Fruit skin is rough and bumpy. Rind is thick and has a very pungent lime aroma. Fruit has little juice, is very bitter and acidic and has many seeds. Fruit and leaves are used to enhance flavour in curries.

## *2.5 Rangpur lime (*Citrus x limonia *Osbeck)*

This variety is commonly known as mandarin lime and originated in the Indo-Pakistan region. It is known under different names in various parts of the world including Brazil (Cravo lemon), the United States (Tahitian orange), Japan (Hime lemon), China (Canton lemon) and India (Marmalade lime). It is neither related to PL nor ML. Rangpur lime (RL) was considered as a hybrid of lime and mandarin as proposed by Webber in 1943. Marker analysis of the organelle DNA identified mandarin as its maternal parent (Gulsen and Roose, 2001) while genome analysis suggested more similarity to citron than mandarin (Barkley et al., 2006). Sequence analysis confirmed Cleopatra mandarin as the maternal parent of RL (Carbonell-Caballero et al., 2015; Curk et al., 2016). Comprehensive gene analysis confirmed RL as a hybrid of mandarin and citron (Curk et al., 2015, 2016; Wang et al., 2017).

Its fruit is like mandarin, loose skinned, flesh is orange coloured, juicy, highly acidic and has many polyembryonic seeds. The fruit is used for making marmalade. It has been mainly used as rootstock (80%) in Brazil for oranges; however, it is not used commercially elsewhere (Bove and Ayres, 2007). It was replaced due to susceptibility to citrus sudden death virus which destroyed millions of trees in Brazil (Muller et al., 2002). Scion varieties on RL are as productive and vigorous as on Rough lemon; however, the fruit size may be smaller and of moderate quality. The fruit bearing starts early, grafted plants show salt and drought tolerance and cold sensitivity. Generally, limes are susceptible to Citrus Tristeza Virus (CTV) and are used as indicator plants to get disease symptoms. RL is susceptible to Phytophthora and resistant to CTV.

## *2.6 Other species*

Other species which are used as lime and are not true limes include Calamondin (Musk lime) recently named as *Citrofortunella mitis* Blanco (Morton, 1987). Other common names include Calamondin orange, Scarlet lime, Golden lime and Kalamansi. Its fruit is used to flavour food and beverages. Kaffir lime (*C. hystrix*) is also commonly known as Magrood, Makrut or Combavas lime and belongs to papeda. It is widely grown in south east Asian countries including Sri Lanka and the Philippines. Kaffir lime fruit juice is very sour and used in medicine.

### 2.7 Lime hybrids

Lime hybrids include Limetta (*C. limetta* Risso) commonly known as sweet limetta and Tunisian SL. Molecular analysis revealed Limetta as a hybrid of (*C. maxima* × *C. reticuleta*) × *C. medica* (Curk et al., 2016). Their fruit is small and oval shaped like kumquats and the entire fruit is edible including the rind. These are more cold tolerant than ML. Limequat (*Citrofortunella floridana* Ingram and Moore) is a bigeneric hybrid of Citrus and Fortunella which originated first in China. Main varieties include Eustis, Lakeland and Tavares. Eustis limequat and Lakeland are hybrids of West Indian lime and round kumquat (*Fortunella japonica*). Eustis lime is commonly known as Chinese lime in Pakistan and is widely cultivated. Tavares is a hybrid of West Indian lime and oval kumquat (*F. margarita*). Blood lime is a hybrid of finger lime and Ellendale mandarin or RL developed in Australia. It has blood-red colour flesh, juice and rind. The fruit is small and sweet, used for juices and beverages. Lime hybrids released by the United States include 'C4-5-20' (seedless triploid lime), 'C4-5-27' (Key lime × Tetraploid lemon) and 'C4-5-33' (seedless triploid lime). Pigmented lime has also been developed by introducing different transcription factors like *Ruby* from blood oranges and *VvmybA1* from grapes (Grosser et al., 2015; Dutt et al., 2016; Ali and Nawaz, 2017). Sydney hybrid is a cross between finger lime and round lime and grown in Australia since the 1940s.

## 3 Economic importance and distribution of limes

Limes are commercially grown throughout tropical, subtropical and Mediterranean climatic conditions. Lime and lemons are the third largest cultivated group in citrus with about 1.6 million tons traded annually (FAOSTAT, 2014). Limes are marketed as fresh fruit and processed products including juices, sauces and marmalades. Since the 1980s, global consumption of limes has increased rapidly, with growing demand and competition (Duportal et al., 2013). Since 2013, global exports of limes and lemons have increased up to 37% (FAOSTAT, 2016). Despite limes have a large genetic diversity, only a few cultivars are commercially cultivated.

Citrus is commercially grown on 8.01 million ha area and produces 132.40 million tons of fresh fruit annually. The share of lime and lemons in the global citrus industry is 13% in terms of area and production with an average global yield of 16 metric tons (MT) $ha^{-1}$. Citrus is widely cultivated in both the northern (NH) and southern hemispheres (SH); however, the NH share of total area and production is over 50%. In lime and lemons, the major producers in NH contribute 66% of the total global production compared with 20% production in the SH (FAOSTAT, 2016). In NH, major lime and lemon producers include India and Mexico which have 30% share of lime and lemons in their domestic citrus production (Table 1). It is noticeable that that lime exports from India are negligible, while Spain, Mexico and Turkey are the largest exporters. Mexico is one of the largest producers of TL which is mainly exported to the United States. Other large producers in NH include China, Spain and Turkey. Turkey has the largest share of lime and lemon production (63%) in relation to overall citrus production. The highest per hectare yield of lime and lemon is 37 MT $ha^{-1}$ produced in the United States, followed by Turkey and Egypt. Pakistan's share in the area and production of lime lemon is merely 3–4% of its total citrus industry. Pakistan produces only 85 000 tons per annum, hence there is an urgent need to enhance production and quality of the produce to boost exports to the international markets (FAOSTAT Database, 2016). Furthermore, there is need to explore the production potential of TL and other promising lime cultivars and its cultivation needs to be improved to increase lime exports to the western world.

In the SH, Argentina is the largest lime and lemon producer yielding 1.67 million tons per annum followed by Brazil. The share of lime and lemon production in the Argentinian citrus industry is more than 50% indicating that after Turkey, Argentina has the largest share of limes in its domestic citrus industry. Brazil is the largest citrus producing country in the SH; however, its share in lime is merely 6% of total citrus production (FAOSTAT, 2016). Exports of TL from Brazil increased up to 118% during 2016. Among lime and lemon-importing countries, the United States, Russia and the Netherlands are the largest importers. In Mexico, ML trees are planted as mix plantations with mangoes, palm and papaya. In Egypt, lime is mainly grown in southern parts of Cairo. In the United States, there is no commercial production of ML, rather TL is preferred. In Brazil it is known as Galego lime and in Iran as Shirazi or Torsh. The triploid TL is mostly produced for the export market. Other triploids such as 'Tanepao' 'Coppenrhad', 'Tanepao', 'Mohtasseb' and 'Ambilobe' produce seedy fruits, have little commercial significance and area under cultivation.

In Pakistan, there has not been much increase in the area under lime and lemon cultivation during the last 10 years. Despite this, lime production has significantly increased in this period by 26161 MT (FAOSTAT, 2016). Per hectare yield also increased with a net rise of 3.3 MT $ha^{-1}$ of fresh fruit, though this needs significant further improvement. Currently, ML acid lime, Eustis (Chinese) limequat and PL are widely cultivated in Punjab, Sindh and other provinces.

**Table 1** An overview of leading citrus, lime and lemon producers in the northern and southern hemispheres

| | Total citrus | | Limes and lemons | | | Limes and lemons share (%) in total citrus | |
|---|---|---|---|---|---|---|---|
| Location | Area (000, ha) | Prod. (000, tons) | Area (000, ha) | Prod. (000, tons) | Yield (tons/ha) | Area (%) | Production (%) |
| World | 8011.95 | 132401.43 | 1078.85 | 17347.15 | 16.08 | 13.47 | 13.10 |
| *Northern hemisphere* | | | | | | | |
| India | 852.85 | 10871.50 | 258.00 | 2978.00 | 11.54 | 30.25 | 27.39 |
| Mexico | 528.18 | 7938.60 | 163.47 | 2429.84 | 14.86 | 30.95 | 30.61 |
| China | 2440.34 | 32698.23 | 106.84 | 2329.86 | 21.81 | 4.38 | 7.13 |
| Spain | 361.06 | 7008.59 | 43.29 | 857.75 | 19.81 | 11.99 | 12.24 |
| Turkey | 465.69 | 1337.03 | 30.03 | 850.60 | 28.32 | 6.45 | 63.62 |
| United States | 298.18 | 7489.78 | 22.06 | 822.00 | 37.27 | 7.40 | 10.97 |
| Iran | 187.88 | 3127.57 | 29.02 | 457.27 | 15.76 | 15.44 | 14.62 |
| Italy | 145.41 | 2528.35 | 25.31 | 379.28 | 14.98 | 17.41 | 15.00 |
| Egypt | 195.96 | 4802.11 | 14.98 | 341.45 | 22.79 | 7.64 | 7.11 |
| Pakistan | 197.91 | 2270.136 | 7.78 | 85.06 | 10.93 | 3.93 | 3.75 |
| *Southern hemisphere* | | | | | | | |
| Argentina | 145.66 | 3281.32 | 52.39 | 1678.34 | 32.03 | 35.97 | 51.15 |
| Brazil | 759.95 | 19591.62 | 47.28 | 1262.35 | 26.70 | 6.22 | 6.44 |
| South Africa | 66.72 | 2230.82 | 15.99 | 323.44 | 20.22 | 23.97 | 14.50 |
| Thailand | 72.77 | 970.63 | 13.32 | 115.04 | 8.63 | 18.31 | 11.85 |
| Australia | 28.87 | 578.82 | 2.05 | 46.79 | 22.85 | 7.09 | 8.08 |

*Source*: FAO production year book (2017), www.faostat.org.

## 4 Reasons for low productivity

Variable pollen functionality and meiotic abnormalities are reported in different lime cultivars including SL with variable frequency particularly at low temperatures. Incompatibility and pollen sterility could also adversely impact fruit setting and yield. In ML, during the rainy season, flowering intensity is reduced, and a greater number of reproductive buds will be available in the spring. Water stress plays a critical role in flower induction and heavy irrigation or higher humidity may reduce flowering. Fruit holding capacity is low in ML and drop rate can be higher. PL is more heat tolerant and less cold sensitive than ML. There is a lack of germplasm resources and crop improvement programmes. Non-availability of certified disease-free plant material is another issue in lime-producing Asian countries, particularly Pakistan. Local nurseries can be a carrier of many diseases if the budwood used is not certified and sanitized or not raised on proper rootstock (Rough lemon in Pakistan). Low-planting density per unit area and lack of good agricultural practices further reduces productivity. Higher susceptibility of lime varieties to citrus canker and other viral diseases including CTV and Witch's broom also adversely affect plant production. Collectively, these factors affect plant life and productivity and need to be addressed.

## 5 Germplasm collections, biodiversity and conservation

Germplasm is any genetic material that could be utilized for further plant propagation and multiplication. It includes seeds, stem, leaf tissues, pollen grains or cell cultures. Any heterogenous material that could be used as heredity stocks and ecotypes collected from different ecological zones could be termed as germplasm. Germplasm collections, their identification, documentation and conservation are the basic requirements to initiate crop improvement programmes and help in minimizing the effect of genetic erosion due to monocropping and introduction of elite cultivars. *Ex-situ* collections offer higher accessibility to researchers and academia across the world. Further, such collections are easy to manage and control under changing climatic conditions (Kruger and Navarro, 2007). University of California Riverside (UCR) and National Clonal Germplasm Repository of Citrus and Date (NCGRCD), USA, have one of the largest citrus (1328 accessions) and lime collections (62 accessions) (Volk et al., 2018). These are categorized as large-fruited acid limes (5), lime hybrids (6), small-fruited acid limes (14), SL (8), limetta types (6), RL and their hybrids (23). Other leading countries include Brazil having 56 accessions followed by France with 40 accessions (Fig. 1). Australia has about 500 accessions of citrus including ten accessions of lime. Australian wild limes including Mount White Lime (*C. garrawayi* FM Bailey)

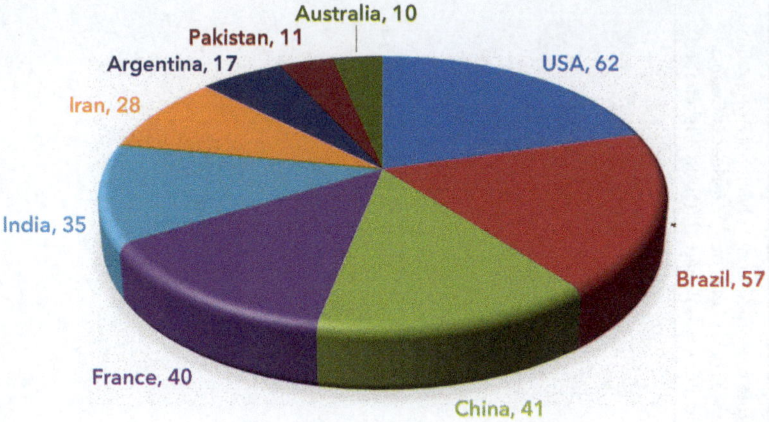


**Figure 1** Global genetic resources of limes.

and Russell River Lime (*C. inodora* FM Bailey) are also vulnerable to rapid extinction that need to be collected and conserved. These are also a source of wide genetic variation such as tolerance to drought and salt, early bearing, disease resistance, dwarfness and coloured fruit (Skyes, 1997). Pakistan is also rich in citrus genetic resource having 150 accessions of citrus at different genetic resource centres. However, it has a limited number of lime species and rootstocks available (Table 2; Figs. 2 and 3). There is rapid erosion of genetic material and thus a major need to identify elite plant material, as well as create local collections to characterise and conserve material at multiple sites for future use.

Extensive characterization of lime germplasm and its diversification using molecular breeding tools, transgenics including genome editing, bioinformatics and next-generation sequencing (NGS) technologies is also needed. Development of core collections is also important along with detailed genetic diversity information for utilization in breeding programmes (Butelli et al., 2017).

International cooperation among R&D and genetic resources centres should be enhanced to make the best use of the available germplasm resources for crop improvement programmes (Hummer et al., 2015). Development of germplasm units at multiple sites is particularly critical in developing countries such as Pakistan for germplasm conservation and better utilization. Mere introduction of exotic material is not enough without detailed study of genetics and breeding behaviour, phenology and physiology to make full use of the available genetic resources. The current focus on fruit quality also needs to be extended to other desirable traits including breeding behaviour, resistance for biotic and abiotic stress. No major achievements in lime

**Table 2** List of acid and sweet lime varieties, hybrids and rootstocks available at different research institutes/universities in Pakistan

| Sr. no. | Varieties | Species | Citrus Research Institute (CRI), Sargodha | Horticulture Research Station (HRS), Sahiwal | Ayub Agriculture Research Institute (AARI), Faisalabad | Institute of Horticultural Sciences (IHS), Univ. of Agriculture, Faisalabad | National Agriculture Research Council (NARC), Islamabad |
|---|---|---|---|---|---|---|---|
| *Acid limes* | | | | | | | |
| 1 | Mexican lime | *C. aurantifolia* | √ | √ | √ | √ | √ |
| 2 | Mexican lime thornless | *C. aurantifolia* | √ | – | – | – | – |
| 3 | Desi lime (seedling populations of ML) | *C. aurantifolia* | √ | √ | √ | √ | √ |
| 4 | Tahiti lime | *C. latifolia* | √ | √ | – | √ | – |
| 5 | Persian lime SPB 7 | *C. latifolia* | √ | – | – | – | – |
| *Sweet limes* | | | | | | | |
| 6 | Palestine/Indian sweet lime | *C. limettioides* | √ | √ | – | √ | – |
| 7 | Local mitha | *C. limettioides* | √ | – | √ | √ | – |
| 8 | Peshawari mitha | *C. limettioides* | √ | – | √ | – | √ |
| *Lime hybrids* | | | | | | | |
| 1 | Eustis limequat/ Chinese lime | *X Citrofortunella* sp. | – | √ | √ | √ | √ |
| *Rootstocks* | | | | | | | |
| 1 | Rangpur lime | *C. limonia* Osbeck | √ | – | – | √ | – |
| 2 | Rangpur lime × Troyer citrange | *C. limonia* × *X Citroncirus* sp. *RUTACEAE* | – | – | – | √ | – |
| *Total accessions* | | *11* | *9* | *5* | *5* | *8* | *4* |

**Figure 2** Acid lime germplasm in Pakistan. (a) Mexican lime thornless, (b) Kaghzi, (c) Tahiti and (d) Eustis.

**Figure 3** Sweet lime germplasm in Pakistan. (a) Palestine, (b) Local and (c) Peshawari.

breeding have been reported in Pakistan. There are many seedling strains available that need to be collected, characterized and conserved. Recently, a research project on development and characterization of polyploids in acid limes has been funded by the Higher Education Commission, Pakistan. The genetically diverse plant material produced will be useful for further selection of elite parental material for breeding programmes.

## 6 Breeding systems

Methods for citrus improvement include conventional breeding, somatic hybridization, somaclonal variation, induced mutations, ploidy manipulations, marker-assisted breeding, genomics and transgenics (Khan and Kender, 2007; Usman and Fatima, 2018). Both rootstock and scion varieties have different criteria for improvement. Better rootstocks should have more adaptation to the local environmental conditions, early bearing, plant uniformity, good seed characteristics, polyembyony, resistance against biotic and abiotic stress, cold hardiness and production of commercially important oils (Grosser and Gmitter, 2005, 2011). Similarly, scion varieties should be early and heavy bearers, less seeded to seedless, monoembryonic, with better fruit size and quality, higher juice content and TSS, flavour and colour, resistance to biotic and abiotic stress conditions including cold hardiness, long harvest span and better storage life (Grosser and Gmitter, 2005; Khan and Kender, 2007).

### *6.1 Conventional and mutation breeding*

Classical breeding approaches have many limitations in lime similar to other citrus species including tree size, long juvenility, nucellar embryony and apomixis, pollen and ovule sterility and higher heterozygosity (Grosser et al., 2000; Singh and Rajam, 2009; Fatima et al., 2015). The commercial release of a new variety using conventional breeding strategies may take 20–30 years. Utilization of modern breeding tools helps to shorten the breeding cycle and target the trait of interest more precisely compared to traditional genetic combinations and screening. Cultivar ML has been widely used as a model species in citrus for somatic hybridization, mutation breeding and transgenics.

Mutants could be developed using physical mutagens including gamma rays and chemical mutagens such as ethylmethane sulphonate (EMS) and colchicine. Little is known about the development of irradiation and EMS-induced variability in lime species. Use of gamma rays to induce variability in both ML and Eustis limequat varieties showed restricted plant growth patterns at higher doses, that is 100 Gy (Usman et al., 2018, unpublished data). Such studies could develop heterozygous material having novel traits which may be characterized using modern breeding and NGS approaches.

### *6.2 Polyembryony, apomixis and candidate gene(s)*

Polyembryony can be gametophytic and sporophytic. In sporophytic embryony, embryos may develop from nucellar tissue, endosperm, integuments and by zygotic cleavage (Batygina and Vinogradova, 2007). Citrus species show both apomictic and non-apomictic behaviour. The phenomenon of apomixis offers fixation of heterozygosity and other economically valuable traits in citrus

(Conner et al., 2015). During the evolutionary process, chances of selection and domestication of apomictic lines have been much higher. Hence, most of the today's citrus species are apomictic in nature except *C. medica*, *C. maxima* and few mandarin hybrids which are monoembryonic. Apomictic genotypes have one zygotic and multiple nucellar embryos which are genetically true to the mother parent (Aleza et al., 2009). The higher frequency of apomixis and nucellar embryony have also reduced the genetic diversity in varieties, hence most of the elite cultivars are highly vulnerable to emerging biotic and abiotic stress (Curk et al., 2015).

Most of the lime species are polyembryonic and utilization of nucellar embryony has been quite useful (Cimen and Yesiloglu, 2016). All the lime species exhibit partial apomixis and several cultivars have originated from a single clonal parent by a series of mutations (Snoussi et al., 2012). Though lime accessions were highly heterozygous, they showed limited diversity within subgroups having a common allele (Ollitrault et al., 2012). Such diversity patterns are less due from sexual recombinations than as the outcome of sporadic mutations and mobility of the transposable elements (Ollitrault et al., 2012; Garcia-Lor et al., 2013) in the form of chromosomal deletions on different chromosomes leading to variability (Curk et al., 2016). Other factors involved in reduced heterozygosity in the subgroups are man-directed selection for desired phenotypes and subsequent clonal propagation (Ollitrault et al., 2003).

Researchers have been trying to identify the genetic mechanism involved and candidate genes regulating apomixis in citrus. Recently, 80 kb of more precise genetic regions containing 11 genes related to apomixis has been identified in a previously proposed 380 kb region (Nakano et al., 2012). In polyembryonic species, hyper expression of candidate gene *Cg4g018970*, encodes for RWP-RK domain referred as *CitRWP*, was observed in ovules while no expression was found in the monoembryonic species (Wang et al., 2017) suggesting its role in regulation of polyembryony. Identification of SNPs and insertion of miniature inverted repeat transposable elements (MITE) in promotor regions only in polyembryonic species further confirmed its association with the nucellar embryony. There is an urgent need to further confirm the functionality of key candidate gene *CitRWP* regulating polyembryony by gene knockout studies in primary and secondary citrus species including limes and its utilization in future crop improvement programmes. Transcriptomic analysis of development of asexual and sexual embryos in citrus revealed that sexual reproduction and nucellar embryogenesis may utilize similar genetic pathways. Six genes associated to nucellar embryos were stress regulated (Wang et al., 2017) which could be further studied across different lime species. There is need to identify and conserve primitive and wild lime land races that have been neglected previously because of presence of higher genetic heterozygosity and disease resistance in these wild taxa. Aleza et al. (2010) confirmed the cleavage of zygotic embryos in the non-apomictic citrus genotypes during interploid crossing and suggested reduced requirement of recovery from underdeveloped seeds due to more probability of embryos having the same genetic origin. Thus, cost-efficiency of the breeding programmes could be enhanced, and the breeder's workload could be reduced.

## 6.3 Cytogenetics

Like other citrus species, chromosome size in lime is also small, making cytogenetic studies relatively difficult (Barba-Gonzalez et al., 2005; Fatima et al., 2015). Hence, there is little work reported on cytogenetics in lime cultivars. Cytogenetic studies were conducted to compare RL with Chironja and Rough lemon. In RL, both genes and chromosomes were involved in regulation of pollen sterility while in Chironja and Rough lemon only genes were involved (Agarwal, 1987). Cytological studies based on chromomycin A3 (CMA) and 4′,6-diamidino-2-phenylindole banding patterns of the chromosomes of ML and SL revealed higher heterozygosity at rDNA sites of both species indicating less genetic similarity between the two species as presumed previously (Carvalho et al., 2005). Among 12 citrus species studied, *Citrus aurantifolia* had different CMA-stained chromosomal banding patterns compared with most of the other citrus species. It had nine chromosomes with one telomeric band, seven chromosomes without bands and two chromosomes with one proximal and one telomeric band (Yamamoto, 2007). Cytogenetic studies have also proposed citron as the true parental species of limes based on its complete chromosomal type's representation in limes (Carvalho et al., 2005). None of the lime progeny having direct relationship with citron showed its role as a female parent. This could be attributed to the reproductive behaviour of citron which is highly homozygous due to cleistogamy, monoembryonic having no apomixis and produce only hybrid embryos.

Citrus species has been widely utilized to study meiotic behaviour, unreduced gamete formation and polyploidization (Aleza et al., 2016a; Cuenca et al., 2015; Fatima et al., 2015; Rouiss et al., 2017a). Advances in the cytogenetics, genomics and transcriptomics have enhanced efficiency of the citrus breeding programmes. Better understanding of the unreduced gamete formation will enable to enhance their production and development of genetically diverse polyploid plant material that could be used for future breeding programmes.

## 6.4 Ploidy manipulations and seedlessness

Diploidy is a general rule in citrus and species are maintained through nucellar embryony and apomixis (Khan and Kender, 2007). Utilization of polyploidization has been a very valuable tool in citrus improvement programmes (Fatima et al., 2002; Usman et al., 2008; Cuenca et al., 2015) and the role of polyploids in the evolutionary history of citrus is still being explored (De Storme and Geelen, 2013). Evolutionary processes could be further enhanced using polyploidization and somatic hybridization in lime. Polyploids provide useful germplasm for future breeding programmes, and have developed plant material with better fruit size and quality attributes. Polyploids may have differential gene expression patterns including circadian genes compared with their diploid progenitors. Polyploids could be spontaneous, induced artificially using colchicine and other such mutagenic agents. In citrus allopolyploids have shown better growth, vigour and better control of the physiological and metabolic pathways regulated by circadian genes due to genomic hybridization (Auger et al., 2005). In polyploids, transcriptomic responses depend upon the chromosome origin rather than chromosome number (Yu et al., 2010). Polyploids are also known to exhibit higher tolerance to abiotic stress conditions including temperature, salinity and drought. Ploidy-based modulation of gene expression has been reported in autotetraploid RL rootstock grafted with scion of diploid sweet orange cultivar Valencia via abscisic acid signalling. The tetraploid-diploid grafted plants showed differential expression of water stress-related genes, reduced stomatal conductance and enhanced drought tolerance (Allario et al., 2013). Triploid TL is one of the least susceptible lime variety that has least threats to its production due to biotic or abiotic stress factors. The lime polyploids also showed plant dwarfism, dark green and thick leaves with more leaf size compared with the diploid plants (Allario et al., 2011). Parentage and ploidy status of different lime species and hybrids is presented in Table 3.

### 6.4.1 Spontaneous polyploids

Limes have wide diversity in spontaneous polyploids, lacking in most of other citrus species except lemon including TL and Triphasia desert lime (Bachi, 1940; Esen and Soost, 1972). The underdeveloped seeds have been a major source of development of triploids and tetraploids in Kaghzi lime and other citrus species (Usman et al., 2006; Fatima et al., 2010). Availability of unreduced gametes play a key role in the development of spontaneous polyploids. In populations of diploid polyembryonic limes, spontaneous doubled diploids (tetraploids) including 'Giant Key' lime have been reported due to frequent availability of 2n gametes (Ollitrault et al., 2008; Aleza et al., 2011). Frequency of spontaneous triploids in breeding programmes was higher in lime species compared with other citrus species (Usman et al., 2006; Ollitrault et al., 2008). Triploids produced also show variation in seedlessness like *C. latifolia* is a seedless triploid while 'Tanepao' Coppenrhad', 'Tanepao', 'Mohtasseb' and 'Ambilobe' are seedy triploids (Ollitrault et al., 2008).

**Table 3** Parentage and ploidy status of lime species and hybrids

| Parentage | | | |
|---|---|---|---|
| ♀ Parent (gametes) | ♂ Parent (gametes) | Proposed hybrid species | Ploidy status |
| *C. mircrantha* (Papeda) | *C. medica* (citron) | *C. aurantifolia* (Mexican lime) | 2n |
| | | *C. macrophylla* (Alemow) | |
| | | *C. aurata* (Pomme d Adam) | |
| | | *C. excelsa* (Nestour lime) | |
| *C. reticuleta* (Mandarin) | | *C. limonia* (Rangpur lime) | 2n |
| | | *C. karna* (Khatta lime) | 2n |
| *C. aurantifolia* (ML, 2n) | | *C. aurantifolia* (Tanepao lime) | 3n |
| | | *C. aurantifolia* (Copperrath lime) | |
| | | *C. aurantifolia* (Mohtasseb lime) | |
| | | *C. aurantifolia* (Ambilobe lime) | |
| *C. aurantium* (sour orange) | *C. aurantifolia* | *C. limetta* (Tunisian sweet lime) | 2n |
| *C. limon* (lemon) | *C. aurantium* | *C. limettioides* (Palestine sweet lime) | 2n |
| *C. aurantifolia* (ML) | *C. aurantifolia* | *C. aurantifolia* (New Caledonia lime) | 2n |
| | | *C. aurantifolia* (Kaghzi lime) | 2n |
| *C. limon* (1n) | *C. aurantifolia* (2n) | *C. latifolia* (Tahiti lime) | 3n |
| | | *C. latifolia* (Persian lime) | 3n |
| | | *C. latifolia* (Bears lime) | 3n |
| | | *C. latifolia* (El Kseur lime) | 3n |
| | | *C. latifolia* (IAC-5 lime) | 3n |
| *C. aurantifolia* (2n) | *C. aurantifolia* (2n) | *C. aurantifolia* (Giant Key lime) | 4n |
| *C. limonia* (2n) | *C. limonia* (2n) | *C. limonia* (Rangpur lime) | 4n |

*Sources*: Curk et al. (2016), Wang et al. (2017), Wu et al. (2018).

### 6.4.2 Somatic hybridization

Genetic variation available in cytoplasmic factors is also very important in breeding programmes (Wang et al., 2010). Cytoplasmic factors control many economically important agronomic traits in citrus (Satpute et al., 2015). During the last three decades, enormous success has been achieved using somatic hybridization (Wang et al., 2013; Satpute et al., 2015). It offers solutions to the problems faced during conventional hybridization including sexual incompatibility, gametic sterility and polyembryony (Grosser and Gmitter, 2011). Somatic hybridization has been widely used for the development of polyploids in lime (Grosser and Gmitter, 2005). Triploid and tetraploid hybrids were developed using electrofusion-based somatic hybridization of protoplasts of ML and clementine (Ollitrault et al., 2000). The potential of citrus polyploids and their development using somatic hybridization, and the use of polyploids in lime crop improvement, have been reviewed (Grosser and Gmitter, 2011). Among fruit quality attributes in polyploids and diploids, mostly the juice contents were affected compared with acidity, solubles and taste. Spontaneous tetraploids recovered from diploid ML had large-sized fruits with dark green leaves and fruit flavour like diploids (Barrett, 1992).

Citrus and its relatives are rich in genes controlling resistance to different biotic and abiotic stress factors. Intergeneric somatic hybrids could be produced using allotetraploidy (Grosser et al., 1996). Protoplasts of ML were fused with leaf protoplasts of sweet orange cv. Valencia to develop allotetraploid hybrid having higher heterosis and vigour (Grosser et al., 1990). These allotetraploid hybrids using various parental combinations with improved traits have revolutionized the lime industry. Somatic hybrids developed from mandarin and RL cultivars with sour orange showed disease tolerance against CTV and bacterial blight (Mendes et al., 2001). Drought-tolerant rootstock was developed by PEG fusion in acid lime cultivars (Mendes-da-Gloria et al., 2000). Rate of embryogenesis was enhanced in somatic hybrids of ML using electrochemical fusion to enhance plant recovery (Olivares-Fuster et al., 2005). A list of recent cybridization and somatic hybridization studies in limes with altered traits is provided in Table 4.

### 6.4.3 Seedlessness

Seedlessness has been one of the most desired traits in citrus including lime. Different approaches have been employed to develop seedless cultivars by interploid hybridization (Usman et al., 2002; Fatima et al., 2002) and somatic hybridization (Grosser and Gmitter, 2009). Odd polyploids (triploids) are usually sterile due to uneven chromosomal distribution in the gametes and could be developed by both methods. In lime triploids, seedlessness has been

**Table 4** Cybridization and somatic hybridization in lime cultivars for crop improvement and seedlessness

| Type of hybridization | Parents | Progeny/traits | References |
|---|---|---|---|
| Cybridization | *C. sinensis* cv. Salustiana + *C. aurantifolia* cv. Mexican lime | - | Olivares-Fuster et al. (2005) |
| Somatic hybridization | Caipira sweet orange + Rangpur lime | Tetraploid/ drought tolerant | Mendes-da-Glória et al. (2000) |
| | Rangpur lime + Cleopatra mandarin | Rootstock improvement | Latado et al. (2002) |
| | Mexican lime + Lakeland limequat, Mexican lime + Valencia sweet orange | Seedless triploid progeny | Grosser and Gmitter (2009) |

attributed to male or female sterility. Triploids also offer higher fruit size and more juice content with a smaller number of seeds. TL is a widely cultivated spontaneous triploid and has larger fruit size than ML (Froelicher, 1999). In TL, parthenocarpic seedlessness is associated with the meiotic chromosomal irregularities indicating triploid origin of the variety (Soost and Cameron, 1980; Vardi et al., 2008). Interploid hybridization of Kinnow mandarin and SL produced less number of developed seeds and higher number of underdeveloped or shrivelled seeds. Use of SL as a maternal parent produced higher number of underdeveloped seeds and triploids (Fatima et al., 2002). A lime breeding programme was initiated in Florida in 2000 to develop hybrids and seedless triploids having enhanced cold tolerance, disease resistance and other novel traits. Out of 650 hybrids developed, 70–80% were triploids. This population could play a major role in future lime crop improvement programmes (Viloria and Grosser, 2005). Seedlessness could also be induced by transforming genes controlling decreased seed setting (SDSL) in to elite lime cultivars (Koltunow et al., 2000).

### 6.4.4 Origin of triploid limes and inheritance patterns

Cytoplasmic pattern of *C. latifolia* was proposed as (*C. limon* × *C. aurantium*) by Bayer et al. (2009) and Froelicher et al. (2011). Curk et al. (2016) analysed different accessions of PL based on nuclear and cytoplasmic patterns and proposed fusion of a 1n lemon ovule by 2n gamete of 2n ML. The diploid gamete might have originated from the Giant Key lime (4n) or as an unreduced gamete (2n). TL, PL and Bears limes are believed as different varieties of a clonal group that was derived from the same hybrid. Details of domestication of PL are provided in previous sections. Cytoplasmic studies revealed that TL has shared cytoplasm of lemon and sour orange (Froelicher et al., 2011). Other groups of seedy triploid limes include 'Coppenrhad', 'Tanepao', 'Mohtasseb' and 'Ambilobe'. These were found to be hybrids of unreduced gamete (2n) from ML fused with haploid (1n) gamete from citron leading to development of these triploids. These varieties were derived from the same hybrid through epigenetic variability and share cytoplasm of ML (Curk et al., 2016). Cytoplasmic and nuclear data analysis demonstrated that both triploid limes including 'Tahiti' (2n *C. aurantifolia* × 1n *C. medica*) and 'Tanepao' (1n *C. limon* × 2n *C. aurantifolia*) are interspecific hybrids involving a common unreduced gamete of ML. The source of unreduced 2n gametes could be doubled-diploid (tetraploid) ML or diploid ML. Since frequency of parental heterozygosity restitution (PHR) was much higher in double diploid ML, triploid TL and Tanepao limes compared with diploid ML, hence it is concluded that the 2n gamete, presumably developed by FDR associated with asynapsis, has more probability to be originated from doubled-diploid ML (Rouiss et al., 2017c). Further studies are suggested that an understanding of the mechanism of development of 2n gametes in diploid ML will demonstrate the definitive origin of spontaneous triploid limes in *C. aurantifolia* and *C. latifolia* developed by natural interploid hybridization.

### 6.4.5 Breeding manipulations of triploid limes: a way forward

The enormous potential of the doubled-diploid ML could be used, and similar parents could be developed to produce unreduced (2n) gametes for the development of triploids such as Tahiti and Tanepao for further utilization as parental materials. Such populations could be screened for novel traits of interest including resistance for disease and insects, improved secondary metabolites and better plant phenology. There is also enormous potential for the development of mutational variants expressing novel traits in ML. Chromosomal doubling restored pollen viability and fertility in tetraploid ML which could be used for developing triploid hybrids. Induced tetraploidy and subsequent hybridization for the development of triploids would be more efficient than searching for spontaneous triploids. Partial apomixis and nucellar embryony in parents and low frequency of natural unreduced gametes are problems in lime breeding for the development of triploids (Cuenca et al., 2015). Selection of non-apomictic female parents could be another option for efficient breeding. Unreduced diploid gametes could transfer 90% of the heterozygosity of ML to the triploid progeny and because of disomic inheritance patterns this would not involve inbreeding depression commonly observed in autotetraploids with tetrasomic inheritance (Gallais, 2003).

### 6.4.6 Meiotic behaviour of tetraploid ML and inheritance patterns

Low temperature could have an important role in polyploidization in citrus as frequency of polyploids have been higher in the NH compared with the SH (Ramsey and Shemske, 1998). Higher polyploidization has also been associated with the presence of apomixis (Horndal, 2006). Low temperature-dependent asynapsis has been shown previously in diploid ML; however, it was absent in tetraploid ML (Iwamasa and Iwasaki, 1963; Rouiss et al., 2017c). Autotetraploids could be distinguished from allotetraploids by determination of tetravalent frequency. In meiosis of tetraploid ML, frequency of bivalents was higher (65%) compared with tetravalents as reported previously in somatic hybrids of citrus (Xie et al., 2015; Rouiss et al., 2017c). Multivalent formation frequency is related to pairing affinity, chromosomal structural variations including length and centromeric position; and may be under genetic control in few species (Jeredi et al., 2012; Cifuentes et al., 2010). Gametophytic sterility reported in diploid ML was attributed to large heterozygous inversion (Iwamasa and Nito, 1988). Pollen viability of diploid ML was low at <10% while it was much higher at 30–64% in tetraploid ML (Pons et al., 2011; Aleza et al., 2012; Rouiss et al., 2017c). Sterility in interspecific diploid hybrids due to lack of chromosomal homology could be fixed by developing allotetraploids which facilitates homologous chromosomal pairing leading to the production of fertile gametes (Contreas et al., 2007). The doubled inversion of arms of the same chromosome in diploid ML occurred due to the structural variability in the parental chromosomes (*C. micrantha* × *C. medica*) of ML which lead to gametic sterility (Carvalho et al., 2005; Curk et al., 2016; Rouiss et al., 2017c).

There are two models for the diploid gametes produced from the tetraploids including disomic inheritance in allotetraploids and tetrasomic in autotetraploids (Sybenga, 2012). In allotetraploids, there were two sets of chromosomes from genetically different genomes. Each chromosome paired only with its homologous counterpart and only bivalents were formed (Sybenga, 2012). This led to disomic inheritance patterns and higher interspecific heterozygosity (Stift et al., 2008). In autotetraploids, the multivalents were formed due to the presence of four homologous chromosomes (Sybenga, 1996). Strict disomic inheritance was reported in tetraploid *Clausena excavata* (Froelicher et al., 2000) while it had preferential patterns in tetraploid ML with 90% PHR which was much higher (Rouiss et al., 2017c) compared with previous reports in different citrus somatic hybrids (Xie et al., 2015; Aleza et al., 2016b).

Most of the interspecific hybrids in citrus have higher gametic fertility compared with ML indicating higher chromosomal pairing affinity and little variation (Ollitrault and Navarro, 2012). In tetraploids, chromosomal pairing options were available which were absent in diploids hence there were chances of having preferential pairing. Two distant parental species were involved in ML and each locus had higher heterozygosity. Molecular studies have revealed a distant relationship and higher genetic diversity in *C. medica* compared with other primary species (Curk et al., 2015; Cabonell-Caballero et al., 2015). Hence, it was presumed that preferential pairing in tetraploid ML may be regulated by variation in genomic sequence and chromosomal structure in distantly related parents (*C. micrantha* × *C. medica*) leading to preferential disomic inheritance. Furthermore, it was evident that overall chromosomal differentiation has been more effective in regulation of pairing than local structural variations like inversion as reported in diploid ML (Rouiss et al., 2017c). Nuclear and chloroplast genome phylogenetic analysis indicated admixture of round lime with the Australian finger limes and the nearest relative to Australian citrus was Fortunella (Wu et al., 2018). Further investigation is required for more understanding of the phylogenetic origin in major lime varieties for better exploitation of the biodiversity for the development of new scion and rootstock varieties.

In conclusion, the lime group is highly genetically polymorphic, has diploid and polyploid varieties, has maternal traits from four out of six cytoplasm types in citrus and nuclear genome contribution from all the four different primary species. Citron has been a common taxon in all the lime varieties; however, it has never been a female parent and is highly homozygous due to cleistogamy. Origin of the triploid groups in lime was not due to the sexual combinations, rather by fusion of haploid gamete from one parent and diploid gamete from the other parent. The low subgroup diversity has been attributed to mutations, epigenetic variation, transposable elements and chromosomal deletions.

## 7 Molecular approaches in crop improvement

Advances in molecular techniques including NGS tools have made it possible to shorten the breeding cycle in fruit trees and identify promising lines much earlier for further use. Modern molecular techniques that have been widely used in citrus include multiomic approaches (genomics, proteomics, transcriptomics), marker-assisted breeding, linkage mapping, bioinformatics and transgenics (Guo et al., 2015; Shiratake and Suzuki, 2016; Rouiss et al., 2017b). These tools help in the identification of candidate genes, proteins and different metabolites that could be used for future crop improvement programmes.

### *7.1 Molecular markers and applications*

Marker-assisted selection and breeding are very useful tools to gear up the process of varietal selection and development for crop improvement programmes in lime. Polymorphic DNA markers linked to genes of interest have been identified in segregating populations (Fang and Roose, 1997). Molecular markers help in more precise selection of the plant material for the desired trait, reduce volume of work and enhance efficiency of the breeding programme. Genetic markers have been used in different lime species for characterization. Molecular markers are being widely utilized in citrus for genetic diversity assessment, identification of zygotic and nucellar seedlings, classification and genetic organization, identification of 2n gametes and polyploids (Usman and Fatima, 2018).

PCR-based markers including simple sequence repeats (SSRs), inter-simple sequence repeats (ISSRs), amplified fragment length polymorphism (AFLPs) and cleaved amplified polymorphisms have been used in citrus (Fang et al., 1997). However, in lime species, reports of marker-assisted selection and breeding have been relatively limited (Ali and Nawaz, 2017). Use of molecular markers for varietal characterization has been reported (Munankarmi et al., 2014; Curk et al., 2016). Among different molecular markers, AFLP, SSRs and ISSRs have been the most commonly used for characterization of lime cultivars compared with other marker types (Khaivi et al., 2015; Sharafi et al., 2016; Munankarmi et al., 2018; Rouiss et al., 2017c). A detailed overview of the markers used in different lime species are provided in Table 5.

**Table 5** Molecular markers used in lime species for genetic diversity assessment and inheritance patterns in polyploids

| Cultivars/Species | Markers | Primer series/ sequence | No of primers used | Optimum primers | Total No of bands | No of polymorphic bands/loci | Polymor-phism (%) | Purpose | Country | No of accessions studied | References |
|---|---|---|---|---|---|---|---|---|---|---|---|
| Limes | AFLP | FAM-6-labelled EcoRI-AXX | 17 | 4 | 193 | 191 | 98.9 | Genetic diversity for Witch's broom disease | Oman | 143 | Al-Sadi et al. (2012) |
| | RAPD | Operon series | 10 | 3 | 76 | 35 | 46 | Genetic diversity | India | 6 | Kumar et al. (2013) |
| Acid lime | SSR | (Barkley et al., 2006; Jannati et al., 2009) | 12 | 11 | 33 | 15 | 50 | Genetic diversity | Nepal | 62 | Shrestha et al. (2012) |
| | ISSR | (Shahsawar et al., 2007) | 49 | 21 | 234 | 204 | 87.18 | Genetic diversity | Nepal | 60 | Munankarmi et al. (2018) |
| Acid lime landraces | RAPD | UBC primers | 40 | 9 | 79 | 75 | 94.94 | Genetic diversity | Nepal | 60 | Munankarmi et al. (2014) |
| Tahiti lime, Kaghzi lime and sweet lime | RAPD | Invitrogen | 25 | 15 | 190 | 155 | 81.5 | Genetic variation | Pakistan | 17 | Naz et al. (2014) |
| Mexican lime | AFLP | - | - | - | - | 126 | 69.84 | Genetic diversity | Iran | 36 | Khaivi et al. (2015) |
| | SSR | CMS (Ahmad et al., 2003; Kijas et al., 1997) | (29), | 9 | 49 | 49 | 100 | To evaluate molecular variation and genetic relatedness | Iran | 28 | Sharafi et al. (2016) |
| | ISSR | (Awasthi et al., 2008) | 9 | 9 | 148 | 131 | 88,2 | Molecular variation and genetic relatedness | Iran | 28 | Sharafi et al. (2016) |
| | A2/J5 primer pairs | (Hocquellet et al., 1999) | - | - | - | - | 71.8% disease | Identify disease incidence (Huanglongbing) | | 116 | Al Fahdi et al. (2018) |
| Persian lime | ISSR | (Tripolitsiotis et al., 2013) | 14 | 8 | 141 | 139 | 99 | Genetic diversity in Persian lime | Iran | 26 | Khankahdani et al. (2017) |
| ClemMex tetraploid hybrids (Clementine × doubled-diploid Mexican lime) | SSRSNP | (Curk et al., 2015; Ollitrault et al., 2012) | 278 | 2 | - | - | - | Chromosomal inheritance patterns | France | 85 | Rouiss et al. (2017c) |

### *7.1.1 Phylogenetic origin of limes and molecular markers*

Identifying the origin of lime taxa has been complex and conflicts in classification have been due to higher sexual compatibility with other citrus species and apomixis. These factors led towards development of interspecific hybrids fixed by clonal propagation or apomixis as separate species (Barrett and Rhodes, 1976; Curk et al., 2016). Molecular marker studies have revealed that all citrus species are the outcome of interspecific hybridization between four groups including mandarin, pummelo, citron and papeda (Garcia-Lor et al., 2013; Ollitrault et al., 2012; Curk et al., 2015). Four ancestral species having diploid and polyploid sections were also involved in limes making this a genetically complex group compared with other citrus species. Four out of six types of different cytoplasm present in citrus have been identified in limes indicating the involvement of four maternal ancestors in the phylogeny. In lime species, *C. latifolia*, *C. limetta*, and lemons CT was like *C. aurantium* indicating the presence of same mitotype in limes and lemons (Froelicher et al., 2011; Carbonell-Caballero et al., 2015; Curk et al., 2015, 2016). Both mitochondrial and chloroplast DNA analysis showed maternal phylogeny of *C. maxima* similar to *C. limettioides*. Similarly, CT found in *C. micrantha* was like all accessions of *C. aurantifolia* and *C. macrophylla* as proposed previously based on organelle DNA (Froelicher et al., 2011; Penjor et al., 2013; Curk et al., 2015).

Molecular marker (SSR) and chloroplast analysis revealed equal contribution of papeda and citron in ML (Ramadugu et al., 2013, 2017; Carbonell-Caballero et al., 2015). Sequence information of the nuclear and organelle genomes also revealed ML as hybrid of papeda and citron (Curk et al., 2016; Wang et al., 2017). Sequencing data of nuclear and organelle DNA revealed sour orange as the maternal parent with other species such as *C. medica*, *C. maxima*, *C. reticulata* and *C. micrantha*. Variety PL have been developed by fertilization of unreduced gamete of ML with haploid ovule of lemon (Curk et al., 2015, 2016; Wang et al., 2017). In SL, marker-based admixture analysis revealed it as citron (60%) and papeda (28%) hybrid (Barkley et al., 2006). SNP analysis of nuclear genes revealed it as citron-mandarin hybrid (Ramadugu et al., 2013) while chloroplast (CPT) sequences were similar to pummelo (Curk et al., 2016). In RL, mandarin was its maternal parent as revealed by RFLP analysis of CPT DNA (Gulsen and Roose, 2001). SSR analysis revealed its higher genomic similarity to citron (Barkley et al., 2006). Complete CPT sequence analysis confirmed Cleopatra mandarin as the maternal parent of RL (Carbonell-Caballero et al., 2015; Curk et al., 2016). Analysis of nuclear and organelle genes revealed RL as a hybrid of mandarin and citron (Curk et al., 2015, 2016; Wang et al., 2017).

Nuclear analysis of limes based on SNPs, indels and SSRs revealed higher heterozygosity compared with other basic taxa and this diversity was the outcome of different sexual recombinations that occurred during the evolutionary process (Gulsen and Roose, 2001; Barkley et al., 2006; Curk et al., 2016). All lime accessions showed contribution from *C. Medica*; however, there was no maternal phylogeny indicating role of citron as a male parent (Froelicher et al., 2011; Luro et al., 2012; Curk et al., 2015, 2016).

The origin of citron was thought to be India (Jena et al., 2009); however, chloroplast phylogenetic analysis revealed the Australasian region as the origin of citron. Nevertheless, India has been a major centre of diversity for citron (Malik et al., 1974). Curk et al. (2016) has proposed *C. micrantha* x *C. medica* model for *C. aurata*, *C. macrophylla*, *C. excelsa* and *C. aurantifolia* (ML) supported by previous reports based on cytogenetic data (Carvalho et al., 2005) and molecular markers (Yamamoto, 2007; Ollitrault et al., 2012; Garcia-Lor et al., 2013). The origin of RL is proposed to be mandarin x citron. Half of the RL karyotypes were similar to haploid chromosome set of *C. medica* while the other half was contributed from Cleopatra mandarin (Curk et al., 2016). *C. limettioides* is proposed to be a hybrid of *C. limon* and *C. aurantium*.

## 7.2 Bioinformatics and genomics

Advances in molecular biology tools and NGS approaches have facilitated the work of molecular biologists, reduced time and effort involved in genomics, proteomics and transcriptomic investigations. In citrus the genome of *C. sinensis* was published first (Xu et al., 2013) followed by the genome of several other citrus species including *C. medica* (Wang et al., 2017), male parent of diploid (ML, Excelsa lime, RL) and triploid (Tanepao, Copperrath, Mohtasseb and Ambilobe) limes. The availability of the genome sequence of citron has opened new horizons and will facilitate genome sequencing of lime species which are not available yet. Chloroplast genome of many citrus species have been studied including Omani lime (*C. aurantifolia*) which was found to be 159 kbp long (Su et al., 2014). Efficient utilization of the available biodiversity in emerging innovative breeding programmes will require a better understanding of the phylogeny and genomic structures of the lime and related species. Details of available genomic databases and other information are provided in Table 6.

## 7.3 Gene expression and transgenics

### 7.3.1 Selection of reference genes

Gene expression analysis using quantitative PCR is a very useful technology to quantify gene expression patterns. For data normalization, reference genes are required which remain stable across variable treatments, growing conditions and stressful environment. More than one stable reference gene is used for relative expression studies. In *C. aurantifolia* and other citrus species' gene

**Table 6** Lime species genomic database and related information

| Species | EST | Genes | Proteins | Nucleotides | *No of sequences | Genome size | Database web links/reference |
|---|---|---|---|---|---|---|---|
| *C. aurantifolia* | 8274 | 138 | 40, 394 | 2786 | 868 | – | https://www.ncbi.nlm.nih.gov/search/?term=citrus+aurantifolia*https://www.citrusgenomedb.org/search/species/list |
| *C. latifolia* | 8756 | 0 | 173 | 266 | 42 | – | https://www.ncbi.nlm.nih.gov/search/?term=citrus+latifolia |
| *C. limettioides* | 604, 879 | 80182 | 617, 672 | 457, 863 | 130 | – | https://www.ncbi.nlm.nih.gov/search/?term=citrus+limettoides |
| *C. limetta* | 0 | 0 | 8 | 19 | 27 | – | https://www.ncbi.nlm.nih.gov/search/?term=citrus+limetta |
| *C. micrantha* | | | 3 | 8 | 39 | – | https://www.ncbi.nlm.nih.gov/search/?term=citrus+micrantha |
| *C. medica* | 1137 | – | 312 | – | – | 406.058 Mb | Genome sequence (https://www.ncbi.nlm.nih.gov/nuccore/MKYO00000000.2)Wang et al. (2017) |

* https://www.citrusgenomedb.org/search/species/list

expression studies, only one gene *18S rRNA* has been consistently used as a reference gene to normalize the gene expression data of several genes (Mardi et al., 2015; Nejat et al., 2015). Little information is available about the expression stability of the selected reference genes. Available citrus genome sequences of *C. sinensis, C. clementina* and *C. medica* are quite helpful for functional genomics in lime species. Among the most damaging diseases in acid limes, 'Witches Broom Disease' and 'Sudden Decline' are caused by *Candidates* Phytoplasma aurantifolia (Silva et al., 2015). Alves et al. (2018) analysed the expression patterns of 14 different genes in ML (*C. aurantifolia*) infected Phytoplasma and used different data normalization algorithms to study the expression stability of the reference genes. No marked differences were observed in the top five across different algorithms including NormFinder, GeNormPlus and BestKeeper. In ML against Phytoplasma infection *PAT1* and *ALPHA-TUBULIN* or *UBI1* could be used as reference genes for stable gene expression and data normalization (Alves et al., 2018).

### *7.3.2* Ruby *gene diversity and anthocyanin production in limes*

Phenolic compounds are produced in most of the angiosperm species and are responsible for developing red, purple and blue colours in different plant organs including leaves, flower petals, fruit peel and flesh (Winkle-Shirley, 2001). Anthocyanin biosynthesis is regulated by various genes including *Ruby* which encodes for transcription factor required for the anthocyanin accumulation in citrus. *Ruby* locus contains MYB transcription factors including two regulators (*Ruby* and *Ruby2*) that regulate anthocyanin biosynthesis (Koes et al., 2005; Tanaka et al., 2008). Any mutation in the MYB regulatory genes alters expression of the biosynthesis genes and anthocyanin pigment development. Functionality of only MYB factor in citrus indicates lack of stringent selection pressure to produce anthocyanins. In limes and lemons, purple pigment is produced in young leaves and flower petals while other citrus species did not produce anthocyanins in these organs. Butelli et al. (2017) reported great variation in *Ruby* in citrus and related species including large deletions, insertions of transposons, translocation and other critical mutations that have altered the anthocyanin accumulation in different tissues.

Citron is involved in parentage of limes including both ML and PL. The *Ruby* allelic composition of ML was similar to the female parent Papeda (Nicolosi et al., 2000). Similarly, *Ruby* allele of lemon was found in PL confirming its parentage (Lemon × ML) as proposed by Curk et al. (2016). The *Ruby* allelic composition of SL is contrary to the parental contribution of orange (Nicolosi et al., 2000) and agrees with the parentage (pummelo × mandarin) proposed by Curk et al. (2016) based on molecular marker analysis. Sour orange and citron are parents of *C. limetta* and sour orange *Ruby* allelic composition was noticed in *C. limetta* (Curk et al., 2016). Parentage of RL suggested by *Ruby* allelic composition was mandarin × citron (Butelli et al., 2017).

Transposons also play a great role in the genomic alterations and gene expression changes during the evolutionary process. In *C. micrantha*, Tcm7 transposon presence associated with a deletion eliminated anthocyanin accumulation in fresh vegetative growth. Anthocyanins play an important role in protection from high light and UV irradiation and the expression of *Ruby* is also associated with light. The *Ruby* expression has been changed in different citrus species according to the climatic conditions of that area (Steyn et al., 2002). The *Ruby* gene expression was different in different citrus species and the changes could be attributed to mutations during the process of evolution.

### *7.3.3 Transgenics*

Tissue regeneration and plant transformation efficiency have been relatively low in most of the citrus species compared with other fruit crops for a long time; however, in lime species regeneration and transformation efficiency have been higher and genes have been transformed against abiotic and biotic stress (Dominguez et al., 2000; Ghorbel et al., 2001; Pena et al., 2004). Further, short juvenility period (2–3 years) of the ML compared with other citrus species (4–10 years) is another advantage facilitating early bearing to test the efficiency of transgene expression (Yang et al., 2007).

Genes controlling decreased seed setting (SDSL-1) were introduced into West Indian lime and fruit with low seed set were obtained. However, the transformation efficiency remained low at 3% (Koltunow et al., 2000) which needs to be enhanced. Transformation stability was enhanced with three extra copies of the transgenes pTiBo542, EHA 105 and virG in lime plants (Pena et al., 2004). Different antioxidants including lipoic acid, glycine betane, glutathione and poly vinylpyrrolidone were tested to enhance efficiency of *Agrobacterium*-mediated transformation in ML. Addition of lipoic acid in the media enhanced callus induction and shoot growth in epicotyl explants while other antioxidants did not show any marked effect on tissue regeneration (Dutt et al., 2011). Recently, cv. ML has been biofortified by transforming anthocyanin synthesis regulating genes including *VvmybA1* (grapes) and *Ruby* (grapefruit). This research venture has opened new horizons to produce pigmented lime and other citrus varieties which may be more helpful in controlling cardiac diseases and cancer due to higher accumulation of anthocyanins and other related pigments. The transcription factors such as CitVHA-c4 and CitERF13 are responsible for citric acid accumulation in citrus and other model crops and could be used as candidate genes for enhanced citric acid accumulation in limes and other acidic citrus species. A detailed overview of progress in transgenics is provided in Table 7.

**Table 7** Overview of genetic transformation studies in lime species

| Specie/variety | Method of transformation | Genes involved | Progeny/traits | References |
|---|---|---|---|---|
| Mexican lime (*C. aurantifolia*) | *Agrobacterium tumefaciens* | GUS | Abiotic stress tolerance | Pena et al. (1997) |
| | *A. tumefaciens* | *CTV-CP* | CTV disease and insect resistance | Dominguez et al. (2000) |
| | RNAi | *p25* | CTV disease and insect resistance | Dominguez et al. (2002a,b) |
| | *A. tumefaciens* | *rolC*, *AtSUC2*, *RsS1*, RTBV | Citrus greening disease | Dutt et al. (2012) |
| | *A. tumefaciens* | *VvmybA1* or *Ruby* | Flower and fruit pigmentation (anthocyanin) | Dutt et al. (2016) |
| West Indian lime (*C. aurantifolia*) | *A. tumefaciens* | *SDSL-1* | Decreased seed setting | Koltunow et al. (2000) |
| Rangpur lime (*C. limonia*) | *A. tumefaciens* | *bO* (Bacterio-opsin) | Resistance against *Phytophthora* | Azevedo et al. (2006) |

In RL and ML species, bacterio-opsin and coat protein gene of CTV were transformed to provide disease and insect resistance (Dominguez et al., 2002a,b; Azevedo et al., 2006). Acid limes, *P. trifoliata* and pummelos are known to have better resistance against citrus greening disease (CGD; Knapp et al., 2004; Ali and Nawaz, 2017). These species could be used as parental material for breeding programmes against CGD.

## 8 Future trends and conclusion

Available land races of ML in the Indo-Pakistan region and elsewhere should be collected, characterized, conserved and superior lines such as Kaghzi lime should be propagated. Interplantation of SL with grapefruit and sweet orange cultivars may enhance fruit size and quality. Old and less productive cultivars should be replaced with new varieties having better yield, seedlessness and higher-disease and insect pest resistance including hybrids recently released by the United States and introduction of coloured ML. New promising rootstocks should be used for propagation. The available genomic database should be used to develop early-bearing and high-yielding scion varieties with higher resistance to biotic and abiotic stress. International germplasm exchange programmes should be further strengthened. Gene knockout and expression-related studies using *CitRWP* and other related genes should be undertaken under different agroclimatic conditions for the development of better cultivars. Development of tetraploid plant material using somatic doubling and somatic hybridization strategies could be useful for the development of seedless varieties. Induced parthenogenesis could be used for the development of haploids along with pollen culture *in vitro*. Mutation breeding followed by NGS technologies could be quite effective in shortening breeding cycles and help in early detection of desired mutants having novel traits. Use of DNA markers associated with economically important horticultural traits could be highly effective in varietal characterization and marker-assisted breeding. Availability of efficient *in vitro* regeneration systems in elite cultivars could be helpful in the development of transgenic plant material against CTV and CGD. Cutting-edge biotechnology tools should be utilized for crop improvement in limes and other related citrus species.

## 9 Where to look for further information

*Book sources*

- Khan, M. M., Al-Yahyai, R. and Al-Said, F. 2017. *The Lime: Botany, Production and Uses*. CAB International, Wallingford, UK.
- Khan, I. A. 2007. *Citrus Genetics, Breeding and Biotechnology*. CAB International, Wallingford, UK.

- Cuenca, J., Garcia-Lor, A., Navarro, L. and Aleza, P. 2018. Citrus genetics and breeding. In: Al-Khayri, J. M., et al. (Eds), *Advances in Plant Breeding Strategies: Fruits*. Springer, Cham, pp. 403–36.
- Ali, M. A. and Nawaz, M. A. 2017. Advances in lime breeding and genetics. In: Khan, M. M., Al-Yahyai, R. and Al-Said, F. (Eds), *The Lime Botany, Production and Uses*. CAB International, Wallingford, UK, pp. 37–53. Chapter 3.


*Review and research articles*


- Eng, W. H. and Ho, W. S. 2019. Polyploidization using colchicine in horticultural plants: a review. *Sci. Hortic.* 246, 604–17.
- Wu, G. A., Terol, J., Ibanez, V., López-García, A., Pérez-Román, E., Borredá, C., Domingo, C., Tadeo, F. R., Carbonell-Caballero, J., Alonso, R. and Curk, F. 2018. Genomics of the origin and evolution of Citrus. *Nature* 554(7692), 311.
- Rouiss, H., Bakry, F., Froelicher, Y., Navarro, L., Aleza, P. and Ollitrault, P. 2017a. Origin of *C. latifolia* and *C. aurantiifolia* triploid limes: the preferential disomic inheritance of doubled-diploid 'Mexican' lime is consistent with an interploid hybridization hypothesis. *Ann. Bot.* 121(3), 571–85.
- Curk, F., Ollitrault, F., Garcia-Lor, A., Luro, F., Navarro, L. and Ollitrault, P. 2016. Phylogenetic origin of limes and lemons revealed by cytoplasmic and nuclear markers. *Ann. Bot.* 117(4), 565–83.


*Expertise and genetic resource centres*

- Fruit Research Institute, Fujian Academy of Agricultural Sciences, China (http://www.cgris.net/China/CHINA%20ex%20situ.htm)
- Citrus variety collections, University of California, Riverside (http://www.citrusvariety.ucr.edu/citrus/limes.html)
- USDA-ARS National Clonal Germplasm Repository (NCGRCD) (https://www.ars.usda.gov/pacific-west-area/riverside-ca/national-clonal-germplasm-repository-for-citrus), (https://www.ars.usda.gov/oc/np/phenolics/comp-1/lime/)
- CRA, Consiglio per La Ricerca e La Sperimentazioni In Agricoltura, Italy (http://sito.entecra.it/portale/cra_dati_istituto.php?id=209)
- Russian Research Institute of Floriculture and sub-tropical fruit crops (RRIFSC) (https://www.greencom.ru/en/firm_info.html/fID/4277)
- NARO Institute of Fruit Tree Science, Tsukuba, Japan (https://www.naro.affrc.go.jp/org/fruit/eng/index-e.html)
- INRA-CIRAD citrus germplasm collection, Corsica, France (https://urgi.versailles.inra.fr/siregal/siregal/welcome.do)

## 10 References


Agarwal, P. K. 1987. Cytogenetical investigation in Rutaceae. I. Meiotic studies in four citrus species of hybrid origin. *Cytologia* 52(4), 753–6. doi:10.1508/cytologia.52.753.

Ahmad, R., Struss, D. and Southwick, S. M. 2003. Development and characterization of microsatellite markers in *Citrus*. *J. Am. Soc. Hortic. Sci.* 128(4), 584–90. doi:10.21273/JASHS.128.4.0584.

Aleza, P., Juárez, J., Ollitrault, P. and Navarro, L. 2009. Production of tetraploid plants of non-apomictic citrus genotypes. *Plant Cell Rep.* 28(12), 1837–46. doi:10.1007/s00299-009-0783-2.

Aleza, P., Juárez, J., Cuenca, J., Ollitrault, P. and Navarro, L. 2010. Recovery of citrus triploid hybrids by embryo rescue and flow cytometry from 2x × 2x sexual hybridisation and its application to extensive breeding programs. *Plant Cell Rep.* 29(9), 1023–34. doi:10.1007/s00299-010-0888-7.

Aleza, P., Froelicher, Y., Schwarz, S., Agustí, M., Hernández, M., Juárez, J., Luro, F., Morillon, R., Navarro, L. and Ollitrault, P. 2011. Tetraploidization events by chromosome doubling of nucellar cells are frequent in apomictic citrus and are dependent on genotype and environment. *Ann. Bot.* 108(1), 37–50. doi:10.1093/aob/mcr099.

Aleza, P., Juárez, J., Cuenca, J., Ollitrault, P. and Navarro, L. 2012. Extensive citrus triploid hybrid production by 2x x 4x sexual hybridizations and parent-effect on the length of the juvenile phase. *Plant Cell Rep.* 31(9), 1723–35. doi:10.1007/s00299-012-1286-0.

Aleza, P., Cuenca, J., Juárez, J., Navarro, L. and Ollitrault, P. 2016a. Inheritance in doubled-diploid clementine and comparative study with SDR unreduced gametes of diploid clementine. *Plant Cell Rep.* 35(8), 1573–86. doi:10.1007/s00299-016-1972-4.

Aleza, P., Garcia-Lor, A., Juárez, J. and Navarro, L. 2016b. Recovery of citrus cybrid plants with diverse mitochondrial and chloroplastic genome combinations by protoplast fusion followed by *in vitro* shoot, root, or embryo micro grafting. *Plant Cell Tissue Organ Cult.* 126(2), 205–17. doi:10.1007/s11240-016-0991-8.

Al Fahdi, A., Al-Mamari, A., Shahid, M. S., Sajeewa, S. S. N., Carvalho, C. M., Elliot, S. L. and Al-Sadi, A. M. 2018. Characterization of Huanglongbing disease associated with acid lime (*Citrus aurantifolia* Swingle) in Oman. *J. Plant Pathol.* 100(3), 419–27. doi:10.1007/s42161-018-0088-9.

Ali, M. A. and Nawaz, M. A. 2017. Advances in lime breeding and genetics. In: Khan, M. M., Al-Yahyai, R. and Al-Said, F. (Eds), *The Lime Botany, Production and Uses*. CAB International, Wallingford, Oxfordshire, UK, pp. 37–53. Chapter 3.

Allario, T., Brumos, J., Colmenero-Flores, J. M., Tadeo, F., Froelicher, Y., Talon, M., Navarro, L., Ollitrault, P. and Morillon, R. 2011. Large changes in anatomy and physiology between diploid Rangpur lime (*Citrus limonia*) and its autotetraploid are not associated with large changes in leaf gene expression. *J. Exp. Bot.* 62(8), 2507–19. doi:10.1093/jxb/erq467.

Allario, T., Brumos, J., Colmenero-Flores, J. M., Iglesias, D. J., Pina, J. A., Navarro, L., Talon, M., Ollitrault, P. and Morillon, R. 2013. Tetraploid Rangpur lime rootstock increases drought tolerance via enhanced constitutive root abscisic acid production. *Plant Cell Environ.* 36(4), 856–68. doi:10.1111/pce.12021.

Almaguer-Vargas, G., Espinoza-Espinoza, J. R. and Quirós-García, J. L. 2011. Desfasamiento de cosecha de limón persa. *Rev. Chapingo. Ser. Hortic.* 17(3), 197–205.

Al-Sadi, A. M., Al-Moqbali, H. S., Al-Yahyai, R. A. and Al-Said, F. A. 2012. AFLP data suggest a potential role for the low genetic diversity of acid lime (*Citrus aurantifolia* Swingle) in Oman in the outbreak of witches' broom disease of lime. *Euphytica* 188(2), 285-97. doi:10.1007/s10681-012-0728-7.

Alves, M. S., Al-Sadi, A. M. and Carvalho, C. M. 2018. Selection of reference genes for quantitative PCR analysis in *Citrus aurantifolia* during phytoplasma infection. *Trop. Plant Pathol.* 43(5), 402–12. doi:10.1007/s40858-018-0224-2.

Auger, D. L., Gray, A. D., Ream, T. S., Kato, A., Coe, E. H. and Birchler, J. A. 2005. Non additive gene expression in diploid and triploid hybrids of citrus. *Genetics* 169(1), 389-97. doi:10.1534/genetics.104.032987.

Awasthi, A. K., Kar, P. K., Srivastava, P. P., Rawat, N., Vijayan, K., Pradeep, A. R. and Urs, S. R. 2008. Molecular evaluation of bivoltine, polyvoltine and mutant silkworm (*Bombyx mori* L.) with RAPD, ISSR and RFLPSTS markers. *Indian J. Biotechnol.* 7, 188–94.

Azevedo, F. A., Filho, F. M., Mendes, B. M. J., Almeida, W. A. B., Schinor, E. H., Pio, R., Barbosa, J. M., Gonzalez, S. G., Carrer, H. and Lam, E. 2006. Genetic transformation of Rangpur lime (*Citrus limonia* Osbeck) with the bO (*bacterio-opsin*) gene and its initial evaluation for *Phytopthora nicotianae* resistance. *Plant Mol. Biol. Rep.* 24, 185–96.

Bachi, O. 1940. Observaceous Citologicalem *Citrus*. I. Numero de cromosomas de algunas especies y variedades. *J. Agron.* 3, 249–58.

Barba-Gonzalez, R., Lim, K. -B., Ramanna, M. S., Visser, R. G. F. and Van Tuyl, J. M. 2005. Occurrence of 2n gametes in the F1 hybrids of Oriental x Asiatic lilies (*Lilium*): relevance to intergenomic recombination and backcrossing. *Euphytica* 143(1–2), 67–73. doi:10.1007/s10681-005-2657-1.

Barkley, N. A., Roose, M. L., Krueger, R. R. and Federici, C. T. 2006. Assessing genetic diversity and population structure in a citrus germplasm collection utilizing simple sequence repeat markers (SSRs). *Theor. Appl. Genet.* 112(8), 1519- 31. doi:10.1007/s00122-006-0255-9.

Barrett, H. 1992. An autotetraploid of the key lime, *Citrus aurantifolia*. *Fruit Var. J.* 46, 166–70.

Barrett, H. C. and Rhodes, A. M. 1976. A numerical study of affinity relationships in cultivated Citrus and its close relatives. *Syst. Bot.* 1, 105–36. doi:10.2307/2418763.

Batygina, T. B. and Vinogradova, G. Y. 2007. Phenomenon of polyembryony. Genetic heterogenity of seeds. *Russ. J. Dev. Biol.* 38(3), 126–51. doi:10.1134/S1062360407030022.

Bayer, R. J., Mabberley, D. J., Morton, C., Miller, C. H., Sharma, I. K., Pfeil, B. E., Rich, S., Hitchcock, R. and Sykes, S. 2009. A molecular phylogeny of the orange subfamily (Rutaceae: Aurantioideae) using nine cpDNA sequence. *Am. J. Bot.* 96(3), 668–85. doi:10.3732/ajb.0800341.

Bove, J. M. and Ayres, A. J. 2007. Etiology of three recent diseases of citrus in Sao Paulo State: sudden death, variegated chlorosis and huanglongbing. *IUBMB Life* 59(4–5), 346–54. doi:10.1080/15216540701299326.

Butelli, E., Garcia-Lor, A., Licciardello, C., Casas, G. L., Hill, L., Recupero, G. R., Keremane, M. L., Ramadugu, C., Krueger, R., Xu, Q., et al. 2017. Changes in anthocyanin production during domestication of Citrus. *Plant Physiol.*, 173(4), 2225–42. doi:10.1104/pp.16.01701.

Cabonell-Caballero, J., Alonso, R., Ibanez, V., Terol, J., Talon, M. and Dopazo, J. 2015. Aphylogenetic analysis of 34 chloroplast genomes elucidates the relationship between wild and domestic species within the genus Citrus. *Mol. Biol. Evol.* 32(8), 2015–35. doi:10.1093/molbev/msv082.

Carvalho, R., Filho, W. S. S., Brasileiro-Vidal, A. C. and Guerra, M. 2005. The relationship among lemons, limes and citron: a chromosomal comparison. *Cytogenet. Genome Res.* 109(1–3), 276–82. doi:10.1159/000082410.

Cifuentes, M., Grandont, L., Moore, G., Chèvre, A. M. and Jenczewski, E. 2010. Genetic regulation of meiosis in polyploid species, new insights into an old question. *New Phytol.* 186(1), 29–36. doi:10.1111/j.1469-8137.2009.03084.x.

Cimen, B. and Yesiloglu, T. 2016. Rootstock breeding for abiotic stress tolerance in citrus. In: Shanker, A. K. and Shanker, C. (Eds), *Abiotic and Biotic Stress in Plants - Recent Advances and Future Perspectives*. InTech Publishers, Rijeka, Croatia, pp. 527–63. Chapter 23.

Conner, J. A., Mookkan, M., Huo, H., Chae, K. and Ozias-Akins, P. 2015. A parthenogenesis gene of apomict origin elicits embryo formation from unfertilized eggs in a sexual plant. *Proc. Natl. Acad. Sci. U. S. A.* 112(36), 11205–10. doi:10.1073/pnas.1505856112.

Contreras, R. N., Ranney, T. and Tallury, S. P. 2007. Reproductive behavior of diploid and allotetraploid *Rhododendron* L., 'fragrant affinity',. *HortScience*. 42, 31–4.

Crane, J. H. and Osborne, J. L. 1991. *Growing 'Tahiti' Limes in the Home Landscape*. IFAS, UF, Gainesville, FL.

Cuenca, J., Aleza, P., Jua´rez, J., Froelicher, Y., Garcı´a-Lor, A., Navarro, L. and Ollitrault, P. 2015. Maximum-likelihood method based on parental heterozygosity restitution of centromeric loci identifies Second Division Restitution (SDR) as the predominant mechanism leading to 2n megagametophytes in *Citrus reticulata*. *Sci. Rep.* 5, 9897. doi:10.1038/srep09897.

Curk, F., Garcia-Lor, A., Luro, F., Snoussi-Trifa, H., Froelicher, Y., Ancillo, G., Navarro, L. and Ollitrault, P. 2015. New insights on limes and lemons origin from nuclear and cytoplasmic markers genotyping and targeted nuclear gene sequencing. *Acta Hortic.* 1065, 135–46.

Curk, F., Ollitrault, F., Garcia-Lor, A., Luro, F., Navarro, L. and Ollitrault, P. 2016. Phylogenetic origin of times and lemons revealed by cytoplasmic and nuclear markers. *Ann. Bot.* 117(4), 565–83. doi:10.1093/aob/mcw005.

Davenport, T. L. 1990. Citrus flowering. *Hortic. Rev.* 12, 349–408.

Davies, F. S. and Albrigo, L. G. 1994. *Citrus*. CAB International, Wallingford, UK.

De Storme, N. and Geelen, D. 2013. Sexual polyploidization in plants–cytological mechanisms and molecular regulation. *New Phytol.* 198(3), 670–84. doi:10.1111/nph.12184.

Dominguez, A., Guerri, J., Cambra, M., Navarro, L., Moreno, P. and Pena, L. 2000. Efficient production of transgenic citrus plants expression the coat protein gene of citrus tristeza virus. *Plant Cell Rep.* 19(4), 427–33. doi:10.1007/s002990050751.

Dominguez, A., de Mendoza, A. H., Guerri, J., Cambra, M., Navarro, L., Moreno, P. and Pena, L. 2002a. Pathogen derived resistance to citrus tristeza virus (CTV) in transgenic Mexican lime (*Citrus aurantifolia* (Christ.) Swing.) plants expressing its p25 coat protein gene. *Mol. Breed.* 10, 1–10.

Dominguez, A., Fagoaga, C., Navarro, L., Moreno, P. and Pena, L. 2002b. Regeneration of transgenic citrus plants under non selective conditions results in high frequency recovery of plants with silenced transgenes. *Mol. Genet. Genomics* 267(4), 544–56. doi:10.1007/s00438-002-0688-z.

Duportal, M., Jorda, E., Sanchez, C., Imbert, É., Loeillet, D. and Vannière, H. 2013. FruiTrop FOCUS Citron. *FruiTrop Cirad* 2013, 140.

Dutt, M., Vasconcellos, M. and Grosser, J. W. 2011. Effects of antioxidants on *Agrobacterium*-mediated transformation and accelerated production of transgenic plants of Mexican lime (*Citrus aurantifolia* Swingle). *Plant Cell Tissue Organ Cult.* 107(1), 79–89. doi:10.1007/s11240-011-9959-x.

Dutt, M., Ananthakrishnan, G., Jaromin, M. K., Brlansky, R. H. and Grosser, J. W. 2012. Evaluation of four phloem-specific promoters in vegetative tissues of transgenic citrus plants. *Tree Physiol.* 32(1), 83–93. doi:10.1093/treephys/tpr130.

Dutt, M., Stanton, D. and Grosser, J. W. 2016. Ornacitrus: development of genetically modified anthocyanin-expressing citrus with both ornamental and fresh fruit potential. *J. Am. Soc. Hortic. Sci.* 141(1), 54–61. doi:10.21273/JASHS.141.1.54.

Esen, A. and Soost, R. K. 1972. Tetraploid progenies from 2x x 4x crosses of *Citrus* and their origin. *J. Am. Soc. Hortic. Sci.* 97, 410–4.

Fang, D. Q. and Roose, M. L. 1997. Identification of closely related Citrus cultivars with inter-simple sequence repeat markers. *Theor. Appl. Genet.* 95(3), 408–17. doi:10.1007/s001220050577.

Fang, D. Q., Federici, C. T. and Roose, M. L. 1997. Development of molecular markers linked to a gene controlling fruit acidity in *Citrus*. *Genome* 40(6), 841–9. doi:10.1139/g97-809.

FAOSTAT Database. 2014. Available at: http://faostat.fao.org/default.aspx (accessed 14 October 2015).

FAOSTAT Database. 2016. Available at: http://faostat.fao.org/default.aspx (accessed 10 July 2016).

Fatima, B., Usman, M., Ramzan, M., Khan, M. M. and Khan, I. A. 2002. Interploid hybridization of kinnow and sweet lime. *Pak. J. Agri. Sci.* 39, 132–4.

Fatima, B., Usman, M., Khan, I. A., Khan, M. S. and Khan, M. M. 2010. Exploring citrus cultivars for underdeveloped and shrivelled seeds: a valuable resource for spontaneous polyploidy. *Pak. J. Bot.* 42(1), 189–200.

Fatima, B., Usman, M., Khan, M. S., Khan, I. A. and Khan, M. M. 2015. Identification of citrus polyploids using chromosome counts, morphological and SSR markers. *Pak. J. Agri. Sci.* 52(1), 107–14.

Froelicher, Y. 1999. Organisation de la diversite dans le genre clausena burm. F. (Rutaceae) Doctorat, Paris, France.

Froelicher, Y., Luro, F. and Ollitrault, P. 2000. Analysis of meiotic behavior of tetraploid *Clausena excavata* species by molecular marker segregation studies. In: *9th Int. Soc. of Citriculture Congress*, South Africa, p. 116.

Froelicher, Y., Mouhaya, W., Bassene, J. B., Costantino, G., Kamiri, M., Luro, F., Morillon, R. and Ollitrault, P. 2011. New universal mitochondrial PCR markers reveal new information on maternal citrus phylogeny. *Tree Genet. Genomes* 7(1), 49–61. doi:10.1007/s11295-010-0314-x.

Frost, H. B. and Soost, R. K. 1968. Seed reproduction, development of gametes and embryos. In: Reuther, W., Batchelor, L. D. and Webber, H. B. (Eds). *The Citrus Industry* (vol. 2). University of California, Barkley, pp. 290–324.

Gallais, A. 2003. *Quantitative Genetics and Breeding Methods in Autopolyploids Plants.* INRA, Paris.

Garcia-Lor, A., Curk, F., Snoussi-Trifa, H., Morillon, R., Ancillo, G., Luro, F., Navarro, L. and Ollitrault, P. 2013. A nuclear phylogenetic analysis: SNPs, indels and SSRs delivers new insights into the relationship. *Ann. Bot.* 111(1), 1–19. doi:10.1093/aob/mcs227.

Ghorbel, R., López, C., Fagoaga, C., Moreno, P., Navarro, L., Flores, R. and Peña, L. 2001. Transgenic citrus plants expressing the citrus tristeza virus p23 protein exhibit viral-like symptoms. *Mol. Plant Pathol.* 2(1), 27–36. doi:10.1046/j.1364-3703.2001.00047.x.

Grosser, J. W. and Gmitter Jr, F. G. 2005. 2004 SIVB congress symposium proceedings 'thinking outside the cell': applications of somatic hybridization and cybridization in crop improvement, with Citrus as a model. *In Vitro Cell Dev. Biol. Plant.* 41, 220–5.

Grosser, J. W. and Gmitter Jr, F. G. 2009. *In vitro* breeding provides new and unique opportunities for conventional breeding. *Acta Hortic.* 829(829), 65–72. doi:10.17660/ActaHortic.2009.829.8.

Grosser, J. W. and Gmitter Jr, F. G. 2011. Protoplast fusion for production of tetraploids and triploids: applications for scion and rootstock breeding in citrus. *Plant Cell Tissue Organ Cult.* 104(3), 343–57. doi:10.1007/s11240-010-9823-4.

Grosser, J. W., Gmitter, F. G., Tusa, N. and Chandler, J. L. 1990. Somatic hybrid plants from sexually incompatible woody species: *Citrus reticulata* and *Citropsis gilletiana*. *Plant Cell Rep.* 8(11), 656–9. doi:10.1007/BF00269986.

Grosser, J. W., Mourao, F. A. A., Gmitter, F. G.Jr., Louzada, E. S., Jiang, J., Baergen, K., Quiros, A., Cabasson, C., Schell, J. L. and Chandler, J. L. 1996. Allotetraploid hybrids between Citrus and seven related genera produced by somatic hybridization. *Theor. Appl. Genet.* 92(5), 577–82. doi:10.1007/BF00224561.

Grosser, J. W., Ollitrault, P. and Olivares-Fuster, O. 2000. Invited review– somatic hybridization in citrus: an effective tool to facilitate variety improvement. *In Vitro Cell. Dev.Biol.-Plant* 36(6), 434–49. doi:10.1007/s11627-000-0080-9.

Grosser, J. W., Viloria, Z. and Dutt, M. 2015. New acid citrus selection of Florida. *Citrus Ind.* 2015, 30–3.

Gulsen, O. and Roose, M. L. 2001. Chloroplast and nuclear genome analysis of the parentage of lemons. *J. Am. Soc. Hortic. Sci.* 126(2), 210–5. doi:10.21273/JASHS.126.2.210.

Guo, F., Yu, H., Xu, Q. and Deng, X. 2015. Transcriptomic analysis of differentially expressed genes in an orange-pericarp mutant and wild type in pummelo (*Citrus grandis*). *BMC Plant Biol.* 15, 44. doi:10.1186/s12870-015-0435-3.

Hocquellet, A., Toorawa, P., Bové, J. M. and Garnier, M. 1999. Detection and identification of the two *Candidatus* Liberobacter species associated with citrus huanglongbing by PCR amplification of ribosomal protein genes of the b operon. *Mol. Cell. Probes* 13(5), 373–9. doi:10.1006/mcpr.1999.0263.

Hodgson, R. W. 1955. Origin of citrus fruits grown in California. *Calif. Citrograph.* 49, 132–6.

Hörandl, E. 2006. The complex causality of geographical parthenogenesis. *New Phytol.* 171(3), 525–38. doi:10.1111/j.1469-8137.2006.01769.x.

Hummer, K. E., Dempewolf, H., Bramel, P., Markham, R. and Stover, E. 2015. Status of global strategies for horticultural fruit crops. *Acta Hortic.* 1101(1101), 147–52. doi:10.17660/ActaHortic.2015.1101.22.

Iwamasa, M. and Iwasaki, T. 1963. On the sterility phenomenon caused by low temperatures in the Mexican lime (*Citrus aurantifolia* Swing.). *Bull. Hort. Res. Stn of Jpn* B2, 25–45.

Iwamasa, M. and Nito, N. 1988. Cytogenetics and the evolution of modern cultivated Citrus. In: *Citriculture: Proceedings of the Sixth International Citrus Congress*, Israel.

Jannati, M. R., Fotouhi, A., Abad, P. and Salehi, Z. 2009. Genetic diversity analysis of Iranian citrus varieties using micro satellite (SSR) based markers. *J. Hortic. For.* 7(1), 120–5.

Jena, S., Kumar, K. and Nair, N. 2009. Molecular phylogeny in Indian *Citrus* L. (Rutaceae) inferred through PCR-RFLP and trnL-trnF sequence data of chloroplast DNA. *Sci. Hortic.* 119(4), 403–16.

Jeridi, M., Perrier, X., Rodier-Goud, M., Ferchichi, A., D'Hont, A. and Bakry, F. 2012. Cytogenetic evidence of mixed disomic and polysomic inheritance in an allotetraploid AABB musa genotype. *Ann. Bot.* 1108, 1593–606.

Khankahdani, H. H., Rastegar, S., Golein, B., Golmohammadi, M. and Jahromi, A. A. 2017. Genetic diversity in Persian lime (*Citrus latifolia* Tanaka) accessions using morphological and molecular markers. *Agric. For.* 63(3), 221–31.

Khiavi, S. J., Hamidoghli, Y., Golein, B. and Sabouri, A. 2015. Evaluation of genetic diversity in acid lime (*Citrus aurantifolia* swingle) genotypes using AFLP markers. *Aust. J. Crop Sci.* 9(10), 996–1002.

Kijas, J. M. H., Thomas, M. R., Fowler, J. C. S. and Roose, M. L. 1997. Integration of trinucleotide microsatellites into a linkage map of *Citrus*. *Theor. Appl. Genet.* 94(5), 701–6. doi:10.1007/s001220050468.

Kahn, T. L., Krueger, R. R., Gumpt, D. J., Roose, M. L., Arpaia, M. L., et al. 2001. *Genetic Resources Conservation Program*. Division of Agriculture and Natural Resources, University of California, Davis, CA, pp. 1–58.

Khan, I. A. and Kender, W. J. 2007. Citrus breeding. In: Khan, I. A. (Ed.), *Citrus Genetics, Breeding and Biotechnology*. CAB International, Wallingford, UK, pp. 1–8. Chapter 1.

Knapp, J. L., Halbert, S., Lee, R., Hoy, M., Clark, R. and Kesinger, M. 2004. *The Asian Citrus Psyllid and Citrus Greening Disease. Agricultural IPM: Fruit (citrus) Florida*. IFAS, UF, Gainesville, FL.

Koltunow, A. M., Vivian-Smith, A. and Sykes, S. R. 2000. Molecular and conventional breeding strategies for seedless citrus. In: *First International Citrus Biotechnology Symposium* (vol. 535), pp. 169–74.

Krueger, R. and Navarro, L. 2007. Citrus germplasm resources. In: Khan, I. A. (Ed.), *Citrus Genetics, Breeding and Biotechnology*. CAB International, Wallingford, UK, pp. 45–140. Chapter 4.

Koes, R., Verweij, W. and Quattrocchio, F. 2005. Flavonoids: a colorful model for the regulation and evolution of biochemical pathways. *Trends Plant Sci.* 10(5), 236–42. doi:10.1016/j.tplants.2005.03.002.

Kumar, M., Parthiban, S., Devi, D. S. and Ponnuswami, V. 2013. Genetic diversity analysis of acid lime (*Citrus aurantifolia* Swingle) cultivars. *The Bioscan* 8(2), 481–4.

Latado, R. R., Derbyshire, M. T. VdC., Tsai, S. M. and Neto, A. T. 2002. Obtenção de hibridos somaticos de limão 'Cravo' e tangerina 'Cleopatra'. *Pesq. Agropec. Bras.* 37(12), 1735–41. doi:10.1590/S0100-204X2002001200009.

Luro, F., Venturini, N., Costantino, G., Paolini, J., Ollitrault, P. and Costa, J. 2012. Genetic and chemical diversity of citron (*Citrus medica* L.) based on nuclear and cytoplasmic markers and leaf essential oil composition. *Phytochemistry* 77, 186–96. doi:10.1016/j.phytochem.2011.12.013.

Mabberley, D. J. 1998. Australia Citreae with notes on Aurantioideae (Rutaceae). *Telopea* 7(4), 333–44. doi:10.7751/telopea19982004.

Malik, M. N., Scora, R. W. and Soost, R. K. 1974. Studies on the origin of the lemon. *Hilgardia* 42(9), 361–82. doi:10.3733/hilg.v42n09p361.

Mardi, M., Karimi, F. L., Gharechahi, J. and Salekdeh, G. H. 2015. In-depth transcriptome sequencing of Mexican lime trees infected with *Candidatus* Phytoplasma aurantifolia. *PLOS ONE* 10, e0130425.

Martinez, G. S. 2002. *Diagnostico del sistema producto limon mexicano en Colima*. Ed. Consejo Estatal del Limon Mexicano del Colima (Coelim-Col). Colima, Colima, Mexico, p. 145.

Mendes, B. M. J., Filho, FdA. A. M., Farias, P. C. D. M. and Benedito, V. A. 2001. Citrus somatic hybridization with potential for improved blight and CTV resistance. *In Vitro Cell Dev. Biol.-Plant* 37(4), 490–5. doi:10.1007/s11627-001-0086-y.

Mendes-da-Glória, F. J., Filho, F.d.e. A. A. M., Camargo, L. E. A. and Mendes, B. M. J. 2000. Caipira sweet orange + Rangpur lime: a somatic hybrid with potential for use as rootstock in the Brazilian citrus industry. *Genet. Mol. Biol.* 23(3), 661–5. doi:10.1590/S1415-47572000000300026.

Moore, G. A. 2001. Oranges and lemons: clues to the taxonomy of citrus from molecular markers. *Trends Genet.* 17(9), 536–40. doi:10.1016/S0168-9525(01)02442-8.

Morton, J. F. 1987. Mexican lime. In: Morton, J. F. (Ed.), *Fruits of Warm Climates*. Florida Flair Books, Florida, pp. 168–72.

Muller, G. W., De Negri, J. D., Vildoso, C. I. A., Mattos, D.Jr., Pompeu, J.Jr., Sobrinho, J. T., Machado, M. A., Carvahlo, S. A. and Girotto, L. F. 2002. Citrus sudden death: a new citrus disease in Brazil. In: *Proceeding of Fifteenth IOCV Conference*. University of California, Riverside, CA, pp. 405–7.

Munankarmi, N. N., Shrestha, R. L., Rana, N., Shrestha, J. K. C., Shrestha, S., Koirala, R. and Shrestha, S. 2014. Genetic diversity assessment of acid lime (*Citrus aurantifolia* Swingle) landraces of eastern Nepal using RAPD markers. *Int. J. Appl. Sci. Biotechnol.* 2(3), 315–27. doi:10.3126/ijasbt.v2i3.10950.

Munankarmi, N. N., Rana, N., Bhattarai, T., Shrestha, R. L., Joshi, B. K., Baral, B. and Shrestha, S. 2018. Characterization of the genetic diversity of acid lime (*Citrus aurantifolia* (Christm.) Swingle) cultivars of Eastern Nepal using inter-simple sequence repeat markers. *Plants* 7(2), 46. doi:10.3390/plants7020046.

Nakano, M., Shimada, T., Endo, T., Fujii, H., Nesumi, H., Kita, M., Ebina, M., Shimizu, T. and Omura, M. 2012. Characterization of genomic sequence showing strong association with polyembryony among diverse *Citrus* species and cultivars, and its synteny with *Vitis* and *Populus*. *Plant Sci.* 183, 131–42. doi:10.1016/j.plantsci.2011.08.002.

Naz, S., Shahzadi, K., Rashid, S., Saleem, F., Zafarullah, A. and Ahmad, S. 2014. Molecular characterization and phylogenetic relationship of different citrus varieties of Pakistan. *J. Anim. Plant Sci.* 24(1), 315–20.

Nejat, N., Cahill, D. M., Vadamalai, G., Ziemann, M., Rookes, J. and Naderali, N. 2015. Transcriptomics-based analysis using RNA-Seq of the coconut (*Cocos nucifera*) leaf in response to yellow decline phytoplasma infection. *Mol. Genet. Genomics* 290(5), 1899–910. doi:10.1007/s00438-015-1046-2.

Nicolosi, E., Deng, Z. N., Gentile, A., La Malfa, S., Continella, G. and Tribulato, E. 2000. Citrus phylogeny and genetic origin of important species as investigated by molecular markers. *Theor. Appl. Genet.* 100(8), 1155–66. doi:10.1007/s001220051419.

Olivares-Fuster, O., Duran-Vila, N. and Navarro, L. 2005. Electrochemical protoplast fusion in citrus. *Plant Cell Rep.* 24(2), 112–9. doi:10.1007/s00299-005-0916-1.

Ollitrault, P. and Navarro, L. 2012. Citrus. In: *Fruit Breeding*. Springer, Boston, MA, pp. 623–62.

Ollitrault, P., Vanel, F., Froelicher, Y. and Dambier, D. 2000. Creation of triploid Citrus hybrids by electrofusion of haploid and diploid protoplasts. *Acta Hortic.* 535(535), 191–8. doi:10.17660/ActaHortic.2000.535.23.

Ollitrault, P., Jacquemond, C. and Dubois, C. 2003. Citrus. In: *Genetic Diversity of Cultivated Tropical Plants*. CRC Press, pp. 209–34.

Ollitrault, P., Dambier, D., Luro, F. and Froelicher, Y. 2008. Ploidy manipulation for breeding seedless triploid citrus. *Plant Breed. Rev.* 30, 323–52.

Ollitrault, P., Terol, J., Garcia-Lor, A., Bérard, A., Chauveau, A., Froelicher, Y., Belzile, C., Morillon, R., Navarro, L., Brunel, D., et al. 2012. SNP mining in *C. clementina* BAC end sequences; transferability in the *Citrus* genus (Rutaceae), phylogenetic inferences and perspectives for genetic mapping. *BMC Genomics* 13, 13. doi:10.1186/1471-2164-13-13.

Pena, L., Cervera, M., Juarez, J., Navarro, A., Pina, J. A. and Navarro, L. 1997. Genetic transformation of lime (*Citrus aurantifolia* Swing): factors affecting transformation and regeneration. *Plant Cell Rep.* 16(11), 731–7. doi:10.1007/s002990050311.

Pena, L., Perez, R. M., Cervera, M., Juarez, J. A. and Navarro, L. 2004. Early events in *Agrobacterium*-mediated genetic transformation of citrus explants. *Ann. Bot.* 94(1), 67–74. doi:10.1093/aob/mch117.

Penjor, T., Yamamoto, M., Uehara, M., Ide, M., Matsumoto, N., Matsumoto, R. and Nagano, Y. 2013. Phylogenetic relationships of *Citrus* and its relatives based on *matK* gene sequences. *PLOS ONE* 8(4), e62574. doi:10.1371/journal.pone.0062574.

Pons, E., Navarro, A., Ollitrault, P. and Peña, L. 2011. Pollen competition as a reproductive isolation barrier represses transgene flow between Compatible and Co-flowering citrus genotypes. *PLOS ONE* 6(10), e25810. doi:10.1371/journal.pone.0025810.

Ramsey, J. and Schemske, D. W. 1998. Pathways, mechanisms, and rates of polyploid formation in flowering plants. *Annu. Rev. Ecol. Syst.* 29(1), 467-501. doi:10.1146/annurev.ecolsys.29.1.467.

Ramadugu, C., Pfeil, B. E., Keremane, M. L., Lee, R. F., Maureira-Butler, I. J. and Roose, M. L. 2013. A six nuclear gene phylogeny of *Citrus* (Rutaceae) taking into account hybridization and lineage sorting. *PLOS ONE* 8(7), e68410. doi:10.1371/journal.pone.0068410.

Ramadugu, C., Razi, M. F., Keremane, M. L., Scora, R. W. and Roose, M. L. 2017. Systematic classification, distribution and botany. In: Khan, M. M., Al-Yahyai, R., Al-Said, F. (Eds), *The Lime Botany, Production and Uses*. CAB International, Wallingford, UK, pp. 12–36.

Reece, P. C. and Childs, J. F. L. 1962. Character differences among seedlings of Persian lime. *P. FL. St. Hortic. Soc.* 75, 110–6.

Rouiss, H., Cuenca, J., Navarro, L., Ollitrault, P. and Aleza, P. 2017a. Tetraploid citrus progenies arising from FDR and SDR unreduced pollen in 4x x 2x hybridizations. *Tree Genet. Genomes* 13(1), 10. doi:10.1007/s11295-016-1094-8.

Rouiss, H., Cuenca, J., Navarro, L., Ollitrault, P. and Aleza, P. 2017b. Unreduced megagametophyte production in lemon occurs via three meiotic mechanisms, predominantly second-division restitution. *Front. Plant Sci.* 8, 1211. doi:10.3389/fpls.2017.01211.

Rouiss, H., Bakry, F., Froelicher, Y., Navarro, L., Aleza, P. and Ollitrault, P. 2017c. Origin of *C. latifolia* and *C. aurantiifolia* triploid limes: the preferential disomic inheritance of doubled-diploid 'Mexican' lime is consistent with an interploid hybridization hypothesis. *Ann. Bot.* 121(3), 571–85. doi:10.1093/aob/mcx179.

Satpute, A. D., Chen, C., Gmitter, F. G., Ling, P., Yu, Q., Grosser, M. R., Grosser, J. W. and Chase, C. D. 2015. Cybridization of grapefruit with 'Dancy' mandarin leads to improved fruit characteristics. *J. Am. Soc. Hort. Sci.* 140(5), 427–35. doi:10.21273/JASHS.140.5.427.

Saunt, J. 1990. *Citrus Varieties of the World*. Sinclair, Norwich, England.

Shahsavar, A. R., Izadpanah, K., Tafazoli, E. and Tabatabaei, B. E. S. 2007. Characterization of *Citrus* germplasm including unknown variants by Inter-Simple Sequence Repeat (ISSR) markers. *Sci. Hortic.* 112(3), 310–4. doi:10.1016/j.scienta.2006.12.039.

Sharafi, A. A., Abkenar, A. A., Sharafi, A. and Masaeli, M. 2016. Genetic variation assessment of acid lime accessions collected from south of Iran using SSR and ISSR molecular markers. *Physiol. Mol. Biol. Plants* 22(1), 87–95. doi:10.1007/s12298-016-0336-4.

Shiratake, K. and Suzuki, M. 2016. Omics studies of citrus, grape and Rosaceae fruit trees. *Breed. Sci.* 66(1), 122–38. doi:10.1270/jsbbs.66.122.

Shrestha, R. L., Dhakal, D. D., Gautum, D. M., Paudyal, K. P. and Shrestha, S. 2012. Genetic diversity assessment of acid lime (*Citrus aurantifolia* Swingle) landraces in Nepal, using SSR markers. *Am. J. Plant Sci.* 3, 1674–81.

Silva, F. N., Souza, A. N., Al-Mahmooli, I., Al-Sadi, A. M. and Carvalho, C. M. 2015. A new disease in *Citrus aurantifolia* in Oman, 'sudden decline', is associated with a pathogen complex including a 16SrII group Phytoplasma. *Phytopathogenic Mollicutes* 5(1), 101–2.

Singh, S. and Rajam, M. V. 2009. Citrus biotechnology: achievements, limitations and future directions. *Physiol. Mol. Biol. Plants* 15(1), 3–22. doi:10.1007/s12298-009-0001-2.

Skyes, S. R. 1997. Australian native limes (Eremocitrus and Microcitrus): a citrus breeder's viewpoint. *Aust. Bushfood Mag.* 3, 12–5.

Snoussi, H., Duval, M. F., Garcia-Lor, A., Belfalah, Z., Froelicher, Y., Risterucci, A. M., Perrier, X., Jacquemoud-Collet, J. P., Navarro, L., Harrabi, M. and Ollitrault, P. 2012. Assessment of the genetic diversity of the Tunisian citrus rootstock germplasm. *BMC Genet.* 13(1), 16. doi:10.1186/1471-2156-13-16.

Snowball, A. M., Halligan, E. A. and Mullins, M. G. 1988. Studies on juvenility in Citrus. In: *Proceedings of the Sixth International Citrus Congress*, Israel, pp. 467–73.

Soost, R. K. and Cameron, J. W. 1980. 'Oroblanco', a triploid pummelo grapefruit hybrid. *Hort. Sci.* 15, 667–9.

Steyn, W. J., Wand, S. J. E., Holcroft, D. M. and Jacobs, G. 2002. Anthocyanins in vegetative tissues: a proposed unified function in photo protection. *New Phytol.* 155(3), 349-61. doi:10.1046/j.1469-8137.2002.00482.x.

Stift, M., Berenos, C., Kuperus, P. and van Tienderen, P. H. 2008. Segregation models for disomic, tetrasomic and intermediate inheritance in tetraploids: a general procedure applied to Rorippa (yellow cress) microsatellite data. *Genetics* 179(4), 2113–23. doi:10.1534/genetics.107.085027.

Su, H. J., Hogenhout, S. A., Al-Sadi, A. M. and Kuo, C. H. 2014. Complete chloroplast genome sequence of Omani lime (*Citrus aurantiifolia*) and comparative analysis within the Rosids. *PLoS ONE* 9(11), e113049. doi:10.1371/journal.pone.0113049.

Swingle, W. T. and Reece, P. C. 1967. The botany of citrus and its wild relatives. In: Reuther, W., Webber, H. J. and Batchelor, L. D. (Eds), *The Citrus Industry*. University of California, Berkeley, CA, pp. 190–430.

Sybenga, J. 1996. Chromosome pairing affinity and quadrivalent formation in polyploids: do segmental allopolyploids exist? *Genome* 39(6), 1176–84. doi:10.1139/g96-148.

Sybenga, J. 2012. *Cytogenetics in Plant Breeding*. Springer Science and Business Media, Berlin, Germany.

Tanaka, Y., Sasaki, N. and Ohmiya, A. 2008. Biosynthesis of plant pigments: anthocyanins, betalains and carotenoids. *Plant J.* 54(4), 733–49. doi:10.1111/j.1365-313X.2008.03447.x.

Tripolitsiotis, C., Nikoloudakis, N., Linos, A. and Hagidimitriou, M. 2013. Molecular characterization and analysis of the Greek citrus germplasm. *Not. Bot. Hort. Agrobot. Cluj* 41(2), 463–71. doi:10.15835/nbha4129302.

Usman, M. and Fatima, B. 2018. Mandarin (*Citrus reticulata* Blanco) breeding. In: Al-Khayri, J. M. et al. (Eds), *Advances in Plant Breeding Strategies: Fruits* . Springer, Cham, pp. 465–533. Chapter 12.

Usman, M., Ramzan, M., Fatima, B., Jaskani, M. J. and Khan, M. M. 2002. *Citrus* germplasm enhancement by interploid hybridization 1. Reciprocal crosses of Kinnow and Succari. *Int. J. Agri. Bio.* 4, 208–10.

Usman, M., Saeed, T., Khan, M. M. and Fatima, B. 2006. Occurrence of spontaneous polyploids in *Citrus*. *Hort. Sci. (Prague)* 33(3), 124–9. doi:10.17221/3751.

Usman, M., Fatima, B., Gillani, K. A., Khan, M. S. and Khan, M. M. 2008. Exploitation of potential target tissues to develop polyploids in citrus. *Pak. J. Bot.* 40(4), 1755–66.

Vardi, A., Levin, I. and Carmi, N. 2008. Induction of seedlessness in citrus: from classical techniques to emerging biotechnological approaches. *J. Am. Soc. Hortic. Sci.* 133(1), 117–26. doi:10.21273/JASHS.133.1.117.

Viloria, Z. and Grosser, J. W. 2005. Acid citrus fruit improvement via interploid hybridization using allotetraploid somatic hybrid and autotetraploid breeding parents. *J. Am. Soc. Hortic. Sci.* 130(3), 392–402. doi:10.21273/JASHS.130.3.392.

Volk, G., Samarina, L., Kulyan, R., Gorshkov, V., Malyarovskaya, V., Ryndin, A., Polek, M., Krueger, R. and Stover, E. 2018. Citrus GeneBank collections: international collaboration opportunities between the US and Russia. *Genet. Resour. Crop Evol.* 65(2), 433–47. doi:10.1007/s10722-017-0543-z.

Wang, L., Pan, Z. Y. and Guo, W. W. 2010. Proteomic analysis of leaves from a diploid cybrid produced by protoplast fusion between Satsuma mandarin and pummelo. *Plant Cell Tissue Organ Cult.* 103(2), 165–74. doi:10.1007/s11240-010-9764-y.

Wang, J., Jiang, J. and Wang, Y. 2013. Protoplast fusion for crop improvement and breeding in China. *Plant Cell Tissue Organ Cult.* 112(2), 131–42. doi:10.1007/s11240-012-0221-y.

Wang, X., Xu, Y., Zhang, S., Cao, L., Huang, Y., Cheng, J., Wu, G., Tian, S., Chen, C., Liu, Y., et al. 2017. Genomic analyses of primitive, wild and cultivated citrus provide insights into asexual reproduction. *Nat. Genet.* 49(5), 765–72. doi:10.1038/ng.3839.

Winkle-Shirley, B. 2001. Flavonoid biosynthesis. A colorful model for genetics, biochemistry, cell biology, and biotechnology. *Plant Physiol.* 126(2), 485–93. doi:10.1104/pp.126.2.485.

Wu, G. A., Terol, J., Ibanez, V., López-García, A., Pérez-Román, E., Borredá, C., Domingo, C., Tadeo, F. R., Carbonell-Caballero, J., Alonso, R., et al. 2018. Genomics of the origin and evolution of *Citrus*. *Nature* 554(7692), 311–6. doi:10.1038/nature25447.

Xie, K., Xia, Q., Wang, X., Liang, W., Wu, X., Grosser, J. W. and Guo, W. 2015. Cytogenetic and SSR-marker evidence of mixed disomic, tetrasomic, and intermediate inheritance in a citrus allotetraploid somatic hybrid between 'Nova'tangelo and 'HB'pummelo. *Tree Genet. Genome.* 11, 112.

Xu, Q., Chen, L. L., Ruan, X., Chen, D., Zhu, A., Chen, C., Bertrand, D., Jiao, W. B., Hao, B. H., Lyon, M. P., et al. 2013. The draft genome of sweet orange (*Citrus sinensis*). *Nat. Genet.* 45(1), 59-66. doi:10.1038/ng.2472.

Yamamoto, M. 2007. Application of fluorescent staining of chromosomes to genetic studies in citrus. *Jpn. J. Plant Sci.* 1, 12-9.

Yang, L., Xu, C. -J., Hu, G. -B. and Chen, K. S. 2007. Establishment of an Agrobacterium-mediated transformation system for *Fortunella crassifolia*. *Biol. Plant.* 51(3), 541-5. doi:10.1007/s10535-007-0117-6.

Yelenosky, G. 1985. Environmental factors affecting citrus. *Fruit Var. J.* 39, 51-7.

Yu, Z., Haberer, G., Matthes, M., Rattei, T., Mayer, K. F. X., Gierl, A. and Torres-Ruiz, R. A. 2010. Impact of natural genetic variation on the transcriptome of autotetraploid *Arabidopsis thaliana*. *Proc. Natl Acad. Sci. U. S. A.* 107(41), 17809-14. doi:10.1073/pnas.1000852107.

# Chapter 7

## Advances in mandarin breeding

*José Cuenca, Luis Navarro and Pablo Aleza, Instituto Valenciano de Investigaciones Agrarias (IVIA), Spain*



### 1 Introduction

Citrus is a leading fruit crop worldwide, with a production of more than 130 million tons in 2016, including oranges (55%), mandarins (25%), lemons/limes (13%), and grapefruits (7%) (FAO, 2018). The main mandarin producers are China (17 million tons) and the Mediterranean area (Spain, 3 million tons; Turkey, 1.3 million tons; Morocco, 1 million tons, and Egypt, 1 million tons). About two-thirds of the total citrus production is destined for the fresh market, particularly mandarin with exports of 32 million tons. The Mediterranean area is the most important region for fresh mandarin fruit exports, representing 60% of the total volume of exports. Spain is the main exporter of fresh mandarins, accounting for 25% of the total exports.

The botanical classification of mandarins is controversial, although efforts have been made in recent years to clarify their origin, phylogeny, and evolution, particularly the importance of introgressions of ancestral genomes in mandarin germplasm (Wang et al., 2018; Wu et al., 2018). The *Citrus* genus belongs to the *Aurantioideae* subfamily within the *Rutaceae* family. The subtribe *Citrinae* (tribe *Citreae*) comprises six genera, which are considered as the true citrus (Swingle and Reece, 1967), namely, *Eremocitrus, Microcitrus, Clymenia, Fortunella, Poncirus,* and *Citrus*. Cultivated citrus arose from natural interspecific hybridizations between four ancestors: citron (*Citrus medica* L.), mandarin (*C. reticulata* Blanco), pummelo (*C. maxima* (L.) Osb.), and a wild papeda citrus (*C. micrantha* Wester)

http://dx.doi.org/10.19103/AS.2019.0054.10

(Scora, 1975; Barrett and Rhodes, 1976; Herrero et al., 1996; Federici et al., 1998; Nicolosi et al., 2000; Gulsen and Roose, 2001; Barkley et al., 2006; Li et al., 2010; Garcia-Lor et al., 2013; Penjor et al., 2013; Ouestlati et al., 2017; Wu et al., 2018).

Swingle and Reece (1967) and Mabberley (1997) classified all mandarin germplasm as *C. reticulata*. In contrast, Tanaka (1977) divided mandarins into 5 groups, including 36 species. More recently, Garcia-Lor et al. (2015) divided mandarins into 7 groups based on their nuclear genomes: *C. deliciosa*, *C. nobilis*, *C. unshiu*, *C. reticulata*, *C. tangerina*, a group comprising mandarin hybrids, and a last group that includes *C. depressa*, *C. reshni*, *C. sunki,* and *C. tachibana*. These groups represent important commercial varieties such as Satsumas (*C. unshiu*), Clementines (*C. deliciosa* X *C. sinensis* (sweet orange)), and many hybrids with *C. reticulata* and *C. tangerine* in their pedigrees.

Recent studies conclude that two independent mandarin domestication events occurred in South China, with interspecific hybridization playing an important role in mandarin diversification (Wu et al., 2018; Wang et al., 2018). Introgressions of other ancestral genomes in most of the available mandarin germplasm have been revealed by several studies (Curk et al., 2014, 2015; Wu et al., 2014, 2018; Oueslati et al., 2017). These ancestral genome introgressions were used by Wu et al. (2018) to classify mandarins into three groups: pure ancestral nonintrogressed mandarins, a second group including varieties with a low percentage of pummelo introgression in small segments, and a last group including mandarin genotypes with high pummelo introgression in long segments. In addition, *C. reticulata* × *C. maxima* reticulation and introgression processes were important in edible mandarin domestication, reflected in the fact that acid mandarins (nonedible) share the same mandarin mitotype (Froelicher et al., 2011; Garcia-Lor et al., 2015), while edible mandarins share the mitotype introgressed by *C. maxima*. It is noticeable that citric acid was dramatically reduced during mandarin domestication, while the sugar content did not increase significantly (Wang et al., 2018).

In general, the varietal group of mandarins shows high genetic variability, making it possible to obtain segregating populations for the development of breeding programs (Garcia-Lor et al., 2015). Mandarin breeding programs are mainly focused on new variety development, although some mandarin genotypes are also used as parents in citrus rootstock breeding programs. For scions, breeding programs are mainly aimed at obtaining seedless, easy-peeling varieties with high-quality fruits characterized by attractive fruit color and good flavor. More recently, obtaining fruits with high anthocyanin content has been included as an objective in some mandarin breeding programs (Rapisarda et al., 2008; Navarro et al., 2015; Russo et al., 2015). In addition, other agronomical traits of importance are disease resistance, early- and late-maturing varieties, high yields, and good postharvest characteristics, such as improved shelf-life.

Citrus growers are also confronted worldwide by increasing biotic and abiotic constraints. Breeding programs are therefore also focused on obtaining new cultivars resistant to diseases and better adapted to climate change. Important pests and diseases affecting citrus are Huanglongbing (HLB), caused by the bacteria *Candidatus liberibacter* spp., which is the most devastating citrus disease threatening citrus cultivation in the world (Bové, 2006; Da Graça et al., 2016; Wang et al., 2017a), Tristeza, caused by the *Citrus tristeza* virus (Cambra and Moreno, 2000), citrus canker, caused by the bacterium *Xanthomonas axonopodis* (Gottwald et al., 2002), citrus variegated chlorosis, caused by the bacterium *Xylella fastidiosa* (Hartung et al., 1994), citrus black spot, caused by the fungus *Phyllosticta citricarpa* (Rodríguez et al., 2018), Mal secco, caused by the fungus *Phoma tracheiphila* (in the Mediterranean Basin; Nigro et al., 2011) and the fruit fly *Ceratitis capitata* (Rodríguez et al., 2011). In addition, some mandarin genotypes are severely affected by Alternaria Brown Spot disease (ABS), produced by the fungus *Alternaria alternata* (Cuenca et al., 2013, 2016).

This chapter reviews the latest advances in mandarin breeding programs, including classical sexual hybridizations and mutation breeding, polyploidy manipulation, and the most recent genetic, molecular, biotechnological, and genomic tools used in breeding.

## 2 Conventional mandarin breeding programs

### 2.1 Nucellar selections

Most citrus species present adventitious embryony, a form of apomixis where somatic embryos arise from the diploid nucellus tissue surrounding the embryo sac. Nucellar embryony is very efficient for the elimination of pathogens. However, since nucellar plants are juvenile, and nonapomitic citrus species are also found, the shoot-tip grafting in-vitro technique has been adopted as a routine protocol for obtaining disease-free plants (Navarro et al., 1975).

Although nucellar plants are normally identical to the mother plant, some selections done with both in-vitro and in-vivo techniques can differ from the original genotype in a number of agronomic characteristics. This makes it possible to use nucellar selection as a breeding strategy. This method has been routinely exploited in citrus for sweet orange (Russo et al., 1981), lemon (Gentile et al., 1992), and 'Satsuma' mandarin breeding, especially in Japan (Omura and Shimada, 2016). In fact, nucellar seedlings of 'Satsuma' mandarin account for many of the registered selections from the Japanese breeding program, which vary in fruit maturity season, fruit shape, rind color and texture, reduced degree of rind puffing, and sugar content. More recently, new selections developed by the Italian program from 'Tarocco' and 'Moro' sweet oranges have been registered, which show differences related to ripening time, anthocyanin content in pulp and peel, fruit shape and size, and fruit firmness (Licciardello et al., 2015) .

## 2.2 Mutation breeding

Some mandarin groups such as Clementines and Satsumas present high heterozygosity and very low genetic diversity, which hamper the development of breeding programs based on sexual hybridization. As a consequence, detecting natural mutations or inducing mutations in elite cultivars are the main sources of variability (Navarro et al., 2015). Selection of spontaneous mutations is an efficient breeding method for these groups. In fact, the frequency of spontaneous mutations in the field is relatively high and most of the Clementines and Satsumas varieties cultivated worldwide have arisen from spontaneous mutations identified on adult trees (Vardi et al., 2008; Ollitrault and Navarro, 2012; Aleza, 2015). These new cultivars usually retain the general characteristics of the original genotype, with some specific changes regarding fruit traits such as fruit color or ripening season. In the Mediterranean region, many new 'Clementine' varieties have been selected in Spain and Morocco, while new mutations of 'Satsumas' have been selected in Japan (Fujii et al., 2016; Omura and Shimada, 2016; Shimizu et al., 2016).

Induced mutation breeding has been used in citrus since 1935, when the first report on the use of irradiation was published (Haskins and Moore, 1935). Induced mutagenesis is a breeding strategy of choice when it is essential to maintain the characteristics of an existing cultivar and promote the modification of one or a small number of traits. In citrus the most widely used mutagen is radiation, particularly X-rays and more recently, $\gamma$-rays. These types of ionizing radiations can induce a wide range of mutation types like chromosome breaks, rearrangements, and point mutations and, as a consequence of such chromosome abnormalities, irradiated citrus cultivars may have few seeds or even be seedless (Roose and Williams, 2007a). Appropriate irradiation doses are critical to success. Treatment of budwoods has been widely used in citrus with good results (Hearn, 1986; Vardi et al., 2003; Gulsen et al., 2007; Bermejo et al., 2011) and a wide range of doses have been applied, with an exposure of 50 Gy/min found to be ideal for seed reduction in citrus (Roose and Williams, 2007a).

Technology required for mutant induction by irradiation is simple and, in many countries, there are companies that provide irradiation services at a low cost. It is not necessary to have previous knowledge of gene control traits, and the resulting trees will not have a juvenile phase. Irradiated plants can be screened more quickly than hybrids obtained by conventional methods. The main disadvantages are the need to screen large populations to find desirable mutations that are produced at random and the potential instability (epigenetics and chimeras) of regenerated plants that requires evaluation through several cycles of propagation to ascertain that the new varieties are stable.

The 'Star Ruby' grapefruit is a classic example of irradiation in citrus (Hensz, 1971). More recently new seedless mandarin cultivars have been obtained by irradiation with excellent results. The 'Orri' mandarin variety was obtained by Vardi et al. (2003) from irradiated budwood of the 'Orah' mandarin and produces zero or few seeds, whereas 'Tango' mandarin was recovered by Roose and Williams (2007b) from irradiated budwood of 'Nadorcott' and also produces very few seeds. Other examples of low seed varieties obtained by irradiation are 'Mor' (Vardi et al., 1993) and 'Murina' (Bermejo et al., 2012) derived from 'Murcott' tangor and 'Nulessin' (Asins et al., 2002) and 'Nero' (Lopez-García et al., 2015) derived from the 'Clemenules' clementine variety.

Somaclonal variation, i.e. the genetic variation found in plants regenerated from tissue cultures, has been used as a spontaneous mutation breeding strategy in citrus, mainly in sweet orange (Grosser et al., 2015). This technique was used in our group at the Instituto Valenciano de Investigaciones Agrarias (IVIA, Spain) for generating 460 somaclones from 'Clementine', but none of them showed improved agronomic characters compared to the original genotype in terms of reduced seed number or fruit maturation date.

## 2.3 Sexual hybridization

Citrus breeding based on sexual hybridization is hampered by the complex reproductive biology present in citrus, such as male/female sterility, self- and cross-incompatibility, and apomixis (Ollitrault and Navarro, 2012; Navarro et al., 2015), thus limiting the range of potential parents that can be used. Mandarins show high genetic variability, allowing the establishment of segregating populations for breeding purposes (Garcia-Lor et al., 2015). New hybrids are characterized by a long juvenile phase, with the presence of thorns and poor flowering. An additional limitation of the sexual hybridization strategy is that diploid hybrids are usually fertile and contain seeds in their fruits. Seedlessness remains the main breeding objective in mandarin breeding, since highly seeded fruits are not accepted for the fresh market (Navarro et al., 2015). Some diploid mandarin varieties are usually seedless due to either male/female sterility or self-incompatibility coupled with parthenocarpy, such as the 'Clementine' and 'Satsuma' groups and other mandarin hybrids; however, some diploid hybrids produce seedy fruits when cross-pollinated. For this reason, the cultivation of self-incompatible parthenocarpic genotypes in isolated blocks results in seedless fruit production (Vardi et al., 2008). This implies that, for diversification programs, male and female sterility must be selected, particularly in areas such as the Mediterranean Basin, where the main seedless easy-peeling cultivars are self-incompatible, but there are fertile male and female 'Clementine' varieties (Ollitrault et al., 2008).

Some important mandarin diploid hybrids have arisen from chance seedlings, i.e. 'Satsumas', 'Clementines', 'Ponkan', 'Ortanique', 'Imperial', 'Ellendale', 'Murcott', and 'Nadorcott'. In some cases, one or both parents have been identified by molecular markers or genomic analysis. Several important diploid mandarin hybrids have been also obtained under controlled pollinations in breeding programs such as 'Fortune', 'Kinnow', 'Minneola', 'Nova', 'Kara', 'Primosole', 'Gold Nugget', 'Kiyomi', and 'Orah' (Cuenca et al., 2018). Interestingly, analyses performed using molecular markers indicated that the male parent was originally mislabeled for some of these hybrids (Barry et al., 2015).

Despite the fact that citrus fruits are generally diploid (2n=2x =18) (Krug, 1943), ploidy manipulation is a typical strategy in mandarin breeding programs, especially for obtaining seedless triploid cultivars and tetraploid genotypes to be used as parents in triploid breeding programs (Aleza et al., 2010a, 2012a,b). Triploid hybrids are generally sterile male and female plants. This means that obtaining triploid hybrids is one of the main breeding strategies in developing new seedless mandarin cultivars (Ollitrault et al., 2008; Ollitrault and Navarro, 2012). Indeed, many triploid mandarins have been released worldwide, such as 'Winola' (Spiegel-Roy and Vardi, 1992), 'Shasta Gold', 'Tahoe Gold', and 'Yosemite Gold', by the US program (Williams and Roose, 2004); 'Tacle', 'Clara', 'Mandared', 'Mandalate', 'Early Sicily', and 'Sweet Sicily' by the Italian program (Recupero et al., 2005; Russo et al., 2015); and 'Garbí' (Aleza et al., 2010b) and 'Safor' (Cuenca et al., 2010) by the triploid breeding program in Spain (Navarro et al., 2015). A patent has been recently issued for the first triploid hybrid recovered using a somatic hybrid as male parent, 'C4-15-19' released in Florida (USPP26086P3). The Spanish breeding program is expected to release two new triploid hybrids in the near future, 'Alborea' mandarin (Cuenca et al., 2015a) and 'IVIA-TRI707'. The triploid breeding strategy is discussed further in the section 'Case study: Achieving sustainability using triploid mandarin cultivars resistant to Alternaria Brown Spot disease' of this chapter.

## 3 Genetic, genomic, and biotechnological tools for mandarin breeding

Advances in molecular tools have enabled the development of new breeding techniques in crops, allowing the development of rapid methods to incorporate traits of interest into new varieties. These rapid methods include genetic maps, marker-assisted selection, genome sequencing, and biotechnological tools. Different techniques are used in mandarin breeding programs. The development of molecular markers in citrus allowed the creation of the

'Clementine' reference genetic map (Ollitrault et al., 2012) as well as the use of marker-assisted selection for some traits (Cuenca et al., 2016; Wang et al., 2017b). Obtaining a haploid line of 'Clemenules' clementine (Aleza et al., 2009a) and the genome sequencing of many citrus species (Xu et al., 2013; Wu et al., 2014; Shimizu et al., 2017; Wang et al., 2017b, 2018) have facilitated studies on phylogenomics, transcriptomics, and functional genomics, among others. In addition, these tools have facilitated developments in genetic transformation, somatic hybridization and, more recently, genome-editing technologies.

### *3.1 Molecular breeding*

Molecular markers have been widely used in citrus for germplasm characterization, phylogenetic analysis, assessment of the genetic origin of polyploid mandarins, genetic mapping, and marker-assisted selection (Cuenca et al., 2011, 2016; Ollitrault et al., 2012; Garcia-Lor et al., 2015; Aleza et al., 2016a; Curk et al., 2016; Yu et al., 2016). More recently, high-throughput sequencing techniques, such as genotyping-by-sequencing or DArT markers, have been used in citrus for phylogenetic analysis and genome wide association studies (GWAS) of fruit quality traits (Gois et al., 2016; Yu et al., 2016; Oueslati et al., 2017; Minamikawa et al., 2017).

The development of molecular markers in citrus, such as AFLPs, RFLPs, RAPDs, SSRs, and SNPs, has enabled the construction and integration of genetic maps of important traits (Cheng and Roose, 1995; Fang et al., 1997; Yang et al., 2003; Asins et al., 2004; Kepiro and Roose, 2010; Butelli et al., 2012, 2017; Cuenca et al., 2013, 2016; Wang et al., 2017b). In 2012, Ollitrault et al. (2012) published the 'Clementine' reference genetic map using 961 SNP, SSR, and Indel markers, spanning 1084.1 cM in 9 linkage groups. More recently, Aleza et al. (2015) located the centromere positions for all linkage groups. This information is of great importance for genetic studies based on recombination patterns, such as polyploid genetics.

Some of these markers are currently integrated in citrus breeding programs to assist in the selection of desired genotypes. Diverse molecular markers, such as the RAPD, SSR, and SNP linked to Alternaria Brown Spot resistance in mandarins, have been developed both at diploid (Dalkilic et al., 2005; Gulsen et al., 2010) and triploid levels (Cuenca et al., 2013, 2016). The locus linked to ABS resistance is located near the centromere in chromosome III and an SNP marker mapped at 0.4 cM from the Alternaria locus is used routinely for marker-assisted selection in mandarin breeding programs (Cuenca et al., 2016). The utility of this selection marker is detailed further in Section 4 'Case study: Achieving sustainability using triploid mandarin cultivars resistant to Alternaria Brown Spot disease' of this chapter. The presence of anthocyanin in the pulp of blood oranges is due to a transposable element insertion in the 5′ extremity of the *RUBY* gene (Butelli et al., 2012, 2017). This trait is being transferred to the mandarin genetic background through sexual hybridizations in order to obtain blood mandarins with the aid of molecular markers linked to the *RUBY* gene. Another trait that is subject to marker-assisted selection in some breeding programs is the apomixis (Wang et al., 2017b), linked to an *RWP-RK* gene that could be related with this character.

## 3.2 Genomic tools

The availability of a saturated genetic map of 'Clementine' was an essential prerequisite in helping to assemble the whole genome sequence (Wu et al., 2014) of a haploid 'Clementine' (Aleza et al., 2009a). Citrus, with a basic chromosome number of 9, has a relatively small genome size, estimated between 370 and 380 Mb (Arumuganathan and Earle, 1991).

As a first step toward genome sequencing, a low-coverage (1.2X) shotgun sequence of the sweet orange genome was carried out, revealing the difficulties related to high heterozygosity. Xu et al. (2013) published the draft sequence of dihaploid sweet orange and resequenced two pummelos and two mandarins in order to investigate the contribution of these two species in the hybrid origin of the sweet orange.

The International Consortium for Citrus Genomics, resulting from collaboration between research centers, private companies, and researchers from France, Italy, USA, Spain, and Brazil, has released the whole genome sequence assembly of a haploid 'Clementine' (Aleza et al., 2009a; Wu et al., 2014). This genome assembly is currently used as a template to organize sequences of highly heterozygous species such as *C. sinensis, C. paradisi*, and *C. limon*. Recently, Wang et al. (2017b), Wu et al. (2018), and Wang et al. (2018) published data on resequencing of several primitive, wild, and cultivated species, including some mandarins like 'Satsuma' (Shimizu et al., 2017). The complete chloroplast genome sequence of sweet orange was published by Bausher et al. (2006). Subsequently, Carbonell-Caballero et al. (2015) sequenced 34 chloroplast genomes, including several mandarin genotypes, with the aim to elucidate the relationships between wild and cultivated citrus species.

The analysis of the transcriptome allows the potential identification of candidate genes related to agronomic traits. Resources for genomic and transcriptomic studies have been developed over the past few years, such as expressed sequence tags (ESTs) collections and microarrays (Forment et al., 2005; Martinez-Godoy et al., 2008). These first studies have resulted in the characterization of gene expression patterns in different tissues or developmental stages in several genotypes (Ancillo et al., 2007; Cercós et al., 2006; Gandía et al., 2007; Alós et al., 2008; Huerta et al., 2008; Gimeno et al.,

2009; Aprile et al., 2011; Bassene et al., 2011; Boava et al., 2011). More recently, RNA-seq technology has resulted in many transcriptomic studies in several citrus species (Shalom et al., 2014; Yu et al., 2012; Sun et al., 2014; Patel et al., 2014; Long et al., 2016; Wang et al., 2017b; Zheng et al., 2012; Zhang et al., 2017). In the case of mandarins, RNA-seq and microarray technology were also used to study the effect of the ploidy level on gene expression (Jiao et al., 2013; Niñoles et al., 2015) and the response to *X. fastidiosa* bacterium infection in mandarins (Rodrigues et al., 2013). In addition, Terol et al. (2016) published an RNA-seq-based reference transcriptome for citrus from the analysis of flower, leaf, root, and phloem organs in 12 citrus species, including mandarins, providing a comprehensive framework for the Citrus transcriptome.

### 3.3 Biotechnological tools

Citrus normally present a juvenile phase in which new hybrids do not flower, and progressive transition to the adult phase can take around 4–6 years, depending on the genotype (Ollitrault and Navarro, 2012), thus hampering breeding based on sexual hybridization. The time needed to obtain a new cultivar with improved characteristics can be potentially reduced by efficient and reliable genetic transformation protocols (Peña et al., 2007), although the results are genotype-dependent.

A strategy for shortening the juvenile period is to produce transgenic plants overexpressing *LEAFY*, *APETALA1*, and *FLOWERING LOCUS T (FT)* genes (Peña et al., 2001; Nishikawa et al., 2010). Another strategy for shortening the juvenile phase is to use mature explants for transformation. The first successful transformation from citrus adult plants was achieved in the sweet range (Cervera et al., 1998) and, subsequently, protocols for transformation from the adult material have been established for other genotypes, such as 'Clementine mandarin' (Cervera et al., 2008).

Other successful transformation events have been performed both to confer stress tolerance and improve fruit quality (Peña et al., 2001; Alquézar et al., 2008; Pons et al., 2014). Many of these have been done specifically to confer disease resistance (Dominguez et al., 2000; Fagoaga et al., 2001; Rodriguez et al., 2011; Dutt et al., 2015; Hao et al., 2016; Riera et al., 2017; Rodriguez et al., 2018). Transgenic citrus varieties expressing genes for antibacterial peptides have already been obtained and are currently being tested for HLB-resistance in the field (Hao et al., 2016). Chemical volatiles repelling the HLB vector have been identified, which can be an efficient way to avoid HLB infection (Alquézar et al., 2017).

Somatic hybridization by protoplast fusion of two nonsexual cells is an important tool in citrus breeding programs, including mandarins. Somatic hybrids are allotetraploid plants possessing the entire nuclear genetic configuration of their parents (Grosser and Gmitter, 2005; Olivares-Fuster et al., 2005; Ollitrault et al., 2008; Dambier et al., 2011; Grosser et al., 2010; Grosser and Gmitter, 2011; Aleza et al., 2016b; Navarro et al., 2015). Somatic hybridization has been used for mandarin improvement using different approaches: the production of allotetraploid parents to be used in interploid crosses to generate triploid mandarins (Grosser and Chandler, 2004; Grosser et al., 2010); the production of triploids by fusing haploid and diploid cells (Kobayashi et al., 1997; Ollitrault et al., 2000); and the production of cybrid plants (diploid plants with the cytoplasm from one parent and the nucleus from the other) for developing seedless cultivars (Xu et al., 2006; Cai et al., 2009; Aleza et al., 2016b) or with improved disease resistance (Omar et al., 2017).

The use of viral vectors can be a very helpful tool to study the function of host genes by both forward and reverse genetics (known as virus-induced gene silencing – VIGS) (Burch-Smith et al., 2004). The use of viral vectors to evaluate plant gene functions is particularly attractive for woody species such as citrus, in which analysis of certain traits like flowering and fruiting by conventional breeding is hampered by their long juvenile period, though genetic transformation of adult plants can be difficult for some genotypes (Cervera et al., 2008). A viral vector based on the citrus leaf blotch virus (CLBV) was developed to be used in functional analyses (Agüero et al., 2012, 2013, 2014), proving its usefulness in inducing precocious flowering in juvenile plants by expressing the *FT* gene. This engineered vector is currently used in the Spanish mandarin breeding program to induce early flowering in the recently obtained diploid and tetraploid hybrids, which are used as parents to obtain triploid mandarin populations (Velázquez et al., 2016). These diploid and tetraploid hybrids are usually obtained from 2x X 2x and 4x X 2x seedlings, respectively, and selected for some traits using molecular markers. Genotypes arising from seeds generally require a minimum period of 4–6 years for flowering under normal conditions. However, using the CLBV–FT vector reduces this period drastically to approximately 6–9 months, allowing their use as parents in new combinations for mandarin breeding.

The VIGS technique has also been used in citrus to induce gene silencing in the HLB-vector *Diaphorina citri* (Hajeri et al., 2014) using the CTV-based RNAi technology (Folimonov et al., 2007). The use of viral vectors, together with the new genetic and genomic tools available, facilitates genetic analysis, allowing the development of markers for focused introgression and marker-assisted early selection.

More recently, several tools for precise plant genome editing have been developed. The use of the clustered regularly interspaced short palindromic repeats/CRISP-associated proteins (CRISPR/Cas) technology (Jinek et al., 2012) has been successfully applied for gene mutation or regulation in many plant species, including tree crops (Nishitani et al., 2016; Ren et al., 2016;

Song et al., 2016). This system recognizes a genome region defined by base complementarity based on the interaction between a single-guide RNA (sgRNA) and the DNA sequence of interest. In citrus, this strategy has been used for modifying the CsPDS gene expression (Jia and Wang, 2014) and for mutating the CsLOB1 gene, conferring some degree of resistance to citrus canker in sweet orange (Peng et al., 2017) and grapefruit (Jia et al., 2017). However, no results for mandarins have been published yet. The use of genome-editing technologies could overcome some limitations inherent to plant transformation and also legal issues in some countries. The USDA (USA) has deregulated the commercialization of two crops obtained through the CRISPR/Cas9 (Waltz, 2016) and has no plans to subject most gene-edited crops to the same regulatory process used for transgenic crops. In addition, other countries are also in favor of CRISPR-obtained plants to be less heavily regulated than transgenic crops (Whelan and Lema, 2015). By contrast, a recent decision of the Court of Justice of the European Union concluded that organisms obtained using CRISPR should be considered as genetic-modified (GM) plants, and therefore, subject to the GM regulations in the European Directive 2001/18/EG (Callaway, 2018), which means they cannot be differentiated from plants obtained through other mutation methods (Abbott, 2015; Cyranoski, 2015).

## 4 Case study: Achieving sustainability using triploid mandarin cultivars resistant to Alternaria Brown Spot disease

Seedlessness is one of the main quality attributes for mandarins. Commercially, a mandarin variety is considered as seedless when less than 5 seeds are found in a sample of 25 fruits. However, in the last few years, the new marketing criteria require no more than 4 seeds in 100 fruits without any one fruit containing more than 2 seeds (Goldenberg et al., 2018). Among mandarins, varieties of the 'Satsuma' group are generally sterile male and female; thus they do not produce seeds in their fruits. However, most commercial mandarin cultivars are characterized by male and female fertility, such as the 'Clementine' group and other important hybrids ('Nadorcott', 'Nova', 'Fortune', and 'Murcott'). These varieties are usually self-incompatible and thus, although they do not produce seeds when grown in isolated blocks, they can produce seeds and induce seed formation under cross-pollination conditions, i.e. in the presence of bees and when cultivated near sexually compatible cultivars (Navarro et al., 2015).

It is estimated that bees and other pollinators pollinate almost 90% of flowering plants and 70% of crops; without bees, these numbers would begin to rapidly decline (Shrestha et al., 2018). Several government agencies regulate the presence of hives to prevent citrus flowers from cross-pollinations, such as the State of California (USA) (California Code of Regulations – Title 3) and the Valencian region (Spain) (DOGV 24/03/2017), two of the main mandarin producing areas for the fresh market. These regulations discourage beekeepers from producing orange-blossom honey in these citrus producing areas, which diminishes bee populations. As a result, achieving genetic seedlessness, which would make it possible to obtain seedless fruits even in the presence of bees, has become an essential objective in most mandarin breeding programs.

There are several breeding strategies to produce new seedless mandarin varieties such as ploidy manipulation for obtaining triploid varieties, the management of male and female sterility in diploid genotypes, inducing mutagenesis, and the use of biotechnological approaches. Among them, ploidy manipulation is currently one of the most used strategies in mandarin breeding in obtaining new seedless cultivars (Ollitrault et al., 2008; Aleza et al., 2010a, 2012a,b). As noted, triploid hybrids are generally male and female sterile and, thus, they do not produce seeds in their fruits nor do they induce seed formation in other cultivars by cross-pollination.

The most promising strategies for triploid citrus breeding are the exploitation of the natural formation of unreduced gametes in diploid X diploid crosses through sexual polyploidization and the utilization of tetraploid parents in interploid (diploid X tetraploid and tetraploid X diploid) crosses (Cuenca et al., 2011; Aleza et al., 2012a,b; Cuenca et al., 2015b). In sexual polyploidization, mandarin triploid embryos are produced in a frequency of 1–20 triploids per 100 fruits, depending on the genotype (Aleza et al., 2010a). Triploid hybrids arise usually from the union of an unreduced (2n) megagametophyte with haploid pollen (Esen and Soost, 1971, 1973; Geraci et al., 1975; Luro et al., 2004; Cameron and Frost, 1968; Cuenca et al., 2011, 2015b). Unreduced diploid pollen has also been reported in a hybrid mandarin X orange (Rouiss et al., 2017). The main mechanism leading to unreduced megagametophyte formation in mandarins is a failure in the second meiotic division, namely second-division restitution (SDR) (Cuenca et al., 2011, 2015b; Aleza et al., 2016a). As a result, the sister chromatids are transmitted in the same diploid gamete and, subsequently, the parental heterozygosity level is reduced (Bastiaanssen et al., 1998; Cuenca et al., 2011; De Storme and Geelen, 2013; Park et al., 2007). By contrast, first-division restitution (FDR) has been described as the main mechanism underlying unreduced pollen formation in citrus (Rouiss et al., 2017). The different mechanisms of unreduced gamete formation have different genetic consequences, particularly affecting the transmission of parental heterozygosity, which, in turn, determines the genetic structure of the resulting triploid populations. The importance of the mechanism of origin is reflected in the particular transmission of the parental heterozygosity in relation to the distance to the centromere. Under FDR, the nonsister chromatids transmit 100% of the parental heterozygosity at the centromere, decreasing to 50% at the telomeres. In the case of SDR, the two sister chromatids transmit no-heterozygosity at the centromere, increasing to 100% of the parental heterozygosity transmission at the telomeres. As a result, the SDR-2n gametes have lower levels of global heterozygosity than the FDR ones (Bretagnolle and Thompson, 1995; Cai and Xu, 2007; Cuenca et al., 2011).

Another useful strategy to obtain mandarin triploid hybrids more efficiently is the use of tetraploid parents crossed with diploid parents (interploid crosses) (Cameron and Bernett, 1978; Esen et al., 1979; Aleza et al., 2012a,b). These tetraploid parents are usually a doubled diploid (DD) obtained by nucellar doubling or colchicine treatments of diploid genotypes (Aleza et al., 2009b, 2011) or tetraploid hybrids, including those obtained by protoplast fusion, as indicated previously. Most of the embryos obtained in interploid crosses are triploid, thus constituting a highly efficient strategy for triploid breeding (Aleza et al., 2012a,b). The genetic outcomes in the segregating populations for this triploid breeding strategy depend on the rate of preferential chromosome pairing (ratio disomic/tetrasomic segregations) and on the double reduction rate. The analysis of the genetic structure of diploid gametes arising from DD 'Clementine' revealed tetrasomic segregation, with some level of disomic and intermediate segregation for some linkage groups; significant double reduction rates were also observed in six of the nine LGs (Aleza et al., 2016a). In addition, the comparison between this population and a population of unreduced gametes from diploid 'Clementine' formed through SDR revealed that both strategies are complementary in terms of genotypic variability. Interploid hybridizations are useful for developing cultivars similar to the tetraploid parent, while sexual hybridization through SDR may potentially create different varieties with new characteristics.

The triploid mandarin breeding strategy is widely exploited in several countries, such as Spain, Italy, France, and the United States (Williams and Roose, 2004; Recupero et al., 2005; Ollitrault et al., 2008; Aleza et al., 2010a, 2012a,b; Grosser et al., 2015; Navarro et al., 2015). As a result, several triploid hybrids have been released. At IVIA in Spain, the triploid breeding program started in 1996 (Navarro et al., 2015) with the aim to obtain new high-quality, easy-peeling, seedless mandarin cultivars, based on both 2x × 2x and interploid sexual hybridizations followed by embryo rescue and flow cytometry. Within this program, more than 6500 triploid hybrids from 209 parental combinations by 2x × 2x pollinations, more than 5300 triploid hybrids from 123 parental combinations by 2x × 4x pollinations, and more than 5800 triploid hybrids from 121 parental combinations by 4x × 2x pollinations have been obtained.

Due to constraints in the citrus reproductive system, such as apomixis in most mandarins, desirable characteristics, or the capacity to produce triploid hybrids (Aleza et al., 2010b), the range of usable parents in mandarin breeding programs is limited. In this sense, many cultivars that are susceptible to ABS disease are being used as parents in mandarin breeding programs, both at diploid and tetraploid levels (McCollum, 2007; Aleza et al., 2010a,b, 2012a,b; Cuenca et al., 2010; Grosser et al., 2010). Some examples of the ABS-susceptible cultivars used as parents in mandarin breeding programs worldwide are 'Fortune', 'Murcott', 'Ponkan', 'Dancy', 'Minneola', 'Nova' or 'Orlando', among others.

ABS is caused by the fungus *Alternaria alternata* and is an important disease affecting mandarins worldwide (Vicent et al., 2000; Timmer et al., 2003; Golmohammadi et al., 2006; Wang et al., 2010). ABS control is primarily based on fungicide application, with up to 10 sprays per year needed to produce quality fruit for the fresh market (Vicent et al., 2007). Genetic resistance represents the best option for controlling the disease (Bhatia et al., 2003; Peres and Timmer, 2006). It offers the possibility of significantly reducing or eliminating the need for pesticides and, at the same time, creating an environmentally safer and more sustainable agriculture.

Inheritance of ABS resistance in citrus is controlled by a single recessive allele (*a*), demonstrated from both diploid (Dalkilic et al., 2005; Gulsen et al., 2010) and triploid progeny analyses (Cuenca et al., 2013, 2016). In this context, only homozygous *aa/aaa* genotypes are resistant to the disease, whereas other allelic configurations (*AA/Aa/AAA/AAa/Aaa*) lead to susceptibility. Hybridizations between resistant cultivars produce only resistant hybrids; nevertheless, the use of susceptible cultivars as parents widens the potential diversity within the breeding programs, but selection of the ABS-resistant hybrids is needed in this case.

Resistance to ABS can be evaluated through *in vitro* inoculation of young leaves with spores of *A. alternata* and selection can be done at a relatively early stage of plantlet development (Vicent et al., 2000). However, it requires the use of fungus spores and young leaves present in a specific developmental stage, which are not always available. In addition, this method is very aggressive and susceptibility to disease can eventually be observed when testing resistant genotypes.

An alternative is selection of ABS-resistant genotypes through phenotyping/biological assays, using marker-assisted selection (MAS) through the identification of genomic regions involved in the ABS response. In a first step, Cuenca et al. (2013) identified a 3.3 Mb region near the centromere in chromosome III, which is linked to the ABS response in mandarins. Later, the same authors performed a fine-mapping of this region and narrowed the linked region to 366 kb and developed SNP markers for an efficient selection of ABS-resistant genotypes (Cuenca et al., 2016). The closest SNP was mapped at 0.4 cM from the ABS resistance locus, allowing the selection of resistant genotypes in very early developmental stages, thus avoiding growing and evaluating susceptible genotypes in the greenhouse and field. The close correlation between the SNP genotyping and the actual allele configuration of the resistance locus was confirmed in the mandarin germplasm. This marker is therefore very useful not only for discriminating between resistant and susceptible cultivars, but also for inferring the allele configuration of the ABS resistance locus. It allows selecting susceptible heterozygous genotypes that can be used as parents in breeding programs and conversely discarding homozygous susceptible varieties.

This SNP marker is currently used for the assisted selection of ABS-resistant genotypes in several mandarin breeding programs worldwide both at diploid and triploid levels. At the Spanish triploid breeding program, it has been successfully applied to select more than 4800 resistant hybrids from more than 9700 hybrids arising from 100 different parental combinations. This analysis is performed at a very early stage of plantlet development, including in-vitro plants, with close to 5000 susceptible genotypes removed from the breeding program. This elimination avoids the need to grow these noninteresting genotypes, which means saving a considerable amount of time, greenhouse and field space, personnel, and money.

In summary, the mandarin breeding program carried out at IVIA in Spain is contributing to sustainable cultivation through applying biotechnological approaches. On the one hand, triploid hybrids provide a solution for seedlessness without diminishing bee populations since no treatments are needed against pollinators, and hives can be maintained in the cultivation area. On the other hand, the application of MAS for selecting ABS-resistant genotypes is environmentally friendly since it ensures that no susceptible hybrids would be grown and thus, no fungicide applications will be needed.

## 5 Conclusions

Recent advances on citrus molecular genetics, genomics and biotechnology allow developing more efficient breeding programs aiming to improve sustainability by obtaining new mandarin varieties with high fruit quality, enriched with health-promoting compounds, seedless and disease resistant. In this regard, triploid breeding has been shown as a useful strategy, resulting in very promising outcomes. The efficiency of mandarin breeding programs will be continuously improved by acquiring new knowledge on gene-trait associations and gene regulations, especially from new genome editing experiments. New gene functions discovery and directed mutation to improve elite mandarin cultivars will be reached through CRISPR/Cas system which has positioned as the most promising strategy for genome editing.

## 6 Where to look for further information

These new results are expected to result from the work developed in several research centers worldwide. The Spanish mandarin breeding program is led by the 'Instituto Valenciano de Investigaciones Agrarias' (www.ivia.es). The French program is developed by the 'Institut National de la Recherche Agronomique/La Recherche Agronomique pour le Développement' (INRA/CIRAD) (www.cirad.fr). The 'Consiglio per la ricerca in agricoltura e l'analisi dell'economia agraria (CREA)' (http://sito.entecra.it) develops mandarin breeding programs in Italy. The 'Agricultural Research Organization (ARO)' at the Volcani Center (www.agri.gov.il) develops breeding programs in Israel. The 'Citrus Research and Education Center' (CREC) in Florida (http://www.crec.ifas.ufl.edu) and the 'Agricultural Experiment Station-Citrus Research Center' in California (www.citrusvariety.ucr.edu) develop mandarin breeding programs in the United States. The 'Institute of Citrus Science (ICS), HZAU' (http://ics.hzau.edu.cn) leads mandarin programs in China and the 'NARO Institute of Fruit Tree Science Citrus Research Station' (www.naro.affrc.go.jp) in Japan. In Australia, mandarin breeding programs are developed by the 'Commonwealth Scientific and Industrial Research Organization (CSIRO)' (www.csiro.au).

## 7 References

Abbott, A. 2015. Europe's genetically edited plants stuck in legal limbo. *Nature* 528(7582), 319–20. doi:10.1038/528319a.

Agüero, J., Ruiz-Ruiz, S., Vives, M. C., Velázquez, K., Navarro, L., Peña, L., Moreno, P. and Guerri, J. 2012. Development of viral vectors based on Citrus leaf blotch virus to express foreign proteins or analyze gene function in citrus plants. *Mol. Plant Microbe Interact.* 25(10), 1326–37. doi:10.1094/MPMI-02-12-0048-R.

Agüero, J., Vives, M. C., Velázquez, K., Ruiz-Ruiz, S., Juárez, J., Navarro, L., Moreno, P. and Guerri, J. 2013. Citrus leaf blotch virus invades meristematic regions in *Nicotiana benthamiana* and citrus. *Mol. Plant Pathol.* 14(6), 610–6. doi:10.1111/mpp.12031.

Agüero, J., Vives, Mdel C., Velázquez, K., Pina, J. A., Navarro, L., Moreno, P. and Guerri, J. 2014. Effectiveness of gene silencing induced by viral vectors based on citrus leaf blotch virus is different in *Nicotiana benthamiana* and citrus plants. *Virology* 460-461, 154–64. doi:10.1016/j.virol.2014.04.017.

Aleza, P. 2015. Workshop on new mandarin varieties. *Acta Hortic.* 1065, 193–9.

Aleza, P., Juárez, J., Hernández, M., Pina, J. A., Ollitrault, P. and Navarro, L. 2009a. Recovery and characterization of a *Citrus clementina* Hort. ex Tan. 'Clemenules' haploid plant selected to establish the reference whole Citrus genome sequence. *BMC Plant Biol.* 9(1), 110. doi:10.1186/1471-2229-9-110.

Aleza, P., Juárez, J., Ollitrault, P. and Navarro, L. 2009b. Production of tetraploid plants of non apomictic citrus genotypes. *Plant Cell Rep.* 28(12), 1837–46. doi:10.1007/s00299-009-0783-2.

Aleza, P., Juárez, J., Cuenca, J., Ollitrault, P. and Navarro, L. 2010a. Recovery of citrus triploid hybrids by embryo rescue and flow cytometry from 2x x 2x sexual hybridisation and its application to extensive breeding programs. *Plant Cell Rep.* 29(9), 1023–34. doi:10.1007/s00299-010-0888-7.

Aleza, P., Cuenca, J., Juárez, J., Pina, J. A. and Navarro, L. 2010b. 'Garbi' mandarin: a new late-maturing triploid hybrid. *HortScience* 45(1), 139–41.

Aleza, P., Froelicher, Y., Schwarz, S., Agustí, M., Hernández, M., Juárez, J., Luro, F., Morillon, R., Navarro, L. and Ollitrault, P. 2011. Tetraploidization events by chromosome doubling of nucellar cells are frequent in apomictic citrus and are dependent on genotype and environment. *Ann. Bot.* 108(1), 37–50. doi:10.1093/aob/mcr099.

Aleza, P., Juárez, J., Cuenca, J., Ollitrault, P. and Navarro, L. 2012a. Extensive citrus triploid hybrid production by 2x x 4x sexual hybridizations and parent-effect on the length of the juvenile phase. *Plant Cell Rep.* 31(9), 1723–35. doi:10.1007/s00299-012-1286-0.

Aleza, P., Juárez, J., Hernández, M., Ollitrault, P. and Navarro, L. 2012b. Implementation of extensive citrus triploid breeding programs based on 4x × 2x sexual hybridisations. *Tree Genet. Genom.* 8(6), 1293–306. doi:10.1007/s11295-012-0515-6.

Aleza, P., Cuenca, J., Hernández, M., Juárez, J., Navarro, L. and Ollitrault, P. 2015. Genetic mapping of centromeres of the nine *Citrus clementina* chromosomes using half-tetrad analysis and recombination patterns in unreduced and haploid gametes. *BMC Plant Biol.* 15, 80. doi:10.1186/s12870-015-0464-y.

Aleza, P., Cuenca, J., Juárez, J., Navarro, L. and Ollitrault, P. 2016a. Inheritance in doubled-diploid clementine and comparative study with SDR unreduced gametes of diploid clementine. *Plant Cell Rep.* 35(8), 1573–86. doi:10.1007/s00299-016-1972-4.

Aleza, P., Garcia-Lor, A., Juárez, J. and Navarro, L. 2016b. Recovery of citrus cybrid plants with diverse mitochondrial and chloroplastic genome combinations by protoplast fusion followed by *in vitro* shoot, root and embryo micrografting. *Plant Cell Tissue Organ Cult.* 126(2), 205–17. doi:10.1007/s11240-016-0991-8.

Alós, E., Roca, M., Iglesias, D. J., Mínguez-Mosquera, M. I., Damasceno, C. M., Thannhauser, T. W., Rose, J. K., Talón, M. and Cercós, M. 2008. An evaluation of the basis and consequences of a stay-green mutation in the navel negra citrus mutant using transcriptomic and proteomic profiling and metabolite analysis. *Plant Physiol.* 147(3), 1300–15. doi:10.1104/pp.108.119917.

Alquézar, B., Rodrigo, M. J. and Zacarías, L. 2008. Regulation of carotenoid biosynthesis during fruit maturation in the red-fleshed orange mutant Cara Cara. *Phytochemistry* 69(10), 1997–2007. doi:10.1016/j.phytochem.2008.04.020.

Alquézar, B., Rodríguez, A., de la Peña, M. and Peña, L. 2017. Genomic analysis of terpene synthase family and functional characterization of seven sesquiterpene synthases from *Citrus sinensis*. *Front. Plant Sci.* 8, 1481. doi:10.3389/fpls.2017.01481.

Ancillo, G., Gadea, J., Forment, J., Guerri, J. and Navarro, L. 2007. Class prediction of closely related plant varieties using gene expression profiling. *J. Exp. Bot.* 58(8), 1927–33. doi:10.1093/jxb/erm054.

Aprile, A., Federici, C., Close, T. , De Bellis, L., Cattivelli, L. and Roose, M. L. 2011. Expression of the H+-ATPase AHA10 proton pump is associated with citric acid accumulation in lemon juice sac cells. *Funct. Integr. Genom.* 11, 551–63. doi:10.1007/s10142-011-0226-3.

Arumuganathan, K. and Earle, E. D. 1991. Nuclear DNA content of some important plant species. *Plant Mol. Biol. Rep.* 9(3), 208–18. doi:10.1007/BF02672069.

Asins, M. J., Juárez, J., Pina, J. A., Puchades, J., Carbonell, E. A. and Navarro, L. 2002. Nulessín, una nueva clementina. *Levante Agric.* 359, 36–40.

Asins, M. J., Bernet, G. P., Ruiz, C., Cambra, M., Guerri, J. and Carbonell, E. A. 2004. QTL analysis of *Citrus tristeza* virus-citradia interaction. *Theor. Appl. Genet.* 108(4), 603–11. doi:10.1007/s00122-003-1486-7.

Barkley, N. A., Roose, M. L., Krueger, R. R. and Federici, C. T. 2006. Assessing genetic diversity and population structure in a citrus germplasm collection utilizing simple sequence repeat markers (SSRs). *Theor. Appl. Genet.* 112(8), 1519–31. doi:10.1007/s00122-006-0255-9.

Barrett, H. C. and Rhodes, A. M. 1976. A numerical taxonomic study of affinity relationships in cultivated Citrus and its close relatives. *Syst. Bot.* 1(2), 105–36. doi:10.2307/2418763.

Barry, G. H., Gmitter Jr., F. G., Chen, C., Roose, M., Federici, C. T. and McCollum, G. T. 2015. Investigating the parentage of 'Orri' and 'Fortune' mandarin hybrids. *Acta Hortic.* 1065, 449–56.

Bassene, J. B., Froelicher, Y., Navarro, L., Ollitrault, P. and Ancillo, G. 2011. Influence of mitochondria on gene expression in a citrus cybrid. *Plant Cell Rep.* 30(6), 1077–85. doi:10.1007/s00299-011-1014-1.

Bastiaanssen, H. J. M., Van Den Berg, P. M. M. M., Lindhout, P., Jacobsen, E. and Ramanna, M. S. 1998. Postmeiotic restitution in 2n-egg formation of diploid potato. *Heredity* 81(1), 20–7. doi:10.1046/j.1365-2540.1998.00370.x.

Bausher, M. G., Singh, N. D., Lee, S. B., Jansen, R. K. and Daniell, H. 2006. The complete chloroplast genome sequence of *Citrus sinensis* (L.) Osbeck var 'Ridge Pineapple': organization and phylogenetic relationships to other angiosperms. *BMC Plant Biol.* 6, 21. doi:10.1186/1471-2229-6-21.

Bermejo, A., Pardo, J. and Cano, A. 2011. Influence of gamma irradiation on seedless citrus production: pollen germination and fruit quality. *Food Nutr. Sci.* 02(3), 169–80. doi:10.4236/fns.2011.23024.

Bermejo, A., Pardo, J. and Cano, A. 2012. Murcott seedless: influence of gamma irradiation on citrus production and fruit quality. *Span. J. Agric. Res.* 10(3), 768–77. doi:10.5424/sjar/2012103-460-11.

Bhatia, A., Roberts, P. D. and Timmer, L. W. 2003. Evaluation of the alter-rater model for timing of fungicide applications for control of Alternaria brown spot of citrus. *Plant Dis.* 87(9), 1089-93. doi:10.1094/PDIS.2003.87.9.1089.

Boava, L. P., Cristofani-Yaly, M., Mafra, V. S., Kubo, K., Kishi, L. T., Takita, M. A., Ribeiro-Alves, M. and Machado, M. A. 2011. Global gene expression of *Poncirus trifoliata*, *Citrus sunki* and their hybrids under infection of *Phytophthora parasitica*. *BMC Genomics* 12, 39. doi:10.1186/1471-2164-12-39.

Bové, J. M. 2006. Huanglongbing: a destructive, newly-emerging, century-old disease of citrus. *J. Plant Pathol.* 88, 7–37.

Bretagnolle, F. and Thompson, J. D. 1995. Tansley Review no-78 - Gametes with the somatic chromosome-number - mechanisms of their formation and role in the evolution of autopolyploid plants. *New Phytol.* 129(1), 1–22. doi:10.1111/j.1469-8137.1995.tb03005.x.

Burch-Smith, T. M., Anderson, J. C., Martin, G. B. and Dinesh-Kumar, S. P. 2004. Applications and advantages of virus-induced gene silencing for gene function studies in plants. *Plant J.* 39(5), 734–46. doi:10.1111/j.1365-313X.2004.02158.x.

Butelli, E., Licciardello, C., Zhang, Y., Liu, J., Mackay, S., Bailey, P., Reforgiato-Recupero, G. and Martin, C. 2012. Retrotransposons control fruit-specific, cold-dependent accumulation of anthocyanins in blood oranges. *Plant Cell* 24(3), 1242–55. doi:10.1105/tpc.111.095232.

Butelli, E., Garcia-Lor, A., Licciardello, C., Las Casas, G., Hill, L., Recupero, G. R., Keremane, M. L., Ramadugu, C., Krueger, R., Xu, Q., Deng, X., Fanciullino, A. L., Froelicher, Y., Navarro, L. and Martin, C. 2017. Changes in anthocyanin production during domestication of citrus. *Plant Physiol.* 173(4), 2225–42. doi:10.1104/pp.16.01701.

Cai, X. and Xu, S. S. 2007. Meiosis-driven genome variation in plants. *Curr. Genom.* 8(3), 151–61. doi:10.2174/138920207780833847.

Cai, X., Fu, J., Chen, C. and Guo, W. 2009. Cybrid/hybrid plants regenerated from somatic fusions between male sterile 'Satsuma' mandarin and seedy tangelos. *Sci. Hortic.* 122(2), 323–7. doi:10.1016/j.scienta.2009.05.017.

Callaway, E. 2018. CRISPR plants now subject to tough GM laws in European Union. *Nature* 560(7716), 16. doi:10.1038/d41586-018-05814-6.

Cambra, M. and Moreno, P. 2000. Tristeza. En: Dran-Vila, N. and Moreno, P. (Eds), *Enfermedades de los cítricos. Parte V: Enfermedades producidas por virus, viroides y agentes similares*. Monografía de la Sociedad Española de Fitopatología N° 2. SEF-Mundi Prensa, pp. 77–81.

Cameron, J. W. and Bernett, R. H. 1978. Use of sexual tetraploid seed parents for production of triploid citrus hybrids. *HortScience* 13, 167–9.

Cameron, J. W. and Frost, H. B. 1968. Genetics, breeding and nucellar embryony. In: Reuther, W., Batcherlor, L. D. and Webber, H. J. (Eds), *The Citrus Industry* (vol. II). University of California Press, Berkeley, pp. 325–70.

Carbonell-Caballero, J., Alonso, R., Ibañez, V., Terol, J., Talon, M. and Dopazo, J. 2015. A phylogenetic analysis of 34 chloroplast genomes elucidates the relationships between wild and domestic species within the genus citrus. *Mol. Biol. Evol.* 32(8), 2015–35. doi:10.1093/molbev/msv082.

Cercós, M., Soler, G., Iglesias, D. J., Gadea, J., Forment, J. and Talón, M. 2006. Global analysis of gene expression during development and ripening of citrus fruit flesh. A proposed mechanism for citric acid utilization. *Plant Mol. Biol.* 62(4–5), 513–27. doi:10.1007/s11103-006-9037-7.

Cervera, M., Juárez, J., Navarro, A., Pina, J. A., Duran-Vila, N., Navarro, L. and Pena, L. 1998. Genetic transformation and regeneration of mature tissues of woody fruit plants bypassing the juvenile stage. *Transgenic Res.* 7, 51–9.

Cervera, M., Navarro, A., Navarro, L. and Peña, L. 2008. Production of transgenic adult plants from clementine mandarin by enhancing cell competence for transformation and regeneration. *Tree Physiol.* 28(1), 55–66. doi:10.1093/treephys/28.1.55.

Cheng, F. S. and Roose, M. L. 1995. Origin and inheritance of dwarfing by the citrus rootstock *Poncirus trifoliate* 'Flying Dragon'. *J. Amer. Soc. Hortic. Sci.* 120(2), 286–91. doi:10.21273/JASHS.120.2.286.

Cuenca, J., Aleza, P., Juárez, J., Pina, J. A. and Navarro, L. 2010. 'Safor' mandarin: a new citrus midlate triploid hybrid. *HortScience* 45, 977–80.

Cuenca, J., Froelicher, Y., Aleza, P., Juárez, J., Navarro, L. and Ollitrault, P. 2011. Multilocus half-tetrad analysis and centromere mapping in citrus: evidence of SDR mechanism for 2n megagametophyte production and partial chiasma interference in mandarin cv. 'Fortune'. *Heredity* 107(5), 462–70. doi:10.1038/hdy.2011.33.

Cuenca, J., Aleza, P., Vicent, A., Brunel, D., Ollitrault, P. and Navarro, L. 2013. Genetically based location from triploid populations and gene ontology of a 3.3-Mb genome region linked to Alternaria brown spot resistance in citrus reveal clusters of resistance genes. *PLoS ONE* 8(10), e76755. doi:10.1371/journal.pone.0076755.

Cuenca, J., Aleza, P., Juárez, J., Pina, J. A. and Navarro, L. 2015a. Two new IVIA triploid mandarin hybrids: 'Alborea' and 'Albir'. *ISHS Acta Hortic.* 1065, 22. International Citrus Congress - International Society of Citriculture, Valencia, Spain, 2012.

Cuenca, J., Aleza, P., Juárez, J., García-Lor, A., Froelicher, Y., Navarro, L. and Ollitrault, P. 2015b. Maximum-likelihood method identifies meiotic restitution mechanism from heterozygosity transmission of centromeric loci: application in citrus. *Sci. Rep.* 5, 9897. doi:10.1038/srep09897.

Cuenca, J., Aleza, P., Garcia-Lor, A., Ollitrault, P. and Navarro, L. 2016. Fine mapping for identification of citrus Alternaria brown spot candidate resistance genes and development of new SNP markers for marker-assisted selection. *Front. Plant Sci.* 7, 1948. doi:10.3389/fpls.2016.01948.

Cuenca, J., Garcia-Lor, A., Navarro, L. and Aleza, P. 2018. Citrus genetics and breeding. In: Al-Khayri, J., Jain, S. M. and Johnson, D. V. (Eds), *Advances in Plant Breeding Strategies: Fruits*. Springer International Publishing, Berlin, Germany, pp. 403–36.

Curk, F., Ancillo, G., Garcia-Lor, A., Luro, F., Perrier, X., Jacquemoud-Collet, J. P., Navarro, L. and Ollitrault, P. 2014. Next generation haplotyping to decipher nuclear genomic interspecific admixture in Citrus species: analysis of chromosome 2. *BMC Genet.* 15, 152. doi:10.1186/s12863-014-0152-1.

Curk, F., Ancillo, G., Ollitrault, F., Perrier, X., Jacquemoud-Collet, J. P., Garcia-Lor, A., Navarro, L. and Ollitrault, P. 2015. Nuclear species-diagnostic SNP markers mined from 454 amplicon sequencing reveal admixture genomic structure of modern citrus varieties. *PLoS ONE* 10(5), e0125628. doi:10.1371/journal.pone.0125628.

Curk, F., Ollitrault, F., Garcia-Lor, A., Luro, F., Navarro, L. and Ollitrault, P. 2016. Phylogenetic origin of limes and lemons revealed by cytoplasmic and nuclear markers. *Ann. Bot.* 117(4), 565–83. doi:10.1093/aob/mcw005.

Cyranoski, D. 2015. CRISPR tweak may help gene-edited crops bypass biosafety regulation. *Nature* 18590. doi:10.1038/nature.2015.18590.

Da Graça, J. V., Douhan, G. W., Halbert, S. E., Keremane, M. L., Lee, R. F., Vidalakis, G. and Zhao, H. 2016. Huanglongbing: an overview of a complex pathosystem ravaging the world's citrus. *J. Integr. Plant Biol.* 58(4), 373–87. doi:10.1111/jipb.12437.

Dalkilic, Z., Timmer, L. W. and Gmitter, F. G. 2005. Linkage of an Alternaria disease resistance gene in mandarin hybrids with RAPD fragments. *J. Amer. Soc. Hortic. Sci.* 130(2), 191–5. doi:10.21273/JASHS.130.2.191.

Dambier, D., Benyahia, H., Pensabene-Bellavia, G., Aka Kaçar, Y., Froelicher, Y., Belfalah, Z., Lhou, B., Handaji, N., Printz, B., Morillon, R., Yesiloglu, T., Navarro, L. and Ollitrault, P. 2011. Somatic hybridization for citrus rootstock breeding: an effective tool to solve some important issues of the Mediterranean citrus industry. *Plant Cell Rep.* 30(5), 883–900. doi:10.1007/s00299-010-1000-z.

De Storme, N. and Geelen, D. 2013. Sexual polyploidization in plants – cytological mechanisms and molecular regulation. *New Phytol.* 198(3), 670–84. doi:10.1111/nph.12184.

Dominguez, A., Guerri, J., Cambra, M., Navarro, L., Moreno, P. and Peña, L. 2000. Efficient production of transgenic citrus plants expressing the coat protein gene of *Citrus tristeza* virus. *Plant Cell Rep.* 19(4), 427–33. doi:10.1007/s002990050751.

Dutt, M., Barthe, G., Irey, M. and Grosser, J. 2015. Transgenic citrus expressing an *Arabidopsis* NPR1 gene exhibit enhanced resistance against Huanglongbing (HLB; citrus greening). *PLoS ONE* 10(9), e0137134. doi:10.1371/journal.pone.0137134.

Esen, A. and Soost, R. K. 1971. Unexpected triploids in Citrus: their origin, identification, and possible use. *J. Hered.* 62(6), 329–33. doi:10.1093/oxfordjournals.jhered.a108186.

Esen, A. and Soost, R. K. 1973. Precocious development and germination of spontaneous triploid seeds in Citrus. *J. Hered.* 64(3), 147–54. doi:10.1093/oxfordjournals.jhered.a108373.

Esen, A., Soost, R. K. and Geraci, G. 1979. Genetic evidence for the origin of diploid megagametophytes in Citrus. *J. Hered.* 70(1), 5–8. doi:10.1093/oxfordjournals.jhered.a109188.

Fagoaga, C., Rodrigo, I., Conejero, V., Hinarejos, C., Tuset, J. J., Arnau, J., Pina, J. A., Navarro, L. and Peña, L. 2001. Increased tolerance to *Phytophthora citrophthora* in transgenic orange plants constitutively expressing a tomato pathogenesis related protein PR-5. *Mol. Breed.* 7(2), 175–85. doi:10.1023/A:1011358005054.

Fang, D. Q., Roose, M. L., Krueger, R. R. and Federici, C. T. 1997. Fingerprinting trifoliate orange germ plasm accessions with isozymes, RFLPs, and inter-simple sequence repeat markers. *Theor. Appl. Genet.* 95(1–2), 211–9. doi:10.1007/s001220050550.

FAO. 2018. Statistics. Available at: http://www.faostat.org.

Federici, C. T., Fang, D. Q., Scora, R. W. and Roose, M. L. 1998. Phylogenetic relationships within the genus Citrus (Rutaceae) and related genera as revealed by RFLP and RAPD analysis. *Theor. Appl. Genet.* 96(6–7), 812–22. doi:10.1007/s001220050807.

Folimonov, A. S., Folimonova, S. Y., Bar-Joseph, M. and Dawson, W. O. 2007. A stable RNA virus-based vector for citrus trees. *Virology* 368(1), 205–16. doi:10.1016/j.virol.2007.06.038.

Forment, J., Gadea, J., Huerta, L., Abizanda, L., Agusti, J., Alamar, S., Alos, E., Andres, F., Arribas, R., Beltran, J. P., Berbel, A., Blazquez, M. A., Brumos, J., Canas, L. A., Cercos, M., Colmenero-Flores, J. M., Conesa, A., Estables, B., Gandia, M., Garcia-Martinez, J. L., Gimeno, J., Gisbert, A., Gomez, G., Gonzalez-Candelas, L., Granell, A., Guerri, J., Lafuente, M. T., Madueno, F., Marcos, J. F., Marques, M. C., Martinez, F., Martinez-Godoy, M. A., Miralles, S., Moreno, P., Navarro, L., Pallas, V., Perez-Amador, M. A., Perez-Valle, J., Pons, C., Rodrigo, I., Rodriguez, P. L., Royo, C., Serrano, R., Soler, G., Tadeo, F., Talon, M., Terol, J., Trenor, M., Vaello, L., Vicente, O., Vidal, Ch, Zacarias, L. and Conejero, V. 2005. Development of a citrus genome-wide EST collection and cDNA microarray as resources for genomic studies. *Plant Mol. Biol.* 57(3), 375–91. doi:10.1007/s11103-004-7926-1.

Froelicher, Y., Mouhaya, W., Bassene, J. B., Costantino, G., Kamiri, M., Luro, F., Morillon, R. and Ollitrault, P. 2011. New universal mitochondrial PCR markers reveal new information on maternal citrus phylogeny. *Tree Genet. Genom.* 7(1), 49–61. doi:10.1007/s11295-010-0314-x.

Fujii, H., Ohta, S., Nonaka, K., Katayose, Y., Matsumoto, T., Endo, T., Yoshioka, T., Omura, M. and Shimada, T. 2016. Parental diagnosis of Satsuma mandarin (*Citrus unshiu* Marc.) revealed by nuclear and cytoplasmic markers. *Breed. Sci.* 66(5), 683–91. doi:10.1270/jsbbs.16060.

Gandía, M., Conesa, A., Ancillo, G., Gadea, J., Forment, J., Pallás, V., Flores, R., Duran-Vila, N., Moreno, P. and Guerri, J. 2007. Transcriptional response of *Citrus aurantifolia* to infection by *Citrus tristeza* virus. *Virology* 367(2), 298–306. doi:10.1016/j.virol.2007.05.025.

Garcia-Lor, A., Curk, F., Snoussi-Trifa, H., Morillon, R., Ancillo, G., Luro, F., Navarro, L. and Ollitrault, P. 2013. A nuclear phylogenetic analysis; SNPs, indels and SSRs deliver new insights into the relationships in the 'true citrus fruit trees' group (Citrinae, Rutaceae) and the origin of cultivated species. *Ann. Bot.* 111(1), 1–19. doi:10.1093/aob/mcs227.

Garcia-Lor, A., Luro, F., Ollitrault, P. and Navarro, L. 2015. Genetic diversity and population structure analysis of mandarin germplasm by nuclear, chloroplastic and mitochondrial markers. *Tree Genet. Genom.* 11(6), e123. doi:10.1007/s11295-015-0951-1.

Gentile, A., Tribulato, E., Deng, Z. N. and Vardi, A. 1992. *In vitro* selection of nucellar lemon callus and regeneration of plants tolerant to *Phoma tracheiphila* toxin. *Adv. Hortic. Sci.* 6, 151–4.

Geraci, G., Esen, A. and Soost, R. K. 1975. Triploid progenies from 2x x 2x crosses of Citrus cultivars. *J. Hered.* 66(3), 177–8. doi:10.1093/oxfordjournals.jhered.a108607.

Gimeno, J., Gadea, J., Forment, J., Pérez-Valle, J., Santiago, J., Martínez-Godoy, M. A., Yenush, L., Bellés, J. M., Brumós, J., Colmenero-Flores, J. M., Talón, M. and Serrano, R. 2009. Shared and novel molecular responses of mandarin to drought. *Plant Mol. Biol.* 70(4), 403–20. doi:10.1007/s11103-009-9481-2.

Gois, I. B., Borém, A., Cristofani-Yaly, M., de Resende, M. D., Azevedo, C. F., Bastianel, M., Novelli, V. M. and Machado, M. A. 2016. Genome wide selection in Citrus breeding. *Genet. Mol. Res.* 15(4), gmr15048863. doi:10.4238/gmr15048863.

Goldenberg, L., Yaniv, Y., Porat, R. and Carmi, N. 2018. Mandarin fruit quality: a review. *J. Sci. Food Agric.* 98(1), 18–26. doi:10.1002/jsfa.8495.

Golmohammadi, M., Andrew, M., Peever, T. L., Peres, N. A. and Timmer, L. W. 2006. Brown spot of tangerine hybrid cultivars Minneola, Page and Fortune caused by *Alternaria alternata* in Iran. *Plant Pathol. J.* 55(4), 578. doi:10.1111/j.1365-3059.2006.01427.x.

Gottwald, T. R., Graham, J. H. and Schubert, T. S. 2002. Citrus canker: the pathogen and its impact. Online. *Plant Health Prog.* 3(1). doi:10.1094/PHP-2002-0812-01-RV.

Grosser, J. W. and Chandler, J. L. 2004. Production of twelve new allotetraploid somatic hybrid citrus breeding parents with emphasis on late maturity and cold-hardiness. *J. Amer. Pomol. Soc.* 58(1), 21–8.

Grosser, J. W. and Gmitter Jr., F. G. 2005. 2004 SIVB congress symposium proceedings 'Thinking outside the cell': applications of somatic hybridization and cybridization in crop improvement, with citrus as a model. *In Vitro Cell. Dev. Biol. Plant* 41, 220–5.

Grosser, J. W. and Gmitter Jr., F. G. 2011. Protoplast fusion for production of tetraploids and triploids: applications for scion and rootstock breeding in citrus. *Plant Cell Tiss. Organ Cult.* 104(3), 343–57. doi:10.1007/s11240-010-9823-4.

Grosser, J. W., An, H. J., Calovic, M., Lee, D. H., Chen, C., Vasconcellos, M. and Gmitter, F. G. 2010. Production of new allotetraploid and autotetraploid Citrus breeding parents: focus on zipper skin mandarins. *HortScience* 45(8), 1160–3. doi:10.21273/HORTSCI.45.8.1160.

Grosser, J. W., Gmitter Jr., F. G., Dutt, M., Calovic, M., Ling, P. and Castle, B. 2015. Highlights of the University of Florida, Citrus Research and Education Center's comprehensive citrus breeding and genetics program. *Acta Hortic.* 1065, 405–13.

Gulsen, O. and Roose, M. L. 2001. Chloroplast and nuclear genome analysis of the parentage of lemons. *J. Amer. Soc. Hortic. Sci.* 126(2), 210–5. doi:10.21273/JASHS.126.2.210.

Gulsen, O., Uzun, A., Pala, H., Canihos, E. and Kafa, G. 2007. Development of seedless and Mal Secco tolerant mutant lemons through budwood irradiation. *Sci. Hortic.* 112(2), 184–90. doi:10.1016/j.scienta.2006.12.040.

Gulsen, O., Uzun, A., Canan, I., Seday, U. and Canihos, E. 2010. A new citrus linkage map based on SRAP, SSR, ISSR, POGP, RGA and RAPD markers. *Euphytica* 173(2), 265–77. doi:10.1007/s10681-010-0146-7.

Hajeri, S., Killiny, N., El-Mohtar, C., Dawson, W. O. and Gowda, S. 2014. *Citrus tristeza* virus-based RNAi in citrus plants induces gene silencing in *Diaphorina citri*, a phloem-sap sucking insect vector of citrus greening disease (Huanglongbing). *J. Biotechnol.* 176, 42–9. doi:10.1016/j.jbiotec.2014.02.010.

Hao, G., Stover, E. and Gupta, G. 2016. Overexpression of a modified plant thionin enhances disease resistance to Citrus canker and Huanglongbing (HLB). *Front. Plant Sci.* 7, 1078. doi:10.3389/fpls.2016.01078.

Hartung, J. S., Beretta, J., Brlansky, R. H., Spisso, J. and Lee, R. 1994. Citrus variegated chlorosis bacterium: axenic culture, pathogenicity, and serological relationships with other strains of *Xylella fastidiosa*. *Phytopathology* 84(6), 591–7. doi:10.1094/Phyto-84-591.

Haskins, C. P. and Moore, C. N. 1935. Growth modifications in citrus seedlings grown from X-rayed seed. *Plant Physiol.* 10(1), 179–85. doi:10.1104/pp.10.1.179.

Hearn, C. J. 1986. Development of seedless grapefruit cultivars through budwood Irradiation. *J. Amer. Soc. Hortic. Sci.* 111(2), 304–6.

Hensz, R. A. 1971. 'Star Ruby', a new deep-red-fleshed grapefruit variety with distinct tree characteristics. *J. Rio Grande Valley Hortic. Soc.* 25, 54–8.

Herrero, R., Asins, M. J., Carbonell, E. A. and Navarro, L. 1996. Genetic diversity in the orange subfamily Aurantioideae. I. Intraspecifies and intragenus genetic variability. *Theor. Appl. Genet.* 92(5), 599–609. doi:10.1007/BF00224564.

Huerta, L., Forment, J., Gadea, J., Fagoaga, C., Peña, L., Pérez-Amador, M. A. and García-Martínez, J. L. 2008. Gene expression analysis in citrus reveals the role of gibberellins on photosynthesis and stress. *Plant Cell Environ.* 31(11), 1620–33. doi:10.1111/j.1365-3040.2008.01870.x.

Jia, H. and Wang, N. 2014. Targeted genome editing of sweet orange using Cas9/sgRNA. *PLoS ONE* 9(4), e93806. doi:10.1371/journal.pone.0093806.

Jia, H., Zhang, Y., Orbović, V., Xu, J., White, F. F., Jones, J. B. and Wang, N. 2017. Genome editing of the disease susceptibility gene CsLOB1 in citrus confers resistance to Citrus canker. *Plant Biotechnol. J.* 15(7), 817–23. doi:10.1111/pbi.12677.

Jiao, W. B., Huang, D., Xing, F., Hu, Y., Deng, X. X., Xu, Q. and Chen, L. L. 2013. Genome-wide characterization and expression analysis of genetic variants in sweet orange. *Plant J.* 75(6), 954–64. doi:10.1111/tpj.12254.

Jinek, M., Chylinski, K., Fonfara, I., Hauer, M., Doudna, J. A. and Charpentier, E. 2012. A programmable dual-RNA-guided DNA endonuclease in adaptive bacterial immunity. *Science* 337(6096), 816–21. doi:10.1126/science.1225829.

Kepiro, J. L. and Roose, M. L. 2010. AFLP markers closely linked to a major gene essential for nucellar embryony (apomixis) in *Citrus maxima* x *Poncirus trifoliata*. *Tree Genet. Genom.* 6(1), 1–11. doi:10.1007/s11295-009-0223-z.

Kobayashi, S., Ohgawara, T., Saito, W., Nakamura, Y. and Omura, M. 1997. Production of triploid somatic hybrids in Citrus. *Jpn. Soc. Hortic. Sci.* 66(3), 453–8. doi:10.2503/jjshs.66.453.

Krug, C. A. 1943. Chromosome number in the subfamily Aurantioideae with special reference to the genus Citrus. *Bot. Gazz.* 104(4), 602–11. doi:10.1086/335173.

Li, X., Xie, R., Lu, Z. and Zhou, Z. 2010. The origin of cultivated Citrus as inferred from internal transcribed spacer and chloroplast DNA sequence and amplified fragment length polymorphism fingerprints. *J. Amer. Soc. Hortic. Sci.* 135(4), 341–50. doi:10.21273/JASHS.135.4.341.

Licciardello, C., Russo, G., Caruso, M. and Caruso, P. 2015. New challanges in citrus research: from varietal innovation to genome. Available at: http://carta.milano.it/wp-content/uploads/2015/11/109_new:challanges.pdf (accessed on 30 October 2018).

Long, J. M., Liu, Z., Wu, X. M., Fang, Y. N., Jia, H. H., Xie, Z. Z., Deng, X. X. and Guo, W. W. 2016. Genome-scale mRNA and small RNA transcriptomic insights into initiation of citrus apomixis. *J. Exp. Bot.* 67(19), 5743–56. doi:10.1093/jxb/erw338.

López-García, A., Terol, J., Tadeo, F. R., Herrero-Ortega, A., Ibañez, V. and Talón, M. 2015. Three new cultivars of Clementine: 'Clemenverd', 'Nero' and 'Neufina'. *Acta Hortic.* 1065, 239–43. doi:10.17660/ActaHortic.2015.1065.26.

Luro, F., Maddy, F., Jacquemond, C., Froelicher, Y., Morillon, R., Rist, D. and Ollitrault, P. 2004. Identification and evaluation of diplogyny in clementine (*Citrus clementina*) for use in breeding. *Acta Hortic.* 663(663), 841–8. doi:10.17660/ActaHortic.2004.663.152.

Mabberley, D. J. 1997. A classification for edible Citrus. *Telopea* 7, 167–72.

Martinez-Godoy, M. A., Mauri, N., Juarez, J., Marques, M. C., Santiago, J., Forment, J. and Gadea, J. 2008. A genome-wide 20 K citrus microarray for gene expression analysis. *BMC Genom.* 9(1), 318. doi:10.1186/1471-2164-9-318.

McCollum, T. G. 2007. Update on the USDA, ARS citrus scion improvement project. *Proc. Fla. State Hortic. Soc.* 120, 285–7.

Minamikawa, M. F., Nonaka, K., Kaminuma, E., Kajiya-Kanegae, H., Onogi, A., Goto, S., Yoshioka, T., Imai, A., Hamada, H., Hayashi, T., Matsumoto, S., Katayose, Y., Toyoda, A., Fujiyama, A., Nakamura, Y., Shimizu, T. and Iwata, H. 2017. Genome-wide association study and genomic prediction in citrus: potential of genomics-assisted breeding for fruit quality traits. *Sci. Rep.* 7(1), 4721. doi:10.1038/s41598-017-05100-x.

Navarro, L., Roistacher, C. N. and Murashige, T. 1975. Improvement of shoot-tip grafting *in vitro* for virus-free citrus. *J. Amer. Soc. Hortic. Sci.* 100, 471–9.

Navarro, L., Aleza, P., Cuenca, J., Juárez, J., Pina, J. A., Ortega, C., Navarro, A. and Ortega, V. 2015. The triploid mandarin breeding program in Spain. *Acta Hortic.* 1065, 389–95.

Nicolosi, E., Deng, Z. N., Gentile, A., La Malfa, S., Continella, G. and Tribulato, E. 2000. Citrus phylogeny and genetic origin of important species as investigated by molecular markers. *Theor. Appl. Genet.* 100(8), 1155–66. doi:10.1007/s001220051419.

Nigro, F., Ippolito, A. and Salerno, M. G. 2011. Invited Review: Mal Secco disease of Citrus: a journey through a century of research. *J. Plant Pathol.* 93(3), 523–60.

Niñoles, R., Aleza, P., Castillo, M. C., Navarro, L. and Ancillo, G. 2015. Ploidy and gene expression in clementine. *Acta Hortic.* 1065, 605–11.

Nishikawa, F., Endo, T., Shimada, T., Fujii, H., Shimizu, T., Kobayashi, Y., Araki, T. and Omura, M. 2010. Transcriptional changes in CiFT introduced transgenic trifoliate orange (*Poncirus trifoliata* L. Raf.). *Tree Physiol.* 30(3), 431–9. doi:10.1093/treephys/tpp122.

Nishitani, C., Hirai, N., Komori, S., Wada, M., Okada, K., Osakabe, K., Yamamoto, T. and Osakabe, Y. 2016. Efficient genome editing in apple using a CRISPR/Cas9 system. *Sci. Rep.* 6, 31481. doi:10.1038/srep31481.

Olivares-Fuster, O., Duran-Vila, N. and Navarro, L. 2005. Electrochemical protoplast fusion in citrus. *Plant Cell Rep.* 24(2), 112–9. doi:10.1007/s00299-005-0916-1.

Ollitrault, P. and Navarro, L. 2012. Citrus. In: Badenes, M. L. and Byrne, D. H. (Eds), *Fruit Breeding: Handbook of Plant Breeding 8*. Springer, New York, pp. 623–62.

Ollitrault, P., Vanel, F., Froelicher, Y. and Dambier, D. 2000. Creation of triploid citrus hybrids by electrofusion of haploid and diploid protoplasts. *Acta Hortic.* 535(535), 191–8. doi:10.17660/ActaHortic.2000.535.23.

Ollitrault, P., Dambier, D., Luro, F. and Froelicher, Y. 2008. Ploidy manipulation for breeding seedless triploid citrus. *Plant Breed. Rev.* 30, 323–52.

Ollitrault, P., Terol, J., Chen, C., Federici, C. T., Lotfy, S., Hippolyte, I., Ollitrault, F., Bérard, A., Chauveau, A., Cuenca, J., Costantino, G., Kacar, Y., Mu, L., Garcia-Lor, A., Froelicher, Y., Aleza, P., Boland, A., Billot, C., Navarro, L., Luro, F., Roose, M. L., Gmitter, F. G., Talon, M. and Brunel, D. 2012. A reference genetic map of *C. clementina* Hort. ex Tan.; citrus evolution inferences from comparative mapping. *BMC Genom.* 13, 593. doi:10.1186/1471-2164-13-593.

Omar, A. A., Murata, M., Yu, Q., Gmitter, F. G., Chase, C. D., Graham, J. H. and Grosser, J. W. 2017. Production of three new grapefruit cybrids with potential for improved citrus canker resistance. *In Vitro Cell. Dev. Biol. Plant* 53(3), 256–69. doi:10.1007/s11627-017-9816-7.

Omura, M. and Shimada, T. 2016. Citrus breeding, genetics and genomics in Japan. *Breed. Sci.* 66(1), 3–17. doi:10.1270/jsbbs.66.3.

Oueslati, A., Salhi-Hannachi, A., Luro, F., Vignes, H., Mournet, P. and Ollitrault, P. 2017. Genotyping by sequencing reveals the interspecific *C. maxima/C. reticulata* admixture along the genomes of modern citrus varieties of mandarins, tangors, tangelos, orangelos and grapefruits. *PLoS ONE* 12(10), e0185618. doi:10.1371/journal.pone.0185618.

Park, T. H., Kim, J. B., Hutten, R. C. B., van Eck, H. J., Jacobsen, E. and Visser, R. G. 2007. Genetic positioning of centromeres using half-tetrad analysis in a 4x-2x cross population of potato. *Genetics* 176(1), 85–94. doi:10.1534/genetics.107.070870.

Patel, M., Manvar, T., Apurwa, S., Ghosh, A., Tiwari, T. and Chikara, S. K. 2014. Comparative de novo transcriptome analysis and metabolic pathway studies of *Citrus paradisi flavedo* from naive stage to ripened stage. *Mol. Biol. Rep.* 41(5), 3071–80. doi:10.1007/s11033-014-3166-x.

Peña, L., Martín-Trillo, M., Juárez, J., Pina, J. A., Navarro, L. and Martínez-Zapater, J. M. 2001. Constitutive expression of Arabidopsis LEAFY or APETALA1 genes in citrus reduces their generation time. *Nat. Biotechnol.* 19(3), 263–7. doi:10.1038/85719.

Peña, L., Cervera, M., Fagoaga, C., Romero, J., Juárez, J., Pina, J. A. and Navarro, L. 2007. Citrus. In: Pua, E. C. and Davey, M. (Eds), *Transgenic Crops V. Biotechnology in Agriculture and Forestry* (vol. 60). Springer, Berlin, Heidelberg, pp. 35–50.

Peng, A., Chen, S., Lei, T., Xu, L., He, Y., Wu, L., Yao, L. and Zou, X. 2017. Engineering canker-resistant plants through CRISPR/Cas9-targeted editing of the susceptibility gene CsLOB1 promoter in citrus. *Plant Biotechnol. J.* 15(12), 1509–19. doi:10.1111/pbi.12733.

Penjor, T., Yamamoto, M., Uehara, M., Ide, M., Matsumoto, N., Matsumoto, R. and Nagano, Y. 2013. Phylogenetic relationships of Citrus and its relatives based on matK gene sequences. *PLoS ONE* 8(4), e62574. doi:10.1371/journal.pone.0062574.

Peres, N. A. and Timmer, L. W. 2006. Alternaria brown spot of citrus in Brazil: evaluation of the Alter-Rater for spray timing and effects on yield of *Murcott tangor*. *Crop Prot.* 25(5), 454–60. doi:10.1016/j.cropro.2005.07.010.

Pons, E., Alquézar, B., Rodríguez, A., Martorell, P., Genovés, S., Ramón, D., Rodrigo, M. J., Zacarías, L. and Peña, L. 2014. Metabolic engineering of β-carotene in orange fruit increases its *in vivo* antioxidant properties. *Plant Biotechnol. J.* 12(1), 17–27. doi:10.1111/pbi.12112.

Rapisarda, P., Bellomo, S. E., Fabroni, S. and Russo, G. 2008. Juice quality of two new mandarin-like hybrids (*Citrus clementina* Hort. ex Tan x *Citrus sinensis* L. Osbeck) containing anthocyanins. *J. Agric. Food Chem.* 56(6), 2074–8. doi:10.1021/jf072616e.

Recupero, G. R., Russo, G. and Recupero, S. 2005. New promising citrus triploid hybrids selected from crosses between monoembryonic diploid female ant tetraploid male parents. *HortScience* 40(3), 516–20. doi:10.21273/HORTSCI.40.3.516.

Ren, C., Liu, X., Zhang, Z., Wang, Y., Duan, W., Li, S. and Liang, Z. 2016. CRISPR/Cas9-mediated efficient targeted mutagenesis in Chardonnay (*Vitisvinifera* L.). *Sci. Rep.* 6, 32289. doi:10.1038/srep32289.

Riera, N., Handique, U., Zhang, Y., Dewdney, M. M. and Wang, N. 2017. Characterization of antimicrobial-producing beneficial bacteria isolated from Huanglongbing escape citrus trees. *Front. Microbiol.* 8, 2415. doi:10.3389/fmicb.2017.02415.

Rodrigues, C. M., de Souza, A. A., Takita, M. A., Kishi, L. T. and Machado, M. A. 2013. RNA-Seq analysis of *Citrus reticulata* in the early stages of *Xylella fastidiosa* infection reveals auxin-related genes as a defence response. *BMC Genom.* 14(1), 676. doi:10.1186/1471-2164-14-676.

Rodríguez, A., San Andrés, V., Cervera, M., Redondo, A., Alquézar, B., Shimada, T., Gadea, J., Rodrigo, M. J., Zacarías, L., Palou, L., López, M. M., Castañera, P. and Peña, L. 2011. Terpene down-regulation in orange reveals the role of fruit aromas in mediating interactions with insect herbivores and pathogens. *Plant Physiol.* 156(2), 793–802. doi:10.1104/pp.111.176545.

Rodríguez, A., Kava, V., Latorre-García, L., da Silva, G. J., Pereira, R. G., Glienke, C., Ferreira-Maba, L. S., Vicent, A., Shimada, T. and Peña, L. 2018. Engineering d-limonene synthase down-regulation in orange fruit induces resistance against the fungus *Phyllosticta citricarpa* through enhanced accumulation of monoterpene alcohols and activation of defence. *Mol. Plant Pathol.* 19(9), 2077–93. doi:10.1111/mpp.12681.

Roose, M. L. and Williams, T. E. 2007a. Mutation breeding in Citrus. In: Khan, I. A. (Ed.), *Citrus Genetics, Breeding and Biotechnology*. CAB International, Wallingford, UK, pp. 345–52.

Roose, M. L. and Williams, T. E. 2007b. Mandarin variety named Tango. US Plant Patent (2007)/0056064 P1.

Rouiss, H., Cuenca, J., Navarro, L., Ollitrault, P. and Aleza, P. 2017. Tetraploid citrus progenies arising from FDR and SDR unreduced pollen in 4x x 2x hybridizations. *Tree Genet. Genom.* 13(1), 10. doi:10.1007/s11295-016-1094-8.

Russo, F., Starrantino, A. and Spina, P. 1981. La selezione clonale dell'arancio e del mandarino: ricerche e risultati ottenuti presso l'Istituto Sperimentale per l'Agrumicoltura di Acireale. *Rivista di ortoflorofrutticoltura italiana* 65)(1), 1–10.

Russo, G., Reforgiato-Recupero, G., Recupero, S. and Paolo, D. P. 2015. 'Sweet Sicily' and 'Early Sicily', two new triploids from the program of CRA-research Centre of

Citriculture and Mediterranean crops. *Acta Hortic.* 1065, 215–21. doi:10.17660/ActaHortic.2015.1065.23.

Scora, R. W. 1975. On the history and origin of citrus. *Bull. Torrey Bot. Club* 102(6), 369–75. doi:10.2307/2484763.

Shalom, L., Samuels, S., Zur, N., Shlizerman, L., Doron-Faigenboim, A., Blumwald, E. and Sadka, A. 2014. Fruit load induces changes in global gene expression and in abscisic acid (ABA) and indole acetic acid (IAA) homeostasis in citrus buds. *J. Exp. Bot.* 65(12), 3029–44. doi:10.1093/jxb/eru148.

Shimizu, T., Kitajima, A., Nonaka, K., Yoshioka, T., Ohta, S., Goto, S., Toyoda, A., Fujiyama, A., Mochizuki, T., Nagasaki, H., Kaminuma, E. and Nakamura, Y. 2016. Hybrid origins of citrus varieties inferred from DNA marker analysis of nuclear and organelle genomes. *PLoS ONE* 11(11), e0166969. doi:10.1371/journal.pone.0166969.

Shimizu, T., Tanizawa, Y., Mochizuki, T., Nagasaki, H., Yoshioka, T., Toyoda, A., Fujiyama, A., Kaminuma, E. and Nakamura, Y. 2017. Draft sequencing of the heterozygous diploid genome of Satsuma (*Citrus unshiu* Marc.) using a hybrid assembly approach. *Front. Genet.* 8, 180. doi:10.3389/fgene.2017.00180.

Shrestha, M., Garcia, J. E., Bukovac, Z., Dorin, A. and Dyer, A. G. 2018. Pollination in a new climate: assessing the potential influence of flower temperature variation on insect pollinator behaviour. *PLoS ONE* 13(8), e0200549. doi:10.1371/journal.pone.0200549.

Song, G., Jia, M., Chen, K., Kong, X., Khattak, B., Xie, C., Li, A. and Mao, L. 2016. CRISPR/Cas9: a powerful tool for crop genome editing. *Crop J.* 4(2), 75–82. doi:10.1016/j.cj.2015.12.002.

Spiegel-Roy, P. and Vardi, A. 1992. 'Shani,' 'Orah' and 'Winola': three new selections from our breeding program. *Proceedings of the 7th International Citrus Congress of the International Society Citriculture*, Acireale, Italy, pp. 72–3.

Sun, Q., Zheng, L., Wan-Peng, X., Shao-Lan, H., Shi-Lai, Y., Qiang, L., Yong-Qiang, Z., Deng, L. and Rang-Jin, X. 2014. Transcriptome analysis of blood orange (*Citrus sinensis*) following fruit bagging treatment by digital gene expression profiling. *J. Hortic. Sci. Biotech.* 89(4), 397–407. doi:10.1080/14620316.2014.11513098.

Swingle, W. T. and Reece, P. C. 1967. The botany of Citrus and its wild relatives. In: Reuther, W., Webber, H. and Batchelor, L. (Eds), *The Citrus Industry* (vol. 1). University of California, Berkeley, USA, pp. 190–430.

Tanaka, T. 1977. Fundamental discussion of Citrus classification. *Stud. Citrol* 14, 1–6.

Terol, J., Tadeo, F. R., Ventimilla, D. and Talon, M. 2016. An RNA-Seq based reference transcriptome for Citrus. *Plant Biotechnol. J.* 14(3), 938–50. doi:10.1111/pbi.12447.

Timmer, L. W., Peever, T. L., Solel, Z. and Akimitsu, K. 2003. Alternaria diseases of citrus – novel pathosystems. *Phytopathol. Mediterr.* 42, 3–16.

Vardi, A., Spiegel-Roy, P. and Elchanati, A. 1993. Mandarin tree named Mor. US Patent PP8, 378.

Vardi, A., Spiegel-Roy, P., Frydman-Shani, A., Elchanati, A. and Neumann, H. 2003. Citrus tree named Orri. US patent PP13616P2.

Vardi, A., Levin, I. and Carmi, N. 2008. Induction of seedlessness in Citrus: from classical techniques to emerging biotechnological approaches. *J. Amer. Soc. Hortic. Sci.* 133(1), 117–26. doi:10.21273/JASHS.133.1.117.

Velázquez, K., Agüero, J., Vives, M. C., Aleza, P., Pina, J. A., Moreno, P., Navarro, L. and Guerri, J. 2016. Precocious flowering of juvenile citrus induced by a viral vector based on citrus leaf blotch virus: a new tool for genetics and breeding. *Plant Biotechnol. J.* 14(10), 1976–85. doi:10.1111/pbi.12555.

Vicent, A., Armengol, J., Sales, R., García-Jiménez, J. and Alfaro-Lassala, F. 2000. First report of Alternaria brown spot of citrus in Spain. *Plant Dis.* 84(9), 1044. doi:10.1094/PDIS.2000.84.9.1044B.

Vicent, A., Armengol, J. and García-Jimenez, J. 2007. Rain fastness and persistence of fungicides for control of Alternaria brown spot of citrus. *Plant Dis.* 91(4), 393–9. doi:10.1094/PDIS-91-4-0393.

Waltz, E. 2016. Gene-edited CRISPR mushroom escapes US regulation. *Nature* 532(7599), 293. doi:10.1038/nature.2016.19754.

Wang, X. F., Li, Z. A., Tang, K. Z., Zhou, C. Y. and Yi, L. 2010. First report of Alternaria brown spot of citrus caused by *Alternaria alternata* in Yunnan Province, China. *Plant Dis.* 94(3), 375. doi:10.1094/PDIS-94-3-0375C.

Wang, N., Pierson, E. A., Setubal, J. C., Xu, J., Levy, J. G., Zhang, Y., Li, J., Rangel, L. T. and Martins, J. 2017a. The *candidatus liberibacter*-host interface: insights into pathogenesis mechanisms and disease control. *Annu. Rev. Phytopathol.* 55, 451–82. doi:10.1146/annurev-phyto-080516-035513.

Wang, X., Xu, Y., Zhang, S., Cao, L., Huang, Y., Cheng, J., Wu, G., Tian, S., Chen, C., Liu, Y., Yu, H., Yang, X., Lan, H., Wang, N., Wang, L., Xu, J., Jiang, X., Xie, Z., Tan, M., Larkin, R. M., Chen, L. L., Ma, B. G., Ruan, Y., Deng, X. and Xu, Q. 2017b. Genomic analyses of primitive, wild and cultivated citrus provide insights into asexual reproduction. *Nat. Genet.* 49(5), 765–72. doi:10.1038/ng.3839.

Wang, L., He, F., Huang, Y., He, J., Yang, S., Zeng, J., Deng, C., Jiang, X., Fang, Y., Wen, S., Xu, R., Yu, H., Yang, X., Zhong, G., Chen, C., Yan, X., Zhou, C., Zhang, H., Xie, Z., Larkin, R. M., Deng, X. and Xu, Q. 2018. Genome of wild mandarin and domestication history of mandarin. *Mol. Plant* 11(8), 1024–37. doi:10.1016/j.molp.2018.06.001.

Whelan, A. I. and Lema, M. A. 2015. Regulatory framework for gene editing and other new breeding techniques (NBTs) in Argentina. *GM Crops Food* 6(4), 253–65. doi:10.1080/21645698.2015.1114698.

Williams, T. E. and Roose, M. L. 2004. 'TDE2' Mandarin hybrid (Shasta Gold(R) Mandarin), 'TDE3' Mandarin hybrid (Tahoe gold(R) Mandarin) and 'TDE4' Mandarin hybrid (Yosemite Gold(R) Mandarin): three new mid and late-season triploid seedless mandarin hybrids from California. In: *Proceedings of 10th International Citrus Congress* (vol. 1). International Society of Citriculture, Agadir, Morocco, pp. 394–8.

Wu, G. A., Prochnik, S., Jenkins, J., Salse, J., Hellsten, U., Murat, F., Perrier, X., Ruiz, M., Scalabrin, S., Terol, J., Takita, M. A., Labadie, K., Poulain, J., Couloux, A., Jabbari, K., Cattonaro, F., Del Fabbro, C., Pinosio, S., Zuccolo, A., Chapman, J., Grimwood, J., Tadeo, F. R., Estornell, L. H., Muñoz-Sanz, J. V., Ibanez, V., Herrero-Ortega, A., Aleza, P., Pérez-Pérez, J., Ramón, D., Brunel, D., Luro, F., Chen, C., Farmerie, W. G., Desany, B., Kodira, C., Mohiuddin, M., Harkins, T., Fredrikson, K., Burns, P., Lomsadze, A., Borodovsky, M., Reforgiato, G., Freitas-Astúa, J., Quetier, F., Navarro, L., Roose, M., Wincker, P., Schmutz, J., Morgante, M., Machado, M. A., Talon, M., Jaillon, O., Ollitrault, P., Gmitter, F. and Rokhsar, D. 2014. Sequencing of diverse mandarin, pummelo and orange genomes reveals complex history of admixture during citrus domestication. *Nat. Biotechnol.* 32(7), 656–62. doi:10.1038/nbt.2906.

Wu, G. A., Terol, J., Ibáñez, V., López-García, A., Pérez-Román, E., Borredá, C., Domingo, C., Tadeo, F. R., Carbonell-Caballero, J., Alonso, R., Curk, F., Du, D., Ollitrault, P., Roose, M. L., Dopazo, J., Gmitter, F. G., Rokhsar, D. S. and Talon, M. 2018. Genomics of the origin and evolution of Citrus. *Nature* 554(7692), 311–6. doi:10.1038/nature25447.

Xu, X. Y., Liu, J. H. and Deng, X. X. 2006. Isolation of cytoplasts from Satsuma mandarin (*Citrus unshiu* Marc.) and production of alloplasmic hybrid calluses via cytoplast-protoplast fusion. *Plant Cell Rep.* 25(6), 533–9. doi:10.1007/s00299-005-0057-6.

Xu, Q., Chen, L. L., Ruan, X., Chen, D., Zhu, A., Chen, C., Bertrand, D., Jiao, W. B., Hao, B. H., Lyon, M. P., Chen, J., Gao, S., Xing, F., Lan, H., Chang, J. W., Ge, X., Lei, Y., Hu, Q., Miao, Y., Wang, L., Xiao, S., Biswas, M. K., Zeng, W., Guo, F., Cao, H., Yang, X., Xu, X. W., Cheng, Y. J., Xu, J., Liu, J. H., Luo, O. J., Tang, Z., Guo, W. W., Kuang, H., Zhang, H. Y., Roose, M. L., Nagarajan, N., Deng, X. X. and Ruan, Y. 2013. The draft genome of sweet orange (*Citrus sinensis*). *Nat. Genet.* 45(1), 59–66. doi:10.1038/ng.2472.

Yang, Z. N., Ye, X. R., Molina, J., Roose, M. L. and Mirkov, T. E. 2003. Sequence analysis of a 282-kilobase region surrounding the *Citrus tristeza* virus resistance gene (Ctv) locus in *Poncirus trifoliata* L. Raf. *Plant Physiol.* 131(2), 482–92. doi:10.1104/pp.011262.

Yu, K., Xu, Q., Da, X., Guo, F., Ding, Y. and Deng, X. 2012. Transcriptome changes during fruit development and ripening of sweet orange (*Citrus sinensis*). *BMC Genom.* 13, 10. doi:10.1186/1471-2164-13-10.

Yu, Y., Chen, C. and Gmitter, F. G. 2016. QTL mapping of mandarin (*Citrus reticulata*) fruit characters using high-throughput SNP markers. *Tree Genet. Genom.* 12(4), 77. doi:10.1007/s11295-016-1034-7.

Zhang, S., Shi, Q., Albrecht, U., Shatters, R. G., Stange, R., McCollum, G., Zhang, S., Fan, C. and Stover, E. 2017. Comparative transcriptome analysis during early fruit development between three seedy citrus genotypes and their seedless mutants. *Hortic. Res.* 4, 17041. doi:10.1038/hortres.2017.41.

Zheng, B. B., Wu, X. M., Ge, X. X., Deng, X. X., Grosser, J. W. and Guo, W. W. 2012. Comparative transcript profiling of a male sterile cybrid pummelo and its fertile type revealed altered gene expression related to flower development. *PLoS ONE* 7(8), e43758. doi:10.1371/journal.pone.0043758.

# Part 2

## Other tropical fruits

# Chapter 8

# Advances in avocado propagation for the sustainable supply of planting materials

*Jayeni Hiti-Bandaralage, Alice Hayward, Chris O'Brien, Madeleine Gleeson, William Nak and Neena Mitter, The University of Queensland, Australia*



## 1 Introduction

Avocado (*Persea americana* Mill) is an ancient tropical fruit species belonging to the plant family, Lauraceae, commonly known as laurels for their distinctive foliage fragrance (Bergh and Ellstrand, 1986). It is the only important fruit species in the family, with other economically important species including cinnamon (*Cinnamomum zeylancium*), camphor (*Cinnamomum camphora*) and bay leaf (*Laurus nobilis*) (Litz et al., 2007). Originating in central Mexico, avocado is one of the earliest domesticated tree crops, with evidence supporting cultivation as early as 4500–2000 BC (Galindo-Tovar et al., 2008). Today, avocado has gained huge commercial momentum as a high-value, highly nutritious food. World production exceeds 5.3 million tonnes/year (2015), with a production increase of over 2 million tonnes in the last decade (2005–15) (FAOSTAT, 2018). In the United States alone, which is the world's biggest consumer of avocado, per capita consumption of avocados has increased 443% over the last two decades (from 1.6 to 7.1 pounds in 2015) (Darnton, 2017). The majority of the world's crop is produced in North and South America (70%), with Mexico contributing

http://dx.doi.org/10.19103/AS.2019.0054.19

one-fifth of the total world production (1.1 million tonnes – 2016) (FAOSTAT, 2018). Concurrent to the increased production of avocado, the land under avocado cultivation has also increased rapidly (5.6 million ha, 2015) (FAOSTAT, 2018).

The increase in avocado consumption is encouraged by the nutritional and health benefits of avocado. It is a fruit packed with a large amount of vitamins A, B, C, E, K, monounsaturated fatty acids, dietary fibre, mineral nutrients (Mg, K, Na) and low levels of saturated fatty acids (Unlu et al., 2005; Dreher and Davenport, 2013). Avocado consumption has also been proven to increase nutrient intake (Fulgoni et al., 2013), for example carotenoid absorption (Unlu et al., 2005), and have positive effects on cardiovascular disease risk (Chong et al., 2006), diabetes and weight management, and cancer risk (Ding et al., 2007), making this buttery fruit even more appealing.

With all of this popularity growth over the last decade, the global production growth of avocado (5% per year) has not been able to meet the increasing trade growth of 12.5% per year (Rijiswick, 2018). Two major barriers to avocado industry growth are: (1) the difficulty in avocado breeding for improved productivity due to its long vegetative phase and lack of advanced molecular breeding and genetic engineering tools to expedite field selection and (2) the difficulty in propagation of avocado plants, limiting global supply.

## 2 Avocado cultivation

Many woody perennial crops are commercially grown as grafted trees, combining important agronomical traits of rootstock and scion cultivars (Warschefsky et al., 2016). Rootstocks are often selected for dwarfing (Vega, 1989), salt or alkalinity tolerance (Pliego-Alfaro and Murashige, 1987; Vega, 1989) and pest and disease resistance (Pliego-Alfaro and Murashige, 1987), while the scion confers fruit set, post-harvest fruit quality and taste characteristics (Ahmed et al., 2001). In the case of avocado, a rootstock is grafted with budwood of a quality fruiting cultivar, most commonly 'Hass' (Platt, 1976), derived from a clonal tree of known ancestry. This ensures conservation of highly desirable productivity, post-harvest and fruit traits, something that is otherwise almost impossible in this open pollinating crop. Grafting also allows for faster maturation and fruiting of commercial trees, yielding within 2–3 years relative to up to 10 years for seedlings.

Avocado rootstocks may be either seedling rootstocks propagated from seed or clonal rootstocks. As an open pollinated plant, avocado seedlings are highly heterozygous, leading to a heterozygous population of orchard plants with respect to growth habit, productivity and resistance to pests and diseases (Sedgley, 1985). Clonal rootstocks have the advantage of being genetically identical, conserving superior traits of the mother tree. This supports optimal productivity and improved fruit yield per tree as well as uniform growth habit for efficient on-farm management (Janick, 2010). In fact, clonal rootstocks are thought to be the only rootstocks for achieving sustainable productivity gains in the future (Ben-Ya'acov, 1987; Wolstenholme, 1987, 2003). This is because avocado is highly sensitive to abiotic conditions; drought, extreme temperatures and soil stress (soil alkalinity, salinity and aeration) (Silva and Ledesma, 2014) as well as to the devastating avocado root-rot fungus *Phytophthora cinnamomi*, although some varieties exhibit some level of tolerance (Silva and Ledesma, 2014). These biotic and abiotic factors influence the total productivity of the plant in terms of yield and health (Ben-Ya'acov, 1987; Menge et al., 2001; Janick, 2010). Additionally, as the world moves towards high-density planting to meet the exponential demand for fruit, uniformity in growth habit is considered essential, especially in regions with high labour costs. Therefore, breeding for improved characteristics and maintaining genetic uniformity through clonal plants could reduce the gap between supply and demand.

Despite the advantages of clonal rootstocks, however, they are notoriously difficult and resource-intensive to produce. Seedlings are much more accessible and hence cheaper than clonal plants. As investment in orchard establishment is mainly governed by plant supply and the cost of plants, in many countries, the majority of avocado plants are grown on seedling rootstocks (Coffey and Guillemet, 1987; Roe et al., 1995). Herein lies a major bottleneck to sustainable avocado propagation, performance and breeding.

## 3 The problem of clonal avocado propagation

Clonal rootstocks can be derived from one of two possible methods: (1) vegetative propagation from cuttings of a high-performing 'mother' tree or (2) production of seeds obtained from controlled pollination (Jain and Ishii, 2003). Controlled pollination of avocado results in very high fruit drop (Barceló-Muñoz and Pliego-Alfaro, 2003) making this commercially unviable for clonal avocado propagation. Thus, in commercial practice, clonal rootstocks are propagated from cuttings induced to form roots.

Clonal propagation of rootstocks depends on the capacity of mature tree cuttings to form post-embryonic, or adventitious roots. As for many woody species, avocado loses competency to form adventitious roots from cuttings as it matures, making regeneration of clonal plants from field-tested rootstock selections very difficult (Platt, 1976; Zulfiqar et al., 2009; Brondani et al., 2012). Different methods including budding, marcottage/air layering (Young, 1961), rooting of cuttings under mist spray and grafting have been attempted to promote rooting with limited success. Additionally, attempts to pre-treat mother plants (e.g. injecting of gibberellic acid, etiolation, ring barking) prior to taking cuttings, and different incubation conditions after rooting treatment, have failed to improve rooting (Frolich, 1951; Reuveni and Raviv, 1980; Barrientos-Priego et al., 1986; Cutting and Van Vuuren, 1988). In all cases, success has been poor due to a very low percentage of root induction, inconsistent rooting, long time for root induction or difficulty in managing experiments under field conditions (Eggers and Halma, 1937; Leal and Krezdorn, 1964; Kadman and Ben-Ya'acov, 1965; Frolich and Platt, 1972; Reuveni and Raviv, 1980).

At present, the most effective industry practice for clonal rootstock production remains a 40-year-old technique based on the Frolich and Platt (1972) double grafting method (Ernst, 1999). Here, budwood from a mature rootstock plant is grafted onto a seedling 'nurse' plant and subjected to etiolation. The base of the etiolated shoot is then treated with a rooting trigger such as auxin. Once the roots are established, the desired fruiting scion (e.g. Hass) is grafted on the rooted rootstock to create a new field-ready plant. In total this process can take up to 18 months to produce a clonal plant, making it time consuming and laborious. This decades-old practice has been improved by Ernst (1999) using a microcloning technique that has slightly increased the efficiency of production. However, despite improvements to this protocol it often results in poor and inconsistent rooting for many rootstock cultivars (Hendry and Staden, 1982; Kadman and Ben-Ya'acov, 1965).

In addition to the technical complexity of this industry process, it is plagued by a number of rate-limiting factors including the germination rate of the nurse-seed, scion and rootstock budwood quality and graft-take success (both grafting steps), rooting efficiency (sometimes as low as 30%) and field-hardening success. Moreover, it depends on sourcing wood from field-maintained scion and rootstock trees, as well as seed from the 'nurse' variety, to produce a field-ready plant. Both seed and budwood availability is highly seasonal and there are often yearly variations in availability due to biennial bearing (for seeds) or climactic conditions. Moreover, production and maintenance of these trees requires a large amount of natural resources, including land, water, fertilizer and pesticides, detracting from food production. This makes it especially resource-intensive compared to seedling rootstocks. On average, a mature avocado tree can potentially yield up to 1000 cuttings and 100–150 fruit over a good season (Graham Anderson pers. comm.; Anderson Horticulture). Thus, to supply a plant order of 10000 plants, at least 100 budwood trees are required plus 1000 seed trees for seedling-based plants, or 200 budwood trees plus 1000 seed trees for clonal plants, assuming 100% propagation efficiency. To sustain the industry of the future, high-density planting (>800 trees/ha) will require double the number of plants as a standard avocado orchard (400 plants/ha) (Köhne and Kremer-Köhne, 1991), and now ultra-high-density planting is being pioneered with 1000–2000 plants/ha depending on cultivar.

Overall, the time, labour and material inputs to current avocado propagation far outweigh the resulting output in tree numbers (Ernst, 1999; Martinez-Pacheco et al., 2010). This is reflected in plant availability and price, especially in developed countries such as the United States, Israel and Australia, where labour costs are high. On a global scale, this has led to severe shortages in elite clonal plants, restricting high-density planting and rapid expansion of commercial orchards. The severity of this situation is exemplified by current waiting lists of over 6 years experienced by farmers to acquire plants, even from the largest propagation nurseries in the world (Hiti-Bandaralage et al., 2017).

An effective and efficient alternative propagation method would potentially revolutionize the avocado industry by meeting the propagule demand globally. Improvements to clonal propagation would facilitate wide uptake of elite varieties with improved yields. This could finally translate the many years of current breeding programmes into millions of dollars for industry. Not only this but in an age where food security is balancing precariously with resource use and environmental impacts, it is important that new technologies can increase food security in a sustainable manner.

## 4 Tissue culture for plant propagation

Plant tissue culture, or *in vitro* micropropagation, is a resource-efficient, cost-effective alternative for clonal propagation (Thorpe, 2007; García-Gonzáles et al., 2010). It is the aseptic culture of tissues under controlled nutritional and environmental conditions *in vitro* (Smith and Drew, 1990; Hussain et al., 2012). The feasibility of tissue culture has been demonstrated repeatedly for herbaceous species (Schroeder, 1980) and has also been commercially successful with tree crops including apple, apricot, cherry and peach (Mohamed-Yasseen, 1993; Dobránszki and da Silva, 2010; Balla and Mansvelt, 2013; Carrasco et al., 2013).

The process of plant tissue culture ideally begins with preparation and maintenance of a small number of donor plants in disease-controlled conditions (e.g. glasshouse) to promote production of healthy young tissues with minimal contamination (Garcia-Gonzáles et al., 2010). The next step is sterilization of explants such as shoot tips, nodal stem segments and leaf pieces, followed by initiation onto an aseptic nutrient medium containing mineral nutrients, a carbon source, plant hormones and various other additives (Murashige and Skoog, 1962). In general, these media components are first tailored to stimulate shoot proliferation from the explants, which can be maintained and multiplied (ideally indefinitely) by subculturing onto fresh medium. A subsequent media of altered composition is then employed to stimulate the shoots to form roots. Once rooted, the plants can be acclimatized to glasshouse conditions before final hardening for field readiness (Hiti-Bandaralage et al., 2018a).

The advantages of plant tissue culture supporting sustainable agriculture are as follows:

1 Small amount of starting material can produce large numbers of clonal plants.

2 No dependence on season or climate allowing year-round, on-demand plant supply.
3 Free from seed and field plantations and associated resource use and transport logistics.
4 Less time and space required than traditional propagation techniques reducing resource use and production costs.
5 Allows for elimination of viruses and disease without use of harmful pesticides or anti-microbials supporting industry health, sustainability and germplasm exchange.
6 The plants or explants can be maintained, moved and stored as demand dictates.
7 Required for development of genetic engineering methods for crop improvement.
8 Allows *in vitro* germplasm conservation and recovery of rare or endangered species.
9 Allows for industrial production of natural plant products with reduced resource use.

## 5 Tissue culture as a sustainable alternative for avocado propagation

The current soil-based clonal propagation practice for avocado is value constrained by its seed dependence, large space requirement, long propagation time, labour requirements, and soil-related quarantine and trade issues. As for other species, tissue culture of avocado could potentially produce a large number of true-to-type plantlets from few explants in a space-efficient, sterile, soil-free and stress-free environment (Hiti-Bandaralage et al., 2017, 2018b). This will allow more rapid and cost-effective translation of breeding programmes to large scale field validation, industry uptake and follow-on productivity gains. It could also facilitate germplasm exchange across normally quarantine restricted borders to allow more expansive uptake of new and improved cultivars. The result could be an expedited ability to increase production at a lower cost. In addition, the development of tissue culture methodologies for avocado is essential if we are to develop *in vitro* germplasm conservation methods (e.g. Encina et al., 2014; O'Brien et al., 2016, 2018) or additional biotechnological tools such as CRISPR (Jinek et al., 2012) for crop improvement. Such tools have been successful for introducing biotic and abiotic stress resistances in other crops (Gao, 2018; Jaganathan et al., 2018) but remain a significant research and industry challenge for avocado. And finally, as tissue culture is seed- and budwood-free, it removes the need for resource-intensive maintenance of thousands of field trees for propagules; each of which require significant investment of labour and precious water and natural land resources, in addition to intensive use of chemical fertilizers and pest/disease-control agents proven to damage human health and the environment. Moreover, such trees use arable land that could otherwise be used for fruit production for human consumption. In all, a successful tissue culture platform for avocado, a crop with limited breeding and improvement to date, could have far-reaching implications for the sustainability of the industry globally.

## 6 Advances in micropropagation of avocado

Despite the potential gains of developing a high-throughput tissue culture system for avocado, similarly to other woody species, avocado is highly recalcitrant to *in vitro* regeneration (Zulfiqar et al., 2009; Castro et al., 1995; Barceló-Muñoz and Pliego-Alfaro, 2003; Pliego-Alfaro and Murashige, 1987). The regeneration capacity of any cultured material is dependent on endogenous factors including genotype, mother plant maturity and explant type, as well as the *in vitro* environment (media components and incubation conditions) (Zirari and Lionakis, 1994). An array of different explant tissues of varying maturity (embryo, shoot tip, stem nodes with axillary buds, leaf, leaf petiole, root pith, stem pith, flower, fruit mesocarp, peduncle, pollen, cotyledon and protoplasts) have been trialled for avocado over the last 45 years. As for other woody species, juvenile material is much more amenable to culture, and protocols optimized for juvenile material are not readily transferable to mature explants (Barceló-Muñoz and Pliego-Alfaro, 2003). The major impediments include: (1) inter-varietal inconsistency, (2) poor shoot proliferation from mature tree explants and (3) poor induction of roots on successfully cultured mature shoots (Pliego-Alfaro and Murashige, 1987). This is a huge bottleneck for clonal avocado propagation, as rootstock and scion selections of known performance exist only as mature field stocks. A comprehensive tissue culture protocol for avocado will require careful optimization of explant selection, explant and mother plant pre-treatment conditions, culture conditions, basal media and hormone regime, though to cultivar choice. This chapter will focus on progress in nodal and also meristem-tip culture as options for avocado propagation using mature material.

### 6.1 Nodal culture for avocado propagation

Of all explant types, nodal culture is generally the most amenable for woody plant species. As such, nodal culture is most widely applied to *in vitro* propagation of avocado (Schroeder, 1973; Harty, 1985; Ahmed et al., 2001). Nodal explants comprise a section of stem with a leaf node containing a regenerable axillary bud. For avocado, nodal sections are derived from the young flush/budwood stem cuttings of avocado shoots and are a very reliable source for preserving genetic stability of elite cultivars (Barceló-Muñoz and Pliego-Alfaro, 2003). For this, a minimal number of donor trees or plants can be maintained for explants in glasshouse conditions. Both juvenile and mature material has been applied successfully to the nodal culture of a number of avocado cultivars (Schroeder, 1973; Martinez-Pacheco et al., 2010).

## 6.2 Culture initiation

The success of *in vitro* regeneration of nodal explants depends on the ontogeny and health of the mother plant, explant source (apical or axillary), position on the mother plant, metabolic status of the bud and genotype (Zulfiqar et al., 2009). The first stage of nodal culture is preconditioning of the nodal explants. This is to minimize bacterial and fungal contamination, promote bud activation rate and reduce browning due to oxidizing exudates (Barceló-Muñoz and Pliego-Alfaro, 2003). The process begins by selection of healthy, stress-free, mother plants, which are prepared by pruning and applying fungicides (Cooper, 1987). Explants obtained from field-grown trees have repeatedly been shown to be highly contaminated with bacteria and fungi, and thus best practice is to maintain a selection of mother trees in glasshouse or controlled conditions. Where this is not possible, antibiotics and fungicides have been used as control measures (Barrera-Guerra et al., 1998; Martinez-Pacheco et al., 2010).

Following explant collection, they are surface sterilized, generally involving washing under running tap water, treatments with a sterilizing agent (e.g. ethanol/sodium hypochlorite) and several washes with sterile distilled water (Cooper, 1987; Nhut et al., 2008). Pre-treatment of explants with antioxidants, such as ascorbic or citric acid, to avoid browning has also been suggested as essential to establish quality cultures (Castro et al., 1995; Barringer et al., 1996). In some cases, an activation step has been employed to expedite growth of dormant apical and axillary buds. Methods of activation include roller drum incubation in liquid medium (Barceló-Muñoz and Pliego-Alfaro, 2003) or etiolation of mother plants prior to explant initiation (Schroeder, 1980; Cooper, 1987; Vega, 1989).

## 6.3 Shoot induction

Once treated for culture, explants are initiated onto a basal media composed of macronutrients (N, K, P, Ca, Mg, S), micronutrients (Fe, Ni, CL, Mn, Cu, Zn, B, Mo) and vitamins, and often hormones essential for plant growth (George et al., 2008). Shoots are then promoted to elongate and proliferate on an optimized media with exogenous hormones (Zulfiqar et al., 2009). Cytokinins (CK) are the primary hormones involved in cell division and shoot regeneration in plants

(Dan and Reichert, 2001). Optimized CK levels are necessary for promoting shoot growth while avoiding excessive callus production, shoot tip necrosis and shoot stunting (Cautín and Vardja, 2001; Muhammad et al., 2007). Unlike juvenile material, which can show 100% shoot regeneration irrespective of cultivar (Pliego-Alfaro, 1988; Barringer et al., 1996; Nhut et al., 2008) and even hormones (Harty, 1985), mature material requires cultivar-specific hormone regimes for shoot regeneration (e.g. Hiti-Bandaralage et al., 2015; Martinez-Pacheco et al., 2010; Nhut et al., 2008; Barceló-Muñoz et al., 1999; Castro et al., 1995; Cooper, 1987). Without hormones, mature nodal shoots show poor vigour and stem elongation, heavy callusing, retarded enlargement, shoot tip dieback, leaf defoliation and leaf abscission (Pliego-Alfaro and Murashige, 1987). Thus determination of the optimum nutritional and hormonal environment specific to a cultivar of choice is a major challenge for successful tissue culture of mature explants (Schroeder, 1968).

In addition to nutrient and hormone regime, various medium consistencies (solid, liquid or double phase (liquid layer on solid media)) and incorporation of high $CO_2$ level in culture vessels have been tested to improve shoot health and proliferation of avocado (Zirari and Lionakis, 1994; Barceló-Muñoz et al., 1999; De la Viña et al., 1999; Nel et al., 1983; Barrera-Guerra et al., 1998). Often these are incorporated intermittently or as pre-treatments, adding to the complexity of avocado tissue culture systems. For best commercial applicability, an optimized solid media without a requirement for additional media phases or complex $CO_2$ treatments is key. We have previously published a promising method for achieving this by varying the hormones and basal nutrients in a cultivar-specific manner (Hiti-Bandaralage et al., 2015). We were able to achieve 100% regeneration of healthy shoots from mature nodes using synergistic hormones to prevent necrosis, defoliation and shoot die back, taking us one step closer to establishing a viable nodal culture method for micropropagation of avocado rootstocks (Hiti-Bandaralage et al., 2015; Fig. 1a).

**Figure 1** Avocado shoots (a), rooted plant (b) and acclimatized plant (c) derived from mature nodal culture (Mitter Lab, University of Queensland, Australia).

## 6.4 Rooting

Even once shoot proliferation is optimized, the biggest bottleneck for mature avocado nodal culture, similarly to traditional clonal propagation, is root induction (Zulfiqar et al., 2009; Pliego-Alfaro and Murashige, 1987; Vidal et al., 1992; Nel and Kotzé, 1984). The major prerequisite for root induction is the use of high-quality, vigorous shoots (Ahmed et al., 2001; Barceló-Muñoz and Pliego-Alfaro, 2003; Nel et al., 1983). However, shoot health must be balanced with hormone use during shooting phases, as the type, quantity and duration of shooting hormone exposure can negatively influence root induction of many woody plants. Most important for rooting is the essential role of the hormone auxin, which again must be optimized cultivar specifically in terms of auxin type, concentration, exposure time and mode of application (Hiti-Bandaralage et al., 2017). Even then, the position of the bud on the nodal section and tree (apical vs. axillary) and the juvenility gradient of the explants can affect root induction capacity (Zulfiqar et al., 2009; Rasmussen et al., 2015). Rooting efficiencies of up to 100% can be achieved for juvenile nodal explants of avocado, however the rooting potential diminishes with increasing plant maturity (Pliego-Alfaro and Murashige, 1987; Rasmussen et al., 2015). Zirari and Lionakis (1994) attempted to root mature avocado shoots (cvs. Duke, Topa-topa, Hass, Fuerte) but the result was leaf defoliation and 100% mortality within four weeks. Similarly, Barrera-Guerra et.al. (1998) were not successful in rooting adult avocado shoots using an array of 23 treatments varying culture environment, basal media and additives, and hormone treatment.

Methods to induce juvenility in mature plant material *in vitro* have been attempted to improve rooting. This kind of phase reversal has been somewhat successful using: (1) continuous subculturing and (2) grafting mature material to juvenile material *in vitro* (Hiti-Bandaralage et al., 2018a). Pliego-Alfaro and Murashige (1987) achieved an increase from 0% to 50% rooting on mature cv. Duke 7 shoots after grafting to juvenile rootstock plus auxin treatment. Phase rejuvenation by pruning of mature plants prior obtaining explants for culturing has also been tested for avocado (Barceló-Muñoz et al., 1999). Pruning plus a 3–5-day incubation in auxin-containing liquid media resulted in the highest rooting percentage ever reported for mature avocado (90%) (Barceló-Muñoz et al., 1999). More recently, however, over 50% rooting was achieved on mature avocado shoots of a Mexican cultivar, Drymifolia, through optimized culture and hormone conditions without the requirement for an additional liquid stage (Martinez-Pacheco et al., 2010). Similarly, in our hands, careful optimization of numerous factors (substrate, macro and micro nutrient, type of auxin and concentration, auxin treatment mode and duration, other additives) has seen rooting efficiency of 50–85% on mature nodal shoots of two cultivars without a requirement for additional media phases (Hiti-Bandaralage pers. comm.; Fig. 1b).

### *6.5 Acclimatization*

Anatomical and physiological changes occur in plants under tissue culture as they adjust to favourable mineral nutrition, moisture, light and external environmental conditions. *In vitro*-grown plants lack proper leaf and root structures to withstand the dynamic *ex vitro* environment and thus require a hardening phase to acclimatize them to the external environment (Hiti-Bandaralage et al., 2018a).

Even once rooted, avocado survival during acclimatization has been reported to be very poor. Prior studies recommend transfer of intact plants to a greenhouse under 100% humidity followed by gradual daily exposure to ambient humidity (Barceló-Muñoz and Pliego-Alfaro, 2003). Additional strategies such as use of different potting media, application of anti-transpirants, inoculation with mycorrhizal symbionts, gradual increase of light conditions, use of mist chambers as well as preconditioning strategies such as supplementing medium with high sucrose doses, paclobutrazol and desiccants during the rooting stage have all been tested (Hazarika, 2003, 2006; Martinez-Pacheco et al., 2010; Cooper, 1987; Vidal et al., 1992). We have achieved 100% plant acclimation success of mature avocado nodal shoots without the need for complex and difficult strategies (Fig. 1c).

Overall, five decades of research has highlighted the issues of cultivar specificity, poor multiplication capacity, poor rooting, leaf defoliation and necrotic die back during extended subculture, as major bottlenecks to commercial application of avocado nodal culture (Castro et al., 1995; Ahmed et al., 2001; Barceló-Muñoz and Pliego-Alfaro, 2003). Despite this, we have had success with high-efficiency shoot induction, rooting and acclimatization of clonal plants from nodal culture. However, the problems of poor inter-cultivar transferability, limited subculture longevity and multiplication rate from nodal cultures still restricts its ability to replace conventional propagation practices. Alternative culture options with increased multiplication rate and longevity will provide the best options for a commercial possibility for avocado. The most promising option for this is meristem culture.

## 7 Meristem culture for the future of sustainable avocado propagation

Meristem culture is a method of culture where an excised meristem dome (0.2-1.0 mm) with one or two leaf primordia is induced to *de novo* organogenesis with the use of suitable tissue culture media and culture conditions (George et al., 2008). Due to the meristematic totipotency of the material, it has the advantage of maximizing multiplication and regeneration rate (Lane, 1978; Rani and Raina, 2000), which is essential for mass propagation. Apical meristem cells from the shoot tip have exorbitant genetic stability, thus are the best source of meristem tissues for clonal propagation. Moreover, due to the small size of tissue it can eliminate viruses, in addition to other microbes, from the plants (Schroeder, 1973, 1976; Romberger, 1980; Yakuwa and Oka, 1988; Grout, 1999). Indeed, meristem culture was first introduced in the 1950s by Morel and Martin (1955) to multiply *Dahlia* and *Cymbidium* species with the objective of pathogen elimination. Due to the potential for mass clonal propagation of elite cultivars that can be guaranteed disease-free; meristem culture is a golden standard for mass commercial propagation in tissue culture. As such, meristem culture has been applied to a wide variety of economically important plants including ornamental, medicinal, forest and fruit species (Walkey, 1972; Persley et al., 1987; Aitken-Christie et al., 1988; Quiroz et al., 2017). The potential of propagation using this technique has been well demonstrated in banana and strawberry (Venkatachalam et al., 2007; Quiroz et al., 2017).

## 7.1 Meristem culture of woody plants

There are reports of meristem culture of several woody species including the economically important fruits apple (*Malus domestica*), pear (*Pyrus* sp.), peach (*Prunus persica* L.) (Walkey, 1972; Postman and Hadidi, 1995), blueberry (*Vaccinium* sp.), blackberry (*Rubus* sp.) (Vasil, 2012) and mulberry (*Morus alba* L.) (Yakuwa and Oka, 1988; Anis et al., 2003). However, despite the advantages of meristem culture over nodal culture, the survival and multiplication of meristem explants is far more challenging than stimulating the growth of an existing nodal axillary bud. This is particularly true for woody species. Moreover, once again, the difficulty of establishing meristem culture in the case of woody plants increases with maturity (Monteuuis, 1991).

For avocado, meristem culture was first explored by Schroeder (1973) using juvenile material. This required the use of large shoot tips (3–8 mm) with well-developed bracts and leaf primordia. Even so, this study demonstrated only very limited long-term survival and multiplication of shoot tips in culture. Moreover, shoot regeneration was hindered by massive callus formed at the base of the tissue resulting in poor shoot elongation and failure in root induction. Hendry and Staden (1982) and Harty (1985) also suggested the difficulty in multiplication of juvenile avocado tips confirming Schroeder's findings.

Culture of shoot tips has also been attempted for avocado disease elimination. Typically elimination of viruses requires meristem explants of 1 mm or less with no visible leaf primordia (Nel and Kotzé, 1984). With avocado, one or two leaf primordia must be left on the meristematic dome in order to overcome mortality and lower regeneration capacity while still minimizing the percentage of virus infected plants (Nel and Kotzé, 1984). The study by Nel and Kotzé (1984) generated only shoots from avocado meristems *in vitro* and then grafted these on seedling rootstocks for recovery of whole plants, achieving a 50% virus elimination rate. However, while this allowed for disease eradication of parent stock, without root induction this is not amenable to a mass clonal propagation technique for elite rootstocks.

To date, no meristem culture system has been established to produce intact avocado plantlets, using either juvenile or mature explants, even with the use of large shoot tips (Schroeder, 1973; Hendry and Staden, 1982; Harty, 1985). Here we present our own advances in this technique for sustainable supply of avocado clonal planting material.

## 7.2 Advances in meristem culture for avocado propagation

In our lab at the University of Queensland we have developed a meristem-tip culture platform for clonal propagation of mature avocado trees. Our complete high-throughput meristem-tip culture system has been successfully optimized to clonally propagate mature material of cultivar 'Reed' but has also been applied to additional avocado cultivars with cultivar-specific modifications. To our knowledge, this is the world's first success with meristem culture of avocado. Importantly for commercial application, this pipeline can produce clonal, rooting-ready shoots in a matter of months. Moreover, at present, a single 'Reed' meristem can reliably give rise to over 500 clonal shoots; easily orders of magnitude more than can be generated by the current nursery clonal propagation technique, or even by nodal culture. In addition to having all the resource-saving benefits inherent to tissue culture, this propagation platform therefore has foreseeable potential as a commercially viable, high-throughput, system for clonal avocado production.

Similar to nodal culture, our meristem-tip culture process is composed of a highly specific stepwise procedure starting from material sterilization, clean culture establishment, shoot multiplication and root induction followed by acclimatization (Fig. 2). Root induction once again was the rate-limiting step of the meristem-tip culture process. However, 100% rooting success has been achieved for cv. Reed after optimizing a number of parameters. In addition, plantlet survival with optimized acclimatization practice is 96–97%.

In addition to cost-effective, rapid clonal propagation of elite cultivars, a successful meristem-tip culture method for avocado could potentially eliminate diseases in propagated materials through efficient and reliable screening starting from mother plants, resulting in reduced losses due to disease at both the propagation and planting stage. This would also eliminate the requirement for harmful chemical treatments, supporting a more sustainable cost-effective propagation practice and industry. Major diseases of avocado trees include *Phytophthora* root rot (caused by the fungus *P. cinnamomi*), Anthracnose (*Colletotrichum gloeosporioides*), bacterial canker and sunblotch – caused

**Figure 2** Avocado mature meristem culture pipeline showing shoot multiplication (a), elongation (b), single shoot (c), rooted and potted plantlet (d) and finally acclimatized plants transferred to nursery conditions for grow out (e).

by a systemic viroid. In the traditional nursery-based propagation method for avocado, vegetative propagation from seeds or cuttings of an infected mother plant produces infected progeny. Such diseases are currently a major constraint to propagation and production requiring vigilant and intensive quarantine and hygiene practices, often involving harsh chemical anti-microbials and pesticides. The problem is especially true for viruses and viroids, which are not responsive to such treatments and extremely difficult to eliminate. The successful meristem-tip culture protocol developed in our hands represents a major breakthrough to address the industry challenge of clonal, disease-free propagation of avocado. Field evaluation of the plants for performance and productivity in coming years is necessary to support commercial adoption of these plants.

## 8 Micrografting for complete field-ready disease-free planting material

Developed in the 1980s, micrografting is a technique involving the placement of a scion (meristem or a shoot tip) on a decapitated rootstock under aseptic conditions, to obtain an *in vitro* grafted plant (Rehmen and Gill, 2015). Micrografting can be used as an integral part of a micropropagation system for the production of disease-free plants (Navarro, 1988) and rejuvenation of mature shoots (Pliego-Alfaro and Murashige, 1987; Revilla et al., 1996; Yıldırım et al., 2010).

Application of *in vitro* micrografting to produce grafted plants, using *in vitro* propagated rootstock and scion material, as a mass propagation practice for woody plants is very rare. This is due to the difficulty of *in vitro* intact plant establishment from mature explant material (Navarro, 1988). Also, their field behaviour and productivity must be validated prior to adoption of micrografted plants in commercial orchards (Hussain et al., 2014), which is time consuming for woody perennials. These limitations can be justified however, as micrografting can potentially eliminate two main problems encountered with traditional grafting failures: microbial contamination and desiccation of scion (Estrada-Luna et al., 2002). *In vitro* micrografting is performed under sterile conditions and incubated under higher relative humidity and thus is advantageous for higher graft success. This technique has been attempted with many woody species, such as sandalwood (*Santalum album* L.) (Sanjaya et al., 2006), citrus (*Citrus* sp.) (Navarro, 1992), cashew (*Anacardium occidentale* L.) (Moneney and Mantell, 2001), almond (*Prunus dulcis* Mill.) (Yıldırım et al., 2010), pistachio (*Pistacia vera*) (Onay et al., 2004), olive (*Olea euroeae*) (Revilla et al., 1996), mulberry (*Morus alba* L.), pear (*Pyrus* spp.) (Hassanen, 2013), apple (*Malus domestica*) (Bhatt et al., 2013) and even with avocado (Pliego-Alfaro and Murashige, 1987; Raharjo and Litz, 2005) for different reasons.

The complexity of this technique is compounded by many factors influencing success. Graft compatibility between rootstock and scion being the most decisive. Other determinants reported to be important include scion origin (actively growing, dormant, *in vivo* forced tips, field plants, *in vitro* shoot tips) (Deogratias et al., 1991; Bhatt et al., 2013), different pre-treatments of scion and rootstock (Hussain et al., 2014), method of grafting (Hassanen, 2013), scion size (Onay et al., 2004; Sanjaya et al., 2006; Hassanen, 2013; Rehmen and Gill, 2015), rootstock age (Deogratias et al., 1991; Sanjaya et al., 2006; Amiri, 2007), media composition (e.g. sucrose concentration, consistency) (Deogratias et al., 1991; Bhatt et al., 2013; Rehmen and Gill, 2015), light conditions and even incubation temperature (Dziedzic and Malodobry, 2006).

The significance of this technique for avocado relates to applications in both research and production. It has been trialled as a rejuvenation method for mature avocado material to restore rooting ability during micropropagation (Pliego-Alfaro and Murashige, 1987), sunblotch viroid elimination (Suarez et al., 2006) and for somatic embryo rescue (Raharjo and Litz, 2005). To date however, no study has focused on *in vitro* micrografting to mass produce field-ready plants as an alternative to nursery grafting. This is likely to predominately be due to unavailability of an efficient *in vitro* micropropagation system for rootstock and scion cultivars.

The traditional nursery grafting technique uses either seedling or clonal rootstocks with budwood collected from mature scion cultivars. This requires large amounts of budwood material from field-grown plants, normally sourced from multiple locations. Availability of budwood is seasonal, with different scion cultivars producing appropriate budwood in different seasons. The quality of the budwood, with respect to the pest and disease and growth stage, is critical for grafting success and mostly not under the control of budwood collectors. The field-grown plants are often infested with non-pathogenic bacteria and fungi that can cause opportunistic infections with poor handling practices. Also, material transport and storage are important aspects in controlling contamination and preserving viability. In the case of rootstocks, seed availability to produce seedlings becomes a limiting factor in propagation due to the seasonal nature of fruit set. This limitation is also valid for clonal rootstocks, as clonal rootstock propagation relies on a nurse seedling (Platt, 1976; Ernst, 1999). Moreover, if a clonal rootstock is used, the same challenges of sourcing material and quality control of rootstock budwood influence the propagation process. Not only acquiring material but also the grafting success itself depends on the season and it is highly variable for some cultivars (Rodrigues et al., 1960). The complex logistics involved in managing the traditional grafting practice of avocado places immense pressure on a nursery's capacity of efficient and timely supply of grafted plants.

In future, integration of a micrografting protocol with the *in vitro* multiplication of avocado scion and rootstock materials could completely abolish the seasonal dependencies of avocado propagation and improve efficiency. Micrografting material under *in vitro* conditions would eliminate disease problems and undesired graft failure due to desiccation. Moreover, graft take would not be influenced by the growing season or *ex vitro* nursery environment. It could also facilitate production of field-ready, disease-free plants to bridge boundaries for plant exports. This would enable plant exchange between nurseries and production regions globally.

## 9 Conclusion

The inefficient propagation of avocado plants for commercial production currently represents a huge bottleneck for intensification of avocado production globally. Current practice is highly resource-intensive and unsustainable if we wish to make the move to lower-footprint, higher-density plantings and meet the growing demand for this high-value food.

For the first time, the long sought-after ability to sustainably mass produce avocado clonal plants from mature material in culture is becoming a feasible reality. Not only does tissue culture significantly reduce the environmental footprint of plant propagation, but it also facilitates development of protocols for genetic improvement of plant performance/resilience and for sustainable conservation of germplasm through cryopreservation. Over the last 45 years, many of the avocado micropropagation studies have been focused on optimizing protocols for material obtained from juvenile plants with an objective of smooth transfer of techniques to mature plants. However, so far this has not been possible with avocado. The limited protocols that have been developed for mature avocado propagation are highly cultivar-specific and do not meet requirements to adopt it as an industrial practice.

We have optimized a successful *in vitro* meristem-tip culture platform that represents a major breakthrough to address the industry challenge of clonal propagation of avocado. Mass-scale propagation of clonal plants of mature origin could address the current and future demand for plants, and support the industries move to higher-density more resource-efficient planting systems. This system has the potential to revolutionize the global avocado industry through sustainable supply of disease-free planting materials year-round.

## 10 Where to look for further information

Plant tissue culture has been a research challenge for avocado and other difficult species for many years. Our tissue culture system for avocado is a world first, and interested readers are welcome to contact us directly at QAAFI, the University of Queensland, to learn of our present and future research plans in this field. For prior work on avocado, we direct readers to seminal works from the laboratory of Fernando Pliego-Alfaro in our References list, who pioneered tissue culture attempts for various avocado cultivars through nodal culture.

For those interested in avocado cultivation and research in general we direct readers to the comprehensive online repository for all things avocado found at http://www.avocadosource.com/. Readers may also refer to Editions of the book *The Avocado: Botany, Production and Uses* edited by Bruce Schaffer, B. Nigel Wolstenholme and Anthony W. Whiley. For those interested in the process and challenges of tissue culture across woody plant species in general we suggest George (2008) and Thorpe (2007) in the References list.

## 11 References

Ahmed, M. F., Kantharajah, A. S. and Holford, P. 2001. Factors affecting adventitious shoot and root formation from the avocado cultivar 'Hass' (*Persea americana* Mill.) *in vitro*. *Gartenbauwissenschaft* 66, 293–7.

Aitken-Christie, J., Singh, A. P. and Davies, H. 1988. *Multiplication of Meristematic Tissue: A New Tissue Culture System for Radiata Pine. In: Genetic Manipulation of Woody Plants*. Springer, Boston, MA, pp. 413–32.

Amiri, M. E. 2007. Special micrografting technique for cherry (*Prunus avium* L.). *Acta Hortic.* 764(764), 151–4. doi:10.17660/ActaHortic.2007.764.18.

Anis, M., Faisal, M. and Singh, S. K. 2003. Micropropagation of Mulberry (*Morus alba* L.) through *in vitro* culture of shoot tip and nodal explants. *Plant Tissue Cult.* 13, 47–51.

Balla, I. and Mansvelt, L. 2013. Micropropagation of peach rootstocks and cultivars. *Methods Mol. Biol.* 11013, 137–48. doi:10.1007/978-1-62703-074-8_10.

Barceló-Muñoz, A. and Pliego-Alfaro, F. 2003. Micropropagation of avocado (*Persea americana* Mill.). In: Jain, S. M. and Ishii, K. (Eds), *Micropropagation of Woody Trees and Fruits*. Springer, pp. 519–42. doi:10.1007/978-94-010-0125-0_17.

Barceló-Muñoz, A., Encina, C. L., Simón-Pérez, E. and Pliego-Alfaro, F. 1999. Micropropagation of adult avocado. *Plant Cell Tissue Organ Cult.* 58(1), 11–7. doi:10.1023/A:1006305716426.

Barrera-Guerra, J. L., Ramirez-Malagon, R. and Martinez-Jaime, O. A. 1998. *In vitro* propagation of avocado (*Persea drymifolia* Ness.). In: *Induced Mutations in Connection with Biotechnology for Crop Improvement in Latin America*. Joint FAO/IAEA Division of Nuclear Techniques in Food and Agriculture, Lima, Peru, pp. 63-9.

Barrientos-Priego, A., Borys, M. W. and Barrientos-Perez, F. 1986. Rooting of avocado cuttings (*Persea americana* Mill.) cvs. 'Fuerte' and 'Colin V-33'. *Calif. Avocado Soc Yearb.* 70, 157–64.

Barringer, S. A., Mohamed-Yasseen, Y. and Splittstoesser, W. E. 1996. *In vitro* multiplication and plantlet establishment of avocado. *In Vitro Cell Dev. Biol. Plant* 32, 119–21.

Ben-Ya'acov, A. 1987. Avocado rootstock-scion relationships. *South Afr. Avocado Grow Assoc. Yearb.* 10, 30–2.

Bergh, B. and Ellstrand, N. 1986. Taxonomy of the avocado. *Calif. Avocado Soc. Yearb.* 70, 135–46.

Bhatt, K. M., Banday, F. A., Mir, M. A., Rather, Z. A. and Hussain, G. 2013. *In vitro* grafting in apple (*Malus domestica*. Borkh) cv. Lal Ambri. *Karnataka J. Agric. Sci.* 26, 399–402.

Brondani, G. E., Baccarin, F. J. B., de Wit Ondas, H. W., Stape, J. L., Gonçalves, A. N. and de Almeida, M. 2012. Low temperature, IBA concentrations and optimal time for adventitious rooting of *Eucalyptus benthamii* mini-cuttings. *J. For. Res.* 23, 583-92. doi:10.1007/s11676-012-0298-5.

Carrasco, B., Meisel, L., Gebauer, M., Garcia-Gonzáles, R. and Silva, H. 2013. Breeding in peach, cherry and plum: from a tissue culture, genetic, transcriptomic and genomic perspective. *Biol. Res.* 46(3), 219–30. doi:10.4067/S0716-97602013000300001.

Castro, M., Oyanedel, E. and Cautín, R. 1995. *In vitro* shoot proliferation in avocado (*Persea americana* Mill.) induced by CPPU. In: *Proceedings of the World Avocado Congress III*, pp. 223–6.

Cautín, R. and Vardja, T. 2001. The effect of cytokinin type and concentration and the number of subcultures on the multiplication rate of some decorative plants. *Proc. Est. Acad. Sci. Biol. Ecol.* 50 22–32.

Chong, E. W.-T., Sinclair, A. J. and Guymer, R. H. 2006. Facts on fats. *Clin. Exp. Ophthalmol.* 34(5), 464–71. doi:10.1111/j.1442-9071.2006.01250.x.

Coffey, M. D. and Guillemet, F. B. 1987. Avocado rootstocks. *Calif. Avocado Soc. Yearb.* 71, 173–9.

Cooper, P. A. 1987. Advances in the micropropagation of avocado (*Persea americana* Mill.). *Acta Hortic.* 212(212), 571–6. doi:10.17660/ActaHortic.1987.212.92.

Cutting, J. G. and Van Vuuren, S. P. 1988. Rooting leafy non-etiolated avocado cuttings from gibberellin-injected trees. *Sci Hortic* 37, 171–6.

Dan, Y. and Reichert, N. A. 2001. Cytokinin metabloism and action. *Annu. Rev. Plant Physiol. Plant Mol. Biol.* 52, 89–118.

Darnton, J. 2017. Avocado consumption: environmental and social considerations. Mitchelton State University Extension. Mitchelton State University.

De la Viña, G., Pliego-Alfaro, F., Driscoll, S. P., Mitchell, V. J., Parry, M. A. and Lawlor, D. W. 1999. Effects of $CO_2$ and sugars on photosynthesis and composition of avocado leaves grown *in vitro*. *Plant Physiol. Biochem.* 37, 587–95. doi:10.1016/S0981-9428(00)80111-4.

Deograties, T. M., Castellani, V., Dosba, F., Juarez, J., Arregui, J. M., Ortega, C., Ortega, V., Llacer, G. and Navarro, L. 1991. Study of growth parameters on apricot shoot tip grafting *in vitro* (STG). *Acta Hortic.* 283, 361–71.

Ding, H., Chin, Y. W., Kinghorn, A. D. and D'Ambrosio, S. M. 2007. Chemopreventive characteristics of avocado fruit. *Semin. Cancer Biol.* 17(5), 386–94. doi:10.1016/j.semcancer.2007.04.003.

Dobránszki, J. and da Silva, J. A. 2010. Micropropagation of apple – a review. *Biotechnol. Adv.* 2(4), 462–88. Chapter 1. doi:10.1016/j.biotechadv.2010.02.008.

Dreher, M. L. and Davenport, A. J. 2013. Hass avocado composition and potential health effects. *Crit. Rev. Food Sci. Nutr.* 53(7), 738–50. doi:10.1080/10408398.2011.556759.

Dziedzic, E. and Malodobry, M. 2006. Vegetative cherry rootstocks in tissue culture. *Sodinink Ir. Daržinink* 25, 77–84.

Eggers, E. R. and Halma, F. F. 1937. Rooting avocado cuttings. *Calif. Avocado Soc. Yearb.* 21, 121–5.

Encina, C. L., Parisi, A., O'Brien, C. and Mitter, N. 2014. Enhancing somatic embryogenesis in avocado (*Persea americana* Mill.) using a two-step culture system and including glutamine in the culture medium. *Sci. Hortic.* 165, 44–50. doi:10.1016/j.scienta.2013.10.019.

Ernst, A. A. 1999. Micro cloning: a multiple cloning technique for avocados using micro containers. *Rev. Chapingo Ser. Hortic.* 5, 217–20.

Estrada-Luna, A. A., López-Peralta, C. and Cárdenas-Soriano, E. 2002. *In vitro* micrografting and the histology of graft union formation of selected species of prickly pear cactus (*Opuntia* spp.). *Sci. Hortic.* 92(3–4), 317–27. doi:10.1016/S0304-4238(01)00296-5.

FAOSTAT. 2018. FAOSTAT. Available at: http://www.fao.org/faostat/en/#data/QC/visualize (Accessed on 19 June 2018).

Frolich, E. F. 1951. Rooting Guatemalan avocado cuttings. *Calif. Avocado Soc. Yearb.* 36, 136–8.

Frolich, E. F. and Platt, R. G. 1972. Use of the etiolation technique in rooting avocado cuttings. *Calif. Avocado Soc. Yearb.* 55, 97–109.

Fulgoni, V. L., Dreher, M. and Davenport, A. J. 2013. Avocado consumption is associated with better diet quality and nutrient intake, and lower metabolic syndrome risk in US adults: results from the National Health and Nutrition Examination Survey (NHANES) 2001–2008. *Nutr. J.* 12, 1. doi:10.1186/1475-2891-12-1.

Galindo-Tovar, M. E., Ogata-Aguilar, N. and Arzate-Fernández, A. M. 2008. Some aspects of avocado (*Persea americana* Mill.) diversity and domestication in Mesoamerica. *Genet. Resour. Crop Evol.* 55(3), 441–50. doi:10.1007/s10722-007-9250-5.

Gao, C. 2018. The future of CRISPR technologies in agriculture. *Nat. Rev. Mol. Cell Biol.* 19(5), 275–6. doi:10.1038/nrm.2018.2.

García-Gonzáles, R., Quiroz, K., Carrasco, B. and Caligari, P. 2010. Plant tissue culture: current status, opportunities and challenges. *Cienc. E Investig. Agrar.* 37, 5–30. doi:10.7764/rcia.v37i3.145.

George, E. F., Hall, M. A. and DeKlerk, G.-J. 2008. *Plant Propagation by Tissue Culture: Volume 1. The Background*. 3rd edn, Springer Science & Business Media, Dordrecht.

Grout, B. W. W. 1999. Meristem-tip culture for propagation and virus elimination. In: Hall, R. D. (Ed.), *Plant Cell Culture Protocols. Methods in Molecular Biology* (vol. 111). Humana Press, Totowa, NJ, pp. 115–25.

Harty, P. A. 1985. Propagation of avocado by tissue culture: development of a culture medium for multiplication of shoots. *South Afr. Avocado Grow Assoc. Yearb.* 8, 70–1.

Hassanen, S. A. 2013. *In vitro* grafting of pear (*Pyrus* spp.). *World Appl. Sci. J.* 21, 705–9.

Hazarika, B. N. 2003. Acclimatisation of tissue-cultured plants. *Curr. Sci.* 85, 1704–12.

Hazarika, B. N. 2006. Morpho-physiological disorders in *in vitro* culture of plants. *Sci. Hortic.* 108, 105–20.

Hendry, N. S. and Staden, J. V. 1982. Attempts at propagation of avocado using various techniques. *South Afr. Avocado Grow Assoc. Yearb.* 5, 71–3.

Hiti-Bandaralage, J., Hayward, A., O'Brien, C. and Mitter, N. 2015. Gibberellin and cytokinin in synergy for a rapid nodal multiplication system of avocado. In: *Actas VIII congreso mundial de la palta*. VIII World Avocado Congress, Lima, Peru, pp. 95–98. 13–18 September 2015.

Hiti-Bandaralage, J. C. A., Hayward, A. and Mitter, N. 2017. Micropropagation of avocado (*Persea americana* Mill.). *Am. J. Plant Sci.* 08, 2898. doi:10.4236/ajps.2017.811197.

Hiti-Bandaralage, J. C. A., Hayward, A., O'Brien, C., Beaveridge, C. and Mitter, N. 2018a. Acclimatisation of micropropagated mature avocado. Acclimatization of micropropagated mature avocado. *Acta Hortic.* 1224, 13–20. doi:10.1766/ActaHortic.2018.1224.3.

Hiti-Bandaralage, J. C. A., Hayward, A., O'Brien, C. and Mitter, N. 2018b. Temporal profile of juvenility-associated microRNAs during tissue culture of avocado. *Acta Hortic.* 1205(1205), 779–84. doi:10.17660/ActaHortic.2018.1205.97.

Hussain, A., Ahmed, I., Nazir, H. and Ullah, I. 2012. Plant tissue culture: current status and opportunities. In: Leva, A. (Ed.), *Recent Advances in Plant In Vitro Culture*. InTech.

Hussain, G., Wani, M. S., Mir, M. A., Rather, Z. A. and Bhat, K. M. 2014. Micrografting for fruit crop improvement. *Afr. J. Biotechnol.* 13(25), 2474–83. doi:10.5897/AJB2013.13602.

Jaganathan, D., Ramasamy, K., Sellamuthu, G., Jayabalan, S. and Venkataraman, G. 2018. CRISPR for crop improvement: an update review. *Front. Plant Sci.* 9, 985. doi:10.3389/fpls.2018.00985.

Jain, S. M. and Ishii, K. 2003. *Micro Propagation of Woody Trees and Fruits*. Springer Science & Business Media.

Janick, J. 2010. *Horticultural Reviews*. John Wiley & Sons.

Jinek, M., Chylinski, K., Fonfara, I., Hauer, M., Doudna, J. A. and Charpentier, E. 2012. A programmable dual-RNA-guided DNA endonuclease in adaptive bacterial immunity. *Science* 337(6096), 816–21. doi:10.1126/science.1225829.

Kadman, A. and Ben-Ya'acov, A. 1965. A review of experiments on some factors influencing the rooting of avocado cuttings. *Calif. Avocado Soc. Yearb.* 49, 67–72.

Köhne, J. S. and Kremer-Köhne, S. 1991. Avocado High Density Planting – A Progress Report, pp. 42–43.

Lane, W. D. 1978. Regeneration of apple plants from shoot meristem-tips. *Plant Sci. Lett.* 13, 281–5.

Leal, F. J. and Krezdorn, A. H. 1964. Rooting avocado cuttings. *Proc. Fla. State Hort. Soc.* 77, 358–62.

Litz, R. E., Raharoj, S. H. T. and Gomez Lim, M. A. 2007. Avocado. In: *Biotechnology in Agriculture and Forestry*. Springer-Verlag, Berlin, Heidelberg, pp. 167–87.

Martinez-Pacheco, M. M., Suarez-Rodríguez, L. M., Hernández-García, A., et al. 2010. *In vitro* propagation of Mexican race avocado (*Persea americana* Mill. Var Drymifolia). In: *International Symposium on Micro and Macro Technologies for Plant Propagation. Acta Hortic.*, pp. 47–52.

Menge, J. A., Personnel, C., Zentmyer, G. A., et al. 2001. Screening and evaluation of new rootstocks with resistance to *Phytophthora cinnamomi*. In: *Proceedings of the California Avocado Research Symposium*, Riverside, CA, pp. 49–53.

Mohamed-Yasseen, Y. 1993. Morphogenesis of avocado *in vitro*. A review. *Calif. Avocado Soc. Yearb.* 77, 101–5.

Moneney, E. E. and Mantell, S. H. 2001. *In vitro* micrografting of cashew. *Plant Cell Tissue Organ Cult.* 66(1), 49–58. doi:10.1023/A:1010687629985.

Monteuuis, O. 1991. Rejuvenation of a 100-year-old *Sequoiadendron giganteum* through *in vitro* meristem culture. I. Organogenic and morphological arguments. *Physiol. Plant* 81, 111–15. doi:10.1111/j.1399-3054.1991.tb01721.x.

Morel, G. et Martin, C. 1955. Guerison de Plantes Attiendes de Maladies A Virus par Culture de Meristemes Apicaux. Rep XIVth Int. Hort. Congress, Netherlands.

Muhammad, A., Rashid, H., Hussain, I. and Naqvi, S. M. S. 2007. Proliferation-rate effects of BAP and kinetin on banana (*Musa* spp. AAA Group) 'Basrai'. *HortScience* 42, 1253–5.

Murashige, T. and Skoog, F. 1962. A revised medium for rapid growth and bio assays with tobacco tissue cultures. *Physiol. Plant.* 15(3), 473–97. doi:10.1111/j.1399-3054.1962.tb08052.x.

Navarro, L. 1988. Application of shoot tip grafting *in vitro* to woody species. *Acta Hortic.* 227, 43–55.

Navarro, L. 1992. Citrus shoot tip grafting *in vitro*. In: Bajaj, Y. P. S. (Ed.), *High-Tech and Micropropagation II. Biotechnology in Agriculture and Forestry* (vol. 18). Springer, Berlin, Heidelberg, pp. 327–38.

Nel, D. D. and Kotzé, J. M. 1984. The role of tissue culture in the avocado plant improvement scheme. *South Afr. Avocado Grow Assoc. Yearb.* 7, 25–6.

Nel, D. D., Kotzé, J. M. and Snyman, C. P. 1983. Progress in tissue culture of avocado. *South Afr. Avocado Grow Assoc. Yearb.* 6, 90–1.

Nhut, D. T., Thi, N. N., Khiet, B. L. T. and Luan, V. Q. 2008. Peptone stimulates *in vitro* shoot and root regeneration of avocado (*Persea americana* Mill.). *Sci. Hortic.* 115(2), 124–8. doi:10.1016/j.scienta.2007.08.011.

O'Brien, C., Constantin, M., Walia, A., Yiing, J. L. Y. and Mitter, N. 2016. Cryopreservation of somatic embryos for avocado germplasm conservation. *Sci. Hort.* 211(1), 328–35. doi:10.1016/j.scienta.2016.09.008.

O'Brien, C., Hiti-Bandaralage, J. C. A., Hayward, A. and Mitter, N. 2018. Avocado (*Persea americana* Mill). In: Jain, S. M. and Gupta, P. (Eds), *Step Wise Protocols for Somatic Embryogenesis of Important Woody Plants*. Springer, Cham, pp. 305–28. doi:10.1007/978-3-319-79087-9_24.

Onay, A., Pirinç, V., Yıldırım, H. and Basaran, D. 2004. *In vitro* micrografting of mature pistachio (*Pistacia vera* var. Siirt). *Plant Cell Tissue Organ Cult.* 77, 215–9.

Persley, G. J., Langhe, E. A. and Langhe, E. F. 1987. Banana and plantain breeding strategies. In: *Proceedings of an International Workshop*, Canberra, Australia.

Platt, R. G. 1976. Current techniques of avocado propagation. In: *Proceedings of the First International Tropical Fruit Short Course: the Avocado*. Florida Cooperative Extension Service and Institute of Food and Agricultural Services, University of Florida, Gainesville, FL, pp. 92–5.

Pliego-Alfaro, F. 1988. Development of an *in vitro* rooting bioassay using juvenile-phase stem cuttings of *Persea americana* Mill. *J. Hortic. Sci.* 63(2), 295–301. doi:10.1080/14620316.1988.11515862.

Pliego-Alfaro, F. and Murashige, T. 1987. Possible rejuvenation of adult avocado by graftage onto juvenile rootstocks *in vitro*. *HortScience* 22, 1321–4.

Postman, J. D. and Hadidi, A. 1995. Elimination of apple scar skin viroid from pears by *in vitro* thermo-therapy and Apical meristem culture. *Acta Hortic.* 221(386), 536–43. doi:10.17660/ActaHortic.1995.386.77.

Quiroz, K. A., Berríos, M., Carrasco, B., Retamales, J. B., Caligari, P. D. S. and García-Gonzáles, R. 2017. Meristem culture and subsequent micropropagation of Chilean strawberry (*Fragaria chiloensis* (L.) Duch.). *Biol. Res.* 50(1), 20. doi:10.1186/s40659-017-0125-8.

Raharjo, S. H. T. and Litz, R. E. 2005. Micrografting and ex vitro grafting for somatic embryo rescue and plant recovery in avocado (*Persea americana*). *Plant Cell Tissue Organ Cult.* 82(1), 1–9. doi:10.1007/s11240-004-5486-3.

Rani, V. and Raina, S. A. 2000. Genetic fidelity of organized meristem-derived micropropagated plants: a critical reappraisal. *Vitro Cell Dev. Biol. - Plant* 36, 319–30.

Rasmussen, A., Hosseini, S. A., Hajirezaei, M. R., Druege, U. and Geelen, D. 2015. Adventitious rooting declines with the vegetative to reproductive switch and involves a changed auxin homeostasis. *J. Exp. Bot.* 66(5), 1437–52. doi:10.1093/jxb/eru499.

Rehmen, H. U. and Gill, M. I. S. 2015. Micrografting of fruit crops-A review. *J. Hortic.* 2, 1–7. doi:10.4172/2376-0354.1000151.

Reuveni, O. and Raviv, M. 1980. Importance of leaf retention to rooting of avocado cuttings. *J. Amer. Soc. Hort. Sci.* 106, 127–30.

Revilla, M. A., Pacheco, J., Casares, A. and Rodríguez, R. 1996. *In vitro* reinvigoration of mature olive trees (*Olea europaea* L.) through micrografting. *In Vitro Cell Dev. Biol.-Plant* 32(4), 257–61. doi:10.1007/BF02822697.

Rijiswick, C. van 2018. World fruit map 2018. In: *RaboReserch Food Agri Bus.* Available at: https://research.rabobank.com/far/en/sectors/regional-food-agri/world_fruit_map_2018.html.

Rodrigues, J., Ryan, G. F. and Frolich, E. F. 1960. Some factors influencing grafting success with avocados. *Calif. Avocado Soc. Yearb.* 44, 89–92.

Roe, D. J., Kremer-Köhne, S. and Köhne, J. S. 1995. Local and imported avocado rootstocks in South Africa. In: *Proceedings of The World Avocado Congress III*, 1995, pp. 132–9.

Romberger, J. A. 1980. Meristems, growth, and development in woody plants: an analytical review of anatomical, physiological, and morphogenic aspects. Department of Agriculture and Forest Services, The United States.

Sanjaya, M. B., Rathore, T. S. and Rai, V. R. 2006. Factors influencing in vivo and *in vitro* micrografting of sandalwood (*Santalum album* L.): an endangered tree species. *J. For. Res.* 11, 147–51. doi:10.1007/s10310-005-0208-1.

Schroeder, C. A. 1968. The longevity of avocado tissue *in vitro*. *Calif. Avocado Soc. Yearb.* 52, 128–30.

Schroeder, C. A. 1973. Apical and other responses of tissues of avocado in aseptic culture. *Calif. Avocado Soc. Yearb.* 57, 138–41.

Schroeder, C. A. 1976. Responses of avocado stem pieces in tissue culture. *Calif. Avocado Soc. Yearb.* 60, 160–3.

Schroeder, C. A. 1980. Avocado tissue *in vitro*. *Calif. Avocado Soc. Yearb.* 64, 139–41.

Sedgley, M. 1985. Some effects of daylength and flower manipulation on the floral cycle of two cultivars of avocado (*Persea americana* Mill., Lauraceae), a species showing protogynous dichogamy. *J. Exp. Bot.* 36(5), 823–32. doi:10.1093/jxb/36.5.823.

Silva, T. A. and Ledesma, N. 2014. Avocado history, biodiversity and production. In: *Sustainable Horticultural Systems*. Springer International Publishing, Cham, pp. 157–205.

Smith, M. K. and Drew, R. A. 1990. Current applications of tissue culture in plant propagation and improvement. *Funct. Plant Biol.* 17(3), 267–89. doi:10.1071/PP9900267.

Suarez, I. E., Schnell, R. A., Kuhn, D. N. and Litz, R. E. 2006. Recovery and indexing of avocado plants (*Persea americana*) from embryogenic nucellar cultures of an avocado sunblotch viroid-infected tree. *Plant Cell Tissue Organ Cult.* 84(1), 27–37. doi:10.1007/s11240-005-7532-1.

Thorpe, T. A. 2007. History of plant tissue culture. *Mol. Biotechnol.* 37(2), 169–80.

Unlu, N. Z., Bohn, T., Clinton, S. K. and Schwartz, S. J. 2005. Carotenoid absorption from salad and salsa by humans is enhanced by the addition of avocado or avocado oil. *J. Nutr.* 135(3), 431–6. doi:10.1093/jn/135.3.431.

Vasil, I. 2012. *Laboratory Procedures and Their Applications*. Elsevier. doi:10.1016/B978-0-12-715001-7.X5001-X.

Vega, D. E. S. 1989. Propagation *in vitro* of rootstocks of avocado. *Calif. Avocado Soc. Yearb.* 73, 149–51.

Venkatachalam, L., Sreedhar, R. V. and Bhagyalakshmi, N. 2007. Genetic analyses of micropropagated and regenerated plantlets of banana as assessed by RAPD and ISSR markers. *In Vitro Cell Dev. Biol.-Plant* 43(3), 267–74. doi:10.1007/s11627-007-9028-7.

Vidal, M. T., Azcón-Aguilar, C., Barea, J. M. and Pliego-Alfaro, F. 1992. Mycorrhizal inoculation enhances growth and development of micropropagated plants of avocado. *HortScience* 27(7), 785–7. doi:10.21273/HORTSCI.27.7.785.

Walkey, D. G. 1972. Production of apple plantlets from axillary-bud meristems. *Can. J. Plant Sci.* 52(6), 1085–7. doi:10.4141/cjps72-186.

Warschefsky, E. J., Klein, L. L., Frank, M. H., Chitwood, D. H., Londo, J. P., von Wettberg, E. J. B. and Miller, A. J. 2016. Rootstocks: diversity, domestication, and impacts on shoot phenotypes. *Trends Plant Sci.* 21(5), 418–37. doi:10.1016/j.tplants.2015.11.008.

Wolstenholme, B. N. 1987. Some aspects of avocado research world-wide. *South Afr. Avocado Grow Assoc. Yearb.* 10, 8–12.

Wolstenholme, B. N. 2003. Avocado rootstocks: what do we know; are we doing enough research. *South Afr. Avocado Grow Assoc. Yearb.* 2003, 106–12.

Yakuwa, H. and Oka, S. 1988. Plant regeneration through meristem culture from vegetative buds of mulberry (*Morus bombycis* Koidz.) stored in liquid nitrogen. *Ann. Bot.* 62(1), 79–82. doi:10.1093/oxfordjournals.aob.a087638.

Yıldırım, H., Onay, A., Süzerer, V., Tilkat, E., Ozden-Tokatli, Y. and Akdemir, H. 2010. Micrografting of almond (*Prunus dulcis* Mill.) cultivars 'Ferragnes' and 'Ferraduel'. *Sci. Hortic.* 125(3), 361–7. doi:10.1016/j.scienta.2010.04.022.

Young, L. B. 1961. Vegetative propagation in avocados by means of marcottage and the rooting of cuttings. *Calif. Avocado Soc. Yearb.* 45, 63–6 227.

Zirari, A. and Lionakis, S. M. 1994. Effect of cultivar, explant type, etiolation pretreatment and the age of plant material on the *in vitro* regeneration ability of avocado (*Persea americana*). *Acta Hortic.* 365(365), 69–76. doi:10.17660/ActaHortic.1994.365.6.

Zulfiqar, B., Abbasi, N. A., Ahmad, T. and Hafiz, I. A. 2009. Effect of explant sources and different concentrations of plant growth regulators on *in vitro* shoot proliferation and rooting of avocado (*Persea americana* Mill.) cv. 'Fuerte'. *Pak J. Bot.* 41, 2333–46.

# Chapter 9

# Advances in cultivation of avocado

*Gary S. Bender, University of California Cooperative Extension, USA*



## 1 Introduction

Avocado is a high-value crop and per capita consumption is still growing (Anon., 2013). In the United States, per capita consumption increased from 1.6 lbs in the 1990s to 7.1 lbs during the years 2013–2017 (Ambrozek et al., 2018). There is interest by growers in planting more land but also in increasing production in the land that is already planted. In response to increasing consumption both in countries that import avocados and in the countries that grow avocados, increasing land area has been planted over the last 20 years (Fig. 1).

Unfortunately, a few countries have been losing land for avocado production due to a variety of factors, including the spread of avocado root rot caused by *Phytophthora cinnamomi* (Zentmyer, 1980), the high prices of water in California (Bender, 2012), and water supply issues in Chile (Milne, 2019) (Fig. 2).

There has always been the potential of losing farmland due to the build-up of salt in the soil, especially along the coastal plain (Gazit and Kadman, 1976). In addition, it is obvious that good avocado land is disappearing near urban areas, especially in California (Fig. 2). Other avocado-growing areas in the world may experience the same trend in the future. With an expanding demand for

http://dx.doi.org/10.19103/AS.2019.0054.20

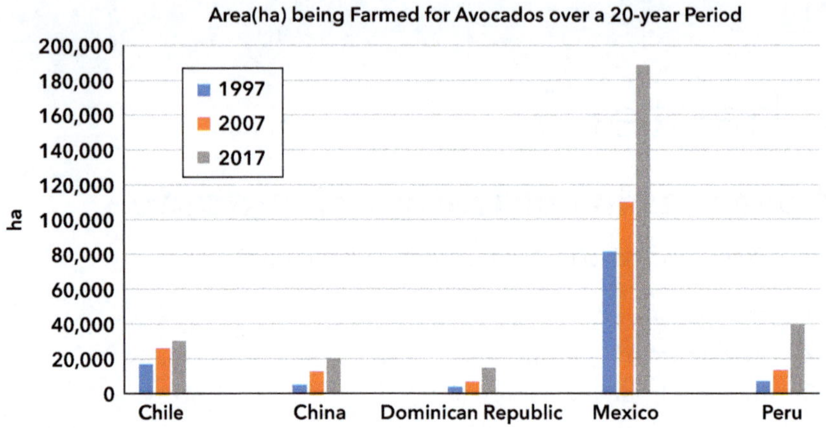


**Figure 1** Increasing area being farmed for avocados 1997–2017. Source: adapted from FAOStat (2019).

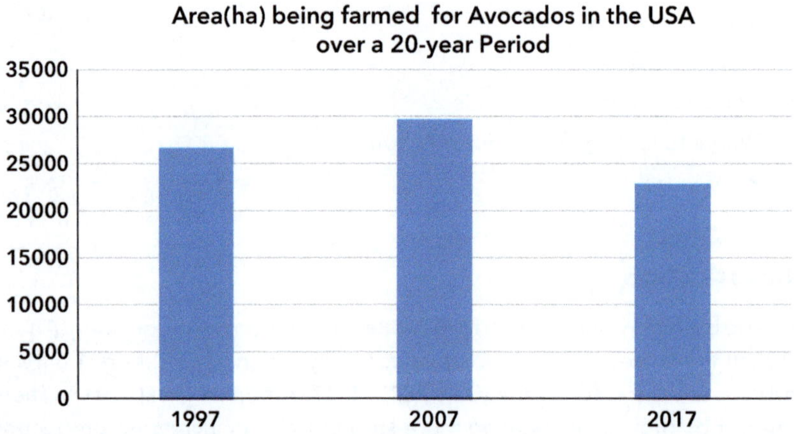


**Figure 2** Decreasing area being farmed for avocados in the United States over a 20-year period.

avocados, growers are challenged to grow enough fruit to meet the needs of the population but not ruin the soil or pollute the groundwater so that production in future generations is sacrificed. The recognition of these problems and the drive to be more sustainable for future production includes many factors that fall under three umbrella categories which include a 'healthy environment' for the trees as well as the people that farm them, 'economic viability' and 'social and economic equity' for farmworkers and others involved in the industry (UC SAREP, 2018). This chapter will address some of the advances in avocado production that address keeping a healthy environment for avocado trees in order to have a more sustainable industry for the future.

Cultural advances in the avocado industry to direct it toward a more sustainable future will be illustrated. But what does a 'sustainable production practice' mean exactly? It is not the same as *organic farming*, but rather a 'reduced chemical approach to farming' (Rice and Rice, 2011). For instance, rather than applying only chemical fertilizers on the soil or through the irrigation system, there is more of an emphasis on building up the soil fertility and beneficial soil microorganisms with the application of composts and mulches and possibly growing legumes between the younger trees to build up soil nitrogen.

Other aspects of sustainable production include using an integrated pest management scheme to control pests. This includes substituting or including biological control agents in pest control practices. Chemicals can be used if the pest can't be controlled, but chemicals should be chosen that are least damaging to parasites and predators of the damaging pest.

Sustainable production also includes managing dense canopies better for increased production and for better air flow to reduce disease pressure. An ideal view of growing sustainably would also have intercropping with other types of crops to slow down the progression of a new pest, but that crop also should be profitable in order for the grower to adopt the practice. Unfortunately intercropping is unlikely because avocados shade the ground so thoroughly that there usually isn't enough light for an understory crop.

## 2 Advances in soil management

Growing sustainably starts with the soil. Avocados are grown on a wide range of soils and many of these soils can pose problems that make it difficult for trees to be optimally productive. In southern California, for instance, soils vary from sands to decomposed granite to heavy clays (Crowley, 2007). Sands and granite have poor water retention which makes irrigation scheduling critical in the dry season. Clays have poor drainage which creates a better environment for zoospores of the avocado root rot fungus *Phytophthora cinnamomi* to move through the soil. Each soil type can pose challenges for growers who are trying to improve their soils. What can growers do to affect water retention, nutrient status, drainage, weed control, and the population of microorganisms that might reduce the population of root rot organisms such as *P. cinnamomi?*

Probably the most important activity for growers is to apply organic matter to their soils, either from agricultural waste, chipped trees and branches from their own trees during pruning operations, or from mulches and compost from landscape prunings in urban areas. There have been indications through the years that mulches and composts can improve soils in tree crops. Researchers have shown that organic matter can reduce water use (Downer et al., 2002), improve water infiltration (Gregoriou and Rajkumar, 1984), reduce weed growth (Lanini et al., 1988), increase soil fertility (Weeks et al., 1950), and control soilborne diseases (Hoitink et al., 1991).

Avocado groves are not generally mulched but it might be time to take a serious look at mulching. There has been a significant amount of research on the application of greenwaste mulches to avocados in southern California during the 1990s and early 2000s (Faber et al., 2003; Menge et al., 1999). They targeted greenwaste mulches because avocado is a high-value crop and could afford hauling and spreading of the materials in groves, and they were near urban areas that had a ready supply of chipped landscape materials. There has also been research on mulching in avocado in Australia as part of a program to develop sustainable orchard management practices (Leonardi, 2012).

In southern California, the researchers worked in two avocado groves and both were on clay/loam soils. The Vanoni grove had 'Hass' on three rootstocks ('Duke 7', 'Thomas' and 'Toro Canyon'). Sixteen trees of each rootstock (trees were 3 years-old) were treated with either no mulch or 1/3 cubic meter/tree of freshly ground *Eucalyptus globulus*. This amount was applied annually for 3 years. The Sprinkling grove was similar in set-up but instead of Eucalyptus they applied mixed yard trimmings at the same rate. This grove also had 'UC 2011' rootstocks instead of 'Toro Canyon'. *P. cinnamomi* was present in the soil at both sites.

After 3 years of mulching, the most pronounced effect in the avocado groves was the improvement in soil nutritional levels at the Vanoni grove, but it did not relate to a significant increase in leaf nutritional levels (Table 1). This could mean that mulching is beneficial for a long-term improvement in the soil nutritional levels, but not for the short term.

**Table 1** Effect of mulch on selected soil and avocado leaf nutritional levels after 4 years of mulching at the Vanoni grove

| Treatment | N | P | K | Ca | Mg | Zn | Mn | pH |
|---|---|---|---|---|---|---|---|---|
| *Soil* | *(%)* | *(ppm)* | *(ppm)* | *(meq/L)* | *(meq/L)* | *(ppm)* | *(ppm)* | |
| Mulch | 1.59 A* | 17.6 A | 284.6 A | 26.7 A | 9.9 A | 7.3 A | 30.7 A | 7.19 B |
| Control | 1.22 B | 12.2 B | 218.3 B | 26.8 A | 8.1 B | 6.7 B | 23.7 B | 7.26 A |
| *Leaf* | *(%)* | *(%)* | *(%)* | *(%)* | *(%)* | *(ppm)* | *(ppm)* | |
| Mulch | 2.54 A | 0.49 A | 1.15 A | 1.43 A | 0.38 A | 20.0 A | 35.5 A | |
| Control | 2.80 A | 0.47 A | 1.12 A | 1.45 A | 0.38 A | 19.3 B | 36.7 A | |

* Mean values followed by identical letters are not statistically different according to ANOVA at $P = 0.05$.
*Source*: adapted from Faber et al. (2003).

**Table 2** Effect of mulch on root length of avocado at two trial sites

| Treatment | Sprinkling ranch | Vanoni ranch |
|---|---|---|
| Control | 13.3 B* | 5.1 B |
| Mulch | 19.0 A | 14.5 A |

* Mean values in each column followed by identical letters are not statistically different according to Walker's k-test.
Root length is measured in cm/cc soil.
*Source*: adapted from Faber et al. (2003).

**Table 3** Effect of mulch on avocado root length, root infection, and Phytophthora populations from samples taken at various soil depths of mulched trees at Sprinkling Ranch (Faber et al., 2003)

| Location | Root length (cm) | Root infection (#/root) | Phytophthora (MPN) |
|---|---|---|---|
| *Mulch* | | | |
| Mulch surface | 0.00 B* | 0.00 B | 0.00 B |
| Mid-mulch | 8.34 AB | 0.00 B | 0.0 B |
| Interface | 22.71 AB | 0.40 B | 0.0 B |
| Soil 7.5 cm | 6.70 AB | 1.20 A | 1.2 A |
| Soil 15 cm | 8.80 AB | 0.80 AB | 0.6 AB |
| *Un-mulched* | | | |
| Soil surface | 0.00 B | 0.0 B | 0.0 B |
| Soil 7.5 cm | 37.34 A | 0.4 B | 1.9 A |
| Soil 15 cm | 6.95 B | 1.0 A | 0.4 AB |

* Mean values in each column followed by the same letter are not significantly different according to two-way ANOVA and LSD ($P < 0.05$) using five replications of mulched and un-mulched tree pairs. Root length is measured in cm/cc soil. Phytophthora is measured by the Most Probable Number method (Alexander, 1982).

Mulching did not affect the size of the canopy volumes at the Vanoni Ranch compared to the controls. However, at the Sprinkling Ranch mulching increased the canopy volumes slightly (but not statistically significant) with the UC2001 and Thomas rootstocks, but it did increase the canopy volume substantially (and significantly) with the 'Duke 7' rootstocks. The authors noted that 'all of the mulched rootstocks had an improved visual rating, But the 'Duke 7' had the greatest growth response'. Of particular interest was the finding that, at both groves, the mulched rootstocks had a change in root architecture (Table 2). There was an increase in root length in the mulched trees at both sites, but a closer examination of root length at the Sprinkling site showed that roots had grown into the mulch and there were lower population of *P. cinnamoni* propagules and less root infection in the mulch (Table 3).

Working on this same project, Menge et al. (1999) explained that the better growth of mulched avocados was due to the relocation of roots into the mulch and into the interface of the mulch, and these roots were healthy (Menge et al., 1999). These researchers found enzymes (cellulose and laminarinase) in the mulch produced by wood-rotting fungi were responsible for destroying hyphae produced by *P. cinnamomi* (Downer et al., 2001). Microorganism activity was also found to be very high in the mulch. They noted that, 'at one site mulching was as effective at reducing the symptoms of avocado root rot as the leading fungicide used to control the disease' (Downer et al., 2001).

In the Australian work, 4 ½-year-old Hass' avocados were mulched in 2011 with filter press, avocado woodchips, and cane-tops. This treatment was compared to a grower-applied mulch of inter-row slashings with a thinly applied filter press. A 3-year cumulative yield showed an average of 167.6 kg/tree for the mulched trees versus 137.9 kg/tree for the grower-applied treatment (Leonardi, 2012).

## 3 Advances in fertility management

Growing avocados sustainably means that the grower must strive to produce optimum crops in order to remain profitable, but at the same time reducing harm to the environment by reducing practices such as over-fertilization and leaching nutrients into the ground water. All plants require 17 nutrients, three of which are supplied by air and water. The other 14 are supplied by soil fertility and some (especially nitrogen) must be supplied by the grower. If any one of these nutrients is in short supply, the tree can produce only to the level of that nutrient (Lovatt, 2015).

### *3.1 Nitrogen*

Avocado trees have relatively few mineral deficiencies in most commercial orchards. The shallow feeder roots seem to be very efficient at recycling nutrients from the decomposing leaf mulch under the trees. Growth of the trees does not make heavy demands on the mineral nutrients in the soil (Wolstenholme, 2018). However, in order to maintain optimum production of fruit, nitrogen in some form should be applied to the trees, probably on a yearly basis. In one multiyear trial, it was concluded that Hass avocados required between 0.68–0.91 kg/tree (1.5–20 lbs/tree) actual nitrogen per year to maintain on optimum yield (Meyer et al., 1992). However, this could vary greatly depending on many factors including rootstocks, amount of rainfall, applied irrigation, or leachability of soils (Lahav et al., 2013). Nitrogen can come from organic mulches and composts, synthetic fertilizers, or from the irrigation water. Cover crops are generally not used because plowing the vegetation into the soil will cut the shallow feeder roots of the avocado. It is important to do a yearly leaf analysis to check the levels of nitrogen, and occasionally zinc, iron, or manganese should be applied if the trees show deficiency symptoms. Phosphorous, potassium, and other nutrients may be desirable for long-term improvement in yield on some soils.

Injecting water-soluble fertilizers through the irrigation system is economical and it places the fertilizer where the roots are located (in the water pattern). It also allows the grower to time the fertilizer application to support the tree in a particular phenological stage of growth. In an earlier experiment, researchers applied nitrogen fertilizer in split applications twice a year, four times a year and eight times a year (Yates et al., 1992). The total amount of actual nitrogen for the year was the same. There were no yield differences between the application methods, but significantly more nitrogen leached below the root system when the fertilizer was applied in greater amounts (such as in the twice a year program) (Yates et al., 1992). Lovatt attempted to refine this by applying nitrogen fertilizer during six equal applications of 0.113 kg (0.25 lb) actual N per tree through the year, but then doubled the amount during one of five of the selected times throughout the year. The yields were the same except when the amounts were doubled in the spring (anthesis, fruit set and initiation of spring flush) and in the fall (end of summer flush and beginning of flower initiation within the buds). Doubling the nitrogen at these two times both gave significantly higher yields than doubling at other times (Lovatt, 2001).

However, in order to move toward the goal of growing sustainably we should reduce our reliance on fossil fuels that are used to make synthetic fertilizers, and instead shift our practices to using sources of nitrogen that will last indefinitely. Practically, this means using organic materials for fertility instead of synthetic fertilizers. Conventional agriculture commonly uses nitrogen fertilizers made by the Haber-Bosch process of converting $N_2$ gas from the air into ammonia by using high temperature and very high pressure. During the reaction, natural gas is burned to create the hydrogen necessary for the reaction. It should be noted that this reaction is often called the most important invention of the twentieth century (Smil, 1999). This reaction is responsible for creating enough fertilizer for agriculture to support an expanding population from 1.6 billion in 1900 to 8 billion today. However, this reaction is usually not considered sustainable because the hydrogen is made from natural gas, and there is a limit to that resource. It might be possible in the future to create hydrogen from electrolysis of water using electricity from solar cells or windmills, but for now it is natural gas.

As described in Section 2 organic compost from yardwaste and crop waste are useful for supplying nutrients to the soil (Table 1). Organic waste on the soil also suppresses weeds, aids in erosion control, reduces evaporative loss, improves soil tilth, and aids in root rot control (Faber, 2008; Matava, 2009; Downer et al., 2002). As can be seen in Table 5, composted yardwaste has a low percentage of nitrogen. Yardwaste should be amended with a composted manure or bloodmeal to bring up the nitrogen content. Yardwaste is useful because it has a high carbon/nitrogen ratio so it breaks down slowly in the orchard, but without extra nitrogen it may cause trees to become nitrogen deficient for a few years. Yardwaste, crop waste, and manures should always be composted in order to reduce weed seeds, reduce ammonia gas, and eliminate the possibility of introducing diseases such as avocado root rot (caused by *Phytophthora cinnamomi*). Care must be taken to make sure these materials are composted properly so that the sides of the piles are turned inward several times for heating throughout the piles (Matava, 2009).

The main problem with applying most organic materials is that they are bulky, low in nitrogen, and expensive. The cost for organic fertilizers is anywhere from 10 to 100 times the cost of conventional fertilizers (Faber, 2007). Due to the low nitrogen content, a significant amount (depending on the organic material) must be applied to each tree. And they are generally not water soluble so they cannot be applied through the irrigation system. They can be applied rather efficiently using tractors pulling side-dump trailers on flatter ground, but most of the avocado orchards in California are on slopes with a lot of rocks. Chile is now the second largest world producer of avocados and exporter of avocados, second to Mexico, and they also have most of their avocados grown on steep slopes (Gardiazabal and Mena, 2011). This means that if mulches, composts, and manures are applied for fertilizers, they have to be applied through a blower tube or by hand, significantly increasing the labor cost in the orchard. In the words of Mary Matava, who has many years of producing and applying composts to avocado, 'in some steep orchards it is simply impractical' (Matava, 2009).

In many crops growing sustainably also means using nitrogen-fixing plants such as legumes to supply some of the nitrogen and other nutrients. Growing legumes around young avocado trees is possible but mature groves have too much shading for these plants to grow. Also, legumes should be tilled into the soil in order to provide the maximum amount of nitrogen into the rootzone, but tilling is not recommended due to the shallow nature of avocado roots (Meyer et al., 1992). Some of the organic sources of nutrients available to growers are presented in Table 4.

Fertilization using organic materials is also different than applying synthetic fertilizers because they break down slowly and supply nitrogen almost like a slow-release fertilizer. This can be an advantage in that there is always some nitrogen available for the trees and it reduces the amount of nitrogen leaching into the groundwater, but can also be a disadvantage because they are slow to correct a deficiency.

As to how much to apply, it was noted by Lahav et al. that 'there is a dearth of scientific literature for avocados' for organic production (Lahav et al., 2013).

**Table 4** Percentage of nutrients in some commonly applied organic materials

| Material | N | $P_2O_5$ | $K_2O$ |
|---|---|---|---|
| Alfalfa meal | 3 | 1 | 2 |
| Blood | 12 | 1.5 | 0.6 |
| Bone meal (steamed) | 0.7–4.0 | 11.0–34.0 | 0 |
| Compost (without manure) | 1 | 1 | 1 |
| Chicken manure | 3 | 3 | 3 |
| Cottonseed meal | 6 | 2.5 | 1.7 |
| Feather meal | 11.0–15.0 | 0 | 0 |
| Fish meal | 10 | 4 | 0 |
| Fish emulsion | 5 | 2 | 2 |
| Grape pomace | 3 | 0 | 0 |
| Kelp | 0.9 | 0.5 | 1.0–4.0 |
| Soluble soy bean meal | 7 | 0 | 0 |

*Source*: adapted from Faber (2008).

However, Crowley tried to work through this difficulty by using the N removal calculator on http://www.avocadosource.com/ (Crowley, 2007). He assumed a yield of 100 kg (220 lbs) of fruit per tree with 100 trees per acre (about 22,000 lbs/acre or 10,000 kg) removed 27.7 kg (61.7 lbs) of nitrogen per acre. He assumed a 50% nitrogen-use efficiency and concluded this crop would require 137.9 kg/ha (123 lbs/ac) of actual nitrogen. He noted that Lovatt found that this is generally the amount of N fertilizer used in California, which ranges 89–178 kg/ha (80–160 lbs/ac) actual N (Lovatt, 1996).

Typical levels of nitrogen in composted organic materials are 5–7% by weight (Crowley, 2007). The mineralization process (conversion of organic N into the ammonium ion by microbes) of 454 kg (1000 lbs) of organic material would therefore yield 23–32 kg (50–70 lbs) actual N. Nitrification (the conversion of ammonium into nitrate) makes this form of N more available to avocado roots, but nitrate is also more leachable (Zicah et al., 1999; Yates et al., 1992). This followed by denitrification (biological reduction of nitrate into $N_2$ gas) makes the use of organic fertilizers rather inefficient. The use of 1969–2959 kg/ha (1757–2640 lbs/ac) of composted organic matter in order to get 137.9 kg/ha (123 lbs/ac) of N should probably be doubled to 3939–5246 kg/ha (3514–4680 lbs/ac) of organic matter in order to get the required amount of N for the avocado trees. An accurate means of predicting the application rate does not exist because these conversion processes are driven by microbes in their environment and there is a great deal of variability in temperature, moisture, humidity, soil types, soil pH, and organic matter remaining in the soil from applications in previous years. Therefore, it is almost mandatory for growers to do a yearly leaf analysis.

### 3.2 Phosphorus

In general, phosphorus deficiency does not appear to be a problem in avocado production (Lahav et al., 2013). Despite the lack of deficiency symptoms, most conventional avocado growers apply some sort of mixed fertilizer solutions containing nitrogen with phosphate. Organic fertilization, especially those that contain manure, have plenty of phosphorus (Crowley, 2007).

Even though phosphorus deficiency should not be a concern in most soils, there could be a potential for deficiency in acid sandy soils with low P reserves. Some of these soils are found in Florida, coastal areas of Queensland and New South Wales in Australia, and some of California's sandy loam soils (Crowley, 2007).

### 3.3 Minor elements

Whether the grower is growing organically, conventionally or a combination, minor element deficiencies do occur and it is important to do a yearly complete leaf analysis in order to correct these problems. In California, certified organic growers may be allowed to correct minor element deficiencies with a synthetically produced minor element if the need is documented with leaf analysis, and the certifying agency gives their written permission. In California, this usually means using zinc sulfate and/or iron sulfate and chelates to correct these deficiencies.

## 4 Advances in irrigation management

Growing avocados in a Mediterranean dry climate such as southern California has certain benefits. For one growers don't have to treat the foliage and fruit with fungicides to reduce fungal diseases. There is also a significant reduction in soil erosion from disturbed land where clearing the hillsides for planting has removed the native plants. However, due to the lack of rain during the dry-summer months, avocado plantings must be irrigated. In dry climate areas, irrigation water usually contains a significant amount of salt which can gradually salinize the soil if not well managed with leaching (Gazit and Kadman, 1976).

A significant advance in the cultivation of avocado is the development, and usage, of soil moisture monitoring equipment. Avocado is a shallow rooted tree, and drying of the soil where the feeder roots are located can occur rather quickly. If the trees are put under too much moisture, stress salts can accumulate and this can lead to reductions in tree growth, lowered leaf chlorophyll, decreased photosynthesis, poor root growth, and leaf scorching (Crowley, 2008). On the other hand, if roots are exposed to too much water with poor drainage, it makes a good environment for *P. cinnamomi* zoospores and root infection increases (Zentmyer, 1980; Menge et al., 1999). In short, especially in drought-prone regions, irrigation scheduling is probably the most important grove operation if growers want to produce optimum yields.

### 4.1 Scheduling irrigations

In California, farm advisors teach the method of knowing 'when to irrigate' and 'how much to irrigate' (Hanson et al., 2000; Bender and Faber, 2009). In order to know 'when' to irrigate, monitoring the soil frequently with an auger or a shovel is necessary, or the grower can install soil moisture monitoring equipment. Tensiometers have been used for this purpose since the 1950s but they have frequent maintenance needs. These are gradually being replaced by Watermarks® (an electrical resistance meter in a stabilized gypsum block) and other devices. These can be read individually, hard wired to a box or have a sending unit send the information by infrared to a receiving unit so that the grower can read his/her soil moisture in many areas in the grove from a computer or smart phone. When using these devices, through experience, the farm advisors in California have determined that an irrigation should occur when the soil moisture reaches 30% depletion in sandy loam soils. This is equivalent to 22–25 cb when using tensiometers and Watermarks®. It is estimated for clay soils with 50% depletion in moisture that an irrigation should occur at about 40–50 cb (Mission RCD, 2000). Other types of soil moisture monitoring meters include conductance and capacitance (measures the di-electric constant in the soil) meters and time domain reflectometers. Each type of meter has different qualities and they are summarized in Table 5.

Knowing *when* to irrigate, as described above, does not tell a person *how much* to irrigate (unless the grower is using a deep tensiometer to know when to turn off the water). In California, we use California Irrigation Management Information System (CIMIS), a system of weather stations that measures evapotranspiration

**Table 5** Characteristics of soil moisture monitoring equipment

| Method | Cost | Ease of use | Accuracy | Reliability | Salt-affected | Stationary |
|---|---|---|---|---|---|---|
| Gypsum block | L | H | H | H | L | YES |
| Tensiometer | L | M | H | M | L | YES |
| Portable tensiometer | M | M | H | M | L | NO |
| Solid-state tensiometer | M | H | H | H | L | YES |
| Time domain reflectometer | H | M | H | H | M | BOTH |
| Neutron probe | H | L | H | H | L | YES |
| Feel (soil probe) | L | H | H | H | L | NO |
| Gravimetric (oven) | L | M | H | H | L | NO |
| Conductance | L | H | M | M | H | BOTH |
| Capacitance | M | H | M | H | M | BOTH |

H, high; M, medium; L, low.
*Source*: adapted from Faber (2017).

**Table 6** Example conversion of ETo to irrigation run times for a sprinkler irrigated field

| | | | |
|---|---|---|---|
| A. | Measured atmometer ETo for 1 week | 2 | Inches |
| B. | Crop coefficient (Kc) | 0.8 | |
| C. | Calculated ETc for the week (=AxB) | 1.6 | Inches |
| D. | Sprinkler application rate | 0.13 | In/hr |
| E. | Hours of irrigation required (=C/D) | 12.3 | Hours |

Note: To convert gallons to inches: gallons ÷ area (square feet) ÷ 0.6234 = inches.
To convert inches to gallons: inches * area (square feet) ÷ 1.604 = gallons.
*Source*: adapted from Battany (2008).

(ETo) by recording solar radiation, humidity, air and soil temperature, wind speed and wind duration, and entering these data into a modified Penman equation. Each weather station is surrounded by eight-inch tall grass, therefore the ETo value equals the amount of water (in inches) used by the grass each day. Crop coefficients that relate ETo to a particular crop have been developed for most of the crops grown in California, and that includes avocado. The latest avocado irrigation trial in California ended in 2007 and the researchers concluded that the crop coefficient for mature avocados is 0.86, although they did make a comment that the crop coefficient for non-shaded younger trees is probably closer to 0.7 (Oster and Arpaia, 2007). A very simple device the growers can use to measure ETo in their grove is the atmometer (Battany, 2008). Atmometers are water-filled devices that measure the loss of water from evaporation. An example of how to convert ETo read from an atmometer is presented in Table 6.

### 4.2 Good-quality water

Sustainability problems with water vary throughout the avocado-producing regions in the world. In order to maintain avocado as a sustainable crop for the future (especially in drought-prone areas such as California, Israel, and parts of Australia), plenty of good-quality (low salinity) water for irrigation is perhaps at the top of the list. Tropical areas that experience frequent rainfall may not have problems with water supply or salt, but they have to apply foliar fungicides on a regular basis to reduce fruit diseases. This is a real problem when there is a wish to reduce the amount of pesticides applied as a sustainability goal.

## 5 Advances in cultivars and rootstocks

### 5.1 Cultivars

Avocados are grown in a wide variety of climates, from a dry-summer Mediterranean climate to more tropical climates where rain is abundant. In the early 1900s, hundreds of seedlings that were thought to be superior cultivars were selected and named, then the 'Hass' cultivar was selected from a chance seedling in 1926 in California and patented by Rudolph Hass in 1935 (Crane et al., 2013). 'Hass' has risen to the top in the commercial market because of its better pulp quality, higher yield, and later maturity than 'Fuerte'. 'Hass' (a black skin cultivar) replaced 'Fuerte' (a green skin cultivar) in the 1960s as the dominant cultivar and is considered the 'standard of excellence' in avocados (Crane et al., 2013). In 2016, it was estimated that 'Hass' is grown on 95.3% of the avocado acreage in California (California Avocado Commission, 2018).

'Hass' is grown primarily in cool subtropical regions and at the higher elevations in some tropical countries (Dominican Republic for instance). 'Hass' generally does not fruit well in low tropical regions that are humid and hot year-round. West Indian selections are mostly grown in these more tropical regions.

An industry that grows only one cultivar is possibly subject to risk. Problems associated with 'Hass' include alternate bearing and producing a large percentage of undersized fruit which are not valuable in the commercial market (Bruwer and Van Rooyen, 2007). Other problems include generally low yields, sensitivity to cold, drought, salinity, insect damage, and certain diseases such as blackstreak (Bergh, 1984). For instance, the cultivars 'Hass' and 'Gwen' are especially sensitive to Persea mite *Oligonychus perseae*, 'Esther' and 'Pinkerton' are moderately sensitive and 'Lamb Hass', 'Fuerte' and 'Reed' are resistant (Kerguelen and Hoddle, 2000).

Fortunately, the industry is making good strides in breeding better cultivars, but still keeping the 'black skin' which the consumer has deemed the proper color for an avocado. 'Hass'-like cultivars bred at the University of California, Riverside include 'GEM', 'Lamb Hass', and 'Harvest'. 'GEM' is increasing in popularity because trees are compact and better suited to high-density plantings, and the fruit quality is excellent (Crane et al., 2013). 'GEM' fruit are larger than 'Hass' and are borne more on the inside of the tree making it easier for pruning. A selection from Mexico called 'Carmen' (also known as 'Hass Carmen') may bloom 1–7 months earlier than 'Hass' and reach physiological maturity in 9–10 months. 'Carmen' may also have two (sometime three) blooms per year resulting in a longer harvesting season.

A new 'Hass'-like selection from South Africa, 'Maluma', is destined for the commercial market because of its superior qualities. It is precocious with a high fairly constant bearing pattern. It is a more compact tree and more suitable than 'Hass' for high-density production. The tree only requires a light pruning after harvest to maintain the growth in high density. The tree is being introduced and evaluated in various avocado-producing countries around the world (Ernst et al., 2015).

### *5.2 Rootstocks*

An advance in the culture of avocados, and possibly the salvation of the avocado industry, has been the improvement in rootstocks. Rootstock research has been a priority in the industry since the discovery of Phytophthora root rot (PRR) caused by *Phytophthora cinnamomi* in the 1930s (Crane et al., 2013). In 1942, Dr. George Zentmyer at the University of California began a program of rootstock selection for this disease. The work has been slow because there was almost no resistance to this fungus because avocado, native to southern Mexico and Central America, had not evolved any resistance to a fungus that was native to New Guinea. But by collecting seed from wild avocados in Central America, screening thousands of seedlings in disease beds, and keeping chance mutations, a degree of resistance slowly developed. 'Duke 7', grown as a clonal rootstock, was the first release by 'Zentmyer' in the mid-1970s (Crane et al., 2013). 'Duke 7' is a productive rootstock on soils without *P. cinnamomi*, but still partially susceptible to the disease in infested soils, especially in water-logged soils.

Improvements have been made in root rot tolerance and now there are three new rootstocks from the University of California, Riverside: 'Zentmyer', 'Uzi', and 'Steddom' (Menge et al., 2012). Unfortunately, 'Zentmyer' is highly susceptible to salinity but could be useful in soils infested with PRR that have low salinity irrigation water. The release of 'Dusa' from the South African avocado breeding program at Westfalia Estate has been a major step forward for the industry because this rootstock has an improved tolerance to PRR and salinity, and does fairly well in cold winters (Menge, 2002; Crowley, 2004). The Ashdot series (predominantly West Indian) has been grown very successfully in Israel and the Ashdon 17 is now recommended for all cultivars (Crane et al., 2013).

Research in resistance to PRR is shifting to elucidating the molecular and genetic basis for the resistance to oomycetes with emphasis on *P. cinnamomi* in avocado. The lab of Dr. Patricia Monosalva at UC Riverside is working to identify these factors and combine them with markers to assist in the UCR Avocado Rootstock Breeding Program. If this is successful, the future avocado rootstocks may eventually include genes for PRR immunity from other species of wild Persea that have immunity to PRR, but are now unusable because they will not take a graft from commercial cultivars of *Persea americana*.

## 6 High-density production

Water prices in southern California continue to increase and there is no end in sight. At the same time, other costs, labor, fertilizer, and taxes also continue to increase. The response from growers has been to turn the water off and discontinue growing avocados (Bender, 2012). In California, this has led to a decline in acreage from 60,756 ha (76,000 ac) in 1988 to 20,234 ha (50,000 ac) in 2016 (California Avocado Commission, 2018). Other countries are experiencing increasing costs, although perhaps not the steep increase in water costs. For avocado production to be sustainable, either avocado fruit prices have to increase or growers must increase the yield.

As for prices, they rise and fall with demand by consumers, alternate bearing in our groves and interference with foreign competition. It is possible to increase the yield by planting trees with a higher density and keeping the trees pruned low. The normal spacing in California is 6.1 m × 6.1 m (20 ft × 20 ft), but one grower in California demonstrated that on a 3 m × 3 m (10 ft × 10 ft) spacing he could produce 33626 kg/ha (30000 lbs/ac) in the sixth year of growth and another grower produced 26900 kg/ha (24000 lbs/ac) in the fifth year of growth (Bender, 2012). The average yield per ha (acre) during the previous 5 years was, at that time, 5604 kg/ha (5000 lbs/ac) and the income with that yield was less than the water costs. It is clear that growers in California needed to show a *dramatic* increase in yield in order to survive and compete in the world market (Bender, 2012).

As an example of what can be accomplished with Hass in a high-density setting, 'Hass' trees planted in 2012 with 'Zutano' as pollinizer trees at a higher rate than normal (one 'Zutano' for every eight 'Hass' trees), 'Hass' yield from the third year to the sixth year of growth averaged 18208 kg/ha (16245 lbs/ac) from 2015 to 2018 (Fig. 3) (Bender et al., 2018). During the same period

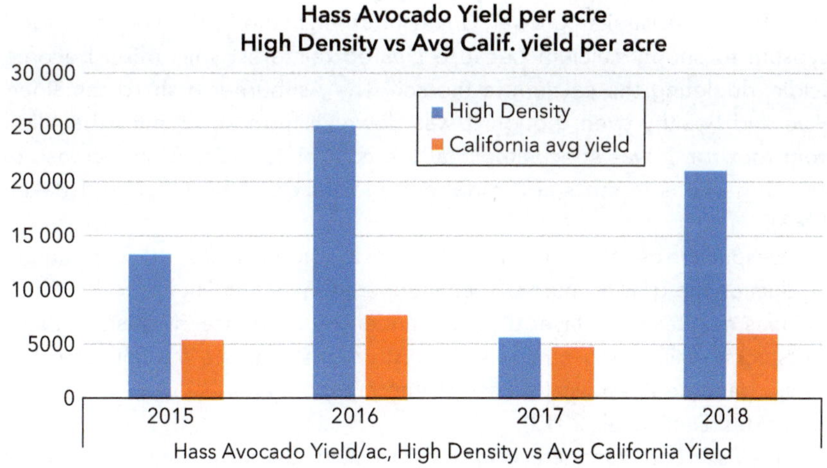


**Figure 3** Comparison of the yield from a high-density avocado trial in California (956 Hass trees/ha) (387 Hass trees/ac) versus California average yield/ac from 2015 to 2018. (Average grove in California has 269 trees/ha, or 109 trees/ac). Source: adapted from Bender et al. (2018).

from 2015 to 2018, the average yield in California was 6749 kg/ha (6021 lbs/ac) (California Avocado Commission, 2018).

Other avocado-growing regions in the world are also taking a close look at high-density production. In Mexico it was thought that avocados should be planted at a spacing of 10–12 meters (32.8–39.4 ft) but the trees reached a height of 12.5 meters (41 ft). This hindered all cultural operations and caused a decline in quantity and quality of fruit (Carrillo-Gallegos and Calderon, 2015). These researchers set up high-density trials throughout the state of Michoacan; the trees were planted between 400 and 555 trees/ha (162–225 trees/ac) instead of the normal rate of 100–69 trees/ha (40–28 trees/ac). They reported an average yield of 18.5 tons/ha (16,512 lbs/ac) during the first five crops, with as high as 30 tons/ha (26,764 lbs/ac) one year.

In Chile, researchers tried very high-density spacings of 2.5 × 1.25m (8.2 × 4.1 ft) and 1.25 × 1.25m (4.1 × 4.1 ft). They reported that 'Hass' achieved a yield of 40 tons/ha (35687 lbs/ac) in the first year of production. An economic analysis showed that 'it is possible to pay back all the initial investment and indirect and production costs with the first crop' (Gardiazabal et al., 2015).

## 7 Root rot management

The control of PRR has been discussed in Section 2 (Advances in Soil Management) and Section 5.2 (Rootstocks). It should be mentioned that the idea of trying to recreate a rainforest type soil under avocado trees in order to control PRR by mulching, applying chicken manure, and cover crops came from an Australian grower, Guy Ashburner, in the 1940s. He also added gypsum to supply calcium because cleared rainforest soils often become acidic, depleting the calcium in the soil. The Ashburner orchard remained alive and healthy even though it was downhill from an orchard that died from root rot. It was speculated that the control came from an increase in microorganisms in the soil (Broadbent and Baker, 1974; Cook and Baker, 1983).

Despite the benefits from mulch, in order to control this disease for optimum production the grower must take an integrated management approach which includes maintaining a layer of mulch under the trees, use of resistant clonal rootstocks, maintaining proper soil moisture (making sure the soil does not remain saturated), application of calcium nutrition, and use of phosphonate fungicides (Dann et al., 2013).

The use of phosphonate fungicides is a major cultural advance for growers trying to control this disease, but it is not allowed in a registered organic program. It is a systemic chemical that moves in the xylem and phloem and it stimulates a defense response in the roots that inhibits colonization by the pathogen. It appears to be best applied by trunk injection in the spring and late summer when growing roots are sinks for the material. It suppresses the infection but rarely directly kills propagules of Phytophthora and does not rid of the soil of the pathogen (Dann et al., 2013; Guest et al., 1995).

Equally important in the integrated control program are quarantines to keep the disease out of disease-free countries and clean nursery stock. If the disease is brought into an area on the roots of nursery trees, the grower cannot get rid of this disease. Therefore, the nursery must be heat-treating their seed, steaming their soil mix, using clean containers, and keeping animals and runoff water out of the nursery. Workers must have clean shoes (Zentmyer, 1980).

## 8 Pest management

The tropical areas in the Americas where avocado originated have a number of economically damaging insects and mites, but most of these remain below damaging levels due to biological control. Problems arise when these pests are introduced into foreign locations where avocados are being grown, but they enter without the biological agents that control them naturally (Pena et al., 2013). Sometimes the natural enemies are suppressed by pesticide applications and then the pest has a rebounding effect to become a more important problem.

Out of necessity, the avocado industry has been a leader in maintaining, and in some cases introducing, biological control to control pests and diseases. Avocado trees are tall, have dense foliage, and are often located on slopes. This makes it difficult, or impossible, for spray equipment to be moved through the orchard. In California, there are five introduced pests, all of which came from Mexico (Pena et al., 2013). Three of these pests have caused significant economic damage and necessitated the application of pesticides applied by helicopters. These pests include Persea mite (*Oligonychus perseae*), avocado thrips (*Scirtothrips perseae*), and *Neohydatothrips burungae*. The industry has focused spraying only those chemicals that are least damaging to beneficial insects, and the problems are slowly coming back under biological control.

Hoddle et al. in California tried to control Persea mite with predator mites and found that *Neoseiulus californicus* McGregor provided control comparable to pesticides but this method was 13–14 times more expensive than application of pesticides (Hoddle et al., 1999, 2000; Kerguelen and Hoddle, 1999).

Avocado thrips were introduced into California in 1996 and more than 20 years later growers still must use pesticides applied by helicopters in selected areas, but they also use licensed private pest control advisers to identify 'hot spots' of insect development. These advisers then make a recommendation to treat if the insect levels appear to be reaching damaging levels. The chemical most commonly used for Persea mite and avocado thrips is abamectin combined with horticultural oil. Fears that these pests may develop resistance to these materials is pushing growers to include other chemicals even though they may be more expensive. Morse et al. recommend that abamectin be applied only every third year to avoid the chance of resistance to the chemical (Pena et al., 2013; Morse et al., 2018).

In the observation of this author, there has developed enough biological control within the avocado trees in California (in most years) to now provide good natural control, but it took a long time to reach this point.

## 9 Conclusion and future trends

The avocado industry has a vibrant growing market, but the industry is having difficulty remaining sustainable in some areas. The water supply in drought-prone areas may be out of our ability to do much about, but we can do a better job of scheduling irrigations and leaching salts. Other factors that growers are capable of influencing are critical in order to produce the highest yields of commercial-quality fruit sustainably, and these are common to all avocado-growing areas in the world. These include maintaining and improving soils, tree fertility, canopy management, breeding programs for new cultivars and rootstocks, root rot management, and pest management. If one factor is deficient, it is possible that avocado fruit production may not be sustainable for the future.

## 10 Where to look for further information

A valuable source of avocado information is available on the website www.avocadosource.com under the listing Journals, Books and Proceedings. In this section of the website there is a complete collection of the California Avocado Society Yearbooks from 1915–2013. Also available are the Proceedings from the World Avocado Congress from 1987–2015, probably the most up-to-date summaries of worldwide avocado research. Selected yearbooks and reports from the Australia Avocado Industry Reports, CICTAMEX Mexico, New Zealand Avocado Growers Association and the South African Avocado Growers Association are also available on this site.

A fairly recent summary of avocado information is found in the book published in 2013 *The Avocado: Botany, Production and Uses – 2nd Edition,* edited by Shaffer, B., Wolstenholme, N. and Whiley, A. (CABI International, Boston MA).

## 11 References

Alexander, M. 1982. Most probable number method for microbial populations. In: *Methods of Soil Analysis, Part 2, Chemical and Microbiological Properties* (2nd edn.). A. L. Page (Ed.). American Society of Agronomy, Madison, WI, pp. 815–20.

Ambrozek, C., Saitone, T. L. and Sexton, R. J. 2018. *Economic Impacts of the California Avocado Commission's Advertising and Promotion Program: 2013-2017*, 25pp.

Anon. 2013. Avocado consumption: Still on the rise? *The Packer*, 24 June 2013. Retrieved on 17 October 2018 from https://www.thepacker.com/article/avocado-consumption-still-rise.

Battany, M. 2008. Atmometers for irrigation management. *Topics in Subtropics* 6(3): 2–4.

Bender, G. S. 2012. Avocado farming with high priced water. *Topics in Subtropics* 10(1): 4–7.

Bender, G. and Faber, B. 2009. Chapter 1: Irrigation. In: *Avocado Production in California – Cultural Care*. G. S. Bender (Ed.). University of California Cooperative Extension, San Diego, pp. 2–5. Retrieved October 2018 from https://ucanr.edu/sites/alternativefruits/files/228975.pdf.

Bender, G. S., Rios, S. and Tanizaki, G. 2018. Improvement of yield per acre by close spacing, pruning of close-spacing 'Hass' and 'Lamb Hass' – A final report. Retrieved on 15 October 2018 from https://www.californiaavocadogrowers.com/sites/default/files/Avocado-High-Density-research-trial-FINAL%20Report.pdf.

Bergh, B. 1984. Avocado varieties for California. *California Avocado Society Yearbook* 68: 75–94.

Broadbent, P. and Baker, K. F. 1974. Behavior of *Phytophthora cinnamomi* in soils suppressive and conducive to root rot. *Australian Journal of Agricultural Research* 25: 121–37

Bruwer, T. and Van Rooyen, Z. 2007. Performance and market acceptability of the 'Hass'-like cultivars 'GEM' and 'Harvest' in South Africa. *Proceedings of the VI World Avocado Congress 2007*. Retrieved on 20 October 2018 from www.avocadosource.com

California Avocado Commission. 2018. Retrieved on 3 March 2019 from https://www.californiaavocadogrowers.com/sites/default/files/documents/2015-Acreage-Inventory-Survey-Results.pdf.

Carrillo-Gallegos, G. and Calderon, V. 2015. Ten years of experience on high-density management in Michoacan, Mexico (abstract). *VIII World Avocado Congress, 2015*, p. 52. Retrieved on 20 October 2018 from www.avocadosource.com.

Cook, R. J. and Baker, K. F. 1983. *The Nature and Practice of Biological Control of Plant Pathogens*. American Phytopathological Society Press, St. Paul, MN.

Crane, J. H., Douhan, G., Faber, B. A., Arpaia, M. L., Bender, G. S., Balerdi, C. F., Barrientos-Priego, A. F. 2013. Cultivars and rootstocks. In: *Avocado, Botany, Production and Uses* (2nd edn.). B. Schaffer, B. N. Wolstenholme and A. W. Whiley (Eds). CABI, Boston, MA, pp. 200–33.

Crowley, D. 2004. Salinity tolerance in avocado. In: *Proceedings of the California Avocado Research Symposium*, 30 October 2004. University of California, Riverside and California Avocado Commission, Irvine, CA, pp. 108–11.

Crowley, D. 2007. Managing soils for avocado production and root health. *California Avocado Society Yearbook* 90: 107–30.

Crowley, D. 2008. Salinity management in avocado orchards. *California Avocado Society 2008 Yearbook* 91: 83–104.

Dann, E. K., Ploetz, R. C., Coates, L. M. and Pegg, K.G. 2013. Foliar, fruit and soilborne diseases. In: *The Avocado, Botany, Production and Uses* (2nd edn.). B. Schaffer, B. N. Wolstenholme and A. W. Whiley (Eds). CABI, Boston, MA, pp. 397–405.

Doll, D. 2015. Soil moisture sensors. Retrieved on 15 October 2018 from http://thealmonddoctor.com/2015/07/10/soil-moisture-sensing-systems/.

Downer, A. J., Menge, J. A. and Pond, E. 2001. Effects of cellulytic enzymes on *Phytophthora cinnamomi*. *Phytopathology* 91: 839–46.

Downer, J., Faber, B. and Menge, J. 2002. Factors affecting root rot control in mulched avocado orchards. *HortTechnology* 12(4): 601–5.

Ernst, A. A., Ernst, Z. R. and Ernst, E. D. 2015. 'Maluma': Establishing a new generation avocado cultivar commercially. *VIII World Avocado Congress Proceedings 2015*, pp. 53–9. Retrieved on 20 October 2018 from www.avocadosource.com.

Faber, B. 2007. So you want to be an organic avocado grower. *California Avocado Society Yearbook* 90: 55–66.

Faber, B. 2008. Organic avocado production. *California Avocado Society Yearbook* 91: 35–43.

Faber, B. 2017. So you want to install soil moisture monitoring meters? *Topics in Subtropics* 15(Fall): 3-5.

Faber, B.A., Downer, A.J. and Menge, J.A. 2003. Effects of mulch on avocado and citrus. *Proceedings V World Avocado Congress 2003*, pp. 719–24. Retrieved on 2 October 2018 from www.avocadosource.com.

FAOStat. 2019. Retrieved on 10 July 2019 from www.fao.org/faostat.

Gardiazabal, F. and Mena, F. 2011. The avocado industry in Chile and its evolution. *California Avocado Society 2011 Yearbook* 94: 52–68.

Gardiazabal, F., Mena, F., Torres, J. and Pinto, A. 2015. Very high density (2.5 × 1.25m and 1.25 × 1.25m) plantings of avocados (*Persea Americana* Mill.) cv. Hass achieved over 40 t/ha in the first year of production (abstract). *VIII World Avocado Congress 2015*, p. 55. Retrieved on 10 October 2018 from www.avocadosource.com

Gazit, S. and Kadman, A. 1976. Growing avocados in areas of high salinity. In: *Proceeding of the First International Tropical Short Course: The Avocado*. J. W. Sauls, R. L. Phillips and L. K. Jackson (Eds). Gainesville Fruit Crops Department, Florida Cooperative Extension Service, University of Florida, pp. 58–60

Gregoriou, C. and Rajkumar, D. 1984. Effects of irrigation and mulching on shoot and root growth of avocado (*Persea americana* Mill.) and mango (*Mangifera indica* L.). *Journal of Horticultural Science* 59(1): 109–17.

Guest, D. I., Pegg, K. G. and Whiley, A. W. 1995. Control of Phytophthora diseases of tree crops using trunk-injected phosphonates. *Horticultural Reviews* 17: 299–330.

Hanson, B., Orloff, S. and Peters, D. 2000. Monitoring soil moisture helps define irrigation management. *California Agriculture* 54(3): 38–42.

Hoddle, M. S., Aponte, O., Kerguelen, V., Heraty, J. 1999. Biological control of *Oligonychus perseae* (Acari: Tetranycheidae) on avocado. I. Evaluating release timings, recovery and efficacy of six commercially available phytoseiids. *International Journal of Acarology* 25: 211–19.

Hoddle M. S, Robinson, L. and Virzi, J. 2000. Biological control of *Oligonychus perseae* (Acari: Tetranycheidae) on avocado. III. Evaluating the efficacy of varying release rates and release frequency of *Neoseiulus californicus* (Acari: Phytoseiidae). *International Journal of Acarology* 26: 203–14.

Hoitink, H. A., Inbar, Y. and Boehm, M. J. 1991. Status of compost-amended potting mixes naturally suppressive to soilborne diseases of floriculture crops. *Plant Disease* 75(9): 869–73.

Kerguelen, V. and Hoddle, M. S., 1999. Biological control of *Oligonychus perseae* (Acari: Tetranycheidae) on avocado. II. Evaluating the efficacy of *Galendromus helveolus* and *Neoseiulus californicus* (Acari: Phytoseiidae). *International Journal of Acarology* 25: 221–9.

Kerguelen, V. and Hoddle, M. S. 2000. Comparison of the susceptibility of several cultivars of avocado to the persea mite, *Oligonychus perseae* (Acori: Tetranychidae). *Scientia Horticulturae* 84: 101–14.

Lahav, E., Whiley, A. W. and Turner, D. W. 2013. Irrigation and mineral nutrition. In: *The Avocado, Botany, Production and Uses*. B. Schaffer, B. N. Wolstenholme and A. W. Whiley (Eds). CAB International, Wallingford, UK, pp. 301–41.

Lanini, W. T., Shribbs, J. M. and Elmore, C. E. 1988. Orchard floor mulching trials in the USA. *Le Fruit Belgium* 56(3): 228–49.

Leonardi, J. 2012. Developing sustainable orchard management practices. Australia Avocado Industry Report 2011–2012, p. 4.

Lovatt, C. J. 1996. Nitrogen allocation within the 'Hass' avocado. *California Avocado Society Yearbook* 80: 75–83.

Lovatt, C. J. 2001. Properly timed soil-applied nitrogen fertilizer increases yield of 'Hass' avocado. *Journal of the American Society for Horticultural Science* 126: 555–9.

Lovatt, C. J. 2015. Optimizing 'Hass' avocado tree nutrient status to increase grower profit – an overview. *Proceedings of the VIII* World *Avocado Congress*, Lima, Peru, pp. 17–23. Retrieved on 25 October 2018 from www.avocadosource.com

Matava, M. 2009. The benefits of mulching avocados. *California Avocado Society 2009 Yearbook* 92: 65–72.

Menge, J. A. 2002. Screening and evaluation of new rootstocks with resistance to *Phytophthora cinnamomi*. *Proceedings of the California Avocado Research Symposium 2002*, pp. 55–9.

Menge, J., Faber, B., Downer, J. and Crohn, D. 1999. Compost demonstration project, Southern California: use of yard trimmings and compost on citrus and avocado. September 1999. Integrated Waste Management Board, Publication #443-99-010. Retrieved on 5 August 2018 from https://www.calrecycle.ca.gov/organics/farming/agdemos/ucriver.

Menge, J. A., Douhan, G. W., McKee, B., Pond, E., Bender, G. S. and Faber, B. 2012. Three new avocado rootstock cultivars tolerant to Phytophthora root rot: 'Zentmyer', 'Uzi', and 'Steddom'. *HortScience* 47(8): 1191–4.

Meyer, J. L., Yates, M. V., Stottlemyer, D. E., Takele, E., Arpaia, M. L., Bender, G. S. and Witney, G. W. 1992. Irrigation and fertilization management of avocado. *Second World Avocado Congress Proceedings*, pp. 281–8. Retrieved on 20 August 2018 from www.avocadosource.com

Milne, N. 2019. Chile's booming avocado business blamed for water shortages. *Global Citizen*. Retrieved on 2 August 2019 from https//:www.globalcitizen.org.

Mission Resource Conservation District. 2000. *Estimating Orchard Water Use with CIMIS*, 20p.

Morse, J., Faber, B. and Hoddle, M. 2018. Avocado thrips. *UC IPM Online*. Retrieved on 30 August 2018 from http://ipm.ucanr.edu/PMG/r8300311.html.

Oster, J. D. and Arpaia, M. L. 2007. Comments about crop coefficients for Hass avocado on clonal rootstocks. Retrieved on 2 June 2018 from http://www.avocadosource.com/.

Pena, J. E., Hoddle, M. S., Aluja, M., Palesky, E., Ripa, R. and Wysoki, M. 2013. Insect and mite pests. In: *Avocado, Botany, Production and Uses* (2nd edn.). B. Schaffer, B. N. Wolstenholme and A. W. Whiley (Eds). CABI, Boston, MA, pp. 423–88.

Rice, L. W. and Rice, R. R. 2011. Outdoor soils and fertility. In: *Practical Horticulture* (7th edn.). Rice, L.W. and Rice R.R. (Eds). Prentice Hall, Upper Saddle River, NJ, p. 92.

Smil, V. 1999. Detonator of the population explosion. *Nature* 400(6743): 415.

UC SAREP. 2018. Retrieved on 27 September 2018 from http://asi.ucdavis.edu/programs/sarep/about/what-is-sustainable-agriculture.

Weeks, W. D., Smith, C. T. and Drake, M. 1950. Residual effects of heavy mulching in a bearing apple orchard on soil nutrients. *Proceedings of the American Society for Horticultural Science* 56: 1–4.

Wolstenholme, B. N. 2018. Avocado growing: planning for the future – keeping the big picture in mind. *Plenary Address: Avocado Brainstorming Meeting*, South Africa.

Yates, M. V., Stottlemyer, D. E., Meyer, J. L. and Arpaia, M. L. 1992. Irrigation and fertilizer management to minimize nitrate leaching in avocado production. *Proceeding of Second World Avocado Congress 1992*, pp. 331–5. Retrieved on 1 August 2018 from www.avocadosource.com.

Zentmyer, G. A. 1980. *Phytopthora cinnamomi* and the diseases it causes. Monograph no. 10. The American Phytopathological Society, St. Paul, MI.

Zicah, S., David, I., Yeselson, Y. and Klein, I. 1999. Seasonal uptake and distribution of 15N-labelled ammonia and nitrate in young avocado trees. *Alon Hanotea* 53, 416–25 (In Hebrew with English summary).

# Chapter 10

# Ecological intensification of banana production systems

*Charles Staver, Bioversity International, France*



## 1 Introduction

In a world racing to address climate change, limits in plant nutrient supplies and water contamination with agricultural pesticides, nutrients and plastics, among others, while feeding and employing a growing and healthier human population, bananas have a bright future. They are a low-cost fruit with

http://dx.doi.org/10.19103/AS.2019.0054.11

recognized health properties. A high diversity of cultivars in taste, aroma and texture offers potential to stimulate mass and niche markets and new uses in cooking. The crop offers good soil cover with a low frequency of soil tillage and can be managed to achieve high yields with year-round or extended production which generates regular cash flow for smallholder households. Increasing average temperatures associated with climate change are improving the suitability of higher altitude tropical zones and the subtropics for cultivation, with only small pockets suffering excessively high temperatures above 35–37°C limiting production (Calberto et al. 2015). Banana is also an extremely flexible crop adapted as an overstory or understory crop in mixed systems as well as annual and perennial stands. It is therefore highly compatible with more diversified and agroecologically based production systems.

Banana production, however, also faces numerous challenges in becoming more resource efficient, profitable to growers and credible with consumers. Unlike many other crops, only slow progress has been made through breeding to develop more productive and resistant cultivars. Yield-threatening diseases such as Fusarium Tropical Race 4 and Banana Bunchy Top Virus are spreading in Asia and Africa. Fusarium Tropical Race 4 has recently been reported in Colombia, the first outbreak in Latin America. Increasing temperatures associated with climate change increase the frequency of weather extremes which, depending on the region, may disrupt routine banana production schedules and increase costs (Calberto et al. 2018). Diverse certification frameworks are required for export and are becoming increasingly strict even in domestic markets (Staver 2018). A growing consumer awareness of food miles (the distance a product moves from field to consumer) may also put pressure on bananas as the most globally traded fruit (Bockel et al. 2018). At the farm level, these issues are beyond the control of growers (or even the associations which represent them). They can only be addressed in terms of crop and input management decisions which growers make. How can yields be increased and timed for optimum returns through the conversion of light, water and nutrients into banana in the most efficient way in terms of land, labor and capital and with minimal environmental impact?

This chapter focuses on immediately available options to growers to be more efficient, productive and competitive, while maintaining and improving the environment. These include strategies for reducing bunch variability, timing production for periods of favorable prices, improving the efficiency of nutrient uptake and recycling, soil health and management for variable weather. Many of these practices are not covered in the codes of good agricultural practices which are oriented to assure marketers and consumers rather than to guide growers toward more efficient production and higher productivity. The chapter first reviews types of banana production systems and then discusses measurement and monitoring of improvements in crop production. The second half of the chapter summarizes and illustrates specific practices that can be used to improve production and productivity from the mat to the field and farm ecosystem.

## 2 Types of banana production systems

Bananas are considered a tropical crop. They are widely grown from sea level to over 2000 meters above sea level in the highlands of Africa, Asia and Latin America and from desert zones like Peru and Sudan to the wet tropics in the Philippines, Central Congo and Costa Rica. However, bananas are also a subtropical crop in both the northern and southern hemisphere from Turkey, Pakistan and China to Argentina, South Africa and Australia (Calberto et al. 2015; Lescot 2018). This wide range of zones represents many different cultivars and production systems. While FAOSTAT collects statistics only for banana and plantain, Fruitrop presents statistics for four distinct groups:

- Cavendish
- other AAA dessert
- ABB plantains
- other cooking bananas (AAA and ABB).

In a survey of constraints on banana yield and profitability, participants were asked to identify their predominant cultivar of interest among eight different banana cultivar groups (Pemsl et al. 2014) (Table 1). Cavendish represented 35% of respondents, plantain 25%, East African Highland bananas 15% and ABB cooking bananas 9%, with the other four groups only representing 16% of responses. This same survey proposed a gradient of banana mat density and associated crops as the basis for six different production systems (Table 1) such that any system can be characterized both by the cultivar(s) present and diverse production system factors.

Other authors have proposed different categories of banana production systems which can be cross-referenced with Table 1. For example, Akinyemi et al. (2010) identified five production systems in Nigeria – plantain as temporary shade for newly established cocoa, bush plantain, taunya cropping in forestry plantations, household compound and plantation. For Africa, Tushemereirwe and Kubiriba (2018) proposed monocrops, backyard and banana intercrop, while recognizing that intercrops can be highly varied from perennial crops to short cycle annuals. For Asia, Molina and Valmayor (1999) highlighted four categories – two monocrops by size of production unit, small holder and corporate monocrops, and two diversified, backyard and mixed-crop covering both temporary and permanent banana planting and both under- and overstory associated crops. Finally, for Latin America and the Caribbean, Dita et al. (2013) proposed five monocrops (conventional and organic Cavendish, subtropical

**Table 1** Two way classification system for banana production systems based on cultivar group and production system characteristics. Any system can be characterized by one or more cultivar groups and a production system

| Cultivar groups | Production systems - mat density, presence of associated crops and duration of banana | Mat density/ha |
|---|---|---|
| Cavendish AAA | 1-Ongoing association with established perennial crops like coffee or coconut | 150–600 |
| Gros Michel AAA<br>other AAA dessert types | 2-Annual mixed food crop association intercropped with Musa (2–5 years) | 400–700 |
| East African Highland AAA | 3-Perennial Musa planting mixed with annual crops | 1000–1400 |
| Plantain AAB<br>Other AAB, including South Pacific plantains | 4-Musa field (2–5 years) with short-term annual crop intercropped during Musa establishment or as temporary shade for a more permanent crop (rubber, cocoa, oil palm) | 1200–1500 |
| ABB cooking bananas | 5-Perennial Musa monocrop (1500–2400 mats/ha) | 1500–2400 |
| Diploid types | 6-Frequently replanted Musa monocrop | 2000–4000 |

Cavendish, plantain, nationally preferred cultivars like Prata and Isla), bananas with perennial crops like coffee and cocoa and small-scale plantings.

Table 2 illustrates how these production systems from other authors align with Table 1. The ongoing association of bananas with shaded coffee is common throughout Latin America, the Caribbean and in areas of East and West Africa (System 1 in Table 1). Coconuts are the other perennial crop in which banana can be cropped throughout the coconut life cycle. Many backyard or compound gardens are also bananas mixed with diverse fruit and ornamental trees such as the 'jardin creole' in Haiti (Jean-Denis et al. 2014). These permanent mixtures can be contrasted with the use of bananas as a temporary crop for the establishment of perennial crops like cocoa, rubber, oil palm and sometimes coconut (System 4 in Table 1). Plant densities are often higher when banana is a temporary crop than for the permanent associations in System 1.

The second system in Table 1 is well illustrated by the bush plantain of Nigeria, a highly diverse mix of annual grain and legume crops, roots and tubers as well as plantain and other banana cultivars which returns to fallow after several plantain harvests, while perennial banana with understory annual crops is found through the East and Central Great Lakes Highlands (System 3 in Table 1). Systems 4, 5 and 6 in Table 1 vary in density and duration of production depending on cultivars and also abiotic conditions and stress. Cavendish bananas for export are often permanent fields (System 5), although China and India use low-cost tissue culture plants for annual replanting at higher densities (System 6). Plantains under certain conditions are permanent, but more generally are replanted from annually to every three years, depending on grower resources and markets. Other cultivars like Silk in Brazil are frequently replanted (System 6), generally after one harvest due to Fusarium wilt problems.

**Table 2** Examples of banana production systems illustrating the different production systems in Table 1

| Production system following Table 1 | Specific characteristics | Cultivar(s) | Zone, country | Banana mat density |
|---|---|---|---|---|
| Associated with perennial crops | Multi-strata Arabica coffee | AAA - Gros Michel, Red/ Green, | Nicaragua, Honduras, Peru | 150-400 |
| | Sun-grown Arabica and Robusta coffee | East African highland AAA | Uganda, Burundi | 800-1000 |
| | Coconut underplanting | Local market | The Philippines, Sri Lanka | |
| | Fruit and timber trees, root and tuber crops - Jardin Creole | Plantain, bluggoe, others | Haiti - The Caribbean | Variable |
| Mixed food crops intercropped with Musa (2-5 years) | Diverse roots tubers, grains and legumes | Plantain | West Africa | 100-500 |
| | Cassava, maize, peanuts, beans | Plantain | Amazon | 500-800 |
| Perennial Musa with annual crops | Beans, pumpkins, leafy vegetables, tomatoes | East African Highland AAA | Uganda, Burundi, Rwanda | 1000-2000 |
| Musa (2-5 years) monocrop or associated | Short cycle annual crops in first 3-4 months | AAB plantain | Central America | 1500-2000 |
| | Fusarium and weevil build-up results in 2-4 harvests only | AAB Isla | Peru | 900-1200 |
| | Temporary shade to cocoa | AAB plantain | West Africa, Peru | 900-1200 |
| Perennial Musa monocrop | Infrequent replanting | Cavendish AAA | Global | 1500-2400 |
| | Infrequent replanting | Prata | Brazil | 1300-1600 |
| | Planted in protected environment | Cavendish | Turkey, Canary Islands, Morocco | 1500-2000 |
| Frequent replant monocrop | Timing for market or subtropical weather | Cavendish | China, Taiwan, India | 1800-2500 |
| | Yield increase and market timing | Plantain | Dominican Republic | 2000-3000 |
| | Fusarium disease build-up results in only 1-2 harvests | Silk | Brazil | 1400-1600 |

This high diversity of cultivars, systems and climates presents considerable challenges to scientists, field technicians and growers to develop strategies for scouting and data collection, recommendations and decision support tools which are broadly applicable. In the next section, we will contrast different frameworks and indicators which are used to identify opportunities for improvement of crop performance.

## 3 Measuring the potential for improvement in banana production systems

Classically, agronomists have focused on closing yield gaps, measured as yield/ha/crop cycle (van Ittersum et al. 2013), and the identification of causal factors explaining these gaps. For crop associations either as crop rotation or as mixed cropping, the expression of yield and yield gap is complicated considerably by diverse factors. The length of the time period to be analyzed may be difficult to determine. For banana, the establishment year may be different from a perennial plantation in production. In addition, certain systems show rapid yield collapse from the first harvest through the second to fourth ratoon crop. In many systems, bananas are only one component of the yield. Can a yield gap be defined without changing the proportion of crop resources captured in commonly practiced crop combinations?

In yield gaps, efficiency is measured for land resources (yield/ha), but crop production demands many additional resource types which provide the basis for other measures of the potential for improvement. These include water, nutrients and other production inputs, labor and capital (Keating et al. 2010). Where labor costs are high, for example, such as Australia or the French West Indies, the efficiency of labor may be equally as important as yield/ha. These analyses are based on an input assembly model of crop production with each factor considered independently. A crop field is also an ecosystem and the numerous ecological processes may offer additional opportunities to address banana production system improvement beyond yield gap. These analyses on the efficiency in the use of factors of production (land, labor, capital and other production resources) and agroecosystems focus primarily at the field.

Certification frameworks broaden the analysis beyond just production. Environmental and social life cycle analysis expand even further the diagnostic to cover the manufacture and disposal of production and marketing inputs in a cradle to grave perspective on the process from farm to market. Each of these themes will be reviewed before we focus on specific practices and principles applicable to the different production systems.

## 4 Using yield gap analysis

Yield gap analysis has been widely applied in annual crops to highlight the potential for yield improvement in different agroclimatic environments (see Global Yield Gap Atlas). Crop models are used to establish potential yield which is then compared with current average grower yields. Analysis of practices among growers with above and below average yields can be used to identify key factors in the yield gap.

Few examples of yield gap analysis are available for banana. Wairegi et al. (2010) analyzed farmer field data and on-farm demonstration plots in three regions in Uganda to establish achievable yields and identify primarily biotic factors in Central Uganda and nutrients and drought in South and Southwest Uganda which reduced yields. Farmer yields ranged from 10–20 t/ha/year contrasted with 60 t/ha considered attainable. In areas of Uganda with <1100 mm annual precipitation, yield reductions may be from 20–65% compared to wetter regions (Van Asten et al. 2011). Norgrove and Hauser (2014) used crop statistics and a review of experiments on plantain to determine that attainable yields are 24–35 t/ha with best practices, while on-farm yields are 4.5–6 t/ha. Dépigny et al. (2017) achieved yields of 35–50 t/ha from nine different plantain cultivars in the first cycle under well-fertilized, non-water limiting conditions. Staver et al. (2016) measured maximum yields of over 50 tons/cycle in Nicaragua, although national average yields are only 16 tons/ha/year over the period 2015–2017 (FAOSTAT). Yields/ha calculated based on bunch weights and mat density can overestimate yields, since up to 10% or more of stems may not reach harvest (Hauser and van Asten 2010).

Although Cavendish bananas are the most widely cultivated cultivar, few analyses of yield gap are available. In Martinique, potential yield was established at 70 t/ha/year from the literature, while the average yield for 329 farms quantified through grower interviews was 28 t/ha/year with a coefficient of variation of 47%. Reported farm yields varied from 7 to 67 t/ha/year explained by variables such as use of tissue culture plants, longer fallows free of banana, chemical destruction of plantations prior to fallow and labor use intensity in crop management. Growers with low yields were also limited by steep slopes of their fields and low capital availability to invest in yield increasing practices such as tissue culture, chemical field destruction and longer crop rotations (Chopin and Blazy 2013). The authors point out that in their regressions, intensity of pesticide and fertilizer use was not a significant factor to explain yield differences, but also noted that low $r^2$ values indicated numerous other factors to be identified with more in-depth studies.

In Kenya smallholder growers produce Cavendish for urban markets. The status of 15 mats at three distances from the homestead was monitored and farmers interviewed monthly for over one year in 10 fields from farmers of three different resource levels (Okumu et al. 2011). Average yields were 28 t/ha/year, well above the national average, with a marked gradient by distance from homestead from 32 to 21 t/ha/year. Total plant biomass did not differ by distance, but harvest index was greater closer to the homestead. The interval between bunches averaged 13 months with no reference to the importance of distance. Pest and disease pressure was low. Soil pH and available N, P and K were highlighted as important explanatory factors. Poor resource farmers had smaller fields with higher levels of nutrients resulting in higher yields. Potential yields were not estimated, but authors postulated that the bimodal rainfall with two dry seasons limited Cavendish yields, although rainfall was 1560 mm/yr.

A study in the Philippines based on interviews with 80 farmers in Mindoro Oriental found average yields of 6.6 t/ha/year with a range from 0.3 to 28 t/ha/year (Bathan and Lantican 2010). Yield variability was linked to factors like soil type, fertilizer (over 75% used no fertilizers), labor effort/ha and distance from residence. Farms with mixed varieties had higher yields. Farmers cited scarce and high cost tissue culture plants and pests and diseases as important factors, although these were not captured in the regression analysis.

These studies of crop yield gap highlight four challenges in using this approach to prioritize steps in banana productivity improvement:

- Firstly, the estimation of potential yield for annual crops is carried out using crop models. Crop models for banana are progressing rapidly (Tixier et al. 2008; Nyombi 2010; Taulya 2015), but are only beginning to address yield potential and cultivar differences. Yields among elite farmers or for experiments are often used to estimate attainable yields which are demonstrated here with the results for plantains. For Cavendish all sites are far below the yield of 70 t/ha/year mentioned in the Martinique study. Yields for Cavendish approaching 100 t/ha/year have also been reported (Canary Islands).
- Secondly, methods for data collection to identify limiting factors vary. Interviews are faster and greater farm to farm variability can be covered, but farmer estimates of even simple elements like yields may be incomplete and other factors like disease presence may be difficult to capture. Field measurements, as shown in the study from Kenya, offer more grounded information, but take longer to execute.
- Thirdly, the interplay of yield components is often not addressed. Growers for export often record boxes/ha rather than bunch size. National markets may operate based on number of fingers or bunch size categories rather than weight. Mats often have more stems than the classic mother-daughter-granddaughter and the time interval between bunches is often not calculated, even though in fields harvested over multiple years, it is the factor most closely linked to yields.
- Fourthly, timing of production during the year is not often cited. Depending on markets and prices, total yields may be less important than the timing of production. Growers in Southeast Asia supplying the Chinese market manage planting dates, sucker uniformity and desuckering to ensure that both the first and second harvests are closely synchronized to peak prices during celebrations of the new year. Under rain-fed conditions in West

Africa, plantain growers who reduce the yield gap may end up with higher production during the glut period when prices collapse. Strategies for off-season production may provide better returns, even if yields are lower or production costs are higher.

The adaptability of banana to mixed cropping (Tables 1 and 2) highlight yet another challenge to building grounded strategies for banana improvement which is not addressed by yield gap analysis for monocrops. Bananas are found in the understory of agroforestry perennial cropping systems and as the overstory for annual crops. How can the potential for improvement be characterized when banana is only one component of the system?

The plantain fields around Njombé, Cameroon studied by Dépigny et al. (2019) illustrate the great variability in mixed systems. The 25 fields varied in initial planted population from less the 300 to almost 3000 corms and average field bunch weight from 5 to 26 kg. They identified key factors positively correlated to higher average field bunch weight – use of nitrogen fertilizer and herbicides, farmer participation in training and farmer specialization in plantain production. Their sampling strategy of bunches in three contrasting seasons during the year did not allow a meaningful estimate of field yield due to several factors. They found a big gap between the population of harvestable stems and their estimate of total bunches harvested due to toppled stems before harvest, theft and possible lapses in farmer data collection. Bunch weights and numbers per season were highly variable with double the number of bunches in the rainy season. Their yield estimate taking into account the different uncertainties was 6.8 t/ha/year varying from 1 to 18 t for the 25 fields. Their estimate of field productive life based on twice yearly counts of stems per mat did not receive much emphasis in their discussion, although this is also a major component of cropping system productivity. They did not report the production of other crops from the same fields.

In a similar cropping system with maize, cassava and plantains in Amazonian Peru, Staver (1989a) identified the cropping system cycle to include both the crop cycle starting at clearing of bush fallow and the re-establishment of bush fallow until the next clearing and cropping cycle. Practices to improve crop performance such as more frequent weeding slowed bush fallow recovery and resulted in longer fallow times. On the other hand, reduced weeding frequency and shorter cropping periods result in more rapid bush fallow recovery, although plantain yields were reduced and time to flowering increased (Staver 1991). Legume trees and shrubs interplanted simultaneously with crops to accelerate bush fallow recovery suppressed herbaceous weed growth and increased woody biomass accumulation while increasing plantain yields (Staver 1989b). In these studies, the within-field variability of plantain yield was not documented.

Bananas are also grown intercropped in shade-grown coffee as a month-by-month source of income and food to complement the annual coffee harvest. To develop criteria for optimization of banana production in this mixed system, Siles et al. (2013a) analyzed both light partitioning and nutrient import-export as the basis for guidelines to achieve increasing banana production, while not sacrificing coffee yields. They proposed a banana density resulting in leaf overlap between neighboring mats of no more than 25% and tree shade not greater than 25–35% to ensure that understory coffee received 50% of available light. Mat to mat variability in bunch size was also recognized as a potential strategy to improve yields. The study did not propose the potential yield or attempt to quantify a yield gap, instead focusing on the identification of improvements which could contribute to increased returns from bananas in shaded coffee. Decisions about whether to invest additional labor or capital in coffee or banana improvement are relevant in a cropping system analysis. Cash income and distribution of labor are different for coffee and banana which may complicate the analysis (Staver et al. 2012).

## 5 Assessing gaps in resource efficiency

The crop yield gap or even cropping system yield gap measured in tons/ha focuses on the productivity of land, but land is not the only production input in agriculture. Eco-efficiency measured by the output of food and fiber per unit of ecological resources such solar radiation, water, nutrients, fossil energy or biodiversity (Keating et al. 2010) goes beyond yield/ha. The 'eco' may also be interpreted to include economic measures of efficiency such as capital and labor – cost/box, input or labor cost/box. Even for a single category of resource, numerous possible calculations are possible – kg applied nutrient/kg total plant biomass or harvested bunches or boxes of fruit. Efficiencies can also be measured at different dimensions from cellular to plant to field and farm. Here we will focus on the plant and field.

Radiation efficiency has been addressed most frequently with recommended leaf area indices for optimum yield. For banana monocrops, leaf area indices of 4.5–5 capture around 90% of available sunlight and any additional leaves will not contribute additional photosynthate (Turner et al. 2007), although the useful LAI may be lower on poorer, lighter textured soils. These are references as a guide, but in practice leaf distribution is uneven and may depend on crop arrangement and also crop phenology. Zones of the soil surface were proposed in a model of cover crops and nitrogen with different levels of light available for banana planted in double rows (Damour et al. 2012). Cover crop growth was variable according to the zone. In high density frequently replanted bananas or plantains, LAI declines with highly synchronized harvest with radiation going uncaptured and falling onto bare soil. Each cropping system shown in

Table 1 has instances in the crop cycle with underutilized radiation. Rodrigo et al. (2001) documented the increase in radiation capture as immature rubber was associated with one, two or three rows of banana. Rubber alone and rubber with one row of banana captured only 40 and 50% of available radiation. Banana with three rows of banana had 23 and 73% greater interception of radiation than banana with two or one rows. In shaded coffee, in studies in Latin America, less than half the coffee plants were rated as highly productive with many plants having limited leaf area and poor vigor. Certain shade trees age rapidly and ten years after planting, growth is slow and leaf production is reduced (Siles et al. 2013a). Such systems in which banana is only one component appear to offer abundant opportunity for increased conversion of radiation into biomass.

The measurement of water use efficiency in banana has been summarized by Carr (2009). In irrigated monocrops, between 40–80 kg fruit/ha were produced for each mm of water. Van Asten et al. (2011) calculated accumulated rainfall for individual stems in experiments in different localities in Uganda compared with their bunch weight to establish that over the range of 500–1750 mm, for every 100 mm decline in rainfall, bunch size declined 1.5–3.1 kg. They found this similar to effects reported in experiments in Puerto Rico and Hawaii with Cavendish. The precipitation use efficiency ranged from 13–40 kg/ha/mm depending on production inputs. Van Asten et al. (2011) highlighted the role of different management practices to address this effect of drought on yields. Potassium fertilization doubled yields, while in other trials cited in their article, the use of mulch increased yields up to 50%. The value of intercropped bananas with immature rubber to increase water use was shown by Rodrigo et al. (2005). Increasing the banana population from 500, 1000 and 1500 mats increased water use linked closely to the increased LAI. Soil water levels were stable across treatments perhaps due to the presence of a kudzu cover crop which had greater (unmeasured) biomass in treatments with lower LAI. In this wet humid site with 8000 mm of rainfall during the experimental period of 800 days, crop transpiration was only 2–11% of total rainfall. Under such conditions, the authors conclude that cropping systems to convert more water to biomass are more important than water use efficiency.

Across the spectrum of banana production systems (Table 2), application of nutrients from external sources is often absent. Under these conditions, a first question to address is the extent of soil mining when nutrients are exported in bananas bunches which are marketed. In three sites in Rwanda, increased banana plant density only increased the loss of macronutrients. The six kg of cattle manure at planting compensated the loss of P, Ca and Mg in low yielding plots, but 4–176 kg N and up to 1000 kg K were exported with all site × mat densities negative, except 2 (Ndabamenye et al. 2013). Even the use of cow manure at planting is not widely practiced among smallholders in the Highlands of East Africa (Mpiira et al. 2013). Bananas interplanted in shaded coffee are rarely fertilized and fertilizer is applied based on the needs of the coffee crop (Siles et al. 2013a).

However, in many intensive production systems, the soil nutrient mining is not the challenge. The fertilizer response trial with Cavendish by Irizarry et al. (2002) illustrates the high levels of fertilizer to cover nutrient exports in harvested bunches which is only half what is being applied. The best treatment produced 57 tons/ha/year containing 180 and 390 kg of N and K respectively with fertilizer application of 240 and 690 kg of N and K applied. This represents an efficiency of 50–75%, although commercially fertilization rates can be upward of 600 kg/ha/year for N and 1000 kg/ha/year for K (Pattison 2006).

Addressing nutrient efficiency across the different production systems faces at least three challenges. First, the possibilities to improve nutrient efficiency and close the yield gap in yield/ha at the same time are debatable (van Noordwijk and Brussaard 2014). Bellamy et al. (2018) profiled three banana systems in Costa Rica -high input conventional export, reduced input convention export and organic agroforestry. The banana yields for the three systems were 59, 41 and 4 tons/ha/year with annual fertilizer applications of 2787, 1924 and 155 kg/ha. This converts to 21, 21 and 25 kg banana per kilogram of fertilizer with efficiencies of less than 50% for the two conventional systems and a high efficiency for the agroforestry system, more nutrients taken off than applied for the organic system.

Second, annual or frequently replanted monocrops require applications to establish plant biomass as well as to ensure expected yields. For such situations, the management of crop residues for nutrient conservation and uptake in the following crop may be relevant. For established plantations, the standing biomass and crop residue represent an important nutrient buffer to maintain plantation growth. Nutrients lost in harvested bunches can readily be calculated annually and compensated.

Third, nutrient efficiency may interact with other factors. Pattison (2006) found that application of lower amounts of nitrogen was part of a bundle of practices which contributed to both soil health and increased production. The highest yielding farms applied less fertilizer but used other practices to ensure soil and root health.

## 6 Assessing the ecological intensification gap

The use efficiency of other production inputs such as pesticides, loss reduction, labor and capital (Table 3) can also be analyzed to orient banana production system improvement. We will not expand on these dimensions here. The approach addressing individual factor productivity and efficiency which is the basis of yield gap can be complemented and expanded building on the concepts of ecological intensification. This approach addresses the weaknesses identified in the previous section on nutrient efficiencies. Ecological intensification has been used as a synonym for sustainable intensification - more yield per unit of resource, while minimizing the impact on the environment (Hochman et al. 2013) which may serve to orient broader agriculture policy. However, this definition misses the fundamental role of ecological processes underlying crop productivity.

Staver et al. (2018) proposed a step by step procedure to improve cropping system performance synchronizing ecological and production processes oriented to increase returns from sunlight, water and nutrients, but also external inputs and labor, to reduce losses to abiotic and biotic stress and reduce off-field impacts. They tested the approach using available data from seven production systems representing five of the production systems in Table 1. In their approach, they first examined whether the production system was threatened by a catastrophic biotic stress like banana bunchy top virus or Fusarium Tropical Race 4 which would nullify any other production constraints. Table 3 summarizes the themes addressed in sequence. The resource efficiency analysis already described above - radiation, water and nutrients - is complemented by indicators of soil protection, nutrient, biomass and organic matter recycling, water cycling, carbon status and a food web perspective on pests and diseases above and below ground (see presentation at http://www.promusa.org/tiki-download_file.php?fileId=2833&display=y). Following work by Calberto et al. (2018), they also highlighted the role of climate and other variability such as product or input prices. How well are practices adapted to addressing normal seasonal changes? Are practices available and being used for moderate variations? Are practices available and being used for extreme events? Finally, indicators are proposed for the impact off-site in lost nutrients, pesticides and air contaminants and the potential of the crop field as habitat for off-site fauna and flora.

They proposed that for each theme criteria could be developed to characterize the potential to intensify ecologically. Among the seven systems studied, two had poor radiation capture, three had unprotected soil, four had underuse of internal nutrient cycling and four had over- and misuse of pesticides. The authors proposed that ecological integration of processes could be achieved by testing practices which address multiple limitations with a single practice such as green manure and residue redistribution. However, much opportunity remains to improve indicators, develop more effective field methods and calibrate the criteria for the potential for ecological intensification.

De Barros et al. (2009) characterized seven contrasting export banana production systems based on slope, altitude, size and intensity of input use using economic and 'emergy' analysis. Emergy analysis calculates the energy from different sources, both natural and fossil, used to generate a product, in this case banana. The natural energy used in the calculations included sunlight, evapotranspiration, wind, organic wastes and top soil loss, while purchased inputs

**Table 3** Criteria for the identification of banana production system improvement

| Analytical tool | Possible calculations |
|---|---|
| *Resource efficiency* | |
| Crop production/ha | Yield unit (bunches, boxes, kg)/unit area/time |
| Yield gap | Potential yield/exploitable yield/range of grower yields |
| Radiation capture | Kg harvested fruit or Kg total plant biomass/unit incoming radiation |
| Water use | Kg harvested fruit or plant biomass/mm irrigation water<br>Kg harvested fruit or plant biomass/mm transpiration<br>% rainfall or total water/ha converted to plant biomass |
| Nutrient conversion | Kg harvested fruit or plant biomass/kg nutrient applied |
| Losses/unit resource | Lodged stems/ha/year<br>Reject bananas (size, bruises, cuts, deformation, pest damage)/total harvest |
| Pesticide use efficiency | Losses or incidence/unit biocide<br>Terrestrial ecotoxicity potential, Human toxicity potential |
| Profitability | Net earnings/box, hectare, returns on management and fixed/variable costs |
| Labor use (family, men/women) | Days labor/ha or box, Labor hours/practice |
| *Indirect Ecological efficiency on-site* | |
| Soil protection | % bare soil |
| Nutrient balance | Nutrients applied/nutrients taken off in harvest |
| Crop and other biomass recycling | Kg crop biomass returned/m2/year<br>% crop biomass strategically placed to protect soil and facilitate nutrient uptake |
| Organic matter cycling | Available carbon<br>Measures of bacteria/fungi ratio |
| Water cycling | Infiltration<br>Water holding capacity |
| Soil as habitat for roots | Soil depth, soil drainage and water table, Soil aggregate stability |
| Biological population regulation | Free-living nematode indices |
| Above and below ground carbon fixing | Accumulated carbon/ha<br>% soil carbon by soil depth |
| Pests and diseases losses | Attributable yield losses/ha, Losses or incidence/unit biocide<br>Measures of incidence and severity |
| Natural control potential (Predator, parasitoid capacity) | Availability of habitat for predators, Extent of use of toxic pesticides |

**Table 3** (Continued)

| Analytical tool | Possible calculations |
|---|---|
| *Climate resilience* | |
| Seasonality | Practices applied to address seasonal fluctuations |
| Moderate variation in production and profits | Scenarios and practices applicable to profile of moderate variability in weather, costs and banana prices |
| Extreme effects on production and profits | As for moderate, but applied to extreme events both in abiotic conditions and input and output prices |
| *Ecological efficiency off-site* | |
| Nutrient leaching/run-off | Kg lost/ha/year, Eutrophication potential |
| Pesticide leaching/run-off | Kg lost/ha/year |
| Aerial contamination | Pesticide drift, greenhouse gas emissions, acidification potential |
| Cropping system as habitat for off-site biodiversity | Insect fauna diversity - parasitic wasps; bird diversity; plant diversity as habitat |

used in banana production were also converted into standard emergy units - solar emjoules. The natural energy used in production was quite similar for all production systems which are all banana monocrops, while both the emergy value of purchased inputs and the yield varied three fold. Emergy indices to compare across systems relate natural energy (both renewable and non-renewable) and purchased inputs. The emergy value of a unit of banana was similar across the seven systems with higher yield based on greater purchased input use on one end and reduced input use and lower yield on the other end. Environmental loading was higher for the high yield-high input systems, while the renewable energy component of production was higher for the low yield-low input systems. The better performing systems environmentally were more poorly performing economically. The authors also pointed out that in economic terms labor was much more expensive than purchased inputs, resulting in increased purchased input use to substitute labor with negative environmental consequences. This situation represents well the challenge of optimizing multiple and often contradictory objectives in seeking the improvement of banana production systems.

## 7 Using life cycle assessment (LCA) and other certification-based measurements to assess performance

Thus far we have taken a producer's or agronomist's view of opportunities to improve banana production systems. Consumers, traders and supermarkets also have an interest in the way food is produced. These concerns are captured in environmental and social life cycle assessment (LCA) and certification frameworks (Table 4). Here we will summarize and illustrate the field, while recognizing the

**Table 4** Indicators/certifications summarizing opportunities for banana production system improvement

| Dimension | Measure or calculation |
|---|---|
| Global warming potential | Quantity of greenhouse gases emitted or $CO_2$ equivalents/kg banana |
| Water footprint | Use of blue, green and grey water/box or ha/year |
| Acidification potential | Measure of generation of atmospheric acids in sulfur dioxide equivalents (1 kg $NH_3$ = 1.88 kg $SO_2$ eq) |
| Eutrophication potential | Measure of nutrient enrichment of water in nitrate equivalents (1 kg $NH_3$ = 1 kg $NO_3$ eq; 1 kg $PO_4$ = 10.45 kg $NO_3$ eq) |
| Primary energy demand | Direct and indirect use of non-renewable and renewable energy in MJ/ha or Kg banana |
| Abiotic depletion potential - fossil | Measure of decreasing availability of resource |
| Human toxicity potential | Conversion of all pesticides into standard diclorobenzene equivalents |
| Environmental ecotoxicity potential | Conversion of all pesticides into standard diclorobenzene equivalents |
| Land use | Hectares occupied in production, packaging and ripening |
| Losses as sub-products or waste | Accounts for larger amount produced than consumed and costs linked to disposal of food waster |
| Social life cycle | Impacts on five groups of people: workers, consumer, local community, society and value chain actors including health and safety, human rights, working conditions, socioeconomic repercussions, cultural heritage and governance |
| *Certification frameworks* | |
| Global Gap | 220 critical control points based often on national standards to ensure food innocuity<br>Risk reduction for supermarkets and intermediaries - label not displayed to consumer |
| Fair Trade | 35 critical control points on fair price, labor safety and ethics;<br>Consumer concern for smallholders and workers |
| Organic | 52 critical control points primarily focusing on natural or naturally derived inputs:<br>Consumer interest in production effects on environment, workers and food contamination |
| Rainforest | 126 critical control points oriented to environment, worker safety, reduced use of more toxic pesticides:<br>Consumer concern for environment |

need for a more complete review. Export bananas are the primary focus of most studies, since banana for local and national production and among neighboring countries is still largely traded informally outside the realm of certification. However, the increasing role of supermarkets in food supply chains in Asia, Africa and Latin America suggests that food safety and certification will play an increasing role (Reardon et al. 2007). Country plans for adaptation to climate change will also influence policies and field practice (Scherr et al. 2012).

A meta-analysis of studies on greenhouse gas emissions for food items placed banana among the highest of the field-grown fruits with 0.79 kg $CO_2$ eq/kg produce (Clune et al. 2016) compared to the average of all field-grown fruits of 0.50. The amount covers from the farm to a regional distribution center. Individual studies of banana varied both in total $CO_2$ and distribution along the production and supply chain. For Central America Craig and Blanco (2009) estimated greenhouse gas emissions in carbon units for a banana at 0.12 kg $CO_2$ eq from the farm to the supermarket with 29% on-farm, 9% in packaging, 36% in transport, 16% in distribution center and retail and 15% in waste disposal. Roibás et al. (2015) found that Ecuadorian conventional bananas generated 0.302 kg $CO_2$ eq/kg banana, while organic bananas required, 0.249 kg $CO_2$ eq/kg banana from production to farm gate. The higher rate for conventional banana was due to nitrogen use, while the organic bananas had higher water use due to lower yields. Across the studies, the major contributions at farm level were linked to fertilizer - emissions and energy costs for manufacture - and post-farm - cardboard box manufacture and fruit transport from field to consumer. The unit of analysis is of primary importance in LCA. All the studies above focused on established plantations for a year-long cycle. New field establishment, plantation renovation and land improvement costs are not included. Bessou et al. (2013) review the challenges associated with LCA in perennial crops. A second major area of uncertainty is the cost of fruit transport depending on end point (port, distribution center, retail) and backhaul cargo in ships. When ships return empty, energy costs are much higher (Lescot 2012). The location of the point of consumption and the modes of transportation in the supply chain also influence this component with Boston and San Francisco intermediate in cost compared to New Orleans at the low end and Minneapolis at the higher end (Craig and Blanco 2009).

Greenhouse gas (GHG) emissions are only one dimension of the life cycle approach (Table 4). Indicators are calculated for nutrients, water, land, pesticides and energy among others from production to consumer, including the processes to manufacture inputs for production, transport and marketing. A comparison of Prata and Cavendish banana production in Brazil by Coltro and Karaski (2019) quantifies for each cultivar 10 indicators commonly used in life cycle analysis (Table 5). Both cultivars are grown in monoculture with abundant fertilizer, fuel and pesticide inputs and marketed side by side in supermarkets. For indicators such as eutrophication, fresh and blue water use, terrestrial and human toxicity potential and land use, over 80% of the value of the indicator occurs on-farm. For global warming potential, acidification, primary energy demand and abiotic depletion potential (fossil), on-farm is only charged with about 50% of value followed by packing and transport.

**Table 5** Life cycle calculations for production of Prata bananas for national markets in Brazil

| Environmental indicator | Unit | Total | On farm (%) | Packaging (%) | Ripen (%) | Transport (%) | Retail (%) |
|---|---|---|---|---|---|---|---|
| $GWP_{100}$ | kg $CO_2$-eq | 0.423 | 51.2 | 7.7 | 1.0 | 28.2 | 12 |
| Acidification | kg $SO_2$-eq | 3.97E-3 | 56.7 | 92. | 6.2 | 22.8 | 0 |
| Eutrophication | kg $PO_4^{3-}$-eq | 2.27E-3 | 82.3 | 4.3 | 2.7 | 10.7 | 0 |
| Total use freshwater | Kg | 1979 | 96.0 | 0.1 | 3.9 | 0 | 0 |
| Blue water use | Kg | 971 | 91.4 | 0.5 | 8.1 | 0 | 0 |
| PED | MJ | 9.18 | 52.7 | 22.6 | 5.4 | 19.4 | 0 |
| ADP fossil | MJ | 4.67 | 44.9 | 17.4 | 1.7 | 36 | 0 |
| TETP | kg DCB-eq | 1.18E-3 | 95.5 | 4.4 | .01 | .04 | 0 |
| HTP | kg DCB-eq | 0.13 | 94.4 | 3.3 | 0.5 | 1.8 | 0 |
| Land use | $m^2$.yr | 0.855 | 97.7 | 2.3 | .01 | 0 | 0 |

GWP = global warming potential; PED = primary energy demand; ADP = abiotic depletion; TETP = terrestrial ecotoxicity potential; HTP = human toxicity potential.
*Source*: reprinted from Coltro and Karaske (2019) with permission from Elsevier.

Calculations of indicators using overall sector values as cited in the previous paragraphs and in Table 5 are useful to contrast banana production and marketing with other commodities but are less useful to identify alternatives for improving indicators for banana. Poore and Nemecek (2018), using a metadata base of 38 000 farms in 119 countries for 40 products, highlight the heterogeneity across and within crops for five environmental indicators. Top performing regions and producers show a threefold or greater difference compared with poorly performing producers or regions. They conclude that this heterogeneity provides opportunities for rapid gains, but also means that alternatives must be custom fit for each region and in many cases for each producer. Among the five indicators they deployed (GHG emissions, acidification, eutrophication, land, scarce water) certain combinations were more difficult to improve simultaneously. Reducing land use, for example, often involves increased nutrient inputs or energy use.

Pattison's (2006) study of 21 export banana farms in Costa Rica illustrates the heterogeneity among growers and the potential to improve efficiency. Growers using certain practices contributing to soil health such as strategic residue management and organic matter application applied less nitrogen but achieved higher yields. Tixier et al. (2008) used a crop model to test scenarios for yield, profitability and environmental indicators in the French West Indies and found that replanting every 3–4 years generated higher yields with production more concentrated in high price periods, but was less efficient in nematicide and nutrient use, thereby generating lower indicators for environmental impact. Chopin et al. (2016) rated 18 technologies for their contribution to sustainability in three different types of banana farms in the French West Indies - smallholder farms and large plantations on flat lands and smallholder hill farms. Their rating was based on the opinions of contrasting stakeholder groups with differing sustainability priorities. Integrated and organic production packages were seen as more applicable than specific individual practices with greater applicability among smallholders than large plantations.

These examples did not include the in-depth calculations required by life cycle analysis, but do illustrate three challenges:

1 improvement in certain life cycle indicators may be contradictory with other indicators as well as profitability and operational routines.
2 life cycle priorities may vary among stakeholders, although there is no common framework for comparison across indicators.
3 region and farm specific-diagnostics are needed to guide plans for improvement.

The approaches described so far do not consider the labor-intensive nature of banana production. Mats in export banana fields are tended at least twice weekly and in bunches at least three or more times from flower emergence to bunch harvest. Labor use is less intensive in the other banana production systems described in Table 2. Part of the environmental life cycle analysis addresses the impacts of pesticide use, for example, on humans. The impacts of pesticides on banana workers in Central America and the French West Indies are one such example. From the early 1990s onwards banana workers in Costa Rica, Philippines and Nicaragua sued Dow Chemical and Dole for damages resulting from the use of the pesticide DBCP in bananas for a decade after its use was prohibited in the United States (Bingham and Monforton 2013). Similarly banana workers in Guadalupe and Martinique sued for damages from problems of sterility resulting from the use of the nematicide chlordecone in banana production. Significant ground water contamination and ongoing health risks continue even to the current day (Multigner et al. 2007).

Social LCA methodologies propose addressing the socioeconomic conditions particularly of workers in production and packing, the local community, consumers and the society at large (Table 4). While environmental impacts are largely negative, social benefits can be both positive and negative. The compliance with labor laws - working conditions, hours, wage levels - is taken into account as well as racial, migratory and gender issues. Two approaches are taken. The performance reference point approach uses national and international standards often captured in international labor codes, while the impact pathway links groups of people with characteristics of the banana production, packing and marketing system (Sala et al. 2015). An example is organic banana replacing cotton production on the northern coast of Peru with numerous socioeconomic improvements.

Certain aspects of LCA like carbon and water footprint are deployed by value chain actors on a stand-alone basis (Lescot 2012; Bockel et al. 2018), while others are used more in global comparisons of food production as we have seen (Poore and Nemecek 2018). Certification frameworks driven by concerns from consumers and marketers about the way food is produced also address many elements from life cycle analysis, although with differing emphasis depending on the specific framework. A study by True Price/TruCost (n.d.) of the external costs of banana production looked at both social and environmental costs, establishing that a box of bananas an additional $6.70 above on-farm costs of production. Sixty percent were social costs such as insufficient earnings to laborers and small growers, social security and health and safety, while 40% were environmental costs such as land use, water depletion, emissions linked to climate change and pollution of water, air and land. Large differences were found among the four countries in the study - Peru, Ecuador, Colombia and Dominican Republic. FairTrade had much lower social costs. The study concluded that labor and input efficiency and improved yields with current inputs could be achieved by benchmarking the better producers as a guide to overall sector improvement.

The International Trade Center (ITC), operating with a mandate from the United Nations and the World Trade Organization, maintains a website which details over 200 different certification schemes (www.standardsmap.org) with a descriptive framework organized under five categories - environment, social, management, quality and ethics. Seventy-one of those listed by the ITC apply to banana with the five categories being divided into 24 subheadings and 268 specific themes. These 268 specific themes are 40% environmental, 32% social, 15% quality, 11% management and 3% ethics related. Staver (2018) reviewed five of the most common certifications - GlobalGAP, EU Organic, FairTrade, Rainforest Alliance/Sustainable Agriculture Network and SA8000. He concluded that, while certification is essential to access many export markets, including multiple labels for a single market, the data collected to show compliance are not designed to support grower decision-making on production and cost management. The requirements for certification generally create awareness about aspects of production which growers may not address in the absence of certification. However, the routine of documentation and recordkeeping is often viewed as an outside imposition rather than a useful management tool. Staver (2018) argues for refocusing the huge effort in recordkeeping for certification toward grower enterprise and ecosystem management to improve productivity, efficiency, profitability and resilience.

## 8 Identifying priorities for improving banana production systems

In this and the previous sections we have undertaken a broad review of possible themes and perspectives to identify and prioritize the potential for banana production system improvement. Yield/hectare/year continues to be a standard metric. However, we have illustrated that numerous other production resources are equally important. The banana field from an ecosystem perspective adds further potential entry points for improvement, while the LCA facilitate an integrated analysis beyond the field and farm. Across these different optics on banana system improvement, we have a wealth of indicators which can be deployed to both characterize the current situation in any of the production systems in Table 1 and facilitate monitoring of improvement in response to the implementation of alternatives. Such efforts in benchmarking and the development of strategies for system redesign and/or continual improvement are fundamental to production system improvement. Although the definition of checklists of good agricultural practices are at the heart of certification frameworks and climate change adaptation programs (Staver 2018), they offer little guidance to individual growers facing their own set of constraints and opportunities.

In the following sections, we refocus our discussion on options for banana production system improvement applicable across the cultivars and production systems covered in Table 1. With the availability of new cultivars still highly limited, we propose six potential entry points which may provide short-term alternatives for more productive, more efficient, more resilient or more profitable systems (Table 6). In line with a focus on grower ecosystem and enterprise management, we will identify indicators in each case for characterizing the current status of fields and for monitoring the effectiveness of alternative practices being implemented.

## 9 Ecological intensification as a basis for improving banana production systems

Agricultural systems have been intensified primarily by the increased use of off-farm inputs. Growers have focused their efforts for higher yields and profits on more and more effective use of these inputs. However, in all cropping systems, both high and low input, ecological processes continue to operate, whether the grower and his or her technical support take them into account or not (Phelan 2009). Here we focus on five levels in the ecological hierarchy for improved banana production, productivity and resilience (Fig. 1). Making agriculture more sustainable has become the major goal with wide consensus. Nobody proposes to sacrifice the future for short-term gains. The consensus

**Table 6** Potential entry points for banana cropping system improvements and a brief summary of possible indicators

| Improvement approach | Ecological or economic process | Possible indicators for diagnostics and monitoring |
|---|---|---|
| Positive selection - suckers from superior mother plants | Variability in traits from plant to plant under similar conditions | Plant performance tracking for plant height, bunch size, finger size and height of follower sucker at flowering and bunch interval |
| Timing and distribution of bunches during year | Time of year and variability or synchronization of flowering and bunch harvest mat to mat in a field due to planting date, type and uniformity of planting material, management of suckers, bunch interval and replanting frequency | Quality, size and uniformity of planting material; Length of bunch interval; Month-by-month bunch production under current practice; Seasonality of price fluctuation and weather phenomena |
| Mat density and associated understory vegetation | Conversion of light and water into different types of plant biomass with different benefits which contribute to overall field performance | Light and water capture by banana and current associated crops; Benefits from associated vegetation - soil cover, biomass production, food and income, habitat role and populations of beneficial organisms |
| Overhead tree shade | Balance between additional resource use for tree growth and tree microenvironment effects on banana plant performance, diseases and contributions in organic matter | Effect of tree resource use on resource availability, banana crop cycle length, microenvironment and disease levels; Seasonal contribution of trees to leaf litter and pruned material |
| Integrated management - nutrient cycling and soil health | Nutrient input and output, plant uptake, decomposition pathways and nature of losses, impact on soil parameters linked to productivity, sustainability and resilience | Input-output calculations, crop residue placement and soil cover, indicators of soil structure, soil biology and carbon |
| Integrated management - weather variability | Timing and nature of crop system design and management practices to provide resistance, buffering and recovery options for crop performance under increasingly variable weather | Diagram of annual management practices and possible weather events; Inventory of weather events with estimation of crop productivity effects; Regional mapping of weather events affecting crop performance |

breaks down when specific technologies are advocated with few indicators and the long-time horizon which sustainability implies. We propose that the improved synchronization of production and ecological processes which underlies ecological intensification can be addressed in a single or several cropping cycles in an ongoing process that starts in the short term. In addition, grower-friendly indicators can be identified to track progress. Our overall goal (multiple integrated objectives) is to increase yields, resource efficiency and resilience of banana production systems by more effectively integrating ecological and production processes.

Here we will focus spatially on the crop field and temporally on the cropping cycle from banana planting to next banana planting in the same field. We will examine approaches at five levels - individual banana plant, population of banana plants, community of different species and their interactions such as plant-plant and food web and ecosystem processes reflecting abiotic factors and their interaction with the community of organisms (Fig. 1). Ecosystem scales beyond the field will be highlighted in the final entry point on moderate and extreme weather events. For each level we will characterize current

| Ecological level | Individual plant: single mat | Population of banana plants in field | Community: all organisms in field | Field ecosystem: | Multi-scale ecosystems beyond the field |
|---|---|---|---|---|---|
| **Improvement approach** | Superior plant selection for suckers | Planting material quality and uniformity; Planting date and replanting frequency Plant density Desuckering to time harvest | Crop interplanting Tree overstory Plant habitat for beneficial organisms | Ecosystem process integration - soil health Nutrient cycling | Ecosystem process integration - weather variability of many types of weather and climate change and variability |

**Figure 1** Contrasting spatial dimensions of ecological approaches to banana system improvement. Photo CIRAD: Endure Guide Number 3: From Science to Field Banana Case Study.

practice with examples, and then discuss one or several options for banana cropping systems improvement, highlighting the relevance, the experimental and practical evidence and possible indicators for use in benchmarking and continual improvement. Numerous reviews are available on banana pest and disease management which we will not discuss here, although there is great potential for ecological intensification.

Before looking at specific practices in ecological intensification, we need to highlight the abiotic contexts of banana production. Although bananas originated as a food crop in the wet tropical lowlands, with appropriate management, they prosper in a wide range of climatic zones (Table 7). Zones such as Kisangani, Democratic Republic of Congo, with uniform year-round temperature of 25°C with no dry or extremely rainy months require few special management practices and produce a relatively stable month-by-month production. Higher altitude sites such as Armenia, Colombia have similar uniformity of conditions throughout the year, although the average annual temperature is 4–5°C lower than Kisangani due to the higher altitude. This temperature difference generates a slower leaf emission rate for Armenia, 21 leaves/year compared to 37 for Kisangani. Other sites in both the tropics and subtropics have seasonal variation due to rainfall and/or temperature which affects leaf emission rate and therefore the seasonality of production. During the winter season in the subtropics such as Uttar Pradesh and Corupá, leaf emission and rates and bunch development decline, while in Uttar Pradesh, the high temperatures in the months prior to the start of the rains may also slow down plant development.

Behind the averages in Table 7 is also year to year variability in temperature and rainfall. For example, Pichilingue, Ecuador, under the effect of El Niño and La Niña varies in annual rainfall more than 1000 mm both above and below the average (Fig. 2). Average annual temperatures also increase or decrease

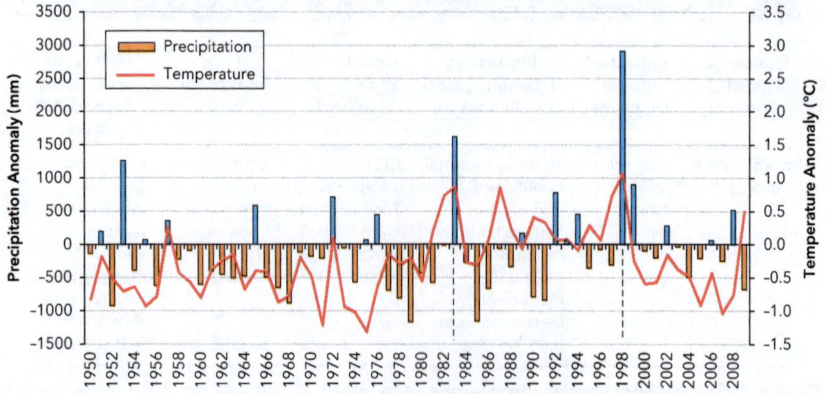


**Figure 2** Variability in average annual rainfall and temperature for Pichilingue, Ecuador. The average annual values for the period in the graph are 2050 mm and 25°C.

**Table 7** Climate parameters for six important production zones for banana

| Site | Total annual precipitation (mm) | # dry months <60 mm | # wet months >300 mm | Annual average temperature (°C) | Temperature coldest month (°C) | Temperature warmest month (°C) | Annual water deficit (mm) |
|---|---|---|---|---|---|---|---|
| Kisangani, DRC | 1800 | 0 | 0 | 25 | 25 | 26 | 794 |
| Pichilingue, Ecuador | 2050 | 6 | 4 | 25 | 24 | 26 | 844 |
| Armenia, Colombia | 2040 | 0 | 0 | 21 | 20 | 22 | 370 |
| Bujumbura, Burundi | 1080 | 4 | 0 | 19 | 19 | 20 | 996 |
| Uttar Pradesh, India | 1210 | 7 | 1 | 25 | 15 | 33 | 1300 |
| Corupá, Brazil | 1520 | 0 | 0 | 18 | 16 | 25 | 358 |

*Source*: adapted from Calberto et al. (2015).

with rainfall. Banana growing areas of Ecuador also experience increased cloud cover and cool temperatures at the start of the dry season which varies in intensity from year to year. Growers use special management practices in an effort to maintain plant growth during this period each year. As we illustrate strategies to improve banana production systems in the next sections, we will reference seasonality of climate which may be important in certain outcomes.

## 10 Improvement options: planting material selection, planting date, replanting frequency and desuckering

The individual plant ecologically in a banana field is the mat originating from a planted sucker made up of successive generations of suckers which develop into new flowering stems. Suckers from existing mats are also extracted for new plantings. This vegetative propagation is considered to generate high uniformity from plant to plant. However, data from different contexts on mat to mat performance suggest that variability is also common, variability which may offer options for banana cropping system improvement. For example, Dépigny et al. (2019) weighed bunches from plantain fields around Njombe, Cameroon, and found high between-field variation which was attributed to differences in field history, use of purchased inputs and farmer skill (Fig. 3).

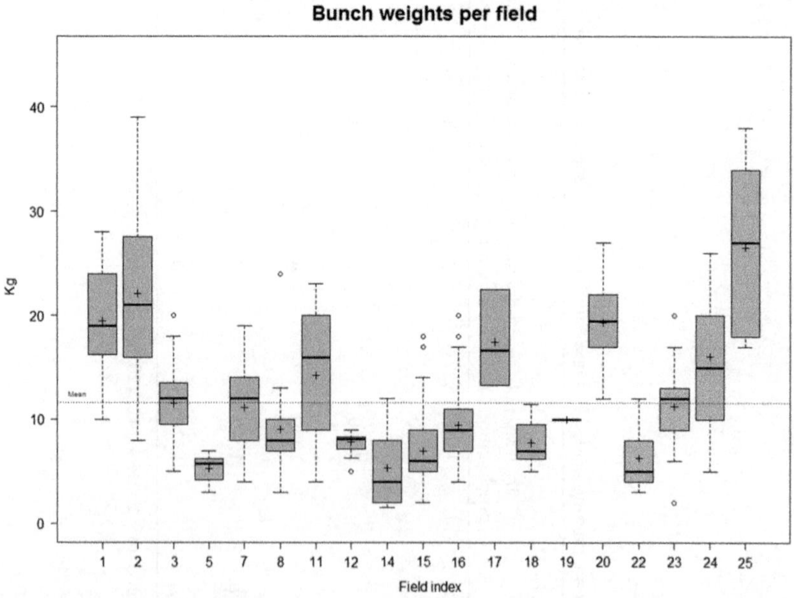


**Figure 3** Variability in plantain bunch size for 25 plantain fields in Njombe, Cameroon. Average bunch weight is marked with (+) in the boxplot. The average bunch weight of all harvested bunches is the horizontal dotted line. Outlier values are represented by small circles. Reprinted from Dépigny et al. (2019) with permission from Elsevier.

Within-field variability is also higher in some fields than others, probably due to different proportions of plantain cultivars, underlying soil variability and uneven distribution of associated trees and associated annual crops (Fig. 3).

In Ecuador, bunch weight was documented on four farms producing organic export bananas (Fig. 4) (Staver 2018). Again field to field differences in average bunch weight are evident. Within fields, a two to threefold difference can be observed between the largest and smallest bunch, in this case with relatively high cultivar uniformity and within-field management. For plantains from three test plots in Nicaragua, within-field variability for finger number (Fig. 5) and bunch weight was high (Staver et al. 2016). Similar variability was found in other plantain cultivars in the same study in Dominican Republic and Ghana. These experimental plots were planted with uniform planting material

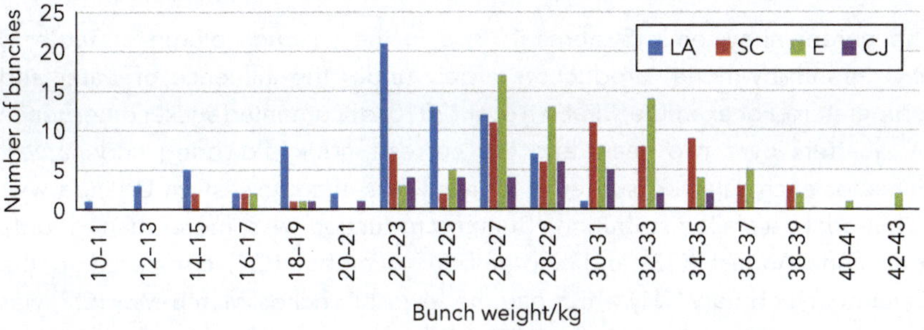


**Figure 4** Bunch size distribution for four banana farms in Ecuador with perennial Cavendish banana for export.

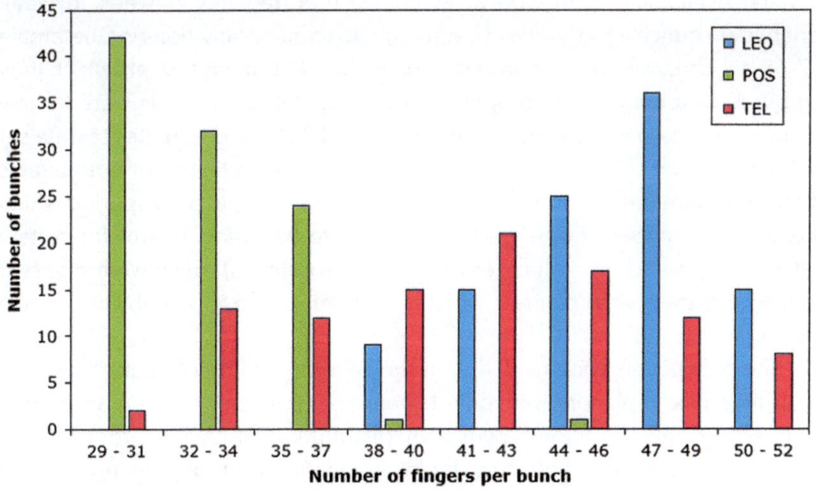


**Figure 5** Frequency of plantain bunch size for three plantain fields in Occidente, Nicaragua. Source: adapted from Staver et al. (2016).

of a single plantain cultivar based only on the size and apparent health of the sucker. This variability in bunch size simultaneously is responsible for average yields below the potential for a given management routine, but is also the source of potential improvement in banana cropping system performance through bigger bunches and higher total production.

Bananas are frequently characterized as producing year-round, but rarely in a single locality are bunches produced in equal proportions month-by-month. The pattern of month-by-month distribution of bunch production for the different systems (Table 1) can be separated into two groups:

- perennial plantings only replanted infrequently.
- replanting with a variable frequency from annually to every 2–4 ratoon crops.

The perennial systems (Systems 1, 3, 5 in the second column of Table 1) have month-by-month production largely under the influence of water and temperature. For example, Birabwa et al. (2010) documented sucker emergence by quarters over two years and the corresponding flowering and harvest dates for each selected sucker in farmer fields in southwestern Uganda with a bimodal rainfall distribution. Bunch production was more than double from June–August (Q3) and September–November (Q4) compared to the December–February (Q1) which had the fewest bunches. March–May (Q2) was an intermediate quarter in terms of numbers of bunches harvested with a third to a half higher more bunches than Q1. Suckers emerging in Q3 and Q4 had shorter total time from emergence to bunch production, primarily a function of water availability. Fortescue et al. (2011) reported higher week to week variability in bunch emergence in perennial Cavendish plantations in Guatemala (37–65 bunches/week) compared to Panama (37–45 bunches/week) which they attributed to a long day photoperiod response, since the fields were largely under adequate water conditions. Turner et al. (2016) used a multi-site data set for local plantain cultivars grown at different altitudes to tease out both effects of cool temperatures during the mid-vegetative period and photoperiod which affected the number of emerged leaves prior to flowering. Corms for certain cultivars originated from altitudes (temperature regimes) which were different from where they were planted which also contributed to variability in cycle length.

The production systems 2, 4, 6 in the second column of Table 1 operate through regular replanting with only 1–4 harvest cycles prior to field uprooting. In these systems the month-by-month distribution of bunch production is influenced by the date of planting, the type and uniformity of planting material and the month-by-month variation in abiotic factors such as water, temperature and day length. In selected farms in Nicaragua and the Dominican Republic highly synchronized bunch emergence and harvest are achieved by the use of microcorms grown out in nurseries where they are sorted by plant size (see Fig. 6a). Blocks of very similar-sized plants result in nearly simultaneous bunch emergence (Fig. 6b). When planting material is less uniform both in size and physiological status, the period of bunch emergence lengthens. No published studies actually report the size and quality of planting material used by farmers as measured at the time of planting, although poor quality planting material is often cited as one of the major limitations on banana production, especially the effect on bunch size and the number of ratoon crops for each mat (Tenkouano et al. 2006; Robinson 2000). Experimental data for plantain in upland Colombia showed that the time to flowering was reduced progressively by a maximum of two months from 18 to 16 months when corm size increased from 0.5 to 4 kg (Belalcazar et al. 1998). Suckers may also show differing degrees of infestation with nematodes or weevils affecting plant vigor, survival and bunch size (Jacobsen et al. 2019). High variability in sucker size and quality leads to dispersed bunch emergence even in the first harvest, which may be more appropriate for home consumption or local marketing.

Depending on local climate and production technology, the choice of planting date may be determined by different factors. In rain-fed areas with long dry seasons, farmers plant under the control of nature at the start of the rainy season. Throughout West Africa, this natural synchronization of the production cycle results in a glut period with very high supply and low prices and an off-season of low supply and high prices (NIHORT 2008; Lopatka et al. 2008). In other zones, growers choose planting dates to target high prices or to ensure that their fields are in less vulnerable vegetative growth during the periods of high winds, typhoons or hurricanes. In the Red River basin around Hanoi, Vietnam, growers use highly uniform planting material of Pisang Awak planted at a time of the year to ensure that bunches are ready to harvest for

**Figure 6** (a) row of microcorms in nursery sorted by size; (b) high-density annual planting with simultaneous flowering resulting from the use of highly uniform planting material.

New Year celebrations in China when prices spike (Fig. 7). In Taiwan, growers adopted annual replanting with tissue culture plants in certain zones to ensure that banana fields are in vegetative growth during the typhoon season. Such plants tolerate high winds better than mats with bunches. In zones of annual replanting in the Dominican Republic, growers choose planting dates both to avoid bunch production during the hurricane season and to target periods of higher prices.

In these production systems lasting only 2–4 harvests, the period of bunch emergence becomes increasingly dispersed with each successive ratoon cycle. Sucker selection is used in certain production systems to maintain a concentrated period of bunch emergence. In the Red River basin around Hanoi, Vietnam, growers of Pisang Awak mentioned in the previous paragraph desucker completely until the appropriate moment that suckers will again produce bunches to coincide with the next New Year price spike (Fig. 7b and c). Plantain growers in the Dominican Republic, using production system 6 (Table 1), desucker for sucker uniformity to achieve a more concentrated crop in the first ratoon crop. In both Vietnam and the Dominican Republic fields are often uprooted and replanted after two harvests, primarily to again target harvests for certain periods during the year.

**Figure 7** (a) highly uniform stand of cultivar Pisang Awak in Red River Delta, Vietnam planted for highly concentrated bunch production for Chinese New Year market; (b and c) complete desuckering used to lengthen production cycle to schedule ratoon bunch production for one year later.

The number of harvest cycles prior to uprooting and replanting depends on many different factors. Disease build-up is a common cause in production zones of Fusarium-susceptible cultivars like maca or silk in Brazil, Gros Michel in Colombia, Iholena in Peru and Pisang Awak in Asia. Depending on disease inoculum in the soil of the field to be planted, the disease presence in corms to be planted and crop management factors, a field may produce up to three harvests. Dita et al. (2018) highlights the challenges of designing production systems of Fusarium-susceptible cultivars in Fusarium-susceptible soils. In West and Central Africa under rain-fed conditions, yield decline in plantain after the second harvest due to multiple factors including nematode and weevil buildup and fertility decline lead to field abandonment back to bush fallow or crop abandonment with a focus primarily on a cocoa perennial crop. Belalcazar et al. (1998) documented yield decline in an experiment with plantain. The percentage of plants harvested declined from 90% in the first harvest to 60% in the third harvest, a major factor in yield decline. In more intensified systems, replanting after only one or two cycles may be motivated by maintaining bunch production targeted to a period of high price or to avoid specific catastrophic weather events.

This brief overview of the current situation focused on mat performance provides a basis to propose four opportunities to address the improvement of banana production systems at the individual plant level which we will describe briefly – selection of superior mother plants as a source of suckers, quality and uniformity of planting material and planting date/replanting frequency and sucker selection.

## *10.1 Selection of mother plants*

The first opportunity to improve cropping system performance at the level of the plant is the selection of superior mother plants as a source of planting material. A plant with a small bunch or a long interval between bunches uses a similar amount of sunlight and water as a plant producing bigger bunches at shorter intervals. The strategy has been harnessed in Cavendish tissue culture. Labs have selected individual plants with superior traits as their source material. The sucker from any mat can be multiplied into 1000 quite uniform plants, but these plants may not represent the best performing mats of the cultivar.

The strategy carried out to develop superior performing Cavendish cultivars for growing conditions in the Canary Islands is a variant of this approach bringing together the efforts of growers, scientists from ICIA and a tissue culture company CULTESA. Growers selected and multiplied interesting plants from among their fields over numerous cycles, largely Dwarf Cavendish. These were then converted into agronomic trials carried out by ICIA scientists. The best performing lines became varieties like Laja, Palmerita, Ricasa and

Brier which were multiplied through tissue culture (Maria Grajal, ICIA, personal communication) and have become the basis for yields near 100 t/ha, among the highest in the world.

A selection process was tested in plantain to determine if the bunches produced by suckers from mother plants with an above average number of fingers had more fingers than similar suckers chosen only for sucker visual quality (Staver et al. 2016). In only one round of selection in nine separate trials with five different cultivars, bunches from suckers selected based on finger number always had more fingers (Table 8). The gains were only a couple fingers for cultivars like FHIA 20 or 21 or Cemsa ¾ which have been introduced relatively recently. Local cultivars, macho × hembra in the Dominican Republic, Apem and Apantu in Ghana and dwarf plantain in Nicaragua showed bigger gains of 5–15 fingers or more. To establish the plots, local scientists proposed the criteria for a superior plant and then visited well managed fields to identify superior plants. For additional rounds of selection, the authors propose to select suckers from plants with the biggest bunches from the first 50% of the field to flower in test plots. In this way, both bunch size and earliness are deployed in superior plant selection. Other criteria such as plant height or height of sucker at flowering could also be integrated into the selection process. The approach may provide the basis to improve productivity of cultivars such as red banana, Gros Michel, bluggoe and Pisang Awak which are not as widely planted, but are still of commercial interest.

## 10.2 Planting material

The second area for improved banana cropping systems is the quality and uniformity of planting material. The technical specifications for different multiplication techniques have been summarized by Lescot and Staver (2010) with an illustrated guide in four languages (Staver and Lescot 2015). See also Tenkouano et al. (2006). The problem of degeneration of the quality of planting material due to the build up of pests and diseases is particularly important, since infected planting material is a primary pathway for pest and disease movement from the field which is the source of the planting material to the new field being planted. Although tissue culture laboratories have become the source of planting materials for Cavendish production for national markets in Asia and large-scale export Cavendish production globally, small growers across all continents still depend primarily on suckers taken from fields planted for production to plant new fields. Jacobsen et al. (2019) suggest that such growers could improve sucker quality by managing their production fields not only for bunches, but also for high quality suckers. Alternative useful practices will depend on cultivars, production techniques and the pests and diseases present. Once suckers have been extracted

**Table 8** Improvement in number of fingers with selection for bunch size

| Site | Cultivar | Selected based on finger # | | | | No selection criteria | | | |
|---|---|---|---|---|---|---|---|---|---|
| | | Days to flower | # fingers | Bunch weight | Sucker ht at flower | Days to flower | # fingers | Bunch weight | Sucker ht at flower |
| Rivas, N* | Cemsa 3/4 | 303 | 50.3 | 18.7 | 78 | 298 | 48.6 | 18.8 | 66 |
| Leon, N | Dwarf | 315 | 52 | 23.1 | ND | 337 | 46 | 20.2 | ND |
| Telica, N | Dwarf | 246 | 44 | 14 | 100 | 272 | 41 | 12.5 | 84 |
| Posoltega,N | Dwarf | ND | 48 | 16.9 | ND | ND | 33 | 11.2 | ND |
| Cibao Central, DR | False horn | 279 | 45 | 15.5 | 79 | 281 | 38 | 13.1 | 81 |
| Cibao Central, DR | FHIA 21 | 331 | 70 | 21 | 94 | 311 | 68 | 20 | 96 |
| Assin Adaw., G | Apantu | 274 | 40 | 11 | 75 | 292 | 34 | 10 | 52 |
| Fumesua, G | Apantu | 272 | 42 | 13 | 85 | 279 | 31 | 10 | 75 |
| Fumesua, G | Apem | 359 | 125 | 19.9 | 148 | 359 | 63 | 10.9 | 47 |

Countries where sites are located: N = Nicaragua; DR = Dominican Republic; G = Ghana.
ND = no data

from well managed fields, diverse practices such as peeling can be used to reduce risk of pest and disease transmission to new fields. To increase the uniformity of planting material, suckers can be sorted by size. As shown in Fig. 6a, microcorms can be grown out in nurseries and sorted prior to planting to improve within row uniformity. The time from planting in the field to flowering is also shorter compared to field planting of normal-sized suckers. Boiling suckers has also been tested (Tenkouano et al. 2006). A relatively simple strategy to improve sucker quality is to use recently established fields as a source of suckers before pests and diseases accumulate in mats rather than the oldest fields.

## 10.3 Planting/replanting dates

The third opportunity to improve banana cropping systems which may be very site specific is the planting date and the replanting frequency. In each specific site, information must be mobilized to review the feasibility of improving cropping system performance based on these entry points. Useful information includes documentation of seasonal price fluctuations and a calendar of weather events with destructive potential for established plantations. Tixier et al. (2007) used the SIMBA model and price fluctuations in the European market to determine the appropriate planting date and replanting frequency for growers in the French West Indies to maximize income. The optimum planting months generated 10–36% increased income compared to worst months.

## 10.4 Sucker selection

The final management approach at the plant level is sucker selection to manipulate timing of bunch production. In a focus on yield maximization/ha/year, the largest or most advanced follower sucker would be selected early to shorten the time interval between bunches in a single mat (Eckstein et al. 2000). However, Birabwa et al. (2010), based on their survey of the month of sucker emergence and the corresponding month of bunch harvest in perennial matooke fields in Southwest Uganda proposed that a delay in desuckering and the selection of smaller less vigorous suckers in late January and February would lead to increased bunch harvest during the high price Christmas season. Similarly Robinson and Nel (1986) working with Cavendish in subtropical South Africa, highlighted that choosing suckers at 10 months rather than earlier could be used to increase bunch harvest in the fall when harvest is normally low and prices higher. We have also seen examples earlier that the timing and the criteria of sucker selection can be used to concentrate bunch production in a shorter period by choosing suckers of a similar size as early as possible rather than leaving the biggest sucker on each mat.

To finalize this discussion of improvement options at the level of individual mats and suckers, we suggest the specific indicators to apply first to determine the current status of an existing banana cropping system and then to measure in the testing of possible improvement (Table 6). The method applied by Birabwa et al. (2010) which focuses on the seasonality of bunch production tracing the date of sucker emergence and the date of flowering and bunch harvest provides information both about bunch variability and the interval between bunches. Such a study could be applied in both perennial plantations as well as newly planted fields and should be paired with detailed weather data from the same sites, including irrigation rates and frequencies. Van Asten et al. (2011) correlated accumulated rainfall over the cycle of a sucker to the size of the resulting bunch, an analysis which could be combined with temperature as a useful diagnostic approach. A second indicator is the characterization of suckers at the moment of planting – primarily size and presence of evidence of pests and diseases. Jacobsen et al. (2019) propose that the quality of planting material used to establish a field should be compared with the quality of the suckers extracted from the field for planting elsewhere as an indicator of the trend in seed health or degeneration. In addition, seasonal fluctuations in bunch prices can provide an orientation about the potential for targeting better prices by shifting planting date, replanting frequency, desuckering routines or type and quality of planting material.

## 11 Improvement options: mat density and associated crops

### *11.1 Mat density*

The mat density across the diverse categories of production systems in Table 1 varies from less than 300 to nearly 3000 plants/hectare. Within each production system high variability has also been documented. Across three zones of highland banana production in Uganda (System 3 in Table 1), Wairegi et al. (2010) found that plant density was concentrated in the range from 1100–1400 mats/ha, but densities from 500 to above 2000 mats/hectare were found. Dépigny et al. (2019) counted a range of 400 up to 1600 mats/ha in plantains with mixed cropping in the Nyombe region of Cameroon, while in the plots set up in collaboration with growers by Staver et al. (2016), plantain densities varied from 1700–2800 mats/ha. For export organic banana in 13 fields in the three main exporting countries, Dominican Republic, Peru and Ecuador (System 5 in Table 1), densities varied from 1400 to 2800 mats/ha (data not published).

The mat densities in the previous paragraph are found with diverse crop associations and ground covers, although not always taken into account in the study. Dépigny et al. (2019) included the presence of annual and perennial crops in their field survey, finding from 0 to 15 with an average of 5 for annual intercrops and from 0 to 4 with an average of 1 for perennial crops. Wairegi et al. (2010) did not include associated crops in their survey, although Ndabamenye et al. (2012) studying a similar system in Rwanda noted that the presence of intercrops was a factor in mat density. The plantain and organic banana systems were market-oriented without intercrops and primarily bare soil, although leguminous cover crops, primarily tropical kudzu, are being promoted in Ecuador. The predominance of flood irrigation in smallholder organic banana in Peru and the Dominican Republic limit the testing of cover crops into existing fields.

Three issues need to be considered in characterizing the potential for cropping system improvement at the level of the banana mat density:

1 the interaction of banana density with components of banana yield.
2 the interaction of associated vegetation with banana yield.
3 the potential of associated vegetation as a host for beneficial and pest or disease organisms.

Banana density in monoculture is highest for fields replanted annually, intermediate when the field will be harvested 3–4 times and lowest in perennial fields. As densities increase past a critical threshold in each of these three production systems, the length of the time to flowering and the interval between bunches increases. At extremely high densities in annual planting above 3300–4000 mats, the percentage of plants producing bunches also declines. Plants which are smaller or suffer a setback are quickly overtopped by neighboring plants and are further delayed. Belalcazar and Cayón (1998), summarizing studies on high-density planting, show that time to flowering may increase from 10–30% and % harvested plants may decline from above 90% to only 80% or less at higher densities in the first harvest. Rosales et al. (n.d.a) summarized experience from Latin America and the Caribbean taking into account both cultivar height and luminosity of the zone. For dwarf cultivars in zones with abundant sunlight, densities up to 4000 plants/ha are viable, although for more average sites, the maximum density should only be 3200 plants/ha. For tall varieties the range in densities is only 2500–3300 plants/ha. They emphasize the importance of uniform planting material and caution that at high densities finger size may be reduced.

For perennial plantations, Turner (1998) suggests that any leaf area index above 5 in export Cavendish would not lead to additional production under conditions of abundant growth resources. Robinson and Nel (1989) compared 1000, 1250, 1666 and 2222 plants ha in a subtropical site. Although yields/ha were highest at the highest density, bunch production was delayed and more spread out, indicating more plant to plant competition. On an annual yield basis, the authors suggested that 1800–2000 mats/ha with a leaf area index of 6 would be the extrapolated optimum density. However, citing diverse practical factors, they recommended the intermediate density of 1666 mats/ha with a leaf area index of 5. The leaf area index of 5 cited by Turner is the minimum leaf area to achieve maximum light capture to convert to yield under conditions of abundant growth resources. However, Robinson (2000) pointed out that if other growth resources are less than optimum, then lower leaf areas may be recommended. On low vigor plots, densities of greater than 3000 mats were needed to achieve LAI of 5, but excessive plant to plant competition reduced individual plant growth and bunch quality. Unless the cause of low vigor can be corrected, density should be reduced to give each mat more space and access to resources. This is well illustrated by the field survey of farmer fields of perennial banana in Rwanda (Ndabamenye et al. 2012) along a rainfall and soil quality gradient. Farmers use only limited amounts of organic amendments in these sites. The authors found that farmers deployed higher densities (>2300 mats/ha) in higher rainfall areas and on better soils. Plant densities were lower in drier study sites with more weathered soils (1000–1200 mats/ha). They found that bunch size was highest at low densities (>20 kg), while intermediate bunches (15 kg) were harvested in more favorable sites even at high densities. Farmers using intercropping managed their perennial banana at lower densities. This brings us to the second issue on intercrops and other vegetation.

## 11.2 Growing bananas with other crops

Bananas are grown with a wide diversity of associated vegetation which contribute to household income and diet, reduce costs and conserve soil resources. Here we will illustrate the potential for improvement with examples focusing on two windows of opportunity - during the initial establishment period of 4–5 months after planting and in established fields.

Belalcazar et al. (1998) tested numerous combinations of maize, beans and cassava during the establishment of plantain in an important coffee zone of Colombia. They concluded that bunch size was not reduced by six rows of *Phaseolus* beans or two rows of maize or cassava planted simultaneously with plantain, but the time to flowering for the first bunch was delayed by two to three weeks with the presence of associated crops and the time to the flowering of the ratoon crop was longer by 1–3 months, indicating that the associated vegetation delayed sucker development. Yields of the associated crops were only a third to a half in association compared to monocrop. Yields of associated crops were lower if planted into plantain established a month earlier and yields dropped to near zero if associated crops were planted into two-month old plantains. Weed control practices are timed to limit weed competition with associated crops and plantain is established with little weed control effort.

Cover crops which are intended to persist in the established plantation can also be planted into new banana plantations. Achard et al. (2018) found that the two species of grass covers, *Brachiaria decumbens* and *Cynodon dactylon*, mowed at two-month intervals delayed first banana flowering compared to a chemical herbicide-based control, and *Cynodon*, although it had lower biomass accumulation, also reduced finger size. Pattison et al. (2018) showed a similar delay in flowering and significantly smaller bunches for *Arachis pintoi* combined with mechanical weed control every two months compared with herbicide-based weed control. Damour et al. (2014) screened 28 broadleaf and grass cover crop species, including annuals and perennials of diverse growth habits and nitrogen-fixing capacity for use as fallow species and intercrops with banana. They characterize the species in small field plots and through literature in terms of nine functions for use in four different cropping situations in banana rotations with cover crop fallows. For their multi-purpose approach combining both open sun growth and intercropping with banana, they identified three promising species – *Dolichus lablab*, *Neonotonia wightii* and *Pueraria phaseoloides*.

Vegetation can also be planted into perennial plantations. Here we will consider two cases in which banana is the dominant crop (systems 3 and 5 in Table 1). In the Highlands of East and Central Great Lakes Africa, perennial bananas are the staple food, but are often undercropped with bush or climbing beans, especially in areas with land scarcity and high human population density. Ntamwira et al. (2014) tested improved and local cultivars of beans in farmer fields under two levels of banana leaf pruning. They found that pruning to 7 leaves did not affect banana bunch size, but improved associated bean crops yields. In two sites with poorer soils, bush beans reduced banana bunch size compared to climbing beans. In another study in Uganda (Ronner et al. 2019), leaf pruning increased available light by 15%, but did not increase climbing bean yields. However, eight leaves were left and the authors calculated that beans in less shaded areas of the fields had higher yields than in more shaded areas. High variability in banana mat density and distribution resulted in high data variability for small plot work. Extrapolation to field scale impacts was not recommended. Blomme et al. (2018) reviewed the potential for interplanting drought- and shade-tolerant crops in established banana fields in Central Africa, including root crops like elephant yam, leguminous ground covers like tropical kudzu and grain legumes like chick peas.

Cover crops in established plantations are intended to provide services such as weed control, soil protection and additional organic matter and possibly nitrogen with either no effect or a positive effect on banana productivity (Damour et al. 2015). Here we will consider some of the evidence and experience on the role of cover crops in established banana. This corresponds to one of four possible uses of cover crops in banana rotations following the framework proposed by Damour et al. (2015), although these authors do not consider the establishment of cover crops in already existing perennial banana fields.

The experiments cited above by Achard et al. (2018) and Pattison et al. (2018) both concluded that in the first ratoon harvest the banana bunch size and interval between bunches were not significantly lower in treatments with the grass or *Arachis* cover crops, although for the *Arachis* cover, bunch size on average was smaller. The overhead banana shade reduced the vigor of the cover crop. Grass cover crop biomass dropped from 0.5–1.0 kg biomass dry matter/m2 to around on 0.2 kg. Tixier et al. (2011) used data from small plots of cover crops under four shade levels and radiation measurements under banana to validate a model of available radiation under banana for cover crops. They rated 11 species on their ability to compete with weeds, to maintain viable biomass long term and to not compete with banana for nutrients. Four species were pre-selected as useful for the French West Indies – *Cynodon dactylon*, *Paspalum notatum* cv. Pensacola, *Alysicarpus ovalifolius*, *Chamaecrista rotundifolia*, but three additional species were potentially useful in radiation profiles – *Neonotonia wightii*, *Pueraria phaseoloides*, *Stylosanthes guianensis* cv. Oxley. They question whether shaded legume cover crops still are able to fix nitrogen. A number of other variables deserve further study – estimates of cut biomass after mowing, the nitrogen contribution with decomposition and the interaction between banana mat density and ground cover growth.

In the case used by Tixier et al. (2011), banana harvest is seasonal based on replanting every several years with periods of more available radiation after the concentration of bunches is harvested. In perennial plantations with only infrequent replanting, radiation conditions are more stable. In Ecuador, a minority of banana growers are managing *Pueraria phaseoloides* in established plantations (Corozo et al. 2016). Using small plants established from seed in nursery bags, growers facilitate the vegetative spread of cover from a few initial plants. Snail herbivory of recently germinated seeds limited the use of direct seeding. Those growers maintaining the cover crop suggest that weeding costs are reduced, although occasional mowing is needed to keep vines off banana stems. One grower suggested that during bunch harvest, workers needed to move more carefully, therefore reducing bruising to fruits. The interaction between cover crops and labor practices is an important point for additional practical analysis.

The third issue for banana cropping system improvement – associated vegetation as a host for other organisms – is illustrated by a study on the impact of cover crops in Cavendish monoculture on nematode communities (Djigal et al. 2012). Compared to bare soil, both grass and legume cover crops for 18 months increased the abundance of different groups of free-living nematodes – bacterivores, carnivores and root-hair feeders. This was reflected in an increase for all cover treatments in the Shannon diversity index, the structural index, indicating positive effects on the length of the food chain and the enrichment index linked to an overall increase in food resources. Plant-feeding nematodes were least abundant in grass cover crop plots. The authors concluded that in grass cover plots, the increased resources stimulated the nematode community and in particular omnivores which acted as predators to reduce the plant-feeding nematode population. Omnivores were not stimulated in legume cover plots. Certain legume covers also had significantly higher plant-feeding nematodes compared to bare soil. For example, *Helycotylenchus multicinctus* levels were higher in *Pueraria*, although *Pratylenchus coffea* was present in both *Paspalum* and *Pueraria* and absent in the other species.

The possible contradictory role of plant diversity in increasing plant parasitic nematodes, while also increasing food resources for soil organisms in general was also proposed by Poeydebat et al. (2017). They sampled 85 plots in Costa Rica and Martinique from monocrops to highly diversified agroforestry. Increased diversity in the low growing plant stratum improved soil biological properties. However, the damage from certain parasitic nematodes to the roots of a susceptible banana cultivar planted as a standardized indicator plant also increased with the diversity in low growing plant stratum and with diversity in the *Musa* cultivars present. The four different plant parasitic nematodes had different responses to diversity, attributed by the authors to their different life cycles. They recommended that plant diversity should be promoted, but for generalist pests, care must be taken to avoid the incorporation of alternate hosts which contribute to the increase of species like *Radopholus similis*, the principal parasitic nematode of banana.

The role of associated vegetation may also be hypothesized in predator and parasitoid diversity of banana bunch pests like flower and red rust thrips, scale and mealy bugs. Arias et al. (2018) documented the presence of parasitoids, predatory mites, predatory stink bugs and entomopathenogenic fungi and their effect on population levels of *Chaetanaphothrips signipennis* or red rust thrips. Habitat augmentation has been proposed in other crops such as sugar cane to increase the availability of nectar or pollen as food, alternative prey or hosts and shelter (Roy et al. 2019). The use of insecticides or fungicides, even organically certified, can also have detrimental effects on these organisms of natural biological control offsetting any positive habitat effect.

To finalize this discussion on mat density and associated crops, we highlight indicators which can be used to characterize current banana cropping systems at the level of mat density and the role of associated vegetation in diverse services (Table 6). A full cropping system cycle should be analyzed from the moment new banana planting is undertaken until when banana is again replanted in the same field. The challenge is to analyze the effectiveness of established leaf area in light and water capture and conversion to biomass (Staver et al. 2018). Periods when light and water are available and temperature is favorable, but leaf area and crop growth are below potential can be observed initially by taking monthly transects at noon on a sunny day in banana fields. Visual estimates can be made of sunlight captured by the banana leaf canopy versus sunlight reaching lower story vegetation or bare soil. Banana leaf area index, time from sucker emergence to flowering and interval between bunches can also be measured and contrasted with growth models (Tixier et al. 2004; Fortescue et al. 2011). The characterization of associated vegetation as habitat or alternate hosts for beneficial organism or pests and diseases is more specialized and are beyond the scope of the chapter, but is an important dimension to consider in banana cropping system improvement.

## 12 Improvement options: banana cropping systems with trees

Bananas are grown with trees overhead primarily in system 1 (Table 1). Here we will look at improvement of the banana productivity in shaded coffee to highlight broader approaches to the use of trees to improve banana cropping systems. Bananas, trees and coffee are found throughout Latin America and the Caribbean, in East Africa and in Asia, a relatively low input approach to cash income and food security. Siles et al. (2013a) in a study of 120 smallholder shaded coffee fields in Honduras, Nicaragua, Costa Rica and Peru found banana mat density from 300 to 600 mats/ha with 950–1200 pseudostems/ha. Tree density varied from 150 to 550 trees/ha with available light for banana ranging from 50 to 70%, and for coffee from 35 to 45% of incident radiation. Rahn et al. (2018) surveyed 144 plots coffee plots with banana and trees along an altitudinal gradient on Mt Elgon, Uganda. Banana density was higher, while coffee and tree densities were lower than in the study in Latin America. The authors formed three classes of fields – open coffee with 29 banana mats/ha, 63 trees/ha and 21% shade, coffee/banana with 1496 banana mats, 48 trees and 28% shade and coffee/tree with 278 banana mats, 146 trees and 48% shade. The level of tree shade on banana was not reported. Mpiira et al. (2013) concluded that trees and bananas in Central Uganda coexisted in the same fields, but did not often overlap.

Trees or shade have been shown to have multiple impacts, both positive and negative, on banana productivity. Early studies such as Murray (1961) found that the time to bunch harvest was longer as light availability decreased – 10.4 months at 100% light and 13.5 months at 20%. Eckstein (1994) found that banana plants along a windbreak received 69% less radiation with a 27% decline in photosynthesis and a 38% decline in transpiration rate. The leaf emission rate was 1–2 leaves/month slower in shade than in sun for the plant and ratoon cycle. Several studies show that bunch size is not greatly affected by some light reduction, and in some cases bunches are larger (Vicente-Chandler et al. 1966; Murray 1961; Eckstein 1994; Torquebiau and Akyeampong 1994). Often this may be due to other micro-environmental factors such as wind protection and reduced temperature, which decrease leaf folding or reduce disease levels (Galan et al. 1997; Vicente-Chandler et al. 1966; Eckstein 1994). In the cases where bunch size was greater or unaffected, banana plant density was low (600–1200) with low light competition among banana plants (Vicente-Chandler et al. 1966; Murray 1961; Torquebiau and Akyeampong 1994), while in Israel planting densities over 2000 flowering stems/ha resulted in yield reductions in the plant crop at shade levels at 70% and in the first ratoon at 20, 40 and 70% (Israelí et al. 1994). Turner (1998) indicated that the increased leaf area and longevity and reduced leaf folding under stress associated with trees contributed to greater light capture which was sufficient to compensate lower light availability under shade.

To develop the basis for recommendations on light partitioning among the tree, banana and coffee strata in these systems, an experiment was set up with four light levels (100, 75, 50, 25%) with six different cultivars in Turrialba, Costa Rica (Valdivia 2012). Natural shade was provided by *Erythrina poeppigiana*. Initial data on growth showed that only Gros Michel was able to maintain leaf area and stem girth even at lower light levels. Net photosynthesis by banana leaves at full sun and 75% full sun exposed to increasing levels of photosynthetically active radiation showed similar response curves, while leaves at 50 and 25% light had diminished response capacity, suggesting that bananas could tolerate up to 25% shade with little impact on productivity (Fig. 8) (Siles et al. 2013b). Data on bunch weights followed the trend for photosynthesis (Fig. 9) with little difference

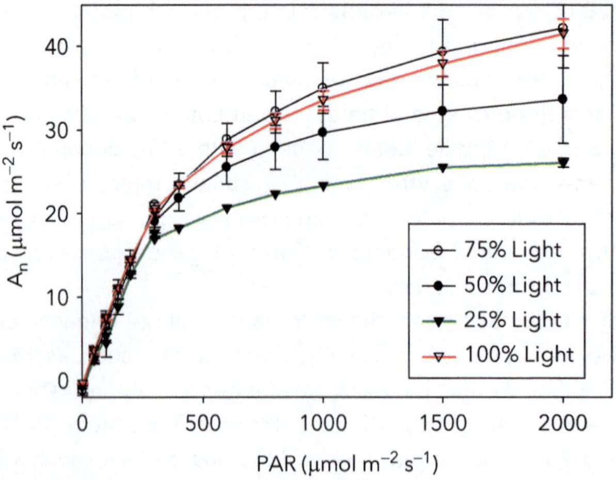


**Figure 8** Photosynthetic response of banana leaves under different levels of light to a burst of radiation. Source: adapted from Siles et al. (2013b).

between full sun and 75% sun. Gros Michel had larger bunches with partial shade and produced bunches in deep shade for a second harvest, not the case for other cultivars (Staver et al. 2013 and unpublished data). The results of this experiment provided the basis to propose an optimum density of Gros Michel of about 250 mats/ha, leaf area to capture 25% of the light, leaving 50% of the incident radiation for coffee, the frequently cited minimum. Dold (2010) estimated that a banana mat density up to 350 mats/ha represents a leaf area index of about 0.5.

Additional design elements should be addressed to complement this rule of thumb for light availability to agroforestry banana. Vicente-Chandler et al. (1966) showed that with 50% shade plantain had fewer damaged leaves from leaf diseases than in full sun – 0 compared to 20%, although bunch weights were similar. For banana, 50% of leaves were heavily infected in full sun compared to only 25% damaged leaves in the 50% shade plots with double the weight and number of hands for bunches from the 50% shade plot. The microclimate effect on leaf disease results from reduced leaf wetting from dew formation at night due to reduced day – night temperature fluctuation. In the Turrialba experiment, the day-night fluctuation in degrees centigrade was greatest a full sun and declined with increasing shade – 3.5, 3.1, 3.0 and 2.7 (Valdivia 2012). Other improvements may also be possible in the tree component. For example, the plots in the Turrialba experiment had *Erythrina* as an overhead shade species, but this species can also be managed at or below the banana canopy through frequent pruning. *Erythrina* trees planted at 6 × 4 m generated over 10 t/ha/year of leaf and small branch biomass with up to 300 kg nitrogen (Montenegro 2005). In decomposition trials lasting 17 weeks, 125 kg nitrogen/ha was released. In other coffee zones, other leguminous tree species from the genus *Inga* are used, although less productive in leaf matter. In a design with frequently pruned trees at or below the banana canopy, timber trees could be used for overhead shade.

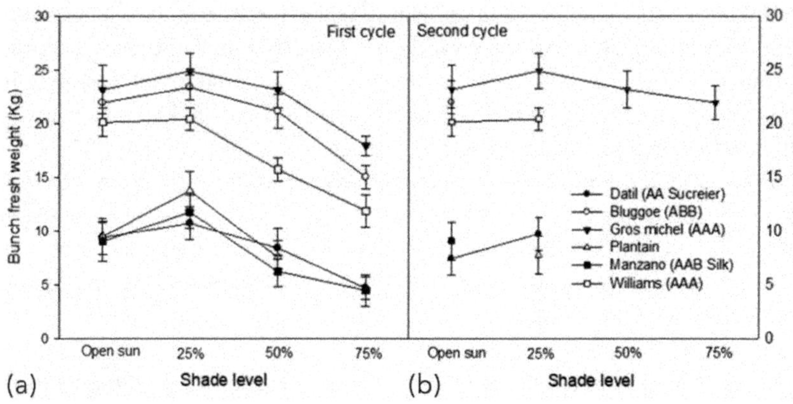


**Figure 9** Bunch size from six cultivars grown under four light levels for first harvest (a) and second harvest (b).

To assess the potential for improvement and identify specific alternatives, diverse measurements can be used as indicators (Table 6). Dold (2010) used a one-time sampling of plots of 25 × 25 m – an inventory of trees, banana mats and coffee present and their characteristics, an estimate of available light at each stratum and an assessment of ground cover. Additional measurements of biomass contributions from leaf fall and pruning may be useful. Data could also be taken following Birabwa et al. (2010) to calibrate current productivity in terms of cycle length from sucker to bunch harvest month-by-month. Similar methods could provide the basis to monitor improvements, drawing not only from shade level and biomass contributions, but also selection of superior mother plants to source planting material and sucker selection to alter month-by-month bunch production.

## 13 Improvement options: nutrient cycling/balance and soil health

We can approximate that each banana pseudostem weighs from 65 to 120 kg at harvest with a harvest index of 25–40% for bunches weighing from 15 to 40 kg. Per hectare this converts into 100–200 tons of total fresh biomass/year, although only 10–30 tons on dry matter basis. Depending on the planting density and the bunch interval, these totals may vary for the different systems in Table 1. Of this total, 15–70 tons/hectare are taken off as bunches, while 50–130 tons remain in the field. A range of nitrogen from 40–150 kg/ha is exported in whole bunches, while for potassium the range is from 150–500 kg/ha/year. Even larger amounts remain behind in crop residue. Approaches to improved banana cropping systems can be found not only in how to compensate the large amounts of nutrients taken off, but how to leverage the large amounts of biomass and nutrients remaining in the field for improved crop performance and soil health.

Currently many smallholders apply few external nutrients, resulting in soil nutrient depletion. The case of Rwanda has already been referenced above (Ndabamenye et al. 2013). On the other hand, in commercial production, nutrients are applied in excess of nutrients taken off in harvest. Pattison et al. (2018) documented increased efficiency of nitrogen use at 180 kg/ha compared with 350 kg N/ha. Approximately 250 kg of fruit per kg of nitrogen were produced at the lower rate, while only 170 kg of fruit per kg of nitrogen were produced at the higher rate. Even at the lower rate more nitrogen was being applied than being taken off. However, the authors also completed an economic analysis of the treatments and show that growers applying the lower rates would face a 17% reduction in return/ha.

Dorel et al. (2008) used results from a fertilizer trial with treatments from 20–420 kg nitrogen/ha to calibrate the nitrogen sub-model to their SIMBA and found that only at the lowest rate was banana growth and development delayed.

They concluded that while the highest rates of nitrogen application served to keep soil N levels in the desired range, much of the nitrogen is leached. The authors also highlighted the incomplete knowledge about the fate of nutrients in the banana stem and corm left in the field after harvest.

A comparison of amounts of N, P and K applied in organic banana with the amounts taken off in harvest among 13 growers in Peru, Ecuador and the Dominican Republic showed that more frequently nitrogen was overapplied, while potassium applications were more often in deficit with potassium taken off in banana. An overview of amounts applied and taken off concluded that small growers were not tracking either nutrients or expenses in their fertilization investments (Rengifo et al. 2017b). Residue distribution was assessed in these same plots to detect whether growers were using crop residue to protect the soil and to ensure rapid uptake of nutrients released from decomposing crop residue. In the extremely dry climate of the northern Peruvian coast, residues dried out before they decomposed. Flood-irrigated fields commonly had a thick layer of undecomposed residues representing both unavailable nutrients and potential substrate for soil biological activity. Bare soil near the base of mats was common in the other two countries, since fertilizers were usually applied right at the base of the stem. Leaves and the stem whorl were often found intact where they had fallen after bunch harvest (Rengifo et al. 2017a). These permanent plantations maintain a large standing biomass of 100–200 t/ha, much of which is recycled internally each harvest cycle. Fertilizer applications can be planned to compensate losses. On the other hand, for frequently replanted fields, nutrients must be mobilized to build a large standing biomass each time bananas are planted. When the field is uprooted nutrients are released from the biomass into the soil. Rosales et al. (n.d.b) emphasize the soil building potential of this production system, but it also has a high potential for nutrient leaching.

These cases of current management of nutrients and residues suggest that the efficiency in the management of nutrients applied and of crop residues that remains offer opportunities for improved banana cropping systems. In the next paragraphs we will illustrate three approaches which are discussed below:

1. covering deficits by mobilizing on-farm resources.
2. introducing off-farm nutrients in proportions more closely linked to amounts lost.
3. accelerating the return of nutrients in crop residues into the banana plant (residue management).

## 13.1 Using on-farm resources

Looking at the first approach, the use of trees to cover banana nutritional needs has already been highlighted in Section 4.3. Leaf and small branch prunings of

*Erythrina* planted with coffee and bananas reached 10 t/ha/year and released 125 kg nitrogen and 95 kg potassium/ha during a 17-week decomposition trial, nutrients available for uptake by associated crops (Montenegro 2005). The use of fodder shrubs, primarily leguminous, was tested in Central Uganda where banana farmers depend on manure to improve banana yields. Unfortunately most farmers have limited resources and access to manure (Mpiira et al. 2013). Seventy-five farm households in three communities installed fodder shrubs and zero-grazed goats to generate additional manure for newly planted banana plots. Over the course of two years, households supplemented on-farm fodder with on average over 100 meters of planted shrubs along field borders. They tripled both their goat population from the initial single pregnant female provided in the test and their resulting monthly manure supply. Banana production in control plots of 10 plants totaled 98 kg, while production from the 10 plants in plots with mulch and manure averaged 340 kg, based both on 50–75% increase in bunch weight and a reduced interval between the first and second harvest. Recycled banana peels also made up an important component of goat diets. Traditionally peels are used to feed animals, but manure is not collected systematically. Farmers concluded that although all the components of the improved technology were already present in their communities, they learned to use and link them for greater productivity and household well-being (Staver et al. 2015). Off-farm nutrients are not introduced in these two approaches. The only new nutrients come from possible nitrogen fixation. Other nutrients originate from field borders or deeper soil layers within the farm. If increased production is sold to market, farm nutrient depletion may be accelerated, suggesting that these strategies should include additional farmer training on farm nutrient budgets.

## 13.2 Using off-farm resources

A second approach to improving nutrient cycling and balance focuses on more efficient introduction of N, P and K from off-farm supplies. In work with small growers of organic export banana, Rengifo et al. (2017c) developed a spread sheet to calculate nutrients taken off compared to nutrients applied to highlight possible imbalances. While such a tool may be of limited value for frequently replanted bananas, in established banana fields in production, a large nutrient reserve and living biomass are already invested. The balancing of inputs and outputs of individual nutrients provides a simple decision tool which can be applied annually. The calculations plus technical assistance to propose modifications reduced imbalances, both deficits and over-application (unpublished data). Growers and their associations need to build such tools into their annual planning to ensure that fertilizers purchased in bulk address imbalances and that resources are budgeted for timely application of the appropriate fertilizer.

## 13.3 Residue management

The third approach focusing on the management of crop residues can be illustrated by a practice tested with smallholder organic export banana growers through a Fontagro-financed research project (Rengifo et al. 2017b). Traditional grower residue management was contrasted with a combination of two altered routines. First, the banana leaves generated by weekly and biweekly deleafing are targeted to cover the closest bare soil compared to common practice in which leaves fall randomly without any redirection. Second, at bunch harvest, the leaves and leaf whorl are stacked about 75 cm in front of the follower stem. The harvested stem is commonly left standing, although each time an additional stem section is cut, the material is also added to the stack. Visual observation suggests that sites of thick residue accumulation generate a proliferation of banana roots and soil biological activity. Fertilizers were also applied under the stack of residue where roots were concentrated. Three treatments in plots of 100 mats were tracked over two bunch harvests:

1 grower practice.
2 strategic placement of residues and fertilizers.
3 adjustment of fertilizer applications using the nutrient balance and strategic replacement of residues and fertilizers.

Indicators of soil health were applied prior to and at the end of the test period (Rengifo et al. 2017a,d). The proposed practices improved the indicators linked to banana productivity. Bunch size increased slightly, although not statistically significant, while the number of bunches/mat/year was increased from 1.14 for the control to 1.23 and 1.31 for the two improved practices respectively, indicating a shortened interval between bunches. This represents 200–400 additional bunches/year with only minor changes in residue management and fertilizer placement. Changes in the other indicators were less consistent and not significant across all growers. Banana residue distribution was improved, but only a minority of growers achieved the target distribution. Indicators like banana root proliferation, bulk density, total soil organic matter, infiltration rates and free- living nematodes improved among some growers, but not others.

The management of crop residue for greater nutrient efficiency and cycling has different challenges when bananas are frequently replanted. A new stand of high-density mats at harvest represents over 200 kg nitrogen and 700 kg potassium. Fertilizer application is even higher to account for inefficiency of nutrient uptake. When the stand is uprooted after one or several harvests, similar amounts may be liberated pending field preparation and replanting which may delay up to several months or more. The university farm in Leon, Nicaragua, has been testing the use of cover crops to take up available nutrients during this interval in their plantain cropping system. Mung bean is planted even before the harvest is complete, although after the leaf canopy has opened up. When stems of all sizes are chopped down and cut up, the living cover of mung bean is already in place. Over several months, stems, leaves and corms decompose beneath the vigorous growth of the cover crop. Replanting is carried out by opening a furrow into the cover crop. The cover crop is only tilled once the new plantation is established. The effectiveness of this approach is still being documented. Other practices and species may be used in other contexts, but overall the strategy of using living biomass as a nutrient sponge or bridge merits practical testing to make banana cropping systems more efficient and productive.

Indicators to characterize current status of nutrient cycling and soil health and to monitor improvements are an expanding field (Stewart et al. 2018; Bünemann et al. 2018) with practical guides also proliferating (e.g., Kittredge n.d.). For banana two useful methods guides are available (Rosales et al. n.d.b; Pattison and Lindsay 2006). Methods for total soil organic matter and sub-components, free-living nematodes and soil physical properties can also be found in specialized publications. Banana-specific procedures are needed to track crop yield and residue management, since harvest may be spread over the twelve months of the year depending on the cropping system. In export banana, extensive data are collected on week by week bunch production and bunch to box ratios by field, but few additional indicators are calculated. Equations to calculate yield and crop biomass based on non-destructive measurements have been developed for different cultivars (Nyombi et al. 2009; Damour et al. 2012; Dold 2010; Ripoche et al. 2012). These equations may be useful in production systems oriented to local and national markets.

## 14 Improvement options: addressing climate and weather variability in banana production

Climate change, measured as the increase in average annual temperatures, is projected to expand the area where banana can be grown, primarily into higher altitude tropical zones and into the subtropics (Calberto et al. 2015). However, the increase in weather variability associated with increasing temperatures will disrupt banana production management routines that have been developed in response to regular seasonal changes which characterize each zone. Calberto et al. (2018) surveyed banana experts globally on the frequency and importance of 16 different weather events. Droughts and excess rain were the general categories most frequently cited followed by winds. Many of the specific events are often moderate in their impact, but more frequent in their occurrence. Variations in rainfall distribution are practically a yearly phenomenon, since the average year almost never occurs. Extreme events are less frequent, but more damaging, often associated with hurricanes or typhoons which bring floods, high winds and saturated soils in different combinations.

The profile of the frequency of moderate and extreme weather events of banana growing areas around the world is not well documented. Some areas are less important for banana growing due to certain risk factors. Export banana production is centered around Davao further south in the Philippines at least in part to the lower frequency of typhoons which strike the northern islands with much greater frequency. The export banana growing area in northwestern Dominican Republic has recently suffered droughts, floods and winds over a two-year period which has made meeting export contracts difficult. The survey of frequency of weather events by Calberto et al. (2018) did not find significant regional differences for the types of events. Practical experience suggests that variability is more localized. However, for whatever the local combination of weather phenomena, the management of variability will be a major challenge to small and large banana growers in the next decades.

The practices and strategies to modify and redesign cropping systems to address weather variability can be organized in two dimensions (1) approach and (2) spatial scale. In terms of the approach, certain practices can help to make the cropping system more resistant. Irrigation to address increasing drought or produce off-season directly offsets the problem of water stress. Levees to avoid floods are also a structure of resistance. Other practices increase cropping system resilience by allowing the production to be maintained or less affected in spite of the weather event. Improved soil organic matter or surface mulch increases water storage capacity and reduces evapotranspiration, thereby reducing impacts of minor droughts. Fertilization can also be varied depending on weather conditions to ensure ongoing plant vigor. Improved drainage and outlet canals evacuate flood waters more quickly to reduce potential damage to plants. Other practices are oriented to recovery in the case that damage is heavy. Practices after floods, winds or low temperatures are needed to bring banana fields back into production. A menu of practices covering all three approaches – resistance, resilience and recovery – needs to be developed for the different moderate and extreme events in each locality.

The different spatial scales for addressing banana cropping system improvement (Fig. 1) offer a second dimension to address weather variability. Table 9 illustrates the extent to which different types of weather events are more usefully addressed at different spatial scales. Certain variability can be addressed by cultivar selection, in particular tolerance to temperature extremes, drought tolerance and plant height, while other events and the accompanying practices are applicable at the field scale, either mat management and distribution or field ecosystem. Here we have expanded our view beyond the field which has been the focus for the other sections. Weather variability options at the farm, multi-farm and larger watershed address many of the events with more serious

**Table 9** Spatial scales and corresponding weather events and possible practices for adaptation

| Scale | Moderate and extreme weather events | Possible practices |
|---|---|---|
| Plant | Low and high temperatures, strong winds, water stress | Cultivar selection – temperature extremes and drought tolerance, height |
| Field of banana mats | Frosts, droughts, winds | Sucker proliferation as buffer, Lower mat density Planting date to reduce risk of wind damage |
| Field scale ecosystem | Droughts due to irregular rains, cloudiness, extended rainy periods or high humidity | Ditches to evacuate water; Soil health practices for water holding capacity; infiltration; Residue management and ground cover; Plant nutrition to ensure vigor; Leaf pruning |
| Farm scale ecosystem | Floods with strong currents, floods standing water | Controlled access drainage ditches; Windbreaks and multi-service field boundary vegetation; Crop rotations |
| Multi-farm ecosystem | Floods, droughts, prolonged dry season winds | Water storage structures; Controlled access irrigation and drainage system; Windbreaks; Protected stream banks; Weather stations |
| Larger watershed ecosystem | Floods, droughts, hurricanes, typhoons | Protected watersheds and river banks; Flood sacrifice zones; Climate-smart roads and bridges; Stream flow records; Flood and damage mapping |

impacts on production. The implementation of practices at the different scales depends on different stakeholders. Plot scale practices are addressed through agronomic research, farmer experimentation and technical assistance, while addressing floods and increasing droughts may be a regional or territorial problem beyond banana depending on government.

To close this section, we highlight indicators for characterization of the current situation. Certain of these (Table 6), we address the importance of registering the intensity and frequency of moderate and extreme weather events and their impact on banana production. Weather, stream flow and banana growth and productivity need to be routinely taken, while the extent of flooding and areas damaged by wind need to be documented sporadically as they occur. An even greater challenge, once data taking infrastructures and routines are established and big data bases are being generated, is data analysis. Teasing out relationships in weather and productivity to guide improved management is a research frontier which we will highlight briefly in the next section.

## 15 Conclusion and future trends

Many research issues on the frontiers of advanced science have been detailed by Staver (2018). Addressing the more immediate challenge of banana cropping system improvement as summarized in this chapter, three points are important to highlight.

First, data-driven agronomy, while a wide topic with advanced science implications (e.g. Wolfert et al. 2017), has very concrete and specific applications as we have seen in the chapter. To provide tools to technical advisors and growers, practical strategies are needed for the collection, storing and analysis of data (Manfre and Laytham 2018). Staver (2018) highlighted the importance of data in strategies for continual improvement and benchmarking applied in banana and proposed that data collected for purposes of certification provide a starting point, but needs to be broadened. From this chapter several topics have emerged:

- selection of superior plants described previously based on bunch size and time to flowering. Other aspects such as sucker and plant height may also be useful to consider as well as selection based on the first harvest versus two or three harvests. The role of mat to mat variability in improving yields and resource efficiency can only advance with extensive data and analysis.
- timing of flowering and bunch production and interval between bunches. This data provide insight on within-field variability and also facilitate the analysis of the role of weather variability on plant productivity and the value of targeted practices in nutrition or sucker management.
- the challenge of moderate and extreme weather events to banana growers. Much remains to be quantified about the frequency, duration and intensity of weather phenomena and their impact on production. Improved weather data accumulated multi-year in parallel with banana performance and losses is a starting point to quantify critical levels and type and degree of impact on production. Once patterns begin to emerge, this also provides the basis for predictive recommendations which guide growers on practices to reduce impacts or to take advantage of improved weather due to variability. Real-time monitoring can also provide the basis for early warning systems.

Second, indicators, both general topics and specific measurements, have been identified to guide a diagnostic and monitoring strategy to apply ecological intensification status to banana cropping system improvement (Table 6). Work remains to validate and simplify indicators already available and to develop more grower-friendly methods. Three topics merit specific mention:

- light and water capture and partitioning are the basis for optimizing primary productivity, yet few practical, field tools are available. Simple, cropping system-specific indicators for seasonal observations are needed to highlight opportunities for improvement and to calculate possible trade-offs.

- nutrient recommendations and more recently soil health have been formulated for banana systems, but have not been integrated with residue management and nutrient cycling which can be guided by an ecological intensification framework. Indicators on the quantification of banana and other crop residues, their nutrient content and decomposition rates and their placement could provide the basis more efficient nutrient use and improved integrated banana system health. While abundant indicators of soil health are on the books, a parsimonious group which serves to guide the assembly of improved practices still needs to be validated.
- The monitoring of pest damage, particularly to the fruit in the packing shed, has been addressed through indicators, most recently by Rojas et al. (2017). However, the status of the biodiversity in the banana field and the types of practices currently used is needed to characterize the potential for habitat enhancement to promote naturally occurring control of insect pests, primarily of the bunch.

Third, conventional intensification of banana production has deployed specific inputs for specific impacts. In an ecological framework, interactions need to be understood among the different components of the system, both planned and unplanned, opening the way for integration efficiencies (Phelan 2009). Well screened practices may affect different aspects of the production system. One important example is associated vegetation which we have already seen can affect light and water capture and partitioning, volume and type of organic matter recycled, soil quality related to water holding and infiltration, soil biology and nematodes and habitat for generalist predators. A second example is crop residue management which interacts with soil biology and root health on the soil surface and the soil profile, the efficiency of nutrient cycling and the severity of certain pests and diseases, including banana weevil and black leaf streak. The harnessing of these simple and complex interactions calls for a research agenda bringing together simple field trials, long-term systems comparisons, farmer participatory approaches, crop system models and big data strategies which link back to the first and second points mentioned above. Such research should address the priorities of small holders as well as export production.

## 16 Where to find further information

- CIRAD: research on the ecological intensification of banana production in the French West Indies
- http://www.endure-network.eu/fr/nouvelles/dernieres_nouvelles/new_guides_for_better_banana_production

- https://www.cirad.fr/en/our-research/tropical-value-chains/banana-and-plantain/context-and-issues
- Australia – National Banana Development and Extension Program https://betterbananas.com.au/
- CLAC: approaches to greater productivity among small growers – organic and fair trade: http://clac-comerciojusto.org/redes-y-productos/redes-de-producto/red-banano/
- Promusa: Musapedia pages in English, Spanish and French: http://www.promusa.org/Pages+in+English, http://www.promusa.org/Páginas+en+español, http://www.promusa.org/Pages+en+français
- Musalac: website for research projects from the regional banana network for Latin America and the Caribbean: http://banana-networks.org/musalac/regional-research-projects/
- FAO Good Agricultural Practices/Pesticide reduction practices: http://www.fao.org/world-banana-forum/projects/good-practices/good agricultural-practices/en/.

## 17 References

Achard, R., Tixier, P., Dorel, M. and Estrade, J. 2018. Intercropping of grass cover crops in banana plantations. *Tropical Agriculture and Development*, 62(1), 1–8.

Akinyemi, S. O. S., Aiyelaagbe, I. O. O. and Akyeampong, E. 2010. Plantain (*Musa* spp.) cultivation in Nigeria: a review of its production, marketing and research in the last two decades. *Acta Horticulturae*, 879, 211–18.

Bathan, B. M. and Lantican, F. A. 2010. Factors affecting yield performance of banana farms in Oriental Mindoro, Philippines. *Journal of International Society for Southeast Asian Agricultural Sciences*, 16(1), 110–20.

Belalcazar, S. and Cayón, G. 1998. Altas Densidades de Siembra. In M. Giraldo, S. Belalcazar, G. Cayón and R. Botero (Eds), *Seminario Internacional sobre producción de plátano*. Corpoica, Colombia.

Belalcazar, S., Cayón, G. and Valencia, J. 1998. Establecimiento del Cultivo. In M. Giraldo, S. Belalcazar, G. Cayón and R. Botero (Eds), *Seminario Internacional sobre producción de plátano*. Corpoica, Colombia.

Bellamy, A. S., Svensson, O., van Den Brink, P. J., Gunnarsson, J. and Tedengren, M. 2018. Insect community composition and functional roles along a tropical agricultural production gradient. *Environmental Science and Pollution Research*, 25(14), 13426–38.

Bessou, C., Basset-Mens, C., Tran, T. and Benoist, A. 2013. LCA applied to perennial cropping systems: a review focused on the farm stage. *The International Journal of Life Cycle Assessment*, 18(2), 340–61.

Bingham, E. and Monforton, C. 2013. The pesticide DBCP and male infertility. In S. F. Hansen (Ed.), *Late Lessons from Early Warnings: Science, Precaution, Innovation*. European Environmental Agency Report. Available at: https://www.eea.europa.eu/publications/late-lessons-2.

Birabwa, R., van Asten, P., Alou, I. N. and Taulya, G. 2010. Got matooke (*Musa* spp.) for Christmas? *Acta Horticulturae*, 879, 113–21.

Blomme, G., Ocimati, W., Groot, J., Ntamwira, J., Bahati, L., Kantungeko, D., Remans, R. and Tittonell, P. 2018. Agroecological integration of shade-and drought-tolerant food/feed crops for year-round productivity in banana-based systems under rain-fed conditions in Central Africa. *Acta Horticulturae*, 1196, 41–54.

Bockel, L., Schiettecatte, L. and Debrune, O. 2018. Life cycle assessment and carbon footprint of banana cultivation Agriculture Organization (FAO), Italy. In G. Kema and A. Drenth (Eds), *Achieving Sustainable Cultivation of Bananas* (pp. 301–22). Burleigh Dodds Science Publishing.

Bünemann, E., Bongiorno, G., Bai, Z., Creamer, R., De Deyn, G., de Goede, R., Fleskens, L., Geissen, V., Kuyper, T., Mäder, P. and Pulleman, M. 2018. Soil quality–A critical review. *Soil Biology and Biochemistry*, 120, 105–25.

Calberto, G., Staver, C. and Siles, P. 2015. An assessment of global banana production and suitability under climate change scenarios. In Aziz Elbehri (Ed.), *Climate Change and Food Systems: Global Assessments and Implications for Food Security and Trade*. Food Agriculture Organization of the United Nations (FAO), Rome.

Calberto, G., Blake, D., Staver, C., Carvajal, M. and Brown, D. 2018. The frequency and effects of weather events on banana productivity-results of a global survey. *Acta Horticulturae*, 1196, 179–86.

Carr, M. K. V. 2009. The water relations and irrigation requirements of banana (*Musa* spp.). *Experimental Agriculture*, 45(3), 333–71.

Chopin, P. and Blazy, J. 2013. Assessment of regional variability in crop yields with spatial autocorrelation: Banana farms and policy implications in Martinique. *Agriculture, Ecosystems & Environment*, 181, 12–21.

Chopin, P., Tirolien, J. and Blazy, J. M. 2016. Ex-ante sustainability assessment of cleaner banana production systems. *Journal of Cleaner Production*, 139, 15–24.

Clune, S., Crossin, E. and Verghese, K. 2016. Systematic review of greenhouse gas emissions for different fresh food categories, *Journal of Cleaner Production* 140, 766–83.doi:10.1016/j.jclepro.2016.04.082.

Coltro, L. and Karaski, T. 2019. Environmental indicators of banana production in Brazil: Cavendish and Prata varieties. *Journal of Cleaner Production*, 207, 363–78.

Corozo, R. E., Arias, M., Durango, W., Staver, C. and Turmel, M. 2016. Cover crops in organic-export banana systems: Farmer experiences and soil health in Ecuador. *Promusa Symposium 2016 -Agroecological Approaches to Promote Innovative Banana Production Systems*, 10–16 October, Montpellier, France. Available at: http://www.promusa.org/article142-2016-France-symposium#poster.

Craig, A. and Blanco, E. 2009. *The Banana Carbon Footprint Case Study*. MIT Center for Transportation and Logistics. Available at: http://ctl.mit.edu/research.

Damour, G., Ozier-Lafontaine, H. and Dorel, M. 2012. Simulation of the growth of banana (*Musa* spp.) cultivated on cover-crop with simplified indicators of soil water and nitrogen availability and integrated plant traits. *Field crops research*, 130, 99–108.

Damour, G., Dorel, M., Tran Quoc, H., Meynard, C. and Risede, J.M. 2014. A trait-based characterization of cover plants to assess their potential to provide a set of ecological services in banana cropping systems. *European Journal of Agronomy*, 52, 218–28.

Damour, G., Garnier, E., Navas, M. L., Dorel, M. and Risede, J. M. 2015. Using functional traits to assess the services provided by cover plants: a review of potentialities in banana cropping systems. *Advances in Agronomy*, 134, 81–133.

De Barros, I., Blazy, J. M., Rodrigues, G. S., Tournebize, R. and Cinna, J. 2009. Emergy evaluation and economic performance of banana cropping systems in Guadeloupe (French West Indies). *Agriculture, Ecosystems and Environment*, 129(4), 437–49.

Dépigny, S., Lescot, T., Achard, R., Diouf, O., Côte, F., Fonbah, C., Sadom, L. and Tixier, P. 2017. Model-based benchmarking of the production potential of plantains (*Musa* spp., AAB): application to five real plantain and four plantain-like hybrid varieties in Cameroon. *The Journal of Agricultural Science*, 155(6), 888–901.

Dépigny, S., Wils, E. D., Tixier, P., Keng, M. N., Cilas, C., Lescot, T. and Jagoret, P. 2019. Plantain productivity: Insights from Cameroonian cropping systems. *Agricultural Systems*, 168, 1–10.

Dita, M., Garming, H., Van den Bergh, I., Staver, C. and Lescot, T. 2013. Banana in Latin America and the Caribbean: current state, challenges and perspectives. *Acta Horticulturae*, 986, 365–80.

Dita, M., Barquero, M., Heck, D., Mizubuti, E. S. and Staver, C. 2018. Fusarium wilt of banana: current knowledge on epidemiology and research needs toward sustainable disease management. *Frontiers in Plant Science*, 9.

Djigal, D., Chabrier, C., Duyck, P. F., Achard, R., Quénéhervé, P. and Tixier, P. 2012. Cover crops alter the soil nematode food web in banana agroecosystems. *Soil Biology and Biochemistry*, 48, 142–50.

Dold, C. 2010. Leaf Area index and measurements of light transmission as management tools in banana production in Central American coffee agroforestry systems. M.Sc. Thesis. Faculty of Agriculture, Rheinische Friedrich-Wilhelms- Universität zu Bonn, Bonn.

Dorel, M., Achard, R. and Tixier, P. 2008. SIMBA-N: modeling nitrogen dynamics in banana populations in wet tropical climate. Application to fertilization management in the Caribbean. *European Journal of Agronomy*, 29(1), 38–45.

Eckstein, K. 1994. *Physiological Responses of Banana in the Subtropics*. DrAGR University of Bonn.

Eckstein, K., Robinson, J.C. and Fraser, C. 2000. The timing of first ratoon sucker selection using tissue culture banana plants. *Acta Horticulturae*, 540, 267–72.

Fortescue, J. A., Turner, D. W. and Romero, R. 2011. Evidence that banana (*Musa* spp.), a tropical monocotyledon, has a facultative long-day response to photoperiod. *Functional Plant Biology*, 38(11), 867–78.

Galan, V., Cabrera, J., Hernandez, P. and Rodriguez M. 1997. Comparison of protected and open sun cultivation of Grande Nain and Dwarf Cavendish bananas. 1. Cultivar evaluation and phonological and production responses. In V. Galan and P. Hernandez (Eds), *International Symposium on Bananas in the Sub Tropics*. Book of Abstracts, p. 153.

Hauser, S. and van Asten, P. 2010. Methodological consideration on banana (*Musa* spp.) yield determinations. *Acta Horticulturae*, 879, 433–44.

Hochman, Z., Carberry, P., Robertson, M., Gaydon, D., Bell, L. and McIntosh, P. 2013. Prospects for ecological intensification of Australian agriculture. *European Journal of Agronomy*, 44, 109–23.

Irizarry, H., Goenaga, R. and Chardón, U. 2002. Nitrogen fertilization in banana grown on a highly weathered soil of the humid mountain region of Puerto Rico. *Journal of Agriculture - University of Puerto Rico*, 86(1/2), 15–26.

Israelí, Y., Plaut, Z. and Schwartz, A. 1994. *Observations on the Effect of Shade on Bananas*. Memorias XI ACORBAT, San Jose, Costa Rica, pp. 409–27.

Jacobsen, K., Omondi, B. A., Almekinders, C., Alvarez, E., Blomme, G., Dita, M., Iskra-Caruana, M. L., Ocimati, W., Tinzaara, W., Kumar, P. L. and Staver, C. 2019. Seed degeneration of banana planting materials: Strategies for improved farmer access to healthy seed. *Plant Pathology*, 68(2), 207–28.

Jean-Denis, S., Jean-Pierre, D., Mutel, M., Duchaufour, H., Langlais, C., Fernandes, P., Alphonse, M. and Malézieux É. 2014. Evolution de la structure d'un système agroforestier en relation avec le cycle de vie familial: cas du jardin de case en Haïti. *Bois et forêts des tropiques*, 321, 7–20.

Keating, B. A., Carberry, P. S., Bindraban, P. S., Asseng, S., Meinke, H. and Dixon, J. 2010. Eco-efficient agriculture: concepts, challenges, and opportunities. *Crop Science*, 50(Suppl. 1), S-109–S-119.

Kittredge, J. N. D. Soil carbon grower on-site test protocols & data sheets V. 4. Northeast Organic Farming Association. Available at: https://www.nofamass.org/sites/default/files/GrowerCarbonPTestManualV.4.pdf.

Lescot, T. 2012. Carbon footprint analysis in banana production. *Second Conference of the World Banana Forum*, Guayaquil, 28–29 Ecuador-February 2012. Available at: http://agritrop.cirad.fr/572587/1/document_572587.pdf

Lescot, T. 2018. Banana genetic diversity. *Fruitrop*, 256, 92–95.

Lescot, T. and Staver, C. 2010. Banana, plantain and other species of *Musacea*. In J. Fajardo, N. Lutaladio, M. Larinde, C Rosell, I. Barker, W. Roca and E. Chujoy (Eds), *Quality Declared Planting Material: Protocols and Standards for Vegetatively Propagated Crops*. FAO Plant Production and Protection Paper 195. Available at: http://www.fao.org/3/a-i1195e.pdf.

Lopatka, J., Topel, J. and de Vasconcellos, P. 2008. *Food Staples in the Ghana School Feeding Program: Analysis of Markets, Value Chains, and Menus*. UC Berkeley – Haas School of Business. International Business Development Program.

Manfre, C. and Laytham, W. 2018. *Putting Data at the Service of Agriculture – A Case Study of CIAT*. Feed the Future, US AID.

Molina, A. and Valmayor, R. 1999. Banana production systems in South East Asia. In C. Pica, E. Foure and E.A. Frison (Eds), *Bananas and Food Security*, pp. 423–36.

Montenegro, E. 2005. Efecto del aporte de nutrientes de la biomasa de tres tipos de árboles de sombra en sistemas de manejo de café orgánico y convencional. M.Sc. Thesis. CATIE Escuela de Posgrado, Turrialba, Costa Rica.

Mpiira, S., Staver, C., Kagezi, G., Wesiga, J., Nakyeyune, C., Ssebulime, G., Kabirizi, J., Nowakunda, K., Karamura, E. and Tushemereirwe, W. 2013. The use of trees and shrubs to improve banana productivity and production in Central Uganda: an analysis of the current situation. In *Banana Systems in the Humid Highlands of Sub-Saharan Africa*.

Multigner, L., Cordier, S., Kadhel, P., Huc-Terki, F., Blanchet, P., Bataille, H. and Janky, E. 2007. Pollution par le chlordécone aux Antilles Quel impact sur la santé de la population? *Environnement, Risques & Santé*, 6(6), 405–7.

Murray, D. 1961. Shade and fertilizer relations in the banana. *Tropical Agriculture*, 38, 123–32.

Ndabamenye, T., Van Asten, P. J. A., Vanhoudt, N., Blomme, G., Swennen, R., Annandale, J. G. and Barnard, R. O. 2012. Ecological characteristics influence farmer selection of on-farm plant density and bunch mass of low input East African highland banana (*Musa* spp.) cropping systems. *Field Crops Research*, 135, 126–36.

Ndabamenye T., Vanlauwe B., Van Asten, P., Blomme, G., Swennen, R., Uzayisenga, B., Annandale, J. and Barnard, R. 2013. Influence of plant density on variability of soil fertility and nutrient budgets in low input East African highland banana (*Musa* spp. AAA-EA) cropping systems. *Nutrient Cycling in Agroecosystems*, 95(2), 187–202.

NIHORT. 2008. Market price survey of some horticultural crops in Nigeria. Unit Report, 15pp.

Norgrove, L. and Hauser, S. 2014. Improving plantain (*Musa* spp. AAB) yields on smallholder farms in West and Central Africa. *Food Security*, 6(4), 501–14.

Ntamwira, J., Pypers, P., van Asten, P., Vanlauwe, B., Ruhigwa, B., Lepoint, P., Dheda, B., Monde, T., Kamira, M. and Blomme, G. 2014. Effect of banana leaf pruning on banana and legume yield under intercropping in farmers' fields in eastern Democratic Republic of Congo. *Journal of Horticulture and Agroforestry*, 6(9), 72–80.

Nyombi, K. 2010. Understanding growth of East Africa highland banana: experiments and simulation. Ph.D. Thesis. Wageingen University.

Nyombi, K., Van Asten, P., Leffelaar, P., Corbeels, M., Kaizzi, C. K. and Giller, K. 2009. Allometric growth relationships of East Africa highland bananas (*Musa* AAA-EAHB) cv. Kisansa and Mbwazirume. *Annals of Applied Biology*, 155(3), 403–18.

Okumu, M., Van Asten, P., Kahangi, E., Okech, S., Jefwa, J. and Vanlauwe, B. 2011. Production gradients in smallholder banana (cv. Giant Cavendish) farms in Central Kenya. *Scientia Horticulturae*, 127(4), 475-81.

Pattison, T. 2006. Banana farm management effects on soil health and plant-parasitic nematodes in Costa Rica. Final report. Queensland Department of Primary Industries and Fisheries, University of Southern Queensland, Australia.

Pattison, T. and Lindsay, S. 2006. *Banana Root and Soil Health User's Manual*. Department of Primary Industries and Fisheries, Queensland, Australia, 63p.

Pattison, A., East, D., Ferro, K. and Dickinson, G. 2018. Agronomic consequences of vegetative groundcovers and reduced nitrogen applications for banana production systems. *Acta Horticulturae*, 1196, 155–62.

Pemsl, D., Staver, C. Creamer, B., Tahirou, A., Arega A. and Rusike, J. 2014. Results of a global online expert survey: major constraints, opportunities and trends for banana production and marketing and priorities for future. RTB banana research. CRP RTB, 52p.

Phelan, P. 2009. Ecology-based agriculture and the next green revolution: is modern agriculture exempt from the laws of ecology? In P. Bohlen and G. House (Eds), *Sustainable Agroecosystem Management*, pp. 97–135.

Poeydebat, C., Tixier, P., Chabrier, C., de Bellaire, L., Vargas, R., Daribo, M. O. and Carval, D. 2017. Does plant richness alter multitrophic soil food web and promote plant-parasitic nematode regulation in banana agroecosystems? *Applied Soil Ecology*, 117, 137–46.

Poore, J. and Nemecek, T. 2018. Reducing food's environmental impacts through producers and consumers. *Science*, 360(6392), 987–92.

Rahn, E., Liebig, T., Ghazoul, J., van Asten, P., Läderach, P., Vaast, P., Sarmiento, A., Garcia, C. and Jassogne, L. 2018. Opportunities for sustainable intensification of coffee agro-ecosystems along an altitudinal gradient on Mt. Elgon, Uganda. *Agriculture, Ecosystems and Environment*, 263, 31–40.

Reardon, T., Henson, S. and Berdegué, J. 2007. 'Proactive fast-tracking' diffusion of supermarkets in developing countries: implications for market institutions and trade. *Journal of Economic Geography*, 7(4), 399–431.

Rengifo, D., Garcia, S., Maldonado, E., Rojas, J., Navia, D., Delgado, R., Staver, C. and Sanchez, S. 2017a. Los Nematodos Como Indicadores Para Orientar el Manejo de Suelos en Sistemas de Producción de Banano Orgánico de Exportación. *III Congreso Latinoamericano y del Caribe de Banano y Platano*, 29–30 November, Colima, Mexico. Available at: http://banana-networks.org/musalac/files/2018/09/NVL-indicador-salud-suelos-banano-organico-Musalac-2017.pdf.

Rengifo, D., Perez, A., Durango, W., Delgado, R., Corozo, R., Rojas, J., Vegas, U. and Staver, C. 2017b. Distribución Estratégica de Residuos y Ajustes en la Zona y Cantidad de Fertilizantes para Mejorar la Nutrición del Banano Orgánico. *III Congreso Latinoamericano y del Caribe de Banano y Platano*, 29–30 November, Colima, Mexico. Available at: http://banana-networks.org/musalac/files/2018/09/redistribucion-residuos-banano-organico-Musalac-2017.pdf.

Rengifo, D., Perez, A., Durango, W., Delgado, R., Corozo, R., Rojas, J., Vegas, U., Staver, C. and Turmel, M. 2017c. Un Cálculo de Balance de Nutrientes para Orientar la Fertilización de Banano Orgánico de Exportación. *III Congreso Latinoamericano y del Caribe de Banano y Platano*, 29–30 November, Colima, Mexico. Available at: http://banana-networks.org/musalac/files/2018/09/Balance-nutrientes-banano-organico-Musalac-2017.pdf.

Rengifo, D., Perez, A., Durango, W., Delgado, R., Corozo, R., Rojas, J., Vegas, U., Staver, C. and Turmel, M. 2017d. Indicadores de Salud de Suelo como Herramienta para Mejorar la Productividad del Banano Orgánico. *III Congreso Latinoamericano y del Caribe de Banano y Platano*, 29–30 November, Colima, Mexico. Available at: http://banana-networks.org/musalac/files/2018/09/Indicadores-Salud-Suelo-Banano-Organico-Musalac-2017.pdf.

Ripoche, A., Achard, R., Laurens, A. and Tixier, P. 2012. Modeling spatial partitioning of light and nitrogen resources in banana cover-cropping systems. *European Journal of Agronomy*, 41, 81–91.

Robinson, J. 2000. Banana productivity-The impact of agronomic practices. *Acta Horticulturae*, 540, 247–58.

Robinson, J. and Nel, D. 1986. The influence of planting date, sucker selection and density on yield and crop-timing of bananas (cultivar 'Williams') in the Eastern Transvaal. *Scientia Horticulturae*, 29(4), 347–58.

Robinson, J. and Nel, D. 1989. Plant density studies with banana (cv. Williams) in a subtropical climate. II. Components of yield and seasonal distribution of yield. *Journal of Horticultural Science*, 64(2), 211–22.

Rodrigo, V., Stirling, C., Teklehaimanot, Z. and Nugawela, A. 2001. Intercropping with banana to improve fractional interception and radiation-use efficiency of immature rubber plantations. *Field Crops Research*, 69(3), 237–49.

Rodrigo, V., Stirling, C., Teklehaimanot, Z., Samarasekera, R. and Pathirana, P. 2005. Interplanting banana at high densities with immature rubber crop for improved water use. *Agronomy for Sustainable Development*, 25(1), 45–54.

Roibás, L., Elbehri, A. and Hospido, A. 2015. Evaluating the sustainability of Ecuadorian bananas: Carbon footprint, water usage and wealth distribution along the supply chain. *Sustainable Production and Consumption*, 2, 3–16.

Rojas, J., Vegas, U., Alburqueque, D., Rengifo, D., Suarez, P., Delgado, R., Corozo, R., Arias, M., Osorio, B., Moyon, D. and Staver, C. 2017. Identificación de las Causas de Pérdida de Fruta en la Etapa de Cosecha y Pos cosecha de Banano Orgánico de Exportación. *III Congreso Latinoamericano y del Caribe de Banano y Platano*, 29–30 November, Colima, Mexico. Available at: http://banana-networks.org/musalac/files/2018/09/Identificaci%C3%B3n-de-las-Causas-de-P%C3%A9rdida-de-Fruta-Musalac-2017.pdf.

Ronner, E., Thuijsman, E., Ebanyat, P. and Giller, K. 2019. Climbing bean x highland banana intercropping in the Ugandan highlands, 24p. Available at: www.N2Africa.org.

Rosales, F., Alvarez, J. and Vargas, A. n.d.a. *Guia Práctica para la Producción de Plátano con Altas Densidades - Experiencias de America Latina y el Caribe*. Bioversity International/MUSALAC. Available at: http://banana-networks.org/musalac/files/2013/01/altas_densidades-1.pdf.

Rosales, F., Pocasangre, L., Trejos, J., Serrano, E. and Peña, W. n.d.b. *Guia del diagnostic de la calidad y salud de suelos bananeros*. Bioversity International. Available at: http://banana-networks.org/musalac/files/2013/01/guia_diagnosticosuelos.pdf.

Roy, S., Jaiswal, A., Sushil, S., Baitha, A. and Roy, M. 2019. Landscape-based habitat engineering for sugarcane ecosystem: a green technological option for pest management. *Sugar Tech* 21 (2), 1–14.

Sala, S., Vasta, A., Mancini, L., Dewulf, J., Rosenbaum, E. (Eds). 2015. *Social Life Cycle Assessment - State of the Art and Challenges for Supporting Product Policies*. European Commission, Joint Research Centre, Institute for Environment and Sustainability, Publications Office of the European Union, Luxemburg.

Scherr, S. J., Shames, S. and Friedman, R. 2012. From climate-smart agriculture to climate-smart landscapes. *Agriculture & Food Security*, 1(1), 12.

Siles, P., Aguilar, C., Quinde, K., Castellón, J., Somarriba, F., Tapia, A., Brenes, S., Deras, M., Matute, N., Staver, C. and Bustamante, O. 2013a. Intercropping bananas with coffee and trees: prototyping agroecological intensification by farmers and scientists. *Acta Horticulturae*, 986, 79–86.

Siles, P., Bustamante, O., Valdivia, E., Burkhardt, J. and Staver, C. 2013b. Photosynthetic performance of banana ('Gros Michel') under a natural shade gradient. *Acta Horticulturae*, 986, 71–77.

Staver, C. 1989a. Why farmers rotate fields in maize-cassava-plantain bush fallow agriculture in the wet Peruvian Amazon. *Human Ecology*, 17(4), 401–26.

Staver, C. 1989b. Shortened bush fallow rotations with relay-cropped Inga edulis and Desmodium ovalifolium in wet central Amazonian Peru. *Agroforestry Systems*, 8(2), 173–96.

Staver, C. 1991. The role of weeds in the productivity of Amazonian bush fallow agriculture. *Experimental Agriculture*, 27(3), 287–304.

Staver, C. 2018. Good agricultural practices: an end point or a starting point for more sustainable banana production? In G. Kema and A. Drenth (Eds), *Achieving Sustainable Cultivation of Bananas* (pp. 179–214). Burleigh Dodds Science Publishing.

Staver, C and Lescot, T. 2015. *Propagating Quality Planting Material to Improve Plant Health and Crop Performance, Key Practices for Dessert Banana, Plantain and Cooking Banana: Illustrated Guide*. Bioversity International, Montpellier, France. Available at: http://banana-networks.org/innovate-plantain/files/2016/02/multiplication_guide_English_Bookmarked_opt.pdf (English version); http://banana-networks.org/innovate-plantain/files/2016/02/multiplication_guide_French_bookmarked_opt.pdf (French version); http://banana-networks.org/musalac/files/2015/02/multiplication_guide_Spanish_bookmarked_opt.pdf (Spanish version); http://banana-networks.org/barnesa/files/2015/08/Guide_Illustre_Arabe.pdf (Arabic version).

Staver, C., Siles, P., Bustamante, O., Garming, H., Castellon, N. and Garcia, J. 2012. Bananas in coffee Agroforestry in Latin America: Assessing ecological and socio-economic benefits. *24th International Conference on Coffee Science*, San Jose, Costa Rica. Available at: https://www.asic-cafe.org/conference/24th-international-conference-coffee-science.

Staver, C., Siles, P., Bustamante, O. and Castellon, N. 2013. Banano y plátano en cafetales agroforestales: Aspectos prácticos para la produccion sostenible. *II Congreso Latinoamericano y del Caribe de Banano y Platano*, 29–30 Mayo, Armenia, Colombia. Available at: http://banana-networks.org/musalac/files/2015/09/003-Charles-Staver-Banano-y-Platano-en-Cafetales-Agroforestales.pdf.

Staver, C., Siles, P., Bustamante, O., Castellon, N. and Mpiira, S. 2015. Improving banana agroecosystem productivity: Lessons for integrated systems intensification. *International Conference on Integrated Systems*, 3–6 March. IITA, Ibadan, Nigeria. Available at: https://fr.slideshare.net/humidtropics/improving-banana-agroecosystem-productivity-lessons-for-integrated-systems-integration-by-charles-staver-et-al.

Staver, C., Turmel, M., Siles, P., Mpiira, S., Bustamante, O., Calberto,, G., Dita,, M., Omondi, A., Alvarez, E. and Zheng, S. 2018. Step by step tools to identify agroecological intensification alternatives for banana cropping systems. *Acta Horticulturae*, 1196, 9–18 (see original presentation at http://www.promusa.org/tiki-download_file.php?fileId=2833&display=y).

Stewart, R., Jian, J., Gyawali, A., Thomason, W., Badgley, B., Reiter, M. and Strickland, M. 2018. What we talk about when we talk about soil health. *Agricultural and Environmental Letters*, 3(1), 1–5.

Taulya, G. 2015. *Ky'osimba Onaanya: Understanding productivity of East African highland banana*. Ph.D. Thesis. Wageningen University.

Tenkouano, A., Hauser, S., Coyne, D. and Coulibaly, O. 2006. Clean planting materials and management practices for sustained production of banana and plantain in Africa. *Chronica Horticulturae*, 46(2), 14–18.

Tixier, P., Malezieux, E. and Dorel, M. 2004. SIMBA-POP: a cohort population model for long-term simulation of banana crop harvest. *Ecological Modelling*, 180(2–3), 407–17.

Tixier, P., Malézieux, E. and Dorel, M. 2007. A model-based approach to maximise gross income by selection of banana planting date. *Biosystems Engineering*, 6, 471–6.

Tixier, P., Malézieux, E., Dorel, M. and Wery, J. 2008. SIMBA, a model for designing sustainable banana-based cropping systems. *Agricultural Systems*, 97(3), 139–50.

Tixier, P., Lavigne, C., Alvarez, S., Gauquier, A., Blanchard, M., Ripoche, A. and Achard, R. 2011. Model evaluation of cover crops, application to eleven species for banana cropping systems. *European Journal of Agronomy*, 34(2), 53–61.

Torquebiau, E. and Akyeampong, E. 1994. Mieux comprendre l'effets de l'ombre -ses effets sur le haricot, le mais et la banane. *L'Agroforesterie Aujourdhui*, 6(4), 17–19.

True Price/TruCost. n.d. *The External Costs of Banana Production - A Global Study*. Available at: https://www.fairtrade.net/fileadmin/user_upload/content/2009/resources/2018_CostsBananaProduction.pdf.

Turner, D. 1998. Ecophysiology of bananas: the generation and functioning of the leaf canopy. *Acta Horticulturae*, 490, 211–22.

Turner, D. W., Fortescue, J. and Thomas, D. 2007. Environmental physiology of the bananas (*Musa* spp.). *Brazilian Journal of Plant Physiology*, 19(4), 463–84.

Turner, D. W., Fortescue, J., Ocimati, W. and Blomme, G. 2016. Plantain cultivars (*Musa* spp. AAB) grown at different altitudes demonstrate cool temperature and photoperiod responses relevant to genetic improvement. *Field Crops Research*, 194, 103–11.

Tushemereirwe, W. and Kubiriba, J. 2018. Banana cultivation in Africa. In G. Kema and A. Drenth (Eds), *Achieving Sustainable Cultivation of Bananas* (pp. 33–64). Burleigh Dodds Science Publishing.

Valdivia, E. 2012. Respuesta de nueve cultivares de musáceas en la etapa vegetativa a cuatro niveles de sombra agroforestal. M.Sc. Thesis. CATIE Escuela de Posgrado, Turrialba, Costa Rica.

van Asten, P., Fermont, A. and Taulya, G. 2011. Drought is a major yield loss factor for rainfed East African highland banana. *Agricultural Water Management*, 98(4), 541–52.

Van Ittersum, M., Cassman, K., Grassini, P., Wolf, J., Tittonell, P. and Hochman, Z. 2013. Yield gap analysis with local to global relevance—a review. *Field Crops Research*, 143, 4–17.

van Noordwijk, M. and Brussaard, L. 2014. Minimizing the ecological footprint of food: closing yield and efficiency gaps simultaneously? *Current Opinion in Environmental Sustainability*, 8, 62–70.

Vicente-Chandler, J., Abruña, F. and Silva, S. 1966. Effect of shade trees on yields of five crops in the humid mountain region of Puerto Rico. *The Journal of Agriculture of the University of Puerto Rico* 50 (3), 218–25.

Wairegi, L., van Asten, P., Tenywa, M. and Bekunda, M. 2010. Abiotic constraints override biotic constraints in East African highland banana systems. *Field Crops Research*, 117(1), 146–53.

Wolfert, S., Ge, L., Verdouw, C. and Bogaardt, M. 2017. Big data in smart farming–a review. *Agricultural Systems*, 153, 69–80.

# Chapter 11

# Advances in breeding of coconut

*L. Perera, Coconut Research Institute, Sri Lanka; R. Manimekalai, Sugarcane Breeding Institute, India; and S. Sudarsono, Bogor Agricultural University, Indonesia*



## 1 Introduction

Coconut, *Cocos nucifera* L., was once the most important palm crop in the humid tropics. Coconut oil trade was at its height around 1908 and coconut oil remained as the most important edible oil worldwide until the 1960s (Nampoothiri et al., 2018). Coconut oil trade subsequently declined in the face of competition from cheaper sources such as oil palm as well as the perception that the high saturated fatty acid content of coconut oil might increase the risk of heart disease. This led to the loss of large coconut plantations with the result that 96% of current total world coconut production is by smallholders (Persley, 1990, 1992).

While guidelines from organizations such as the American Heart Association continue to advise against high intake of processed coconut oil as a potential risk factor in cardiovascular disease (Sacks et al., 2017), some research has highlighted the potential health benefits of the medium chain saturated fatty acids (MCT) present in virgin coconut oil. MCT have been shown to increase levels of high-density lipoprotein (HDL) which reduces the risk of cardiovascular disease (Cardoso et al., 2015). Now, there is a growing body of evidence to suggest that coconut oil may have a role in reducing the risk of cardiovascular disease (Eyres et al., 2016; Wallace, 2019). Coconut contains also lauric acid,

http://dx.doi.org/10.19103/AS.2019.0054.21

which is linked to an improved immune response (Peedikayil et al., 2016). There is also limited evidence that coconut oil may reduce the risk of Alzheimer's disease, though more research is needed to substantiate this claim (Dofy, 2012; Swaminathan and Jicha, 2014). As a result, there is an increasing consumer interest in products such as coconut water and virgin coconut oil among western consumers in particular, which may help boost future production.

After an introduction to the current extent of coconut cultivation and uses, the chapter discusses the origins and botany of coconut. It then reviews available genetic resources. The chapter goes on to review conventional breeding techniques such as mass selection and hybridization before discussing the developments in molecular breeding techniques for coconut.

## 2 Coconut cultivation and uses

Coconut has a pan-tropical distribution, occurring in the area between latitudes 20°N and 20°S of the equator and at altitudes from sea level to 1200 m. Coconut is currently grown in 89 tropical countries on 12 million ha (APCC, 2014). Coconut grows best under conditions of high humidity and at temperatures between 27°C and 30°C on moderately to well-aerated soils. These conditions mean that most cultivation is concentrated in the Asia-Pacific region. About 65% of world coconut production comes from Southeast Asia, mainly from the Philippines and Indonesia (which account for 7.19 million ha of production). Other important Asian producers are India and Sri Lanka. The remaining coconut production comes almost equally from the Pacific, American and African regions. Mexico and Brazil are the major coconut producers in the American region, while Ivory Coast, Mozambique and Tanzania are the major producers in the African region (https://coconutcommunity.org/statistics).

The International Coconut Community (ICC), formally the Asian and Pacific Coconut Community (APCC), is the only international organization specifically representing coconut-producing countries established in 1969 under the United Nations Economic and Social Commission for Asia and the Pacific (UN-ESCAP). The ICC has 18 member countries accounting for over 90% of the world's coconut production and exports of coconut products. The ICC member countries include the following Asian and Pacific countries:

- Asia: India, Indonesia, Malaysia, Sri Lanka, Thailand, Vietnam and Timor Leste.
- Pacific: Federated States of Micronesia, Fiji, Kiribati, Marshall Islands, Papua New Guinea, Philippines, Samoa, Solomon Islands and Tonga.

Jamaica and Kenya are associate member countries of the ICC. Membership to the ICC is open to all coconut-producing countries with the unanimous consent of the existing members and by acceding to the agreement establishing the ICC.

While coconut is mainly harvested for its oil, almost every part of the palm has a use (Burkill, 1966; Harries, 1995). Botanically, coconut is a fibrous drupe. The endosperm has a solid component (kernel) and a liquid component (coconut milk), protected by a hard endocarp (shell) and a fibrous mesocarp (husk). The exocarp, the thin smooth outer layer, is found in varying colours. The mesocarp is thick and composed of a fibre called coir (Fig. 1).

The kernel (or the solid endosperm) of the nut is primarily used to produce copra (dried endosperm) from which coconut oil is extracted. Desiccated coconut (shredded dried endosperm) is another important kernel product that is used in confectionary manufacture. Coconut milk and creams extracted from the kernel are often used in food preparation. In addition to its use as cooking oil, it is also used as a raw material in manufacturing products such as soap, glycerine, margarine, detergents, cosmetics and pharmaceuticals. The fibre extracted from the husk (exocarp) of the nut has versatile uses in traditional products such as doormats, baskets, hats, fans, brooms, mattresses and ropes. In countries such as Sri Lanka, fibre products can be more valuable than kernel-based products. Coir pith, the remainder of the husk after fibre extraction, is used as an organic matter for soil moisture conservation. Coconut fronds are woven into roofing material in poor communities. The stem (trunk) is made into furniture, handicrafts and building materials. The sugary sap obtained by

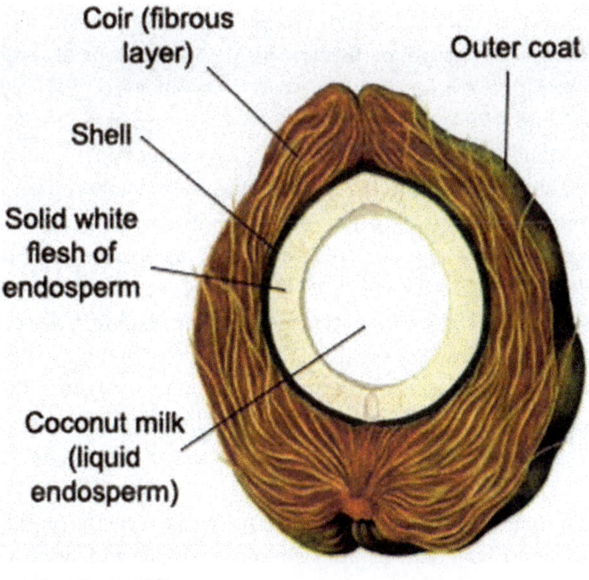


**Figure 1** Coconut components.

tapping unopened inflorescence gives an alcoholic beverage (known as toddy or palm wine). Fermented toddy is the base for the production of arrack (a distilled alcoholic drink), vinegar and jaggery (a concentrated form of sugar). The shell (endocarp) of the nut is used to produce activated charcoal (used for medical purposes) and is polished and decorated to make ornaments. The remainder of the copra after oil extraction is called coconut cake or poonac and is used as a feed supplement for livestock due to its rich protein content.

In addition to these more established uses, there are a wide range of applications, which has increased its market potential (Nayer, 2017). These include the use of coconut fibre in emerging high-value products such as geotextiles and natural-fibre composites, and the use of coir pith as a growing medium for pot plants in hydroponic cultivation. As noted earlier, the increasing interest in the possible health benefits has led to new products such as virgin coconut oil (including in capsule form), coconut water (as a natural energy drink), coconut sap sugar (as a low-calorie sweetener) and coconut milk powder/cream.

## 3 Origin and botany

The broad distribution of coconut in the tropics, the long history of human interaction with coconut in many regions of the world and the lack of wild relatives have made it difficult to determine the origin of coconut. As a result, there have been a number of conflicting theories. Early theories supported a New World origin for coconut with subsequent dispersal to Asia and Polynesia (Guppy, 1906; Cook, 1911; Ridley, 1930). This theory was supported by the presence of the closest relative of coconut, Cocoid palm, in the Northwestern part of South America. However, a range of ethnological, entomological, archaeological and palaeontological evidence suggests a different theory of Polynesian and Asian origin (Child, 1964, 1974; Burkill, 1966; Fremond et al., 1966; Dennis and Gunn, 1971; Mahabale, 1976; Harries, 1978; Purseglove, 1985). Although the origin of coconut is still not fully resolved, the most likely region for coconut domestication is now considered to be Melanesia, on the coasts and islands between Southeast Asia and the Western Pacific, approximately between New Guinea and Fiji (Child, 1964; Harries, 1990; Purseglove, 1985). Gunn et al. (2011) analysed a sample of 1322 coconut accessions representing the geographical and phenotypic diversity of coconut using ten microsatellite loci and proposed two independent geographical origins of coconut cultivation: the islands of Southeast Asia and the southern margins of the Indian subcontinent.

From its putative centre of origin, coconut was disseminated both east and west by plant materials carried by ocean currents (Ohler, 1984) and by human dissemination, particularly by ancient Melanesian and Polynesian sea voyagers, who carried coconut as a source of food and drink on long sea voyages (Harries, 1978). There is evidence that coconuts were on the Pacific

Coast of Panama from the pre-Columbian era (i.e. before 1492), either via early Polynesians carrying them or by ocean currents distributing them naturally (Whitehead, 1976). Coconut reached West Africa only after AD 1500 through Portuguese traders (Purseglove, 1985), possibly from East Africa (perhaps via Malaysian seafarers who reached Madagascar in the first century AD), and then spread to the Caribbean and Atlantic Coast of America (Child, 1974; Purseglove, 1972; Harries, 1977; Zizumbo-Villarreal, 1996). Plantations were developed throughout the tropics by the end of nineteenth century. Coconut populations that are now considered endemic to the Atlantic Coast of Africa, America and around the Caribbean region are derived from coconuts in East Africa, India and Sri Lanka (Harries, 1977), while coconuts in the Pacific Coast of America are related to coconuts from the Pacific islands and Southeast Asian according to fruit component analysis data (Harries, 1978).

Coconut (*Cocos nucifera* L.) is a member of the monocot palm family Arecaceae (alternatively called Palmae) and the subfamily Cocoideae that includes 27 genera and 600 species and is currently the only species of the genus *Cocos*. Coconut possess a diploid genome with 32 chromosomes ($2n=2x=32$). The scientific classification of coconut palm is given below:

Kingdom: Plantae
Order: Arecales
Family: Arecaceae
Subfamily: Arecoideae
Tribe: Cocoeae
Genus: *Cocos*
Species: *Cocos nucifera*
Binomial name: Cocos *nucifera* L.

The major classification of coconut is based on stature and breeding behaviour, which groups coconut into two broad categories (Menon and Pandalai, 1958):

- Tall (referred to as variety Typica)
- Dwarf (referred to as variety Nana).

The major characteristics of tall (known as 'Talls') and dwarf (known as 'Dwarfs') varieties are shown in Table 1. Talls are the most common and commercially important varieties, growing to around 20–30 m. They are largely cross-pollinated and characterized by their late-bearing nature and medium to large-sized nuts. Talls are generally hardy and thrive well in wide range of environmental conditions.

In contrast, Dwarfs grow to a height of about 10–15 m and are predominantly self-pollinating (autogamous). Despite being self-fertilising,

**Table 1** Contrasting features of tall (Typica) and dwarf (Nana) varieties of coconut

| Character | Tall (Typica) | Dwarf (Nana) |
|---|---|---|
| Stature | Tall (about 20–30 m) | Short (about 10–15 m) |
| Enlargement at the base of the stem (Bole production) | Yes | No |
| Time taken to first flower | 6–8 years | 3–4 years |
| Economic life span | Long (about 80–100 years) | Short (about 40 years) |
| Bearing nature | Continuous | Seasonal |
| Nut production/palm/year | Average 40 | Average 80–100 |
| Copra amount and quality | 200 g/nut, good quality | 80–100 g/nut, poor quality |
| Growing conditions | Variable | Sensitive to climatic changes |
| Breeding habit | Out breeding | Inbreeding |
| Genetic constitution | Heterozygous | Homozygous |

Dwarfs cross-pollinate with one another and with Talls. They are early bearing and generally produce small nuts in large numbers (50–60 nuts per bunch) with about 14–16 bunches per year. Their nuts have distinctive colours such as green, red, yellow and brown. There is a disagreement as to whether Dwarfs are the result of inbreeding among Talls or are the result of domestication of coconut by man (, Swaminathan and Nambiar 1961; Raveendranath and Ninan, 1974; Purseglove, 1985; Harries, 1978, 1995). In addition to tall and dwarf varieties, there are intermediate varieties, referred to as semi-tall or semi-dwarf varieties (Liyanage, 1958; Menon and Pandalai, 1958).

Harries (1978) proposed two main types of tall coconuts:

- 'Niu kafa'
- 'Niu vai'

'Niu kafa' produces large, long, angular, thick-husked and slow-germinating nuts with less free water content. This type evolved naturally and was disseminated via ocean currents (the wild type) and is concentrated in South Asia, East and West Africa. The 'Niu vai' type is characterized by spherical nuts with an increased proportion of endosperm and reduced husk thickness, early germination and disease resistance. 'Niu vai' is concentrated in Southeast Asia, the Pacific and the west coast of Panama and is believed to have evolved as a result of selection by humans for characteristics such as increased nut size (Child, 1964). Harries (1978) and Foale (1991) have suggested that introgression of these two types, together with further selection and dissemination by man, are responsible for the wide range of varieties and pan-tropical distribution of coconut seen today. Based on this work, Bahala et al. (1989) has suggested three different categories of coconuts: wild ('Niu kafa'), domestic ('Niu vai') and modern cultivated varieties. It is generally accepted that pure wild-type coconuts no longer exist though their characteristics are present in modern populations. Current coconut varieties are a complex mix of wild and domestic characteristics, depending on the predominant type when cultivation began in each region. Coconut grows to a maximum height of about 30 m in a period of about 60 years. They have large, long fronds about 3.5 m long with 60-90 cm long pinnate configuration. Coconut flowers continuously at the rate of 12 inflorescences on average per year. The coconut palm, like other Palmae species, has a fibrous root system. Roots are produced continuously throughout the lifespan of the tree, typically running 3 m horizontally from the base and to a depth of about 1.5 m, though some roots may penetrate as far as 10 m down in deep loamy soils. New roots emerge and die in a continuous process. The root epidermis hardens over time and becomes less permeable to water and nutrients.

The coconut palm has a single terminal bud (cabbage). Permanent damage to the growing bud causes death of the palm. In the tall variety the base of the trunk is enlarged to form a bole which is about 150 cm in diameter and about 45–60 cm in height. From the bole upwards, the trunk reduces to about 90 cm in diameter. Stem growth is rapid up to 15 years of age at a rate of about 10–150 cm per year, reducing to about 30–40 cm per year between 15 and 40 years of age and thereafter about 10–15 cm per year. The coconut stem has no cambium but up to about 18000–20000 vascular bundles. As a result, although coconut cannot rejuvenate injured tissues, it can tolerate some damage by continuing the translocation of food and water through undamaged tissues.

The coconut palm is monoecious, that is, its inflorescence carries both male and female flowers. In both tall and dwarf coconut varieties male flower opening commences from the date of opening of the inflorescence and generally lasts for about 18–22 days (male phase). The female phase in tall coconuts, in contrast, begins about 3 days after the male phase lapses and lasts only about 2–3 days. The receptivity of the female flowers to the previous inflorescence lapses when a new inflorescence opens. Tall varieties of coconuts are thus largely cross-pollinated. However, some cultivars (e.g. 'Rennel tall', 'Polynesian tall', 'Malayan tall' and 'New Hebrides tall') produce inflorescences more frequently, allowing inter-spadix pollination within the same palm, thus promoting selfing in usually cross-pollinating Talls. Some cultivars are strictly allogamous (e.g. 'West African tall'). In dwarf varieties, the female phase begins at the same time as the male phase and lasts for 12–15 days. As a result, self-pollination is a common phenomenon in dwarf coconuts.

Pollination is usually via insects, particularly honeybees (*Apisindica*). Pollination mites (*Neocypholae lapsampulula*) are present in large numbers in the coconut inflorescences and assist pollination. Honeybees transport the mites from palm to palm. Wind pollination plays a minor role. With successful natural pollination, 50–70% of female flowers usually abort within the period of first 3 months, depending on the prevailing weather conditions and the competition for carbohydrates from the developing bunches in the palms. The remaining flowers develop into fruits, which mature in about 11–12 months.

## 4 Genetic resources

Coconut, *Cocos nucifera* L., is the only species in the genus *Cocos* with no known closely related wild relatives in contrast to many other cultivated crops. There is significant morphological diversity (Narayana and John, 1949; Gangolly et al., 1957; Menon and Pandalai, 1958; Liyanage, 1958; Bourdeix et al., 2005). There is a great diversity in fruit characteristics within and between populations in size, shape, colour and composition. Fruit shape varies from near-spherical to short- and long-angular. There is a great deal of variation in other characteristics such as frond morphology, orientation of the crown, number and length of bunches, number of female flowers and number of nuts per palm even within populations. Morphological diversity is higher in Southeast Asia than in South Asia, Africa or South America (Benbadis, 1992; Whitehead, 1976). Many different varieties and types (commonly referred to as ecotypes) have been described based on morphology and breeding habits (Narayana and John, 1949).

Coconut genetic resources are currently threatened by high rates of genetic erosion as a result of urbanisation, the spread of more intensive forms of agriculture, extreme weather events associated with climate change and the impact of pests and diseases. Moreover, coconut development programmes involving replanting traditional areas of coconut cultivation focus on planting a small number of high-yielding varieties. This has led to the development of programmes to conserve genetic diversity, for example, through germplasm collections. These germplasm collections are mainly collections of palms in the field (*ex situ* gene banks). Since this is very resource intensive, alternative cryopreservation techniques have also been investigated. A review of key issues in conserving coconut genetic resources is Batugal et al. (2005).

The International Coconut Genetic Resources Network (COGENT) was established in 1992 as a global network to coordinate national conservation programmes. In 1996 COGENT initiated a germplasm conservation programme in 13 Asia-Pacific countries with the funding from the Asian Development Bank (ADB). This programme was extended to Mauritius, Madagascar, Seychelles, Sri Lanka and Vietnam in 1997. The germplasm collection and conservation programmes worldwide now have 1680 accessions, comprising of over 1200 tall and 400 dwarf varieties, recorded in the Coconut Genetic Resources Database (CGRD). Selection of those coconut accessions/collections is mainly based on phenotype and ecological distribution. As a result these collections probably contain some duplications. COGENT has also established large multi-site international coconut gene banks in Indonesia, India, Papua New Guinea and Ivory Coast (IPGRI, 1997, 1998) in addition to existing collections in Tanzania, Ivory Coast, Philippines and India.

Countries with narrow genetic diversity need to strengthen their gene banks with genetic sources from other countries. However, safe importation of genetic materials is essential to prevent accidental introduction of new pests and diseases. Seed nuts are not suitable since any pathogens in the nuts are protected from decontamination techniques such as fumigation. The only safe method of germplasm movement between countries is via embryos. Mature embryos of 10–12-month-old nuts are collected, sterilized and pre-cultured in an appropriate storage medium (Nguyen et al., 2015). Once they arrive at the recipient laboratories the embryos are transferred to a normal embryo germination medium, raised as plants, acclimatized and field-planted. However, the overall success of embryo recovery has been as low as 10% mainly due to poor *in vitro* germination. The success of embryo recovery depends on factors such as storage conditions and duration of storage, as well as the quality of staff and infrastructure in laboratories (Benbadis, 1992).

The genetic diversity of coconut has been assessed by various authors using molecular marker systems (Everard, 1996; Ashburner et al., 1997; Duran et al., 1997; Wadt et al., 1999; Dasanayake et al., 2003; Manimekalai and Nagarajan, 2006; Rajesh et al., 2013; Novarianto et al., 2001; Pandin, 2009a,b; Lebrun et al., 1998, 1999; Teulat et al., 2000; Perera et al., 1998, 1999, 2000, 2001, 2003; Rivera et al., 1999; Meerow et al., 2003; Akuba, 2002). High levels of genetic diversity in tall coconuts has been observed across all regions where they are grown. The genetic diversity of dwarf varieties is lower, mainly owing to their self-pollinating nature and recent divergence from tall coconut varieties. A high level of genetic diversity in coconut has been observed using molecular markers (Perera et al., 2000, 2003). The mean genetic diversity of coconut on average was 0.647 (Nei, 1987; Perera et al., 1999), with much higher genetic diversity in tall compared to dwarf varieties (Perera et al., 2004).

## 5 Conventional breeding techniques: mass selection and hybridization

The traditional aim of coconut breeding has been to improve economically important traits such as earliness in flowering, high nut yield in terms of nut size and numbers as well as kernel and oil percentage. With the threat of climate change, drought has become a more significant constraint to coconut production and breeding for drought tolerance has become an important goal in coconut breeding programmes. In the same context, there are now breeding programmes for cold tolerance in Southern China and for cyclone tolerance in India as well as some Caribbean and Pacific countries. Although many pests of coconut can be controlled effectively with physical, chemical and biological control methods, there are some pests like coconut mite (*Aceria guerreronis*) that are not easy to control with these techniques. As a result breeding for resistance to pests has also become a goal in current breeding programmes. Similarly, breeding for disease resistance has also become a high priority as the only sustainable method to deal with diseases such as lethal yellowing (LY) and its variants caused by phytoplasma.

The long vegetative period and highly heterozygous nature of coconut, the large area of land required for evaluation, the lack of vegetative propagation methods and the small number of seeds produced per year limit the use of traditional breeding methodologies employed for other crops. Due to the long juvenile phase (7–10 years), phenotypic evaluation of desirable reproductive traits and progeny testing is a lengthy process. Unlike rubber, tea and oil palm no clonal propagation method exists in coconut and hence multiplication of elite trees is not possible. Moreover, low seed production (50–80 nuts/palm/year over 12 months period) compared to that of oil palm, coupled with low rate of success (20%) in hand pollination, not only limits the seed production from a particular pair of elite palms, but also limits the size of a full-sib family for evaluation and genetic trials. The development of inbred lines from heterozygous palms is almost impossible because of the long period for seed production. As a result, traditional coconut breeding is confined to mass selection of phenotypically superior mother palms and inter- and intra-varietal hybridisation.

Most of the mature coconuts grown currently are probably derived from mass selection based on criteria developed informally by growers. The range of traditional and improved varieties has been reviewed by Bourdeix et al. (2006). Most large coconut plantations were established from seeds collected based on the preference of the individual grower, primarily based on yield. Mass selection was done using only the female parent and only half of the additive genetic effects were gained as the male parent was not controlled. Research has focussed on estimation of heritability values for a number of useful characteristics in coconut (Liyanage and Sakai, 1960). These studies revealed the effectiveness of selection of parent palms based on husked nut weight, setting of female flowers into fruits, yield of copra (dried kernel) per palm per year, number of female flowers per bunch and nuts per bunch. In addition, characteristics such as short internode length and high endosperm weight have also been considered as useful criteria in selecting seed palms (Liyanage et al., 1988).

Seed production by artificial cross-pollination between high-yielding parent palms (hybridization) is another breeding strategy used in coconut. Dwarf by tall hybrids developed by artificial pollination have shown remarkable gains in terms of combined traits of nut yield and precocity in both parents. Many Dwarfs by tall hybrids have been produced since their first discovery, utilising different tall and different dwarf forms (Nuce De Lamothe and Bernard, 1985; Bourdeix et al., 1990).

In inter-varietal hybridization, different crosses between genetically diverse varieties are made using hand-pollination techniques. The principle of hand pollination focuses on emasculation of the inflorescence followed by isolation of the female flowers by means of a bag. When the female flowers become receptive, selected pollen is artificially introduced. The procedure involves three steps: collection and preparation of the pollen, isolation of female flowers and pollinating the female flowers with selected pollen. The seeds derived from different crosses are evaluated in the field and subsequently suitable cross/crosses are identified. Seed production by artificial pollination is expensive, laborious and results in a limited number of seed nuts. Selected hybrids are multiplied by the isolated seed gardens technique, which allows direct natural pollination between selected parents. The isolation of the seed garden is achieved either by a forest barrier around the seed garden that is thick enough to prevent pollen arriving from outside or by planting 12 guard rows around the seed garden from the same pollen parent to trap pollen coming from outside. Within the seed garden, the female parent is emasculated so that only pollen from the male parent pollinates the female flowers of the female parents. This is the most commonly practised technique for the mass production of hybrid seeds.

Micropropagation by immature ovary culture in Sri Lanka (Perera et al., 2007, 2008a,b) and plumule explants derived from zygotic embryos (Chan et al., 1998) in Mexico have shown promising results in mass propagation of coconut through tissue culture. A disadvantage of micropropagation through zygotic embryos is that the performance of the clones is not known till they are fully grown and display target characteristics in the field. One of the limitations of ovary culture is the limited number of female flowers in inflorescence. This has been overcome by callus multiplication steps after initiation of the callus. Attempts have been made to select elite mother palms that show a good response to *in vitro* culture conditions. Research suggests that there is no somaclonal variation or genetic instability occurring in plantlets that are regenerated from ovary explants, suggesting that large-scale micropropagation is feasible (Bandupriya and Dunwell, 2012; Bandupriya et al., 2017). Recent research has also improved micropropagation through zygotic embryos so that it can be scaled up for commercial application (Oropeza et al., 2016). Anther culture has also shown promising results (Perera et al., 2008a).

## 6 Molecular breeding techniques

Although conventional coconut breeding using standard conventional breeding techniques based on phenotypic selection have been successful in producing high-yielding varieties, the inherent constraints in coconut breeding make the potential use of biotechnologies in coconut breeding highly attractive. The application of molecular genetics in coconut breeding, particularly molecular markers, began in the early 1990s and their application in coconut have been diverse, ranging from assessing genetic diversity to creating high-resolution genetic maps for identification of particular traits.

Initially the studies were aimed at assessment of coconut genetic diversity and genetic relatedness at the DNA level, using universal marker techniques such as RAPD (Everard, 1996), RFLP (Lebrun et al., 1999, 1998), AFLP (Perera et al., 1998), ISTR (Duran et al., 1997; Rohde et al., 2000) and SSR (Perera et al., 2000, 2001, 2003). They have also been used to identify varieties correctly, offsetting the limitations of phenotypic markers which are not always accurate. This technique has been used to identify the Sri Lanka Brown Dwarf variety which was used in the hybrid breeding programme in Sri Lanka with two successful hybrids released for commercial planting in 2012 (Perera et al., 2002, 2012). Perera et al. (2004) reported the successful use of microsatellite molecular markers in identifying and differentiating coconut varieties used as parents in the Sri Lanka coconut breeding programme (Sri Lanka Tall, Sri Lanka Green Dwarf and Sri Lanka Yellow Dwarf varieties) and then in identifying the resulting hybrids (Sri Lanka Green Dwarf × Sri Lanka Tall and Sri Lanka Yellow Dwarf × Sri Lanka Tall). Similarly, Bandaranayake and Kearsey (2005) have used markers to purify the San Ramon variety from mixed populations for planting in seed gardens. The application of microsatellite markers to study somaclonal variation in clonal plants regenerated from various explants through somatic embryogenesis and successfully field planted in Sri Lanka have been reported by Fernando et al. (2004). Microsatellite markers have confirmed the absence of variants within the clones (Perera et al., 2008a,b).

Molecular breeding is the application of molecular tools in breeding programmes to increase the efficiency and effectiveness of breeding programmes. Several attempts have been taken to use molecular tools in coconut breeding programmes through quantitative trait loci (QTL) mapping/gene discovery and marker-assisted selection (Huang et al., 2013; Rajesh et al., 2018). The first genome map of coconut was developed for an $F_1$ population of East African Tall × Laguna Tall based on ISTR markers (Rohde et al., 1999). The Indonesian Palm Genome Project has sequenced the genomes of 'Tenga Tall' and 'Pati Kopyor Green Dwarf' as representative tall and dwarf palm varieties (Lestari et al., 2016). Alsaihati et al. (2014) reported the size of coconut genome at around 2.6 Gb with 50–70% repeat sequences. Building on this work, a draft coconut genome has recently been completed (Xiao et al., 2017). Over the last four years a number of research institutions have contributed to generate coconut transcriptome data including: Academia Sinica (https://www.sinica.edu.tw/); the Central Plantation Crops Research Institute (CPRI, http://cpcri.gov.in/) and University Putra Malaya (UPM, http://www.upm.edu.my/). Much of this data is deposited in publicly available databases such as TROPgeneDB (http://tropgenedb.cirad.fr/tropgene/JSP/index.jsp) and NCBI GenBank Database. The NCBI GenBank database currently has 211 744 entries of coconut nucleotide sequences of either chloroplastic, mitochondrial or nuclear genomes. (http://www.ncbi.nlm.nih.gov/nucleotide, accessed: 10 February 2016, at 10:25).

This work has made it possible to identify QTLs which can be used to target desirable traits for breeding. An example is a mapping population developed in the Philippines from a cross between Malayan Yellow Dwarf × Laguna Tall using AFLP, ISTR, RAPD and ISSR markers. Three hundred and eighty-two markers have been identified resulting in sixteen linkage groups and identifying six QTLs for early germination, early flowering and high yield in coconut (Herran et al., 2000). QTLs for other traits such as leaf production, girth and height have also been identified for the same mapping population (Ritter et al., 2000). Another mapping population in the Ivory Coast derived from Cameroon Red Dwarfs and Rennell Island Tall has been used to map 280 markers in 16 linkage groups resulting in identification of several QTLs related to nut number, number of bunches and traits related to fruit quality (Lebrun et al., 2001; Baudouin et al., 2006). An $F_1$ mapping population derived from East African Tall and Rennell Island Tall developed at Chambezi Research Station (Tanzania) has been used to identify QTL for wax content and wax components (Riedel et al., 2009).

The size of the mapping population and selection of parents for the maximum segregation of traits are critical when producing a genome map of any crop (Bandaranayake and Kearsey, 2005). The size of the mapping population is particularly a crucial issue in coconut because obtaining a reasonable number of seed nuts from a particular crossing combination is a difficult task given very low seed production, that is, about 100 seed nuts per palm per year. Bandaranayake and Kearsey (2005) concluded that 200–400 individuals were needed for a mapping population. Furthermore, genetic relationship studies in coconut suggest that maximum segregation of traits in coconut can only be obtained by crossing between varieties of Southeast Asia and the Pacific group and Indo-Atlantic group. A large mapping population has now been developed in Sri Lanka comprising 350 individuals arising from a cross between 30 genetically identical Sri Lanka Red Dwarf (Southeast Asia and Pacific group) and a single Sri Lanka Tall (Indo-Atlantic group) to obtain maximum segregation of traits and sufficient number of meioses to analyse. An improved linkage map can be derived by generating an $F_2$ population from this $F_1$ population. Another mapping population, particularly segregating for tolerance to *Aceria* mite, has been constructed in Sri Lanka between tolerant Sri Lanka Yellow Dwarf and a highly susceptible Sri Lanka Tall palm (Perera et al., 2007). Attempts are also being undertaken to investigate the possible synteny of the oil palm and the coconut genomes as both palm species are diploid and have the same chromosome number, $2n = 32$. Synteny would possibly speed up research by increasing the marker density on the respective linkage maps by the exchange of DNA markers. Development of number of EST-SSRs from oil palm has been undertaken and their cross-amplicability is under investigation.

Molecular markers are now being used in areas such as resistance breeding. A good example is the identification of WCLWD phytoplasma in coconut palms in the southern province of Sri Lanka (Perera et al., 2012) and its application in reliable screening of resistant parents for resistant breeding programmes. A recent report on cloning and molecular characterization of cDNA of two genes homologous to the NPR family linked to a salicylic acid (SA)-dependent defence response in coconut palm, which improves resistance to LY (Nic-Matosi et al., 2017). Puch-Hau et al. (2015) have also identified candidate genes conferring resistance to LY. These findings offer new opportunities to improve disease resistance breeding programmes.

The development of mapping populations is time-consuming, laborious and costly. Association studies that correlate traits and molecular markers are an alternative to identify QTLs/molecular markers associated with important traits using natural populations. However, association mapping requires larger populations and high-throughput marker systems to be successful (Batugal et al., 2009). Shalini et al. (2007) have identified a combination of molecular markers linked to mite resistance by stepwise multiple regression using coconut genotypes and mite-resistant and susceptible accessions collected from different parts of South India. In recent years, a series of gene-based polymorphic SSR markers have been developed using coconut transcriptome sequences and subsequently have been used in association studies to identify markers associated with the height of coconut trees (Xia et al., 2014). Transcriptome and RNA Seq data generation coupled with identification of differentially expressed (DE) genes should also help in improving resistance to diseases such as root wilt disease (Fan et al., 2013; Huang et al., 2013; Rajesh et al., 2016).

## 7 Conclusion

Coconut is an important plantation crop with versatile uses. Systematic breeding programmes have been carried out since the early 1920s and considerable improvements have been achieved especially in terms of yield and time taken to initiate flowering. Due to inherent constraints in coconut, breeding methods have been largely confined to mass selection and inter and intra-varietal hybridization. Most coconut growing countries are now doing research to develop new molecular techniques to improve the efficiency and effectiveness of breeding programmes. However, multinational research collaborations will be needed to commercialize these new technologies for application in the field.

## 8 Acknowledgements

Some of the information presented in this chapter came from previous book chapters and papers published by the authors and from the information shared with other researchers during the compilation of this chapter. Therefore, the authors would like to thank all the researchers who were involved in the publishing of this chapter, especially Vijitha Vidanarachchi, Dharshani Bandupriya, I. Maskromo, P. Lestari, Kasun Meegahakumbura and Auchithya Dissanayake.

## 9 References

Akuba, R. H. 2002. Breeding and population genetic studies on coconut (*Cocos nucifera* L.) composite variety using morphological and microsatellite markers. PhD thesis. UPLB, Philippines.

Alsaihati, B., Liu, W., Lin, Q. and Al-Mssallem, I. S. 2014. Coconut genome *de novo* sequencing. Poster PO46, Genome sequencing and ESTs session, 13 January 2014. *Plant and Animal Genome XXII (International Plant and Animal Genome Conference)*, San Diego, CA.

APCC. 2014. *Coconut Statistical Year book 2014 of Asian and Pacific Coconut Community*.

Ashburner, G. R., Thompson, W. K. and Halloran, G. M. 1997. RAPD analysis of South Pacific coconut palm populations. *Crop Science* 37(3), 992–7. doi:10.2135/cropsci1997.0011183X003700030048x.

Bahala, R. T., Cano, S. B. and Santos, G. A. 1989. Recommended coconut variety hybrids/cultivars for the national planting/replanting program. Philippine Coconut Authority, Agricultural Research and Development Branch.

Bandaranayake, C. K. and Kearsey, M. J. 2005. Genome mapping, QTL analysis and MAS: importance, principle, constraints and application in coconut. *International Plant Genetic Resources Newsletter* 142, 47–54.

Bandupriya, H. and Dunwell, J. 2012. Overexpression of *CnANT*, coconut *BABYBOOM* homologue alters plant growth and morphology in transgenic *Arabidopsis* plants. *Tropical Agricultural Research* 23(3), 249–60. doi:10.4038/tar.v23i3.4662.

Bandupriya, H. D. D., Iroshini, W. W. M. A., Perera, S. A. C. N., Vidhanaarachchi, V. R. M., Fernando, S. C., Santha, E. S. and Gunathilake, T. R. 2017. Genetic fidelity testing using SSR marker assay confirms trueness to type of micropropagated coconut (*Cocos nucifera* L.) plantlets derived from unfertilized ovaries. *The Open Plant Science Journal* 10(1), 46–54. doi:10.2174/1874294701710010046.

Batugal, P., Rao, R. and Oliver, R. 2005. *Coconut Genetic Resources*. International Plant Genetic Resources Institute (IPGRI), Selangor, Malaysia.

Batugal, P., Bourdeix, R. and Baudouin, L. 2009. Coconut breeding. In: Jain, S. M. and Priyadarshan, P.M. (Eds), *Breeding Plantation Tree Crops: Tropical Species*. Springer, Berlin, pp. 327–76.

Baudouin, L., Lebrun, P., Konan, J. L., Ritter, E., Berger, A. and Billottee, N. 2006. QTL analysis of fruit components in the progeny of a Rennell Island Tall coconut (*Cocos nucifera* L) individual. *Theoretical and Applied Genetics* 112(2), 258–68. doi:10.1007/s00122-005-0123-z.

Benbadis, B. K. 1992. Coconut and date palm. In: Hammerschlag, F. A. and Litz, R. W. (Eds), *Biotechnology of Perennial Fruit Crops*. CAB International, Wallingford, UK, pp. 383–400.

Bourdeix, R., N'Cho, Y. P., Le Saint, J. P. and Sangare, A. 1990. Coconut (*C. nucifera* L.) selection strategies. I. Rundown of achievements. *Oléagineux* 45, 366–71.

Bourdeix, R., Guarino, L., Ramanatha Rao, V. and Baudouin, L. 2005. Status, gaps and strategy in coconut germplasm collecting. In: Batugal, P., Ramanatha Rao, V. and Oliver, J. (Eds), *Coconut Genetic Resources*. International Plant Genetic Resources Institute – Regional Office for Asia, the Pacific and Oceania (IPGRI-APO), Serdang, Selangor DE, Malaysia, pp. 44–64.

Bourdeix, R., Konan, J.-L. and N'Cho, Y. 2006. *Coconut: a Guide to Traditional and Improved Varieties*. CIRAD, Paris, France.

Burkill, I. H. 1966. A dictionary of the economic products of the Malay Peninsula. Ministry of Agriculture and Co-operatives, Kuala Lumpur, Malaysia.

Cardoso, D. A., Moreira, A. S., de Oliveira, G. M., Raggio Luiz, R. and Rosa, G. 2015. A coconut extra virgin oil-rich diet increases HDL cholesterol and decreases waist circumference and body mass in coronary artery disease patients. *Nutricion Hospitalaria* 32(5), 2144–52. doi:10.3305/nh.2015.32.5.9642.

Chan, J. L., Saénz, L., Talavera, C., Hornung, R., Robert, M. and Oropeza, C. 1998. Regeneration of coconut (*Cocos nucifera* L.) from plumule explants through somatic embryogenesis. *Plant Cell Reports* 17(6–7), 515–21. doi:10.1007/s002990050434.

Child, R. 1964. *Coconuts*. Longmans, London, p. 216.

Child, R. 1974. Coconuts (2nd edn.). Longmans, London, 335p.

Cook, O. F. 1911. History of the coconut palm in America. *American Journal of Science* s4-31(183), 221–6. doi:10.2475/ajs.s4-31.183.221.

Dasanayake, P. N., Everard, J. M. D. T., Karunanayake, E. H. and Nandadasa, H. G. 2003. Characterization of coconut germplasm by microsatellite markers. *Tropical Agricultural Research* 15, 51–60.

Dennis, J. V. and Gunn, C. R. 1971. The case against trans-Pacific dispersal of the coconut by ocean currents. *Economic Botany* 25(4), 407–13. doi:10.1007/BF02985208.

Dofy, L. 2012. Coconut oil for Alzheimer's Disease. *Clinical Practice* 1(2), 12–17.

Duran, Y., Rohde, W., Kullaya, A., Goikoetxea, P. and Ritter, E. 1997. Molecular analysis of East African tall coconut genotypes by DNA marker technology. *Journal of Genetics and Breeding* 51, 279–88.

Everard, J. M. D. T. 1996. Use of molecular markers for breeding of the coconut palm (*Cocos nucifera* L.). M.Sc. thesis. University of New England, Arm dale, Australia.

Eyres, L., Eyres, M. F., Chisholm, A. and Brown, R. C. 2016. Coconut oil consumption and cardiovascular risk in humans. *Nutrition Reviews* 74(4), 267–80. doi:10.1093/nutrit/nuw002.

Fan, H., Xiao, Y., Yang, Y., Xia, W., Mason, A. S., Xia, Z., Qiao, F., Zhao, S. and Tang, H. 2013. RNA-Seq Analysis of *Cocos nucifera*: transcriptome sequencing and *de novo* assembly for subsequent functional genomics approaches. *PLoS ONE* 8(3), e59997. doi:10.1371/journal.pone.0059997.

Fernando, S. C., Weerakoon, L. K.,Perera, P. I. P., Bandupriya, H. D. D., Ambagala, I., Gamage, C. K. A., Shantha, E. S., Gunathilake, T. R. and Perera, L. 2004. Genetic fidelity and ex vitro performance of tissue cultured coconut plants. In: Peiris, T. S. G. and Ranasinghe, C. S. (Eds), *Tree of Life – New Trends in the Millennium*. Proceedings of the International Conference to Mark the 75th Anniversary of Coconut Research Institute, Sri Lanka, Part II. Ceylon Printers, Colombo.

Foale, M . A. 1991. Coconut genetic diversity – present knowledge and future research needs. *Proceedings of the International Workshop on Coconut Genetic Resources*, Cipanas, Indonesia, 8–11 October 1991.

Fremond, Y., Ziller, R. and Nuce De Lamothe, M. 1966. *The Coconut Palm*. G.P. Maisonneuveet Larose, Paris.

Gangolly, S. R., Satyabalan, K. and Pandalai, K. M. 1957. Varieties of coconut. *Indian Coconut Journal* 10, 3–28.

Gunn, B. F., Baudouin, L. and Olsen, K. M. 2011. Independent origins of cultivated coconut (*Cocos nucifera* L.) in the old-world tropics. *PLoS ONE* 6(6), e21143. doi:10.1371/journal.pone.0021143.

Guppy, H. B. 1906. *Observation of Naturalists in the Pacific between 1896–1899*. Macmillion, London.

Harries, H. C. 1977. Sixteenth century dissemination of the coconut palm. *Isis* 68(4), 605–6. doi:10.1086/351879.

Harries, H. C. 1978. Evolution, dissemination and classification of *Cocos nucifera* L. *Botanical Review* 44(3), 265–319. doi:10.1007/BF02957852.

Harries, H. C. 1990. Malaysian origin for a domestic *Cocos nucifera*. In: Basset al, P. (Ed.), *The Plant Diversity Malaysia*. Kluwer Academic Publishers, Netherlands, pp. 351–7.

Harries, H. C. 1995. Coconut (*Cocos nucifera* L.). In: Smartt, J. and Simmonds, N. W. (Eds), *Evaluation of Crop Plants* (2nd edn.). Longman, London.

Herran, A., Estioko, L., Becker, D., Rodriguez, M. J. B., Rohde, W. and Ritter, E. 2000. Linkage mapping and QTL analysis in coconut (*Cocos nucifera* L.). *Theoretical and Applied Genetics* 101(1–2), 292–300. doi:10.1007/s001220051482.

Huang, Y. Y., Matzke, A. J. M. and Matzke, M. 2013. Complete sequence and Comparative analysis of the chloroplast genome of coconut palm *Cocos nucifera*. *PLoS ONE* 8(8), e74736. doi:10.1371/journal.pone.0074736.

IPGRI. 1997. *1996 Annual Report of IPGRI*. International Plant Genetic Resources Institute (IPGRI), Rome, Italy.

IPGRI. 1998. *1997 Annual Report of IPGRI*. International Plant Genetic Resources Institute (IPGRI). Rome, Italy.

Jay, M., Bourdeix, R., Potier, F. and Sanslaville, C. 1989. Premiers résultats de l'étude du polymorphisme des polyphénolsfoliaires du cocotier. *Oléagineux* 44(3), 115–27.

Lebrun, P., N'cho, Y. P., Seguin, M., Grivet, L. and Baudouin, L. 1998. Genetic diversity in coconut (*Cocos nucifera* L.) revealed by restriction fragment length polymorphism (RFLP) markers. *Euphytica* 101(1), 103–8. doi:10.1023/A:1018323721803.

Lebrun, P., N'cho, Y. P., Bourdeix, R. and Baudouin, L. 1999. Le cocotier. In: Hamon, P., Seguin, M., Perrier, X. and Gassmann, J. C. (Eds), *Diversité génétique des plantes cultivées*. CIRAD, Montpellier, pp. 219–40.

Lebrun, P., Baudouin, L., Bourdeix, R., Konan, J. L., Barker, J. H., Aldam, C., Herran, A. and Ritter, E. 2001. Construction of a linkage map of the Rennell Island Tall coconut type

(*Cocos nucifera* L.) and QTL analysis for yield characters. *Genome* 44(6), 962–70. doi:10.1139/g01-085.

Lestari, P., et al. 2016. Pemanfaatan Teknologi Genomikadan Bioteknologi untuk Mendukung Percepatan Pemuliaan Palma Penting di Indonesia. KKP3S project annual report. IAARD, Ministry of Agriculture, Republic of Indonesia.

Liyanage, D. V. 1958. Varieties and forms of coconut palms grown in Ceylon. *Ceylon Coconut Quarterly* 9, 1–10.

Liyanage, D. V. and Sakai, K. I. 1960. Heritabilities of certain yield characters of the coconut palm. *Journal of Genetics* 57(2–3), 245–52. doi:10.1007/BF02987231.

Liyanage, D. V., Wickramaratne, M. R. T. and Jayasekara, C. 1988. Coconut breeding in Sri Lanka: a review. *Cocos* 6, 1–26. doi:10.4038/cocos.v6i0.2051.

Mahabale, T. S. 1976. The origin of the coconut. *The Palaeobotanist* 25, 28–248.

Manimekalai, R. and Nagarajan, P. 2006. Inter-relationship among coconut (*Cocosnucifera* L) germplasm accessions using RAPD technique. *Genetic Resources and Crop Evolution* 53(6), 1137–44. doi:10.1007/s10722-005-1303-z.

Meerow, A. W., Wisser, R. J., Brown, J. S., Kuhn, D. N., Schnell, R. J. and Broschat, T. K. 2003. Analysis of genetic diversity and population structure within Florida coconut (*Cocos nucifera* L.) germplasm using microsatellite DNA, with special emphasis on the Fiji Dwarf cultivar. *Theoretical and Applied Genetics* 106(4), 715–26. doi:10.1007/s00122-002-1121-z.

Menon, K. P. V. and Pandalai, K. M. 1958. *The Coconut-A Monograph*. Indian Central Coconut Committee, pp. 86–102.

Nampoothiri, U., Krishnakumar, V., Thampan, P. K. and Nair, M. A. (Eds). 2018. *The Coconut Palm (Cocos nucifera L.)*. Springer Publishers, Heidelberg, Germany.

Narayana, G. V. and John, C. M. 1949. Varieties and forms of coconut. *The Madras Agricultural Journal* 36, 349–66.

Nayer, N. 2017. *The Coconut: Phylogeny, Origins and Spread*. Academic Press, Cambridge, MA.

Nei, M. 1987. *Molecular Evolutionary Genetics*. Columbia University Press, New York, NY.

Nguyen, Q. T., Bandupriya, H. D., López-Villalobos, A., Sisunandar, S., Foale, M. and Adkins, S. W. 2015 Tissue culture and associated biotechnological interventions for the improvement of coconut (*Cocos nucifera* L.): a review. *Planta* 242(5), 1059–76. doi:10.1007/s00425-015-2362-9.

Nic-Matosi, G., Narvaez, M., Peraza-Echeverria, S., Saenz, L. and Oropeza, C. 2017. Molecular cloning of two novel NPR1 homologue genes in coconut palm and analysis of their expression in response to the plant defense hormone salicylic acid. *Genes and Genomics* 39(9), 1007–19 doi:10.1007/s13258-017-0566-z.

Novarianto, H., Kumaunang, J., Tulalo, M. A., Masniawati, A. and Hartana, A. 2001. Keragaman genetic kelapa Dalam Mapangetnomor 32 hasilpenyerbukansendiriberda sarkanpenanda RAPD. *Pemberitaan Puslitbangtri* 7(2), 43–8.

Nuce De Lamothe, M. de and Benard, G. 1985. L'hybride de cocotier PB121 (ouMawa) (NJMxGOA). *Oleagineux* 40, 261–6.

Ohler, J. G. 1984. Coconut, tree of life. Plant production and Protection. Paper 57. Food and Agricultural Organisation of the United Nations, Rome.

Oropeza, C., Saenz, L., Chan, J., et al. 2016. Coconut micro-propagation in Mexico using plumule and floral explants. *Cord* 32(2), 1–6.

Pandin, D. S. 2009a. Keragaman genetic kelapa Dalam Tenga (DTA) danDalam Mapanget (DMT) berdasarkanpenanda RAPD. Buletin Palma No. 16.

Pandin, D. S. 2009b. Keragaman genetic kelapa Dalam Bali (DBI) danDalam Sawarna (DSA) berdasarkanpenanda RAPD. Buletin Palma No.17.

Peedikayil, F. C., Remy, V., John, S., Chandru, T. P., Sreenivasan, P. and Bijapur, G. A. 2016. Comparison of antibacterial efficacy of coconut oil and chlorhexidine on Streptococcus mutants: an in vivo study. *Journal of International Society of Preventive and Community Dentistry* 6(5), 447–52. doi:10.4103/2231-0762.192934.

Persley, G. J. 1990. *The Coconut Palm: Prosperity or Poverty*. ACIAR, Canberra, p. 81.

Persley, G. J. 1992. *Replanting the Tree of Life: Towards an International Agenda for Coconut Palm Research*. CAB International, Wallingford, UK.

Perera, L., Russell, J. R., Provan, J., McNicol, J. W. and Powell, W. 1998. Evaluating genetic relationships between indigenous coconut (*Cocos nucifera* L.) accessions from Sri Lanka by means of AFLP profiling. *Theoretical and Applied Genetics* 96(3–4), 545–50. doi:10.1007/s001220050772.

Perera, L., Russell, J. R., Provan, J. and Powell, W. 1999. Identification and characterization of microsatellite loci in coconut (*Cocos nucifera* L.) and the analysis of coconut populations in Sri Lanka. *Molecular Ecology* 8(2), 344–6. doi:10.1046/j.1365 294X.1999.00534.x.

Perera, L., Russell, J. R., Provan, J. and Powell, W. 2000. Use of microsatellite DNA markers to investigate the level of genetic diversity and population genetic structure of coconut (*Cocos nucifera* L.). *Genome* 43(1), 15–21. doi:10.1139/g99-079.

Perera, L., Russell, J. R., Provan, J. and Powell, W. 2001. Levels and distribution of genetic diversity of coconut (*Cocos nucifera* L., varTypica, form typica) from Sri Lanka assessed by microsatellite markers. *Euphytica* 122(2), 381–9. doi:10.1023/A:1012987224319.

Perera, L., Peries, R. R. A. and Padmasiri, M. H. L. 2002. Sri Lanka Brown Dwarf coconut: a potential coconut variety for future breeding. *Plant Genetic Resources Newsletter* 131, 28–34.

Perera, L., Russell, J. R., Provan, J. and Powell, W. 2003. Studying genetic relationships among coconut varieties/populations using microsatellite markers. *Euphytica* 132(1), 121–8. doi:10.1023/A:1024696303261.

Perera, L., Fernando, W. B. S. F., Hearth, N., Fernando, A., Russell, J., Provan, J. and Powell, W. 2004. Use of microsatellite DNA markers for population analysis, variety identification and for hybridity testing of coconut in Sri Lanka. In: Peiris, T. S. G. and Ranasinghe, C. S. (Eds), *Proceedings of the International Conference to Mark the 75th Anniversary of Coconut Research Institute, Sri Lanka*. Part II, Ceylon Printers, Colombo.

Perera, P. I. P., Hocher, V., Verdeil, J. L., Doulbeau, S., Yakandawala, D. M. D. and Weerakoon, L. K. 2007. Unfertilized ovary, a novel explant for coconut (*Cocos nucifera* L.) somatic embryogenesis. *Plant Cell Reports* 26(1), 21–8. doi:10.1007/s00299-006-0216-4.

Perera, P. I. P., Yakandawala, D. M. D., Verdeil, J.-L., Hocher, V. and Weerakoon, L. K. 2008a. Somatic embryogenesis and plant regeneration in unfertilised ovary explants of coconut (*Cocos nucifera* L.). *Journal of Tropical Agricultural Research* 20, 226–33.

Perera, P. I. P., Perera, L., Hocher, V.,Verdeil, J. L.,Yakandawala, D. M. D. and Weerakoon, L. K. 2008b. Use of SSR markers to determine the anther-derived homozygous lines in coconut. *Plant Cell Reports* 27(11), 1697–703. doi:10.1007/s00299-008-0592-z.

Perera, L., Meegahakumbura, M. K.,Wijesekara, H. R. T., Fernando, W. B. S. and Dickinson,M. J. 2012. A Phytoplasma is associated with the Waligama coconut leaf wilt disease in Sri Lanka. *Journal of Plant Pathology* 94(1), 205–9.

Puch-Hau, C., Oropeza-Salin, C., Peraza-Echeverria, S., et al. 2015. Molecular cloning and characterization of disease-resistant gene candidates of the nucleotide binding sites (NBS) for *Coco nucifera* L. *Physiological and Molecular Plant Pathology* 89, 87–96.

Purseglove, J. W. 1965. The spread of tropical crops. In: Baker, H. G. and Stebbins, G. L. (Eds), *The Genetics of Colonising Species*. Academic Press, New York, pp. 337–89.

Purseglove, J. W. 1968. Origin and distribution of the coconut. *Tropical Science* 10, 191–9.

Purseglove, J. W. 1972. *Tropical Crops: Monocotyledons*. Longmans, London.

Purseglove, J. W. 1985. *Tropical Crops: Monocotyledons* (Rev. and updated edn.). Longman, Harlow, England.

Rajesh, M. K., Jerald, B. A., Reith, P., Regis Jacob Thomas, Fayas, T. P., Rachana, K. E. and Anitha Karun. 2013. Development of a RAPD-derived SCAR marker associated with tall-type palm trait in coconut. *Scientia Horticulturae* 150(4), 312–6.

Rajesh, M. K., Fayas, T. P., Naganeeswaran, S., Rachana, K. E., Bhavyashree, U., Sajini, K. K. and Karun, A. 2016. *De novo* assembly and characterization of global transcriptome of coconut palm (*Cocos nucifera* L.) embryogenic calli using Illumina paired-end sequencing. *Protoplasma* 253(3), 913–28. doi:10.1007/s00709-015-0856-8.

Rajesh, M. K., Rachana, K. E., Kulkarni, K., Sahu, B.B., Thomas, R. J. and Karun, A. 2018. Comparative transcriptome profiling of healthy and diseased Chowghat Green Dwarf coconut palms from root (wilt) disease hot spots. *European Journal of Plant Pathology* 151(1), 173–93.

Raveendranath, T. G. and Ninan, C. A. 1974. A study of the somatic chromosome complements of tall and dwarf coconuts. *Jounal of Plantation Crops* 1, 17–22.

Ridley, H. N. 1930. *The Distribution of Plants throughout the World*. London.

Riedel, M., Riederer, M., Becker, D., Herran, A., Kullaya, A., Arana-López, G., Peña-Rodríguez, L., Billotte, N., Sniady, V., Rohde, W. and Ritter, E. 2009. Cuticular wax composition in *Cocos nucifera* L. physicochemical analysis of wax components and mapping of their QTLs onto the coconut molecular linkage map. *Tree Genetics and Genomes* 5(1), 53–69. doi:10.1007/s11295-008-0168-7.

Ritter, E., Rodriguez, M. J. B., Herran, A., Estioko, L., Becker, D. and Rohde, W. 2000. Analysis of quantitative trait loci (QTL) based on linkage maps in coconut(*Cocos nucifera* L.). In: Arencibia, A. (Ed.), *Plant Genetic Engineering Towards the Third Millennium*. Elsevier Science, Amsterdam, The Netherlands, pp. 42–8.

Rivera, R., Edwards, K. J., Barker, J. H. A., Arnold, G. M., Ayad, G., Hodgkin, T. and Karp, A. 1999. Isolation and characterization of polymorphic microsatellites in *Cocos nucifera* L. *Genome* 42(4), 668–75. doi:10.1139/g98-170.

Rohde, W., Becker, D., Kullaya, A., Rodriguez, M. J. B., Herran, A. and Ritter, E. 1999. Analysis of coconut germplasm biodiversity by DNA marker technologies and construction of a first genetic linkage map. In: Oropeza, C., Verdeil, J. L., Ashburner, G. R., Cardeña, R. and Santamaría, J. M. (Eds), *Current Advances in Coconut Biotechnology*. Current Plant Science and Biotechnology in Agriculture (vol. 35). Kluwer Academic Publishers, Dordrecht, The Netherlands, pp. 99–120.

Rohde, W., Herran, A., Estioko, L., et al. 2000. Mapping of DNA markers, homeotic genes and QTLs in coconut (Cocos nucifera L.). In: Rajanaidu, N., et al (Eds), *International Symposium on Oil Palm Genetic Resources and Utilization*. Malaysian Palm Oil Board, Kuala Lumpur, Malaysia.

Sacks, F. M., Lichtenstein, A. H., Wu, J. H. Y., Appel, L. J., Creager, M. A., Kris-Etherton, P. M., Miller, M., Rimm, E. B., Rudel, L. L., Robinson, J. G., Stone, N. J. and Van Horn, L. V. 2017.

Dietary fat and cardiovascular disease: a presidential advisory from the American Heart Association. *Circulation* 136(3), 1–23. doi:10.1161/CIR.0000000000000510.

Shalini, K. V., Manjunatha, S., Lebrun, P., Berger, A., Baudouin, L., Pirany, N., Ranganath, R. M. and Theertha Prasad, D. T. 2007. Identification of molecular markers associated with mite resistance in coconut (*Cocos nucifera* L.). *Genome* 50(1), 35–42. doi:10.1139/g06-136.

Swaminathan, A. and Jicha, G. A. 2014. Nutrition and the prevention of Alzheimer's dementia. *Frontiers in Aging Neuroscience* 6, 282. doi:10.3389/fnagi.2014.00282.

Swaminathan, M. S. and Nambiar, M. C. 1961. Cytology and origin of the dwarf coconut palm. *Nature* 192(4797), 85–6. doi:10.1038/192085a0.

Teulat, B., Aldam, C., Trehin, R., Lebrun, P., Barker, J. H. A., Arnold, G. M., Karp, A., Baudouin, L. and Rognon, F. 2000. An analysis of genetic diversity in coconut (*Cocos nucifera*) populations from across the geographic range using sequence-tagged microsatellites (SSRs) and AFLPs. *Theoretical and Applied Genetics* 100(5), 764–71. doi:10.1007/s001220051350.

Wadt, L. H. O., Sakiyama, N. S., Pereira, M. G., Tupinamba, E. A., Ribeiro, F. E. and Aragao, W. M. 1999. RAPD markers in the genetic diversity study of the coconut palm. In: Oropeza, C., Verdeil, J. L., Ashburner, G. R., Cardeña, R. and Santamaría, J. M. (Eds), *Current Plant Science and Biotechnology in Agriculture*. Current Plant Science and Biotechnology in Agriculture (vol. 35). Kluwer Academic Publishers, Dordrecht, The Netherlands, 89–97. doi:10.1007/978-94-015-9283-3_7.

Wallace, T. C. 2019. Health effects of coconut oil: a narrative review of current evidence. *Journal of the American College of Nutrition* 38(2), 97–107. doi:10.1080/07315724.2018.1497562.

Whitehead, R. A. 1976. Coconut. In: Simmonds, N. W. (Ed.), *Evolution of Crop Plants*. Longman Publishers, London, UK, pp. 221–5.

Xia, W., Xiao, Y., Liu, Z., Luo, Y., Mason, A. S., Fan, H., Yang, Y., Zhao, S. and Peng, M. 2014. Development of gene-based simple sequence repeat markers for association analysis in *Cocos nucifera*. *Molecular Breeding* 34(2), 525–35.doi:10.1007/s11032-014-0055-x.

Xiao, Y., CSU, P., Fan, H., Baudouin, L., Xia, W., Bocs, S., Xu, J., Li, Q., Guo, A., Zhou, L., Li, J., Wu, Y., Ma, Z., Armero, A., Issali, A. E., Liu, N., Peng, M. and Yang, Y. 2017. The genome draft of coconut (*Cocos nucifera*). *GigaScience* 6(11), 1–11. doi:10.1093/gigascience/gix095.

Zizumbo-Villarreal, D. 1996. History of coconut (*Cocos nucifera* L.) in Mexico: 1539–1810. *Genetic Resources and Crop Evolution* 43(6), 505–15. doi:10.1007/BF00138827.

# Chapter 12

# Advances in cultivation of coconut

*M. A. Foale, Quang Thien Nguyen and Steve W. Adkins, The University of Queensland, Australia*



## 1 Introduction

The coconut palm evolved over millions of years in the coral atoll ecosystem of tropical regions (Harries, 1978; Harries and Harries, 2018), developing a buoyant seed capable of dispersal over thousands of kilometres of ocean. It requires a very high tide or a cyclonic surge to be delivered onto a suitable strand site. The seed possesses a long-lasting endosperm which is the sole support of seedling growth for 4 months, enabling early establishment, then nurturing leaf expansion for several more months until it becomes redundant (Foale, 1968).

It is the oil content of the solid endosperm and the emulsion extracted from it, as well as the useful volume of liquid endosperm (coconut water) that have become the principal primary coconut products of trade in today's world. Traditionally, the human inhabitants of coral atolls and of the coastal zone of high tropical islands have consumed derivatives of the coconut kernel, mostly extracted oil and coconut cream, as a staple source of energy and nutrition.

http://dx.doi.org/10.19103/AS.2019.0054.22

**Figure 1** A very ancient palm with low light interception.

In places where fish and coconut form the foundation of the diet people have enjoyed excellent health and strength (Prior et al., 1981).

During industrialisation the rapid growth of the human population in Europe and North America caused a shortage of edible oil. While this had been traditionally sourced from domestic animals (sheep, cattle and pigs), the shortfall resulted in demand for coconut oil. Besides that, it was found that soap made from coconut oil was of outstanding quality. The oil was transported from the tropical world as copra, a transportable form that was produced by using hot air or sun drying of the kernel of the mature coconut fruit. Such was the price paid for this oil, during the late nineteenth and early twentieth centuries, that significant investment by European capital was made in coconut plantations throughout the relatively humid coastal tropical world. As there were colonial administrations active in the Indian and Pacific Ocean countries, Southeast Asia, the South Pacific (Bennett, 1987), Africa, the Caribbean and tropical South America, coconut plantations flourished in all those regions. The total area occupied by coconut was approximately 10 million ha.

Although the strand is the natural habitat of the coconut it was found that the palm also flourished on many different soil types provided that there was well-distributed annual rainfall amounting to at least 2000 mm. Plantations prospered until the widespread disruption of world trade during the Second World War. The resulting acute shortage of edible oil resulted in a rapid rise, in oil-importing countries, in the cultivation of alternative oil-seed crops including sunflower, rapeseed, soybean and peanut. The price of coconut oil staged a partial comeback once trade resumed. As the price fell further, however, and many countries in the coconut world achieved independence during the post-war decades from the 1940s to 1980s, many external investors declined to regenerate their once profitable assets. Plantations either reverted to indigenous ownership or were subdivided into small tenant holdings during this period of decline in the market value of copra and coconut oil.

In the tropics the oil palm (*Elaeis guineensis*, Jacq.), a forest palm of West Africa, gained investor attention at this time because it was found to be considerably higher yielding than the coconut. The processing of ripe oil palm bunches was readily and cheaply mechanisable, imparting a further advantage over coconut. In spite of efforts to raise the potential yield of coconut oil through selection and hybrid breeding (Bourdeix et al., 2005; Batugal and Rao, 2005), which did achieve more than 3 t/ha with intensive management including irrigation, the oil palm was capable of producing more than 5 t/ha of oil. Corley (1985) notes a record coconut yield of 6.5 t/ha of copra, and for oil palm 10 t/ha of oil. An early plantation scale copra yield from Solomon Tall palms of 3.75 t/ha was recorded in the 1920s by Unilever in the Solomon Islands (Green and Foale, 1961).

This high yield would undoubtedly have been supported by the breakdown of the very large amount of biomass generated from the rain forest felled in order to establish the plantation. As plantations age such initial nutrient supply is bound to dwindle, at which time the fertiliser needs of the palm need to be met as precisely as possible. Experience in the late decades of the twentieth century with a method known as 'foliar diagnosis' led to the definition of critical levels being proposed for both macro and micro nutrients (Taffin, 1993; Foale, 2003). A summary of these critical levels is presented in Table 1.

**Table 1** Coconut foliar analysis critical levels

| | |
|---|---|
| *% of dry matter from 14th youngest frond - IRHO* | |
| Nitrogen | 1.8-2.0 |
| Phosphorus | 0.12 |
| Potassium | 0.8-1.0 |
| Magnesium | 0.2-0.3 (interact with potassium) |
| Calcium | 0.3-0.5 |
| Sodium | Not essential |
| Chlorine | 0.5-0.6 |
| Sulphur | 0.15-0.2 |
| *Trace elements parts per million* | |
| Boron | 10 |
| Manganese | >30 |
| Iron | >30 |
| Copper | 3-7 |
| Molybdenum | 0.15 |
| Zinc | 15 |
| Aluminium | >38 (not essential - higher toxic) |

Besides the challenge to achieve higher yield, the coconut is also beset by limited diversity in its adaptation to a range of pests and diseases. While the original atoll habitat, and seagoing dispersal, generated a very robust seed to support early growth of the seedling, the atoll ecosystem generally has few biohazards in the form of hostile insects and microorganisms. The natural ecosystem on a high island, on the other hand, has been found to host many biohazards absent from atolls, so that a modern atoll population is likely to be vulnerable to them. It is therefore a significant challenge, in developing new plantations, to identify genotypes possessing tolerance or resistance to anticipated hazards such as a leaf beetle, a nutfall bug, a fungus, mycoplasma or virus diseases (Taffin, 1993; Foale, 2003).

## 2 Improving potential yield of the coconut

An understanding of the physiology of yield in the coconut can assist in forming a strategy to raise yield potential. There are two quite distinct groups in the coconut world known as Tall and Dwarf. A characteristic of a Tall population is variability of physical traits, including the anatomical traits that contribute to yield, as they are naturally heterozygous or cross-breeding. Self-pollination, a characteristic of most Dwarf populations, is rare in the Tall palm. As a consequence of regular crosspollination there is obvious variation in the anatomical traits that contribute to the yield of the Tall. These include the following: potential fruit size (this varies also with the number of fruit on a bunch); the number of inflorescences produced per year (also affected by the environment); the number of fruit per bunch (related to the number of female flowers as well as the success of pollination); the proportion of kernel, by weight, in the mature fruit; and the proportion of dry matter accumulation invested in the growth of the trunk (Foale, 1992). Research has not yet been directed into comparing the net assimilation rate (weight of dry matter assimilated per joule of intercepted solar radiation (Foale, 1968)) of different palms, or populations. Breeding for yield must therefore focus on the observable variables listed above, as well as earliness of flowering.

There is a great deal of difference in fruit composition between coconut populations, related to the effect of human selection for either a larger proportion of water and kernel or fibre quality. Whitehead (1966) identified two extreme types which are known as 'niu vai' and 'niu kafa', being the Polynesian names for 'water nut' and 'fibre nut', respectively. Subsequently, Harries (1978) presented data on fruit composition that characterised the high and low extremes of dry matter partitioned to husk. This ranged from around 70% in a wild-type fruit, free of human selection, and 30% in large fruit that had been selected especially to provide a source of drinking water, and fresh kernel, while voyaging at sea. Harries pointed out that many populations found in traditional coastal villages had an intermediate husk proportion due to introgression between the original coconut population that existed prior to any human colonisation and the selected larger fruit type introduced by the settlers. Each extreme type can still be found however, in isolated locations. For example, there is an apparently truly isolated 'wild-type' population on North Keeling Island (Leach et al., 2003) and the clearly domesticated populations of San Ramon (The Philippines), Rennell (Solomons), Rotuma (Fiji) and Samoa (Niu Vai) (Whitehead, 1966).

Selection of Tall palms for high yield offers only limited progress due to their heterozygosity (Child, 1964), but hybridising the Tall and the Dwarf has provided some gain (Bourdeix et al., 2005). Generally the Dwarf flowers a year or two earlier than the Tall, and it also partitions a lesser proportion of dry matter to the trunk, with the result that a DXT hybrid shows intermediate expression of these two traits. Examples of successful DXT hybrids are the PB121 (Malayan Yellow Dwarf by West African Tall) produced in Ivory Coast and the Maren (Malayan Orange Dwarf by Rennell Tall) in Solomon Islands, both of which delivered a yield gain of at least 30%. Other similar examples of successful hybrids abound, while the potential copra yield suggested by Corley (1985) on the basis of optimum light interception and ideal partitioning to kernel is actually double the best yield cited by him of 6.3 t/ha. In the realm of plant management the difficulty of increasing planting density to achieve greater light interception is that more biomass tends to be partitioned to trunk extension, at the expense of a higher share going to yield.

## 3 Protecting the coconut from biohazards

The coconut palm has prospered on atolls through geological time, growing there within an ecosystem much less complex than those adjacent to the strands of high islands and continental land masses. Globally, several acute hazards are active, threatening to destroy the coconut population in particular areas. The lethal yellowing disease, present in Jamaica, Mexico and many Caribbean islands (Whitehead, 1966, 1968) was found to be caused by a mycoplasma which killed the palm within a year or two of the first symptoms. The Tall population of Jamaica is mostly descended from palms introduced during and since the sixteenth century from the Cape Verde islands where the coconut was first established in 1501 by Vasco da Gama with seeds from India (Harries, 1978). Lethal yellowing is currently causing serious damage in both West and East Africa.

Germplasm from the Pacific, including the Panama Tall variety and orange Malayan Dwarf, showed resistance to lethal yellowing for decades, but eventually they also succumbed – due perhaps to a mutation of the mycoplasma. While there has been mention of similar symptoms in isolated locations of Kalimantan and Papua New Guinea, there is no sign in those cases of rapid spread of the disease.

Lethal yellowing evidently conforms with similar examples of harm due to both pathogenic organisms and insects affecting susceptible populations, while not affecting others due to evident resistance. The leaf beetle, *Brontispa longissima,* has recently devastated populations in the Philippines, Vietnam and beyond. At the same time it is almost unnoticed, except on seedlings or on introduced genotypes, in Papua New Guinea and Solomon Islands, where it is endemic. Another interesting case is the virus responsible for foliar decay disease in Vanuatu (Randles et al., 1986; Taffin, 1993). This organism is lethal to exotic germplasm and was only discovered when introductions were made in the context of coconut research in the 1960s. The virus induces no symptoms in the local Tall population but devastates introductions when they are exposed to a species of leaf hopper (*Myndus taffini)* which transfers inoculant from the local Tall.

These and other examples show that some populations have evolved to gain resistance to a threat through natural selection due to the presence of that threat. The challenge to future research is to identify, if possible, the genetic signature of specific resistances, in order to enable immediate selection of germplasm for use in specific locations. Until that is possible it remains important to screen introduced germplasm for resistance to dominant threats.

## 4 Mass production of elite genotypes

Somatic embryogenesis (SE) is a powerful tool for the production of a large number of plants from an elite individual selected for high potential yield and robust resistance to major biohazards. The first recorded success with SE in coconut was described by Branton and Blake (1983), but attempts by others, while yielding healthy plantlets, have been on such a limited scale to date as to be unaffordable. At the time of writing there are several groups worldwide including in Australia, Mexico, the Philippines and Sri Lanka, researching on coconut which have advanced the technique close to being economically viable for mass plantlet production. While much work has been done on the heterozygous embryo by inducing callus formation from a slice of the plumule, the future lies in cloning somatic tissue to generate plantlets that are true-to-type for the selected parent palm. Most success has been achieved with tissue from a very immature inflorescence, which unfortunately involves sacrificing the elite selected palm – a high price at the beginning, but offset by the potential high rate of multiplication now achievable.

Great care is needed in selecting a suitable donor palm for SE. This need is highlighted by experience with high yielding hybrids. Both the PB121 hybrid from Ivory Coast and the Maren hybrid from Solomon Islands have respectively a yellow and orange Dwarf parent, which are more susceptible to particular strains of bud rot disease (*Phytophthora palmivora)* than the Tall. This places these hybrids at risk where the organism is endemic as well as having evolved into a local variant. Where there is a promising natural genotype or hybrid, care should be taken to assess the likely biohazards where new planting or replanting is to take place. Ideally the genetic signature for tolerance for a range of diseases including bud rot, lethal yellowing, *Brontispa* beetle, rhinoceros beetle and so on will be identified to support the use of SE to supply replanting material gains strength.

## 5 The challenge of affordable SE

SE first was achieved by two independent research groups in Germany and the United States, when plantlets were regenerated from cultured carrot (*Daucus carota* L.) 'mother' cells (Steward et al., 1958; Reinert, 1959). Since then, the capacity to produce somatic embryogenic structures and plantlets from undifferentiated cells has become the focus of much research on many species (Fehér et al., 2003; Verdeil et al., 2007; Ikeuchi et al., 2013). Even though SE can be achieved in many species, it has been much more difficult to achieve in others, and this includes the coconut. The first attempts at coconut SE were undertaken four decades ago (Eeuwens and Blake, 1977; Pannetier and Buffard-Morel, 1982). These early studies focussed on several plant somatic tissues as initial explants (i.e. young leaves, stem slices from young seedlings, sections from rachillae of young inflorescences) to produce embryogenic callus (Branton and Blake, 1983a; Gupta et al., 1984). Nevertheless, there has recently been a shift to either somatic tissues (e.g. immature inflorescences, ovaries) or the easier-to-manipulate zygotic tissues (e.g. immature or mature embryo and embryo-derived plumule) to achieve SE in coconut. While the immature embryo was found to be wanting, the responsiveness of the easier-to-obtain mature embryo was dramatically improved by transverse slicing (Adkins et al., 1998; Samosir, 1999), and at a later date by the isolation and culture of the plumular tissue (Chan et al., 1998; López-Villalobos, 2002; Pérez-Núñez et al., 2006). More recently, attention has returned to the harder-to-use somatic tissue explants such as young inflorescence tissue (Antonova, 2009; Sandoval-Cancino et al., 2016).

Regarding culture media, the Y3 (Eeuwens, 1976) and BM72 (Karunaratne and Periyapperuma, 1989) formulations have been the most widely used for the early stages of SE callus culture, while MS (Murashige and Skoog, 1962) and B5 (Gamborg et al., 1968) have been found to be less effective (Branton and Blake, 1983b; Bhallasarin et al., 1986) (Table 2). The addition of sucrose (3–4%) was shown to be important for coconut SE to take place, while AC (0.1–0.3%) has been extensively used to prevent explanted tissue and callus from browning, which is a stress-related response, caused by the undesired production of secondary plant products such as phenols, or ethylene (Samosir, 1999). However, the presence of activated charcoal (AC) in the culture medium dramatically interferes with the activity of the exogenously applied plant-growth regulators and other media supplements, leading to uncertainty about the exact functional concentrations of these additives within the medium (Pan and van Staden, 1998). Differences in particle size, and the potency of the various AC types, have been shown to influence the frequency of somatic embryogenic callus formation (Sáenz et al., 2010). In addition to AC, biochar (BC) is an emerging environmentally friendly form of charcoal that is more affordable (Kochanek et al., 2016). Also, BC may not bind 2,4-D to the same degree while still helping to prevent oxidative browning. Hence, it is worth comparing this type of charcoal with AC in terms of their interaction with 2,4-D. Knowledge gained through this comparison may help improve the efficiency of callus growth, leading to an improved scale-up in clonal propagation.

Another universal toxin-absorbing agent, polyvinylpyrrolidone (PVP), was tested in coconut leaf-derived cell suspension cultures but without any significant effect (Basu et al., 1988). However, polyvinylpolypyrrolidone (PVPP), used in zygotic embryo-derived callus culture, was found to have some positive effect in promoting the rate of SE (Samosir, 1999). The frequent subculturing of the explant tissues and the developing somatic embryogenic callus is another useful approach to reduce exposure to toxic phenols (Fernando and Gamage, 2000; Pérez-Núñez et al., 2006), even though the cultured tissue would encounter further stress during the transfer process. In this context, further study should be focussed on the use of alternative agents by which phenolic oxidation can be reduced. Recently, aminoindan-2-phosphonic acid (AIP), which can inhibit the

**Table 2** Important studies related to somatic embryogenesis in coconut

| Variety/cultivar[a] | Initial explants (age)[b] | Culture media and plant-growth regulators (optimal combinations reported)[c]: Callus induction → proliferation | Maturation → germination (modifications only) | Responses/ results[d] | Reference |
|---|---|---|---|---|---|
| JMD | Seedling stem and rachillae of young inflorescences | Y3 + Sucrose (6.8%) + 2,4-D (0.1 μM) + BAP (5 μM) + $GA_3$ (10 μM) | - | EC | Eeuwens and Blake (1977) |
| MDY × WAT | Young leaves | Y3 + MW Vit + Suc (2%) + 2,4D (n/a) + AC (n/a) | Reducing 2,4-D (n/a) + BAP (n/a) | EC | Pannetier and Buffard-Morel (1982) |
| JMD | Seedling stem and rachillae of young inflorescences | MS macro + Y3 micro + modified Blake vit + Sucrose (5%) + AC (0.25%) + 2,4-D (100 μM) + BAP (5 μM) + 2iP (5 μM) | Reducing 2,4-D (0.1 μM) | EC | Branton and Blake (1983) |
| IWCT | MZE and stems, leaves and rachillae | Modified Y3 + 2,4-D (452 μM) + NAA (27 μM) + BAP (8.88 μM) + Kin (4.65 μM) + AC (0.25%) | Reducing 2,4-D (2.3 μM) | EC | Gupta et al. (1984) |
| IWCT | IZE in enclosing soft endosperm | Y3 + 2,4-D (226 μM) + Kin (9.4 μM) + AC (0.1%) | Reducing 2,4-D (4.5 μM) | Aneuploid callus cells | Kumar et al. (1985) |
| IWCT | MZE (8–10 mpp) | B5 + IAA-asp (7 μM) + IAA-ala (7 μM) | B5 + IAA-asp (7 μM)/IAA-ala (7 μM) + Kin (9.4 μM)/BAP (8.8 μM) ↓ B5 + NAA (2.7 μM) + BAP (9.4 μM) + PVP (0.1%) | EC, SEM, PR | Bhallasarin et al. (1986) |
| var. *typical* | IZE (6–8 mpp) | BM72 + Sucrose (3%) + AC (0.25%) + 2,4-D (12–20 μM) | Reducing 2,4-D (8 μM) + BAP (10 μM) | EC, SEM, PR | Karunaratne and Periyapperuma (1989) |

(*Continued*)

**Table 2** (Continued)

| Variety/cultivar[a] | Initial explants (age)[b] | Culture media and plant-growth regulators (optimal combinations reported)[c]: Callus induction → proliferation | Maturation → germination (modifications only) | Responses/ results[d] | Reference |
|---|---|---|---|---|---|
| MYD × WAT, WAT × MYD, and MYD | Immature inflorescences | Y3 + MW Vit + Sucrose (4%) + 2,4-D (250–300 μM) + AC (0.2%) ↓ Increasing 2,4-D (450–550 μM) | Reducing 2,4-D + incorporating BAP | EC, SEM, PR | Verdeil et al. (1994) |
| BLT | MZE slices | Y3 + Sucrose (3%) + 2,4-D (125 μM) + AC (0.25%) + AVG (1 μM) + STS (2 μM) | Reducing 2,4-D + Putrescine (7.5 μM) + Spermine (1 μM) | EC, SEM | Adkins et al. (1998) |
| MMD | MZE plumules | Y3 + 2,4-D (100 μM) +AC (0.25%) | reducing 2,4-D (1 μM) + BAP (50 μM) | EC, SEM, PR | Chan et al. (1998) |
| MYD and BLT | MZE slices | Y3 + Sucrose (3%) + 2,4-D (125 μM) + AC (0.25%) | Reducing 2,4-D + NAA (10 μM) + ABA (5 μM) | EC, SEM, PR | Samosir (1999) |
| SLT | IZE | BM72 + Sucrose (4%) + 2,4-D (24 μM) + ABA (2.5–5 μM) | Reducing 2,4-D (8–16 μM) + ABA (5 μM) | EC, SEM, PR | Fernando and Gamage (2000) |
| SLT | MZE plumules | BM72 + Sucrose (6%) + AC (0.25%) + 2,4-D (0.1 μM) + BA (5 μM) ↓ Increasing 2,4-D (24 μM) + Sucrose (4%) | Reducing 2,4-D (16 μM) + ABA (5 μM) ↓ PGR-free | EC, SEM, PR | Fernando et al. (2003) |
| MGD | MZE plumules | Y3 + Sucrose (3%) + AC (0.25%) + 2,4-D (600 μM) | Reducing 2,4-D (6 μM) + BA (300 μM) | EC, SEM, PR | Pérez-Núñez et al. (2006) |

| | | | | | |
|---|---|---|---|---|---|
| SLT | Unfertilised ovaries (-4, -5 and -6 stages of ovary maturity) | BM72 + Sucrose (4%) + AC (0.1%) + 2,4-D (100 μM) | Omitting 2,4-D + ABA (5 μM) + $AgNO_3$ (10 μM) ↓ Modified Y3 + 2,4-D (8–16 μM) + ABA (5 μM) | EC, SEM, PR | Perera et al. (2007) |
| SLT | Unfertilised ovaries (-4 stage of ovary maturity) | BM72 + Sucrose (4%) + AC (0.1%) + 2,4-D (100 μM) + TDZ (9 μM) ↓ Reducing 2,4-D (66 μM) | PGR-free ↓ BAP (5 μM) + GA3 (0.45 μM) +2iP (45 μM) | EC, SEM, PR | Perera et al. (2009) |
| MYD | Immature inflorescences | Y3 + Sucrose (3%) + AC (0.25%) + 2,4-D (250 μM) + 2iP (5 μM) + BAP (5 μM) ↓ Adding TDZ (10 μM) | Reducing 2,4-D ↓ PGR-free + Ancymidol (30 μM) | EC, SEM, PR | Antonova (2009) |
| MYD × MXPT | Immature inflorescences | Y3 + Sucrose (3%) + AC (0.25%) + 2,4-D (650 μM) | Reducing 2,4-D (6 μM) + BAP (300 μM) + GA3 (4.6 μM) ↓ Plantlet 2,4-D (6 μM) + BAP (300 μM) | EC, SEM, PR | Sandoval-Cancino et al. (2016) |

[a] *BLT* Batu Layar Tall, *JMD* Jamaican Malayan Dwarf, *IWCT* Indian West Coast Tall, *MGD* Malayan Green Dwarf; *MMD* Mexican Malayan Dwarf, *MXPT* Mexican Pacific Tall, *MYD* Malayan Yellow Dwarf, *SLT* Sri Lanka Tall and *WAT* West African Tall.

[b] *IZE* immature zygotic embryos, *MZE* mature zygotic embryos, *mpp* months post pollination.

[c] *ABA* abscisic acid, *AC* activated charcoal, *B5* Gamborg et al. (1968) medium, *BAP* 6-benzylaminopurine, *Blake* Blake (1972) medium, *BM72* Karunaratne and Periyapperuma (1989) medium, *2,4-D* 2,4-dichlorophenoxyacetic acid, $GA_3$ gibberellic acid, *IAA-ala* indole-3-acetic acid-alanine, *IAA-asp* indole-3-acetic acid-aspartate, *2iP* 2-isopentyl adenine, *Kin* kinetin, *MS* Murashige and Skoog (1962) medium, *MW Vit* Morel and Wetmore (1951) vitamins, *NAA* naphthalene acetic acid, *Nitsch* Nitsch (1969) medium, *PGR* plant-growth regulators, *PVP* polyvinylpyrrolidone, *TDZ* Thidiazuron, *Y3* Eeuwens (1976) medium.

[d] *EC* embryogenic callus; *SEM* somatic embryo maturation, *PR* plantlet regeneration – not mentioned.

activity of phenylalanine ammonia lyase, has successfully prevented phenolic compounds from being produced in the tissues of sweet wormwood (*Artemisia annua* L.), sugar maple (*Acer saccharum* Marsh.) and American elm (*Ulmus americana* L.) (Jones et al., 2012; Jones and Saxena, 2013).

As seen in many other species, the sequential development of SE-derived plantlets is typically divided into three stages: (1) the production of callus and its proliferation; (2) the formation, maturation and germination of the somatic embryo; and (3) the acclimatisation of the plantlet to *ex vitro* conditions. Callus formation is commonly achieved with a high concentration of auxin, usually 2,4-dichlorophenoxyacetic acid (2,4-D). However, the most successful concentration of 2,4-D seems to differ between cultivars and explants, as well as its response to other medium additives, such as AC – of suitable quality, which are also included (Table 2). For instance, while a low 2,4-D (24 µM) treatment was found to be optimal to initiate callus production on the zygotic embryo from Sri Lanka Tall (Fernando and Gamage, 2000), a much higher dose (125 µM) was needed for Malayan Yellow Dwarf and Buta Layar Tall (Adkins et al., 1998; Samosir, 1999). For callus production on immature inflorescence tissue and embryo-derived plumule, an even higher concentration of 2,4-D (up to 600 µM) was required (Verdeil et al., 1994). Complications arise when a high concentration of 2,4-D is used for extended periods as such treatments can induce chromosomal aberrations in the cultured tissue (Blake and Hornung, 1995). In addition, it is now thought that coconut tissue can metabolise 2,4-D into fatty acid analogues, which are subsequently incorporated into triacylglycerol derivatives (López-Villalobos et al., 2004). These latter molecules represent a stable and stored form of 2,4-D that can continue to arrest somatic embryo formation, even when 2,4-D has been removed from the medium. Apart from 2,4-D, other auxins such as NAA (27 µM) in combination with 2,4-D (452 µM) have been used to promote callus formation on rachillae explants (Gupta et al., 1984). In addition, a study of the ultrastructural changes that take place during the acquisition of SE potential suggests that the gametophyte-inducing conditions, produced by 2,4-D are required for the successful transition from the vegetative to the embryogenic state (Verdeil et al., 2001). However, the repeated use of synthetic auxin at a high concentration may lead to somaclonal variation, as reviewed by Krishna et al. (2016). Therefore, there is a need to identify the lowest effective 2,4-D concentration in culture media whenever possible in order to produce true-to-type clones.

The addition, to the callus proliferation and maturation medium, of a cytokinin such as 6-benzylaminopurine, thidiazuron, kinetin or 2-isopentenyl adenine, at 5–10 µM, is also commonplace (Table 2). Callus formation is often best achieved in the dark for at least 30 days, after culture initiation and at 28 ± 1°C (Adkins et al., 1998). However, in one study, dark incubation was successfully extended to 90 days to achieve greater callus production (Pérez-Núñez et al.,

2006). Further improvement in the timely production of SE callus has been achieved by including in the medium one of the multifunctional polyamines, particularly putrescine (7.5 mM) or spermine (1.0 μM), to protect the explanted tissue from ethylene damage and/or to promote the rate of SE (Adkins et al., 1998). Ethylene production inhibitors, such as aminoethoxyvinylglycine (AVG), and ethylene-action inhibitors, such as silver thiosulphate (STS), have also been shown to provide a beneficial environment for callus multiplication, and for the formation of somatic embryos (Adkins et al., 1998). In several studies, the conversion of undifferentiated callus to SE callus was achieved by the reduction or removal of 2,4-D from the culture medium (Table 2). Furthermore, Chan et al. (1998) showed that incubating callus under a 12-h photoperiod (45–60 μmol/$m^2$/s photosynthetic photon flux density) significantly improved the rate of SE, as compared to that produced under darkness. Incorporating or increasing the amount of BAP (from 50 up to 300 μM) could also promote SE, leading to a greater yield of viable plantlets at the end of the culture phase (Pérez-Núñez et al., 2006; Chan et al., 1998).

A moderate application of abscisic acid (*ca.* 5 μM) has been shown to enhance the formation and the maturation of somatic embryos (Samosir et al., 1999; Fernando and Gamage, 2000; Fernando et al., 2003). In addition, the use of osmotically active agents, such as polyethyleneglycol (PEG 3%), in combination with ABA (45 μM) has also been shown to be beneficial, not only for the production of somatic embryos but also for their subsequent maturation and germination (Samosir et al., 1998). In a more recent study using immature inflorescence explants, Antonova (2009) demonstrated the benefit of using a specific growth retardant, ancymidol (30 μM), to elevate the somatic embryo germination frequency from a few per cent to 56%.

It is worth noting that cell suspension culture systems have also been successful in raising the rate of SE for some members of the *Arecaceae* (the palm family), including oil palm (*Elaeis guineensis* Jacq.; Teixeira et al. (1995)) and date palm (*Phoenix dactylifera* L. (Sané et al., 2006)). The de-differentiated cells of suspension cultures are believed to be beneficial, not only for increasing the rate of plantlet regeneration but also for the large-scale production of secondary metabolites (Mustafa et al., 2011). Plantlet regeneration from cell suspension culture has been reported in lemongrass [*Cymbopogon martini* (Roxb.) Wats] (Patnaik et al., 1997), crown flower (*Calotropis gigantea* L.) (Roy et al., 2000) and sugarcane (*Saccharum officinarum* L.) (Thorat et al., 2017). It has also been shown that embryogenic suspension cultures can be cryopreserved for a germplasm conservation purpose in Norway spruce (*Picea abies* L.) and Sitka spruce [*Picea sitchensis* (Bong.) Carr.] (Find et al., 1998). Cell suspension cultures possessing a high degree of homogeneity and rapid cell growth can also be used as a simplified system for further study of plant cell physiology (Moscatiello et al., 2013). However, establishment of cell suspension culture in coconut is yet to be achieved. Apart from an attempt using leaf cells to induce cell division by Basu et al. (1988), there has not been any extensive research reported on this aspect. Therefore, further research is needed to successfully generate embryogenic cell suspension cultures, for improvement in both the SE productivity and additional applications.

Temporary immersion systems have been employed with date palm (*Phoenix dactylifera* L. (Tisserat and Vandercook, 1985)) and peach palm (*Bactris gasipaes* Kunth. Steinmacher et al. (2011)) to raise the rate of plantlet regeneration. These two techniques, 'cell suspension culture' and 'temporary immersion system', when applied to coconut SE cultures, could possibly facilitate the rapid multiplication of robust plantlets, thereby creating a platform for mass clonal propagation. However, the *ex vitro* acclimatisation of the somatic embryo-derived plantlet has yet to be refined, with the present rate of success no better than 50% (Fuentes et al., 2005). Further improvement may come from using a photoautotrophic culture system (Samosir and Adkins, 2014) and/or through the incorporation of fatty acids, notably lauric acid, into the plantlet maturation medium (López-Villalobos et al., 2001, 2011).

The economic demand for elite fruit quality and biohazard-resistant coconut varieties has grown rapidly in recent years. There is now an urgent need to provide a substantial quantity of superior planting material to boost the potential income of the producer. Embryo culture of non-germinable seednuts is already underway on a moderate scale in the Philippines, and to a lesser extent in Indonesia, Thailand and Vietnam. Even though little practical success has been achieved so far, through SE propagation, the cloning approach has great potential for the mass production of uniform, genetically superior, coconut plantlets. Further optimisation should be undertaken to exploit the full capacity of this method, particularly testing other somatic tissues as explants (e.g. the immature inflorescence). In addition, some innovative techniques, such as embryo transplantation and marker-assisted breeding, should be considered. With the recent worthwhile increase in genetic information for the identification of marker genes, molecular screening and identification of candidate cultivars can now be undertaken far more efficiently than in the past. Introduction of an attractive trait, such as aromatic water, into a hybrid combination with Malayan Gold (Orange) Dwarf is currently underway to improve fruit value and harvest index. The identification of genetic markers for resistance to specific biohazards such as *Phytophthora* bud rot and the *Oryctes rhinoceros* beetle would be extremely valuable when selecting a source palm for tissue to initiate SE.

Inefficient plantlet regeneration from *in vitro* culture systems remains a major bottleneck for many coconut research groups around the globe. This relates to the variable success rate of explanted tissues *in vitro*, the subsequent slow growth of tissues *in vitro*, their further lack of vigour when planted *ex vitro*, as well as the need locally to upgrade tissue culture skill. A robust protocol for coconut tissue culture is being achieved, but less rapidly than for many other plant species; experience with those could help drive future improvements in coconut *in vitro* culture. The literature suggests that, with further work, it may be possible to generate a highly efficient embryogenic cell suspension culture, thereby achieving an affordable clonal propagation system for coconut.

## 6 Molecular aspects for SE in coconut

SE is a multi-step process which involves the transition of a single cell into a somatic pro-embryo structure and finally into a somatic embryo. Hence, alterations in the physiological and biochemical characteristics of the cell must occur to create a condition in which somatic embryogenic competence can be acquired (Umehara et al., 2007; Pandey and Chaudhary, 2014). To achieve such alterations, cells can be affected by several factors, including the presence of certain plant-growth regulators, which act to change the existing pattern of gene expression to one that promotes SE. Subsequently, these changes in competence regulate the biosynthesis of certain enzymes, which drive the cell to adopt the new function (Pandey and Chaudhary, 2014; Chugh and Khurana, 2002; Fehér et al., 2003). This process is commonly known as cell specification and is an important genetic event in the formation of somatic embryos (Miyashima et al., 2013; Smertenko and Bozhkov, 2014; Umehara et al., 2007).

Studies on specific gene expression have been used to help unravel the molecular mechanisms which regulate the process of SE in coconut (Pérez-Núñez et al., 2009). It is likely that the dissecting out of the key molecular elements will help to improve the efficiency of the existing clonal propagation protocols. Bandupriya et al. (2013) have been able to isolate a homologous gene (i.e. *CnANT*) to the *Arabidopsis AINTEGUMENTA*-like gene in coconut, which encodes for two APETALA 2 domains and a linker region. The analysis of *CnANT* transcripts demonstrated that this gene is involved in coconut SE, and is at its highest level of expression during the callus induction phase when cells are acquiring SE competence (Bandupriya et al., 2013, 2014). The role of *CnANT* in SE was studied in explants derived from *Arabidopsis* overexpressing lines. The upregulation of the *CnANT* gene caused increased shoot organogenesis even in culture media devoid of plant-growth regulators (Bandupriya and Dunwell, 2012). However, the spontaneous formation of somatic embryos, as reported with other *PL/AIL* genes, was not observed with the *CnANT* gene (Bandupriya and Dunwell, 2012; Boutilier et al., 2002; Tsuwamoto et al., 2010).

Like the *CnANT* gene, the *CnCDKA* and *CnSERK* homologs have also been isolated from coconut and shown to be associated with the induction of SE in this species (Pérez-Núñez et al., 2009). The *CnCDKA* gene encodes a cyclin-dependent kinase, which regulates cell division following its activation by certain cyclins (Montero-Cortes et al., 2010a). The *CnSERK* gene encodes a protein receptor (Pérez-Núñez et al., 2009), which may be a component of a signalling cascade involved in regulating the rate of SE (Hecht et al., 2001; Schmidt et al., 1997; Santos et al., 2005; Thomas et al., 2004). *In situ* hybridisation has shown the transcripts of both genes to be localised in the SE structures that form on callus, and within meristematic centres. The molecular mechanisms of *CnCDKA* and *CnSERK* genes to confer embryogenic competence to somatic cells are still unknown, but experimental results indicate that these genes are reliable molecular markers for this biological process (Montero-Cortes et al., 2010a). One further molecular strategy adopted to improve the rate of coconut SE involved the upregulation of genes that affect the formation of shoot meristem production in somatic embryos of other species. Montero-Cortes and co-workers isolated the coconut *CnKNOX1* gene, a *KNOX* class I gene, which was expressed exclusively in tissue with meristematic activity (Montero-Cortes et al., 2010b). They established that the *CnKNOX1* gene was responsive to the addition of gibberellin during coconut SE, resulting in an increased rate of somatic embryo formation and germination.

Considering the limited understanding of the molecular mechanisms that underlie coconut SE, it is apparent that more research is needed in this area before a further impact upon the rate of coconut SE can be achieved. The isolation and characterisation of genes which regulate the formation of the root apex, such as the *PL/AIL* genes are still in their infancy, while the discovery of genes which specify the shoot apex has not even commenced. The study of these embryogenic genes, as well as other genes, encoding regulatory factors (such as the B3 domain transcription factor family) that are involved in lipid metabolism, represents an important avenue to explore in coconut research (Kim et al., 2013).

## 7 Management from nursery to productive plantation

The coconut seed is enclosed in its husk, a robust fibrous structure which imparts buoyancy, enabling dispersal of the seed to distant habitats. Ideally a batch of seed nuts is placed in a pre-nursery, so that the sprouted seeds can be placed at intervals in groups, to enable comparison of the growth rate within each group. Seedling vigour is considered a good indicator of health and yield potential. When the most vigorous 50% of seedlings were selected their yield performance was 30% higher than unselected palms in the first 5 years of production but fell to 13% after that.

The husk complicates the growth of a viable seedling because it reduces the ease with which the shoot and roots emerge. Additional uncertainty arises when the seeds are collected after falling from the tree because in some cases germination has already taken place and the orientation of the shoot within the husk is unknown. If the shoot is not oriented vertically when the seed is placed in the pre-nursery it usually becomes trapped within the husk and fails to emerge. Germination and emergence therefore depend on the navigation of the shoot through the husk, which itself evolved specifically for oceanic dispersal. The husk has no role in germination except to protect the delicate emerging shoot. Care is needed to avoid the possible negative effect of the husk on seedling success. Seedling growth can be more closely monitored if part or all of the husk is removed (Foale, 1968).

Reduction of the risk of failure, and possibly an aid to stimulating germination by improving the access of moisture to the germ pore, can be achieved by trimming at least some of the husk away from the seed. This is reported by Kenman (1973), who found that slicing a portion of husk away from 'above' the germ pore (a common practice around the coconut world), as well as removing the somewhat water-repellent exocarp from beneath the seed nut, boosted the rates of both germination and growth.

Complete removal of the husk was adopted by Foale (1968) in order to accurately record date of germination, and also to enable measurement of nut diameter when comparing the growth of three varieties, each subdivided into three nut sizes. The delicate new shoot required protection against foraging insects and rats for 3 weeks, after which it had 'hardened' into a robust structure.

Once a healthy seedling has emerged it is best placed in a polybag of 15–20 L volume. If the seednut is bulky the husk should be trimmed further to enable comfortable placement into the polybag. Traditionally, coconut seeds have been placed on the soil surface of the nursery and retained there for 6–8 months, whereupon they are uprooted – exposing and damaging the roots, while about half the endosperm has been extracted by the haustorium (Foale, 1968). During planting in the field most of these roots are dried and broken, so that the seedling becomes reliant for survival on the residual endosperm, which then supports recovery of the roots' ability to absorb water, which is essential to support renewed transpiration and photosynthesis. The polybag seedling on the other hand can be retained in the nursery for up to 12 months, its vigorous growth based on a rapidly expanding leaf area (Foale, 1968). Careful transplanting into a hole that is large enough to accommodate the undisturbed soil volume within the polybag will result in little, if any, transplanting shock. While no replicated comparison has been done, experience has shown that the palm raised in a polybag will flower as much as 1 year sooner than one from an open soil nursery.

As a fully developed palm has a more-or-less spherical crown, the optimum plant arrangement to achieve maximum light interception is the triangular or isometric pattern. Each palm has six equidistant neighbours whereas in a square planting pattern the palm has only four close neighbours. For a given spacing within the row the isometric pattern accommodates 15% more plants per ha calculated using the formula for density: $D = 10000/(S^2 \times 0.866)$ (where D is density of palms per ha and S is the spacing in metres). A spacing of 8.5 m results is an isometric density of 160 palms/ha which, in good growing conditions, can achieve up to 90% light interception by Tall palms with a fully developed crown, between the ages of 15 and 30 years (Taffin, 1993). The density should be 10% greater for a Tall by Dwarf hybrid and 30% greater for the Dwarf. If growing conditions are less than ideal, due to a marked dry season and limited opportunity for irrigation, the density should be reduced by 10–20%.

The period between transplanting into the field and light interception exceeding around 60% at around 7–9 years of age offers the opportunity for intercropping, particularly with annual food crops. Pasture grazing is not possible at this early stage, as the palm is highly palatable to farm animals, thus restricting grazing to those that are tethered, particularly goats. There is an opportunity for research locally, to identify suitable short-term crops, for this early period of transition of the plantation from low to high light interception. A beneficial leguminous cover crop such as *Pueraria* will control weed growth and boost soil nitrogen (Bourke, 1975).

As the length of the frond diminishes steadily, from 6 m at around 30 years of age to 4.5 m at 50 years, light interception inevitably falls to around 50% and then further as the palm becomes senile (Fig. 1). The older plantation therefore offers an opportunity for a range of intercrops, again depending on the local economy, including cacao, coffee, kava, citrus, banana and breadfruit, to pasture. An advantage of intercropping, in addition to potential income, is that weed growth is suppressed to some extent, but there is potential for water competition. This is particularly the case in a farming system where the fruit is left to fall it would perhaps be hard to find among weeds. At the very least, in the absence of an intercrop, bushy weeds that reduce access and use soil water need to be eliminated.

Interestingly the productivity of fruit does not fall with ageing as rapidly as the reduction in light interception, because the palm partitions less and less of new biomass to the trunk as it ages (Foale and Ashburner, 2004). Based on their data for trunk extension between successive fronds and trunk diameter, the calculated volume of trunk biomass associated with a single frond at 20 years of age is around 2.5 L compared to 0.3 L at 60 years of age.

The productivity of a plantation is bound to be limited, at least part of the time, when the water supply is plentiful, by a limiting mineral nutrient. Frequently it is potassium which is exported from the paddock as a major component of the fruit, particularly in the husk and the nut water, as well as being readily leached from the profile during heavy rainfall. Returning the husk to the field is helpful but in many cases low potassium still limits productivity because of the high need of the coconut (Taffin, 1993). There is a method available to identify the most important limiters on growth, known as foliar diagnosis. Research has been done to identify critical levels of many mineral elements in coconut leaf tissue, and the values used as a guide to palm health are shown in Table 1 (Taffin, 1993; Foale, 2003).

Many coconut regions experience a dry season long enough to reduce yield potential. Ideally the palm at all ages does not experience yield-lowering water deficit, which implies well-distributed rainfall of at least 200 mm/month. The mean daily value for evapotranspiration in the humid tropics is around 3–5 mm. This implies that a saturated soil profile with the capacity to store up to 300 mm of water in the root zone will avoid water deficit during rainless periods of up to 6 weeks or more. Where the soil stores less water however, or when rainless periods in excess of 6 weeks are frequent, the coconut will respond well to irrigation unless it has ready access to a water table, as, for example, on an atoll or sandy strand. The feasibility of irrigation will depend on the severity of the need, the availability of suitable water and the affordability of a distribution system. Where rainfall is well in excess of 200 mm in some months, and the soil permeability is low, palms are grown on parallel banks allowing excess water to drain from the depressed inter-row.

The coconut palm is attractive to many pests as well as being susceptible to a number of fatal diseases, as mentioned earlier (Taffin, 1993). Globally, the outstanding threat is from a phytoplasma, spread by leaf hopper insects, which causes lethal yellowing disease (LYD), a fatal disease that leads to the death of mature palms within 3 years of the first symptoms. First described in Jamaica in the 1960s (Whitehead, 1966), variants of LYD have since been reported in other Caribbean islands, Mexico, Florida in the United States, several countries in West Africa and East Africa as well as in isolated reports from Indonesia (Natuna wilt) and Papua New Guinea (Bogia coconut syndrome). There are some varieties that have shown resistance, though this is reported to have broken down in Jamaica, due perhaps to a mutation of the phytoplasma. There is much uncertainty about whether the absence of more virulent forms of LYD in the Asia Pacific region is due to resistance of the dominant palm genotypes or absence of active leaf hopper species. There is no known means to arrest the devastation of LYD once a palm has been infected, as death usually follows within 3 years, so its presence constitutes a major threat to the coconut population wherever it is active. Screening of germplasm from regions where LYD has not established has been done in the search for resistance (Whitehead, 1968).

There is a region of the Philippines affected by Cadang-cadang disease, due to a viroid which is fatal to the palm – sometimes lasting for over 15 years from first symptoms, in which case it does not result in immediate loss of production. It has not been identified in other countries and its mechanism of transmission remains unknown (Randles and Imperial, 1984). There is a virus in Vanuatu which is tolerated by the indigenous palm population but spreads rapidly, by the agency of a plant hopper, to all introduced populations (Randles,

1987). Globally, there is quarantine protection against the introduction of germplasm from Vanuatu to avoid the risk of spread of the foliar decay virus.

The principal fungal pathogen of coconut is *Phytophthora palmivora* which is present in many strains throughout the world. Here again screening is required to assess the resistance of a population of coconut to a specific local strain. An example of serious damage was noted when some populations of a coconut hybrid from West Africa succumbed to bud rot in Indonesia where there was a different strain of *Phytophthora* (Blaha et al., 1994). It is clear that vigilance is needed to check the different strains between source and destination of coconut seeds. There are many other pathogens and harmful insects, summarised well by Taffin (1993).

Another aspect of coconut protection is the development of a strategy to avoid severe wind damage. Smallholdings are often found to have palms of mixed age. Clearly the younger palms would suffer from reduced light interception, but at least they would benefit from the wind-breaking effect of the taller palms in the event of cyclonic wind. A case could be made for planting perhaps three or four rows of palms on adjacent strips of land at intervals of 5 years. Taller strips would shelter the shorter ones, and the combined production would be more stable, in the long term, as the strips would reach their peak yield 5 years apart. This would also enable replanting to be phased over an extended period.

## 8 Harvesting

The collection of fruit, whether immature for coconut water or fully mature, is readily achieved with a long-handled 'hook knife', adequate to reach a bunch height of 4 m or so above the ground. Above that height the immature fruit must be lowered to the ground to avoid splitting when it falls; this requires the use of a ladder when feasible, or climbing in order to attach a lowering rope to the harvested bunch. The mature fruit can be allowed to fall from any height without damage, but there is an increasing level of hook-knife operational skill required, especially when the bunch is more than 8 m above ground. Many farmers prefer climbing, using notches cut into the trunk as footholds. Four or five times a year they will harvest three mature, or near-mature bunches. Dehusking in the field has the advantage of providing mulch, but the husk may be saleable for fibre and cocopeat production, or for burning in some models of copra dryer. Dehusked nuts are transported either to the farmer's own copra dryer, or to the farm gate for sale to a collector from a processing factory. The immature fruit is transported whole to a collection point for sale.

Toddy or coconut sap is harvested, generally from young palms accessed by a bamboo ladder when available, or other climbing aids, by tapping the emerged spathe shortly before it is due to open. The spathe is bound along its length with a cord and the tip is cut to allow the sap to flow. Twice daily the cut is renewed allowing up to 5 L/day of sap to be collected, more if palms are irrigated to avoid periods of water deficit. This procedure is followed for up to 30 days per spathe, which may overlap with tapping on the next spathe which has emerged. No fruit are produced on a spathe which has been tapped for toddy.

Sometimes green fronds are harvested for use as cladding in temporary shelters, while dead fronds will fall or be cut down to be left in the field as mulch, or to be used in fencing for livestock, particularly pigs.

## 9 Products and processing

Primary products on farm include nut water, sap or toddy from the unopened spathe, copra, being dried kernel and husk as well as mature and immature fruit for sale to a processor. Both shell and husk may fuel the copra dryer, but if alternative fuel is available the shell is converted to charcoal, which itself may become domestic fuel or be sold to a manufacturer of activated carbon, a very high-value product. A common model of smallholder copra dryer comprises a rack upon which freshly split half nuts are placed to be dried by fuel burning, either coconut shell or firewood (Fig. 2). Shell is preferred as it produces very little smoke in contrast to firewood, thereby achieving higher-quality copra. After 1 day of drying the kernel has shrunk sufficiently from the shell of the half nut to be easily removed, freeing the shell for fuel to dry the next batch, or for conversion to charcoal in an underground pit for sale or domestic fuel.

Copra ideally has moisture content no higher than 6%, at which it may be stored and transported, free from colonisation by microorganisms or invasion by insects. The oil content of anhydrous copra is around 65–70%, and 95% of this is extractable in a process of grinding, heating and applying high pressure. The residual cake, comprising 38% by weight of the original copra, contains around 8–10% coconut oil, 20% crude protein, 14% crude fibre and 44% carbohydrate (Thieme, 1968). It is valued as a high energy feed supplement for livestock.

The oil pressed from copra has a brownish colour, a notable aroma and significant free fatty acid content. To achieve food-grade quality it is refined, bleached and deodourised (RBD). This oil is suitable for domestic frying applications, for the manufacture of coconut butter by hydrogenation of the fraction of unsaturated fatty acids present, and for use in soap manufacture.

As alternatives to copra, there are three other processes used to extract the oil from the coconut kernel at low temperature, resulting in what has become known as virgin coconut oil (VCO). The traditional fermentation process is based on the coconut milk extracted from shredded fresh kernel, which is left to stand at room temperature for 48 h. While the microbial action involved is

**Figure 2** Burning coconut shell in a copra dryer.

not well understood the milk emulsion breaks down so that the separated oil floats to the surface and is easily collected. Filtration results in a 'water clear' VCO which has a food market value much higher than the RBD oil from copra, and known to be rich in antioxidants (Bawalan, 2011). A second process uses centrifugal separation of the oil from coconut milk.

The third VCO production method is known as direct micro-expelling, which involves partially drying the white kernel, that has been shredded direct from the half nut (avoiding the testa), at a temperature below 60°C. When the moisture content has fallen to around 12% it is subjected to moderate hand-operated pressure which recovers up to 85% of the oil. Simple filtration is then required to produce VCO of water clear quality, comparable to the wet process (Anon., 2018). If the moisture content is significantly less than 12%, the recovery of oil using low pressure is much reduced. The residual cake, containing around 68% fibre, protein and soluble carbohydrate and 32% oil, on an anhydrous basis, has potential value in the human diet, with further processing.

VCO has a melting point around 23°C, marketed for convenience in a screw-top jar to enable access when it is solid. Removal of the longer-chain fatty acids by fractionation, however, results in an oil that has a melting point below 0°C, comparable to olive oil for example, which is widely used in salad dressing. Hydrogenation of the 8% unsaturated fatty acids in coconut oil yields 'coconut butter' - a popular ingredient in recipes for confectionery treats in a warm environment, where its melting point of 33°C is an advantage.

The husk may be processed employing either wet or dry retting, which achieves the separation of the fibres from the pith. Wet retting involves soaking the husk for up to several weeks to achieve separation by bacterial action, while dry retting involves mechanised crushing and combing of the husk. Coconut fibre products include yarn, rope, mat and board. The pith is transformed by drying and pressing into briquettes which expand dramatically in water; it is known as cocopeat, which is very popular in potting mixes and hydroponic systems in horticulture. The husk of immature fruit requires wet retting over several months, producing a high quality white fibre.

A traditional process in coconut communities is the preparation of coconut milk from scraped mature coconut kernel (Fig. 3) which is a daily ingredient in the diet of a high proportion of tropical population worldwide. Coconut milk, processed on a large scale, is an important commercial product marketed with different descriptions related to the oil content of the emulsion, which ranges from 28% in coconut cream down to 8% in coconut milk 'lite'.

Coconut water from immature fruit, harvested from when it reaches full physical size around 6 months after pollination, to around 9 months or more, has become a major export from leading coconut countries as well as being drunk

**Figure 3** Coconut milk extraction by hand.

traditionally. The combined popularity of coconut water and VCO, in particular, have recently generated increased profit resulting in renewed interest in both replanting and new plantings in the coconut world.

Fresh toddy is a solution of up to 16% sugar which can be used as a drink, particularly nutritious for children where fresh fruit is scarce. If it is allowed to ferment for 24 h or longer, natural fermentation results in alcohol production to a concentration depending on the sugar content. Evaporated toddy yields coconut palm sugar, which is highly valued. Distillation of the alcohol and ageing in barrels results in a liqueur type of product, known as coconut arrack.

Coconut water is widely drunk locally either out of the trimmed fruit or from a carton. The water is also the feedstock for *nata de coco* which contains microbial cellulose produced by the addition of a specific culture. It is popular in desserts, particularly in Indonesia. Evaporation of coconut water yields a powder that is rehydrated for consumption as a drink.

## 10 Conclusion and future trends

The coconut has become a valuable resource, progressing from its pivotal role in the sustenance of coastal communities in the humid tropics to a source of food, drink and horticultural materials worldwide. Following a period of high profitability as a source of oil for soap making and explosives, from the late nineteenth century until 1930, the production industry experienced decades of depression in world trade from which it is recovering in recent decades. The development of the product known as VCO, in particular, has led this recovery. The dominant form of the coconut palm, known as the Tall, is heterozygous, which means that there is a high degree of variation in potential productivity. A hybrid form between the Tall and the homozygous Dwarf offers more uniform progeny, but the challenge of achieving diversity of adaptation to biohazards, along with the low reproduction rate due to the small number of seeds produced by an individual plant need to be addressed. Identification of the genetic signature of resistance to biohazards would enable development of more robust varieties for future production.

Nutrient management of the coconut has been well-researched, so that foliar analysis is capable of identifying specific major and trace nutrients that could limit growth when water deficit is not operating, enabling targeted use of specific fertilisers. Methods of seedling management have been developed to minimise the shock of transfer of the seedling from the nursery to the field.

The need for massive replanting of the ageing coconut populations around the globe has identified a major research need, being for the vegetative multiplication of identified superior genotypes. The coconut is considered a recalcitrant species in relation to multiplication signalling that a major investment is needed to overcome this challenge.

Marker genes need to be identified with respect to specific threats, such as lethal yellowing disease and major insect predators such as rhinoceros beetle, the *Brontispa* leaf miner and many others.

## 11 Where to look for further information

The resurgence in the market for coconut food products, especially virgin coconut oil and coconut water, has created great interest in many coconut-producing countries around the world. There is a global organisation known as the International Coconut Community (ICC), formerly known as the Asian and Pacific Coconut Community. Practically all producing countries have joined the ICC which has its headquarters in Jakarta, Indonesia. Funded by contributions from all member countries the ICC provides up-to-date market information for all coconut products and convenes regular conferences on technical issues relating to production, processing and health issues. It also supports the International Conference on Coconut Oil which holds biennial meetings.

A high level of investment in coconut research is found in Mexico, Ivory Coast, India, Sri Lanka, Thailand, Indonesia and Philippines, while several other countries follow developments closely. The Australian Centre for International Agricultural Research (ACIAR) has supported coconut research, especially in the Pacific region, for several decades. Issues of greatest concern to researchers include identifying resistance to devastating diseases including lethal yellowing (mycoplasma) and phytophthera (fungal) bud rot. Reference to this work is readily found by searching the internet.

The demand for renewal of old plantings and expansion of production areas has led to an intense effort in several countries, and in Australia supported by ACIAR, to develop an affordable process for SE. The Technical Advisory Group of the ICC maintains an interest in this research in several different countries, while the ICC's scientific Advisory Committee on Health disseminates information on the exceptionally favourable influence of coconut oil in the human diet.

## 12 References

Adkins, S. W., Samosir, Y. M. S., Ernawati, A., Godwin, I. D. and Drew, R. A. 1998. Control of ethylene and use of polyamines can optimise the conditions for somatic embryogenesis in coconut (*Cocos nucifera* L.) and papaya (*Carica papaya* L.). In: Drew, R. A. (Ed.), *Proceedings of the International Symposium of Biotechnology in Tropical and Subtropical Species*, Brisbane, Australia, pp. 459–66. doi:10.17660/ActaHortic.1998.461.54.

Anon. 2018. Available at: http://www.coconutoil.com.au/dme-solution.

Antonova, I. D. 2009. Somatic embryogenesis for micropropagation of coconut (*Cocos nucifera* L.). PhD Thesis. The University of Queensland, Australia.

Bandupriya, H. and Dunwell, J. 2012. Overexpression of *CnANT*, coconut *BABYBOOM* homologue alters plant growth and morphology in transgenic *Arabidopsis* plants. *Tropical Agricultural Research* 23, 249–60.

Bandupriya, H. D. D., Gibbings, J. G. and Dunwell, J. M. 2013. Isolation and characterization of an *AINTEGUMENTA*-like gene in different coconut (*Cocos nucifera* L.) varieties from Sri Lanka. *Tree Genetics and Genomes* 9(3), 813–27. doi:10.1007/s11295-013-0600-5.

Bandupriya, H. D. D., Gibbings, J. G. and Dunwell, J. M. 2014. Overexpression of coconut *AINTEGUMENTA*-like gene, *CnANT*, promotes in vitro regeneration in transgenic *Arabidopsis*. *Plant Cell, Tissue and Organ Culture* 116(1), 67–79. doi:10.1007/s11240-013-0383-2.

Basu, A., Sethi, U. and Guhamukherjee, S. 1988. Induction of cell division in leaf cells of coconut palm by alteration of pH and its correlation with glyoxalase-I activity. *Journal of Experimental Botany* 39(12), 1735–42. doi:10.1093/jxb/39.12.1735.

Batugal, P., Rao, V. R. And Olivier, J. 2005. *EDS: Coconut Genetic Resources*. International Bureau of Plant Genetic Resources, Regional Office IBPGR-APO, Serdang, Malaysia.

Bawalan, D. D. 2011. *Processing Manual for Virgin Coconut Oil, Its Products and Byproducts for Pacific Island Countries and Territories*. Secretariat of Pacific Communities, Noumea, New Caledonia.

Bennett, J. A. 1987. *Wealth of the Solomons: A History of a Pacific Archipelago*. University of Hawaii Press, Honolulu.

Bhallasarin, N., Bagga, S., Sopory, S. K. and Guhamukherjee, S. 1986. Induction and differentiation of callus from embryos of *Cocos nucifera* L. by IAA-conjugates. *Plant Cell Reports* 5(5), 322–4. doi:10.1007/BF00268591.

Blaha, G., Hall, G., Warokka, J. S., Concibido, E. and Ortiz-Garcia, C. 1994. *Phytophthora* isolates from coconut plantations in Indonesia and Ivory Coast: characterization and identification by morphology and isozyme analysis. *Mycological Research* 98(12), 1379–89. doi:10.1016/S0953-7562(09)81067-8.

Blake, J. 1972. A specific bioassay for inhibition of flowering. *Planta* 103, 126–8. doi:10.1007/bf00387363.

Blake, J. and Hornung, R. 1995. Somatic embryogenesis in coconut (*Cocos nucifera* L.). In: Jain, S., Gupta, P. and Newton, R. (Eds), *Somatic Embryogenesis in Woody Plants*. Kluwer Publishers, Dordrecht, pp. 327–40.

Bourdeix, R., Konan, J. L. and N'Cho, Y. P. 2005. COCONUT – a guide to traditional and improved varieties. Collection Catalys. Cirad – CNRA France, p. 104.

Bourke, R. M. 1975. Evaluation of leguminous cover crops at Keravat, New Britain. *Papua New Guinea Agricultural Journal* 25(1), 1–9.

Boutilier, K., Offringa, R., Sharma, V. K., Kieft, H., Ouellet, T., Zhang, L. M., Hattori, J., Liu, C. M., van Lammeren, A. A. M., Miki, B. L. A., et al. 2002. Ectopic expression of BABY BOOM triggers a conversion from vegetative to embryonic growth. *The Plant Cell* 14(8), 1737–49. doi:10.1005/tpc.001941.

Branton, R. L. and Blake, J. 1983a. A lovely clone of coconuts. *New Scientist* 26, 554–7.

Branton, R. L. and Blake, J. 1983b. Development of organized structures in callus derived from explants of *Cocos nucifera* L. *Annals of Botany* 52(5), 673–8. doi:10.1093/oxfordjournals.aob.a086624.

Chan, J. L., Saenz, L., Talavera, C., Hornung, R., Robert, M. and Oropeza, C. 1998. Regeneration of coconut (*Cocos nucifera* L.) from plumule explants through somatic embryogenesis. *Plant Cell Reports* 17(6-7), 515–21. doi:10.1007/s002990050434.

Child, R. 1964. *Coconuts*. Longmans, London, p. 216.

Chugh, A. and Khurana, P. 2002. Gene expression during somatic embryogenesis - recent advances. *Current Science* 83, 715–30.

Corley, R. H. V. 1985. Yield potentials of plantation crops. In: *Potassium in the Agricultural Systems of the Humid Tropics*. Proceedings of the 19th Colloquium of the International Potash Institute, Bangkok.

Eeuwens, C. J. 1976. Mineral requirements for growth and callus initiation of tissue explants excised from mature coconut palms (*Cocos nucifera*) and cultured *in vitro*. *Physiologia Plantarum* 36(1), 23–8. doi:10.1111/j.1399-3054.1976.tb05022.x.

Eeuwens, C. J. and Blake, J. 1977. Culture of coconut and date palm tissue with a view to vegetative propagation. *Acta Horticulturae* 78(78), 277–86. doi:10.17660/ActaHortic.1977.78.35.

Fehér, A., Pasternak, T. P. and Dudits, D. 2003. Transition of somatic plant cells to an embryogenic state. *Plant Cell, Tissue and Organ Culture* 74(3), 201–28. doi:10.1023/A:1024033216561.

Fernando, S. C. and Gamage, C. K. A. 2000. Abscisic acid induced somatic embryogenesis in immature embryo explants of coconut (*Cocos nucifera* L.). *Plant Science: an International Journal of Experimental Plant Biology* 151(2), 193–8. doi:10.1016/S0168-9452(99)00218-6.

Fernando, S. C., Verdeil, J. L., Hocher, V., Weerakoon, L. K. and Hirimburegama, K. 2003. Histological analysis of plant regeneration from plumule explants of *Cocos nucifera*. *Plant Cell, Tissue and Organ Culture* 72(3), 281–3. doi:10.1023/A:1022345011002.

Find, J. I., Kristensen, M. M. H., Nørgaard, J. V. and Krogstrup, P. 1998. Effect of culture period and cell density on regrowth following cryopreservation of embryogenic suspension cultures of Norway spruce and Sitka spruce. *Plant Cell, Tissue and Organ Culture* 53(1), 27–33. doi:10.1023/A:1006070729489.

Foale, M. A. 1968. The growth of the young coconut palm (*Cocos nucifera* L). 1. The role of the seed and of photosynthesis in seedling growth up to 17 months of age. *Australian Journal of Agricultural Research* 19(5), 781–9. doi:10.1071/AR9680781.

Foale, M. A. 1992. Physiological basis for yield in coconut. In: Nair, M. K., Khanm H. H., Gopalasundaram, P. and Bhaskara Rao, E. V. V. (Eds), *Advances in Coconut Research and Development*. Oxford and IBH Publishing Co Pty Ltd, New Delhi.

Foale, M. A. 2003. *The Coconut Odyssey – the Bounteous Possibilities of the Tree of Life*. Monograph 101. Australian Centre for International Agricultural Research, Canberra, p. 132.

Foale, M. A. and Ashburner, R. 2004. The coconut palm. In: Chopra, V. L. and Peter, K. V. (Eds), *Handbook of Industrial Crops*. Indian Council of Agricultural Research, New Delhi. Chapter 6.

Fuentes, G., Talavera, C., Desjardins, Y. and Santamaria, J. M. 2005. High irradiance can minimize the negative effect of exogenous sucrose on photosynthetic capacity of in vitro grown coconut plantlets. *Biologia Plantarum* 49(1), 7–15. doi:10.1007/s10535-005-7015-6.

Gamborg, O. L., Miller, R. A. and Ojima, K. 1968. Nutrient requirements of suspension cultures of soybean root cells. *Experimental Cell Research* 50(1), 151–8. doi:10.1016/0014-4827(68)90403-5.

Green, A. H. and Foale, M. A. 1961. Improvement of coconut production on the high islands of the Pacific. In: *Proceedings of the 10th Pacific Science Congress*, Honolulu, Hawaii.

Gupta, P. K., Kendurkar, S. V., Kulkarni, V. M., Shirgurkar, M. V. and Mascarenhas, A. F. 1984. Somatic embryogenesis and plants from zygotic embryos of coconut (*Cocos nucifera* L.) *in vitro*. *Plant Cell Reports* 3(6), 222–5. doi:10.1007/BF00269297.

Harries, H. C. 1978. The evolution, dissemination and classification of *Cocos nucifera* L. *The Botanical Review* 44(3), 265–319. doi:10.1007/BF02957852.

Harries, H. C. and Harries, J. 2018. The cradle of the coconut: Emilio Chiovenda's "La Culla del Cocco" (1921–1923) revisited. *Webbia* 73(1), 155–8. doi:10.1080/00837792.2018.1443640.

Hecht, V., Vielle-Calzada, J. P., Hartog, M. V., Schmidt, E. D. L., Boutilier, K., Grossniklaus, U. and de Vries, S. C. 2001. The Arabidopsis *SOMATIC EMBRYOGENESIS RECEPTOR kinase 1* gene is expressed in developing ovules and embryos and enhances embryogenic competence in culture. *Plant Physiology* 127(3), 803–16. doi:10.1104/pp.010324.

Ikeuchi, M., Sugimoto, K. and Iwase, A. 2013. Plant callus: mechanisms of induction and repression. *The Plant Cell* 25(9), 3159–73. doi:10.1105/tpc.113.116053.

Jones, A. M. P. and Saxena, P. K. 2013. Inhibition of phenylpropanoid biosynthesis in *Artemisia annua* L.: a novel approach to reduce oxidative browning in plant tissue culture. *PLoS ONE* 8(10), e76802. doi:10.1371/journal.pone.0076802.

Jones, A. M. P., Chattopadhyay, A., Shukla, M., Zon, J. and Saxena, P. K. 2012. Inhibition of phenylpropanoid biosynthesis increases cell wall digestibility, protoplast isolation, and facilitates sustained cell division in American elm (*Ulmus americana*). *BMC Plant Biology* 12, 75. doi:10.1186/1471-2229-12-75.

Karunaratne, S. and Periyapperuma, K. 1989. Culture of immature embryos of coconut, *Cocos nucifera* L.: callus proliferation and somatic embryogenesis. *Plant Science* 62(2), 247–53. doi:10.1016/0168-9452(89)90087-3.

Kenman, E. 1973. Effect of seednut trimming on the germination and growth of coconut (*Cocos nucifera L.*). *Papua New Guinea Agricultural Journal* 24, 26–9.

Kim, H. U., Jung, S. J., Lee, K. R., Kim, E. H., Lee, S. M., Roh, K. H. and Kim, J. B. 2013. Ectopic overexpression of castor bean *LEAFY COTYLEDON2* (*LEC2*) in *Arabidopsis* triggers the expression of genes that encode regulators of seed maturation and oil body proteins in vegetative tissues. *FEBS Open Bio* 4, 25–32. doi:10.1016/j.fob.2013.11.003.

Kochanek, J., Swift, R. S., Kochanek, M. A., Cox, J. and Flematti, G. R. 2016. A systematic approach to recycling organics for horticulture: comparing emerging and conventional technologies. *Acta Horticulturae* 1112(1112), 327–34. doi:10.17660/ActaHortic.2016.1112.44.

Krishna, H., Alizadeh, M., Singh, D., Singh, U., Chauhan, N., Eftekhari, M. and Sadh, R. K. 2016. Somaclonal variations and their applications in horticultural crops improvement. *3 Biotech* 6(1), 54. doi:10.1007/s13205-016-0389-7.

Kumar, P. P., Raju, C. R., Chandramohan, M. and Iyer, R. D. 1985. Induction and maintenance of friable callus from the cellular endosperm of *Cocos nucifera* L. *Plant Science* 40(3), 203–7. doi:10.1016/0168-9452(85)90204-3.

Leach, B. J., Foale, M. A. and Ashburner, G. R. 2003. Some characteristics of wild and managed coconut palm populations and their environment in the Cocos (Keeling) Islands, Indian Ocean. *Genetic Resources and Crop Evolution* 50(6), 627–38. doi:10.1023/A:1024493030687.

López-Villalobos, A. 2002. Roles of lipids in coconut (*Cocos nucifera* L.) embryogenesis. Thesis/dissertation. University of London.

López-Villalobos, A., Dodds, P. F. and Hornung, R. 2001. Changes in fatty acid composition during development of tissues of coconut (*Cocos nucifera* L.) embryos in the intact nut and *in vitro*. *Journal of Experimental Botany* 52(358), 933–42. doi:10.1093/jexbot/52.358.933.

López-Villalobos, A., Hornung, R. and Dodds, P. F. 2004. Hydrophobic metabolites of 2,4-dichlorophenoxyacetic acid (2,4-D) in cultured coconut tissue. *Phytochemistry* 65(20), 2763–74. doi:10.1016/j.phytochem.2004.08.034.

López-Villalobos, A., Dodds, P. F. and Hornung, R. 2011. Lauric acid improves the growth of zygotic coconut (*Cocos nucifera* L.) embryos *in vitro*. *Plant Cell, Tissue and Organ Culture* 106(2), 317–27. doi:10.1007/s11240-011-9924-8.

Miyashima, S., Honda, M., Hashimoto, K., Tatematsu, K., Hashimoto, T., Sato-Nara, K., Okada, K. and Nakajima, K. 2013. A comprehensive expression analysis of the *Arabidopsis MICRORNA165/6* gene family during embryogenesis reveals a conserved role in meristem specification and a non-cell-autonomous function. *Plant and Cell Physiology* 54(3), 375–84. doi:10.1093/pcp/pcs188.

Montero-Cortes, M., Rodriguez-Paredes, F., Burgeff, C., Perez-Nunez, T., Cordova, I., Oropeza, C., Verdeil, J.-L. and Sáenz, L. 2010a. Characterisation of a cyclin-dependent kinase (*CDKA*) gene expressed during somatic embryogenesis of coconut palm. *Plant Cell, Tissue and Organ Culture* 102(2), 251–8. doi:10.1007/s11240-010-9714-8.

Montero-Cortes, M., Saenz, L., Cordova, I., Quiroz, A., Verdeil, J. L. and Oropeza, C. 2010b. GA3 stimulates the formation and germination of somatic embryos and the expression of a *KNOTTED*-like homeobox gene of *Cocos nucifera* (L.). *Plant Cell Reports* 29(9), 1049–59. doi:10.1007/s00299-010-0890-0.

Morel, G. and Wetmore, R. H. 1951. Fern callus tissue culture. *American Journal of Botany* 38(2), 141–3. doi:10.2307/2437837.

Moscatiello, R., Baldan, B. and Navazio, L. 2013. Plant cell suspension cultures. In: Maathuis, F. J. M. (Ed.), *Plant Mineral Nutrients, vol. 953. Methods in Molecular Biology*. Humana Press, New York, NY, pp. 77–93. doi:10.1007/978-1-62703-152-3_5.

Murashige, T. and Skoog, F. 1962. A revised medium for rapid growth and bio assays with tobacco tissue cultures. *Physiologia Plantarum* 15(3), 473–97. doi:10.1111/j.1399-3054.1962.tb08052.x.

Mustafa, N. R., de Winter, W., van Iren, F. and Verpoorte, R. 2011. Initiation, growth and cryopreservation of plant cell suspension cultures. *Nature Protocols* 6(6), 715–42. doi:10.1038/nprot.2010.144.

Nitsch, J. P. 1969. Experimental androgenesis in *Nicotiana*. *Phytomorphology* 19, 389–404.

Pan, M. J. and van Staden, J. 1998. The use of charcoal in *in vitro* culture – a review. *Plant Growth Regulation* 26(3), 155–63. doi:10.1023/A:1006119015972.

Pandey, D. K. and Chaudhary, B. 2014. Oxidative stress responsive *SERK*1 gene directs the progression of somatic embryogenesis in cotton (*Gossypium hirsutum* L. cv. Coker 310). *American Journal of Plant Sciences* 5, 80–102. doi:10.4236/ajps.2014.51012.

Pannetier, C. and Buffard-Morel, J. 1982. Production of somatic embryos from leaf tissues of coconut, *Cocos nucifera* L. In: *Proceedings of the 5th International Plant Tissue Culture Congress*, Tokyo, Japan.

Patnaik, J., Sahoo, S. and Debata, B. K. 1997. Somatic embryogenesis and plantlet regeneration from cell suspension cultures of Palmarosa grass (*Cymbopogon martinii*). *Plant Cell Reports* 16(6), 430–4. doi:10.1007/BF01146788.

Perera, P. I., Hocher, V., Verdeil, J. L., Doulbeau, S., Yakandawala, D. M. and Weerakoon, L. K. 2007. Unfertilized ovary: a novel explant for coconut (*Cocos nucifera* L.) somatic embryogenesis. *Plant Cell Reports* 26(1), 21–8. doi:10.1007/s00299-006-0216-4.

Perera, P. I. P., Vidhanaarachchi, V. R. M., Gunathilake, T. R., Yakandawala, D. M. D., Hocher, V., Verdeil, J. L. and Weerakoon, L. K. 2009. Effect of plant growth regulators on ovary culture of coconut (*Cocos nucifera* L.). *Plant Cell, Tissue and Organ Culture* 99(1), 73–81. doi:10.1007/s11240-009-9577-z.

Pérez-Núñez, M. T., Chan, J. L., Sáenz, L., González, T., Verdeil, J. L. and Oropeza, C. 2006. Improved somatic embryogenesis from *Cocos nucifera* (L.) plumule explants. *In Vitro Cellular and Developmental Biology - Plant* 42(1), 37–43. doi:10.1079/IVP2005722.

Pérez-Núñez, M. T., Souza, R., Saenz, L., Chan, J. L., Zuniga-Aguilar, J. J. and Oropeza, C. 2009. Detection of a *SERK*-like gene in coconut and analysis of its expression during the formation of embryogenic callus and somatic embryos. *Plant Cell Reports* 28(1), 11–9. doi:10.1007/s00299-008-0616-8.

Prior, I. A., Davidson, F., Salmond, C. E. and Czochanska, Z. 1981. Cholesterol, coconuts and diet on Polynesian atolls; a natural experiment; the Puka Puka and Tokelau Islands Studies. *Am. J. Clin. Nutr.* 34(8), 1551–61.

Randles, J. W. 1987. Small circular single-stranded DNA associated with Foliar Decay disease of Coconut Palm in Vanuatu. *J. Gen. Virol.* 68, 273–80.

Randles, J. W. and Imperial, J. S. 1984. Coconut Cadang Cadang Viroid. CMI/AAB descriptions of plant viruses, No. 287. July 1984.

Randles, J. W., Julia, J. F., Calvez, C. and Dollet, M. 1986. Association of single-stranded DNA with the foliar decay disease of the coconut palm in Vanuatu. *Phytopathology* 76(9), 889–94. doi:10.1094/Phyto-76-889.

Reinert, J. 1959. Über die Kontrolle der Morphogenese und die Induktion von Adventivembryonen an Gewebekulturen aus Karotten. *Planta* 53(4), 318–33. doi:10.1007/BF01881795.

Roy, A. T., Koutoulis, A. and De, D. N. 2000. Cell suspension culture and plant regeneration in the latex-producing plant, *Calotropis gigantea* (Linn.) R. *Plant Cell, Tissue and Organ Culture* 63(1), 15–22. doi:10.1023/A:1006497220647.

Sáenz, L., Herrera-Herrera, G., Uicab-Ballote, F., Chan, J. L. and Oropeza, C. 2010. Influence of form of activated charcoal on embryogenic callus formation in coconut (*Cocos nucifera*). *Plant Cell, Tissue and Organ Culture* 100(3), 301–8. doi:10.1007/s11240-009-9651-6.

Samosir, Y. M. S. 1999. Optimisation of somatic embryogenesis in coconut (*Cocos nucifera* L.). PhD Thesis. The University of Queensland, Australia.

Samosir, Y. M. S. and Adkins, S. W. 2014. Improving acclimatization through the photoautotrophic culture of coconut (*Cocos nucifera*) seedlings: an in vitro system for the efficient exchange of germplasm. *In Vitro Cellular and Developmental Biology - Plant* 50(4), 493–501. doi:10.1007/s11627-014-9599-z.

Samosir, Y. M. S., Godwin, I. D. and Adkins, S. W. 1998. An improved protocol for somatic embryogenesis in coconut (*Cocos nucifera* L.). In: Drew, R. A. (Ed.), *Proceedings of the International Symposium of Biotechnology in Tropical and Subtropical Species* (vol. 461), Brisbane, Australia, pp. 467–76.

Samosir, Y. M. S., Godwin, I. D. and Adkins, S. W. 1999. The use of osmotically active agents and abscisic acid can optimise the maturation of coconut somatic embryos. In: Oropeza, C. (Ed.), *Current Advances in Coconut Biotechnology*. CAB International, Wallingford, UK, pp. 341–54.

Sandoval-Cancino, G., Sáenz, L., Chan, J. L. and Oropeza, C. 2016. Improved formation of embryogenic callus from coconut immature inflorescence explants. *In Vitro Cellular and Developmental Biology - Plant* 52(4), 367–78. doi:10.1007/s11627-016-9780-7.

Sané, D., Aberlenc-Bertossi, F., Gassama-Dia, Y. K., Sagna, M., Trouslot, M. F., Duval, Y. and Borgel, A. 2006. Histocytological analysis of callogenesis and somatic embryogenesis from cell suspensions of date palm (*Phoenix dactylifera*). *Annals of Botany* 98(2), 301–8. doi:10.1093/aob/mcl104.

Santos, M. D., Romano, E., Yotoko, K. S. C., Tinoco, M. L. P., Dias, B. B. A. and Aragao, F. J. L. 2005. Characterisation of the cacao *somatic embryogenesis receptor-like kinase* (*SERK*) gene expressed during somatic embryogenesis. *Plant Science* 168(3), 723–9. doi:10.1016/j.plantsci.2004.10.004.

Schmidt, E. D. L., Guzzo, F., Toonen, M. A. J. and deVries, S. C. 1997. A leucine-rich repeat containing receptor-like kinase marks somatic plant cells competent to form embryos. *Development* 124(10), 2049–62.

Smertenko, A. and Bozhkov, P. V. 2014 Somatic embryogenesis: life and death processes during apical-basal patterning. *Journal of Experimental Botany* 65(5), 1343–60. doi:10.1093/jxb/eru005.

Steinmacher, D. A., Guerra, M. P., Saare-Surminski, K. and Lieberei, R. 2011. A temporary immersion system improves *in vitro* regeneration of peach palm through secondary somatic embryogenesis. *Annals of Botany* 108(8), 1463–75. doi:10.1093/aob/mcr033.

Steward, F. C., Marion, M. O. and Mears, K. 1958. Growth and organized development of cultured cells. II. Organization in cultures grown from freely suspended cells. *American Journal of Botany* 45(10), 705–8. doi:10.2307/2439728.

Taffin, G. de. 1993. *Le* cocotier. LE TECHNICIEN D'AGRICULTURE TROPICALE. Editions Maisonneuve et Larose, Paris.

Teixeira, J. B., Sondahl, M. R., Nakamura, T. and Kirby, E. G. 1995. Establishment of oil palm cell suspensions and plant regeneration. *Plant Cell, Tissue and Organ Culture* 40(2), 105–11. doi:10.1007/BF00037662.

Thieme, J. G. 1968. Coconut oil processing. FAO Agricultural Development Paper No. 89. FAO, Rome.

Thomas, C., Meyer, D., Himber, C. and Steinmetz, A. 2004. Spatial expression of a sunflower *SERK* gene during induction of somatic embryogenesis and shoot organogenesis. *Plant Physiology and Biochemistry: PPB* 42(1), 35–42. doi:10.1016/j.plaphy.2003.10.008.

Thorat, A. S., Sonone, N. A., Choudhari, V. V., Devarumath, R. M. and Babu, K. H. 2017. Plant regeneration from cell suspension culture in *Saccharum officinarum* L. and ascertaining of genetic fidelity through RAPD and ISSR markers. *3 Biotech* 7(1), 16. doi:10.1007/s13205-016-0579-3.

Tisserat, B. and Vandercook, C. E. 1985. Development of an automated plant culture system. *Plant Cell, Tissue and Organ Culture* 5(2), 107–17. doi:10.1007/BF00040307.

Tsuwamoto, R., Yokoi, S. and Takahata, Y. 2010. *Arabidopsis* EMBRYOMAKER encoding an AP2 domain transcription factor plays a key role in developmental change from vegetative to embryonic phase. *Plant Molecular Biology* 73(4–5), 481–92. doi:10.1007/s11103-010-9634-3.

Umehara, M., Ogita, S., Sasamoto, H., Koshino, H., Nakamura, T., Asami, T., Yoshida, S. and Kamada, H. 2007. Identification of a factor that complementarily inhibits somatic embryogenesis with vanillyl benzyl ether in Japanese larch. *In Vitro Cellular and Developmental Biology - Plant* 43(3), 203–8. doi:10.1007/s11627-006-9016-3.

Verdeil, J. L., Huet, C., Grosdemange, F. and Buffard-Morel, J. 1994. Plant regeneration from cultured immature inflorescences of coconut (*Cocos nucifera* L.): evidence for somatic embryogenesis. *Plant Cell Reports* 13(3–4), 218–21. doi:10.1007/BF00239896.

Verdeil, J. L., Hocher, V., Huet, C., Grosdemange, F., Escoute, J., Ferrière, N. and Nicole, M. 2001. Ultrastructural changes in coconut calli associated with the acquisition of embryogenic competence. *Annals of Botany* 88(1), 9–18. doi:10.1006/anbo.2001.1408.

Verdeil, J. L., Alemanno, L., Niemenak, N. and Tranbarger, T. J. 2007. Pluripotent versus totipotent plant stem cells: dependence versus autonomy? *Trends in Plant Science* 12(6), 245–52. doi:10.1016/j.tplants.2007.04.002.

Whitehead, R. A. 1966. *Sample Survey and Collection of Coconut Germ Plasm in the Pacific Islands*. Ministry of Overseas Development. HMSO, London, p. 78.

Whitehead, R. A. 1968. *Collection of Coconut Germ Plasm from the Indian region, Peru and the Seychelle Islands and Testing for Resistance to Lethal Yellowing Disease in Jamaica*. FAO of the United Nations, Rome, p. 40.

# Chapter 13

# Advances in guava breeding

*Pankaj Kumar and K. Usha, Division of Fruits and Horticultural Technology – Indian Agricultural Research Institute, India; and Bhupinder Singh, Centre for Environment Science and Climate Resilient Agriculture (CESCRA), India*



## 1 Introduction

The guava (*Psidium guajava* L.) popularly known as 'superfruit', 'poor man's fruit' or 'apple of tropics' belongs to the Myrtaceae family (Nakasone and Paull, 1998; Tate, 2000). The Myrtaceae family is represented by 133 genera and 3800 species of trees and bushes and is widely distributed over equatorial regions in tropical and subtropical climates (Mitra and Bose, 1985). Guava originated in tropical America (Hayes, 1953), Mexico (De Candolle, 1904) and Brazil (Purseglove, 1968) and is rich in vitamins 'A' and 'C' with seeds that are rich in omega-3, omega-6 polyunsaturated fatty acids, fiber, minerals, potassium, magnesium, essential nutrients and antioxidant pigments (Mitra and Sanyal, 2004). Carotenoids and polyphenols produce the fruit's color and guavas that are red or orange in color have more potential value as antioxidant sources than yellowish-green ones (Singh et al., 2005; Nimisha et al., 2013). In Spanish, the tree is known as *guayabo* or *guayavo*, the fruit *guayaba* or *guava*. The French call it *goyave* or *goyavier*; the Dutch, *guyaba*, *goeajaaba*; the Surinamese, *guave* or *goejaba;* and the Portuguese, *goiaba* or *goaibeira*. Hawaiians, guava or *kuawa,* in Guam, *abas*; in Malaya, guava or *jambu batu*, in India, tropical Africa and the Philippines *bayabas* and Indians of Mexico and Central and South America call it by tribal names– *pichi, posh, enandi, etc.*

http://dx.doi.org/10.19103/AS.2019.0054.23

Guava fruit is commercially important in India, China, Indonesia, South Africa, Florida, Hawaii, Egypt, Yemen, Brazil, Mexico, Colombia, West Indies, Cuba, Venezuela, New Zealand, Philippines, Vietnam and Thailand due to its year-round availability, rich nutritional and medicinal value, and affordable price, suitability for transportation, handling and consumer preference (Adesida and Farombi, 2012; Barbalho et al., 2012; Nimisha et al., 2013). There are more than 160 guava cultivars known globally (Table 1), with huge demand in the national and international market. Brazil, Mexico, Thailand, Peru and the Philippines are the major exporting countries of guava fruits and processed products (Varmudy, 2013).

All the cultivars of Indian guava (Table 2) belong to a single species *Psidium guajava* with basic chromosome number X=11 (haploid). Hayes (1953) reported the genus to contain about 150 species, but only a few were studied in detail. The two species, pyriferum and pomiferum mentioned by Linnaeus are nothing but trees with pear-shaped and round-shaped fruits (Bailey, 1919). In guava, most of the commercial varieties are reported to be diploids with chromosome number 2n =22, except the seedless types which are triploids 2n = 33 (Kumar and Ranade, 1952). The chromosome number of *P. friedrichsthalianum* Niedenzu was reported to be 2n = 22 (Srivastava, 1977). Subsequently, several *Psidium* species like *P. cattleianum, P. molle, P. acutangulam, P. araca, P. chinensis,*

**Table 1** Commercial cultivars of guava in the world

| Country | Cultivars |
|---|---|
| Australia | Allahabad Safeda; Beaumont; Lucknow-49; KaHua Kula |
| Bangladesh | Swarupkathi; Mukundapuri; Kanchannagar; Kazi |
| Brazil | Paluma; Rica; Pedro Sato; Kumagai; Sassaoka; Ogawa; Yamamoto; Século XXI, Cortibel |
| Colombia | Puerto Rico; Trujillo, Rojo Africano; Extranjero |
| Costa Rica | Tai-kuo-bar |
| Cuba | Enana Roja Cubana; EEA 1-23 |
| Egypt | Bassateen El Sabahia; Bassateen Edfina; Allahabad Safeda |
| Malaysia | Kampuchea (Vietnam, GU8); Hong Kong Pink; Jambu Kapri Putih; Maha 65; Bentong Seedless (Malaysian S.); Taiwan Pear |
| Mexico | Media China, Regional de Calvillo, China, la Labor, Acaponeta, Coyame |
| Puerto Rico | Corozal Mixta; Corriente; Seedling 57-6-79 |
| South Africa | Fan Retief, Frank Malherbe |
| Taiwan | Tai-kuo-bar |
| Thailand | Glom Sali; Glom Toon Klau; Khao Boon Soom |
| Vietnam | Glom Toon Klau; Khao Boon Soom |
| USA (Hawaii) | Beaumont; Pink Acid; Ka Hua Kula |

*Source*: adapted from Pommer and Murakami (2009).

**Table 2** Guava cultivars grown in different states of India

| State | Cultivars |
|---|---|
| Andhra Pradesh | Allahabad Safeda, Anakapalli, Banarasi, Chittidar, Hafsi, Sardar, Smooth Green, Smooth White, Dharward, Dholka, Kothrud, Lucknow-24, Sardar |
| Assam | Amsophri, Madhuriam, Safrior Payele |
| Bihar | Anakapalli, Banarasi, Bangalore, Chittidar, Hafsi, Nagpur Seedless, Allahabad Safeda |
| Gujarat | Nasik, Seedless, Sindh |
| Karnataka | Allahabad Safeda, Arka Mridula, Sardar, Navalur |
| Maharashtra | Dharward, Dholka, Kothrud, Lucknow-24, Sardar |
| Punjab | Allahabad Safeda, Shweta, Sardar(L-49), Punjab Pink |
| Tamil Nadu | Anakapalli, Banarasi, Bangalore, Chittidar, Hafsi |
| Uttar Pradesh | Allahabad Safeda, Chittidar, Hafsi (Red Fleshed), Harijha, Seedless, Apple color, Chittidar, Red Fleshed, Banarasi Surkha, Sardar, Mirzapur Seedless |
| West Bengal | Bariampur and cvs. of Uttar Pradesh |

*Source*: adapted from Teaotia et al. (1962) and Pandey (1968).

*P. cujavillis,* and *P. pumilum* were recognized and documented (Table 3). The wild species of guava are of considerable importance in breeding programs.

## 2 Floral biology

The knowledge of flower bud development, time of anther dehiscence and anthesis, the degree of cross pollination, the extent of fruit set in guava is a prerequisite for guava hybridization program (Braganza, 1990). Guava flowers and sets fruits at different times and at different locations depending on the existing climatic conditions where it is cultivated. Three flowering seasons, *viz.*, 'Ambebahar' (March–May), 'Mrig bahar' (July–September) and 'Hatti or Hasta bahar' (October) are observed in the peninsular regions of western and southern India (Cheema et al., 1954) and only two seasons (March–May and again in July–September) are observed in the subtropical North India (Ojha et al., 1986). The duration of flowering in each season lasts from 35 to 45 days.

In guava, two or three flower buds either singly or in cymose and are borne in leaf axil on current season's growth, both on the terminal and lateral shoots. The flowers consist of a superior calyx with five lobes and the corolla consists of 6–10 petals arranged in one or two whorls. The androecium consists of 160–400 thin filaments carrying bilobed anthers, closely packed together. The gynoecium consists of an inferior ovary, syncarpous, with axillary placentation and subulate style. The style is smooth and bearded at the summit.

**Table 3** Guava species and their general characteristics

| S. No. | Species | characteristics |
|---|---|---|
| 1 | *P. Cattleianum var. Cattleianum* (Sabine) Strawberry guava | It is a small tree, resistant to cold. Thin skinned, round, sweet flavored, fragrant fruits (2.5 cm in diameter) with numerous seeds. It is known as 'Strawberry guava' due to its sweet aroma which is reminiscent of strawberry and lacks muskiness of the common guava. Preferred by certain tribals (Normand, 1994). |
| 2 | *P. Cattleianum* (Sabine) *var. Lucidum* Lemon guava | Relatively hardy species with small, globose, juicy, acidic, sulfur yellow colored fruits. |
| 3 | *P. guineense*(Sw)Syn: *P. molle* (Bertol), *P. Araca* (Raddi), *P. Schiedeanum* Berg. Brazilian or Castilian guava | Dwarf, slow growing shrub resistant to drought. Fruits are round with yellow skin, pale yellow pulp surrounding the white central pulp and numerous hard seeds (Correa et al., 2011). |
| 4 | *P. friedrichsthalianum* (Niedenzu) Chinese guava or Costa Rican guava | It is a tall tree about 7–10 m high. Fruits are globose, small, sour, highly acidic fruits, good for jelly making. Wilt and nematode (*M. incognita*) resistant. |
| 5 | *P. montanum* (Swartz) Mountain guava | It is found in the mountains of Jamaica. Poor quality round fruits with white pulp and more seeds. |
| 6 | *P. araucanum (Soares-Silva* and *Proenca)* | A large tree with globose or pear-shaped fruits with thin pericarp, yellowish-green at maturity, angular or lenticulate seeds. |
| 7 | *P. acutangulum DC* | The tree ranges in height from 26 to 40 ft. Fruits are round to pear-shaped, pale yellow to yellowish white acidic pulp but well flavored pulp containing few hard, triangular seeds. |
| 8 | *Psidium guajava* L. Common guava | It is a large dicotyledonous shrub, or small evergreen tree, generally 3–10 m high. Fruits are ovoid or pear-shaped, 4–12 cm long, weighing 500 g; skin yellow when ripe, sometimes flushed with red, creamy-white or creamy-yellow to pink or red; mesocarp, juicy. With numerous cream to brown, kidney-shaped or flattened seeds. The exterior of the fruit is fleshy, and the center consists of a seedy pulp. |

Flowers located between the middle and base of the branch are most likely to bear fruits. When flower buds reach their maximum development, sepals begin to break down at various points, indicating the start of anthesis which may vary depending on the day temperature. The full development of the flower from the flower bud differentiation until flower opening takes about 32–42 days depending on the variety and environmental conditions. The flower buds that open the next morning show a cracking of the calyx on the previous day usually 24 hours before flower opening. The stigma becomes receptive within 2–3 hours after opening and remains so up to 48 hours thereafter. There are reports that the receptivity commences 2 days before anthesis and lasts up to 4 days after anthesis (Singh and Sehgal, 1968). But maximum pollen germination on stigma occurred when pollination was done 2 hours after anthesis. In wild species of *Psidium,* the best period of stigmatal receptivity is within an hour of the opening of the flowers (Teotia et al., 1970; Pereira and Sao Jose, 1987).

The anthesis starts at 4 a.m. and continues until 10 a.m., with peak anthesis between 6 and 8 a.m. (Pereira and Sao Jose, 1987). Anthers dehisce 15 to 30 minutes after anthesis and continue up to 2 hours (Balasubramanyam, 1959). In the case of wild *Psidium* species, the majority of flowers open between 4 and 6 a.m. (Teotia et al., 1970). About 80% of anthers shed their pollens in about 45 minutes after the opening. In *Psidium friedrichsthalianum* peak dehiscence occurs between 7 and 9 a.m., while in *P. cujavillus* and *P. cattleianum* it is between 9 to 11 a.m. (Teotia et al., 1970). The pollen is round with large grains (Srivastava, 1974). At the time of dehiscence viability of pollen varies from 42 to 95% depending on the varieties. Pollen viability of seedless varieties is less than 50%, and in seeded diploid varieties like Chittidar or Allahabad Safeda, it is over 90%. Sucrose solution (4%) was found as the best medium for artificial pollen germination (Balasubramanyam, 1959).

Pollen grains of commercial cultivars 'Chittidar' could be stored for about 5 months at 0°C with 25% RH. Pollen grains of *P. guajava* L., *P. guineese* Swartz, *P. molle* Bertol, *P. chinese* Lodd, and *P. cattleianum* Sabine var. Lucidum remains viable for 1 day under field conditions but is viable for 90 to 135 days at low temperature (0–4.5°C) and RH 0–25%. Pollen grains of cv. seedless could be stored for 7 months in a refrigerator at 5±2°C, but lost viability fully in a month at room temperature (Chezhiyan, 1988).

## 3 Breeding objectives

The major problems facing the guava industry in different parts of the world are the severity of wilt disease, high seed content of diploid commercial varieties and poor yield with small, irregular and misshapen fruits of triploid seedless varieties (Negi and Rajan, 2007). An additional difficulty in guava breeding is the floral structure (epigynous flower, with abundant incurred stamens of various sizes), pollen viability, long juvenile period, self-incompatibility and heterozygous nature, variation in fruit characteristics from one year to another caused by environmental factors limit the scope of conventional breeding methods in guava (Dinesh and Reddy, 2001; Coser et al., 2012; Usman et al., 2013). There is also evidence of variation between fruits on the same tree and, to a lesser extent, among trees of the same genotype (Thaipong and Boonprakob, 2005). This forces the breeder to test the same cultivars for many years before final characterization and use in breeding programs (Nimisha et al., 2013). Today guava breeding is aimed at developing varieties with high yield, high quality, fruits of uniform good size and shape, attractive skin and pulp color, fewer or soft seeds, long storage life, suitable for table and processing purposes and resistant to fruit fly and wilt (Zipori et al., 2007; Nimisha et al., 2013). Tree shape and size with good branch angles, producing a spreading tree with dwarf or semi-dwarf habit, a high fruit: shoot ratio is important characters particularly in high-density planting system to minimize the branch breakage when carrying heavy crops and to reduce the labor input on tree pruning. Further, the breeding for biochemicals and antioxidants in the fruit can be another important line of research, which, however, requires quantitative dissection for these attributes in the available germplasm of guava. Kherwar and Usha (2016), at the Indian Agricultural Research Institute, New Delhi, screened fifty genotypes of guava for fruit quality attributes and reported the existence a wide variation in fruit TSS, lycopene, ascorbic acid and total antioxidants, that is, 14.19–6.60 per cent, 0.51–3.03 mg g-100, 99.0–257.5 mg g-100 and 6.18–12.56 μmol TE g-100 respectively. Further, the character association and path analysis of these fifty genetically diverse indigenous genotypes of guava for eleven major bioactive and antioxidant compounds yielded a significantly positive phenotypic and genotypic association of reducing and non-reducing sugar with the total sugar content. The titrable acidity revealed significant positive association with the reducing sugar and the ascorbic acid. Lycopene, on the other hand, showed highly significant positive correlation with ascorbic acid, acidity and pectin. Overall the path analysis confirmed the direct effect of pectin, anthocyanin, lycopene, TSS, non-reducing sugar; and indirect effect of ascorbic acid, sugar: acid ratio, total sugar, acidity, and reducing sugar for achieving amenability in bioactive and antioxidant compounds of guava (Thonte and Chakrawar, 1981). Enormous natural

**Table 4** Donor parents for improving fruit shape of guava

| S. No. | Fruit shape | Varieties |
|---|---|---|
| 1 | Globose | Allahabad Safeda, Apple Color, Arka Amulya, Arka Mridula, Benaras, Behat Coconut, Chittidar, Dharwad, Hafsi, Local 2, Mirzapur seedling, Nasik, Philippine guava, Red flesh, Sindh, Smooth green, Surka Chitti, Superior Sour Lucidum, Dhareedar, Aneuploid 2, Lalit |
| 2 | Pyriform | Bangalore Local, G-6, White flesh |
| 3 | Ovate | Florida seedling, Surka Chitti Neputani |
| 4 | Oblong | Oblong, Aneuploid-1, Nagpur Seedless, Seedless triploid, Thailand 2, Lucknow 42 |

*Source*: adapted from Dinesh and Vasugi (2010).

**Table 5** Donor parents for improving fruit size of guava

| S. No. | Fruit weight (g) | Varieties |
|---|---|---|
| 1 | Small (16–100) | Aneuploid 1, Apple color, EC 14039, EC 147037, G-6, Hafsi, Local 1, Local 2, Nagpur seedless, *Psidium cattleianum var. Lucidum, P. chinensis, P. friedrichsthalianum, P. molle*, Pati, Philippine guava, Portugal, Seedless (triploid), Sindh, GR1 |
| 2 | Medium (100–140) | Allahabad Safeda, Arka Amulya, Arka Mridula, Bangalore Local, Chittidar, 7-39, EC 147034, 7-12 IC 147036, 9-35 EC 147036, EC 162904, Florida seedling, Lucknow 42, Mirzapur seedling, Nasik, Phili (pink), Red flesh, Sardar, Smooth Green, Spear acid, Surkachitti, Surka Chitti Neputani, White flesh |
| 3 | Large (>140) | Abu Ishakwala, Benaras, Behat Coconut, Chakaiya Ruthmanagar, Dharwad, Kamsari, Karela, *P. quadrangularis*, Superior Sour lucidum, 1 kg guava |

**Table 6** Donor parents for improving fruit color of guava

| S. No. | Peel color | Varieties |
|---|---|---|
| 1 | White | Abu Ishakwala, Allahabad Safeda, Aneupoloid 2, Apple color, Arka Amulya, Arka Mrudiula, Behat coconut, Benaras, Chakaiya Ruthmanagar, Chittidar, Dharwad, Florida seedling, Hafsi, Karela, Local 1, Local 2, Lucknow 42, Mirzapur seedling, Nagpur Seedless, Nasik, *P. chinensis*, Sardar, Seedless (triploid), Singh, Smooth Green, Superior Sour lucidum, Surka Chitti, Surka Chitti Neputani, White Flesh |
| 2 | Shades of red | Aneuploid 1, GR-6, Kamsari, *P. chinensis*, Pati, Phili (pink), Portugal, Red Flesh, GR1 |
| 3 | Shades of yellow | Spear acid, Bangalore Local, *P. cattleianum var. lucidum, P. Quadrangularis* |
| 4 | Purple pericarp | Purple Local, Philippine guava |

genetic variability for fruit shape, weight, color, quality, antioxidants and lycopene content and resistant to pest and diseases available (Tables 4–8) can be exploited through conventional and molecular breeding approaches to improve guava (Giri et al., 2004).

# 4 Methods of guava improvement

## *4.1 Heritability*

Commercially important traits such as yield, fruit size, certain types of disease resistance and quality characteristics (vitamin C, acidity, pectin content etc.) are often low in the heritability category (Golberg and Levy, 1941). The obovoid shape of the fruit is dominant over a round (oblate) and pyriform, bold seeds

**Table 7** Donor parents for improving different horticultural traits

| S. No. | Parameters | Varieties |
|---|---|---|
| 1 | Dwarfness | Apple color, Aneuploid No 82, *P. chinensis* |
| 2 | Good Yielder | Benaras, 7-39 EC147034, EC 162904, Behat Coconut, |
| 3 | Fruit quality | Cortibel selections CI, CIII, CIIIh, CIV, CV, CVI, CVII, and CVIII CXI, and the commercial cultivar PalumaCI, CII and CVII showed better performance |
| 3 | High Vitamin C | Mirzapur seedling, *P. chinensis*, Chakaiya Ruthmanagar, Dhareedar, EC 162904, G-6 |
| 4 | High TSS | Dhareedar, Allahabad Safeda, Arka Mridula, Seedless, Sindh, Hafsi, Bangalore Local, Surka Chitti, Behat Coconut |
| 5 | Processing | 7-12, EC 147036, 7-39 EC 147034 |

**Table 8** Donor parents resistant to pests and diseases in guava

| S. No. | Pest | Resistant variety | References |
|---|---|---|---|
| 1 | Tea mosquito bug | Hafsi, Superior Sour lucidum, EC 147036, EC 147039 | Dinesh and Vasugi (2010) |
| 2 | Spiralling whitefly | Arka Amulya, Benaras, Spear acid, *Psidium chinensis, P. friedrichsthalianum,* EC 147039 | Dinesh and Vasugi (2010) |
| 3 | Fruitfly | *Psidiumchinensi,* Kamsari, Red flesh, Superior Sour lucidum, EC 147037, EC147039 | Dinesh and Vasugi (2010) |
| 4 | Root knot nematode | Costa Rican wild guava (*P. friedrichstalianium*) is resistant and compatible as rootstock with *P. guajava* cv. Paluma, in field conditions<br>*Psidium guajava* x *P. guineense* is resistant to *Meloidogyne enterolobii* | Bezerra et al. (2018) |
| 5 | Guava wilt | *Psidium molle, P. Guineense* and Philippine Guava (Purple type) has shown promise as a wilt-resistant rootstock,<br>*Psidiummolle* x *P. guajava*– is free from wilt<br>TS-G2 rootstock is resistant to guava wilt | CISH, Lucknow<br>Severn-Ellis et al. (2012) |
| 6 | Bark-eating caterpillar | Superior Sour Lucidum and Seedless, Portugal and Seedless, Lalit and Seedless, Spear Acid and Seedless, had contrasting response to bark-eating caterpillar | Naga Chaithanya et al. (2016) |

were dominant over soft seeds and red pulp color is dominant to white pulp color and is governed monogenically (Rajan et al., 2005). Dinesh and Vasugi (2010) studied inheritance pattern using 'Purple guava' and 'Arka Mridula' as parents and observed that hybrids segregate in a ratio of 3:1 for green leaf types and purple leaf types. The attractive pulp color and high yields of 'Beaumont' could be transferred to other white sweet cultivars. Many cultivated red fleshed varieties are heterozygous for this character. A linkage was found between red-flesh color and bold seed size (Srivastava and Srivastava, 1965; Rajan et al., 2005). Dinesh and Yadav (1998) studied the F1 progenies of crosses among 'Apple Colour' and other guava varieties and found that genotypic variance was less than phenotypic variance for fruit weight, length, volume, width and TSS. In another study, the length of fruit showed high values of heritability in the narrow sense, while the other traits presented low and medium values (Marak and Mukanda, 2007; Pérez et al., 2012).

Genomic resources for *Psidium* at NCBI are scarce and currently, for *Psidium gujava* only 58 nucleotides, 22 GSS (Genomic Survey Sequences), 9 protein sequence, 13 EST (Expressed Sequence Tags) and 86 CDS (coding DNA sequences) are available and a few more for other *Psidium* species (Sonah et al., 2011). The genome size of guava is ~400 Mbp and is distributed on 11 chromosomes (Feuillet et al., 2011). There is not a single database which contains the sequence information on guava genomic or transcriptomic sequences. This hinders our understanding of variability among *Psidium* species, linkages, trait linked gene identification and transfer of genes within or across sources to improve guava for commercial success. Availability of reliable genetic transformational protocol is also a constraint. Despite several problems in guava breeding a few successful reports are available (Subramanyam and Iyer, 1982, 1998; Gomez-Lim and Litz, 2004; Ribeiro and Pommer, 2004; Thaipong and Boonprakob, 2005; Dinesh and Iyer, 2005; Patel et al., 2007; Pommer and Murakami, 2009; Rodriguez-Medina et al., 2010).

## 4.2 Plant introduction

Guava is a native of South American countries and is an open-pollinated crop exhibiting high levels of genetic diversity (Antonio et al., 2011) due to the continuous cultivation of heterogeneous seedling trees. Introduction saves time and many clones introduced (Table 9) are being commercially cultivated or used in breeding programs for guava improvement. Many guava clones were introduced to India by the Portuguese from Hawaii, Brazil, Thailand, Australia, the United States, Philippines, China and so on during the seventeenth century (Mehta et al., 2016). Important among them are Beaumont and Indonesian Seedless from Australia, Acerapera from Brazil, Verdie from the United States and wild species like *Psidium sartorianum*, *P. littorale* var. *longipes* and *P. longipes* from the United States. Introductions of Indian cultivars like Allahabad Safeda and Sardar also showed excellent results in other parts of the world (Gonzaga et al., 1999). Utmost care needs to be taken while introducing guava plants since there is a danger of introducing new pests and diseases along with plant

**Table 9** Guava varieties developed through introduction

| S. No. | Cultivar | Introduced from Country |
|---|---|---|
| 1 | Beaumont and Indonesian Seedless | Australia, Hawaii |
| 2 | Acerapera | Brazil |
| 3 | Allahabad Safeda and Sardar | India |
| 4 | Tai-Kuo-bar | Taiwan |
| 5 | Giant Thailand Guava | Thailand |
| 6 | Verdie, *Psidium sartoriaum, Psidium littorale var. longipes, Psidium longipes* | USA |

material. The introduced acidic types like Beaumont were prone to 'Stylar End Rot' caused by *Phomopsis psidii*. This disease was also inherited by developed hybrids. Hence strict plant quarantine measures are necessary at the time of introduction to prevent entry and spread of pests or diseases.

### 4.3 Selection

Guava is a predominantly self-pollinated crop but crosspollination of up to 35% also occurs under natural conditions providing a heterozygous, open-pollinated seedling population. Due to cross-pollinated species, substantial variability exists in seedling populations grown in different regions (Srivastava, 2005). Guava orchards raised from seeds are highly variable due to pollination by insects and have high rates of segregation. High heterozygosis could be a merit if the stringent selection is made and selected plants could be used as parents for future breeding programs. Selection based on phenotypic characters to select superior types from the seedling population for high yield and quality is an age-old practice followed in guava improvement programs. Clonal identifications based on various morphological characters may not be reliable to discriminate closely related genotypes because of the effect of environmental changes, epistatic interactions, pleiotropic effects, and so on (Loh and Rao, 1984; Chandra et al., 2005a,b). Most of the cultivars grown on a commercial scale are however seedling selections from the well-known parent cultivars (Table 10). During the last 15 years, guava improvement work from different parts of the world resulted in the release of more than 150 superior selections.

### 4.4 Hybridisation

Guava breeding through conventional approaches is a long-term and cumbersome process that relies on the arbitrary rearrangement of existing genes between two closely related parent plants (Pommer, 2012). The existence

**Table 10** Guava varieties developed through selection

| S. No. | Varieties | Parentage and place developed | Characters |
|---|---|---|---|
| 1 | Allahabad Safeda | Local selection from Allahabad | It is the most popular variety in India and has acquired large variations due to seed propagation. This is the progenitor of many Indian varieties and occupies the largest area under cultivation. Fruits are round, large in size, skin with smooth, light yellow on ripening, pulp white, firm, excellent in quality with high TSS and Vitamin C, pleasant flavor and a few, soft seeds. Bearing large, uniformly pink fruits with deep pink flesh. |
| 2 | Sardar Guava | Allahabad Safeda from Ganeshkhand Fruit Experimental Station, Pune | It is a selection from Allahabad Safeda made at Poona during 1927. The plant has a spreading nature. The tree is dwarf with open, rounded crown. The fruits are medium to large and contain a crisp, soft and creamy pulp; it is a heavy bearing variety. The fruit has a slightly acidic flavor, attractive aroma, with many seeds, and good keeping quality. |
| 3 | Guava (G-1) | Half-sib population of Apple Color | High TSS (12-15°B), longer shelf life, Red skinned. |
| 4 | CISH-G-1 | Red colored guava from CISH, Lucknow | Deep red colored with soft seeds, high TSS (15° B), long shelf life and higher anthocyanin content. |
| 5 | CISH-G-2 | Red-colored guava from CISH, Lucknow | Crimson colored attractive fruit, stripes in groove, seeds soft. |
| 6 | CISH-G-3 (Lalit) | Half-sib population of Apple color from CISH, Lucknow | Saffron yellow colored fruits with pink flesh, good blend of sugar and acid. Suitable for both table and processing purposes. The pink color in the beverage remains stable for more than a year in storage. 24 % higher yield than AS. |

(*Continued*)

**Table 10** (Continued)

| S. No. | Varieties | Parentage and place developed | Characters |
|---|---|---|---|
| 7 | CISH-G-4 (Shweta) | Half-sib population of Apple color from CISH, Lucknow | Attractive fruits, sub-globose with few soft seeds, high TSS (14° Brix) and attractive pink blush, good yield potent. |
| 8 | CISH-GS-35 | Half-sib progeny of Allahabad Safeda from CISH, Lucknow | High yielder, attractive, fruit surface, color, soft seeds and TSS (14° Brix). |
| 9 | Paluma | Rubi-Supreme from UNESP, Brazil | Light-pink, thick and firm pulp and few seeds. |
| 10 | Rica | Supreme from UNESP, Brazil | Highly productive plants (more than 50 $t.ha^{-1}$), red pulp, very pleasant flavor (11°Brix) and low acidity. Few and small seeds. |
| 11 | Pedro Sato | Red Ogawa No. 1 from UNESP, Brazil | Pink pulp, few seeds. |
| 12 | Bhavanagar selection | From Gujarat | Fruits small and in cluster, heavy bearer, seeds soft and plenty. |
| 13 | Anand selection (Red) | Local selection from Anand | Fruit medium size, flesh reddish and soft, seeds many and soft, pleasant flavor. |
| 14 | Arka Mridula | (Selection-8) from IIHR, Bangalore | From 200 open-pollinated seedlings of variety Allahabad Safeda, one seedling selection, Selection-8, was found to be promising. Plants are dwarf, higher yield and more pectin content. |
| 15 | Dhawal | Allahabad Safeda | The trees are vigorous, medium tall with erect and upright branching and a flat crown. Fruits are medium to large sized, roundish ovate in shape Vitamin C more than 250 mg per 100 g fruit. |
| 16 | Lalima | Half-sib population of Apple Guava | Plants are semi-vigorous and spreading type. Fruits are medium sized (180-200 g), Good TSS, lower acidity. |
| 17 | AS 1, AS 2, AS 3 | Seedlings selection of Allahabad Safeda from NDUAT, Faizabad | Promising with respect to fruit quality and yield. |

**Table 10** (Continued)

| S. No. | Varieties | Parentage and place developed | Characters |
|---|---|---|---|
| 18 | Pant Prabhat | Seedlings selection of Allahabad Safeda from GBPUAT, Pantnagar | Plant growth in this line is upright; with broad leaves, the tree is highly productive (100–125kg). Fruit skin is smooth and light yellow in color, fruits medium sized with average fruit weight of 150-172g, pulp is white, seeds are small and medium soft, the fruit has a sweet taste with pleasant flavor, ascorbic acid content varies from 125mg (rainy season) to 300mg/ 100g fruit weight (winter season). TSS varies from 10.5 to 13.5°B. |
| 19 | Dhareedar | A selection from an old, seedling orchard from Fruit Research Station, Kuthulia, Rewa | The trees are vigorous, medium tall with erect and upright branching and a flat crown. Fruits are medium to large sized, roundish ovate in shape with 5–7 raised lines on the surface of mature fruits, the peel in greenish yellow, the pulp soft and sweet (TSS 11.7° Brix). |
| 20 | Chittidar | Local selection | The fruits are characterized by numerous, red dots on skin. Fruits are sub-globose, with white pulp, high TSS and Vitamin C content of 240 mg /100g pulp. |
| 21 | G. Vilas Pasand | Local selection from Bulakihar (Malihabad), Lucknow | Vigorous, wide spreading with bushy, low growing habit. Fruits are round to ovoid, skin texture is course to smooth, fruit skin pale yellow to golden, color of flesh is creamy-white, soft, very large fruit (400g to 800g), less seeds, high vitamin C content, very productive throughout the year. |

(*Continued*)

**Table 10** (Continued)

| S. No. | Varieties | Parentage and place developed | Characters |
|---|---|---|---|
| 22 | Selection 1, 3, and selection 5 | Pink pulp Navalur guava from UAS, Dharwad | Fruit were big and pear-shaped with mimosa colored skin and light-pink colored pulp. Plants exhibited a high bearing capacity. |

of wide genetic variability lead to the development of guava hybrids through controlled artificial pollinations in India, Cuba, Colombia, Venezuela, the United States, Malaysia and Nigeria. For making crosses in a hybridization program, chosen flowers are emasculated at least one hour before anthesis, a day before opening at 'calyx break stage' and bagged. Fresh pollen from the pollen parent is collected from an unopened flower, at calyx break stage. Pollination is carried out within 2 hours after emasculation between 10 a.m. and 12 noon by gently smearing the stigmatic surface with collected or stored viable pollen and flowers are rebagged. Bags are removed after 5–6 days of pollination. The maximum fruit set occurs when the stigmas are pollinated within 2 hours after anthesis (Xu et al., 2007). Guava improvement work resulted in the development of hybrids such as 'Safed Jam', 'Kohir Safeda' and 'Arka Amulya' in India (Table 11). 'Século XXI' in Brazil, was developed through inter varietal hybridization. Interspecific hybridization between *Psidium molle* and *P. guajava* has yielded hybrids which are resistant to wilt and are grafting compatible (Leslie Landrum et al., 1995).

## 4.5 Polyploidy in guava

Seedlessness is an important commercial attribute in guava (Majumdar and Singh, 1964). Triploid, diploid, tetraploid and aneuploid forms of *Psidium guajava* varieties with distinct characteristics for size and shape of fruits, the color of flesh types ranging from 'seedless' to 'less-seeded' were identified as bud sports, seedling mutants or recombinant forms (Table 12). Cytological examination of the 'seeded', 'less-seeded' or 'seedless' types has shown differences in chromosome number and chromosomal abnormalities, self-incompatibility, gene dissimilarities contribute to this feature (Hirano and Nakasone, 1969). The 'highly seeded' and 'less-seeded' types are diploids while the 'highly-seedless' types are triploids (Raman et al., 1969; Iyer and Subramanyam, 1971). Breeders have been investigating methods that allow sexual recombination that result in progeny with nearly seedless fruit. Seedlessness was believed to be due to autopolyploid and triploids (Raza et al., 2003). Seedless varieties available at IARI and Saharanpur, a seedless type at Poona, India, Java's Seedless Red Guava (Red Jakarta guavas x white Thai), 'Indonesian Seedless' and 'Bangkok

**Table 11** Guava varieties developed through hybridization

| S. No. | Variety | Parentage | Characteristics |
|---|---|---|---|
| 1 | H-1 developed at FRS Anantharajupet (AP) | Red Fleshed x Saharanpur Seedless | Flesh pinkish, moderate seeds |
| 2 | H-2 developed at FRS Anantharajupet (AP) | Smooth Green x Nagpur Seedless | Fruit medium, cream color, less-seeded, soft and easily crushed |
| 3 | H-3 developed at FRS Anantharajupet (AP) | Allahabad x Red Fleshed | Fruit medium, Flesh white, moderate seed content |
| 4 | H-4 developed at FRS Anantharajupet (AP) | Smooth Green x Saharanpur Seedless | Fruit medium, Flesh white, many seeded, soft and easily crushed |
| 5 | H-5 developed at FRS Anantharajupet (AP) | Red Fleshed x Nagpur Seedless | Flesh white, sweet taste, seeds are big but soft and few |
| 6 | H-6 developed at FRS Anantharajupet (AP) | Banrasi x Allahabad | Flesh cream colored, less-seeded, soft and easily crushed |
| 7 | Arka Amulya developed at IIHR, Bangalore | Seedless x Allahabad Safeda | Plants are semi-vigorous and spreading type. Fruits are medium sized (180-200g), weight of 100 seed is 1.8g, pulp is white, with high TSS (12.5° B), fruits have good keeping quality. Pulp white with few soft seeds |
| 8 | Hybrid 16-1 developed at IIHR, Bangalore | Apple Color x Allahabad Safeda | Fruit peel bright red, flesh firm, few seeds, high TSS, good keeping quality |
| 9 | Arka Kiran developed at IIHR, Bangalore | Kamsari × Purple Local | Plants are semi-vigorous, amenable to high-density planting. The fruits are sub-globose, weighing about 200-230g. The pulp is deep pink, thick and has good flavor. The seeds are medium soft (9.0 kg cm-2), with high lycopene content (7.45 mg /100g) and TSS of 12.0 to 12.5°B |
| 10 | Arka Rashmi developed at IIHR, Bangalore | Kamsari x Purple Local | Deep pink pulp, seeds are medium soft, high lycopene content (235 mg/100 g) |
| 11 | Safed Jam developed at FRS, Sangareddy | Allahabad Safeda x Kohir | Fruit size bigger, few seeds |
| 12 | Kohir Safeda developed at FRS, Sangareddy | Kohir x Allahabad Safeda | Large fruits, white flesh, few seeds |

(*Continued*)

**Table 11** (Continued)

| S. No. | Variety | Parentage | Characteristics |
|---|---|---|---|
| 13 | Hisar Safeda developed at C.C.S. HAU, Hisar | Allahabad Safeda x Seedless | Compact crown, seed count low, seeds soft, TSS=13.4%, acidity=0.38%, average yield 114kg/tree/yr |
| 14 | Hisarsurkha developed at C.C.S. HAU, Hisar | Apple Color x Banrasi Surkha | Tree crown broad to compact, skin yellow with red dots in low temperature, pink color flesh, low seed count and TSS=13.6%, acidity=0.46%, average yield 86 kg/tree/yr |
| 15 | RCGH 1 ICAR Research Complex for NEH Region, Umiam, Meghalaya | Sour type x Red fleshed local | White flesh, highest yield (39.05 kg/plant), highest TSS (10.83°B), total sugar (8.07%), lowest acidity (0.50%) |
| 16 | RCGH4 developed at ICAR Research Complex for NEH Region, Umiam, Meghalaya | Red fleshed x Allahabad Safeda | Red flesh |
| 17 | RCGH 7 developed at ICAR Research Complex for NEH Region, Umiam, Meghalaya | Lucknow-49 x Pear-shaped | White flesh, Less number of seeds/100 g fruit weight (111.18) |
| 18 | Punjab Pink developed at PAU, Ludhiana | Portugal x L 49 = F1 x Apple Color | Medium to large fruit size, attractive red colored skin sometimes in summer season and golden yellow in winter season. Red flesh, pleasant flavor |
| 19 | Twenty-first century developed at UNESP, Brazil | Supreme-2 x Paluma | Very productive plant with a short cycle (130 days from bloom to harvest), rosy-red large fruits great flavor |

Apple' are triploids with somatic chromosome number of 2n = 33 (Kumar and Ranade, 1952; Raman et al., 1971). Another seedless cultivar, 'Salithong', has chromosome number equal to 2n=22 (Srisuwan, 2003). A natural autotetraploid in *P. guajava* (Naithani and Srivastava, 1966) and induced tetraploidy in cv. Allahabad Safeda by treating shoot tips with 0.1% colchicine were reported (Janaki Ammal, 1951; Ram Kumar, 1975). Shafaat Mohammed (1975) studied the breeding behavior of the aneuploids in guava (*Psidium guajava* L.) such as trisomic, tetrasomic and higher aneuploids. He observed that reciprocal crosses between aneuploids and diploids indicated less than 100% crossability. The aneuploids, when used as the male parent, crossed less frequently than as female parents and some aneuploids crossed more readily than others. Differences were observed in fruit size, fruit weight, and seed number in

**Table 12** Polyploidy in guava

| S. No. | Ploidy status | Species/ varieties | Chromosome number | References |
|---|---|---|---|---|
| 1 | Diploid | *P. guajava* | 22 | Kumar and Ranade (1952) |
| 2 | Triploid | A natural triploid | 33 | Kumar and Ranade (1952), Raman et al. (1971) |
| 3 | Tetraploid | Tetraploidy in guava has been induced too with colchicine treatment | 44 | Janaki Ammal (1951), Ram Kumar (1975) |
| 4 | Tetrasomic | Promising tetrasomic dwarfing rootstock (Aneuploid No. 82) | 14 | Sharma (1982) |
| 5 | Hexaploid | *P. cattleianum* var. Lucidum | 66 | Hirano (1967) |
| 6 | Tetraploid | *P. cujavillus, P. guineense* | 44 | Hirano (1967) |
| 7 | Tetraploid | *P. friedrichsthalianum* | 44, 66 | Hirano (1967) |

reciprocal crosses. The extra chromosome was found to be transmitted through both the egg cell and the pollen. However, the frequency of transmission was greater through the egg cell than the pollen. As high as 26% transmission of extra chromosomes were obtained through the egg cell. There was no clear-cut difference between trisomics and higher aneuploids with regard to the frequency of transmission of the extra chromosomes. In guava, where a large number of seeds are a disadvantage, aneuploidy breeding appears to be beneficial (Majumder and Mukherjee, 1972a,b). In order to evolve a variety with fewer seeds and high yield, crosses were made between seedless triploid and seeded diploid variety Allahabad Safeda at the Indian Agricultural Research Institute, New Delhi. Out of the 73 $F_1$ hybrid seedlings raised, 26 were diploids, 9 trisomics (2n+1), 5 double trisomics (2n + 1 + 1) and 14 tetrasomics (2n + 2). They showed distinct variation in tree growth habit and leaf and fruit characters. Three tetrasomic plants had a dwarf habit and normal shape and size of fruits with less number of seeds (Mukherjee, 1977). Mohammad (1974) reported obtaining 30 trisomics, 2 double trisomics, 1 tetrasomic and 2 higher aneuploids in progenies of open-pollinated triploid and triploid × diploid crosses. Reduction in growth and the size of leaf distinguished aneuploids from diploids. Aneuploids, particularly trisomics, had promising qualities and may be useful in developing plants with reduced seediness and possibly in providing dwarfing rootstocks. Sharma (1982) identified a promising dwarf rootstock aneuploid No. 82 (named later as Srijan) through a selection made out of 48 different aneuploid seedlings at IARI, New Delhi. The aneuploid No.

82 is a tetrasomic and induced substantial dwarfing of Allahabad Safeda in terms of plant height, plant spread and tree volume (Sharma et al., 1992). A cross between diploid Allahabad Safeda and seedless triploid was named as Arka Amulya and was released from IIHR, Bangalore, India. Fruits are medium in size with white flesh, soft seeds, high TSS and with good keeping quality. In India, a cross between a triploid seedless and diploid 'Allahabad Safeda' was performed to develop a cultivar with fewer seeds and increased yield potential. In the F1 generation, 26 diploid plants, nine trisomics, five double trisomic and 14 tetrasomics were obtained with variations observed in plant growth habit, flower and fruit characters. It is known that trisomic and other aneuploidy can be used to overcome the problem of incompatibility in certain interspecific crosses. Some seedless cultivars like 'Gonzalo No.1', 'Gonzalo No. 2', 'Ibarra' and 'N7' seedless were obtained. Although seedlessness is an important commercial attribute, poor germination of pollen grains and low fruit yields with few viable seeds were also observed.

### 4.6 Mutation

Mutations occur naturally in guava. Many researchers induced mutations through gamma irradiation and chemicals. Buds of guava cv. Sardar treated with Gamma rays (1, 2, 3, 4 and 5 kR) were budded onto Lucknow-49 rootstock (Nalini et al., 1973; Brar and Bal, 2003). Variability for plant height, internodal length, and stem diameter was maximum in 2 kR treatments while, for the number of branches, number of leaves and breadth of leaves, maximum variability was noted in 4, 1 and 3 kR treatments, respectively. Mutagenic treatments had no significant effect on stomatal size. In vitro mutagenesis, followed by micropropagation via axillary bud proliferation of shoot tips in guava, was carried out by Zamir et al. (2003). Optimum shoot proliferation was recorded in MS medium supplemented with 1.0mg BAP and 250mg L-glutamine/liter. Rooting of cultured shoots was observed in half-strength MS medium supplemented with IAA and IBA. LD50 was observed at 45 Gy. Rates of more than 75 Gy were lethal to explants.

## 5 Biotechnological techniques

### 5.1 Molecular approaches

A close genetic relationship among cultivars, somatic mutations, and changes due to environmental alterations can create problems incorrect identification of germplasm. The combination of morpho-agronomic characters and DNA molecular markers are the main working tools for genomics surveys and constitute novel tools of great utility (Table 13). They are useful to characterize

**Table 13** Molecular approaches used in improvement of guava

| Biotechnological techniques | Reference |
|---|---|
| *DAMD Markers for guava improvement* | |
| The genetic diversity and relatedness among 22 guava accessions comprising commercial cultivars, breeding lines, and unimproved cultivars were tested. Similarity between pairs of cultivars calculated by DAMD analysis were in the range of 0.22–0.95, with the maximum similarity (0.95) between CISH-G-4 and CISH-G-5 and least similarity was between *P. acutangula* and G-6 (0.22) | Saxena et al. (2007) |
| Studied the genetic diversity and relatedness among 22 guava accessions. The clustering revealed that most of the cultivars originated from the Indo-Gangetic plains and exotic cultivars formed a separate group. | Bajpai et al. (2008) |
| *RAPD Markers for guava improvement* | |
| Determined the genetic relationship in 13 north Indian cultivars of guava | Dahiya et al. (2002) |
| Analyzed molecular diversity of 41 different genotypes of guava | Prakash et al. (2002) |
| Determined the phylogenetic relationship in 18 cultivars of Taiwan | Chen et al. (2007) |
| Examined the diversity among 22 *Psidium guajava* L. cultivars | Sharma et al. (2007) |
| Identified molecular markers associated with high quercetin accumulation in the leaves of guava | Feria-Romero et al. (2009) |
| Studied cluster analysis in *P. guajava* samples | Ahmed et al. (2011) |
| Analyzed the genetic diversity between 20 guava accessions | Pessanha et al. (2011) |
| *AFLP Markers for guava improvement* | |
| Analyzed a set of 48 guava accessions | Hernandez-Delgado et al. (2007) |
| Molecular characterization of 48 Mexican native cultivars of guava | Hernandez-Delgado et al. (2007) |
| Studied genetic diversity of Mexican guava germplasm | Sánchez-Teyer et al. (2010) |
| *SSR Marker for guava improvement* | |
| Constructed (GA)n and (GT)n microsatellite-enriched library | Risterucci et al. (2005) |
| Assessed genetic variability in the collection | Valdés-Infant et al. (2007) |
| Transferred guava-derived SSR primers to other *Myrtaceae* species | Briceño et al. (2010) |
| Characterized 13 guava accessions | Yadav et al. (2011) |
| Characterized diversity of guava germplasm in the United States | Viji et al. (2010) |
| Developed the guava microsatellite (SSR) markers using the SAT software | Risterucci et al. (2010) |

(*Continued*)

**Table 13** (Continued)

| Biotechnological techniques | Reference |
|---|---|
| Studied genetic diversity of the natural population of *P. guajava* in Venezuela | Briceño et al. (2010) |
| Characterized nine guava cultivars and marker mPg-CIR09 was found to be the most suitable primer combination for guava accession characterization | Dinesh and Reddy (2001) |
| Characterized guava cultivars and generated barcodes | Kanupriya et al. (2011) |
| Characterized the genetic diversity among 28 *P. guajava* genotypes | Coser et al. (2012) |
| Discriminated the wild guava tree | Nogueira et al. (2012) |
| Studied the genetic distance among guava genotypes | Noia et al. (2012a,b) |
| Studied the genetic homogeneity of guava plants derived from somatic embryogenesis | Rai et al. (2012) |
| Studied the genetic diversity of wild guava from southern Espírito Santo and commercial cultivars | Noia et al. (2012a,b) |
| Transferred SSR markers developed in guava (*Psidium guajava* L.) to 4 Myrtaceae species | Rai et al. (2013) |
| *ISSR Markers for guava improvement* | |
| Developed molecular marker for pulp color in guava | Nandgav (2010) |
| Assessed genetic variability among 11 *Psidium* sp. | Mani et al. (2011) |
| Assessed the clonal fidelity of micro-propagated guava | Liu and Yang (2012) |
| *SPAR Markers for guava improvement* | |
| Tried to associate SPAR markers with the pink color of guava | Nandgav. (2010) |
| Two main clusters separating the genus, *Feijoa sellowiana* and *Psidium* sp. with a genetic distance of 0.86 | Mani et al. (2011) |

guava germplasm, estimate diversity level and for parentage relationship among accessions, outcrossing rates, outside pollen contamination in seed orchards and for risk assessment of gene flow from plantation to natural stands and to recommend genotypes with conservation and breeding potential (Bernardo, 2008; Barbour et al., 2010). They can be used from any tissue at any time during the plant growth, and thus expedite the process of variety identification and breeding and help in overcoming the limitations of traditional methods (Azofeifa-Delgado, 2006). DNA markers are the predominant types of genetic markers for marker-assisted breeding (MAB) (Arif et al., 2011). Molecular approaches, such as RAPD for the estimation of genetic diversity (Chen et al., 2007; Feria-Romero et al., 2009), ALFP for genetic characterization of genotypes (Hernandez-Delgado et al., 2007), and simple sequence repeat (SSRs) for cultivar identification and linkage mapping (Risterucci et al., 2005), have been reported in guava (Table 13). Ferreira et al. (2016) carried out a study on molecular detection of *Erwinia psidii* in guava plants under greenhouse and field conditions and found that the PCR method could be very useful tool in a multiphasic approach for diagnosis of bacterial blight.

The report on molecular characterization involving SSR-based survey of guava genome for the development of a molecular linkage map is very limited. In an earlier report of Lepitre et al. (2010), SSR markers have been mapped in combination with AFLP markers for enriching the available maps (Valdés-Infante et al., 2007; Rodriguez et al., 2007). Studies carried out on transferability of microsatellites from *Psidium guajava* to *Eugenia, Myrciaria, Campomanesia* and *Syzygium* Species (Myrtaceae) found that SSR is advantageous for characterization, conservation and breeding of these species (Zucchi et al., 2002, 2003; Santos et al., 2007; Acuña et al., 2012; Rai et al., 2013; Ferreira-Ramos et al., 2014; Nogueira et al., 2016). These studies also selected most suitable SSR for each species and genus, thus contributing for a better comprehension of the relationships between the species, taxonomy, structure and genetic diversity, construction of genetic maps, breeding and conservation programs in the family. Transferability success of any DNA sequence by PCR is inversely related to the evolution and the distance between two species (Barbará et al., 2007; Biffin et al., 2010; Kalia et al., 2011).

Kherwar et al. (2018) used microsatellite markers to discriminate between wild species, commercial cultivars and a new collection from farmer's field by generating a unique molecular fingerprint/profile, which can serve as a foundation for flawless selection and characterization of the germplasm. They showed that the SSR analysis could efficiently and accurately distinguish between commercial, red-flesh, white-flesh guava or uncultivated guavas and that the observed genetic diversity could be exploited in guava breeding initiatives. The genetic diversity analysis further helped in tracking the duplicate and misidentified cultivars and verification of synonyms and homonyms.

Padmakar et al. (2015) carried out a study on genotyping of guava full-sib population with 94 F1 progenies, derived from a two-way pseudo-test cross strategy differing in fruit quality traits such as seed strength (hardness/softness), fruit weight, and TSS, using SSR markers and sequence-related amplified polymorphic (SRAP) primer combinations. SSR's high-throughput genotyping, revealed 64.3% parental polymorphism that generated 321 alleles during the mapping population survey. Twenty percent of parental polymorphism was revealed with SRAPs that generated 126 scorable markers. Two linkage maps were constructed for parents using maternal-specific 143 testcross marker loci and paternal-specific 127 testcross markers along with 60 intercross marker loci. The estimated genome coverage was of 87.32% in Kamsari and 83.74% in Purple Local. These genetic maps will play a pivotal role in identification of complex quantitative trait loci (QTLs) related to fruit quality. From the above literature, it can be amply deduced that SSR, ISSR and SRAP markers can be used in order to specify the gaps in the gene pools and can help to predict nature and strength of correlation between morpho-physiological characters and the data generated by molecular markers.

## 5.2 Genetic mapping

Modern technologies of gene sequencing, microarray experiments, attempts made to understand gene and protein expression within the cell of an organism and information on molecular markers has been extremely helpful in identifying regions on chromosomes (QTL) that bring about variation in a trait, thereby providing tools that can lead to far more accurate selection procedures for genetic improvement (Schulman, 2007; Narain, 2010; Soto-Cerda and Cloutier, 2012; Nimisha et al., 2013). Advances in genomics, the study of the DNA structure of genomes, has led to complete genetic mapping of many fruit species increasing the possibilities to locate useful genes affecting quantitative traits. The combination of advances in informatics, sequencing technology, and genomics makes it theoretically possible to select promising types on the basis of the genotype instead of the phenotype (Li and Homer, 2010; Treangen and Salzberg, 2012). This has potential benefits to guava breeding where the cost of phenotype selection are very high because of the long time necessary for fruiting and the large amounts of land required for each seedling. Genome-wide dissection of quantitative traits requires the construction of complete genetic linkage maps with sufficient genomic coverage and large segregating populations to allow detection of moderate to large effect QTLs. Some genomic resources have been generated within the EU-Project 'Improvement of guava: Linkage mapping and QTL analysis as a basis for marker-assisted selection (GUAVAMAP; FP6-2003-INCO-DEV-2 No. 015111)'. SSR markers for *Psidium guajava* were developed massively. These were used to characterize guava germplasm collections in different countries at the molecular level. Genomic cosmid (cos sites + plasmid = cosmid) library was established in guava (Ritter, 2012). Molecular hybridization was done with heterologous probes for resistance (RGL sequences) and homeotic genes (MADX-box and HOMEO-box genes). PCR primers were designed for these sequences for downstream applications (Grattapaglia et al., 2012). On the other hand, linkage maps were constructed in different genetic backgrounds and QTL analyses for several useful traits were conducted. Individual and combined parental linkage maps have been constructed in three mapping populations (Enana x N6, Enana x Suprema Roja and Enana x Belic L-207) based on AFLP, SSR, and Conserved Ortholog Set (COS) markers. These population maps contain between 500 and 1000 markers and have lengths of 1500 to 2200 cm each. Individual linkage groups vary between 150 and 240 cm in length and contain between 35 and over 100 markers each

(Rodríguez et al., 2010). Also, an integrated guava reference linkage map was established based on common markers map with a very high marker density of over 1700 markers. QTL analyses for plant height, petiole length, leaf length and width, yield, fruit number, and average fruit weight, fruit length and width, internal and external pulp thickness, seed number and weight, vitamin C contents, acidity, and total soluble solids have been performed in these populations. Over 100 QTLs were detected for the different characters in the three populations and varied between 2 and 13 QTLs per trait. Individual QTLs explained between 5 and over 40% of the total variance. Total variance explained by the sum of all detected QTLs varied between 20 and over 50% between traits and populations (Grattapaglia et al., 2012). Some of the QTLs are co-located or closely linked to SSR markers, which allows an efficient selection for MAB in the different genetic background. Several QTLs for the same or for related characters occurred at the same location in different maps, allowing in this way to compare co-locations of QTLs in different genetic backgrounds (Babu et al., 2004; Rodríguez et al., 2007; Valdés-Infante et al., 2007, 2012).

## 5.3 Micropropagation

Generations of new and promising variability through somaclonal variant selection, production of androgenic and gynogenic haploids to achieve homozygous plants with stress-tolerance and free from diseases through shoot-tip culture are a need of the hour. A technique for successful *in vitro* propagation of guava germplasm using shoot-tip explants from mature trees was reported (Jaiswal and Amin, 1987; Papadatou et al., 1990; Jaiswal and Amin, 1992). Addition of activated charcoal to the medium enhanced rooting of explants and vegetative growth of established plantlets. Embryogenesis and plant regeneration from mesocarp of *Psidium guajava* L. was reported (Chandra et al., 2004a,b). Explants were cultured on modified MS medium fortified with 2,4-D (2.0 mg/L), ascorbic acid (100 mg/L), L-glutamine (400 mg/L) and sucrose (6%). Embryogenic proliferating tissue was induced, which was translucent, mucilaginous and differentiated into many small somatic embryos. The somatic embryos were retained in the same medium, where simultaneously differentiation of new somatic embryos and their conversion into plantlets was observed. Organogenesis has been induced *in vitro* both from mature tree explants (Amin and Jaiswal, 1987) and seedling explants of guava (Singh et al., 2002; Shah et al., 2008). Detection of genetic variation is also important for micropropagation and *in vitro* germplasm conservation to eliminate undesirable somaclonal variations. Rai et al. (2012) evaluated the genetic homogeneity of 1-year-old guava (*Psidium guajava* L.) plants developed from *in vitro* somatic embryogenesis with 6 SSR and 10 ISSR primers and the monomorphic banding pattern in micro-propagated plants. The mother plant confirms the genetic homogeneity of the *in vitro* raised plants and demonstrates the reliability of *in vitro* propagation system for guava. Twenty-one ISSR primers were employed to test 30 progenies of clonally propagated guava plants. The similarity indicators between the progenies and the mother ranged from 0.92 to 1.0. Such a similarity indicated a very low polymorphism. The low somaclonal variation rate of 1.65% indicated the genetic stability of micro-propagated guava. Therefore, micropropagation is a safe mode for multiplication of true to type plants of guava (Amin and Jaiswal, 1988). This is also very important for the commercial propagation of elite clones with desirable fruit traits (Liu and Yang, 2012).

## 5.4 Genetic transformation

Genetic engineering shortens the breeding period is a promising production alternative. It has now opened up new possibilities by allowing the transfer of individual genes within or across species and compliments the classical breeding approaches. For this approach to gain wide acceptance an efficient micropropagation and regeneration procedure to produce large numbers of rooted plants from unique plants is a prerequisite (Liu and Yang, 2012). Genetic transformation opens the opportunity for genetic manipulation of plants at the cellular level and provides the means for modifying single horticultural traits without significantly altering other aspects of the phenotype (Singh et al., 2004; Krishna and Singh, 2007). The main target of gene transfer techniques is to produce improved varieties through the incorporation of horticultural important genes into existing cultivars (Singh et al., 2004). An engineered *Agrobacterium tumefaciens* strain LBA 4404 harboring binary vector pBI121 having selectable markers (nptII and GUS) with CaMV 35S promoter gene has been used for transformation of guava (Biswas et al., 2007). Preliminary work on the genetic transformation of guava with cold hardiness genes (CBF1, CBF2, and CBF3) was successfully demonstrated by Biswas et al. (2005, 2007). However, complete regeneration of transformed plants could not be achieved. Guava fruit was identified as a particularly rich source of hydroperoxide lyase (HPL) activity. The enzyme HPL is widely distributed in plants and is involved in the biosynthesis of volatile aldehydes and alcohols that are important constituents in the characteristic flavor of fruits, vegetables, and leaves (Kim and Grosch, 1981). The HPL enzyme purified from guava fruits was cloned by polymerase chain reaction (PCR) using 30 and 50 rapid amplification of cDNA ends (Tijet et al., 2000). The sequence showed approximately 60–70% identity to the known 13-hydroperoxide lyases. One guava HPL gene was sub-cloned into a plant transformation vector, and a transformation protocol was performed in order to analyze the effects of HPL gene silencing in the guava transformants

(Valecillos and Fermin, 2008). Such efforts will ultimately provide the most rapid advances in guava.

## 6 Conclusion and future trends

Guava is a nutritionally valuable crop with large potential in the domestic and export market however its productivity is limited by several pathogens. The advances in biotechnological tools using cell, tissue, and organ culture techniques have considerable potential for the improvement of guava within a limited time frame. Guava triploid production is difficult due to the reproductive complications of the plant. Few seeds/soft seeds preferred by consumers direct the need for aneuploidy breeding, development of autotetraploids and *in vitro* genetic manipulation of somatic cells. Accurate variety identification using molecular tools is a basic requirement for protection of guava germplasm, as well as for answering legal disputes arising from commercialization of cultivars with superior traits. With the development and availability of an array of molecular markers and dense molecular genetics maps in crop plants, marker-assisted selection has become possible for traits both governed by major genes as well as QTLs. Somaclonal variation, recovery of haploid plants from anther/pollen culture, offers advantages in guava breeding to select plants resistant to the wilt disease (Singh et al., 1977; Evans and Sharp, 1988). Tools like MAS can be highly useful for parentage analysis, clonal fingerprinting, genetic diversity analysis, development of genetic linkage maps, in disease and pest resistance breeding and in relation to flavor, color, fragrance, vitamins, fruit softening, and other traits of interest. With the advent of robust sequencing techniques and research focus on developing transformation efficient protocol for guava, the opportunities to increase the rate of genetic gain in guava breeding programs and challenges of improving guava seems highly achievable.

The right technology is pivotal to improving productivity of fruit crops, including guava. It is high time that an all-out effort is focused on the adoption of this technology by the guava growers. Credit availability, micro and easy financing conditions are however desirable for the adoption of key technologies that have emerged over the years and aim at improving the guava production and fruit quality, but can also be popularized among horticulturists dealing with other fruit crops. Further, important strategies have to be identified for enhancing quality guava production to make it competitive in the world market. Engrossed research at the ICAR-IIHR, Bengaluru and ICAR-IARI, New Delhi hold promise in this regard, since these institutes are mandated to develop varieties and farm management technologies aimed at improving quality guava production and to popularize them among farmers through institutional and governmental support at the state and national level. With concerted efforts worldwide, development of guava that is rich in minerals and antioxidants is not impossible.

## 7 References


Acuña, C. V., Fernandez, P., Villalba, P. V., García, M. N., Hopp, H. E., and Poltri, S. N. M., 2012. Discovery, validation, and in silico functional characterization of EST-SSR markers in *Eucalyptus globulus*. *Tree Genet. Genom.*, 8, 289–301.

Adesida, A., and Farombi, E. O., 2012. Free radical scavenging activities of guava extract *in vitro*. *Afr. J. Med. Med. Sci.*,41, 81-90.

Ahmed, B., Mannan, M. A., and Hossain, S. A., 2011. Molecular characterization of guava (*Psidium guajava* L.) germplasm by RAPD analysis. *Int. J. Na. Sci.*, 1(3), 62–7.

Amin, M. N., and Jaiswal, V. S., 1987. Rapid clonal propagation of guava through *in vitro* shoot proliferation on nodal explants of mature trees. *Plant Cell Tiss. Organ. Cult.*,9, 235–43.

Amin, M. N., and Jaiswal, V. S., 1988. Micropropagation as an aid to rapid cloning of a guava cultivar. *Sci. Hortic.*, 36, 89–95.

Antonio, C., Santos, F., Corrêa, L. C., and da Costa, S. R., 2011. Genetic divergence among *Psidium* accessions based on biochemical and agronomic variables. *CRP Breed. Appl. Biotech.*, 11, 149–56.

Arif, I., Khan, H. A., Bahkali, A. H., Al Homaidan, A. A., Al Farhan, A. H., Al Sadoon, M., and Shobrak, M., 2011. DNA marker technology for wildlife conservation. *Saudi J. Biol. Sci.*, 18, 219–25.

Azofeifa-Delgado, A., 2006. Uso de marcadores moleculares en plantas: aplicacionesen frutales del tropico. *Agron. Mesoam.*, 17, 221–42.

Babu, R., Nair, S. K., Prasanna, B. M., and Gupta, H. S., 2004. Integrating marker-assisted selection in crop breeding - prospects and challenges. *Curr. Sci.*, 87, 607–19.

Bailey, L. H., 1919. *Standard Encyclopaedia of Horticulture*. Macmillan Publishers, New York, NY, pp. 2847–9.

Bajpai, A., Chandra, R., Rajan, S., and Srivastava, N., 2008. RAPD and minisatellite markers for genetic diversity and relationship in guava varieties. *Ind. J. Genet.*, 68, 441–5.

Balasubramanyam, V. R., 1959. Studies on blossom biology of guava (*Psidium guajava* L.). *Ind. J. Hortic.*, 16, 69–75.

Barbalho, S. M., Farinazzi-Machado, F. M. V., Alvares-Goulart, R., Brunnati, A. C. S., Ottoboni, A. M. M. B., and Nicolau, C. C. T.,2012. *Psidium guajava* (Guava): A plant of multipurpose medicinal applications. *Med. Aromat. Plt.*, 1, 1-6.

Barbará, T., Palma-Silva, C., Paggi, G. M., Bered, F., Fay, M. F., and Lexer, C., 2007. Cross-species transfer of nuclear microsatellite markers: potential and limitations. *Mol. Ecol.*, 16, 3759–67.

Barbour, R. C., Wise, S. L., McKinnon, G. E., Vaillancourt, R. E., Williamson, G. J., and Potts, B. M., 2010. The potential for gene flow from exotic eucalypt plantations into Australia's rare native eucalypts. *For. Ecol. Manag.*, 260, 2079–87.

Bernardo, R., 2008. Molecular markers and selection for complex traits in plants: Learning from the last 20 years. *Crop Sci.*, 48, 1649–64.

Bezerra, C. F., Rocha, J. E., Nascimento Silva, M. K. D., de Freitas, T. S., de Sousa, A. K., Dos Santos, A. T. L., da Cruz, R. P., Ferreira, M. H., da Silva, J. C. P., Machado, A. J. T., Carneiro, J. N. P., Sales, D. L., Coutinho, H. D. M., Ribeiro, P. R. V., de Brito, E. S., and Morais-Braga, M. F. B., 2018. Analysis by UPLC -MS-QTOF and antifungal activity in Guava. *Food Chem. Toxicol.*, 119, 122–32.

Biffin, E., Lucas, E. J., Craven, L. A., da Costa, I. R., Harrington, M. G., and Crisp, M. D., 2010. Evolution of exceptional species richness among lineages of fleshy-fruited Myrtaceae. *Ann. Bot.*, 106, 79–93.

Biswas, B. K., Yadav, A., Joshee, N., and Yadav, A. K., 2005. *In vitro* plant regeneration and genetic transformation to enhance cold hardiness in guava: A nutraceutical fruit. In: *Proceeding of 1st International Guava Symposium*. CISH, Lucknow, India, pp. 31–2.

Biswas, B. K., Joshee, N., Yadav, A., and Yadav, A. K. 2007. Development and application of biotechnology in guava: a nutraceutical fruit. *Acta Hortic.*, 744, 267–76.

Braganza, M. A., 1990. Floral biology studies and varietal evaluation in genus *Psidium*. M.Sc. thesis submitted to University of Agricultural Sciences, Bangalore.

Brar, H. S., and Bal, J. S., 2003. Studies on the use of gamma rays on the performance of guava budlings. *Ann. AgriBio. Res.*, 8, 213–17.

Briceño, A., Aranguren, Y., and Fermin, G., 2010. Assessment of guava-derived SSR markers for the molecular characterization of *Myrtaceae* from different ecosystems in Venezuela. *Acta Hortic.*, 849, 139–46.

Chandra, R., Bajpai, A., Gupta, S., and Tiwari, R. K., 2004. Embryogenesis and plant regeneration from mesocarp of *Psidium guajava* L. (guava). *Ind. J. Biotech.*, 3, 246–48.

Chandra, R., Mishra, M., Abida, M., and Singh, D. B., 2005a. Paclobutrazol mediated somatic embryogenesis in guava. In: *Abstract of 1st International Guava Symposium*. CISH, Lucknow, India, p. 34.

Chandra, R., Mishra, M., and Bajpai, A., 2005b. Biotechnological interventions for improvement of guava (*Psidium guajava* L.). In: Kishun, R., Mishra, A. K., Singh, G. and Chandra, R. (Eds), *Proceeding of 1st International Guava Symposium*. CISH, Lucknow, India, pp. 19–25.

Cheema, G. S., Bhat, S. S., and Naik, K. C., 1954. *Commercial Fruits of India*. Macmillan & Co., New York, NY.

Chen, T. W., Chang-Chai, N. G., Wang, C. Y., and Shyu, Y. T., 2007. Molecular identification and analysis of *Psidium guajava* L. from indigenous tribes of Taiwan. *J. Food Drug Anal.*, 15, 82–8.

Chezhiyan, N., 1988. Stigma receptivity, flower shedding, flower abnormality and pollination studies in *Psidium* sp. *Madras Agric. J.*, 75(1–2), 29–32.

Correa, L. C., Santos, C. A. F., Lima, G. P. P., Rodrigues, M. A., and Costa, T. P. P., 2011. Genetic similarity among accessions of guava and Brazilian guava aracazeiros based on AFLP markers. *Rev. Bras. Frutic.*, 33, 859–67.

Coser, M. S., Fontes, M. P. M., and Ferreira, F. S. M., 2012. Pollen viability of guava genotypes from different locations. *3rd International Symposium on Guava and Other Myrtaceae*. Petrolina, PE, Brazil. Session III: Myrtaceae Plant Breeding and Post-Harvest Management.

Dahiya, K. K., Archak, S., and Karihaloo, J. L., 2002. DNA fingerprinting of guava (*Psidium guajava* L.) cultivars using RAPD markers. *Ind. J. Plant Genet. Resour.*, 15, 112–15.

De Candolle, A. P., 1904. *Origin of Cultivated Plants*. Kegal Paul, London.

Dinesh, M. R., and Iyer, C. P. A., 2005. Significant research achievement in guava–improvement and future needs. In: Kishun, R., Mishra, A. K., Singh, G., and Chandra, R. (Eds), *Proceedings of the 1st International Guava Symposium*. CISH, Lucknow, India, pp. 7–16.

Dinesh, M. R., and Reddy, B. M. C., 2001. Evaluation of *Psidium guajava* accessions and some other *Psidium* species for fruit characters. *J. Appl. Hortic.*, 3, 41.

Dinesh, M. R., and Vasugi, C., 2010. Guava improvement in India and future needs. *J. Hortic. Sci.*, 5, 94–108.

Dinesh, M. R., and Yadav, I. S., 1998. Half-sib analysis in guava (*Psidium guajava*). *Ind. J. Hortic.*, 55, 20–2.

Evans, D. A., and Sharp, W. R., 1988. Somaclonal variation and its application in plant breeding. Feature article. *IAPTC Newslett.*,54, 2–10.

Feria-Romero, I. A., Astudillo-Dela, H. V., Chavez-Soto, M. A., Rivera-Arce, E., Lopez, M., Serrano, H., and Lozoya, X., 2009. RAPD markers associated with quercetin accumulation in *Psidium guajava*. *Bio. Plant.*, 53, 125–8.

Ferreira, M. Á. S. V., Blum, L. E. B., Silva, C. F., Uesugi, C. H., and Marques, A. S., 2016. Molecular detection of Erwinia psidii in guava plants under greenhouse and field conditions. *Cienc. Rural.*, 46, 9.

Ferreira-Ramos, R., Accoroni, K. A. G., Rossi, A., Guidugli, M. C., Mestriner, M. A., Martinez, C. A., and Alzate-Marin, A. L., 2014. Genetic diversity assessment for *Eugenia uniflora* L., *E. pyriformis* Cambess, *E. brasiliensis* Lam. and *E. francavilleana* O. Berg neotropical tree species (Myrtaceae) with heterologous SSR markers. *Genet. Resour. Crop. Evol.*, 61, 267–72.

Feuillet, C., Leach, J. E., Rogers, J., Schnable, P. S., and Eversole, K., 2011. Crop genome sequencing: Lessons and rationales. *Trends Plant Sci.*, 16, 77–88.

Giri, C., Shyamkumar, B., and Anjaneyulu, C., 2004. Progress in tissue culture, genetic transformation and applications of biotechnology to trees: an overview. *Trees Struct. Funct.*, 18, 115–35.

Golberg, L., and Levy, L., 1941. The vitamin C content of fresh, canned and dried guava. *Nature*, 148, 286. (Cited from Mitra, S. K. and Bose, T. K. 1990. Nature, guava. In: Bose, T. K. and Mitra, S. K. (Eds), *Fruit: Tropical and Sub Tropical*. Nayaprokash, Calcutta, pp. 280–303).

Gomez-Lim, M. A., and Litz, R. E., 2004. Genetic transformation of perennial tropical fruits. *In Vitro Cell Dev. Biol. Plant.*, 40, 442–9.

Gonzaga, N. L., Bezerra, J. E. F., and Montano, J. C., 1999. Introduction and evaluation of Indian varieties of guava in the region of Submedio San Francisco. *Pesquisa em Andamento da Embrapa Semi Arido*, 95, 3.

Grattapaglia, D., René, E., Vaillancourt, R. E., Shepherd, M., Thumma, B. R., Foley, W., Külheim, C., Potts, B. M., and Myburg, A. A., 2012. Progress in Myrtaceae genetics and genomics: *Eucalyptus* as the pivotal genus. *Tree Genet. Genom.*, 8, 463–508.

Hayes, W. B., 1953. *Fruit Growing in India*. Kitabistan, Allahabad.

Hernandez-Delgado, S., Padilla-Ramirez, J. S., Nava-Cedillo, A., and Mayek-Perez, N., 2007. Morphological and genetic diversity of Mexican guava germplasm. *Plant Genet. Resour.*, 5, 131–41.

Hirano, R. T., 1967. Chromosomal and pollination studies as related to intra-specific and interspecific compatibility in the genus *Psidium*. Master's thesis, University of Hawaii, Honolulu.

Hirano, R. T., and Nakasone, H. Y., 1969. Chromosome numbers of ten species and clones in the genus *Psidium*. *J. Am. Soc. Hortic. Sci.*, 94, 83-6.

Iyer, C. P. A., and Subramanyam, M. D., 1971. Problems with triploidy in guava. *SABRAO Newslett.*, 3(1), 31-3.

Jaiswal, V. S., and Amin, M. N., 1987. *In vitro* propagation of guava from shoot culture of mature trees. *J. Plant Physiol.*, 130, 7-12.

Jaiswal, V. S., and Amin, M. N., 1992. Guava and jackfruit. In: Hammerschlag, F. A., and Litz, R. E. (Eds), *Biotechnology of Perennial Fruit Crops, Biotechnology in Agriculture* (vol. 8). CAB International, Wallingford, UK, pp. 421-31.

Janaki Ammal, E. K. J., 1951. Chromosomes and horticulture: Tetraploids in guava. *J. Royal Hortic. Soc.*, 76, 236-9.

Kalia, R. K., Rai, M. K., Kalia, S., Singh, R., and Dhawan, A. K., 2011. Microsatellite markers: An overview of the recent progress in plants. *Euphytica*, 177, 309-34.

Kanupriya, Latha, P. M., Aswath, C., Reddy, L., Padmakar, B., Vasugi, C., and Dinesh, M. R., 2011. Cultivar identification and genetic fingerprinting of guava (*Psidium guajava* L.) using microsatellite markers. *Int. J. Fruit Sci.*, 11, 184-96.

Kherwar, D., and Usha, K., 2016. Genetic variations, character association and path analysis studies in guava (*Psidium guajava* L.) for bioactive and antioxidant attributes. *Ind. J. Plant Physiol.*, 21(3), 355-36.

Kherwar, D., Usha, K., Amitha Mithra, S. V., and Singh, B., 2018. Microsatellite SSR marker assisted assessment of population structure and genetic diversity for morpho-physiological traits in guava (*Psidium guajava* L.). *J. Plant Biochem. Biotechnol.*, 27(3),284-92.

Kim, I. S., and Grosch, W., 1981. Partial-purification and properties of a hydroperoxide lyase from fruits of Pear. *J. Agric. Food Chem.*, 29, 1220-5.

Krishna, H., and Singh, S. K., 2007. Biotechnological advances in mango (*Mangifera indica* L.) and their future implication in crop improvement–a review. *Biotech. Adv.*, 25, 223-43.

Kumar, L. S. S., and Ranade, S. G., 1952. Autotriploid in guava (*Psidium guajava* L.). *Curr. Sci.*, 21, 75-6.

Lepitre, V., Nansot, G., Grangeon, R., Pomies, V., Rivallan, R., Risterucci, A. M., Valdés-Infante, J., Rodríguez-Medina, N. N., Muth, J., Boike, J., Prüfer, D., Becker, D., Rohde, W., Ritter, E., and Billotte, N., 2010. The microsatellite (SSR)/AFLP reference linkage map of guava. *Acta. Hortic.*, 849, 183-91.

Leslie Landrum, R. W., Dennis Clark, W., William, P., and Brendecke, J., 1995. Hybridization between *Psidium guajava* and *P. guineense* (Myrtaceae). *Econ. Bot.*, 49, 153-61.

Li, H., and Homer, N., 2010. A survey of sequence alignment algorithms for next-generation sequencing. *Brief. Bioinform.*, 5, 473-83.

Liu, X., and Yang, G., 2012. Assessment of clonal fidelity of micro-propagated guava (*Psidium guajava*) plants by ISSR markers. *AJCS*, 6(2), 291-5.

Loh, C. S., and Rao, A. N., 1989. Clonal propagation of guava (*Psidium guajava* L.) from seedling and grafted plants and adventitious shoot formation *in vitro*. *Sci. Hortic.*, 39, 31-9.

Majumder, P. K., and Mukherjee, S. K., 1972a. Aneuploidy in guava. I. Mechanism of variation in number of chromosomes. *Cytologia*, 37, 541-8.

Majumder, P. K., and Mukherjee, S. K., 1972b. Aneuploidy in guava. II. The occurrence of trisomics, tetrasomics and higher aneuploids in the progeny of triploid. *Nucleus*, 13, 42-7.

Majumder, P. K., and Singh, R. N., 1964. Seedlessness in guava (*Psidium guajava L.*). *Curr. Sci.*, 33, 24–5.

Mani, A., Mishra, R., and Thomas, G., 2011. Elucidation of diversity among *Psidium* species using morphological and SPAR methods. *J. Phytol.*, 3(8), 53–61.

Marak, J. K., and Mukanda, G. K., 2007. Studies on the performance of open pollinated seedling progenies of guava cv. 'apple colour'. *Acta. Hortic.*, 735, 79–84.

Mehta, S. K., Singh, K. K., Jat, D. K. and Rana, D. K.. 2016. Comparative studies of physico-chemical characteristics of various cultivars of guava (*Psidium guajava* L.) under sub tropical valley condition of Garhwal Himalaya (Uttarakhand), India. *Plant Arch.* 16(1), 361-4.

Mitra, S. K., and Bose, T. K., 1985. Guava. In: Bose, T. K. (Ed.), *Fruits of India - Tropical and Sub-tropical*. Naya udyog, Calcutta, 721pp.

Mitra, S. K., and Sanyal, D., 2004. *Guava*. Indian Council of Agricultural Research, New Delhi, India, p. 87.

Mohammad, S., 1974. Aneuploidy in guava. *Biol. Plant.*, 16, 382–8.

Mukherjee, S. K., 1977. Improvement of mango, grapes and guava. In: Nijjar, G. S. (Ed.), *Fruit Breeding in India*. Oxford & IBH, New Delhi, pp. 15–20.

Naga Chaithanya, M. V., Ramesh, S., Dinesh, M. R., Sailaja, D. and Aswath, C. 2016. Developing mapping populations for identifying genomic regions controlling resistance to bark-eating caterpillar (*Indarbela tetraonis*) in guava. *J. Crop Improv.*, 30(4), 371–7.

Naithani, S. P., and Srivastava, H. C., 1966. Autotetraploidy in *Psidium guajava* L. *Naturwissenchaft.*, 8, 205–6.

Nakasone, H. Y., and Paull, R. E., 1998. *Tropical Fruits*. CAB International, Wallingford, UK, pp. 93–8.

Nalini, A. S., Gubbaiah, Narayana-Reddy, M. A., Farooki, A. A., and Nalawadi, U. G., 1973. Studies on the floral biology of guava (*Psidium guajava* L.) variety Lucknow-49 (Sardar). *Mysore J. Agric. Sci.*, 7(1), 24–37.

Nandgav, S., 2010. Development of molecular marker for pulp colour in guava (*Psidium Guajava* L.). Thesis.

Narain, P., 2010. Statistical genomics and bioinformatics. *J. Hortic. Sci.*, 5, 85–93.

Negi, S. S., and Rajan, S., 2007. Improvement of guava through breeding. *Acta Hortic.*, 735, 31–7.

Nimisha, S., Kherwar, D., Ajay, K. M., Singh, B. and Usha, K. 2013. Molecular breeding to improve guava (*Psidium guajava* L.): current status and future prospective. *Sci. Hortic.*, 164, 578–88.

Nogueira, M. A., Guilhen, H. S. J., Mangaravite, E., Ferreira, A., and Ferreira, F. S. M., 2012. Discriminant analysis of wild guava tree by morphological descriptors and microsatellites. *3rd International Symposium on Guava and Other Myrtaceae*. Petrolina, PE, Brazil. Session IV: Myrtaceae Biotechnologies and Genomic.

Nogueira, A. M., Ferreira, A. and Ferreira, M. F. S., 2016. Transferability of microsatellites from *Psidium guajava* to *Eugenia*, *Myrciaria*, *Campomanesia*, and *Syzygium* species (Myrtaceae). *Plant Mol. Biol. Rep.*, 34, 249–56.

Noia, R. L., Coser, M. S., Guilhen, S. H. J., Adésio Ferreira, A., and Ferreira, S. M. F., 2012a. Genetic distance of guava genotypes from different altitudes by microsatellites. *3rd International Symposium on Guava and Other Myrtaceae*. Petrolina, PE, Brazil. Session IV: Myrtaceae Biotechnologies and Genomic.

Noia, R. L., Nogueira, M. A., Mangaravite, E., Adésio Ferreira, A., and Ferreira, F. S. M., 2012b. Genetic diversity of wild guava from southern Espírito Santo and commercial cultivars by microsatellites. *Proceeding of 3rd International Symposium on Guava and Other Myrtaceae*. Petrolina, PE, Brazil. Session IV: Myrtaceae Biotechnologies and Genomic.

Normand, F., 1994. Strawberry guava, relevance for reunion. *Fruits*, 49, 217-27.

Ojha, A. P., Tiwari, J. P., and Mishra, K. K., 1986. Studies on growth, flowering and yield of guava (*Psidiumguajava* L.) under terai condition of U.P. *Prog. Hortic.*, 8, 205–6.

Padmakar, B., Kanupriya, C., Madhavi Latha, P., Prashant, K. S., Dinesh, M. R., Sailaja, D., and Aswath, C., 2015. Development of SRAP and SSR marker-based genetic linkage maps of guava (*Psidium guajava* L.). *Sci. Hortic.*, 192, 158–65.

Pandey, S. D., 1968. The guava of Uttar Pradesh: A classification. *Hortic. Adv.*, 7, 72-98.

Papadatou, P., Pontikis, C., Ephtimiadou, E., and Lydaki, M., 1990. Rapid multiplication of guava seedlings by *in vitro* shoot-tip culture. *Sci. Hortic.*,45, 99–103.

Patel, R. K., Yadev, D. S., Babu, K. D., Singh, A., and Yadav, R. M., 2007. Growth, yield and quality of various guava (*Psidium guajava*L.) hybrids/cultivars under mid hills of Meghalaya. *Acta Hortic.*, 735, 57–9.

Pereira, F. M., and Sao Joco, A. R., 1987. Study of different methods of pollen collection and pollination of guava (*Psidium guajava* L.) (Portuguese, English abs.). *Cientifica*, 15(1/2), 85–92.

Pérez, L., Sigarroa, A., Rodríguez, N. N., Valdés-Infante, J., Bandera, E., and Pommer, V. C., 2012. Heritability estimates of guava (*Psidium guajava* L.) agricultural important characters evaluated in three populations. *3rd International Symposium on Guava and Other Myrtaceae*. Petrolina, PE, Brazil. Session III: Myrtaceae Plant Breeding and Post-Harvest Management.

Pessanha, P. G. O, Viana, A. P., Amaral Júnior, A. T., Souza, R. M., Texeira, M. C. and Pereira, M. G. 2011. Assessment of genetic diversity in access to Psidum spp. via RAPD markers. *Rev. Bras. Frutic.*, 33, 129–136.

Pommer, C. V., 2012. Guava world-wide breeding: major techniques and varieties and future challenges. *3rd International Symposium on Guava and Other Myrtaceae*. Petrolina, PE, Brazil. Session III: Myrtaceae Plant Breeding and Post-Harvest Management.

Pommer, C. V. and Murakami, K. R. 2009. Breeding Guava (*Psidium guajava* L.) In: *Breeding Plantation Tree Crops: Tropical Species*. Springer, New York, pp. 83–120

Prakash, D. P., Narayanaswamy, P., and Sondur, S. N., 2002. Analysis of molecular diversity in Guava using RAPD markers. *J. Hortic. Sci. Biotechnol.*, 77, 287–93.

Purseglove, J. W., 1968. *Tropical Crops: Dicotyledons*. John Wiley and Sons, Inc., New York, NY.

Rai, M. K., Phulwaria, M., Gupta, K. M., Shekhawat, N. S., and Jaiswal, U., 2012. Genetic homogeneity of guava plants derived from somatic embryogenesis using SSR and ISSR markers. *Plant Cell Tiss. Organ. Cult.*, 111, 259–64.

Rai, M. K., Phulwaria, M., and Shekhawat, N. S., 2013. Transferability of simple sequence repeat (SSR) markers developed in guava (*Psidium guajava* L.) to four Myrtaceae species. *Mol. Biol. Rep.*, 40, 5067–71.

Rajan, S., Yadava, L. P., Kumar, R., and Saxena, S. K., 2005. Selection possibilities for seed content- a determinant of fresh fruit quality in guava (*Psidium guajava* L.). *J. Appl. Hortic.*, 7, 52–4.

Raman, W. M., Manimekalai, G., and Ramalingam, R. S., 1969. Observation on seedlessness, fruit development and cytology of varieties of guava. *Madras Agric. J.*, 56, 255–61.

Raman, V. S., Sri Rangaswamy, S., and Manimekalai, F., 1971. Triploidy and seedlessness in guava (*Psidiumguajava* L). *Cytologia*, 36, 392–9.

Ram Kumar, 1975. Inducing polyploidy and cytological studies in guava (*Psidium guajava* L). *Ind. J. Hortic.*, 32, 128–30.

Raza, H., Khan, M. M., and Khan, A. A., 2003. Review: Seedlessness in Citrus. *Int. J. Agric. Biol.*, 5(3), 388–91.

Ribeiro, I. J. A., and Pommer, C. V., 2004. Breeding guava (*Psidium guajava*) for resistance to rust caused by *Puccinia psidii*. *Acta. Hortic.*, 632, 75–8.

Risterucci, A. M., Duval, M. F., Rohde, W., and Billotte, N., 2005. Isolation and characterization of microsatellite loci from *Psidium guajava*. L. *Mol. Ecol. Notes*, 5, 745748.

Risterucci, A. M., Nansot, G., Grangeon, R., Lepitre, V., de Reeper, A., Argout, X., Ruiz, M., and Billotte, N., 2010. Development of Guava Microsatellite (SSR) markers using the SAT Software. *Acta Hortic.*, 849, 113–20.

Ritter, E. 2012. Guava biotechnologies, genomic achievements and future needs. In: *3rd International Symposium on Guava and Other Myrtaceae*. Petrolina, PE, Brazil. Session IV: Myrtaceae Biotechnologies and Genomic.

Rodríguez-Medina, N., Valdés-Infante, J., Becker, D., Velázquez, B., González, G., Sourd, D., Rodríguez, J., Billotte, N., Risterucci, A. M., Ritter, E., and Rohde, W., 2007. Characterization of guava accessions by SSR markers, extension of the molecular linkage map, and mapping of QTLs for vegetative and reproductive characters. *Acta Hortic.*, 735, 201–15.

Rodríguez-Medina, N. N., Valdés-Infante, J., Rodríguez, A. J., Velásquez, B. J., Rivero, D., Martínez, F., González, G., and Sourd, G. D., 2010. Genetic resources and breeding of guava (*Psidium guajava* L.) in Cuba. *Biotech. Aplic.*, 27(3).

Rodríguez-Medina, N. N., Valdés-Infante, J., Gonzalez, G., Fuentes, V., and Canizares, J., 2010. Genetic resources of Guava (*Psidium guajava* L.) in Cuba: Germplasm characterization and breeding. *Acta. Hortic.*,849, 341–8.

Sánchez-Teyer, L. F., Barraza-Morales, A., Quiroz-moreno, A., Ortiz-Garcia, M. M., Becerril-chi, K., Keb-llanes, M., O'Connor-Sanchez, A., and Padilla-Ramirez, J. S., 2010. Genetic diversity of Mexican guava germ plasm evaluated using AFLP and SSR Markers. *Acta. Hortic.*, 849, 255–60.

Santos, K. L., Welter, L. J., Dantas, A. C. M., Guerra, M. P., Ducroquet, J. P. J., and Nodari, R. O., 2007. Transference of microsatellite markers from *Eucalyptus* spp to *Acca sellowiana* and the successful use of this technique in genetic characterization. *Genet. Mol. Biol.*, 30, 73–9.

Saxena, S. K., Rajan, S., Chandra, R., Srivastava, N., and Bajpai, A., 2007. Molecular characterization of closely related open pollinated seedling selections in guava. *Acta Hortic.*, 735, 49–56.

Schulman, A. H., 2007. Molecular markers to assess genetic diversity. *Euphytica*, 158, 313–21.

Severn-Ellis, A., Schoeman, M. H., Willemse, S., Sippel, A., Rees, J. and de Castro, M. 2012. Towards guava wilt disease resistance in South Africa. In: *3rd International Symposium on Guava and other Myrtaceae*. Petrolina, PE, Brazil, 23–25 April 2012. *Acta Hortic.* 959, 73–9.

Shafaat Mohammed. 1975. Investigations on the breeding behaviour of aneuploids of guava. Thesis submitted to the Division of Fruits and Horticultural Technology, IARI, New Delhi.

Shah, S. T., Zamir, R., Ahmad, J., Ali, H., Lutfullah, G., 2008. *In vitro* regeneration of plantlets from seedling explants of guava (*Psidium guajava* L.) cv. Safeda. *Pak. J. Bot.*, 40, 1195–200.

Sharma, Y. K., 1982. Rootstock investigation in guava (*Psidium guajava* L.). Meerut University, Meerut, India, Thesis submitted for the award of Ph.D. degree to Meerut University.

Sharma, Y. K., Goswami, A. M., and Sharma, R. R., 1992. Effect of dwarfing aneuploid guava rootstock in high density orcharding. *Ind. J. Hortic.*, 49, 31-6.

Sharma, A. S., Sehrawat, S. K., Singhrot, R. S., and Boora, K. S., 2007. Assessment of genetic diversity and relationship among *Psidium* spp. through RAPD analysis. *Acta Hortic.*, 735, 71-8.

Singh, R. and Sehgal, O. P. 1968, Studies on blossom biology of *Psidium guajava* L. (guava). II. Pollen studies, stigma receptivity, pollen and fruit set. *Ind. J. Hortic.*, 25:52-59

Singh, V. R., Dhar, L., and Singh, G., 1977. Note on the performance of guava cultivars and *Psidium* species against wilt disease under natural field conditions. *Haryana J. Hortic. Sci.*,6(3–4), 149–50.

Singh, S. K., Meghwal, P. R., Sharma, H. C., and Singh, S. P., 2002. Direct shoot organogenesis on hypocotyl explants from *in vitro* germinated seedlings of *Psidium guajava* L. cv. Allahabad Safeda. *Sci. Hortic.*, 95, 213–21.

Singh, M., Jaiswal, U., and Jaiswal, V. S., 2004. *In vitro* regeneration and improvement in tropical fruit trees: an assessment. In: Srivastava, P. S., Narula, A., and Srivastava, S. (Eds), *Plant Biotechnology and Molecular Markers*. Anamanya Publishers, New Delhi, pp. 228–43.

Singh, D., Mohammed, S., and Shukla, A. K., 2005. Micropropagation of guava as influenced by season, source and retreatment to explant. In: *Abstract of 1st International Guava Symposium*. CISH, Lucknow, India, p. 33.

Sonah, H., Deshmukh, R. K., Singh, V. P., Gupta, D. K., Singh, N. K., and Sharma, T. R., 2011. Genomic resources in horticultural crops: Status, utility and challenges. *Biotech. Advan.*, 29(2), 199–209.

Soto-Cerda, B. J., and Cloutier, S., 2012. Association mapping in plant genomes. In: Caliskan, M. (Ed.), *Genetic Diversity in Plants*. InTech, Croatia, pp. 1–26.

Srisuwan, A., 2003. Chromosome number of dessert type, processing type guava (*Psidium guajava* L.) and their hybrids. Available at: http://www.cab.kps.ku.ac.th/dbcab/thisis /reports/01-06-02-000047.pdf.

Srivastava, O. P., 1974. Studies on the flowering habit, blooming period, anthesis, dehiscence and pollen grain of *Psidium guajava* L. varieties Apple Colour, Chittidar and Red Flesh. *Prog. Hortic.*, 6, 71–7.

Srivastava, H. C., 1977. Cytological studies in *Psidium friedrichsthalianum* N. *Cytologia*, 42, 395–400.

Srivastava, U., 2005. Genetic resources management in guava. In: Kishun, R, Mishra, A. K., Singh, G., and Chandra, R. (Eds), *Proceeding of 1st International Guava Symposium*. CISH, Lucknow, India, pp. 17–18.

Srivastava, R. P., and Srivastava, R. K., 1965. Physicochemical studies on Safeda Allahabad and Red fleshed guavas. *Punjab Hortic. J.*, 5, 12–15.

Subramanyam, M. D., and Iyer, C. P. A., 1982. Improvement of guava by breeding. Report, Fruit Workshop, Nagpur, pp. 117–18.

Subramanyam, M. D., and Iyer, C. P. A., 1998. *Report, Fruit Research Workshop on Tropical and Subtropical Fruits*. Rajendra Agricultural University, Pusa, India, pp. 81-4.

Tate, D., 2000. *Tropical Fruit of Thailand*. Asia Books Co. Ltd., Bangkok, Thailand.

Teaotia, S. S., Pandey, I. C., and Agnihotri, B. N., 1962. Study of some guava varieties (*Psidium guajava* L.) of Uttar Pradesh. *Ind. Agric.*, 6, 47-53.

Teotia, S. S., Tripathy, R. S., and Bhan, S. 1970. Evaluation of indices for determination of maturity of guava (*Psidium guajava* L.). *Hortic. Adv.*, 7, 97-103.

Thaipong, K., and Boonprakob, U., 2005. Genetic and environmental variance components in guava fruit qualities. *Sci. Hortic.*, 104, 37-47.

Thonte, G. T., and Chakrawar, V. R., 1981. The variablility and correlation studies of guava strains. *National Symposium on Subtropical Fruit Crops*, Bangalore, p. 17.

Tijet, N., Waspi, U., Gaskin, D. J. H., Hunziker, P., Muller, B. L., Vulfson, E. N., Slusarenko, A., Brash, A. R., and Whitehead, I. M., 2000. Purification, molecular cloning, and expression of the gene encoding fatty acid 13-hydroperoxide lyase from guava fruit (*Psidium guajava*). *Lipids*, 35, 709-20.

Treangen, T. J., and Salzberg, S. L., 2012. Repetitive DNA and next generation sequencing: computational challenges and solutions. *Nat. Rev. Genet.*, 13, 36-46.

Usman, M., Samad, W. A., Fatima, B., and Shah, M. H., 2013. Pollen parent enhances fruit size and quality in intervarietal crosses in guava (*Psidium guajava*). *Int. J. Agric. Biol.*, 15, 125.

Valdés-Infante, J., Rodríguez, N. N., Becker, D., Velazquez, B., Sourd, D., Espinosa, D., and Rohde, W., 2007. Microsatellite characterization of guava (*Psidium guajava* L.) germplasm collection in Cuba. *Cultiv. Trop.*, 28, 61-6.

Valdés-Infante, J., Rodríguez, N. N., Barbara, V. J., Gaspar, S. D., Gonzalez, G., Rodriguez, J. A., and Rohde, W., 2012. Tools for a genetic breeding program for guava (*Psidium guajava* L.) in Cuba. *Agron. Costarric.*, 36, 111-29.

Valecillos, C., and Fermin, G., 2008. Cloning and sequencing of the HPL gene and genetic transformation in guava. *Acta. Hortic.*, 849, 245-50.

Varmudy, V., 2013. Guava: potential untapped. facts for you. Available at: http://www.ffymag.com/ admin/ issuepdf/ Guava_FFYApril-13.pdf.

Viji, G., Harris, D. L., Yadav, A. K., and Zee, F. T., 2010. Use of microsatellite markers to characterize genetic diversity of selected accessions of guava (*Psidium guajava*) in the United States. *Acta. Hortic.*, 859, 169-76.

Xu, J. H., Cheng, C. Z., Zhang, L. M., Zhang, Z. H., Xu, Y. J., and Zheng, S. Q., 2007. Pollen parent effects on fruit quality of 'Jiefangzhong' loquat. *Acta. Hortic.*, 750, 361-66.

Yadav, H. C., Yadav, A. L., Yadav, D. K. and Yadav, P. K., 2011. Effect of foliar application of micronutrients and GA3 on fruit yield and quality of rainy season guava (*Psidium guajava* L.) Cv. L-49. *Plant Arch.*, 11, 147-9.

Zamir, R., Khattak, G. S. S., Mohammad, T., Shah, S. A., Khan, A. J. and Ali, N. 2003. *In vitro* mutagenesis in guava (*Psidium guajava* L.). *Pak. J. Bot.*, 35(5), 825-8.

Zipori, I., Shuker, S. Dag, A., and Tomer, E., 2007. Guava breeding in Israel. *Acta. Hortic.*, 735, 39-47.

Zucchi, M. I., Brondani, R. P. V., Pinheiro, J. B., Brondani, C., and Vencovsky, R., 2002. Transferability of microsatellite markers from *Eucalyptus* spp. to *Eugenia dysenterica* (Myrtaceae family). *Mol. Ecol. Notes*, 2, 512-13.

Zucchi, M. I., Brondani, R. P. V., Pinheiro, J. B., Chaves, L. J., Coelho, A. S. G., and Vencovsky, R., 2003. Genetic structure and gene flow in Eugenia dysenterica DC in the Brazilian Cerrado utilizing SSR markers. *Genet. Mol. Biol.*, 26, 449-57.

# Chapter 14

## Advances in guava cultivation

*Noosheen Zahid and Mehdi Maqbool, University of Poonch, Pakistan; and Ajit Singh and Asgar Ali, University of Nottingham Malaysia Campus, Malaysia*



### 1 Introduction

Guava (*Psidium guajava*) belongs to the Myrtaceae family. It is also commonly known as guyava, kuawa or guayba. Guava fruits (Figs. 1 and 2) are used for both fresh consumption of the fresh fruit and processing. In comparison with other fruit tree species, it has exceptional properties in terms of productivity, hardiness, adaptability and nutritional value. Because of its versatility and widespread use, guava has been appropriately called 'the apple of the tropics'. It is one of the most common and major fruits globally and is considered as the fifth most important fruit (in terms of production area and volume) after mango, citrus, banana and apple.

There is a need to enhance more traditional, small-scale methods of production to make guava more competitive in world markets. One modern technique is a shift to high-density planting (HDP), which is widely used for crops like apple in Europe and North America. This includes the adoption of an appropriate planting density, good quality planting material, canopy management and an efficient support and orchard management system with the appropriate inputs. Given the popularity of high-density technology and

http://dx.doi.org/10.19103/AS.2019.0054.24

**Figure 1** Guava flower.

its likely benefits, it is now the right time to encourage the adoption of this technology by guava growers. This chapter explores the ways guava cultivation can be sustainably modernized in this way.

## 2 Crop climatic requirements and distribution

Guava is grown from tropical to subtropical regions and can also grow well in both humid and dry climates. This versatile crop can be grown on a variety of soils such as poor and saline soils and, to a lesser extent, in water-logged conditions. Guava can be grown from sea level to 21000 m altitude, at temperatures between 20 and 30°C with annual rainfall between 1000 and 2000 mm (Paull and Bittenbender, 2006). The crop grows best when the annual rainfall is below 40 inches (1016 mm), and its growth is restricted between June and September. Areas with a distinct winter season improve the yield and quality of the fruit. As noted, guava trees are very hardy and can thrive in all types of soil conditions from alluvial to lateritic, but they are sensitive to waterlogging. The crop can be grown on heavier but well-drained soil. The best soils are deep, friable and well-drained ones. Guava can grow in a soil having a pH of 6.5–8.5. As guava is a surface-rooted plant, the surface soil should be rich in nutrients. All these reasons make it possible for farmers to grow guava plants

**Figure 2** Guava fruit.

on a commercial scale. However, good management practices are needed to get a profitable crop.

Guava is a medium-sized tree of about 30 feet in height. The flower is mainly self-pollinated, but cross-pollination (about 35%) also occurs through bees and insects. The proportion of cross-pollination varies with the environment, genotype and area of production (Martínez de Lara et al., 2004). The guava tree normally produces as many as three crops in a year, a unique phenomenon in tropical and subtropical regions because there is more than one growing season during a year. The flowers (Fig. 1) are hermaphroditic, solitary or grouped in twos and threes in the leaf axils. They are pinkish white and have a pleasant smell. In a mild climate, guava can flower and bear fruits throughout the year, provided water and temperature are not limiting factors. Blooming period varies from 25 to 45 days depending upon the cultivar, the season and the region where it is grown. Initial fruit set in guava is quite high and approximately 80% of the flowers become fruits. During severe fruit drops only 34–56% of the fruits reach maturity. The formation of fruit can first be noticed after 12 days from flowering. The rainy season crop is harvested during August and the winter season crop during January–February.

Guava is currently grown in more than 60 countries. It is cultivated in India, Mexico, Brazil, Thailand, Spain, Portugal, Southern France, Israel, Panama, Costa

**Table 1** Top guava-producing countries in the world

| Rank | Country | Production in (MT), 2011 |
|---|---|---|
| 1 | India | 17 650 000 |
| 2 | China | 4 366 300 |
| 3 | Thailand | 2 550 600 |
| 4 | Pakistan | 1 784 300 |
| 5 | Mexico | 1 632 650 |
| 6 | Indonesia | 1 313 540 |
| 7 | Brazil | 1 188 910 |
| 8 | Bangladesh | 1 047 850 |
| 9 | Philippines | 825 676 |
| 10 | Nigeria | 790 200 |

Rica, Nicaragua, Bolivia, Malaysia, Kenya, the USA (Hawaii, California and Florida), New Zealand, Philippines, China, Indonesia, Cuba, Java, Venezuela, Pakistan, Australia and in some of the African countries. Major producers of guava are India, Brazil and Mexico (Table 1). Other leading producers are South Africa, Jamaica, Kenya, Cuba and the USA (mainly Florida and Hawaii), Egypt, Thailand, Columbia, Pakistan and the Philippines. International trade is largely limited to processed guava products and includes exports to the USA, Japan and Europe.

## 3 Composition, nutritional value and uses

Guava fruit is round and greenish yellow in colour. Its flesh is whitish in colour, containing numerous small seeds. Guava fruit is best eaten when it has perfectly ripened and is freshly plucked from the trees. It emits a sweet aroma and is pleasantly sweet and refreshingly acidic in its flavour. It is wholly edible, including its skin, which is thin and merged with its pulp. Ninety-five per cent of guava fruits are consumed either as fresh or as processed fruits. Only 0.05% of produce is exported (Reetz, 2015).

Guava fruit is normally consumed fresh, but it is also eaten cooked. It is suitable for the preparation of marmalade, jellies and syrup. Peeled and hollowed-out fruit cooked in light syrup strongly resembles the taste of European pears. Fruit jellies and jam are often made from guava. Guava paste, or guava cheese as known in the West Indies, is made by evaporating the pulp with sugar - it is eaten as a sweetmeat. A firm in the Philippines dehydrates slices of the outer, non-seeded part of the fruit to make a similar product. In some Asian countries such as Indonesia, the leaves are used for cooking (Orwa et al., 2009). Eaten green, the fruit is astringent and is considered as a remedy for dysentery (Lekeu, 2001). The dried ripe fruits are recommended

**Table 2** Antioxidants in different fruits

| Fruit | Antioxidants (mg/100 g) | Fruit | Antioxidants (mg/100 g) |
|---|---|---|---|
| Guava | 496 | Chiku (sapodilla) | 55 |
| Plum | 330 | Papaya | 50 |
| Custard apple | 202 | Banana | 30 |
| Mango | 170 | Sweet lemon | 26 |
| Pomegranate | 135 | Orange | 24 |
| Apple | 123 | Watermelon | 23 |
| Grapes | 85 | Pineapple | 22 |

*Source*: adapted from Sreeramulu et al. (2013).

as a remedy for dysentery, while the leaves and fruits are used as a cure for diarrhoea. It exhibits antibacterial action against intestinal pathogens such as *Staphyloccocus* (Milyani, 2012; Kafle et al., 2018).

For every 100 g of fresh edible matter, guava fruit contains 80–85 g water, 1–7 g protein, 0.4 g fat, 11–13 g carbohydrate and 5.5 g fibre. Vitamin C content in guava is five times higher than that of orange. It is also rich in vitamin A, with a content similar to that of avocado and two and seven times higher than that of banana and orange, respectively (Samson, 1986; Lekeu, 2001; Amusa et al., 2006). Amusa et al. (2006) observed significant reduction in the carbohydrate, protein, crude fibre and protein content in anthracnose-infected guava fruit compared to uninfected fruit. A study by the National Institute of Nutrition in Hyderabad, India, has revealed that fruits like guava and apple are found to be rich in antioxidants, which are linked to the prevention of cellular damage associated with cancer, ageing and a variety of degenerative diseases (Sinha, 2011). Antioxidants available in different fruits are shown in Table 2. As Table 2 shows, guava has the highest level of antioxidants of the fruits listed (496 mg/100 g). Beside total antioxidants, the fruit is rich in ascorbic acid (200–300 mg), carotene (200–400 I.U.), vitamin A, thiamine, riboflavin and different minerals like phosphorus, iron and calcium (Mahattanatawee et al., 2006). Guava is particularly rich in pectin and vitamin C (Saini et al., 2018).

## 4 Propagation methods

Different propagation methods for guava production are used in different countries (which are described in Table 3). The main methods are:

- Propagation by seed
- Vegetative propagation: grafting, air layering, shoot cutting, root cutting and budding

**Table 3** Area, production and propagation methods of different guava cultivars

| Country | Area (×1000 ha) | Production (× 1000 t) | Productivity (t ha$^{-1}$) | Main cultivar | Propagation method | Reference |
|---|---|---|---|---|---|---|
| Bangladesh | 18.0 | 200.0 | 11.11 | Kazi Peara (guava), BARI Peara, BAU Guava 1–8, Kazi, Swarupkathi, Mukundapuri and Allahabad | Stem herbaceous cuttings, air layering (>95%) | Rahim et al. (2011), Saha (2015) |
| India | 255.0 | 4102.0 | 16.08 | Allahabad Safeda, Sardar (L49), Lalit, Banarasi and Shweta | Stem herbaceous cuttings, chip budding and grafting (>95%) | NHB (2015–16), Mitra (2014) |
| Indonesia | 9.4 | 208.0 | 22.12 | Indonesian white, Indonesian seedless, ARP9409 and JBT001 | Stem herbaceous cuttings, grafting (>95%) | BPS (2014), Nasution and Hadiati (2014) |
| Malaysia | 4.0 | 65.0 | 16.25 | Red Malaysian, GU 8 (Jambu Kamopochea), GU 11 (Giant guava), GU15 (Jade Seedless), Batu Arang, Subang 6 and Sitiawan | Stem herbaceous cuttings, grafting andair layering (>95%) | Pommer and Murakami (2009), MAM (2013) |
| Pakistan | 66.38 | 496.0 | 7.47 | Sadabahar Gola (Round), Sadabahar Surahi (Pyriform), Large Pyriform, Larkana Pyriform and Safeda | Seed (>80%), cutting and air layering | MNFSR (2014), Usman et al. (2013) |
| Thailand | 7.94 | 168.0 | 21.15 | Klomsali, Salithong and Kimchu | Soft wood cuttings, grafting (>95%) | Siwarungson et al. (2013) |
| Taiwan | 7.12 | 168.0 | 23.59 | Century, Pearl and Crystal | Soft wood cuttings and wedge grafting (>98%) | Agriculture and Food Agency (2014) |

| | | | | | | |
|---|---|---|---|---|---|---|
| Egypt | 17.5 | 315.0 | 18.0 | Bassateen El Sabahia, Bassateen Edfina and Allahabad Safeda | Soft wood cuttings and mound layering (>85%) | Pommer and Murakami (2009), EMALR (2014), El-Shereif (2016) |
| South Africa | 0.5 | 12.5 | 25.0 | TSG2, Fan Retief, Fredene, Dimple, Jonelle, Welheim, Frederika, Frank Marlhebe, Van Zyl, Rousseau and Du Preez | Soft wood cuttings, air layering and grafting (>80%) | DAFF (2015), HORTGRO (2015) |
| Sudan | 22.0 | 473.0 | 21.5 | A-H-180, A-H-132, A-H-157, Pakistani and Beamonal | Air layering, semi-hardwood cuttings | Agriculture in Sudan (1999) |
| Cuba | 7.31 | 47.88 | 6.55 | Enana Roja Cubana | Grafting on seedlings, cutting | Pérez-Rodríguez et al. (2013), Valdés-Infante et al. (2012), Pommer et al. (2006) |
| Mexico | 20.44 | 302.7 | 14.81 | Wild types named 'China' and 'Media China' | Seeds and root cuttings | Martínez de Lara et al. (2004) |
| Brazil | 14.98 | 349.6 | 23.34 | Paluma, Rica, Século XXI, Pedro Sato, Sassaoka, Kumagai and Ogawa | Stem herbaceous cuttings (> 95%) | Agrianual (2016), Pereira and Nachtigal (2009) |
| Colombia | 8.34 | 69.1 | 8.28 | Regional Blanca, Regional Roja, Ráquira Blanca, Guavatá Victoria, Pera and Manzana | Seeds (in home backyards), grafting or cutting for commercial orchards | Ocampo and Núñez (2007), Parra-Coronado (2014) |
| Venezuela | 3.50 | 55.7 | 15.9 | Criolla Roja, San Miguel and Rio Chiquito | Seeds | Aular and Casares (2011) |

- Sexual propagation
- Asexual propagation: *in vivo* propagation and *in vitro* propagation.

This section describes each of these techniques. Seed propagation of guava is restricted to commercial production; however, countries with low levels of technology cultivate guava with seed (Pereira, 1990). Plants produced by seeds have great variability, which is an undesirable characteristic for commercial orchards (Pereira et al., 2016). Seed-propagated plants with higher juvenility period delay fruit production for up to 2-4 years. Propagation with seed is acceptable only in breeding programmes where genetic variability is required. Another drawback of seed propagation is virtual elimination of certain cultivars such as 'Guanabara' (Pereira and Nachtigal, 2009).

Amongst vegetative propagation techniques, grafting is the most successful method in guava production. Sprouting requires 9–12 days for initiation, and, under greenhouse conditions, grafting is successful up to 70–92% (Singh, 2007). Air layering is a simple and efficient method to propagate guava trees. For air layering, physiological condition of branches and application and concentration of growth regulators are important factors. The main disadvantage of air layering is low yield of the mother plant as compared to shoot and root cutting (Urdaneta et al., 2009). Propagation through shoot cutting was not very important in the past but is now gaining in popularity. In this method, shoots of all ages can be used. Shoot cuttings are treated with root promoters and placed in a propagation unit at a temperature of 28°C. Cuttings are ready in 3 months to establish a field. Cuttings from young shoots give best results (Pereira et al., 2016).

Propagation through root cuttings was once common. However, it is now considered responsible for guava wilt disease caused by a soil born fungus, and so it is no longer in use (Pereira et al., 2016; Gupta et al., 2010). Budding in guava can be done before or during the growing season. In some species budding is successful in dormant conditions. For budding, the compatibility of scion and rootstock is essential. 'T' budding and shield budding are successful in most guava varieties.

Sexual propagation is not successful in guava cultivation due to poor seed germination (Usman et al., 2012). The presence of tannins in seed, the hard seed coat and seed dormancy are responsible for low seed germination (Ali et al., 2007). But different seed treatments such as the use of growth regulators ($GA_3$) and seed treatment with 10% HCl can increase seed germination up to 90% (Butt et al., 2013). The main drawback of sexual propagation in guava production is the degradation of variety, but the technique increases heterozygosity and helps in maintaining heterogeneous populations for operative selection.

Asexual propagation techniques include *in vivo* propagation. Conventional *in vivo* techniques delay guava production due to seasonal dependence, the need for more plant material for propagation and a longer juvenile phase (Usman et al., 2014). Herbaceous cuttings are more effective than air layering, grafting and budding techniques (Table 4). Different substrate mixtures are used to increase rooting with or without the use of auxins (Gautam et al., 2010; Cheng et al., 2011; Saha, 2015).

*In vitro* propagation in guava reduces the use of plant material and helps in the establishment of cultivars with better growth (Chandra and Mishra, 2007; Usman et al., 2012). Direct organogenesis has been done in different pyriform and round-shaped cultivars (Usman et al., 2012). Somatic embryogenesis is also undertaken for guava production and multiplication of elite cultivars (Nic-Can et al., 2015). Zygotic embryos are also used as an explant. Internodes, leaf disc monocarps and petals are also used in embryogenesis (Butt et al., 2013).

## 5 Planting and cultivation techniques

A field should be deeply ploughed and properly levelled before the planting operations begin. Guava can be planted in February–March or August–September with earth balls or even bare rooted. In the latter case, plants should be defoliated and roots should be covered with some moist material.

The most innovative method in guava production is high-density plantation rather than traditional, large, widely spaced trees often characterized by declining yields. HDP means a density which gives the maximum crop yield. High-density plant spacing increases production and productivity and improves fruits quality. It helps to improve the efficient use of land, water, light, fertilizers and pesticides compared to traditional planting systems. At present, there is a need to replace the traditional planting system with improved HDP. A study of the impact of high-density plant spacing on fruit yield and quality of guava (Mahesh et al., 2017) found that plant spacing of 2.0 × 1.5 m produced maximum sugars and ascorbic acid but had a negative effect on fruit physical appearance. High-density plantation gave significantly higher yield per hectare.

Fruit size is also affected by rainy and winter seasons when combined with planting density. During the winter season, 2.0 × 1.5 m planting density produced bigger size fruit due to more availability of metabolites and less fruit load on plants (Nautiyal et al., 2016). Early planted guava fields at a spacing of 2.0 × 1.0 m produced better vegetative growth along with improved parameters such as yield, fruit weight and fruit quality (Kumawat et al., 2014). High plant density does not affect leaf nutrient status. With HDP, relative humidity and canopy temperature differ when compared with normal planting density. These microclimatic conditions enhance the penetration of solar radiations, resulting in high productivity and reduced incidence of insect pests (Saini et al., 2018).

The principle of crop regulation is the manipulation of natural flowering and fruiting of guava trees in the desired season to increase fruit yield and

**Table 4** Varieties of guava produced by clonal propagation in the field

| Propagation method | Media | Treatments | Cultivars | Country | Reference |
|---|---|---|---|---|---|
| Cuttings (soft wood and semi-hardwood) 10–15 cm of length | | | Red Malaysian, GU8 (Jambu Kampuchea), GU 11 (Giant guava), GU15 (Jade Seedless), BatuArang, Subang 6 and Sitiawan | Malaysia | Pommer and Murakami (2009) |
| | Vermiculite, soil, sand | Hoagland solution Girdling and ringing of sprout conditioning | - | India | Gautam et al. (2010), Pandey and Bisen (2010) |
| | 1:2:1 (peat moss, perlite, vermiculite or sterilized sand) | IBA, NAA | Century, Pearl and Crystal | Taiwan | Cheng et al. (2011) |
| | Soil | IBA, NAA | Safeda | Pakistan | Kareem et al. (2013) |
| | | | Klomsali, Salithong, Kimchu | Thailand | Siwarungson et al. (2013) |
| | Soil | | Allahabad Safeda | Egypt | El-Shereif (2016) |
| | Soil | | Indonesian white, Indonesian seedless | Indonesia | Nasution and Hadiati (2014) |
| | Sand | | | Iran | Sardoei (2014) |
| Air layering | Sphagnum moss | IBA/NAA | Century, Pearl and Crystal | Taiwan | Cheng et al. (2011) |
| | Soil, vermi compost, chicken manure, Sphagnum peat, coconut peat | IBA | Allahabad Safeda | India | Maurya et al. (2012) |
| | Soil, manure, composting | | BARI Guava-1, 2 e 3 | Bangladesh | Saha (2015) |

| | | | | |
|---|---|---|---|---|
| Mound layering | Soil | Bassateen El Sabahia; Bassateen Edfina; | Egypt | Pommer and Murakami (2009) |
| | Soil | L-49 (Sardar) | India | Dadas et al. (2015) |
| Cleft grafting, veneer grafting | Soil | Kazipiara, Swarupkhati, polypiara | Bangladesh | Rahim et al. (2011) |
| Wedge grafting | 1:1:3 (sand, compost, rich black soil) | Century, Pearl and Crystal | Taiwan | Cheng et al. (2011) |
| | Soil | Arka Amulya Hisar Surkha Hisar Safeda Allahabad Safeda Lalit Pant Guava L-49 and local guava (rootstock) | India | Mamta et al. (2014), Disket et al. (2014) |

quality while reducing reliance on inputs such as fertilizers and pesticides. The concept of crop regulation in guava is based on the phenomenon that guava flowers with new and succulent growth. Different methods are used in crop regulations of guava.

Induction of water stress from December until the onset of monsoon is recommended to produce a good winter yield (Boora et al., 2016). The amount of winter crop can be increased by changing the fertilizer schedule from May to June. This change in the nutrition schedule results in the induction of more vegetative growth for winter cropping, which consequently increases the yield (Boora et al., 2016).

Pruning is the judicious removal of plant parts such as branches, water shoots or terminal buds to maintain the shape of the tree and its productivity. For high-density plantation, pruning is the main factor that is responsible for the maintenance of the microclimate of plant canopy. Guava responses very well to training and pruning as it seems to have a vital role in controlling the number and size of new shoots (Saini et al., 2018). Severe pruning should be avoided. About 25% pruning of lateral shoots gives more promising results in terms of yield as compared to intensively pruned trees (Lotter, 1990). The time of pruning is also important for good quality guava production. Plants pruned during July have a healthy growth, high yield and highly nutritious fruit (Medina-Urrutia and Campbell, 1994). Pruning is also helpful in reducing tree height and increasing canopy and in maintaining tree size (Rajwant and Dhaliwal, 2001). Flower intensity and fruit setting is improved with 25% pruning during June–July (Mishra and Pathak, 1998). Pruning of terminal shoots up to 20–30 cm during the current spring season is helpful in managing the rainy season crop (Tiwari and Lal, 1984).

Deblossoming can be induced by the use of different chemicals such as naphthalene acetic acid, naphthalene acetamide, 2,4-dichlorophenoxy acetic acid, potassium iodide and urea. Different concentrations with different combinations have varying degrees of success (Tiwari et al., 1992). Along with these chemicals, hand thinning is also helpful in the reduction of the rainy season crop and increased production of the winter season crop (Singh and Kumar, 1993).

## 6 Irrigation and fertilization

Young guava plants require irrigation at weekly intervals during the summer months and two to three lots of irrigation during the winter months. In the case of fruit-bearing trees, irrigation should be provided for flowering and better fruit setting at intervals of 2–3 weeks during the summer months and at monthly intervals during the winter months. Irrigation is also essential to check excessive fruit drop during the summer months.

Recent studies have shown that drip irrigation gives remarkable results in terms of yield and physicochemical characteristics of guava in comparison to

**Table 5** Doses of manures and fertilizers required for guava crop

| Age of tree (years) | FYM (kg/plant) | Urea (g/plant) | SSP (kg/plant) | MOP (g/plant) |
|---|---|---|---|---|
| 1-3 | 10-20 | 150-200 | 0.5-1.5 | 100-400 |
| 4-6 | 25-40 | 300-600 | 1.5-2.0 | 600-1000 |
| 7-10 | 40-50 | 750-1000 | 2.0-2.5 | 1100-1500 |
| 10 and above | 50 | 1000 | 2.5 | 1500 |

flood irrigation. As research shows, the main advantage of drip irrigation is promoting the bearing habit of plants starting from the third year, with an average production of 8.31 tons/ha in an orchard planted in 5 × 5 m distance and 12.00 tons/ha with 6 × 6 m planted orchard (Mandal et al., 2007). Drip irrigation has also proved to be successful for guava grown in saline-affected areas (Singh, 2018).

Manure and fertilizer requirements for guava vary depending on varieties, the age of plants, the fertility status of soil, climatic conditions and management practices. Guava fruiting depends on the current season's growth. Manures and fertilizers need to encourage the vegetative growth and therefore fruit bearing. Good fertilization of guava plants not only increases their yield but also improves the quality of fruits (Table 5). Farmyard manure should be applied in May. One half of the inorganic fertilizer used should be applied during May-June and the remaining half during September–October.

Zinc deficiency is a common problem in the growth of guava plants. Affected plants produce small leaves with yellowing or chlorosis in between the veins of plants. Plant growth is suppressed and branches die back. Zinc deficiency could be corrected by spraying trees with zinc sulphate lime mixture (1 kg of zinc sulphate + ½ kg of unslaked lime in 100 litres of water). Two or three such sprays should be given at fortnightly intervals to cure zinc deficiency.

Space between guava plants could be economically utilized by suitable intercrops planted during the pre-bearing period in guava orchards. During the initial 3–4 years, vegetables like brinjal, radish, carrot and okra could be interplanted in vacant spots of land. Leguminous crops like cowpea, grams and beans can be safely sown as intercrops. Intercropping should be stopped when the trees attain the age of full growth. The advantage of intercropping is that the empty land is put into productive use before the trees attain the full growth and become productive themselves. The disadvantage is that the trees can be damaged when cultivating the crops. Annual crops will also compete with guava trees for nutrients and water.

## 7 Pests and diseases

Guavas suffer from a range of insect pests and diseases, with recent increases in the numbers of insect pests such as fruit borer, bark-eating caterpillars, fruit flies, white flies, semi-loppers, stem borers and fruit borers. Among different pests infesting guava, lepidopteron fruit borers such as *Deudorix isocrates* Fab. and *Dichocrosis punctiferalis* Guen., which were previously considered minor pests, are now becoming a significant problem (Yadav and Pandey, 1995a,b; Devi and Jha, 2017). Both the species are important polyphagous pests and can cause losses of up to 24% in guava (Biswas et al., 1995). The incidence of these pests is significantly affected by abiotic factors such as temperature, relative humidity and rainfall. During the monsoon season, adult fruit flies lay eggs on the surface of fruits. On hatching, maggots enter fruits and, in most cases, fruits drop.

Spraying with insecticides such as malathion and demicron can minimize the incidence of these pests. However, indiscriminate use of pesticides in either dust or spray form can leave residues on fruits, which pose a potential hazard to human health. The use of resistant varieties is one of the most cost-effective and environmentally friendly means of protecting crops against insect pests (Devi and Jha, 2017). Other control practices such as bagging of fruits, detection of early infestation by periodic monitoring, regular destruction of infested fruits and use of light traps are helpful in the control of pomegranate butterfly (Sharma et al., 2014).

Fruit fly (Diptera:Tepritidae), which has the ability to adapt to a range of conditions, is one of the main pests of guava (Godoy et al., 2011). It causes significant yield losses as well as additional costs in control measures and reduced trade, (Araujo et al., 2013). The population of fruit flies fluctuates depending on factors such as the availability of primary or alternate hosts and environmental conditions. Lower temperatures favour an increase in the Tephritidae population (Araujo et al., 2008). This pest can be controlled by different cultural practices such as prompt collection and destruction of fallen fruit, removing loose tree barks as well as tillage of tree basins (which disrupts activities such as oviposition by adult beetles). Pupae and larvae can also be destroyed by encouraging natural enemies, including parasitoids such as *Spalangia philippinensis* and *Opius compensates* (Dawabah et al., 2019).

Along with these pests, guava crop is also prone to different diseases such as wilt disease caused by different fungi. Various fungus species have been reported to attack different parts of the plants (Table 6). Wilt can cause losses ranging from 5 to as high as 60% and mainly occurs in plants grown in alkaline soil with a pH range of 7.5–9.0. Clay loam and sandy loam soils are more prone to this disease (Naz et al., 2013). Infected guava plants start showing immediate signs of wilting with the onset of the rainy season in August, with most plants dying in September and October. Plants more than five years of age are prone to this disease. Wilt symptoms varies during different times of the year. Common symptoms are the yellowing of leaves with interveinal chlorosis and the appearance of small, dried fruits on trees.

**Table 6** List of fungal species causing wilt in guava plants

| Fungal species | Part of plant on which it attacks | References |
|---|---|---|
| *Phytopthora parasitica* | Whole plant | Bokhari and Sahi (2011) |
| *Cephalosporium* sp. | Whole plant | Vestal (1941) |
| *Glomerella cingulata* Stonem. | Guava fruit | Kafi and Siddique (1964) |
| *Alternaria alternate* | Bark of guava | Fateh et al. (2011) |
| *Alternaria tenuissima.* (Kunze ex Pers.) | Guava leaves | Ghaffar and Kafi (1968) |
| *Cladosporium* sp. | Guava leaves | Ghafoor and Khan (1976) |
| *Curvularia* sp. | Tree branches | Fateh et al. (2011) |
| *Fusarium solani* f. sp. *psidi ava* | Whole plant | Bokhari and Sahi (2011) |
| *Fusarium oxysporum* | Tree branches | Fateh et al. (2011) |
| *Fusarium chlamydosporum* | - | Gupta and Misra (2012) |
| *Gleosporium psidii* | Tree twigs | Abbas et al. (2014) |
| *Fusarium oxysporum* f. sp. *psidi* | Whole plant | Bokhari and Sahi (2011) |
| *Penicillium* sp. | Bark of guava | Fateh et al. (2011) |
| *Pestalotiopsis brevista* Sacc. | Leaves and fruit | Ghaffar and Kafi (1968), Kafi and Siddique (1964) |
| *Phomopsis* sp. | Whole plant | Ghaffar and Kafi (1968), Kafi and Siddique (1964) |
| *Phoma psidii* | On dead branches | Ahmad (1956, 1968) |
| *Diplodia psidii* | Tree branches | Ahmad (1967, 1969) |
| *Stagnopsis psidii* | Branches | Ahmad (1967, 1969) |
| *Pestalotia psidii* Pat. | Dead twigs | Saleem et al. (1988) |
| *Lasiodiplodia undulata* (Pat) | Dead branches | Bokhari and Sahi (2011) |
| *Polyporus* sp. | Tree branches | Ghafoor and Khan (1976) |
| *Verticillium albo-atrum* | Whole plant | Gupta et al. (2003) |
| *Botyodiplodia theobromae* | - | Pandit and Samajpati (2002) |
| *Sclerotium rolfsii* and *Rhizoctonia solani* | - | Misra et al. (2003) |

Guava wilt can be controlled by a strict sanitation regime in orchards with wilted trees uprooted and burnt. Symptoms of the disease do not appear with green manuring, and disease development is less when organic sources of nitrogen are used. Soil solarization with transparent polyethylene sheet during May–June has been suggested for the control of wilt pathogens (Dwivedi, 1993). Special care should be taken during the transplantation of guava plants. *Trichoderma* spp. *Aspergillus niger* AN27 can be used as a biological control against guava wilt (Naz et al., 2013). Soil amendments with lime, neem and gypsum are helpful in controlling wilt disease. Intercropping with turmeric and marigold is also effective against wilt (Prasad et al., 2003).

## 8 Harvesting and postharvest operations

Guavas are harvested throughout the year around the world. The peak harvesting periods in North India, for example, are August for the rainy season crop, November–December for the winter season crop and March–April for the spring season crop. In the mild climatic conditions in other parts of the country, peak harvesting periods vary significantly. Ripening of guava starts on the tree and continues after harvesting. Ripening is accelerated during the rainy season due to high temperature and slows down in the winter season in low temperature conditions. During ripening, guava shows intense changes in skin colour and firmness due to high respiration and ethylene biosynthesis (Singh, 2011; Teixeira and Durigan, 2010). Firmness in guava fruit declines during ripening, and this decrease is associated with the decrease in pectin tissues due to hydrolytic activity of polygalacturonase enzymes (Chitarra and Chitarra, 2005; Mitra et al., 2012).

Being a delicate fruit, guava requires a very careful handling during harvesting and transport. Being highly perishable, guava fruits are typically marketed immediately after harvest. In countries such as India, fruits are packed in baskets made from locally available packaging materials. If the fruit needs to be transported over longer distances, wooden or corrugated fiberboard boxes are used along with good cushioning materials such as paddy straw, dry grass, guava leaves or rough paper. Good ventilation is necessary to prevent the build-up of heat. Fruits can be stored for a few days at room temperature. Fruits treated with aureofungin and ethylene chlorohydrin plus calcium carbonate can be stored at room temperature for a period of 5 days without spoilage and up to 7 days with only 25% spoilage (Yamashita and Benassi, 2000). Cooling can also extend shelf life. In a study by Silip et al. (2003), guava fruits were hydro-cooled and stored at 5°C, 10°C and 15°C for 6 weeks. Hydro-cooled fruits showed less changes in colour, had a lower incidence of browning and had less weight loss, particularly when stored at 10°C. Research suggests that storing fruits at an optimum temperature range is key to maintaining the fruit quality and minimizing postharvest losses (see Table 7) (Singh, 2011).

Besides cold storage, the ripening of guava can be delayed by other techniques. The use of ionizing radiations on guava cultivars 'Allahabad Safeda' and 'Lucknow-49' has been shown to reduce respiration rates by minimizing ethylene production during storage (Singh and Pal, 2009). Guava treated with 10% $O_2$ and 5% $CO_2$ for 24 h can be stored for 2 weeks at 4°C without chilling (Bautista and Silva, 1997). Low-density polyethylene films retard ripening of guava by slowing down the processes of softening and increasing soluble solids, acidity and ascorbic acid (Lazan and Ali, 1997). The use of a coextruded polyolefin film maintained good sensory characteristics of the cultivar 'Kumagai' for 28 days under refrigerated storage (Jacomino et al., 2001), while up to

**Table 7** Storage temperature and shelf life of guava

| Maturity Stage | Temperature (°C) | Shelf life (days) | Reference |
|---|---|---|---|
| Breaker stage | Ambient | 1–2 | Mamede et al. (2016) |
| Breaker stage | 10 | 14–21 | Teixeira and Durigan (2010) |
| Mature green | 5-8 | 14-21 | Kader (1999) |
| Fully ripe | 8–10 | 7 | Kader (1999) |
| – | 20 | 7 | Paull and Chen (2014) |
| – | 3.5-7 | Several weeks | Vazquez-Ochoa and Colinas-Leon (1990) |
| – | 10 | 25 | Silip and Hajar (2007) |

6 weeks of shelf life was achieved for the cultivar 'Pedro Sato' with the use of modified atmosphere packaging in Cryovac packaging under refrigeration at 8 °C (Yamashita and Benassi, 2000). The use of edible coatings has also been employed in enhancing the shelf life of guava. As compared to non-treated guava, coated fruits exhibited lower weight loss during storage and thus a longer shelf life (Ribeiro et al., 2005). Use of 2% and 4% hydroxypropylcellulose reduced fruit softening up to 35% and 45%, respectively (McGuire and Hallman, 1995), whereas fruits coated with 5% carnauba wax reduced fruit softening up to 30% (McGuire and Hallman, 1995). Guava coated with chitosan and gelatin reduced metabolic activity and thus extended the shelf life (Oshiro et al., 2012).

## 9 Conclusion and future trends

In conclusion, guava is gaining popularity in the whole world and guava growing area has increased approximately 64% since 1990. A large number of varieties have been developed with the features of less seed, coloured varieties along with good keeping quality. But still guava wilt is a cry for farmers. Mother plants infected with wilt are the main source of this disease. Nowadays, guava biotechnology and producing guava varieties through bud and stem culture are becoming famous. The conventional breeding techniques are slow and require long time to grow disease free plants. A large number of research papers and reports have been published on guava production through micropropagation, but these techniques are not working on a commercial scale. Thus there is an urgent need to develop efficient protocols which are applicable for short breeding cycle, rapid seed viability and propagation of rare plants.

## 10 References

Abbas, S. Q., Niaz, M., Iftakhar, T., Perveen, A., Abbas, A. and Riaz, S. 2014. New fungal records on guava (*Psidium guajava*) from Pakistan. *Proceedings of the Pakistan Academy of Sciences* 51, 121–7.

Agrianual. 2016. *Goiaba*. FNP Consultoria & Comércio, São Paulo, pp. 293–6.

Agriculture and Food Agency. 2014. Index of agricultural production: guava. In: *Agricultural Statistics Yearbook*. Council of Agriculture, Taipei, p. 87.

Agriculture in Sudan. 1999. Typed script in press. Ministry of Agriculture and Forestry, Karthoum.

Ahmad, S. 1956. Fungi of Pakistan:. *Biological Society of Pakistan Lahore* 1, 1–126.

Ahmad, S. 1967. Contributions to the fungi of Pakistan: V. *Biologia* 13, 15–42.

Ahmad, S. 1968. Further contributions to the fungi of Pakistan: V. *Biologia* 14, 1–11.

Ahmad, S. 1969. Fungi of Pakistan: monogarph. *Biological Society of Pakistan Lahore* 5, 1–110.

Ali, N., Mulwa, R. M. S., Norton, M. A. and Skirvin, R. M. 2007. Radical disinfestation protocol eliminates *in vitro* contamination in guava (*Psidium guajava* L.) seeds. *Plant Cell, Tissue and Organ Culture, Berlin* 91(3), 295–8. doi:10.1007/s11240-007-9283-7.

Amusa,N. A., Ashaye, O. A., Amadi, J. and Oladapo,O. 2006. Guava fruit anthracnose and the effects on its nutritional and market values in Ibadan, Nigeria. *Journal of Applied Sciences* 6(3), 539–42. doi:10.3923/jas.2006.539.542.

Araujo, E. L., Silva, R. K. B., Guimarães, J. A., Silva, J. G. and Bittencourt, M. A. L. 2008. Levantamento e flutuação populacional de moscas-das-frutas (Diptera: Tephritidae) em goiaba *Psidium guajava* L., no município de Russas (CE). *Revista Caatinga* 21, 138–46.

Araujo, E. L., Ribeiro, J. C., Chagas, M. C. M., Dutra, V. S. and Silva, J. G. 2013. Moscas-das-frutas (Diptera: Tephritidae) em um pomar de goiabeira, no semiárido brasileiro. *Revista Brasileira de Fruticultura* 35(2), 41–51.

Aular, J. and Casares, M. 2011. Consideraciones sobre la produccion de frutas en Venezuela. *Revista Brasileira de Fruticultura, Jaboticabal* 33, 187–98. doi:10.1590/S0100-29452011000500022.

Badan Statistica Indonesia. 2014. Sub-directorate of statistical compilation and publication. Disponível em: http://www.bps.go.id.

Bautista, P. B. and Silva, M. E. 1997. Effects of CA treatments on guava fruit quality. *Proceedings of the 7th International Controlled Atmosphere Conference*. University of California, Davis, CA.

Biswas, G. C., Ahmad, M., Karim, M. A. and Mian, M. Y. 1995. Incidence of fruit borer (*Virachola isocrates* F.) on guava at Khagrachari Hill District. *Bangladesh Journal of Entomology* 5(1/2), 33–9.

Bokhari, A. A. and Sahi, S. T.. 2011. Studies on guava decline and disease management Pakistan. In: *Abstracts "Challenges and Options for Plant Health Management", 8th National Conference of Pakistan Phytopathological Society*, 28–29 November 2011, University of Agriculture, Faisalabad, Pakistan, p. 64.

Boora, R. S., Dhaliwal, H. S. and Arora, N. K. 2016. Crop regulation in guava-a review. *Agricultural Reviews* 37(1), 1–9. doi:10.18805/ar.v37i1.9258.

Butt, M., Usman, M. and Fatima, B. 2013. Enhanced seed germination and callogenesis under long days using leaf disc as explant in guava cultivars. *Biologia, Pakistan* 59, 293–8.

Chandra, R. and Mishra, M. 2007. Biotechnological interventions for improvement of guava (*Psidium guajava* L.). *Acta Horticulturae* 735, 117–25. doi:10.17660/ActaHortic.2007.735.15.

Cheng, M., Lee, S. and Teck, C. 2011. *Taiwan Guava Production Manual*. Horticulture Crop Training and Demonstration Center, Cayo District, pp. 7–15.

Chitarra, M. I. F. and Chitarra, A. B. 2005. *Pós-colheita de frutas e hortaliças: fisiologia e manuseio* [*Post Harvest of Fruits and Vegetables: Physiology and Handling*] (2nd edn.). UFLN, Lavras, p. 783.

Dadas, M. M., Patil, S. D. and Deshmukh, P. L. 2015. Standardization of layering time in guava (*Psidium guajava* L.). *Bioinfolet, New Delhi* 12, 506–7.

Dawabah, A. A. M., Al-Yahya, F. A. and Lafi, H. A. 2019. Integrated management of plant-parasitic nematodes on guava and fig trees under tropical conditions. *Egyptian Journal of Biological Pest Control* 29(1), 29. doi:10.1186/s41938-019-0133-9 .

Department of Agriculture, Forestry and Fisheries. 2015. Local market fruit sales data. Directorate of Agriculture Statistics, Pretoria. Disponível em: www.daff.gov.za.

Devi, A. R. and Jha, S. 2017. Incidence and screening of guava (*Psidium guajava* L.) varieties against fruit borer (*Deudorix Isocrates* F.) in eastern gangetic plains of West Bengal, India. *International Journal of Current Microbiology and Applied Sciences* 6(10), 1689–98. doi:10.20546/ijcmas.2017.610.204.

Disket, D., Bakshi, P., Wali, V. K., Bhushan, B. and Sharma, A. 2014. Growth and yield attributes of commercial guava (*Psidium guajava* L.) cultivars under sub-tropical condition. *Indian Journal of Plant Physiology, New Delhi* 19(1), 79–82. doi:10.1007/s40502-014-0076-9.

Dwivedi, S. K. 1993. Fungitoxicity of *Foeniculum vulgare* seed oil used against a guava wilt pathogen. *National Academy of Science Letters* 16, 207.

Egyptian Ministry of Agriculture and Land Reclamation. 2014. Bulletin of economic affairs sector. Available at: http://sis.gov.eg.

El-Shereif, A. R. 2016. Tropical and subtropical fruits in Egypt. *ISHS Newsletter, The Hague*, no. 14, January 2016.

Fateh, F. S., Kazmi, M. R., Ahmad, I. and Mukhtar, T. 2011. Common fungi found in decline affected Mango and Gauva orchids in Punjab. In: *Abstracts "Challenges and Options for Plant Health Management", 8th National Conference of Pakistan Phytopathological Society*, 28–29 November 2011. University of Agriculture, Faisalabad, Pakistan, p. 30.

Gautam, N. N., Singh, K., Singh, B., Seal, S., Goel, A. and Goel, V. L. 2010. Studies on clonal multiplication of guava (*Psidium guajava* L.) through cutting under controlled conditions. *Australian Journal of Crop Science, Lismore* 4, 666.

Ghaffar, A. and Kafi, A. 1968. Fungi of Karachi. *Pakistan Journal of Science* 20, 5–10.

Ghafoor, A. and Khan, S. A. J. 1976. *List of Diseases of Economic Plants in Pakistan*. Department of Plant Protection, Ministry of Food, Agriculture and Under Developed Area, Government of Pakistan, 85.

Godoy, M. J. S., Pacheco, W. S. P. and Malavasi, A. 2011. Moscas-das-frutas quarentenárias para o Brasil. In: Silva, R. A., Lemos, W. P. and Zucchi, R. A. (Eds), *Moscas-das-frutas na Amazônia brasileira: diversidade, hospedeiros e inimigos naturais*. Embrapa Amapá, Macapá.

Gupta, V. K. and Misra, A. K. 2012. *Fusarium chlamydosporum*, causing wilt disease of guava (*Psidium guajava* L.) in India. *Archives of Phytopathology and Plant Protection* 45(20), 2425–8. doi:10.1080/03235408.2012.728055.

Gupta, P., Dubey, V., David, A. and John, D. 2003. *Verticillium albo-atrum*? a new pathogen associated with guava wilt. *Indian Phytopathology* 56, 503–5.

Gupta, V. K., Misra, A. K., Gaur, R. K., Jain, P. K., Gaur, D. and Saroj, S. 2010. Current status of Fusarium wilt disease of guava (*Psidium guajava* L.) in India. *Biotechnology, Harare* 9, 176–95.

HORTGRO. 2015. Guava production and export data. SATI, Information and Market Intelligence Division, Paarl. Disponível em: www.hortgro.co.za.

Jacomino, A. P., Sarantópoulos, C. I. G. L., Sigrist, J. M. M., Kluge, R. A. and Minami, K. 2001. Sensorial characteristics of "Kumagai" guavas submitted to passive modified atmosphere in plastic packages. *Journal of Plastic Film and Sheeting* 17(1), 6–21. doi:10.1106/R06Q-QRR9-BMPJ-6Y9Y.

Kader, A. 1999. *Guava: Recommendations for Maintaining Postharvest Quality*. Division of Agriculture and Natural Resources, University of California.

Kafi, A. and Siddique, Z. A. 1964. Russetting of guava in Karachi Malir area. Abstracts Pakistan Science Conference A: 56.

Kafle, A., Sangita, S., Reddy, I. and Chapagain, M. 2018. A review on medicinal properties of *Psidium guajava*. *Journal of Medicinal Plant Studies* 6, 44–7.

Kareem, A., Jaskani, M. J., Fatima, B. and Sadia, B. 2013. Clonal multiplication of guava through softwood cuttings under mist conditions. *Pakistan Journal of Agricultural Science* 50, 23-7.

Kumawat, K. L., Sarolia, D. K., Kaushik, R. A. and Jodha, A. S. 2014. Effect of different spacing on newly planted guava cv. L-49 under ultra-high density planting system 9, 3729–35.

Lazan, H. and Ali, Z. M. 1997. Guava. In: Shaw, P. E., Chan Jr., H. T., Nagy, S. and Wardowski, W. F. (Eds), *Tropical and Subtropical Fruits* (vol. 3). Florida Science Publication, Gainesville, FL, pp. 20–41.

Lekeu, J. P. 2001. Guava (*Psidium guajava* L.). In: Raemaekers, R. H. (Ed.), *Crop Production in Tropical Africa*. DGIC, CIP Royal Library Albert I, Brussels, Belgium, pp. 592–5.

Lotter, J. De V. 1990. Vegetative and reproductive habit of guava in relation to pruning methods. *Acta Horticulturae* 275, 229–37.

Mahattanatawee, K., Manthey, J. A., Luzio, G. S., Talcott, S. T., Goodner, K. and Baldwin, E. A. 2006. Total antioxidant activity and fiber content of select Florida-grown tropical fruits. *Journal of Agricultural and Food Chemistry* 54(19), 7355–63. doi:10.1021/jf060566s.

Mahesh, P., Shant, L., Pankaj, N. and Prabhakar, J. 2017. Response of high density spacing on physic-chemical quality and yield of guava (*Psidium guajava* L.) cv. Plant prabhat. *International Journal of Agriculture Sciences* 9, 3962–5.

Mamede, A., Barboza, H., Soares, A. G. and Fonseca, A. J. 2016. Postharvest physiology and technology for fresh guavas. Available at: https://www.resaerchgate.net/publication/303875667.

Mamta, J., Syamal, M. M. and Singh, S. P. 2014. Comparative efficacy of different propagation techniques in guava. *Indian Journal of Horticulture, Bangalore* 71, 315–20.

Mandal, G., Kumar, S., Kumar, R. and Singh, R. 2007. Effect of drip irrigation and plant spacing on yield, quality and economic return of guava (*Psidium guajava* L.) grown in saline soil. *Acta Horticulturae* 735(735), 427–32. doi:10.17660/ActaHortic.2007.735.60.

Martínez de Lara, J., Barrientos-Lara, M. C., Reyes de Anda, A. C., Hernández Delgado, S., Padilla Ramírez, J. S. and Mayek Pérez, N. 2004. Diversidad fenotípica y genética en huertas de Guayabo de Calvillo, Aguascalientes. *Revista Fitotecnia Mexicana, Chapingo* 27, 243–9.

Maurya, R. K., Ray, N. R., Chavda, J. C., Chauhan, V. B. and Patil, A. K. 2012. Evaluation of different organic media and water holding materials with IBA on rooting and survival of air layering in guava (*Psidium guajava* L.) cv. ALLAHABAD SAFEDA. *Asian Journal of Horticulture, Muzaffarnagar* 7, 44–7.

McGuire, R. G. and Hallman, G. J. 1995. Coating guavas with cellulose- or carnauba-based emulsions interferes with postharvest ripening. *HortScience* 30(2), 294–95. doi:10.21273/HORTSCI.30.2.294.

Medina-Urrutia, V. M. and Campbell, R. J. 1994. Pruning and paclobutrazol affect the growth and production of young mango 'Tommy Atkins' trees. *Proceedings of the International Society for Tropical Horticulture* 38, 50–5.

Milyani, R. 2012. Inhibitory effect of some plant extracts on clinical isolates of *Staphylococcus aureus*. *African Journal of Microbiology Research* 6(36), 6517–24. doi:10.5897/AJMR11.119.

Ministry of Agriculture Malaysia. 2013. *Agriculture Statistics Handbook*. Kuala Lumpur.

Ministry of National Food Security and Research. 2014. *Government of Pakistan*.

Mishra, H. K. and Pathak, R. A. 1998. Effect of shoot pruning on crop regulation in guava (*Psidium guavjava* L.) cv. L-49 (Sardar). *Progressive Horticulture* 30(1–2), 78–81.

Misra, A. K., Pandey, B. K., Prasad, B. and Shukla, S. K. 2003. Pathogenic diversity in the cause of wilt disease of guava. *Indian Phytopathology* 56, 314–.

Mitra, S. 2014. Tropical and subtropical fruits in India. *Journal of Tropical Crop Science, Bogor* 1, 1–2.

Mitra, S. K., Devi, H. L., Chakraborty, I. and Pathak, P. K. 2012. Recent development in postharvest physiology and storage of guava. *Acta Horticulturae* 959(959), 89–95. doi:10.17660/ActaHortic.2012.959.10.

Nasution, F. and Hadiati, S. 2014. Characterization and clustering of some guava germplasm collections based on leaf and fruit characters. *Agrivita Journal of Agricultural Science* 36(1), 91–6. doi:10.17503/Agrivita-2014-36-1-p091-100.

National Horticulture Board. 2015. Indian horticulture database. Ministry of Agriculture, Government of India, Gurgaon. Disponível em: http://www.nhb.gov.in.

Nautiyal, P., Lal, S., Dimri, D. C. and Arora, I. 2016. Shoot pruning severity in high density of guava (*Psidium guajava* L.). *International Journal of Agriculture Sciences* 8, 2427–31.

Naz, A., Naz, H. and Ashraf, S. 2013. Management of guava wilty by biological agent. *International Journal of Agriculture and Food Science and Technology* 4, 667–70.

Nic-Can, G. I., Galaz-Ávalos, R. M., De-la-Peña, C., Alcazar-Magaña, A., Wrobel, K. and Loyola-Vargas, V. M. 2015. Somatic embryogenesis: identified factors that lead to embryogenic repression. A case of species of the same genus. *PLoS ONE, San Francisco* 10(6), e0126414. doi:10.1371/journal.pone.0126414.

Ocampo, F. and Nũñez, V. M. 2007. Propagación *in vitro* de *Psidium guajaba* mediante organogénesis directa a partir de segmentos nodales. *Revista Corpoica - Corpoica Ciencia y Tecnología Agropecuaria* 8(1), 22–7. doi:10.21930/rcta.vol8_num1_art:79.

Orwa, C., Mutua, A., Kindt, R., Jamnadass, R. and Simons, A. 2009. *Agroforestree Database: a Tree Reference and Selection Guide* Version 4.0. World Agroforestry Centre, Kenya.

Oshiro, A. M., Daiane, M. D., Scalon, S. and de Paula, Q. 2012. Preservação de Goiabas 'Pedro Sato' Armazenadas Sob Atmosfera Modificada Em Refrigeração. (Preservation of guavas 'Pedro Sato' stored under modified atmosphere in refrigeration). *Revista de Ciências Agrárias (Journal of Agricultural Sciences)* 35(1), 213–21.

Pandey, S. K. and Bisen, A. 2010. Effect of mechanical treatments on rooting in cuttings of guava, lemon and pomegranate. *Journal of Horticulture and Forestry, Nairobi* 2, 95–8.

Pandit, P. K. and Samajpati, N. 2002. Wilt disease of guava (*Psidium guajava* L.) by *Botryodiplodia theobromae* Pat. *Journal of Mycological Research* 40, 71–3.

Parra-Coronado, A. 2014. Maduración y comportamiento poscosecha de la guayaba (*Psidium guajava* L.). Una revisión. *Revista Colombiana de Ciências Hortícolas, Tunja* 8(2), 314–27. doi:10.17584/rcch.2014v8i2.3223.

Paull, R. E. and Bittenbender, H. C. 2006. *Psidium guajava* guava. In: Janick, J. and Paull, R. E. (Eds), *The Encyclopedia of Fruit and Nuts*. Cambridge University Press, Cambridge, pp. 541–9.

Paull, R. E. and Chen, C. C. 2014. Guava: postharvest quality maintenance guidelines. Fruit, Nut, and Beverage Crops F.N. 41 (June).

Pereira, F. 1990. Factors affecting guava production and quality with special reference to Sao Paulo, Brazil. *Acta Horticulturae, The Hague* 275(275), 103–10. doi:10.17660/ActaHortic.1990.275.8.

Pereira, F. M. and Nachtigal, J. C. 2009. Melhoramento genético da goiabeira. In: Natale, W., Rozane, D. E., Souzia, H. A. and Amorim, D. A. (Eds), *Cultura da goiaba do plantio à comercialização*. FCAV, Capeas, CNP, FAPESP, Fundunesp, SBF, Jaboticabal, pp. 371–98.

Pereira, F. M., Usman, M., Mayer, N. A., Nachtigal, J. C., Maphanga, O. R. M. and Willemse, S. 2016. Advances in guava propagation. *Revista Brasileira de Fruticultura* 39(4), e-358. doi:10.1590/0100-29452017.

Pérez-Rodríguez, J., Arangurengonzalez, M., Luzbet-Pascual, R., Reynaldo-Escobar, I. M. and Rodrigueze, D. J. 2013. Aportes a la producción intensiva de plantas de guayabo (*Psidium guajava* l.) a partir de esquejes en los viveros comerciales. *CitriFrut, La Habana* 30, 11–6.

Pommer, C. V. and Murakami, K. R. 2009. Breeding guava (*Psidium guajava* L.). In: Mohan Jain, S. and Priyadarshan, P. M. (Eds), *Breeding Plantation Tree Crops: Tropical Species*. Springer, New York, pp. 83–120.

Pommer, C. V., Murakami, K. R. and Watlington, F. 2006. Goiaba no mundo. *O Agronômico, Campinas* 58, 22–6.

Prasad, D., Shukla, S. K., Prasad, B. and Misra, A. K. 2003. Effect of intercrop and different doses of NPK on the incidence of guava wilt. *Proceedings of the Seminar on Plant Diseases of National Importance with Special Reference to Guava Wilt and Mango Malformation*, 4–5 April. CISH and IISR, Lucknow, pp. 53.

Rahim, M. A., Islam, K. K., Islam, M. S., Alam, M. S., Islam, F. and Naher, N. 2011. Conservation of guava (*Psidium guajava*) germplasm using wilt-resistant rootstock of poly piara (*Psidium cattleianum*). In: Roskaft, E. and Chivers, J. (Eds), *Present State, Problems and Prospects of Its Conservation*. University of Chittagong, Bangladesh, pp. 127–39.

Rajwant, K. and Dhaliwal, G. S. 2001. Effect of time and pruning intensity on tree canopy volume, girth and plant height in Sardar guava. *Haryana Journal of Horticulture Science* 30, 154–6.

Reetz, E. R. 2015. *Anuário Brasileiro Da Fruticultura*. [*The Annual Brazilian Fruit Crops*]. Editora Gazeta Santa Cruz LTDA, Santa Cruz do Sul, RS.

Ribeiro, V. G., Assis, J. Sd, Silva, F. F., Siqueira, P. P. X. and Vilaronga, C. P. P. 2005. Storage of guavas 'Paluma' under refrigeration and at environmental conditions, with coated and uncoated fruits with carnauba wax (in Portuguese). *Revista Brasileira de Fruticultura* 27(2), 203–6. doi:10.1590/S0100-29452005000200005.

Saha, M. G. 2015. Production technology of important fruit crops in agroforestry system. In: Shahjahan, M., Ahmed, S. and Jewell, K. N. A. (Eds), *Training Manual Agroforestry Technologies for Professionals*. Agricultural Research Council Farmgate, Bangladesh.

Saini, H., Baloda, S., Vijay and Saini, P. 2018. Effect of pruning on productivity of guava under high density plantation - a review. *Current Journal of Applied Science and Technology* 27, 1-12.

Saleem, A., Hussain, M. and Ansar, M. 1988. A report on Pestalotia dieback of Guava plant in Pakistan. *Pakistan Journal of Forestry* 41(3-4), 38.

Samson, J. A. 1986. *Tropical Fruits*. Longman Groups, UK, p. 2.

Sardoei, A. S. 2014. Effect of different media of cuttings on rooting of guava (*Psidium guajava* L.). *European Journal of Experimental Biology, London* 4, 88–92.

Sharma, R. R., Reddy, S. V. R. and Jhalegar, M. J. 2014. Pre-harvest fruit bagging: a useful approach for plant protection and improved post-harvest fruit quality-a review. *Journal of Horticultural Science and Biotechnology* 89(2), 101–13. doi:10.1080/14620316.2014.11513055.

Silip, J. J. and Hajar, S. A. 2007. Relationship between precooling storage temperature and storage duration to the quality characteristics of guava (*Psidium guajava* L. Cv. Kampuchea). *Acta Horticulturae* 735, 535–46. doi:10.17660/ActaHortic.2007.735.70.

Silip, J. J., Ahmad, S. H., Russly, A. R., Wan Mohammad, W. A. and Ali, A. 2003. Responses of Guava cv. Kampuchea (*Psidium guajava* Linn.) to hydro-cooling time, storage temperature and storage duration. *Asian Journal of Microbiology, Biotechnology & Environmental Sciences* 5(3), 303–6.

Singh, G. 2007. Recent development in production of guava. *Acta Horticulturae, The Hague* 735, 161–76.

Singh, S. P. 2011. Guava (*Psidium guajava* L.). In: Yahia, E. (Ed.), *Postharvest Biology and Technology of Tropical and Subtropical Fruits – Volume 3 Cocona to Mango*. Elsevier Science, p. 213.

Singh, A. 2018. Soil and water management practices for fruit cultivation in salt affected soils. In *book: Advances in Salinity and Sodicity Management under Different Agro-Climatic Regions for Enhancing Farmers' Income*. ICAR-CSSRI, Karnal, pp. 172–81.

Singh, U. R. and Kumar, R. 1993. Crop regulation in guava. In: Chadha, K. L. and Pareek, O. P. (Eds), *Advances in Horticulture*. Malhotra Publishing House, New Delhi, pp. 1197–1204.

Singh, S. P. and Pal, R. K. 2009. Ionizing radiation treatment to improve postharvest life and maintain quality of fresh guava fruit. *Radiation Physics and Chemistry* 78(2), 135–40. doi:10.1016/j.radphyschem.2008.09.007.

Sinha, K. 2011. Study shows costly fruits are not always the healthiest. *The Times of India (Mumbai edition)*, 12 October 2011.

Siwarungson, N., Ali, I. and Damsud, T. 2013. Comparative analysis of antioxidant and antimelanogenesis properties of three local guava (*Psidium guajava* L.) varieties of Thailand, via different extraction solvents. *Journal of Food Measurement and Characterization, New York* 7(4), 207–14. doi:10.1007/s11694-013-9157-0.

Sreeramulu, D., Reddy, C. V. K., Chauhan, A., Balakrishna, N. and Raghunath, M. 2013. Natural antioxidant activity of commonly consumed plant foods in India: effect of domestic processing. *Oxidative Medicine and Cellular Longevity* 2013, 369479. doi:10.1155/2013/369479.

Teixeira, G. H. A. and Durigan, J. F. 2010. Effect of controlled atmospheres with low oxygen levels on extended storage of guava fruit (*Psidium Guajava* L. Pedro Sato). *Horticultural Science* 45(6), 918–24.

Tiwari, J. P. and Lal, S. 1984. Research report. ICAR Workshop on Fruit Research, Lucknow.

Tiwari, R. B., Tiwari, J. P. and Lal, S. 1992. Effect of shoot pruning, NAA and urea on cropping pattern of guava Indian. *Journal of Horticulture* 49, 305–8.

Urdaneta, A. B. S., SUÁREZ, E., Gonzalez, M. R., Amaya, Y., Colmenares, C. B. and Ortega, J. 2009. Efecto del ácido indolbutírico sobre el enraizamiento de acodos aéreos de guayabo (*Psidium guajava* L.) en el municipio Baralt, Venezuela. Evaluación preliminar. *Revista UDO Agrícola, Cumaná* 9(1), 113–20.

Usman, M., Butt, M. and Fatima, B. 2012. Enhanced in vitro multiple shoot induction in elite Pakistani guava cultivars for efficient clonal plant multiplication. *African Journal of Biotechnology, Nairobi* 11, 10182–9187.

Usman, M., Samad, W. A., Fatima, B. and Shah, M. H. 2013. Pollen parent enhances fruit size and quality in intervarietal crosses in guava (*Psidium guajava*). *International Journal of Agriculture and Biology, Faisalabad* 15, 125–9.

Usman, M., Shah, M. H., Badar, A., Fatima, B., Sabir, M. and Zaman, Q. 2014. Media steaming and coco-coir enhance growth of rough lemon (*Citrus jambhiri* L.) stock. *Pakistan Journal of Agricultural Science, Faisalabad* 51, 615–23.

Valdés-Infante, J., Rodríguez, N. N., Velásquez, J. B., Sourd, D. G., González, G., Rodríguez, J. A. and Rohde, W. 2012. Herramientas para un programa de mejoramiento genético del guayabo (*Psidium guajava* L.) en cuba. *Agronomía Costarricense, San Jose* 36, 111–29.

Vazquez-Ochoa, R. I. and Colinas-Leon, M. T. 1990. Changes in guavas of three maturity stages in response to temperature and relative humidity. *Horticultural Science* 25, 86–7.

Vestal, E. F. 1941. *A Textbook of Plant Pathology*. Kitabistan, Allahabad and Karachi, p. 645.

Yadav, L. B. and Pandey, V. 1995a. Incidence of pomegranate butterfly *Virachola isocrates* in guava orchard and guava based cropping system. *Bulletin of Entomology* 36, 84–91.

Yadav, L. B. and Pandey, V. 1995b. Guava fruit infestation by pomegranate butterfly *Virachola isocrates* in relation to weather factors. *Bulletin of Entomology* 36(1/2), 113–6.

Yamashita, F. and Benassi, M. T. 2000. Influence of modified atmosphere packaging and calcium treatment on the kinetics of ascorbic acid degradation and weight loss in guavas. *Cienciae Tecnologia de Alimentos* 20, 27–31.

# Chapter 15

# Crop improvement in jackfruit

*S. Shyamalamma and S. Priyanka, University of Agricultural Sciences, India; and G. S. K. Swamy and P. M. Sampath, University of Horticultural Sciences, India*



## 1 Introduction

The Moraceae family contains three genera of economic importance:

- Morus (*M. alba*; mulberries)
- Ficus (*F. carica*; figs)
- Artocarpus (*viz.*, *A. heterophyllus, A. altilis, A. hirsutus*; including jackfruit)

The entire genus is characterized by fruits with a high nutrient content (as discussed in Section 4). The family Moraceae comprises more than 1000 species across 67 genera, mostly of tropical shrubs and trees (Bailey, 1942). The genus *Artocarpus* is said to contain around 70 species (Zerega et al., 2010). Among the species found commonly in India are jackfruit (*Artocarpus heterophyllus* Lam.), breadfruit (*Artocarpus altilis*), monkey fruit (*Artocarpus lakoocha*) and wild jack (*Artocarpus hirsutus*). Jackfruit and breadfruit are widely grown by smallholders for family use. Monkey fruit (locally known as 'Vatehuli') is widely known as a substitute for tamarind in cooking in the coastal belt of Karnataka, Maharashtra and Goa. This chapter focuses on jackfruit. It reviews the origin, distribution and varieties of the fruit. It then looks at uses and nutritional composition, before discussing breeding and improved varieties.

http://dx.doi.org/10.19103/AS.2019.0054.25

## 2 Origin, distribution and uses

The lack of knowledge on the precise area of origin relates to the lack of research in this area, although the work of Jarrett (1959) has substantially contributed to clarifying earlier confusion. The tree is reported to be native to rainforests of Malaysia and the Western Ghats of India (Figs. 1 and 2). More recently, molecular research has provided important pointers to which species might form the primary and secondary genepools of jackfruit. However, this work is preliminary because the research has been done in Japan, Thailand and the United States with a limited number of accessions available for analysis (Kanzaki et al., 1997; Schnell et al., 2001; Zerega et al., 2014).

Jackfruit is now widely grown in many Asian countries, especially Bangladesh, Myanmar, Nepal, Sri Lanka, Thailand, Malaysia, Indonesia, India and the Philippines (Fig. 2). It is also grown in Southern China and in the Indo-Chinese region in Laos, Cambodia and Vietnam (Manjunatha, 1948; Morton, 1965; Narasimham, 1990; Gunasena et al., 1996). From Asia, it spread to tropical Africa, especially the eastern parts of the continent - Zanzibar, Kenya,

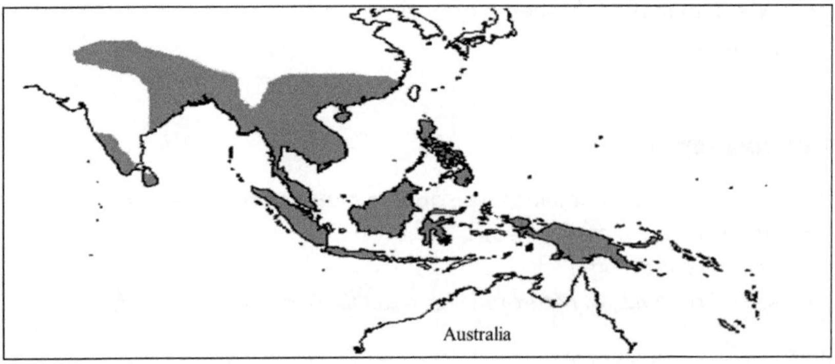


**Figure 1** The native range of the genus *Artocarpus* (indicated by shaded areas). Source: adapted from Zerega et al. (2014).

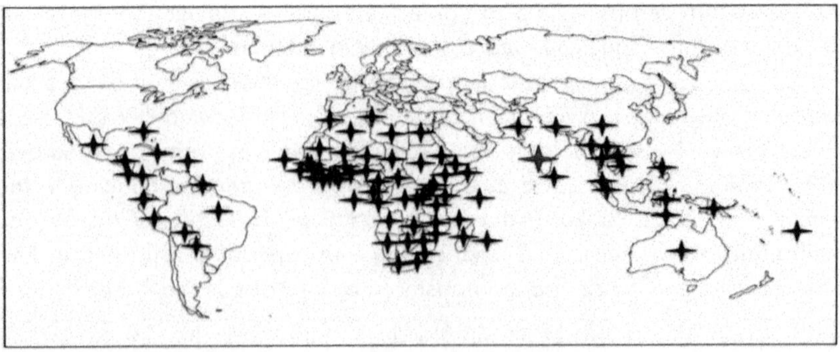

**Figure 2** Areas where jackfruit is cultivated. Source: adapted from Zerega et al. (2014).

Uganda and Madagascar. It is also grown in Mauritius and in some countries in Latin America such as Mexico. Leading producers are India, Bangladesh and Indonesia (Altendorf, 2018).

In India, jackfruit is distributed widely along the west coast. Jackfruit is also now gaining popularity in the southern states, particularly in Kerala, Karnataka, Maharashtra, Tamil Nadu, Andhra Pradesh and the Andaman-Nicobar Islands. It also grows in the wild in the country's north-eastern states, such as Assam, Meghalaya, Tripura Nagaland, as well as in the eastern states such as Bihar, West Bengal and Orissa. Though the crop is gaining importance, its area and production is still small (Table 1).

The genetic diversity of jackfruit can be seen in the range of genotypes found in individual states such as Karnataka. The genotypes found in districts such as Tubagere Hobli and Doddaballapur Tq are famous for their sensory quality, including the pulp colour which varies from orange to red. The genotypes found in the Tumakuru district are unique with thicker bulbs and smaller seeds, the pulp colour being mostly red and yellow. Most of the jackfruit genotypes found in Hassan district are also unique with very thick, round bulbs, with the colour varying from cream to yellow. The genetic diversity of jackfruit found in Tamil Nadu is also large, especially in Panruti, Cuddalore district, with fruits available almost all year round. The fruits found in this belt are also unique, with a typically round shape and uniform weight of 10–15 kgs.

There are several varieties/types of jackfruits which differ widely in size, shape, bearing, maturity, appearance and taste. Wide variations in sweetness, acidity, flavour and taste are also observed. Dutta (1956) described different varieties of jackfruit and grouped them into four main groups:

- **Soft-fleshed type**: When the fruits are ripe, they become soft, juicy and very sweet in taste. The flesh/bulbs can be separated easily.
- **Firm-fleshed type (hard flesh)**: When the fruits are ripe, the pulp is firm and crispy; taste varies from very sweet to less sweet.

**Table 1** Area, production and productivity of jackfruit in different states of India

| States | Area (000′ ha) | Production (000′ mt) | Productivity (t/ha) | Year |
|---|---|---|---|---|
| Arunachal Pradesh | 1.06 | 8.52 | 8.03 | 2005 |
| Jharkhand | 4.10 | 41.00 | 10.01 | 2008 |
| Karnataka | 6.78 | 231.57 | 34.17 | 2005 |
| Tripura | 2.20 | 12.50 | | |
| West Bengal | 10.42 | 143.47 | 13.76 | 2010 |
| Kerala | 97.54 | 348 million fruits | 3568 fruits/ha | 2003-04 |
| Assam | 22.00 | 201.00 | 9.49 | 2003-04 |

*Source*: APAARI (2012).

- **Rudrakshi**: The fruits are very small, weighing 1-5 kg, and round with fewer spikes on the fruit's surface. Quality of fruit is medium and is suitable as a root stock material.
- **Singapore or Ceylon jackfruit**: Introduced from Ceylon, this is a highly precocious type and produces fruit within 2-3 years; trees also start bearing off-season,that is in winter (September–December).

Varieties can be used in several different ways, as can the different parts of the fruit.

Jackfruit is a multipurpose tree species widely regarded as second only to coconut in value. Both the fruit and the wood are widely used. The fruits have several uses, including both tender and ripe fruits. Tender jackfruit is rich in mineral nutrients and fibre and is gaining importance as a meat substitute due to its unique taste and appearance after cooking. It has been reported that the glycaemic index of tender jackfruit is very low and found to be beneficial for people with type 2 diabetes (Devalaraja et al., 2011). Recent products include minimally processed tender (ready-to-cook) fruit with an enhanced shelf life of up to 3 days under ambient temperatures (Ramli et al., 2017).

The mature fruit has several uses. The flakes are used in the preparation of chips, papads (dough), dehydrated flakes for vegetable purpose, and the whole fruit after dehydration is used for animal feed purpose. Among these products, chips and papads are now produced on a commercial scale. Fried chips are the main product used by consumers. Vacuum-fried chips are becoming popular due to the natural taste and flavour the process is able to retain. The shelf life of normal chips ranges from 1–3 months with air-locked packing. However, vacuum-fried chips have a shelf life of over 1 year.

The ripe fruit has several uses. The main use is as fresh fruit. However, ready-to-eat flakes in punnet boxes are becoming more popular among consumers. The ripe fruit is also used in the preparation of several value-added products such as squash, jam, juice, pulp and as dehydrated powder for food and nutraceutical uses. The pulp, packed in retort pouches of 1–5 kgs, is used in preparation of products such as sweets and ice cream in Indian states such as Karnataka, Kerala and Maharashtra.

Jackfruit is comprised of flakes (35–50%), rind (35–40%) and seeds (10–15%). The edible parts are the flakes and seeds. The seeds have a lower nutritive value than the bulbs. The mature seeds are mainly used in cooking and in the food industry for starch. The seeds also contain two glycoproteins: jacalin and artocarpin (Reddy and Sasikala, 2016). Jacalin is a tetrameric two-chain lectin combining a heavy α-chain of 133 amino acid residues with a light β-chain of 20–21 amino acid residues. It has antifungal, antibacterial and antiviral properties. Seed lectins are therefore used as biocontrol agents and in binding of immunoglobulins.

## 3 Biology

The botanical name of jackfruit is *Artocarpus heterophyllus* Lam. It is a member of the family Moraceae. It is tetraploid, with somatic chromosome number 2n = 56, (2n = 4x = 56), (Darlington and Wylie, 1956). The tree is characterized to be monoecious (having separate male and female flowers on the same tree), evergreen and latex-producing, with high-strait trunk and thick branches.

The trees start flowering and bearing fruits in about –8 years after planting as seedlings and 3–4 years in grafted plants. A large number of flowers are borne on club-shaped rachis. The inflorescence is a spike covered by spathe. Female spikes are borne on foot stalks, while the male spikes appear both on the foot stalks and on the terminal branch lets. Foot stalks bearing female spikes are much more vigorous than those of male spikes. Jackfruit has a monoecious type of flowering. Though flowering is common in November to March in South Indian conditions and from February to April in North Indian conditions, flowering may range from September to December in off-season trees. The number of days for spathe shedding is usually around 9 days (8–10 days). Protrusion of stigma is seen from 1to 2 days after opening of spathe. An important feature is a consistent ratio and spread of female and male flowers (1:3).

Vegetable-type jackfruit trees are normally expected to flower and bear fruits throughout the year; however, some trees flower only from November to March. Spathe emergence in these trees is usually around 5 days (4–7 days). Male inflorescences are axillary on leafy twigs or from trunk and branches, but typically above female inflorescences, drooping, oblong, ellipsoidal or clavate, 5–10 × 2–3 cm, green; flowers numerous, densely crowded, with 2-lobed calyx and single stamen. Male spikes are uniformly spread all over the tree. The surface of a young male spike is smooth, while that of the female is granular. In general, male flowers out-compete the female ones in number. Samaddar and Yadav (1982) have reported up to 96% male inflorescences per tree. Female inflorescences are borne on short, stout twigs on the main trunk and primary branches mainly; peduncle stout with green fleshy ring at apex, larger than male inflorescence; flowers small, calyx tubular, ovary oblong and compressed, style terminal and oblique, stigma clavate. The length and diameter of female inflorescences are larger than those of males. Azad (1989) has reported the lowest, 3–5% of female inflorescences, while Sahadevan et al. (1950) reported 3–40%.

The anthers are yellow, grey and ash-grey. After dehiscence, the male spike turns black due to the growth of mould and drops down. Dehiscence occurs usually in the afternoon. Anthesis is at its peak between 3:00pm and 4:00pm. The stigmas are creamy white and protrude. Stigmas remain receptive for about 36 h. Pollination and fertilization are completed within 3–6 days of anthesis. Pollination is of an anemophilous type. Seed is covered by a thin pericarp and perianth is fleshy and edible. Fruits are ready for harvest from

March to July. Fruits take 90–140 days from spike appearance to full maturity. The receptivity of stigma, which is around 36 h, is one of the limitations for incomplete pollination of female fruits (Azad, 1989).

Jackfruit has typical characteristics of anemophilous species. The ripe male inflorescences have a sweet scent of honey and burnt sugar which attracts small flies and beetles and these probably bring about pollination. Wind also acts as a pollinating agent. Under open pollination, the fruit set is about 75%, which can be improved by hand pollination (Samaddar, 1985). Hand pollination is carried out by brushing the male inflorescence ready for pollen shedding over the female inflorescence with stigma in receptive condition. Uneven pollination leads to mis-shaped fruits.

The fruit is a large syncarp with the biggest of all cultivated fruits measuring 30–90 × 25–50 cm the shape varies from round, clavate to oblong, with short sharp hexagonal fleshy spines, fruits at immature stages will posses green rind, at maturity, the rind colour varies from green to light yellow and in some cases it will brown. The perianths of the individual flowers become the fleshy pericarp, surrounding the seeds. Each pericarp with seed is an individual fruit, generally referred to as flake. The pericarp/flake is yellow, or yellow-white, orange, coppery red and firm (Corner, 1938). Seeds have thick gelatinous yellow covering, large, oblong, about 3 × 2 cm.

## 4 Nutritional composition

An overview of the nutritional composition of jackfruit compared to fig and mulberry is provided in Table 2. The nutritional composition of jackfruit bulbs, seeds and rind varies as shown in Table 3.

Jackfruit flakes are normally fibrous and are composed of mono-, di- and polysaccharides. Berry and Kalra (1988) reported that ripe jackfruit flakes contain 77.2% moisture, 18.9% carbohydrate, 1.9% protein, 0.1% fat, 1.1% fibre, 0.8% total mineral matter with a calorific value of 84 k cal per 100 g pulp. Selvaraj and Pal (1989) studied the biochemical changes during ripening of jackfruit and reported that organic acids, amino acids, lipid constituents, aroma characteristics, fruit firmness, starch, tannins, soluble amino acids, phospholipids and free fatty acids decreased during ripening. In contrast, sucrose, glucose, fructose, sugar-to-acid ratio, citric-to-maleic acid ratio, total lipids, total fatty acids and total sterols increased. Reported that glucose (10.94%), fructose (1.42%), xylose (0.18%), rhamnose (0.44%), arabinose (1.51%), galactose (2.47%) and three other unknown sugars were present in ripe jackfruits.

Jackfruit seeds are underutilized in both human and animal nutrition due to lack of information about their nutritional status. Carbohydrates are the main components of the seed in the form of starch. The nuts are reported to contain

**Table 2** Nutritional composition of fig, jackfruit and mulberry

| Composition | Fig | Jackfruit | Mulberry |
|---|---|---|---|
| Protein (g) | 0.75 | 1.72 | 1.44 |
| Fat (g) | 0.3 | 0.64 | 0.39 |
| Carbohydrate (g) | 19.18 | 23.4 | 9.8 |
| Fibre (g) | 2.9 | 1.5 | 1.7 |
| Total sugars (g) | - | 20.6 | - |
| Calcium (mg) | 35 | 34 | 39 |
| Magnesium (mg) | 17 | 27.0 | 18 |
| Phosphorus (mg) | | 36 | |
| Potassium (mg) | 232 | 303 | 194 |
| Sodium (mg) | 1 | 3 | 10 |
| Iron (mg) | 0.370 | 0.6 | 1.85 |
| Vitamin A (IU) | 142 | 110 | 25 |
| Thiamine (mg) | 0.06 | 0.105 | - |
| Riboflavin (mg) | 0.05 | 0.055 | 0.101 |
| Vitamin C (mg) | 2 | 13.7 | 36.4 |
| Carotene β (μg) | 8585 | 61 | 9 |
| Energy (Kj) | 74 | 95 | 43 |

*Source*: USDA National Nutrient database.

25–52% of starch (Bobbio et al., 1978) and forms a highly rigid gel. Although seeds are rich in protein (6–8%), carbohydrates, fibre, vitamins and minerals, and provide about 135 k cal/100 g of energy, they are yet to be exploited commercially.

Jackfruit rind constitutes 35–40% of the fruit bulk and is a good source of pectin and minerals. Bhatia et al. (1959) described a method for the preparation of pectin from different portions of the rind of ripe fruits. The crude protein yield of ripe and raw fruit is 1.22% and 0.47%, respectively. Reported that the inedible portion of ripe jackfruit contains 83% water, 8.23% crude fibre, 1.81% crude protein, 0.89% crude fat, 0.54% total ash and 5.68% carbohydrates.

## 5 Breeding and examples of improved varieties

The objectives of any crop improvement programme are to develop new varieties with the following characteristics:

- high yielding with regular bearing (early or late or year-round bearing);
- a higher proportion of flakes with fewer seeds and less gum;

**Table 3** Composition of jackfruit (on the basis of fresh weight) per 100 g edible portion

| Composition | Young fruit | Ripe fruit | Seed |
|---|---|---|---|
| Water (g) | 76.2-85.2 | 72.0-94.0 | 1.0-64.5 |
| Protein (g) | 2.0-2.6 | 1.2-1.9 | 6.6-7.04 |
| Fat (g) | 0.1-0.6 | 0.1-0.4 | 0.40-0.43 |
| Carbohydrate (g) | 9.4-11.5 | 16.0-25.4 | 25.8-38.4 |
| Fibre (g) | 2.6-3.6 | 1.0-1.5 | 1.0-1.5 |
| Total sugars (g) | - | 20.6 | - |
| Total minerals (g) | 0.9-0.8 | 7-0.9 | 0.9-1.2 |
| Calcium (mg) | 30.0-73.2 | 20.0-37.0 | 50.0 |
| Magnesium (mg) | - | 27.0 | 54.0 |
| Phosphorus (mg) | 20.0-57.2 | 38.0-41.0 | 38.0-97.0 |
| Potassium (mg) | 287-323 | 191-407 | 246 |
| Sodium (mg) | 3.0-35.0 | 2.0-41.0 | 63.2 |
| Iron (mg) | 0.4-1.9 | 0.5-1.1 | 1.5 |
| Vitamin A (IU) | 30 | 175-540 | 10-17 |
| Thiamine (mg) | 0.05-0.15 | 0.03-0.09 | 0.25 |
| Riboflavin (mg) | 0.05-0.2 | 0.05-0.4 | 0.11-0.3 |
| Vitamin C (mg) | 12.0-14.0 | 7.0-10.0 | 11.0 |
| Energy (Kj) | 50-210 | 88-410 | 133-139 |

*Sources*: Narasimham (1990), Soepadmo (1992), Gunasena et al. (1996) and Azad (2000).

- medium-size fruits with a smooth surface and a thin layer of rind;
- orange or red-coloured flakes;
- good taste, crispness with low fibre; and
- resistance to biotic and abiotic stresses.

Detailed requirements for fruit for different end uses are summarized in Table 4.

Jackfruit breeding faces a number of challenges including:

- A long generation cycle;
- A long juvenile period;
- Limited opportunity for recombination; and
- A highly heterozygous character, requiring a large population for selection.

These challenges have limited breeding programmes to date.

There is currently little information on the heritability of desirable traits such as the number of fruits per tree, the size of fruit and the thickness of rind. There is a need for basic research into the mode of inheritance of important characteristics as well as identification of genes for traits such as seedlessness, dwarfing, resistance to biotic and abiotic stresses such as fruit fly, fruit rot or

**Table 4** Selection characteristics of fruits for specific purposes

| | Characteristics to be considered |
|---|---|
| *A. Fresh fruit purpose* | |
| Yield | 120–200 fruits per tree per year |
| Fruit size | Average fruit weight (small to medium) |
| Flake quality | Orange colour, good taste and firm texture |
| Number of flakes/kg fruit | <20 |
| Flakes/fruit ratio | >0.50 |
| Total soluble solids | >25°Brix |
| Fruit rind thickness | <1.0 cm |
| Flake thickness | >0.50 cm |
| Latex exudation | Nil/low |
| Fruit shape | Round to oval |
| Bearing habit and season | Twice or all season |
| Number of seeds per kg fruit | <120 g |
| Vivipary | Absent |
| Shelf life (days) | >3 days in RT |
| *B. Cooking purpose* | |
| Fruit maturity period | 3–4 weeks old |
| Cooking quality | Should not become bitter |
| Tenderness | Tender |
| Flake fibre content | Less |
| Fruit rind thickness | <1 cm |
| Browning | Less phenol content |
| Latex exudation | Nil or low |
| Seedlessness/seed softness | Good |
| Shelf life (days) | 3 days after minimal processing |
| Bearing habit and season | Twice/All round |
| *C. Processing purpose* | |
| Dry matter content | High |
| Flake fibre content | Low |
| Flake quality | Dark yellow to orange colour |
| Carotenoid content | High |
| Total soluble solids | >15–30°Brix |
| Flakes/fruit ratio | >0.50 |
| Fruit size | >15 kg |
| Latex exudation | Nil or low |
| Fruit rind thickness | <1 cm |
| Flake thickness | >0.2 mm |
| Shelf life (days) storability of fruits | >3 days in RT |

soil salinity. Recent research has begun to identify morphological and genetic variation as a basis for breeding (Azad et al., 2007; Shyamalamma et al., 2008; Haq, 2011; Akter and Rahman, 2017).

Limited work has been done on improving the jackfruit through crossing (hybridization). As wide genetic variability exists in the crop, there is a need to exploit heterosis for traits such as improvement in yield and fruit quality. Interspecific hybridization also deserves attention in order to identify species that are compatible with jackfruit and which are the sources of useful traits. An example of the potential of hybridization is the programme at the Indian government Fruit Research Station in Kallar (in Tamil Nadu State) using the Singapore jack variety as female and Velipala variety as male. Twelve hybrid seedlings were planted. After 10 years, plants produced around 30 fruits with an average weight of fruits per tree of 43.3 kg. The hybrids resembled the mother plant. The fruits of hybrids are tastier with bigger carpels than either of the parents (Veeraraghavadattam et al., 1996).

Since grafting is one of the principal methods of propagation, research has focussed on optimizing the technique (Zaher, 2008; Selvi et al., 2008; Sidhu, 2012; Priya et al., 2014). This has made it possible to identify good-quality grafts, with early-bearing characteristics, and bearing twice or throughout the year.

## 5.1 Examples of improved varieties

### 5.1.1 Palur 1 Jack or PLR, 1

Tamil Nadu Agricultural University (TNAU) in India developed this improved variety (Fig. 3) in 1992. It is a clonal selection from Panikkankuppam village of Panruti taluk, in the South Arcot district of Tamil Nadu. It is high yielding with medium height, less spreading and suitable for high-density planting. In addition to regular bearing season (March–June), it also produces fruits off season (October–December). The annual yield per tree is about 80 fruits weighing around 900 kg. The average fruit weight is 12 kg, containing 115–120 flakes. The off-season bearing is a unique trait of this variety and the fruits fetch a premium market price. The fully ripe fruits have a flat stigmatic surface instead of a spiny surface. The fruit quality is good with attractive golden yellow firm flakes. The flakes are very sweet with high consumer appeal and good keeping quality. The seeds are also edible with high palatability. TSS is 19°brix.

### 5.1.2 Pechiparai - I or PPI 1 Jack

TNAU developed this improved variety (Fig. 4) in 1996. It is a clonal selection from Ulagumoodu Local near Pechiparai and produces high-quality crisp carpels with high TSS and ascorbic acid content. The flakes are sweet and tasty with a

**Figure 3** PLR 1 variety of jackfruit.

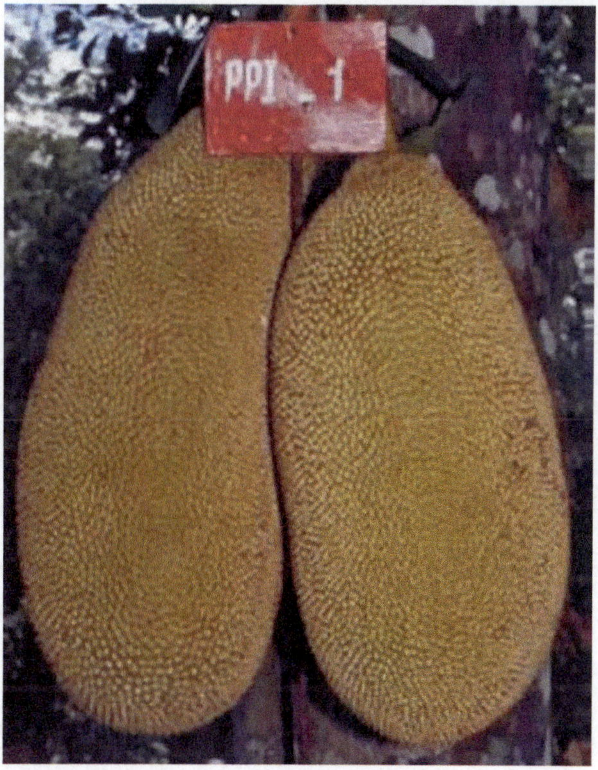


**Figure 4** PPI 1 variety of jackfruit.

pleasant aroma. The maximum bearing takes place on tree trunks. This variety is suitable for both commercial planting and planting in domestic gardens.

### 5.1.3 PalurJack 2 or PLR Jack - 2

TNAU developed this improved variety (Fig. 5) during 2007. It is a clonal selection from Pathirakkotai Local. It has bigger sized fruits and highly palatable and edible flakes, which attract higher prices due to higher keeping and other quality characteristics. The variety is less susceptible to major insect pests and diseases. It can be planted during June–December and is suitable for growing in tropical climates with deep, well-drained loam soils. This variety yields 95–110 fruits/tree/year weighing an average of 1600–1950 kg.

### 5.1.4 Swarna Halasu

Swarna Halasu (Fig. 6), released from UAS Bangalore, is a clonal selection made from a mother plant near the UAS Hebbal Quarters, No. B-15. It is unique with respect to the following characteristics:

- Fruits are borne in clusters on the trunk, primary branches and also on the secondary and tertiary branches.
- Fruits are round and medium-sized (6–8 kg/fruit), with thin rind (<1 cm) suitable for small families and kitchen gardens.
- Pulp is dark yellow, hence the name Swarna, thick and juicy with a TSS of 23–25°Brix, with a pulp recovery of 660 g/kg fruit.
- Thick bulbs are suitable for table use as well as processing into chips, papad and squash.

**Figure 5** PLR (J) 2 variety of jackfruit.

**Figure 6** Swarna variety of jackfruit: (a) fruiting tree, (b)fruit and flakes.

- Fruits are harvested between April and July.
- **Lalbaugh Madhura:** It is an early bearing, high yielding Jackfruit variety released from UAS, Bengaluru and State Dept. of Horticulture, Govt. of Karnataka during 2018. The fruits are big in size, weighing an average of 6-18 kg. The flakes are bigger, weighing an average 75–100 g, thus 9–10 bulbs will weigh one kg. The flake TSS content ranges from 29–30°Brix. The fruits have high flake to fruit ratio of 0.55. Fruits are suitable for table purpose as well as for value addition into Jam, Squash, Pulp, Halwa and Chips. The fruit yield is high, 10–15 years old trees produces 80–100 fruits and total yield per tree would be 650–800 kg/tree/year. Flakes are crisp, thick and good for table purpose and value addition into squash, jam, juice, halwa and so on.

### *5.1.5 Gumless type (pre-released)*

The gumless type (Fig. 7), developed at the Indian Institute of Horticultural Research (IIHR) in Bangalore, has medium-to-low latex with medium fruits (6.4–9.0 kg) with an oblong fruit shape and containing 120–140 flakes/fruit with a flakes:fruit ratio of 0.45–0.50. The flakes are light yellow, sweet-tasting with TSS of 26–30°Brix. The variety will soon be released.

### *5.1.6 Konkan Prolific*

Konkan Prolific (Fig. 8) was released in 2004 by the Regional Fruit Research Station (RFRS) in Dapoli (Maharashtra). It is a seedling selection with a potential yield of 450–550 kg/tree and an average yield of 420.56 kg/tree, producing

**Figure 7** Gumless type jackfruit.

**Figure 8** Konkan Prolific Jackfruit variety.

73 fruits/tree/year. Individual fruits have a weight of 5.70 kg, with a fruit length of 34.4 cm, and a fruit breadth of 15.2 cm. The fruit is medium sized with an oblong shape, firm flesh even in the monsoon season, golden yellow flesh. Fruit bearing is in bunches, with a spiny surface, Flowering is in January–February, bearing after 6–7 years of planting. Fruits have TSS of 25°Brix and seed weight of 132 g.

## *5.2 Genotypes identified by the University of Agricultural Sciences (UAS) in Bangalore (Karnataka)*

Genotypes are classified as:

(A) Seasonal bearing genotypes (March–July);
(B) Twice bearing genotypes;

(C) Off-season bearing genotypes (August–February);
(D) All-season bearing genotypes.

Examples of seasonal bearing genotypes (A) include the following.

### 5.2.1 Janagere

The jackfruit genotype Janagere (Fig. 9) has been identified from a farmer's field near Janagere, Magadi Hobli, by UAS Bangalore. It is unique with respect to the following characteristics:

- The tree is more than 200 years old.
- Around 250–300 fruits are borne in bunches, mostly on trunk and primary branches. Fruits are ellipsoid and medium sized (8–10 kg/fruit), with rind thickness around 1 cm.
- Pulp is yellow, thick and firm with a TSS of 23 to 24°Brix.
- Thick bulbs are suitable for domestic consumption, with pulp content of 560 g/kg fruit.
- The bulbs are firm, free from fibres and easily digestible.
- Fruits are harvested between March and July.

### 5.2.2 Kachahalli Roadside (Kempusiddappa)

Kachahalli Roadside (Kempusiddappa) (Fig. 10) is an elite genotype from Kachahalli village, Tubgere Hobli, Doddaballapura Tq. It is unique with respect to the following characteristics:

- Fruits are borne on the trunk and primary branches.
- Fruits are oblong and large sized (12–15 kg/fruit), with a thick rind (1–2 cm) suitable for commercial cultivation for fresh fruit.

**Figure 9** Janagere Jackfruit tree.

**Figure 10** Kachahalli Roadside Jackfruit tree.

- The pulp is dark orange in colour, thick and firm. Each flake weighs 60-65 g, with a TSS of 21-28°Brix.
- The unique taste and thick bulbs make it more suitable for fresh consumption. The pulp recovery per kg fruit is around 600 grams.
- Fruitsare harvested between April and June.

### 5.2.3 Kachahalli Table type 12

Kachahalli Table type 12 (Fig. 11) is an elite genotype from Kachahalli village, Tubgere Hobli, Doddaballapura Tq. It is unique with respect to the following characteristics:

**Figure 11** Kachahalli Table type 12.

- Fruits are borne on trunk and primary branches, with a yield of 50–80 fruits/tree.
- Fruits are ellipsoid and medium sized (10–15 kg/fruit), with a thicker rind of 2 cm, suitable for commercial cultivation for fresh fruit, with 490 g pulp/kg fruit.
- The pulp is reddish orange in colour, thick and juicy with a TSS of 25–27°Brix.
- Thick bulbs are suitable for the table as well as processing into squash.
- Fruits are harvested between March and June.

### 5.2.4 National Seed Project (NSP)

The National Seed Project (NSP) jackfruit (Fig. 12) genotype is an elite selection, made from the trees near the National Seed Project (NSP) at UAS (Bangalore). It is unique with respect to the following characteristics:

- Fruits are borne on trunk and primary branches.
- Fruits are oblong, very large (>25 to 27 kg/fruit) with rind thickness of 1.7–1.8 cm.
- The tree bears an average of 70–80 fruits/tree/year.
- Pulp is yellow, medium thickness and juicy with a TSS of 23–25°Brix.

**Figure 12** National Seed Project (NSP).

- Medium-size bulbs are suitable for table use and processing into chips and papad.
- Fruits are harvested between April and July.

An example of twice bearing genotypes (B) is the following.

### *5.2.5 Byrachandra*

The jackfruit (Fig. 13) genotype Byrachandra has been identified by Dr. S. V. Hittalmani, Former Additional Director of Horticulture, Lalbaugh, Bangalore, from a local selection near Byrapattana, Ramanagara district. Hort farm. It is unique with respect to the following characteristics:

- Fruits are borne on the trunk and primary branches.
- Fruits are round and large sized (>10 kg/fruit), with thin rind of <1 cm.
- Pulp is orange red in colour, hence the name Chandra, thick and juicy with a TSS of 25–26°Brix.
- Medium-size bulbs each weigh 43–45 g, with a pulp recovery of 650 g/kg fruit.
- Fruit is very suitable for table use and processing into squash.
- Twice bearing type with an average yield of 80–100 fruits/tree/year.

Examples of off-season bearing genotypes (C) include the following:

**Figure 13** Byrachandra.

### 5.2.6 Veerappa Surenahalli

The Veerappa Surenahalli (Fig. 14) genotype is from a farmer in Surenahalli village, Tovinkere, Tumkur district. It is unique with respect to the following characteristics:

- Fruits are borne on the trunk and primary branches.
- Fruits are ellipsoid and small sized (<5 kg/fruit), with rind thickness of 1.2–1.4 cm.
- Pulp is orange red to golden yellow in colour, thick and not juicy with a less TSS of 16–18°Brix.
- Large size bulbs, each weighing 54–58 g, with a pulp recovery of 400 g/kg fruit, highly suitable for table use and processing into value-added products as it is an off-season bearing genotype.

### 5.2.7 Muniraju Ramachandra Hosur

The Muniraju Ramachandra Hosur (Fig. 15) genotype is from a farm in Ramachandra Hosur, Chikkaballapura district. It is unique with respect to the following characteristics:

- Fruits are borne on the trunk and primary, secondary and tertiary branches, with a yield of 120–150 fruits/tree.
- Fruits are ellipsoid and small sized (<5 kg/fruit), withrind thickness of <2 cm, suitable for small families and fresh fruit use, with pulp content of 400 g pulp/kg fruit.

**Figure 14** Veerappa Surenahalli.

**Figure 15** Muniraju Ramachandra Hosur.

- Pulp is light orange in colour, thick and less juicy with a TSS of 20–22°Brix.
- Big, thicker bulbs are suitable for table use as well as preparation of chips.
- Fruits are harvested between August and February.

## 6 Conclusion

Jackfruit has been considered as the 'rice' tree in Sri Lanka, underlining the importance of the fruit as a staple food. As a result, the government has banned felling of jackfruit trees without official permission. In India, shortages of inputs such as water and labour are encouraging many farmers to shift to cultivation of perennial horticulture crops such as mango and jackfruit which are better able to deal with abiotic stresses such as more unpredictable rainfall patterns. There is a huge demand for jackfruit flakes with good quality and colour in the Southern states of India while, in Northern India, many people eat tender jackfruit given its nutritional and wider health benefits. The availability of good-quality grafts, with early-bearing characteristics, and bearing twice or throughout the year, means that farmers are now adopting jackfruit cultivation on a large scale in Southern India. This has increased the opportunities to invest in breeding programmes to produce new varieties for a range of end uses.

## 7 Where to look for further information

For further detailed information on jackfruit research and plant material availability, the interested persons can contact the faculty working in two universities as indicated below.

1 Dr. Shyamalamma, S. Prof. of Horticulture, Dept. of Biotechnology, UAS, GKVK, Bengaluru-560605, Karnataka, India. Email: shyamala_reddyrs@yahoo.co.in
2 Dr. G. S. K. Swamy. Prof. of Horticulture, Dept. of Fruit Science, College of Horticulture, Mysore, Karnataka, India. Email: swamy.gsk@uhsbagalkot.edu.in.

## 8 References

APAARI. 2012. *Jackfruit Improvement in the Asia-Pacific Region – A Status Report*. Asia-Pacific Association of Agricultural Research Institutions, Bangkok, Thailand.

Akter, A. and Rahman, H. 2017. Evaluation of jackfruit (*Artocarpus heterophyllus* Lam.) germplasm. *RRJoB*, 7(1): 38–53.

Altendorf, S. 2018. *Minor Tropical Fruits: Mainstreaming a Niche Market*. FAO, Rome, Italy.

Azad, A. K. 1989. Studies on floral biology, pattern of fruitset and quality of fruits borne in different section, of jackfruit plants. MS Thesis, Bangladesh Agricultural University, Mymensigh, Bangladesh.

Azad, A. K. 2000. Genetic diversity of jackfruit in Bangladesh and development of propagation methods. PhD Thesis, University of Southampton, United Kingdom.

Azad, A. K., Jones, J. and Haq, N. 2007. Assessing morphological and isozyme variation of jackfruit (*Artocarpus heterophyllus* Lam.) in Bangladesh. *Agroforest. Syst.*, 71: 109–25.

Bailey, L. H. 1942. *The Standard Encyclopedia of Horticulture*. The Macmillan Co., New York, pp. 401–2.

Berry, S. K. and Kalra, C. L. 1988. Chemistry and technology of jack fruit (*Artocarpus heterophyllus* Lam.) – a review. *Indian Food Packer*, 8: 63–75

Bhatia, B. S., Siddappa, G. S. and Lal, G. 1959. Preparation of pectin extract and syrup from Jackfruit rind. *Indian J. Agril. Res.*, 29: 75.

Bobbio, F. O., El-Dash, A. A., Bobbio, P. A. and Rodrigues, L. R. 1978. Isolation and characterization of the physicochemical properties of the starch of jackfruit seeds (*Artocarpus heterorphyllus*). *Cereal Chem.*, 55: 505–511.

Corner, E. J. H. 1938. Notes on the systematics and distribution of Malayan phanerogams II. The jack and the chempedak. *Gardener. Bull.*, 19: 56–81.

Darlington, C. D. and Wylie, A. P. 1956. *Chromosome Atlas of Flowering Plants*. George Allen and Unwin Ltd, London, 184p.

Devalaraja, S., Jain, S. and Yadav, H. 2011. Exotic fruits as therapeutic complements for diabetes, obesity and metabolic syndrome. *Food Res. Int.*, 44(7): 1856–65.

Dutta, S. 1956. Cultivation of Jackfruit in Assam. *Indian J. Hortic.*, 13(4): 189–97.

Gunasena, H. P. M., Ariyadasa, K. P., Wikramasinghe, A., Herath, H. M. W., Wikramasinghe, P. and Rajakaruna, S. B. 1996. *Manual of Jack Cultivation in Sri Lanka*. Forest Information Service, Forest Department, 48p.

Haq, N. 2011. The jackfruit. In: Thottappily, G., et al. (Eds), *Genetic Resources and Crop Improvement*. Studium Press LLC, Houston, TX.

Jarret, F. M. 1959. Studies in *Artocarpus* and allied genera, I. General considerations. *J. Arnold Arbor.*, 40(1): 1–29.

Kanzaki, S., Yonemori, K., Sugiura, A. and Subhadrabandu, S. 1997. Estimation of phylogenetic relationship among 11 *Artocarpus* species. *Sci. Hortic.*, 70: 57–66.

Manjunatha, B. L. 1948. *Wealth of India* (vol. 1). CSIR Publications, New Delhi, pp. 1–240.

Morton, J. F. 1965. The jackfruit (*Artocarpus heterophyllus*). Its culture, varieties and utilization. *Proceedings of the Florida State Horticultural Science Society*, 78: 336–44.

Narasimham, P., 1990. Bread fruit and Jackfruit. In: Nagy, S., Shaw, P. E. and Wordowski, W. P. (Eds), *Fruits of Tropical and Sub-Tropical Origin: Composition, Properties and Uses*. Florida Science Source Inc, pp. 193–259.

Priya, D. S., Sunetra, T., Mathala, J. G., Thangam, M. and Singh, N. P. 2014. *A Guide to Jackfruit: Cultivation and Value Addition*. Technical bulletin No. 41. Indian Council for Agricultural Research (ICAR), Goa, India.

Ramli, R., Azmi, A., Johari, N. and Noor, S. 2017. Minimally-processed jackfruit: opportunities for the foodservice industry. *J. Food Res.*, 9(1): 1–7.

Reddy, M. and Sasikala, P. 2016. Partial characterization of lectin from *Artocatpus* sps. *Res. J. Pharm., Biol. Chem. Sci.*, 7(2): 1663–70.

Sahadevan, P. C., Abdul, S. A. and Menon, K. C. P. 1950. Floral abnormalities in Jack. *Madras Agric. J.*, 37: 497–8.

Samaddar, H. N. 1985. Jackfruit. In: Bose, T. K. (Ed.), *Fruits of India: Tropical and Subtropical*. Naya Prokash, Calcutta, pp. 487–97.

Samaddar, H. N. and Yadav, P. S. 1982. Flowering, floral biology, pollination, fruit set and fruit growth of jack fruit. *Hortic. J.*, 1: 54–60.

Schnell, R. J., Olano, C. T., Campbell, R. J. and Brown, J. S. 2001. AFLP analysis of genetic diversity within a jack fruit germplasm collection. *Sci. Hortic.*, 91: 261–74.

Selvaraj, Y. and Pal, D. K. 1989. Biochemical changes during the ripening of jackfruit (*Artocarpus heterophyllus* L.). *J. Food Sci. Tech.* 26(6): 304–7.

Selvi, R., Kumar, N., Selvarajan, M. and Anbu, S. 2008. Effect of environment on grafting success in jackfruit. *Indian. J. Hortic.*, 65: 341–3.

Shyamalamma, S., Hegde, M., Chandra, B. C. and Naryanswamy, P. 2008. Evaluation of genetic diversity in jackfruit (*Artocarpus heterophyllus* Lam.) based on amplified fragment length polymorphism markers. *Genet. Mol. Res.*, 7(3): 645–56.

Sidhu, A. 2012. *Jackfruit Improvement in the Asia-Pacific Region: A Status Report*. Asia-Pacific Association of Agricultural Research Institutions, Bangkok, Thailand.

Soepadmo, E. 1992. *Artocarpus heterophyllus* Lamk. In: Verheij, E. W. and Coronel, R. E. (Eds), *Plant Resources of Southeast Asia, No. 2: Edible Fruits and Nuts*. PROSEA, Wagengingen, The Netherlands, pp. 86–91.

Veeraraghavadattam, D., Jeeva, S. and Rabindran, R. 1996. *Scientific Fruit Culture*. SuriAssociates, Coimbatore, p. 129.

Zaher, A. 2008. Using grafting for propagation of jackfruit and producing rootstocks for grafting. *Am-Eurasian J. Agric. Environ. Sci.*, 3(3): 459–73.

Zerega, N. J. C., Nur Supardi, M. N. and Motley, T. J. 2010. Phylogeny and recircumscription of artocarpeae (Moraceae) with a focus on *Artocarpus*. *Syst. Bot.*, 35(4): 766–82.

Zerega, N. J. C., Ravikanth, G., Gardner, E. M., Laricchia, K., Melhem, T., Wang, M., Witherup, C. E., Setty, S., Zuberi, M. I., Jagdish, M. R. and Ragone, D. 2014. New genomic resources to understand genetic diversity and promote crop improvement of jackfruit (*Artocarpus heterophyllus*), breadfruit (*A. altilis*), and other under utilized *Artocarpus* crops. *Procedings of the International Symposium on Jackfruit and Breadfruit of the Tropics*, Bengaluru, Karnataka, India.

# Chapter 16

# Achieving sustainable cultivation of litchi

*Xuming Huang, South China Agricultural University, China*



## 1 Introduction

Litchi (*Litchi chinenesis* Sonn.) (also known as lychee) cultivation is limited to a very narrow climate range, because the crop is susceptible to freezing damage but requires chilling temperatures for flower induction. It is commercially grown in around 20 countries in two narrow belts in the northern and southern hemispheres characterized by the right subtropical climate for successful cultivation (Fig. 1). The important producers are China (590 000 ha, including 11 000 ha in Taiwan), India (84 000 ha), Vietnam (49 000 ha), Thailand (20 000 ha) and Madagascar (13 000 ha). With around 70% of the total area of litchi production, China produces over 2 500 000 tons (the country's fifth largest fruit in terms of cultivation area), while Taiwan produces around 130 000 tons and India 700 000 tons. The smaller producers include Mexico, Brazil and the United States in the Americas, Bangladesh, Nepal and Pakistan in the Indian sub-continent as well as countries such as South Africa, Australia, Reunion Island and Israel.

Litchi production faces a number of challenges. The production is strongly influenced by climate. The global warming increases the tendency for warm winters, threatening flower differentiation and thus productivity (Kumar, 2014). The extreme climate conditions lead to drought and heat stresses that suppress fruit development and cause fruit drop, cracking and sunburn.

Marketability of litchi fruit is strongly limited by its perishability, as the fruit has a very short shelf life at ambient temperatures (Huang et al., 2005). It must be harvested and marketed within 7–10 days from fruit reaching maturity. Overripe fruit produces off-flavors due to loss of sugars and accumulation of alcohol and

http://dx.doi.org/10.19103/AS.2019.0054.13

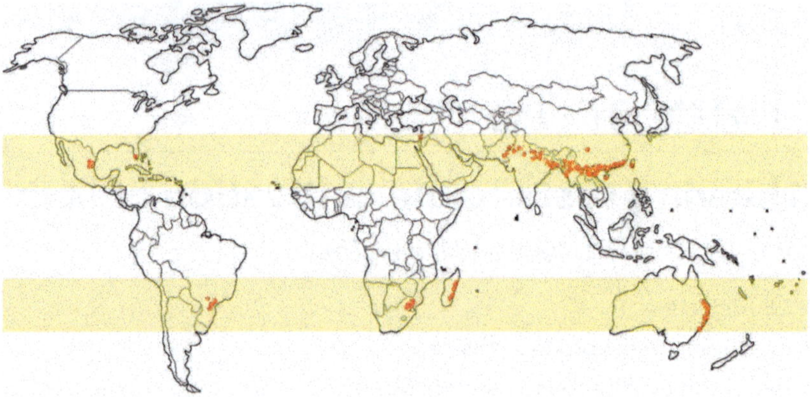

**Figure 1** Global distribution of litchi production.

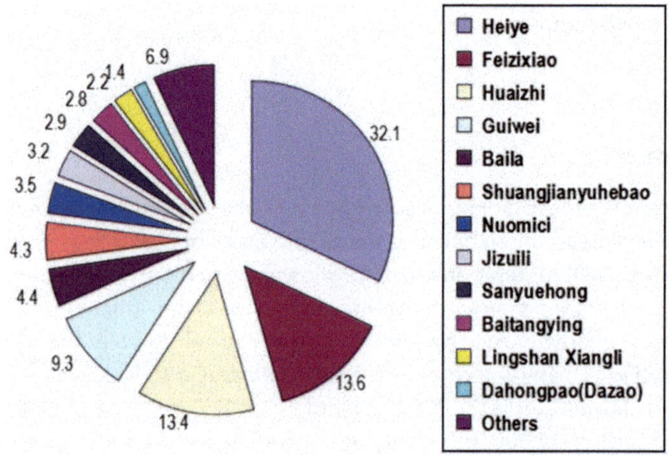


**Figure 2** Key cultivars in the litchi industry in China.

acetaldehyde in the flesh and thus are less marketable (Pesis et al., 2002). In most countries, the litchi industry is also based on narrow cultivar diversity. For instance, the 'Mauritius' ('Dazao' or 'Tia So') variety dominates the litchi industry in southern African countries, Brazil, the United States, Mexico and Israel. Even in China, with over 300 recorded varieties, the litchi industry depends mainly upon a few cultivars, with 'Heiye', 'Huaizhi' and 'Feizixiao' accounting for two-thirds of the total production (Fig. 2). This narrow cultivar diversity, which limits the growing season, as well as a limited period to pick mature fruit, results in a short harvest season. Seasonal overproduction relative to market demand results in significantly lower prices, which affects the profitability of production. Profitability is also influenced by variations in fruit quality as well as high labor and other costs of production.

To improve litchi production, research needs to address issues such as increasing the range of cultivars, particularly to extend the production season, as well as to improve yields and quality. It also needs to ensure that cultivation is more efficient and, therefore, more profitable and sustainable.

## 2 Extension of the production season

The price of fresh litchi is highest at the beginning and toward the end of the season. Higher and more stable prices can be achieved by extending the season to reduce the risk of oversupply. Litchi is a seasonal flowering crop, flowering in response to winter chill and blooming in spring. The extension of the production season can be achieved by developing earlier or later season cultivars and by geographic expansion of litchi cultivation.

In Guangdong, the largest litchi-producing province in China, the litchi season used to start with the 'Sanyuhong' variety in May and end with 'Huaizhi' in late July. The season has been advanced by half a month with hybrid lines such as 9918, 9919, D11 and D13, which mature earlier than 'Sanyuehong' (Fig. 3) (Ouyang et al., 2009). The season has also been extended by another half month with the introduction of the late season cultivar 'Maguili', a chance seedling selection, which matures in mid-August in Gaozhou in west Guangdong (Ouyang et al., 2002). With these new varieties, litchi production season has been extended to three months from late May to late August in Guangzhou (Fig. 4).

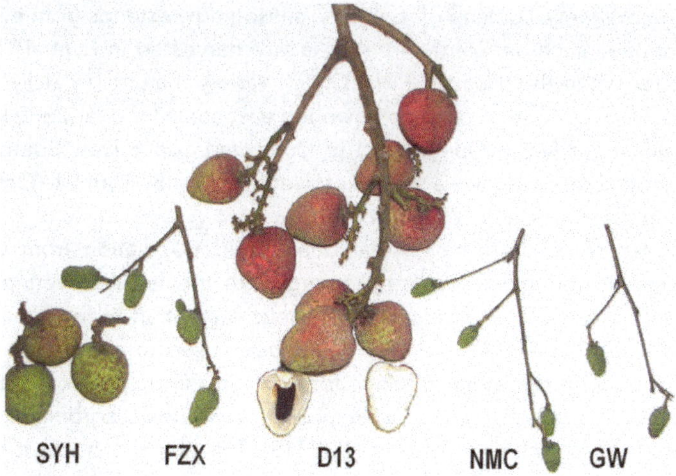


**Figure 3** Comparison of the fruit of an early line D13 with those of referenced litchi varieties. Photo taken on May 2, 2005. SYH=Sanyuehong; FZX=Feizixiao; C=Nuomici; GW=Guiwei.

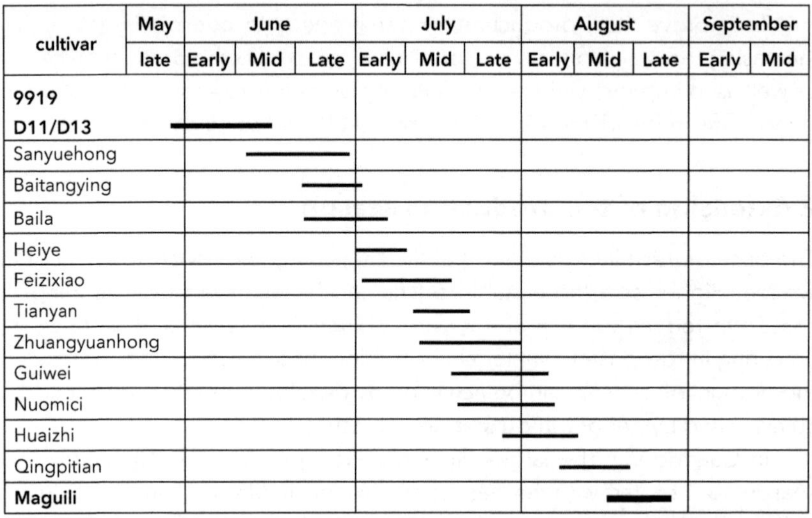


**Figure 4** Extension of litchi production season with different cultivars in central Guangdong, China.

The production season can be further extended by expanding into new geographical areas where seasonal patterns differ. In China, the development of litchi production in the southernmost province of Hainan and in Sichuan province further north, together with the use of new cultivars, has made it possible to extend the litchi season to 6 months from mid-April in Lingshui (19°N) using the 'Sanyuehong' variety to early September in Luzhou (29°N) with the 'Maguili' variety. Looking globally, the season now extends from mid-April, when varieties such as 'Kom' and 'Kaloka' are harvested in central Thailand, to October when the 'Huaizhi' ('Wai Chee') variety matures in Sicily, Italy. In Australia, litchi cultivation extends down the east coast for over 2500 km from Cooktown in Far North Queensland to Coffs Harbour in New South Wales, allowing the country to have a season of over five months from late October to late March (Diczbalis et al., 2010).

Chinese researchers have created a hybrid population from crossing the earliest-flowering variety 'Sanyuehong' with the latest-flowering variety 'Maguili' as a basis to understand the genetic regulation of litchi maturation (Zhao et al., 2010). The F1 population has been used to construct a genetic linkage map with molecular markers for flowering and other key traits. This research has made it possible to identify genetic groups according to early or late-maturing characteristics (Liu et al., 2015). This has also made it possible to identify genotypes with, for example, extremely early maturing (EEM) characteristics (Liu et al., 2018). A series of new lines has been generated from the crosses for further breeding.

## 3 Improving cultivar quality and yield

Quality characteristics of litchi fruit include larger fruit size, aborted seed (which results in higher flesh content), small size seed, good texture and taste. Currently most early cultivars, especially 'Sanyuehong' and its close varieties such as D13 and D11, are of below-average quality, with low sugar/acid ratio and a tough texture. The late cultivar 'Maguili' is also of average quality because of large seed size and soft flesh. The breeders in China have been trying to cross these cultivars with high-quality varieties with the aim of breeding early and late cultivars with better quality (Zhang and Liu, 2015). The current higher-quality varieties also have drawbacks. In China, for example, the small-seeded 'Nuomci' and 'Guiwei' varieties are high-quality, premium-priced varieties but are erratic in bearing (thus requiring more careful cultivation), have a relatively high chilling requirement for flowering, which limits their area of cultivation, and are also more susceptible to diseases. 'Nuomici' is also particularly susceptible to excessive fruit drop and cracking.

These limitations have led to the development of new varieties, which combine quality with greater reliability in cultivation as well as allowing a longer season. A number of these improved varieties are listed in Table 1. For example, 'Nuomici' and 'Guiwei' are being replaced by 'the second generation of "Nuomici"' including 'Lingfengnuo', 'Jingganghongnuo' and 'Xianjinfeng' with comparable quality but better reliability. In South Africa, 'Early Delight' is replacing 'Mariutius' and 'Sanyuehong' due to earliness in flowering and better quality. In Australia, B3 (Kwai May Pink) has gained popularity among growers due to its reliability, wide adaptability and superior quality to 'Tai So' which used to be the dominant cultivar. In the future, conventional breeding approaches will be supplemented by molecular techniques which identify genetic markers related to traits such as fruit size, coloration and shelf life (Lai et al., 2015; Kumar, 2014; Mir, 2018).

## 4 Improving tree management

The growth of a bearing tree in a year can be divided into vegetative growth and reproductive growth (Fig. 5). Since the vegetative flush growth is strongly suppressed by the terminal panicle and fruit, the vegetative growth of litchi generally does not occur until the fruit is harvested. Litchi is a typical recurrent plant with cyclic flush growth (Fu et al., 2014). The flush growth starts with terminal bud break, followed by outgrowth of the rudimentary leaves and shoot elongation. With the expansion of the new leaves, the terminal bud returns to rest status and stays dormant until the new leaves harden and turn green (Zhang et al., 2016). If warm temperatures are maintained, litchi undergoes this vegetative growth cycle continuously. However, as Figure 5 indicates, chilling

**Table 1** New litchi cultivars recently released and their major characteristics

| Cultivar | Origin | Major characters | Reference |
| --- | --- | --- | --- |
| Jiganghongnuo | China, chance seedling selection, possible hybrid of Huaizhi and Nuomici | Small seeded with high flesh recovery (~80%), resistant to cracking and diseases, high productivity, good storability, large fruit size (up to 39 g), late season | Ye et al. (2011) |
| Lingfengnuo | China, chance seedling selection, possible hybrid of Huaizhi and Nuomici | Small seeded with high flesh recovery (~80%), high TSS/acid ratio, resistant to cracking and diseases, long harvest period (up to 20 days) due to development of sweetness and flavor well before the fruit mature, high productivity, good storability, late season; medium fruit size (average 22 g) | Fan et al. (2010) |
| Xianjingfeng | China, chance seedling selection | Small seeded with high flesh recovery (~80%), resistant to cracking, high productivity, late season; medium fruit size (average 25 g) | Qiu et al. (2010) |
| Fengshanhongdenglong | China, chance seedling selection | Small seeded with high flesh recovery (~80%), resistant to cracking, high productivity, late season; medium fruit size (average 25 g) | Ou et al. (2012) |
| B3 (Kwai May Pink) | Australia, chance seedling selection from 'Guiwei' | High productivity, good return flower after heavy pruning, harvest period due to the development of acceptable sweetness and flavor well before the fruit mature, mid-season. Medium fruit size (~20 g), mostly large seeded with a flesh recovery of 70%, mid-season, currently the No.1 cultivar in Australia | Menzel et al. (2005) |
| R1G22 or Early delight | South Africa, selection from open pollinated 'Mauritius' seedlings | Early season, high productivity, good color and size (~26 g), good flesh recovery (70%) | Fronman et al. (2014) |
| DB5-27 and DB17-70 | Israel | Good quality | |

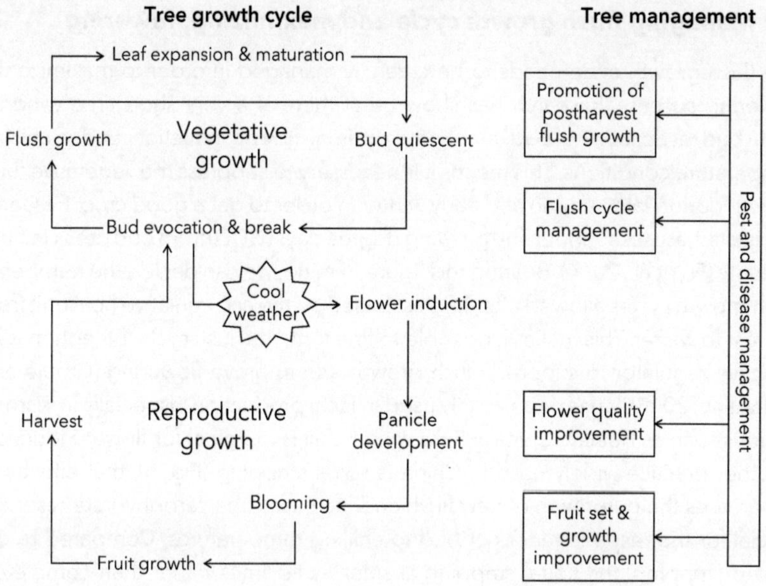


**Figure 5** Tree growth development cycle of litchi and tree management at different developmental phases.

temperatures in winter induce buds to exit this cycle and initiate flowering and the process of reproductive growth, which ends with the maturation of fruit.

During the phase of vegetative growth after harvest, major tree management operations include promotion of postharvest flush growth to foster strong bearing for the next season and regulation of the flush cycle to maximize flower induction during the winter. During the reproductive growth phase, major tree management operations include improvement of flower quality and promotion of fruit set and development.

### *4.1 Postharvest tree recovery*

Fruit harvesting also involves heading i.e. removing the terminal fruit and releasing the axillary buds from para-dormancy. Early harvesting thus results in early flush growth and early tree recovery. A heavy fruit load may delay and weaken the postharvest flush growth (Yuan et al., 2009). The flush growth requires significant amounts of water and nutrients. Irrigation and fertilization are therefore the key practices in encouraging flush growth. The application of fast-acting fertilizer is generally done prior to the outgrowth of flush after harvest, while organic fertilizers can be applied earlier before harvest. It is recommended that about 50% of the annual nitrogen (N) and 25% of the potassium (K) and phosphorous (P) requirements are applied at this point (Zhang et al., 2010).

## 4.2 Managing flush growth cycle and maximizing flowering

The flush growth cycle needs to be carefully managed in order to maximize the flowering success. Research has shown that there is a very short time window, from bud evocation to bud break, for optimal flower induction under chilling temperature conditions. This means it is necessary to suppress the vegetative flush growth during late autumn and early winter in order to get a good crop. Research suggests that it takes about 560 growing degree days (GDD) from bud break to flush maturity (Fu et al., 2014). Bearing this figure in mind, we can decide the number of flush growth cycles allowed after harvest based on the accumulative heat unit from harvest to winter. This makes it possible to fine tune the flush cycle. Ethephon is an effective regulator to suppress flush growth and improve flowering (Cronje and Ratlapane, 2018). It has been widely used in litchi production, especially in warmer regions such as Hainan, China, where winter chill is marginal for flower induction. Another practice widely used in China is spiral stripping (Fig. 6), that effectively suppresses the outgrowth of new flush and increases tree carbohydrate reserves, crucial for the responsiveness of bud to chilling temperatures. Compared to the closed stripping, the spiral stripping is safer as its limits rather than completely cutting off the photosynthate supply to the roots (Zhou et al., 1998). The stripping width and spiral circle number are decided based on the tree vigor and the expected length of time for which flush growth control is needed.

**Figure 6** Spiral girdling on trunk and common tools used.

### 4.3 Improvement of fruit set

Litchi is a shy-bearing species, with a final fruit retention rate of less than 5% due to poor fruit set (Stern and Gazit, 2010). Recent research has helped to improve the understanding of the mechanisms underlying the fruit set, development and maturation (Wang et al., 2011). The problem of poor fruit set limits the use of some high-quality varieties such as 'Feizixiao', an early variety with reliable flowering, large fruit size and excellent fruit quality. Studies have shown that 'Feizixiao' produces large panicles with low-quality flowers, as reflected by low pollen vitality and poor development of the embryo sac, which lead to failure of fertilization, a precondition for fruit set (Jiang et al., 2012). In addition, blooming of the large panicle exhausts the carbohydrate reserve crucial for early fruit development (Yuan et al., 2009; Jiang et al., 2012). Therefore, both poor fertilization and carbohydrate deficiency are involved in poor fruit set. Reducing flower numbers by panicle pruning (tipping plus branch thinning) improves the quality of the remaining flowers, reduces carbohydrate reserve consumption, and thus improves fruit set (Fig. 7). Research shows that timing of pruning is important. Panicle pruning carried out during panicle elongation stimulates panicle branching and fails to reduce flower numbers. The best time for panicle pruning is the early blooming stage when panicle elongation has stopped. Currently, panicle pruning is a routine treatment in 'Feizixiao' cultivation. However, it is a labor-costly practice. 'Chemical pruning' with ethephon plus uniconazole is now replacing hand flower thinning.

### 4.4 Improved postharvest handling

Marketing litchi fruit in fresh condition is critical to meet the market requirements with deteriorated fruit being a major problem (Fig. 8). Improving postharvest handling is a major challenge in improving litchi production. In countries like

**Figure 7** A non-pruned large panicle with poor fruit set (left) and a pruned panicle of 'Feizixiao' with excellent set (right).

**Figure 8** Deteriorated fruit on the shelf in a supermarket in the United States.

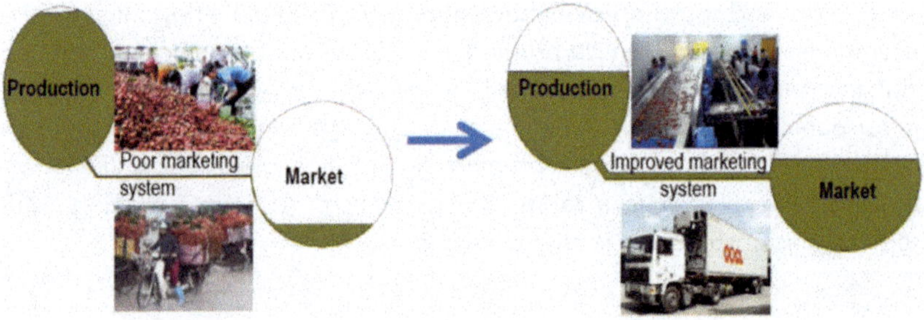


**Figure 9** Widening the marketing channel with better postharvest practices and infrastructure solves the problem of seasonal overproduction of litchi and makes the litchi industry more sustainable.

Australia, harvested litchi is rapidly graded, sorted, packed and marketed via a cold chain that extends from field to shelf. However, in most developing countries with poor infrastructure, postharvest losses of litchi are high due to poor marketing channels and lack of a reliable cold chain (Fig. 9).

Current commercial postharvest handling depends largely on sulfur fumigation for transport to more distant markets. The treatment suppresses fruit browning and maintains the red color associated with quality fruit. However, the $SO_2$ generated during fumigation causes a potential health hazard to workers and leads to residues on the fruit, which result in an undesirable taste and potential food safety problems for consumers. With more strict regulation of sulfur residue in food, there is an urgent need to develop alternative treatments to maintain fruit freshness. Quarantine treatments to eliminate pests and pathogens such as heat and near-freezing temperature treatments are too aggressive for this perishable fruit and cause pericarp browning if sulfur fumigation is not used. Mild alternatives such as γ-ray irradiation has been tested. An irradiation dose of 300–400 Gy (gray) caused 100% mortality of the stem-end borer pest (Hu et al., 1995) and reduced fruit deterioration during storage (Pandey et al., 2013). γ-Ray irradiation at 400 Gy is therefore used for exporting litchi from Australia to New Zealand and the United States. Other techniques to improve postharvest shelf life of litchi include the use of postharvest dips such as in citric acid, chitosan coatings, biocontrol agents such as *Bacillus subtilis*, modified atmosphere packaging and controlled atmosphere storage, though most have negative effects on flavor and other aspects of sensory quality (Sivakumar et al., 2007).

## 5 Conclusion and future trends

Searching through SCOPUS database reveals a constant and rapid increase in publications related to litchi in the past 10 years, suggesting an increasing activity of basic and applied research of litchi across the world. Among the 12180 publications, China contributed the largest number (4782, including 277 from Taiwan), followed by India (1507), the United States (1279) and Thailand (488). In China, significant governmental funding has been and will continue to be invested to support national litchi and longan cultivation. Key ongoing challenges include ageing orchards, a production season which still remains concentrated over a few months, shortage of farm labor and ever increasing labor costs. There has been significant research into litchi genomics and the identification of molecular markers related to important traits. These studies will contribute new knowledge about the biology of litchi, facilitate breeding for improved target traits and generate novel strategies for precise control of plant development.

## 6 Acknowledgement

The author would like to thank Francis Dodds, Editorial Director at Burleigh Dodds Science Publishing, for his help in improving the text in the author's first draft.

## 7 References

Cronje, R. and Ratlapane, I. M. 2018. Effect of full-cover ethephon applications on flowering and yield of 'Mauritius' litchi in South Africa. *Acta Horticulturae*, 1211:71–8.

Diczbalis, Y., Nichools, B., Gorves, I. and Lake, K. 2010. Sapindaceae production and research in Australia. *Acta Horticulturae*, 863:49–58.

Fan, Y., Yin, J. H., Liu, C. M., Luo, S., Zhou, J. and Huang, X. M. 2010. A new high quality litchi cultivar, Lingfengnuo. *Journal of Fruit Science*, 27(5):852–3.

Fronman, I. J., Husselman, J. H., Cronje, R. B., Severn-Ellis, A. and Sippel, A. D. 2014. Plant improvement strategies for litchi in South Africa. *Acta Horticulturae*, 1029:65–71.

Fu, X. Y., Mo, W. P., Zhang, J. Y., Zhou, L. Y., Wang, H. C. and Huang, X. M. 2014. Shoot growth pattern and quantifying flush maturity with SPAD value in litchi (*Litchi chinensis* Sonn.) *Scientia Horticulturae*, 174:29–35.

Hu, M., Yao, Z., Qiu, Y., Liu, X. and Hou, R. 1995. Disinfestation of *Conopomorpha sinensis* Bradley larvae with gamma irradiation as a quarantine treatment. *Journal of Zhongkai Agrotechnical College*, 8(1):46–51.

Huang, X. M., Wang, H. C., Yuan, W. Q., Lu, J. M., Yin, J. H., Luo, S. and Huang, H. B. 2005. A study of rapid senescence of detached litchi: Roles of water loss and calcium. *Postharvest Biology and Technology*, 36:177–89.

Jiang, S. Y., Xu, H. Y., Wang, H. C., Hu, G. B., Li, J. G., Chen, H. B., and Huang, X. M. 2012. A comparison of the costs of flowering in 'Feizixiao' and 'Baitangying' litchi. *Scientia Horticulturae* 148:118–25.

Kumar, R. 2014. Effect of climate change and climate variabiable conditions on litchi productivity and quality. In Cronje (Ed.), *IV International Symposium on Lychee, Longan and Other Sapindeaceae Fruit*, Acta. Hort. 1029, International Society for Horticultural Science (ISHS) (https://www.actahort.org/books/1029/index.htm).

Lai, B., Hu, B., Qin, Y-H., et al. 2015. Transcriptomic analysis of *Litchi chinensis* pericarp during maturation with a focus on chlorophyll degradation and flavonoid biosynthesis. *Genomics*, 16:225–43.

Liu, W., Xiao, Z., Bao, X., et al. 2015. Identifying litchi (*Litchi chinensis* Sonn.) cultivars and their genetic relationships using single nucleotide polymorphism (SNP) markers. *PLoS ONE*, 10(8):e0135390 (doi: 10/1371/journal.pone.0135390).

Liu, W., Xiao, Z., Fan, C., et al. 2018. Cloning and characterization of a flavonol synthase gene from *Litchi chinensis* and its variation among litchi cultivars with different fruit maturation periods. *Frontier in Plant Science*, 9:567–78.

Menzel, C. M., Huang, X. M. and Liu, C. M. 2005. Cultivars and plant improvement. In Menzel, C. M. and Waite, G. K. (Eds), *Litchi and Longan: Botany, Production and Uses*. CABI Publishing.

Mir, H. 2018. Biotechnological advances in litchi: Present scenario and future prospects. In Ahmad, F., et al. (Eds), *V International Symposium on Lychee, Longan and Other Sapindeaceae Fruit, Acta. Hort.* 1211, International Society for Horticultural Science (ISHS) (http://www.actahort.org/books/1211/index.htm).

Ou, L. X., Chen, J. Z., Cai, C. H., Cui, B. G., Xiang, X., Sun, X. M., Ye, L. H., Chen, Q., Qiu, Y. P. and Li, Z. Q. 2012. New high quality litchi cultivar, Fengshanhongdenglong. *Journal of Fruit Science*, 29(1):314–15.

Ouyang, R., Liang, Y. G., Liu, C. M., Wang, Z. H., Hu, G. B., Lin, Z. X. and Chen, H. 2002. Selection of the extremely late litchi cultivar-Maguili. *South China Fruits*, 31(4):42–4.

Ouyang, R, Hu, G. B., Wang, Z. H., Liu, C. M. and Qin, Y. H. 2009. Some extremely early new litchi cultivars selected by South China Agricultural University. *Proceedings of the 20th Symposium on the Development of Litchi and Longan Industry Across the Taiwan Strait*. Guangdong Science and Technology Association, pp. 61–7.

Pandey, N., Joshi, S., Singh. C., et al. 2013. Enhancing shelf life of litchi through integrated surface coating and gamma irradiation. *Radiation Physics and Chemistry*, 85:197–203.

Pesis, E., Dvir, O., Feygenberg, O., Ben Arie, R., Ackerman, M. and Lichter, A. 2002. Production of acetaldehyde and ethanol during maturation and modified atmosphere storage of litchi fruit. *Postharvest Biology and Technology*, 26:157–65.

Qiu, Y. P., Li, Z. Q., Yuan, P. Y., Ou, L. X., Xiang, X., Chen, J. Z., Sun, Q. M. and Cai, C. H. 2010. Trait observation of a new litchi cultivar, Xianjingfeng. *Guangdong Agricultural Science*, (8):65–6.

Sivakumar, D., Korsen, L. and Zeeman, K. 2007. Postharvest management and quality retention of litchi during storage. *Fresh Produce*, 1(1):66–75.

Stern, T. and Gazit, S, 2010. The reproductive biology of litchi. *Horticultural Reviews*, 28: 393–453.

Wang, R., Xu, S., Shen, H., Wu, L. F., Cao, H. L., Ye, W. H. and Cai, C. X. 2011. Influence of experimental warming on yield and fruit quality in Guiwei litchi. *Journal of Anhui Agricultural Science*, 39(24):14566–8.

Ye, Z. X., Hu, G. B., Chen, J. Z., Liu, C. M., Zheng, W. T., Li, B., Lin, C. X. and Liao, J. R. 2011. A new late-maturing litchi cultivar 'Jinggang Hongnuo' resistant to fruit cracking. *Acta Horticulturae Sinica*, 38(1):189–90.

Yuan, W. Q., Huang, X. M., Wang, H. C., Li, J. G., Chen, H. B. and Yin, J. H. 2009. Seasonal changes in carbon nutrition reserve in Nuomici litchi trees and its relation to fruit load. *Acta Horticulturae Sinica*, 36(11):1568–74.

Zhang, Y. F. and Liu, C. M. 2015. Studies on breeding system of litchi in an artificial way. *Tropical Agricultural Science and Technology*, 38(2):20–3.

Zhang, C. L., Chen, H. B. and Zhang, X. M. 2010. Irrigation and fertilization in litchi trees. In Chen, H. B. (Ed.), *Practical Techniques in Litchi Production*. Guangdong Science and Technology Press, Guangzhou, pp. 102–14.

Zhang, H., Li, H., Lai, B., Xia, H. Q., Wang, H. C. and Huang, X. M. 2016. Morphological characterization and gene expression profiling during bud development in a tropical perennial, *Litchi* chinensis Sonn. *Frontier in Plant Science*. doi: 10.3389/fpls.2016.01517.

Zhao, Y. H., Guo, Y. S., Hu, Y. L., Zhang, B., Liu, R., Ouyang, R., Fu, J. X. and Liu, C. M. 2010. Construction of a Genetic Linkage Map of Litchi with RAPD, SRAP and AFLP. *Acta Horticulturae Sinica*, 37(5):697–704.

Zhou, X. J., Wu, D. Y. and Huang, H. B. 1998. Translocation and distribution of 14C-photosynthates in lychee (*Litchi chinensis* Sonn. cv. Nuomici) by closed- and spiral-girdling. *Acta Agriculturae Nucleatae Sinica*, 12(6):365–8.

# Chapter 17

# Advances in breeding of mangoes

*Sunday O. S. Akinyemi, National Horticultural Research Institute, Nigeria*



## 1 Introduction

Mango (*Mangifera indica* L.) and closely related genera (*Mangifera* spp.) belong to the family Anacardiaceae that consists of dicotyledonous trees and shrubs. Mango is distributed widely in the world's tropical and subtropical regions. The cultivated mango is probably a natural hybrid between *M. indica* and *M. sylvatica* that occurred in Southeastern Asia and India. On the basis of ancient accounts, it was believed for many years that mango must have originated from India and spread outward from Southeast Asia. The selection of wild types has occurred for 4000–6000 years, and vegetative propagation has been undertaken for at least 400 years in India. More than 1000 cultivars currently exist around the world (Mukherjee, 1953; Condole, 1984). Measured by production, mango is the fifth largest fruit crop in the world (FAO, 2010), after bananas, grapes, apples and oranges, and the second most important tropical fruit crop. Despite ranking fifth in total world fruit production, its breeding potential has not been as well exploited as that of other major crops.

Mango is considered to be a difficult plant species to handle in breeding programmes because of certain inherent characteristics including long juvenile phase, high level of heterozygosity, heavy fruit drop and the large progeny population and acreage required for a meaningful assessment of hybrids. Despite its economic importance, genomic resources for mango are limited and genetics of useful horticultural traits are still poorly understood.

http://dx.doi.org/10.19103/AS.2019.0054.27

Globally, there are a number of problems that affect mango production. Being highly cross-pollinated and owing to the fact that most of the superior clones of mango including 'Indian' and 'Floridan' cultivars are monoembryonic, propagation through sexual means does not ensure true-to-type plant reproduction. This leads to a long juvenile phase in such plants. Production problems are associated with both scion and rootstock. Scion cultivar problems include biennial bearing habit, large tree size, susceptibility to major diseases and pests, short-postharvest life and fruit disorders like malformation, spongy tissue and so on (Iyer and Degani, 1997; Krishna and Singh, 2007).

Almost all the popular cultivated varieties are chance seedlings resulting from natural cross-pollination. Some hybrids are gaining ground owing to their novel characteristics. However, there is an acute shortage of this material. Most of the important varieties are not amenable to hi-tech cultivation practices and do not meet the requirements of modern horticultural production systems such as dwarfing, precocity and regularity in bearing, high yield, resistance to diseases, pests and physiological disorders and good keeping quality which limits the world mango trade due to the perishable nature of the fruit (Lizada, 1993). Mango is a climacteric fruit and so long-distance transport can be a challenge. There has been a long desire to develop new varieties with these desirable horticultural attributes. It has been suggested that an ideal mango variety should be a dwarf, a regular bearer and with a medium size fruit (250–300 g) (Singh, 1996). One problem in breeding new varieties is different perceptions of quality. As an example, while some European consumers prefer medium-sized fruit with no fibre, and which can be easily halved, African consumers prefer larger varieties with some fibre content.

## 2 Conventional breeding methods

Since the beginning of agriculture from eight to ten thousand years ago, farmers have been altering the genetic makeup of the crops they grow. Farmers would select the best plants from each year's crop to provide the seed for the next generation and sometimes acquire seed from other farmers, engaging in both simple mass selection and introduction of new germplasm (Kloppenburg, 2004). Modern plant breeding came into being when man learned that crop plants could be artificially mated or cross-pollinated to be able to improve the characters of the plant (Poehlman and Sleper, 1999). Desirable characteristics from different parent plants could be combined in the offspring (Stiner et al., 2000; Akinyemi, 2017). The science of genetics revolutionized the selection process, significantly improving the efficiency of breeding (Acquaah, 2012; Briggs and Knowles, 1967). Modern plant breeders follow standard methods to create variability, discriminate among the variability and develop cultivars for release to farmers (Chahal and Gosal, 2002; Acquaah, 2012; Allard, 1960).

Plant breeding can be divided into two basic types - conventional and novel (molecular) (Acquaah, 2012; Chahal and Gosal, 2002). Molecular techniques are used primarily to create new variability, after which the breeder uses conventional breeding methods for cultivar development (Jain and Kharkwal, 2004).

In conventional breeding, breeders assemble desirable traits from different, but usually closely related, plants into the new cultivar using the techniques of crossing (hybridization) (Jain and Kharkwal, 2004; Acquaah, 2012). Consequently, the product of conventional breeding only emphasizes target traits pre-existing in the genetic potential of the species, without introducing new genes (Jakowitsch et al., 1999). Historically, most of the current mango cultivars grown worldwide originated from selection of open-pollinated seedlings as chance seedlings. Conventional varietal improvement methods such as selection from natural seedlings and grafts, hybridization, clonal selection and mutation breeding are the most commonly adopted methods in mango improvement.

Breeding methods include:

- selection from open-pollinated seedlings occurring naturally
- controlled pollinations (using hand pollination of selected flowers on a large number of panicles)
- enclosing self-incompatible female and male parents and cross-pollinating using insects
- maintaining hybrid populations by grafting scions on established plants
- pre-selection of mango hybrids.

Selection from natural mutants for important agronomical traits (such as precocity, yield, regular bearing and resistance to diseases) might be improved by sport selection (Usman et al., 2001). Varieties produced this way may produce good crops in some years but often not in others.

For many years, new varieties have been selected on the basis of size, appearance and eating quality with very little attention given to their production capabilities. Young and Ledin (1954) discussed several methods for improving mango varieties in Florida. Among these methods were selection of chance seedlings, further introductions and use of bud mutations. Of these methods the selection of chance seedlings has produced more new varieties than any of the others. The improvements of Florida mangos to date are largely due to this method. A few of them are of excellent quality and are gradually finding their way into the accepted group of marketable mango varieties. Most of them, while they may be of acceptable quality, are, however, erratic or poor bearers.

Hybridization is a cross between individuals from separate populations that differ in one or more heritable traits (Harrison, 1990; Goulet et al., 2017).

Hybridization has played an important role in the evolution of many lineages. With the growing availability of genomic tools, it is becoming increasingly clear that gene flow between divergent taxa can generate new phenotypic diversity, allow for adaptation to novel environments and contribute to speciation. Hybridization can have immediate phenotypic consequences through the expression of hybrid vigour. On longer evolutionary time scales, hybridization can lead to local adaptation through the introgression of novel alleles and transgressive segregation and, in some cases, result in the formation of new hybrid species. In spite of the difficulties encountered in mango hybridization with controlled pollination, significant progress has been made with regard to release of new hybrid varieties from various breeding centres. In many countries, a large number of hybrids are being intensively screened and it is expected that many more hybrid cultivars will be released in the near future.

The discovery of self-incompatibility in some cultivars has led to further improvement in hybridization known as the caging technique. This involves the enclosure of two desirable parents of synchronous flowering in a screened enclosure or insect proof cage with specially reared pollinators (insects) to provide a method of hybridization, using self-incompatible (female) and male parents (Sharma et al., 1972). This method has been used in Israel (Degani et al., 1993), Brazil and South Africa.

Bud mutants occur frequently and are a valuable source of variation. Albino mutant occurrence is quite frequent in mango seedlings and in certain shoots of mature trees. There are only two cases of somatic mutants yielding new varieties. 'The Davis' sport of the 'Haden' cultivar (Young and Ledin, 1954) and the 'Rosica' bud mutant of the Peruvian cultivar 'Rosado de Ica' (Medina, 1997). The mutants are high-yielding, regular bearing with good seedling. Induced mutations by radiations were first used by Siddiqui et al. (1966). He exposed dormant buds of Langra to gamma radiations and grafted on one-year-old seedlings, resulting in heavier and larger fruits. Singh and Chadha (1981) located four superior clones from orchards of Dashehari while Singh et al. (1985) isolated high-yielding clones from 'Langra' orchards. Sharma and Majumder (1988) irradiated bud sticks of Dashehari with chemical mutagens (EMS and NMU) and top worked them on to 10-year-old seedlings. The mutants showed dwarfness, changes in shape, leaf serration and total soluble solids (TSS). Mutations can be successfully propagated asexually. The techniques used need to be perfected in mango to make mutation breeding more useful.

## 3 Molecular methods in breeding

As noted, conventional breeding of woody perennial fruit crops like mango is difficult owing to a range of factors such as long juvenile phase, self-incompatibility, low fruit set, high fruit drop, single seed per fruit, high degree of cross-pollination, polyembryony, polyploidy and heterozygous nature, meagre information on inheritance of important quantitative traits and so on. Molecular approaches offer an efficient alternative tool for conventional breeding. They are very useful for characterizing the genetic diversity/relatedness among different cultivars or species of mango, for identifying genes of commercial interest, for improvement through gene transfer technologies, overcoming reproductive isolation barrier via protoplast fusion and so on.

Employing biotechnology to correct genetic flaws in existing varieties has great potential. Functional genomics could provide insights into genetic regulation of plant function and novel means of isolating genes for manipulation in transgenic plants. Evaluating the performance of transgenic fruit-free cultivars requires approximately 12 years or until fruiting and flowering have been observed, depending on the species. Molecular breeding thus represents a highly efficient approach for developing improved, perennial tropical fruit cultivars (Gomez-Lim and Litz, 2007).

## 4 In vitro techniques

This section discusses a number of techniques:

- Tissue culture
- Micropropagation
- Somatic embryogenesis
- Organogenesis
- Cryopreservation.

The growth of plant tissues in a suitable culture medium (in vitro), which is often referred to as tissue culture, offers an effective solution to many problems related to the propagation of fruit crops. It has several advantages over conventional methods of propagation such as very high multiplication rate in a short period, germination of uniform true-to-type and disease-free plants through meristem culture, creation of variants and long-term storage of valuable germplasm. Tissue culture techniques now exist for most crop species, although continued optimization is still required, especially for woody plants (Ara and Sharma, 2012). Several efficient protocols have been established for plant regeneration in woody fruit trees like mango, and serious efforts have been made all over the world in the last few decades (Litz and Gomez-Lim, 2005; Gomez-Lim and Litz, 2007; Krishna and Singh, 2007; Litz, 2009). Some of the tissue culture protocols established in mango include:

In vitro plant propagation, better known as 'micropropagation' has become established because the system facilitates rapid multiplication of superior clones within a short timescale, can ensure freedom from diseases and speeds up tree breeding programmes (Ara and Sharma, 2012). Presently, three in vitro methods are commonly employed for micropropagation of woody perennials that include (i) clonal propagation (ii) adventitious bud formation (a form of organogenesis) and (iii) somatic embryogenesis. In mango, the first and the second methods have not proved reliable and only somatic embryogenesis has been in practice for plantlet regeneration. Several cultivars of mango are now being generated by in vitro methods (Krishna and Singh, 2007; Litz, 2009), but this crop is still observed as a relatively difficult crop for tissue culture in comparison to other horticultural crops (Ara and Sharma, 2012).

Somatic embryogenesis in mango depends on induction of nucleolar tissue in ovular halves of fruits to produce callus, somatic embryo formation, proliferation of somatic embryos and their synchronized development into plantlets (Chaturvedi et al., 2004). The plants regenerated from nucellar tissues are true-to-type. Many researchers have been successful in using this technique to multiply mono and polyembryonic mango varieties (Nower, 2013; Malabadi et al., 2011; Mishra et al., 2010; Krishna and Singh, 2007; Chaturvedi et al., 2004). Somaclonal variation could be of great value to the breeders as it is an efficient tool to create variations. Single gene mutations may result in the alteration of a significant horticultural trait.

Relatively, low plant recovery from embryogenic cultures was reported initially. Numerous developmental anomalies, such as poly cotyledony, fasciation, absence of bipolarity and formation of secondary embryos from hypocotyls were described, which contribute to the failure of embryos to develop to maturity in a normal fashion (De Wald et al., 1989). In addition, the problems of precocious germination and progressive necrosis have also prevented the efficient recovery of mango plantlets (De Wald et al., 1989). In an attempt to mitigate these problems, De Wald et al. (1989) investigated in detail the suitable conditions for optimizing the induction and maintenance of embryogenic cultures and for the maturation of somatic embryos. As a result somatic embryogenesis has been a widely used technique (Litz et al., 1982, 1984; Litz, 1984; Patena et al., 2002; Ara et al., 2000; Deore et al., 2000; Rivera-Domínguez et al., 2004; Ermayanti and Rantau, 2009; Ara and Sharma 2012).

Intraspecific somatic hybrids of mango in the form of in vitro plantlets are possible. Somatic hybridization of mango via protoplast fusion has been attempted at cultivar level. Enzymatically isolated protoplasts from leaves of greenhouse-grown seedlings of cultivars 'Tommy Atkins', 'Keitt' and 'Haden' and from proembryonic masses (PEMs) of cv. 'Kensington Pride' were used by Rezazadeh et al. (2011). Protoplast fusion of these mango cultivars achieved an 18% success rate when compared to the fusion frequency of 5–10% obtained in other studies.

The process of initiation and development of an organ, referred to as organogenesis is a promising field for the regeneration of plant from cell and callus culture. With the discovery of cellular totipotency and the development of methods for regeneration of plant from in vitro cell culture, a large number of economically important plant species have been successfully regenerated from callus culture via organogenesis. The production of large numbers of haploids from microspores and the possibility of raising triploids from endosperm cell culture are dramatic examples of the potential role of organogenesis in genetics and plant breeding. Organogenesis for in vitro regeneration of mango has been difficult because of exudation of phenolics from the cut ends of the explants that lead to blackening followed by necrosis and ultimate death of explants. This problem has been addressed by Rao et al. (1982) who removed cotyledons from mango embryos and explanted them in MS medium supplemented with 15% (v/v) coconut water, 26.9 μM NAA and 11.6–23.2 μM kinetin.

It has long been recognized that preserving plant biodiversity is essential for both classical and modern plant-breeding programmes including genetic engineering (Cruz-Cruz et al., 2013). Cryopreservation, the storage of living materials at cryogenic temperatures (below −130°C), is an alternative to conventional field and in vitro germplasm collection which enables plant genetic resources to be conserved safely and cost effectively for decades with minimal requirements of space and routine maintenance (Engelmann, 2004; Pence, 2011; Volk et al., 2014). More importantly, cryopreservation eliminates the need for regular renewal of the collection, thus reducing the risk of genetic erosion caused by pests, diseases, weather conditions, pollution and genetic variations (Panis and Lambardi, 2006; Wang et al., 2014).

Cryopreservation of seeds is the most efficient option for many species due to ease of application and the amount of diversity conserved (Engelmann, 2004; Pritchard and Nadarajan, 2008). The status of seed cryobanking has been reviewed extensively (Pence, 2011; Pritchard, 2007; Pritchard and Nadarajan, 2008; Stanwood and Bass, 1981; Touchell and Dixon, 1994). Following the vitrification revolution in the 1990s (Benson, 2004), the range of cryogenic techniques increased significantly, resulting in successful cryopreservation of many tropical crops and species with recalcitrant seeds (Engelmann and Takagi, 2000; Sakai and Engelmann, 2007).

Mango seeds are highly recalcitrant and cannot be stored. Hence, tissue culture methods can be an ideal approach (Engelmann, 1991). Pliego-Alfaro et al. (1996a,b) cultured somatic embryos on mannitol and ABA-supplemented medium, which suppressed growth so that cultures could be extended up to three months. They suggested that medium-term germplasm storage can be made in mango through in vitro means. Wu et al. (2003) compared three techniques for cryopreserving the embryonic mango cultures. The best result was obtained with vitrification, while the encapsulation/dehydration yielded no recovery.

## 5 The use of molecular markers

Outbreeding, the wide range of climatic conditions prevailing in the different growing areas for mango, and widespread hybridization and recombination of characteristics have contributed to the high genetic diversity in mango (Ravishankar et al., 2000). There has been a considerable confusion in the nomenclature of mango cultivars because of the use of synonyms for many cultivars, which increases the difficulty of identifying cultivars (Krishna and Singh, 2007). To efficiently and effectively utilize mango's diverse genetic resources, it is first necessary to characterize the available germplasm. The use of DNA markers have been found to contribute immensely for the cultivar identification. Breeding projects can be improved by taking advantage of DNA markers for MAS, since this will decrease the number of generations needed for gene introgression and will facilitate the identification of genetically different parents to generate heterosis (Lavi et al., 2004). Molecular approaches are commonly used for taxonomical characterization, to understand the regulation and expression of important traits/genes and so on. They offer an efficient alternative tool to conventional breeding. They are useful for characterizing the genetic diversity/relatedness among different cultivars or species of mango, for identifying genes of commercial interest, improvement through gene transfer technology, creation of variations in existing cultivars, in vitro, overcoming reproductive isolation barrier via protoplast fusion and so on.

The initial molecular markers used for mango were isozyme markers, which were used in a survey of genetic variation (Gan et al., 1981) and for the identification of loci that are polymorphic among cultivars (Degani et al., 1990). Several approaches and makers are now frequently used in mango. The presence of a low level of polymorphism prompted workers to rely more on DNA markers. Different DNA markers have been employed in mango to study monogenic and polygenic traits. As it is a tropical fruit tree, the isolation of a sufficient quality of genomic DNA for using in PCR-based DNA marker technology very often poses severe problems due to the presence of inhibitors such as polysaccharides, which inhibit the enzymatic DNA processing or phenolics as inhibitors of PCR reactions (Ramirez et al., 2004). The well-established CTAB protocol of Doyle and Doyle yielded excellent DNA templates for PCR amplification for mango (Ramirez et al., 2004). Recently, Gomathi et al. (2005) reported a simple and efficient method for isolation of DNA from healthy and malformed floral tissues of mango (*M. indica* L.).

This section reviews the following types of marker that are discussed below:

- Variable number tandem repeat sequence (VNTRS)
- Rapid amplified polymorphic DNA (RAPD)
- Single sequence repeats (SSRs)
- Amplified fragment length polymorphism (AFLP)
- Inter-Simple-Sequence-Repeats (ISSR).

Looking at variable number tandem repeat sequence (VNTRS), Adato et al. (1995) analysed DNA fingerprint patterns of some mango genotypes using minisatellite multilocus probes. They also carried out genetic analysis of the progeny of the cross 'Tommy Atkins × Keitt'. The six bands of the parents showed polymorphism with an average transmission frequency of 65% and 81% for maternal and paternal specific bands.

Rapid amplified polymorphic DNA (RAPD) are DNA markers that are produced by amplification of random DNA segments using single, short primers by means of a polymerase chain reaction (PCR). Schnell et al. (2005) used RAPD genetic markers to identify mango cultivars and to determine genetic relationships among cultivars. In their work on molecular characterization of different mango cultivars, Arogundade et al. (2017) used RAPD genetic markers and reported that the 45 tested-RAPD markers classified all the mango accessions used into four main groups, indicating that many of the mango accessions belong to the same group. RAPD markers were also utilized by Jayasankar (1995) for characterizing fungi-resistant mutants from embryogenic cultures of 'Carabao' and 'Hindi' that had been recurrently selected with a phytoxin isolated from *colletotichum gloeosportoides*.

Simple sequence repeats or SSRs, also known as microsatellites, as DNA markers, have advantages over many other marker types. Microsatellites are reported to be more variable than RFLPs or RAPDs, and have been widely utilized in plant genomic studies (Weber, 1990; He et al., 2005). They are highly polymorphic, have multiple alleles and are co-dominant. SSRs have been widely utilized for the conservation of genetic resources and in population genetics, molecular breeding and paternity-testing studies (Ellegren, 2004). In mango, SSR markers are particularly important in the identification of cultivars, the determination of genetic variability, germplasm conservation and identifying the domestication and movement of germplasm (Viruel et al., 2005). More than 100 SSR markers have been developed by using different mango germplasms (Chiang et al., 2012; Duval et al., 2005, Honsho et al., 2005; Ravishankar et al., 2011; Schnell et al., 2005; Viruel et al., 2005). The true potential of microsatellite over other types of genetic markers lies in their use to track desirable traits in large-scale breeding programmes and as an anchor points for map-based gene cloning strategies.

Amplified fragment length polymorphism (AFLP) is a PCR-based technique which allows inspection of polymorphism at fairly a large number of loci within a very short span of time and, at the same time, requires a very small amount of DNA. AFLP markers are suitable for cultivar identification, estimation of genetic relationship and mapping of QTLs in mango (Kashkush et al., 2001). The information generated by AFLP analysis regarding genetic relatedness and diversity existing in mango genepool are useful for breeding of improved mango varieties (Yamanaka et al., 2006). Eiadthong et al. (1999) has used AFLP analysis to study genetic relationships among mango cultivars in Thailand. Ravishankar et al. (2000) investigated the genetic relatedness of 18 mango cultivars grown in different parts of India through RAPD markers. From their results they concluded that majority of cultivars were evolved from a local mango genepool and later domesticated.

Amplification of inter-simple-sequence-repeats (ISSRs) is a relatively novel technique and has proven to be a powerful, rapid, simple, reproducible and inexpensive way to assess genetic diversity or to identify closely related cultivars in many species, including fruit trees. González et al. (2002) tested a range of ISSR primer sequences in mango and identified those that show clear polymorphisms between cultivars grown in Australia. They suggested the potential use of DNA markers for mango improvement in the areas of variety identification, validation of parentages, estimation of genetic variation in existing populations and characterization of rootstocks.

## 6 Genetic engineering of mango

Isolation, introduction and expression of foreign DNA in plants and animals is called genetic engineering. It involves the introduction of a gene by the techniques of recombinant DNA technology and genetic transformation. It provides an exciting tool to improve specific traits of inbreeds, provided a transgene is available for this purpose. As an example, a suitable version of the cry gene of *Bacillus thuringiensis* has been transferred in maize inbreeds to create insect-resistant commercial hybrids. Similarly, many transgenes have been shown to generate resistance to fungal diseases.

Genetic transformation provides the means for modifying single horticultural traits in perennial plant cultivars without altering their phenotype. This capability is particularly valuable for tree species in which development of new cultivars is often hampered by their long generation time, high levels of heterozygosity and nucellar embryony. Targeting specific gene traits is predicated upon the ability to regenerate elite selections from cell and tissue cultures. The integrity of the clone would thereby remain unchanged except for the altered trait (Gomez-Lim and Litz, 2007). In nature, gene transfer is unpredictable which makes the percentage recovery of desired gene combination subject to efficient screening and selection. Additionally, their range in terms of species involved is dependent on sexual compatibility.

This limits the movement of gene across different taxa. Gene transfer using biotechnology entails targeted manipulation of the genetic material towards a desired end. In mango, gene cloning and genetic transformation techniques are still mostly confined to control overripening to extend the postharvest life of mango fruits, as well in the manipulation of aroma and flavour (Litz, 2009; Bimal and Singh, 2017).

Success has been achieved in obtaining genetically transformed mango embryogenic cultures using *Agrobacterium tumefaciens* (Mathews et al., 1992, 1993). Engineered *A. tumefaciens* strains were used for the transformation of 'Keitt' and 'Hindi' mango, respectively. A preliminary study indicated that 12.5 $\mu gml^{-1}$ kanamycin sulfate was toxic to proembryo mass grown in suspension culture, while 200 $\mu gml^{-1}$ kanamycin sulfate was toxic in semisolid medium (Mathews and Litz, 1990). Cruz et al. (1997) successfully transformed 'Hindi' mango cultures with an engineered disarmed strain of *A. tumefacians* (LBA 4404) which contained ACC oxidase, ACC synthase and alternate oxidase genes in antisense configuration with the aim of expressing anti-ripening behaviour in fruits to extend postharvest life. Martínez et al. (2001) isolated a mango cDNA homologue of ethene receptor gene ETR-1, which expressed transiently during fruit ripening as well as in wounding of tissue. Cruz et al. (2000) optimized the particle bombardment parameters and reported the successful transient expression for the beta-glucuronidase gene in polyembryonic mases. Stable transgene expression was achieved using green fluorescent protein as reporter gene.

## 7 Conclusion

Despite the fact that mango is considered to be a difficult plant species to handle in breeding programmes, because of its inherent characteristics, appreciable success has been achieved in this area. Moving from conventional methods of crop improvement to non-conventional methods, with combination of both in many cases, has brought about some new varieties and potential solutions to some of the inherent problems in the production of the crop. A key problem, however, is that newer varieties or hybrids are not easily accepted because of a highly regionalized preferences. Research may be better directed towards finding varieties and rootstocks resistant to pests, diseases and abiotic stress.

## 8 References

Acquaah, G. 2012. *Principles of Plant Genetics and Breeding* (2nd edn.). John Wiley and Sons Ltd, West Sussex, UK.

Adato, A., Sharon, D., Lavi, U., Hillel, J. and Gazit, S. 1995. Application of DNA fingerprints for identification and genetic analyses of mango (*Mangifera indica*) genotypes. *Journal of the American Society for Horticultural Science* 120(2), 259–64. doi:10.21273/JASHS.120.2.259.

Akinyemi, S. O. S. 2017. Mango production in Nigeria: current situation and future prospects. In: Valavi, S. G., Raj Mohan, K, Govil, J. N., Peter, K. V. and Thottappilly, G. (Eds), *Mango. Volume 2: Cultivations in different Countries*. Studium Press LLC, Houston, TX, pp. 186–207.

Allard, R. W. 1960. *Principles of Plant Breeding*. John Willey and Sons Inc., New York.

Ara, H. and Sharma, A. K. 2012. Tissue culture protocol for regeneration and germplasm conservation of mango. In: Valavi, S. G., Raj Mohan, K, Govil, J. N., Peter, K. V. and Thottappilly, G. (Eds), *Mango. Volume 1: Production and Processing Technology*. Studium Press LLC, Houston, TX.

Ara, H., Jaiswal, U. and Jaiswal, V. S. 2000. Plant regeneration from protoplasts of mango (*Mangifera indica* L.) through somatic embryogenesis. *Plant Cell Reports* 19(6), 622–7. doi:10.1007/s002990050783.

Arias, R. S., Borrone, J. W., Tondo, C. L., Kuhn, D. N., Irish, B. M. and Schnell, R. J. 2012. Genomics of tropical fruit tree crops. In: Shunell, R. J. and Priyadarshan, P. M. (Eds), *Genomics of Tree Crops*. Springer, New York, pp. 209–39.

Arogundade, O., Akinyemi, S. O. S., Akin-Idowu, P. E., Aderonmu, O. I.. and Esuola, C. O. 2017. Molecular characterization of different mango (Mangifera indica L.) cultivars in Southwest Nigeria using Random Amplification of Polymorphic DNA (RAPD) markers. Lambert Academic Publishing.

Benson, E. E. 2004. Cryoconserving algal and plant diversity: historical perspectives and future challenges. In: Fuller, B. J., Lane, N. and Benson, E. (Eds), *Life in the Frozen State*. CRC Press, Boca Raton, pp. 299–328.

Bimal, R. and Singh, A. K. 2017. Biology and biotechnology of mango (*Mangifera indica* L.) with reference to *in vitro* cell and tissue culture in endangered and endemic cultivars: a review. *Journal of Cell and Tissue Research* 17(3), 6285–92.

Briggs, F. N. and Knowles, P. F. 1967. *Introduction to Plant Breeding*. Reinhold Publishing Corp, New York.

Chahal, G. and Gosal, S. S. 2002. *Principles and Procedures of Plant Breeding: Biotechnological and Conventional Approaches* (vol. 21). Narosa Publishing House, New Delhi, pp. 64–89.

Chaturvedi, H. C., Agnihotri, S., Sharma, M., Sharma, A. K., Jain, M., Gupta, P. and Kidwai, N. R. 2004. Induced nucellar embryogenesis *in vitro* for clonal multiplication of *Mangifera indica* L. var. Ambalavi: a dwarfing rootstock. *Indian Journal of Biotechnology* 3(2), 221–8.

Chiang, Y. C., Tsai, C. M., Chen, Y. K. H., Lee, S. R., Chen, C. H., Lin, Y. S. and Tsai, C. C. 2012. Development and characterization of 20 new polymorphic microsatellite markers from *Mangifera indica* (Anacardiaceae). *American Journal of Botany* 99(3), e117–9. doi:10.3732/ajb.1100443.

Condole, A. D. 1984. *Origin of Cultivated Plants*. Vegal Paul Trench and Co., London, pp. 1–67.

Cruz, H. A., Gómez, L. M. A. and Litz, R. E. 1997. Transformation of mango somatic embryos. *Acta Horticulturae* 455, 287–91.

Cruz, H. A., Town, L., Cavallaro, A. and Botella, I. R. 2000. Transient and stable transformation in mango by particle bombardment. *Acta Horticulturae* 509, 237–42.

Cruz-Cruz, C. A., González-Arnao, M. T. and Engelmann, F. 2013. Biotechnology and conservation of plant biodiversity. *Resources* 2(2), 73–95. doi:10.3390/resources2020073.

Degani, C., El-Batsri, R. and Gazit, S. 1990. Enzyme polymorphism in mango. *Journal of the American Society for Horticultural Science* 115(5), 844–7. doi:10.21273/JASHS.115.5.844.

Degani, C., Cohen, M., Reuvani, O., El-Batsri, R. and Gazit, S. 1993. Frequency and characteristic of zygotic seedlings from polyembryonic mango cultivars determined using isozymes as genetic markers. *Acta Horticulturae* 341(341), 78–85. doi:10.17660/ActaHortic.1993.341.6.

Deore, A. C., Chavan, S. S., Ranade, S. S., Deshpande, R. S. and Dhonukshe, B. L. 2000. Factors affecting *in vitro* somatic embryogenesis from nucellar tissues of Alphonso mango. *Annual Review of Plant Physiology* 14(2), 182–6.

De Wald, S. G., Litz, R. E. and Moore, G. A. 1989. *Journal of the American Society for Horticultural Science* 114, 837–41.

Duval, M. F., Bunel, J., Sitbon, C. and Risterucci, A. M. 2005. Development of microsatellite markers for mango (*Mangifera indica* L.). *Molecular Ecology Notes* 5(4), 824–6. doi:10.1111/j.1471-8286.2005.01076.x.

Eiadthong, W., Yonemoni, K., Kanzaki, S., Sugiura, A., Utsunomiya, N. and Subhadrabandhu, S. 1999. Identification of mango cultivars of Thailand and evaluation of their genetic variation using the amplified fragments by simple sequence respect (SSR) anchored primers. *Scientia Horticulturae* 82, 57–66.

Ellegren, H. 2004. Microsatellites: simple sequences with complex evolution. *Nature Reviews. Genetics* 5(6), 435–45. doi:10.1038/nrg1348.

Engelmann, F. 1991. *In vitro* conservation of tropical plant germplasm-a review. *Euphytica* 57(3), 227–43. doi:10.1007/BF00039669.

Engelmann, F. 2004. Plant cryopreservation: progress and prospects. *In Vitro Cellular and Developmental Biology - Plant* 40(5), 427–33. doi:10.1079/IVP2004541.

Engelmann, F. and Takagi, H. 2000. *Cryopreservation of Tropical Plant Germplasm: Current Research Progress and Applications*. JIRCAS/IPGRI, Rome.

Ermayanti, T. M. and Rantau, D. E. 2009. J. Biotechnol. Res. Tropical Reg. 2(2), 1–5.

FAO. 2010. Statistical database. 15 February 2011. Available at: http://faostat.fao.org (accessed on 15 February 2011).

Gan, Y. Y., Zaini, S. and Idris, A. 1981. Genetic variation in the grafted vegetatively propagated mango (*Mangifera indica* L.) L. *Pertanika* 4, 53–63.

Gomathi, K. A., Nailwal, T. K., Gurdeep, B., Shukla, A. and Pant, R. C. 2005. A rapid and efficient procedure for extracting high quality genomic DNA from mango (*Mangifera indica* L.). *Physiology and Molecular Biology of Plants* 11, 169–71.

Gómez, M. A. and Litz, R. E. 2004. Genetic transformation of perennial tropical fruits. *In Vitro Cellular and Developmental Biology - Plant* 40(5), 442–9. doi:10.1079/IVP2004547.

Gomez-Lim, M. A. and Litz, R. E. 2007. *Biotechnology in Agriculture and Forestry* (vol. 60). Springer, Heidelberg, pp. 51–71.

González, A., Coulson, M. and Brettell, R. 2002. Development of DNA markers (ISSRs) in mango. *Acta Horticulturae* 575(575), 139–43. doi:10.17660/ActaHortic.2002.575.13.

Goulet, B. E., Roda, F. and Hopkins, R. 2017. Hybridization in plants: old ideas, new techniques. *Plant Physiology* 173(1), 65–78. doi:10.1104/pp.16.01340.

Harrison, R. G. 1990. Hybrid zones: windows on evolutionary process. *Oxford Surveys in Evolutionary Biology* 7, 69–128.

He, X. H., Li, Y. R., Guo, Y. Z., Tang, Z. P. and Li, R. B. 2005. Inter-simple sequence repeat (ISSR) analysis of different native mango cultivars in Guangxi. *Molecular Plant Breeding* 3, 829–34.

Honsho, C., Nishiyama, K., Eiadthong, W. and Yonemori, K. 2005. Isolation and characterization of new microsatellite markers in mango (*Mangifera indica*). *Molecular Ecology Notes* 5(1), 152–4 doi:10.1111/j.1471-8286.2004.00866.x.

Iyer, C. P. A. and Degani, M. R. 1997. Advances in classical breeding and genetics in mango. *Acta Horticulturae* 455(455), 252–67. doi:10.17660/ActaHortic.1997.455.34.

Jain, H. K. and Kharkwal, M. C. 2004. *Plant Breeding Mendelian to Molecular Approaches*. ISBN 978-94-007-1040-5.

Jakowitsch, J., Mette, M. F., van der Winden, J., Matzke, M. A. and Matzke, A. J. M. 1999. Integrated pararetroviral sequences define a unique class of dispersed repetitive DNA in plants. *Proceedings of the National Academy of Sciences of the United States of America* 96(23), 13241–6.

Jayasankar, S. 1995. *In vitro* selection of mango for anthracnose resistance and characterization of the regenerants. Ph.D. thesis. University of Florida.

Kashkush, K., Jinggui, F., Tomer, E., Hillel, J. and Lavi, U. 2001. Cultivar identification and genetic map of mango (*Mangifera indica* L.). *Euphytica* 122(1), 129–36. doi:10.1023/A:1012646331258.

Kloppenburg, J. R. 2004. *First the Seed: the Political Economy of Plant Biotechnology, 1492-2000* (2nd edn.). University of Wisconsin Press, Madison.

Krishna, H. and Singh, S. K. 2007. Biotechnological advances in mango (*Mangifera indica* L.) and their future implication in crop improvement – a review. *Biotechnology Advances* 25(3), 223–43. doi:10.1016/j.biotechadv.2007.01.001.

Lavi, U., Kashkush, K., Sa'ada, D., Shats, H., Ravid, U. and Tomer, E. 2004. Mango breeding and the potential of modern biology. *Acta Horticulturae* 645(645), 51–9. doi:10.17660/ActaHortic.2004.645.2.

Litz, R. E. 1984. *Horticultural Science* 19, 715–7.

Litz, R. E. 2009. *The Mango: Botany, Production and Uses*. CAB International, Wallingford, UK.

Litz, R. E. and Gomez-Lim, M. A. 2005. *Biotechnology of Fruit and Nut Crops*. CAB International, Wallingford, UK, pp. 40–61.

Litz, R. E., Knight, R. L. and Gazit, S. 1982. Somatic embryos from culture ovules of polyembryonic *Mangifera indica* L. *Plant Cell Reports* 1(6), 264–6. doi:10.1007/BF00272635.

Litz, R. E., Knight Jr., R. J. and Gazit, S. 1984. *In vitro* somatic embryogenesis from *Mangifera indica* L. callus. *Scientia Horticulturae* 22(3), 233–40. doi:10.1016/0304-4238(84)90056-6.

Lizada, M. C. C. 1993. Mango. In: *Biochemistry of Fruit Ripening*. Springer, Dordrecht, pp. 255–71. doi:doi:10.1007/978-94-011-1584-1.

Malabadi, R. B., Teixeira da Silva, J. A., Nataraja, K., Vijaya, K. S. and Mulgund, G. S. 2011. Induction of somatic embryogenesis in mango (*Mangifera indica* L.). *International Journal of Biological Technology* 2(2), 12–8.

Martínez, P. G., Gómez, R. L. and Gómez-Lim, M. A. 2001. Identification of an ETR-1 homologue from mango fruit expressing during fruit ripening and wounding. *Journal of Plant Physiology* 158(1), 101–8. doi:10.1078/0176-1617-00238.

Mathews, H. and Litz, R. E. 1990. Kanamycin sensitivity of mango somatic embryos. *HortScience* 25(8), 965–6. doi:10.21273/HORTSCI.25.8.965.

Mathews, H., Litz, R. E., Wilde, H. D., Merkela, S. A. and Wetzstein, H. Y. 1992. Stable integration and expression of beta-glucuronidase and NPT II genes in mango somatic embryo. *In Vitro Cellular and Developmental Biology - Plant* 28(4), 172–8. doi:10.1007/BF02823312.

Mathews, H., Wilde, H. D., Litz, R. E. and Wetzstein, H. Y. 1993. Genetic transformation of mango. *Acta Horticulturae* 341(341), 93–7. doi:10.17660/ActaHortic.1993.341.8.

Medina, J. P. 1997. 'Rosica'-a new mango variety selected in Ica, Peru. *Fruit Varieties Journal* 31, 88–9.

Mishra, M., Shree, Y., Pati, R., Seal, S., Shukla, N., Kamle, M., Chandra, R. and Srivastava, A. 2010. Micropropagation of *Mangifera indica* L. cv. Kurakkan through somatic embryogenesis. *Indian Journal of Genetics and Plant Breeding* 70(1), 85–90.

Mukherjee, S. K. 1953. The mango–its botany, cultivation, uses and future improvement, especially as observed in India. *Economic Botany* 7(2), 130–62. doi:10.1007/BF02863059.

Nower, A. A. 2013. *In vitro* production of somatic embryos from nucellus of mango (*Manglfera indica* L.). *Life Science Journal* 10(2).

Panis, B. and Lambardi, M. 2006. Status of cryopreservation technologies in plants (crops and forest trees). In: Ruane, J. and Sonnino, A. (Eds), *The Role of Biotechnology in Exploring and Protecting Agricultural Genetic Resources*. Food and Agriculture Organization of the United Nations, Rome, pp. 61–78.

Patena, L. F., Carlos-Refuerzo, L. R. and Barba, R. C. 2002. *In Vitro Cellular and Developmental Biology* 38, 173–7.

Pence, V. C. 2011. Evaluating costs for the in vitro propagation and preservation of endangered plants. *In Vitro Cellular and Developmental Biology - Plant* 47(1), 176–87. doi:10.1007/s11627-010-9323-6.

Pliego-Alfaro, F., Litz, R. E., Moon, P. A. and Grey, D. J. 1996a. Effect of abscisic acid, osmolarity and temperature onin vitro development of recalcitrant mango nucellar embryos. *Plant Cell, Tissue and Organ Culture* 44(1), 53–61. doi:10.1007/BF00045913.

Pliego-Alfaro, F., Monsalud, M. J., Litz, R. E., Gray, D. J. and Moon, P. A. 1996b. *Plant Cell, Tissue and Organ Culture* 44, 63–70.

Poehlman, J. M. and Sleper, D. A. 1999. *Breeding Field Crops* (4th rev. edn.). Iowa State University Press.

Popenoe, F. W. 1920. The mango in southern California. *Journal of Economic Botany* 1, 153–200.

Pritchard, N. 2007. Efficient and effective implementation of people-related projects. *Industrial and Commercial Training* 39(4), 218–21, doi:10.1108/00197850710755159.

Pritchard, H. W. and Nadarajan, J. 2008. Cryopreservation of orthodox (desiccation tolerant) seeds. In: Reed, B. M. (Ed.), *Plant Cryopreservation: a Practical Guide*. Springer, Berlin, pp. 485–501.

Ramirez, I. M., Rodríguez, N. N., Valdes, I. J., Capote, M., Becker, D. and Rohde, W. 2004. Isolation of genomic DNAs from the tropical fruit trees avocado, coconut, guava and mango for PCR-based DNA marker application. *Cultivos Tropicales* 25, 33–8.

Rao, A. N., Sin, Y. M., Kathayoda, N. and Hutchinson, J. F. 1982. *Tissue Culture of Economically Important Plants*. COSTED, Singapore, pp. 124–37.

Ravishankar, K. V., Anand, L. and Dinesh, M. R. 2000. Assessment of genetic relatedness among mango cultivars of India using RAPD markers. *Journal of Horticultural Science and Biotechnology* 75(2), 198–201. doi:10.1080/14620316.2000.11511223.

Ravishankar, K. V., Mani, B. H. R., Anand, L. and Dinesh, M. R. 2011. Development of new microsatellite markers from mango (*Mangifera indica*) and cross-species amplification. *American Journal of Botany* 98(4), e96–9. doi:10.3732/ajb.1000263.

Rezazadeh, R., Harrison, D. K. and Williams, R. R. 2011. Intraspecific somatic hybridization of mango (*Mangifera indica* L.) through protoplast fusion. *Journal of Applied Horticulture* 13(2), 101–7.

Rivera-Domínguez, M., Manzanilla-Ramírez, M. A., Robles-González, M. and Gómez-Lim, M. A. 2004. Induction of somatic embryogenesis and plant regeneration of 'Ataulfo' mango (*Mangifera indica* L.). *Plant Cell, Tissue and Organ Culture* 79(1), 101–4. doi:10.1023/B:TICU.0000049442.54898.6f.

Sakai, A. and Engelmann, F. 2007. Vitrification, encapsulation-vitrification and droplet-vitrification: a review. *Cryo Letters* 28(3), 151–72.

Schnell, R. J., Olano, C. T., Quintanilla, W. E. and Meerow, A. W. 2005. Isolation and characterization of 15 microsatellite loci from mango (*Mangifera indica* L.) and cross-species amplification in closely related taxa. *Molecular Ecology Notes* 5(3), 625–7. doi:10.1111/j.1471-8286.2005.01018.x.

Sharma, D. K. and Majumder, P. K. 1988. Induction of variability in mango through physical and chemical mutagens. *Acta Horticulturae* 231, 112–6.

Sharma, D. K., Majumder, P. K. and Singh, R. N. 1972. Inheritance pattern in Mango (*Mangifera indica* L.). *Proceedings of the Symposium on Recent Advances in Horticultural*. UP Institute of Agricultural Sciences, Kanpur, pp. 66–8.

Siddiqui, S. H., Mujeeb, K. A. and Wasti, S. M. 1966. Evolution of new varieties of mango (*Mangifera indica* L.) through induced somatic mutations by ionizing radiations. *Proceedings of the Ist Agricultural Symposium Atomic Energy Commission*, Dacca, pp. 34–7.

Singh, R. N. 1996. *Mango*. Indian Council of Agricultural Research, New Delhi.

Singh, H. and Chadha, K. L. 1981. Improvement of Dashehari by clonal selection. *National Symposium on Tropical and Sub-tropical Fruit Crops*. Horticultural Society of India, Bangalore, p. 5 (Abstr.).

Singh, R. N., Singh, G., Rao, O. P. and Mishra, J. S. 1985. Improvement of Banarsi Langra through clonal selection. *Progressive Horticulture* 17, 273–7.

Stanwood, P. C. and Bass, L. N. 1981. Seed germplasm preservation using liquid nitrogen. *Seed Science and Technology* 9, 423–37.

Stiner, M. C., Munro, N. D. and Surovell, T. A. 2000. The tortoise and the hare: small-game use, the broad-spectrum revolution, and paleolithic demography. *Current Anthropology* 41(1), 39–79. doi:10.1086/300102.

Touchell, D. H. and Dixon, K. W. 1994. Cryopreservation for seed banking of Australian species. *Annals of Botany* 40, 541–6.

Usman, M., Fatimaand, B. and Jaskani, M. J. 2001. Breeding in mango. *International Journal of Agriculture and Biology* 3, 522–6.

Viruel, M. A., Escribano, P., Barbieri, M., Ferri, M. and Hormaza, J. I. 2005. Fingerprinting, embryo type and geographic differentiation in mango (*Mangifera indica* L., Anacardiaceae) with microsatellites. *Molecular Breeding* 15(4), 383–93. doi:10.1007/s11032-004-7982-x.

Volk, G. M., Shepherd, A. and Borrart, R. M. 2014. Strategies for improved efficiency when implementing plant vitrification techniques. *Acta Horticulturae* 1039(1039), 85–9. doi:10.17660/ActaHortic.2014.1039.9.

Wang, B., Wang, R. R., Cui, Z. H., Bi, W. L., Li, J. W., Li, B. Q., Ozudogru, E. A., Volk, G. M. and Wang, Q. C. 2014. Potential applications of cryogenic technologies to plant genetic improvement and pathogen eradication. *Biotechnology Advances* 32(3), 583–95. doi:10.1016/j.biotechadv.2014.03.003.

Weber, J. L. 1990. Informativeness of human (dC-dA), (dG-dT)n polymorphisms. *Genomics* 7(4), 524-30. doi:10.1016/0888-7543(90)90195-z.

Wu, Y. J., Huang, X. L., Xiao, J. N., Li, X. J., Zhou, M. D. and Engelmann, F. 2003. Cryopreservation of mango (*Mangifera indica* L.) embryogenic cultures. *Cryo Letters* 24(5), 303–14.

Yamanaka, N., Hasran, M., Xu, D. H., Tsunematsu, H., Idris, S. and Ban, T. 2006. Genetic relationship and diversity of four *Mangifera* species revealed through AFLP analysis. *Genetic Resources and Crop Evolution* 53(5), 949–54. doi:10.1007/s10722-004-6695-7.

Young, T. W. and Ledin, R. B. 1954. Mango breeding. *Proceedings of the Florida State Horticultural Society* 67, 241–4.

# Chapter 18

# Advances in mango cultivation

*Víctor Galán Saúco, Instituto Canario de Investigaciones Agrarias, Spain*



## 1 Introduction

The mango (*Mangifera indica* Linn) is the most important species of the Anacardiaceae family, due to its worldwide distribution and economic importance. Mangoes are the third most common tropical fruit after bananas/plantains and pineapples, and are ranked fifth among all fruits worldwide, with apples and grapes as the top two fruits. The mango is commercially cultivated in all continents, both in the tropics and subtropics from 36° north latitude in Málaga, Spain, to 33° south latitude in South Africa. According to FAO statistics, mangoes are produced in 100 countries (105 if including Spain, Portugal, Italy, Morocco and Turkey, which are not listed by the FAO). World mango production has increased steadily from 15700 × $10^3$ tons in1990, 25040 × $10^3$ tons in 2000, 30880 × $10^3$ tons in 2006, 42140 × $10^3$ tons in 2012 to 46509 × $10^3$ tons in 2016, although FAO statistics include mangosteens and guavas under the same heading. Due to its widespread geographical distribution and the development of cultural techniques that enable out of season flowering, it is possible to supply world markets with mangoes throughout the year.

The mango plant has long been a cultural and religious symbol in its area of origin, i.e. South East Asia, but it is also the most popular fruit in many Latin American countries, where it is an important part of local tradition and culture. It has a similar status in many African countries (Galán Saúco, 2016).

http://dx.doi.org/10.19103/AS.2019.0054.28

## 2 Genetic considerations

The mango possesses 40 chromosomes and has been described as allotetraploid (Mukherjee, 1950). Tetraploidy was first identified in 1951 by Roy and Viswewarya (cited in Majumder and Sharma, 1985) and later proven via flow cytometry and chromosome counting (Galán Saúco et al., 2001; Rayner, 2002), and appears to be common in polyembryonic mangoes (Smith et al., 2008; Galán Saúco, 2008). There are two main groups of mango: polyembryonic and monoembryonic. In polyembryonic types, adventitious embryos grow from the nucellar maternal tissue giving rise to genetically identical progenies, while in monoembryonic types, a sexual embryo develops, giving rise to different individual seedlings. Polyembryonic types are generally used as rootstocks, while most commercial global mango cultivars, with a few exceptions - 'Kensington Pride', a polyembryonic mango, being the most well known - are monoembryonic types.

Several mango breeding programmes exist, some of which have been active for more than 50 years in certain countries, e.g. Israel (Tomer et al., 1997), Australia (Bally et al., 2009) South Africa (Human et al., 2009), Brazil (Rossetto et al., 1997) and India (Dinesh et al., 2016). However, the majority of worldwide commercial cultivars are not a product of a breeding programme but have resulted via the selection of chance seedlings. The Floridian cultivars 'Tommy Atkins', 'Kent', 'Keitt' and to a lesser extent 'Osteen', 'Palmer', 'Haden' or 'Irwin', dominate the global fresh mango export market. Other cultivars available on the international market originated through selection of chance seedlings such as 'Ataulfo' from Mexico, 'Madame Francis' from Haiti, 'Nan Doc Mai' from Thailand, 'Carabao' from the Philippines or 'Alphonso', 'Sindhri' and 'Chawsa' from India and Pakistan, respectively (Galán Saúco, 2017). Although many different cultivars have been obtained through breeding programmes, some are protected by Breeders Rights. The only ones that are marketed on a commercial scale outside the country of origin are the Israeli cultivars 'Omer', 'Shelly', 'Naomi' or 'Kastury' and 'Calypso' from Australia (Galán Saúco, 2017; Cohen et al., 2016). A few others cultivars from the Indian breeding programme (Dinesh et al., 2016) merit to be mentioned, particularly 'Mallika', recommended for planting in Florida (Ledesma, 2016; Ledesma and Campbell, 2016) or 'Amrapalli' has long been used for high-density planting (Majunder and Sharma, 1988). As explained by Galán Saúco (2019), only one rootstock, 13-1 from Israel, was obtained through selection work (Gazit and Kadman, 1980).

The development of new mango cultivars via traditional mango breeding programmes is a slow process, and generally takes over 20 years (Tomer et al., 1997; Bally et al., 2009 et passim). However, the development of molecular markers and new generation sequencing technologies will considerably improve the results of worldwide mango breeding programmes by including, as per the Australian breeding programme (Bally, 2013), marker-assisted screening, disease screening, post-harvest evaluation and a genomic gene discovery programme.

The monoembryonic characteristic of many mango plants and the high level of heterozygosity of the species (Iyer and Degani, 1997), in combination with the long period of mango cultivation in different geographical and climatic regions of the world, has given rise to a large number of different cultivars. Until late into the 20th century, mangoes were classified via morphological and phenological characters using descriptors (IPGRI, 2006). These descriptors vary according to the influence of the environment, phenological phase of the plant and the fact that the limited number of discriminating traits (Khan et al., 2015) are not sufficient to accurately identify cultivars. Advances over recent decades in the development of different molecular markers for fruit tree crops (Larrañaga and Hormaza, 2016) are an extremely important tool for future mango breeding, and are an essential and highly reliable complement to morphological and phenological markers. Two excellent reviews (Pérez and Hormaza, 2018; Bally and Dillon, 2018) of advances in mango genomics have recently been published and the most relevant facts are summarized below.

Biochemical markers, based on the separation of allozymes and isozymes, were extensively used from the 1970s to the end of the 20th century. These markers enabled the discrimination of genotypes initially considered to be clones and the study of the paternity of commercial varieties. However, they are not currently widely used due to the large number of DNA markers that have been developed over the last four decades.

Restriction fragment length polymorphism (RFLP) markers are of limited use in mangoes; however, they have been useful for the analysis of interspecific relationships among species of *Mangifera* or, in combination with randomly amplified polymorphic DNA (RAPDs), separating monoembryonic and polyembryonic cultivars. RAPDs have been extensively used in mango genetic diversity studies, particularly to estimate the genetic relationships among cultivars and group them according to geographical origins, in addition to distinguishing sexual and nucellar seedlings in polyembryonic cultivars. However, due to problems in result reproducibility and the development of more powerful molecular markers, their use in plants in general, and in mangoes in particular, has diminished significantly over recent years.

Amplified fragment length polymorphism (AFLPs), another DNA markers, are a valuable approach to study genetic relationships between mango genotypes and for the construction of genetic linkage maps. In addition, they have been used for the analysis of phylogenetic relationships among *Mangifera* species. Inter simple sequence repeats (ISSRs) or simple sequence repeats (SSR) anchored primers have been used to identify mango cultivars and establish geographical groupings. One important recent contribution of

ISSR markers, outlined by Pérez and Hormaza (2018), has been shown through the identification of zygotic and nucellar seedlings, which demonstrated that in the Brazilian cultivar 'Uba', the most vigorous seedling is not always of nucellar origin (Rocha et al., 2014). ISSR markers have also recently been used to establish genetic diversity among *Mangifera* species (Ariffin et al., 2015).

Pérez and Hormaza (2018) highlighted that microsatellites or SSRs have resulted in extensive progress in mango cultivar identification, paternity breeding and germplasm conservation, indicating that the SSRs developed for mangoes can also be used for closely related species of *Mangifera*. They have been widely used for molecular characterization and genetic diversity studies in different mango-producing areas of the world. SSR and expressed sequence (EST)-SSR markers have been used for mapping, parentage assignment and verification, as well as inferring phylogenetic relationships between *Mangifera* species and varieties (Bally and Dillon, 2018). The use of these molecular markers is of particular relevance for mango breeding programmes as they have enabled confirmation of previous results obtained through AFLPs indicating that although selfing can occur, mangoes tend to produce fruits via cross fertilization (Pérez-Barraza et al., 2015, 2016). Other advances in the use of molecular markers for mangoes, also reviewed by Pérez-Barraza et al. (2018), involve obtaining an EST-SSR for several mango cultivars, the use of sequence-tagged microsatellite sites (STMS) or start codon targeted (SCoT) polymorphism markers and other less common molecular markers for genetic diversity analysis of mangoes. The authors also indicated the potential of next-generation sequencing (NGS) technologies for mangoes, e.g., sequencing the whole genome of *Pseudomonas syringae*, pv. *Syringae*, the causal agent of apical necrosis, and the identification of virulence genes (Martínez García et al., 2015). Recently developed cleaved amplified polymorphic sequence (CAPS) markers for mangoes enable the identification of true hybrids in F1 progeny and are a simple, rapid and highly reliable tool for efficient mango breeding (Bally and Dillon, 2018).

Two reviews (Pérez and Hormaza, 2018; Bally and Dillon, 2018) report the considerable advances made in mango 'omics', particularly in the characterization of transcriptomics of different mango tissues which, when combined with NGS, can greatly enhance the existing knowledge of mango gene function and regulation. They conclude that, judging from the work carried out by research groups in Australia, the United States, Israel and India, we may not be far from sequencing the whole mango genome, which will be an extremely powerful tool to facilitate future mango breeding. Using NGS technologies, Sherman et al. (2015) identified 332 016 single nucleotide polymorphisms (SNPs) and used a subset of 293 SNPs to analyse genetic diversity in the Israeli mango collection, whereas Kuhn et al. (2016) reported the development of approximately 400 000 SNPs.

Finally, the recent generation of a mango consensus genetic map, based upon individual maps from seven F1 hybrid populations, will improve the efficiency and help overcome the challenges facing mango breeding programmes (Khun et al., 2017). Markers and regions of the genome associated with important horticultural traits, such as embryo type, branch habit, bloom, ground skin colour, blush intensity, beak shape and pulp colour, have been identified. As indicated by the authors, these findings are a clear demonstration of the application of genomic tools for the breeding and selection of improved mango cultivars, although markers associated with important quantitative traits must also be found to further improve mango breeding efficiency. To tackle this objective, new projects by the same research group are being undertaken in order to find the association of quantitative traits with SNP markers.

Another advance for mango cultivation that may arise in the near future comes from the possible use of closely related *Mangifera species* in breeding. The genus *Mangifera* consists of up to 69 species (Kosterman and Bompard, 1993) which exhibit great genetic diversity and are distributed over a wide geographical area under different environmental conditions. Some of them, such as *Mangifera laurina, M. griffithii* and *M. quadrifida* are tolerant to flooding. *M. laurina* also exhibits resistance to anthracnosis, while others like *M. sylvatica* are resistant to cold conditions (Bompard and Schnell, 1997). Many of the wild *Mangifera* species, particularly *M. casturi*, *M. lalijiwa*, *M. odorata*, *M. zeylanica* (Barbosa Gámez et al., 2017) and *M. caesia* (Ledesma, 2016) have been widely cultivated for fruit in their area of origin. A number of them are graft compatible with the mango (Bompard, 2009; Campbell and Ledesma, 2013; Ledesma, 2017), and others, such as M. *laurina*, *M. lalijiwa* and *M. Kasturi* are already used as rootstocks; the former in Borneo, *M. lalijiwa* in Hawaii and *M. kasturi* both in Indonesia and Hawaii (Galán Saúco, 2019). Recent work by Campbell and Ledesma (2013) and Ledesma (2017) at the Fairchild Tropical Botanical Garden (FTBG) have shown not only the cross compatibility of many of these *Mangifera* species but also some successful crosses with selected mango cultivars. More than 33 accessions of *Mangifera* species are present at the FTBG collection (Ledesma and Campbell, 2016) and although obtaining interspecific hybrids is still in its infancy (Ledesma and Campbell, 2017), the possibility of a new generation of rootstocks and/or commercial cultivars adapted to diseases, abiotic stresses and perhaps climate change, derived from these interspecific crosses is highly promising.

## 3 Physiological aspects

Two reviews focusing on physiological aspects of mangoes have been recently published, one by Normand and Lauri (2018), dealing with tree growth and development, and the other by Pérez-Barraza and Osuna-García (2018) mainly devoted to the flowering process. The more relevant aspects of these two important papers are summarized below.

The mango tree exhibits rhythmic growth, terminal flowering and both monopodial branching - growth and branching of an axis with an active apical bud - and sympodial branching - development of new axes from lateral buds once the apical bud is transformed into an inflorescence. It also has an acrotonic branching pattern with the development of new axes appearing at the end of the parent growth unit and only rarely also appearing at the distal end of the penultimate growth unit. In summary, a mango tree can be described as a succession of growth units connected either through the activity of the apical bud as it forms an axis or through the activity of lateral buds that form ramifications. As indicated by Normand and Lauri (2018), following the previous work of Hallé and Oldeman (1970) and Goguey (1995, 1997), mango tree architecture follows the Scarrone model.

As in many other plant species, temperature is the main factor affecting mango tree growth and fruit development (Galán Saúco, 2008). However, for mangoes, the time needed to complete growth unit development is cultivar dependent with different cultivars responding differently to temperature (Dambreville et al., 2015). As indicated by many researchers (Normand and Lauri, 2018), mango tree phenology is largely determined by environmental factors such as temperature and water availability, and the control of these two factors are of crucial importance to improve mango cultivation. Other factors, such as tree age, also affect the pattern of vegetative growth of young seedlings or mangoes. During the early non-flowering phase, flushes are regular and continuous, independent of environmental conditions, whereas old trees generally produce only one or a very reduced number of vegetative flushes per year. More recent studies have shown that other endogenous factors exert considerable influence upon mango phenology and must be considered in order to improve cultural practices. In fact, as highlighted by Normand et al. (2009) and Dambreville et al. (2013), apical growth units have a higher probability of burst than lateral growth units. However, other factors, such as the burst date of the growth unit and the number of sister growth units, also influence the probability of flushing, which varies between cultivars. Along this line, Dambreville (2013) indicated that within a season favourable for vegetative growth, the burst date of daughter growth units is positively correlated with the burst date of the mother growth unit, although Whiley et al. (1989) reported that the delay between both dates depends on the temperature: the higher the temperature, the shorter the delay.

Normand and Lauri (2018) reported that the vigour of the current-year branch positively influences the number of daughter growth units produced during the following period of vegetative growth and the burst probability of a growth unit. Referring to the work of different researchers, they also indicated the existence of differences among cultivars in relation to the pattern of vegetative growth, in terms of the number and date of vegetative flushes and the number of growth units produced. The variation between cultivars as a result of genetic differences and environment conditions explains why the vegetative growth pattern of the same cultivars behave differently in the tropics compared with the subtropics. There is a close relationship between vegetative growth and reproductive growth, both at the growth unit level, the scaffold branch and the whole tree level, with reproduction having a negative effect on subsequent vegetative growth and heavy fruit load delaying vegetative growth, indicating that this effect is also cultivar dependent (Normand, 2016). The delay in vegetative growth in fruiting terminals explains why, in subtropical areas with a low temperature, especially with a heavy load and late cultivars, there is no vegetative growth after harvest or this only occurs in the non-fruit set terminals and the tree enters an alternate bearing cycle or exhibits phenological asynchronism. As a strategy for cultivation, alternate bearing is also reinforced in areas of the subtropics through pruning. The negative effect of endogenous factors, such as fruit overload and vegetative growth, on the reproductive process is also reported by Pérez-Barraza and Osuna-García (2018), stating that a very heavy load or excessive vegetative growth can inhibit the process of flower induction due to the large quantity of gibberellins present in the tree.

Several authors cited by Normand and Lauri (2018) have highlighted the requirement of a rest period, which is variable and cultivar dependent, for the growth unit to produce flowers; the earlier appearance during the vegetative growth period, or even during the previous season, being more prone to flower than those of recent emergence. This reinforces the previously mentioned asyncrony in flowering and, as indicated by the authors, supports the theory first mentioned by Davenport (2009), of a negative relationship between the age of the growth unit and the requirement for cool temperatures to induce flowering. Other authors, mentioned by Pérez-Barraza and Osuna-García (2018) agreed that under subtropical conditions, the stimulus for floral induction is low temperature, but recent work indicates that in tropical conditions low temperatures are rare or non-existent and that the growth age of the last vegetative flush is the dominant factor regulating induction. Shoots must be in a state of dormancy long enough, (normally from 7 weeks to 5 months), to induce flowering in the absence of low temperatures. Again, there is a clear cultivar effect, with different cultivars requiring different periods of rest. Furthermore, as flowering has been observed to occur in the cultivar 'Ataulfo' under the tropical conditions of Nayarit, Mexico, with temperatures of 27/17°C (day/night), in shoots between 2 and 4 months old (Pérez-Barraza, 2015), they concluded that under tropical conditions, age and temperature are both important to induce flowering.

The position of the growth unit is also relevant for flowering. Normand (2009) stated that terminal growth units have greater potential for flowering than lateral ones and also reported the existence of a quadratic relationship between diameter of the growth unit and flowering probability; the bigger the diameter, the greater the probability of flowering. Dambreville et al. (2013) also indicated that the position of the growth unit also affects the number of inflorescences produced and the date of flowering, but again this is cultivar dependent. They also confirmed the positive effect, reported by Issarakraisila et al. (1997), of current-year branch vigour on flowering, but indicated that this effect varies according to the cultivar. In addition to the interrelationships mentioned above, it is worth noting that Capelli et al. (2016) found a positive and linear relationship between vegetative growth and the number of fruits produced during the same growing cycle, but again this is cultivar dependent.

As indicated by Normand and Lauri (2018), cultural practices for the successful cultivation of mangoes should be based on flowering and fruiting, but consideration should also be given to vegetative growth and its relationship with the reproductive process. The knowledge of these relationships may be useful in pruning, particularly if removing wood to allow more light to the internal canopy (see the section of cultivation practices). However, as they also highlight, other considerations should be taken in account in order to understand the behaviour of mango trees. First, terminal growth units are not independent and the relationships described above are affected by other factors at different hierarchical scales (i.e. branch, scaffold branch or whole-tree scale). As they also state, the level of fruit production affects their physiology, in particular their carbohydrate content, and therefore their ability to grow and flower. Consequently, there is a need for future studies to look at different hierarchical scales, in addition to those carried out mainly at the growth unit level, to better understand mango tree behaviour. In order to obtain the maximum sustainable yield from mango trees, the cultivar dependence relationship, as well as the cultivar environmental relationship, must be considered for cultural practices that also have an important influence on reproductive and vegetative processes.

There have been recent advances in identification, quantification and expression of genes related to flowering, also reviewed by Pérez-Barraza and Osuna-García (2018), which help to understand the reproductive process and can be important in solving the problem of irregular flowering. Nakagawa et al. (2012) identified and characterized the *MiFT* gene in the 'Irwin' mango, which increases its expression under inductive low temperature conditions and in trees with low fruit load in mature leaves. Recent work by Pérez-Barraza (2015) in Mexico also found that the expression of this gene in mature leaves of the 'Ataulfo' mango decreased after the application of gibberellins and increased after the application of gibberellin inhibitors, which corroborates the role of this gene in the flowering process in mangoes.

## 4 Cultivation techniques

Increasing planting density is probably the most important cultural practice that has changed in recent years. As with many other fruit crops, mangoes are currently planted at a much higher density. Compared with traditional plantings at densities of 8 × 8 to 10 × 10 m (Mitra and Bhagwan, 2018), modern mango plantings are established at densities around 7 × 4 m in the tropics and 6 × 2, 3 × 4 or 5 × 1.5 m in the subtropics (Fig. 1) (Galán Saúco, 2008, 2018b; Mitra, 2018). Of particular interest are the ultra-high-density plantings, which were initiated in India after the discovery of the dwarf hybrid cultivar 'Amrapalli' (Majumder et al., 1982) with plantings at 2.5 × 2.5 m (1,600 pl/ha). This was further developed by Oosthuyse (1993, 2005) in South Africa who proposed ultra high-density planting of mangoes at densities of 3 × 1.5 m, with rows oriented in the north–south direction to obtain maximum light. The benefits of this system include: a reduced time from planting until the trees attained optimal canopy cover and maximum fruit production, reduced disease and pest pressure, reduced solar injury and wind damage to trees and fruits, reduced requirement for the control of weeds and grass growth, markedly increased ease of carrying out tree and fruit manipulations and harvesting. Hedge dimension containment is vital for the system's success, which is achieved by selective branch heading shortly after fruit set and general branch heading after fruit harvest. The cost of establishment can be greatly reduced by planting seeds in the land directly upon commencement, which increases cost efficiency. The system is particularly amenable to drip irrigation and fertigation. A report of the excellent results of an orchard of 'Tommy Atkins' planted at this density in 1996 has recently been published (Oosthuyse, 2018).

In addition to plantings in India, ultra-high-density plantings have been successfully established, following Oosthuyse's recommendations, in South Africa, Egypt, Mexico and Peru with the Floridian cultivars 'Tommy Atkins', 'Kent', 'Irwin', 'Keitt', with the South African cultivar 'Heidi' and with the Egyptian cultivars 'Eweisse', 'Saddik' and 'Hindi Bi Sinnara'. The 'Keitt' cultivar is particularly well adapted to this system because its usually heavy load keeps the tree canopy smaller and it can be cultivated on trellises due to its long spreading flushes. Planting trellise systems is common in Malaga, Spain, at planting distances of 2.5 × 2 m, and palmetto systems are found in El Algarve, Portugal, at 2.0–3.0 × 1.5–2.5 m (Galán Saúco, 2015).

Ultra-high-density plantings are more adapted to the subtropics due to the short non-productive phase, the reduced number of vegetative flushes per year and the regularity of mango flowering. This allows closer distances between mango plants than in the tropics where tree growth is much greater due to year-round temperatures suitable for mangoes (Galán Saúco, 2018b).

**Figure 1** Higher density in the suntropics.

As a result, the use of a growth retardant, e.g. paclobutrazol (PBZ), is proposed in warmer areas for successful ultra-high-density planting (Oosthuyse, 2018). However, as PBZ is forbidden in many countries, other growth constraints such as a heavier soil, moderate salinity problems or a dwarfing rootstock may aid in the establishment of ultra-high-density plantings. The success of the system depends on the right application of pruning procedures at specific times during the yearly growth cycle in order to obtain suitably mature terminal shoots when conditions are inductive for flowering. Despite these limitations, for many mango researchers, the ultra-high-density system of producing mangoes is the most appropriate solution to overcome low productivity and the long gestation period for the early return and export of mangoes from convenience and productivity aspects (Banik et al., 2013; Oosthuyse, 2018).

Training and pruning procedures for modern mango plantings have been described by many researchers (Oosthuyse and Jacobs, 1995; Oosthuyse, 1997, 2005; Galán Saúco, 2008; Crane et al., 2009). They are well known by efficient farmers throughout the world and enable a quick well-aerated ramified structure that can fruit from the third year or even in the second year for cultivars that are able to produce additional vegetative shoots concurrently with fruit growth, e.g. 'Heidi' and 'Zill'. Modern mango cultivation, and especially ultra-high-density plantings, require careful training and pruning from planting and under subtropical conditions will require control of flowering even in the nursery (Galán Saúco, 2018b) because the low temperature stimulus in the cool subtropics may enable trees to flower shortly after grafting (Fig. 2), especially when the nursery is placed is in a cooler location and climatization is not provided and if flowering is not avoided or destroyed it can weaken the tree by depleting reserves. Specific measures to avoid this problem include spraying giberellic acid ($GA_3$) during the winter, which is useful in delaying annual flowering also in adult trees in the subtropics (Galán Saúco, 2008). However, the best option is hand-pruning the flowers that appear immediately below the terminal node when fruits reach pea size (stage 701 of the BBCH scale) (Hernández Delgado et al., 2011).

It is worth highlighting the possibility of training by combining bending with pruning, following the procedure detailed in an experiment by Ledesma et al. (2016). This procedure involves tipping each central leader shoot of the grafted trees at 50 cm to induce precocious branching while maintaining all lateral shoots that are tipped after the second vegetative flush. Each branch is then pulled down to the horizontal. Control trees received initial pruning and tipping, but no branch bending (Fig. 3). 'Mallika' mangoes used in this experiment were planted in May 2015. All bent trees started blooming in March 2016, only 10 months after field planting and had a significant increase in trunk diameter as well as lateral growth and spread over the treatment time. In contrast, most control trees did not flower during this season and finished the experiment taller, with a significant reduction in lateral growth and spread, and with a branch system more susceptible to damage by wind or heavy rains. Although more time is needed to properly evaluate this bending procedure, it looks promising when trying to develop and maintain small, mushroom-shaped trees that may be capable of the early production of large crops and high-quality fruits in the tropics. Although the procedure

**Figure 2** Precocious flowering at the nursery.

**Figure 3** Training of tres with combining bending and pruning. Courtesy: N. Ledesma, Fairchild Tropical Botanical Garden.

should be validated for different cultivars and different characteristics should be assessed, e.g. the height of the first tipping in different locations and the long-term effect on the full production phase, it looks promising for mango cultivation.

Cultural practices for flower induction in the tropics, as well as for the control of flowering in the subtropics to avoid coinciding with low temperatures that can lead to embryo abortion, have been well described

(Davenport and Núñez Elisea, 1997; Davenport, 2009; Tongumpai et al., 1997; Galán Saúco, 2008, 2015; Torres et al., 2009; Hermoso González et al., 2018 et passim).

An interesting practice for the subtropics has been developed in Malaga, Spain, (Hermoso González et al., 2018) where medium to late season cultivars, such as 'Osteen' or 'Kent' and late season cultivars such as 'Keitt', do not produce vegetative growth after harvest, even after quick pruning of the terminal, and the trees are committed to alternate bearing, resulting from the short period with good temperatures after harvest. In order to obtain tree uniformity, this alternate bearing tendency is reinforced through pruning by applying the following procedure (Fig. 4):

Year ON (high yield). Total or partial machete pruning of terminals to induce a 2nd flowering with good moderate sized fruits.
Year OFF (without fruit). Spring pruning of productive shoots of the previous year's vegetative shoots for the next 'ON' year.

This practice allows a good crop every 2 years and has the additional advantage of obtaining smaller sized and more coloured fruits, which is commercially

**Figure 4** Alternate bearing pruning system.

beneficial particularly for the cultivar 'Keitt' that otherwise produces green to yellow fruits with sizes too big for the European market.

Growth regulators, e.g. PBZ, are commonly used in mango orchards, on many cultivars, and their positive effect on mango flowering is well documented. There is evidence that PBZ is able to induce flowering in the tropics without low temperatures or PBZ application may complement the need for low temperatures for floral induction (several authors cited by Pérez-Barraza and Osuna-García, 2018). PBZ is still used in many mango-producing countries in the tropics, but if detected, mangoes will be rejected in certain markets. Consequently, other growth regulators that can be approved for mangoes must be identified. Recent studies using the cultivar 'Ataulfo' in Mexico indicated that P-Ca could be a possible substitute to PBZ causing similar effects by reducing vegetative growth and increasing and advancing flowering (Pérez-Barraza et al., 2016)

Work involving various cultivars in different countries has also demonstrated the beneficial impact on fruit set of spraying $GA_3$ at flower bud differentiation or at full bloom (several authors cited by Pérez-Barraza and Osuna-García, 2018), which is a finding of great interest in subtropical areas when flowering occurs under low temperature conditions. However, as Singh (2009) have shown, under the Mediterranean climate of Western Australia, spray applications of $GA_{13}$ or $GA_4/A_7$ resulted in higher mean fruit set compared with $GA_3$ or $GA_7$. $GA_{13}$ (10 ppm) gave the best results regarding final fruit retention, illustrating the need for further studies on the effects of different types of gibberellins in mango cultivation using different cultivars and environments.

Recent work by Pérez-Barraza et al. (2015) indicated that the application of 50 $mg{\cdot}L^{-1}$ of cytokines to 'Ataulfo' at full bloom stage increased the size of seedless fruits with no impact on normal seeded fruits. This could be good practice to reduce the problem of embryo-aborted fruits that is very common in the subtropics when flowering occurs under low temperature conditions (Galán Saúco, 2008). Pérez-Barraza and Osuna-García (2018) also indicated that the use of combined applications of growth regulators such as thidiazuron (TDZ) and $GA_3$ are effective in increasing fruit size of the seedless fruit of the 'Ataulfo' mango.

It is generally accepted that mango flowers are mainly pollinated by many different insects, particularly *Diptera* and *Hymenoptera*, with houseflies and bees (*Apis cerana* and *A. mellifera*) being the best mango pollinators. Consequently, it has been suggested that the reduced use of insecticides and presence of small areas of native flowers close to mango orchards is beneficial. In places like La Mayora in Malaga, artificial structures similar to bee hives (Fig. 5) are also placed on the border of mango experimental plantings to attract common

**Figure 5** Artificial bee hive.

bees, whose role in pollination is proven for many crops and may be good pollinators for mangoes (Nates Parra, 2005).

As indicated in the review by Levin (2018), many studies have focused on mango water use and irrigation including the assessment of pre-flowering irrigation treatments, pre-harvest irrigation cessation and the effect of season-long reduced volume applications on fruit quantity and/or quality, the effects of irrigation at different phenological periods and the effects of pre-flowering irrigation on trees receiving flower-promoting treatments; however, the irrigation needs of mango have not been well investigated. In particular, very few studies have investigated regulated deficit irrigation (RDI) strategies at different phenological stages, with the most recent studies focusing mainly on developing different irrigation strategies to save water and increase water use efficiency, rather than maximizing crop production.

Mangoes are a drought-resistant crop with adaptive features that enable drought tolerance, such as deep and long-life tap roots, coriaceous leaves with thick cuticles and recycling resin ducts to reduce wilting (Bally, 2006). Succesful mango farming depends on the application of effective irrigation and fertilization management, based on modern techniques and proven strategies, because water deficits during the reproductive cycle can adversely affect fruit retention and early fruit growth (Schaffer et al., 2009; Whiley and Schaffer, 1997) and consequently, the final yield.

As in many other crops, studies to determine mango irrigation requirements have been based on two different approaches (Levin, 2018):

1 Those related to determining the crop coefficient (Kc) based on local or regional evapotranspiration (ETP) or pan evaporation.
2 Those based on the use of plant- or soil-based parameters.

Crop coefficients have been established for mango in different locations and are used to determine mango irrigation needs in Australia, Brazil, Israel and many other countries.

In the second approach, the use of sensors to evaluate the plant's response to water deficits rather than sensing the soil moisture status directly have been reported in recent years although, as Levin (2018) indicated, both methods have not yet been widely adopted for mangoes. Very few reports exist on the use of soil sensors to establish mango water requirements, but there are studies investigating the use of plant sensors, including leaf water potential measurements, xylem sap flow in the tree trunk, micro-variation (shrinking and expansion) of branch/trunk diameter (microdendrometry) and canopy temperature as indicators of water needs (several authors cited by Levin (2018). Of particular interest are two works of Lu (2005, 2013). In the first, he developed a 'shrinkage index' to evaluate trunk diameter variations indicating that changes of this value appear to be a good indicator of the onset of water stress. Results from the second study showed a general correlation between whole-tree water use (sap flow) and tree size (trunk diameter) for 'Kensington Pride', 'Irwin' and 'Nam Dok Mai' grown in Darwin and Katherine (Australia), and also indicated that sap flow was not as sensitive as microdendrometry and was also a less sensitive indicator of water stress under field conditions in high water-holding capacity soils. Nevertheless, Lu concluded that the use of plant sensors could be very useful for research methods, but are not well adapted to farm conditions.

The most interesting finding is the possibility of applying a deficit irrigation (DI) strategy to mangoes. The review by Levin (2018) concluded that partial root zone drying (PRD) and regulated deficit irrigation (RDI) strategies can potentially be applied to mango trees without significantly affecting yield parameters but increasing, not always significantly, the water use efficiency. However, as he also mentions, most of the studies on DI in mangoes have been in locations with relatively low yields (15–25 tons/ha) and may not be extrapolated to highly productive farms. In the case of Israel, with much higher yields, experiments by Levin et al. (2015a,b) with the cultivar 'Keitt' identified that contrary to previous thoughts, the post-harvest period is the most sensitive phenological period to irrigation and that reduction of water during this period has a great impact on yield, in terms of quantity and quality; whereas water reduction during the period from fruit set to hardening of the seed will mainly impact the total number of fruits but not necessarily final yield. Water reduction during the period between seed hardening and harvest will also mainly affect final fruit size and yield. These considerations clearly illustrate the interaction between the vegetative and reproductive periods commented upon in the physiological section of this chapter. As Levin (2018) indicates, this has an important impact on optimal Kc values that will depend both on the phenological period and production level and that as a result of this study, growers in Israel have made important water savings of around 25–35% per season.

Finally, it should be mentioned that although drip irrigation may provide the most efficient way to irrigate from a resource (water and fertilizer) and water and fertilizer management point of view, as indicated by Lu (2013) there are some disadvantages. The main constraints for drip irrigation are economic factors such as the high cost of spare parts, high initial expenses for installation of a drip irrigation system, minor repairs, inadequate and uncertainty of the power supply, frequent clogging of drippers and microtubes due to poor water quality (e.g. high level of organic material dissolved in the water). These limitations explain why in Australia, and other parts of the world, micro-sprinkers are commonly used in mango orchards.

As with any crop, appropriate nutrition is essential to enable plant growth and development, better tolerance to pests and diseases, and for maximum yields of fruits of good quality and shelf life. In order to obtain the best results from fertilizer practices, nutrients should be applied at the right time according to plant needs using the most appropriate source.

It is not in the scope of this chapter to discuss mango fertilization practices; however, this subject has been detailed by de Carvalho Genú and de Queiroz Pinto (2002), Litz (2009), Galán Saúco (2008) and Oosthuyse (2006). Furthermore, the latter author gives specific recommendations for fertilization of ultra-high-density mango plantings (Oosthuyse, 2005), and recommendations for organic fertilization have also been published in India (Chadha et al., 1980, cited by Iyer, 2004). The only significant new input on the subject deals with the role of magnesium (Mg) in mango fertilization. Appropriate Mg fertilization improves yield, vigour and general growth of the plant, and influences skin colour, internal fruit breakdown and shelf life as well as improving tolerance to cold conditions, high solar radiation and hot water treatment (Galán Saúco, 2018a).

As Levin (2018) recently highlighted, fertigation (fertilizer applied through the irrigation system) in a 100% automated irrigation system potentially has the best results for mango nutrition. However, he indicates that this may not be a good option in the case of organic cultivation because of the following constraints:

- The limited number of organic liquid fertilizers and the different combination of minerals compared with conventional ones.
- The low mineral concentration of these organic fertilizers, particularly in N (usually < 5%).

Due to these constraints it is not possible to apply a much higher amount of fertilizer per irrigation via the fertigation system in order to incorporate total fertilizer needs. In addition, the alternative of using a solid organic fertilizer requires more labour and the nutrient concentration is not always known. An organic mango orchard is, however, a relatively new production system with advantages and disadvantages and, due to an increased market demand, is being implemented in many mango-producing areas around the world. It is out of the scope of this chapter to discuss cultural practices and recommendations for organic cultivation of mangoes, but the reader can find a recent review in Medina-Urrutia et al. (2018).

Greenhouse cultivation of mangoes is currently only carried out in the subtropics, e.g. Japan (Yonemoto, 2018), Spain (both in the Canary Islands and Malaga) and Portugal (El Algarve) (Galán Saúco, 2015) with around 100 ha in the last two countries, and around 150 ha in Japan. Despite the small area still cultivated under this system, the potential of protected mango cultivation both for the tropics and subtropics, is high and interest in consumption of tropical fruits in places such as The Netherlands is so high that they are already considering producing tropical fruits under greenhouse in this country using technology similar to vegetable or flower production (Bakker, 2018).

It should also be mentioned that effort is focused on cultivating mangoes in heated greenhouses in Hokkaido Islands (Japan) with winter temperatures reaching −20°C. These efforts are driven by the high prices obtained in this country by greenhouse-produced mangoes harvested at the full maturity stage, which can compensate their higher cost of production. The possibility of delivering mangoes to the European Union market in a fresh ready-to-eat condition is also an important incentive for mango greenhouse cultivation in Spain.

The main theoretical advantages for protected mango cultivation, already reported by Galán Saúco (2015), are:

- Shortening of the young, unproductive period without causing premature ageing of the trees.
- Protection against adverse climatic conditions (hail, wind, sunburn, low winter temperature, rain during flowering).
- Increased flowering (not in the tropics) and fruit set.

- Increased foliar surface and consequent increase of photosynthetic capacity.
- Better control of irrigation and the possibility of applying water stress to obtain out of season production.
- Advance of harvesting season due to a temperature increase (not in the tropics).
- Possibility of obtaining good late-harvesting crops (not in the tropics).
- Better fruit presentation.
- Reduction of sunburn damage.
- Better and easier control of pests and diseases.
- Better possibilities for organic cultivation due to lower pest incidence.

Some of these advantages need further explanation. Mangoes cultivated in greenhouses in the subtropics suffer the same problems with low winter temperatures as those cultivated in the open, but the higher diurnal temperatures obtained during the whole year inside the greenhouse, or heating if applied, allow a shortening of the unproductive phase. It is also a well-known fact that to obtain abundant flowering after flower induction, higher temperatures are required to avoid any additional stress that can prevent the triggering of flowering (Albrigo and Galán Saúco, 2004). As a result, the higher diurnal temperatures occurring under greenhouse conditions enable excellent flowering and a consequent increase of fruit set. These advantages can not occur under tropical conditions where temperatures are optimum for mango growth and development. However, protection against climatic factors, i.e., rainfall, during flowering may also prevent the incidence of diseases such as anthracnosis (see below) giving rise to higher fruit set, which also occurs in the tropics.

It has been observed that under subtropical Mediterranean conditions no chemical treatment against Mediterranean fruit fly (*Ceratitis capitata*) is required, which is simply controlled by the use of fly traps in the greenhouses. Treatments against bacteriosis (*Pseudomonas syringae*), transmitted by dew or rainfall, or against powdery mildew (*Oidium mangiferae*), transmitted by the wind, are greatly reduced to 2–3 foliar applications in contrast to the 9–10 required in the open air (E. Guirado, CSIC, La Mayora Experimental Station, Pers. Comm., 2013). The control of anthracnosis is also unnecessary, as it is transmitted through water splashes that do not occur during greenhouse cultivation (Galán Saúco, 2008). These advantages, i.e. better pest and disease control, are found in the subtropics and tropics. In an experiment in Thailand (Juntamanee et al., 2013), the quality of fruits of 'Nam Doc Mai' cultivated under plastic was greatly improved compared with those cultivated in the open due to a significant reduction of anthracnosis incidence (during pre- and post-harvest) and to the complete control of thrips damage on the fruit surface. It is also important to mention that in this experiment, an increase of both net photosynthesis and stomatic conductance was observed as a consequence of the reduced leaf water pressure deficit, which agrees with the above-mentioned theoretical advantages of protected mango cultivation.

The potential to produce water stress under greenhouse conditions could lead to advanced flowering or obtaining out of season flowering, which is also possible in the humid tropics via previously mentioned techniques that include water stress, growth regulator and $KNO_3$ applications (Alburquerque et al., 2002; Tongumpai et al., 1997; Galán Saúco, 2008).

It is obvious that under subtropical climates the increase of diurnal temperature under greenhouse conditions will result in advancing the harvest season. This is the case for early cultivars such as 'Lippens' or 'Torbet', that are not sensitive to embryo abortion (Galán Saúco, 2008) and do not require the removal or delay of terminal flowering, which will result in excellent crops that obtain a higher price. However, even in the case of late cultivars such as 'Keitt' grown in the south of Spain, that are cultivated via forcing alternate bearing, the use of polifibril black nets whitewashed at the time of harvesting, resulted in better coloured fruits than in the open (E. Guirado, CSIC, La Mayora Experimental Station, Pers. Comm., 2013).

It is also possible to reduce sunburn damage by using shading nets of appropriate density (Pavel et al., 2000) and although shading may slightly reduce fruit size this could be an advantage for cultivars of higher weight such as 'Keitt' (Whiley and Schaffer, 1997). As mentioned above, the possibility of obtaining fruits of appropriate size and exceptional quality, both externally and internally, allows mangoes cultivated under greenhouses to reach very high prices in markets like Japan, which compensate for the higher cost of mango production under greenhouse conditions (Galán Saúco, 2017).

The cultivation of mangoes in greenhouses also has disadvantages. Beside the costs derived from establishment, periodical replacement of laterals and covers, and cooling or heating systems if used, there are problems originating from the difficulties of eliminating plastics and nets without damaging the environment. Furthermore, it is vital to introduce pollinating agents. Both the cost and environmental problems are common to the greenhouse cultivation of any crop and are of minimal importance in places like the south east of Spain where mangoes are planted in already existing greenhouses previously cultivated with vegetables or flowers (Galán Saúco, 2015). Mango pollination is essential and is achieved by insects, with the common house fly and bees as the main agents, which necessitates their introduction into the greenhouse. Although it has been reported that the nectar of mango flowers is not especially attractive for bees in the open (Singh, 1988; Jiron and Hestron, 1985; Young and Sauls, 1979), bees play an important role in mango pollination in greenhouses (Galán Saúco, 1996; Pavel et al., 2000). This explains why in Spain, bee hives are placed inside the mango greenhouses or at least within their proximity, opening the lateral windows during the day to allow their entry into the greenhouse. Fish waste to rear flies, which are used as pollinators, was introduced into Japanese mango greenhouses with the associated bad odour inside the greenhouse (Yonemoto, 2018). Pollination with flies is also used in Spain for mangoes due to the recent commercialization by Polyfly of flies that can reach into small flowers, like those of the mango, and are active at temperatures between 14 and 36°C (Anon., n.d.)

As indicated before, mangoes in greenhouses are cultivated in Malaga, Spain, using trellise systems at planting distances of 2.5 × 2 m with 3 horizontal wires, the first at around 1.00 m above the soil and the highest at 2.5 m or in palmette systems like in El Algarve, Portugal, at 2.0–3.0 × 1.5–2.5 m but there is a lack of comparative economic studies regarding these cultivation systems. When cultivated using these systems it is necessary that once the plant reaches the first wire, continuous pinching should be practised to train the mango to the other wires and to fill the horizontal surfaces with terminals able to produce flowers during the next season. It is also necessary to cut back to the last node those terminals that have fructified to favour the growth of vegetative flushes prior to the onset of winter. As in the open, it is convenient to force alternance by pruning at the beginning of the spring of the 'OFF' year which will lead to the growth of vegetative flushes during late spring, summer and autumn to sustain next year's crop (E. Guirado, CSIC, La Mayora Experimental Station, Pers. Comm., 2013).

Trees are trained in an open vase form in Japanese greenhouses where tree spacing is 2.5 × 2.5 m, but they require thinning 5–7 years after planting to avoid crowding. Trees are initially formed by cutting the main stem at a height of 30–40 cm above the ground allowing 3–4 shoots as scaffold branches which are cut later at a length of 30–40 cm, allowing the development of 2–3 shoots from each of the scaffold branches, known as the sub-scaffold branches. The lateral shoots grown from these sub-scaffold branches become the fruiting branches. Tree height is kept at 1.5 m or less and stakes and pipes are used on the ground to tie down the limbs and branches (Yonemoto, 2018)

'Irwin', 'Osteen', 'Tommy Atkins', 'Kent' and 'Keitt', usually as monovarietal plantings, are the common cultivars planted in greenhouses. Their yields in Spain are similar to those obtained in the open, but it would be desirable to study the possibilities of increasing yield in mixed plantings with several cultivars under greenhouse conditions due to the apparent preference of these cultivars for cross pollination, (Pérez-Méndez, 2013); however, this could lead to additional management problems. 'Irwin' is the cultivar planted in Japan, due to its red colour and pleasant taste with maximum yields of 20 tons/ha (Yonemoto, 2018)

Temperatures in the tropics are favourable for growth during the whole year and more growth flushes are produced annually than in the case of the subtropics. This makes it more difficult to control the grow of mangoes. Because of this cultivation of mangoes in greenhouses in the tropics may necessitate the use of dwarfing rootstocks (i.e. 'Kalapady', 'Vella-i-kulumban', 'Kerela Dwarf', 'Manjeera', 'Creeping' or 'Amrapalli' (Singh and Singh, 1976; Iyer and Subramanyan, 1986; Vargas Ramos et al., 2004), slow-growing cultivars (i.e. 'Lippens', 'Irwin', 'Keitt' or 'Sensation' among the Floridian cultivars or others like 'Amrapalli' 'MDCH-1' ('Madhulika') (Singh, 1999) or a combination of both, as in the case of 'Palmer' and 'Irwin', grafted on 'Eldon' or 'Edward' grafted on 'Julie' (Cedeño-Maldonado et al., 1988).

## 5 Conclusion

This chapter reviews the recent advances of different aspects that influence mango cultivation, including genetic and physiological considerations as well as cultural practices.

There have been recent important advances in mango genomics, particularly the shift from the use of biochemical markers to DNA markers that have been increasingly developed over the last four decades. Considerable advances have also been made regarding the knowledge of mango gene function and regulation, especially through the characterization of transcriptomics of different mango tissues combined with NGS technologies. A great advance in this field is the recent generation of a mango consensus genetic map and, although markers associated with important quantitative traits must also be found to further improve mango breeding efficiently, new projects are trying to find this association. Recent studies on graft and sexual compatibility of the mango with other *Mangifera* species show important adaptation or resistance to biotic and abiotic stresses that open the door to a new generation of rootstocks and/or commercial cultivars, perhaps even better adapted to the ongoing global warming.

From a physiological point of view, recent work has confirmed previous findings indicating that in tropical areas of the world, where low temperatures are rare or non-existent, the growth age of the last vegetative flush is the dominant factor regulating flower induction conditions. The recent advances made in identification, quantification and expression of genes related to flowering can help to understand the reproductive process. These studies must continue in the future and will be of great value in solving the problem of irregular flowering in mangoes.

Cultural practices for the successful cultivation of mangoes must not only be based on flowering and fruiting, but should also consider vegetative growth and its relationship with the reproductive process and the cultivar/rootstock planted. Increased planting density is probably the most important cultural practice that has changed over recent years. Furthermore, ultra-high-density plantings require careful training and pruning and have been established with success in different countries with various cultivars. An interesting new training system that includes the possibility of combining bending branches with pruning has been proposed in Florida. A novel practice for the subtropics has been developed for late cultivars in Spain, which reinforces the alternate bearing tendency in order to obtain satisfactory yields, both in quantity and quality, every 2 years.

Some advances have been made regarding finding substitutes for PBZ with recent studies in Mexico indicating that calcium prohexadiona can be used instead of PBZ to reduce vegetative growth and induce flowering in the tropics. Although the beneficial impact on fruit set of spraying $GA_3$ applications at flower bud differentiation or at full bloom was already known, recent studies indicate different effects of various types of gibberellins on the total fruit set of mangoes, illustrating the need for further studies regarding their effect on different cultivars and environments. The beneficial effect reported for the application of cytokinins at the full bloom stage, and thidiazuron and gibberellic acid for increasing the size of seedless fruits is also of great interest for the subtropics in order to reduce the problem of embryo-aborted fruits.

Recent studies on applying deficit irrigation strategies to mangoes without reducing yield parameters have permitted the important reduction of water consumption, which is of special interest for dry areas and could also be of great help to counteract possible drought episodes in the event of the projected global climate change.

Recent advances in the field of mango nutrition have been limited, except perhaps regarding the positive effect of magnesium fertilization on different aspects of tree growth, increased yield, improved fruit quality and increased shelf life and tolerance to cold conditions, high solar radiation and hot water treatment. Although organic mango cultivation is increasingly implemented all over the world, studies regarding efficient fertilization are still needed in order to obtain the maximum yield from this relatively new mango production system, and should be increasingly used in mango cultivation to satisfy the growing demand for organic mangoes.

Finally, cultural techniques for efficient mango cultivation under greenhouse conditions have being implemented in different countries and although this system of cultivation is restricted to the subtropics there are economic reasons for the protected cultivation of mangoes even in temperate countries.

## 6 Where to look for further information

Active research at different levels of various aspects of mango cultivation is carried out in many countries in which this crop is cultivated and some of the most relevant institutions dealing with mango research are listed below:

- Australia: Queensland Department of Agriculture, Horticulture and Forestry Sciences (www.daf.qld.gov.au).
- Brazil: Empresa Brasileña de Pesquisa Agropecuaria (*EMBRAPA*) (www.embrapa.br).
- France: CIRAD (www.cirad.fr).
- India: Central Institute for Subtropical Horticulture (ICAR) (https://icar.org.in).
- Mexico: Instituto Nacional de Investigaciones forestales Agrícolas y Pecuarias (www.inifap.gob.mex).
- South Africa:
    - ARC Institute for Tropical and Subtropical Crops (www. Arc.agric.za)
    - HortResearch (www.hortresearch.co.za).
    - Spain: Instituto de Hortofruticultura Subtropical y Mediterránea Estación Experimental 'La Mayora' (www.ishm.uma-csic.es).
    - Instituto Canario de Investigaciones Agrarias (www.icia.es).
- Thailand: Chiang Mai University (www.cmu.ac.th).
- USA:
    - University of Florida. Institute of Food and Agricultural Sciences (UF/IFAS) (https://trec.ifas.ufl.edu/faculty/tropical-fruits-crop-faculty/).
    - Fairchild Tropical Botanical Garden (www.fairchildgarden.org).

Every 3–4 years the International Society for Horticulture Science (ISHS) organizes an International Mango Symposium with presentations being published in Acta Horticulturae. Other scientific journals also that report advances in mango cultivation include: Fruits, Hortscience and Revista Brasileira de Fruticultura. Several countries in Latin America, mainly, Mexico, Brazil and Peru also organize regular annual meetings in Spanish dealing with mangoes. Annual mango meetings are also organized in several countries of Asia such as India and Australia. Specialized research topics focusing on mangoes are also regularly studied by the National Mango Board of the USA (www.mango.org).

## 7 References

Albrigo, G. L. and Galán Saúco, V. 2004. Flower bud induction, flowering and fruit-set of some tropical and subtropical fruit tree crops with special reference to *Citrus*. *Acta Horticulturae* 632(632), 81–90. doi:10.17660/ActaHortic.2004.632.10.

Alburquerque, J. A. S., Medina, V. D. and Mouco, M. A. C. 2002. Induçao floral. In: Genú, P. J. C. and Pinto, A. C. Q. (Eds), *A Cultura da Mangueira*. Embrapa Informaçao Tecnológica, Brasilia, pp. 259–76.

Anon. n.d. Polyfly. Polinización con moscas. Available at: www.polyfly.es (accessed on 12 December 2018).

Ariffin, Z., Md Sah, M. S., Idris, S. and Hashim, N. 2015. Genetic diversity of selected mangifera species revealed by inter simple sequence repeats markers. *International Journal of Biodiversity* 2015, 1–8. doi:10.1155/2015/458237.

Bakker, S. 2018. Paises Bajos podría producir frutas exóticas en el futuro. Available at: https://www.freshplaza.es/article/9035767 (accessed on 12 December 2018).

Bally, I. S. E. 2006. *Mangifera indica* (mango). In: *Species Profiles for Pacific Island*. Agroforestry, ver. 3.1, pp. 25.

Bally, I. S. E. 2013. New mango hybrids from Australia. *Acta Horticulturae* 975(975), 55–61. doi:10.17660/ActaHortic.2013.975.4.

Bally, I. S. E. and Dillon, N. L. 2018. Mango (*Mangifera indica* L.) breeding. In: Al-Khayri, J. M., Jain, S. M. and Johnson, D. V. (Eds), *Advances in Plant Breeding: Fruits*. Springer International Publishing, Cham, pp. 811–96. Chapter 20.

Bally, I. S. E., Lu, P. and Johnson, P. R. 2009. Mango breeding. In: Jain, S. M. and Pryyadarshan, P. M. (Eds), *Breeding Plantation Tree Crops*. Springer, pp. 51–82.

Banik, B. C., Maitry, P. K., Hasan, M. A. and Ghosh, S. N. 2013. High density orchard systems for 'Himsagar' mango in the alluvial zone of West Bengal. *Acta Horticulturae* 975(975), 241–6. doi:10.17660/ActaHortic.2013.975.26.

Barbosa Gámez, I., Caballero Montoya, K. P., Ledesma, N., Sáyago Ayerdi, S. G., García Magaña, MdL., Bishop von Wettberg, E. J. and Montalvo-González, E. 2017. Changes in the nutritional quality of five Mangifera species harvested at two maturity stages. *Journal of the Science of Food and Agriculture* 97(14), 4987–94. doi:10.1002/jsfa.8377.

Bompard, J. M. 2009. Taxonomy and systematics. In: Litz, R. E. (Ed.), *The Mango, Botany, Production and Use* (2nd edn.). CAB International, Wallingford, pp. 19–41.

Bompard, J. M. and Schnell, R. 1997. Taxonomy and sistematics. In: Litz, R. E. (Ed.), *The Mango, Botany, Production and Uses*. CAB International, Wallingford, pp. 21–48. Chapter 2.

Campbell, R. J. and Ledesma, N. 2013. Update on new Mangifera species in Florida, USA. *Acta Horticulturae* 992(992), 95–8. doi:10.17660/ActaHortic.2013.992.10.

Capelli, M., Lauri, P. É. and Normand, F. 2016. Deciphering the costs of reproduction in mango – vegetative growth matters. *Frontiers in Plant Science* 7, 1531. doi:10.3389/fpls.2016.01531.

Cedeño-Maldonado, A., Pérez, A. and Reyes-Soto, I. 1988. Effect of dwarfing rootstocks on tree size and yield of selected mango varieties. *The Journal of Agriculture of the University of Puerto Rico* 72(1), 1–8.

Cohen, Y., Vaknin, M. and Mauch-Mani, B. 2016. BABA-induced resistance: milestones along a 55-year journey. *Phytoparasitica* 44(4), 513–38. doi:10.1007/s12600-016-0546-x.

Crane, J. H., Salazar García, S., Lin, T. S., de Queiroz Pinto, A. C. and Ziu, Z. H. 2009. Crop production management. In: Litz, R. E. (Ed.), *The Mango, Botany, Production and Uses* (2nd edn.). CAB International, Wallingford, pp. 432–83.

Dambreville, A., Lauri, P. É, Trottier, C., Guédon, Y. and Normand, F. 2013. Deciphering structural and temporal interplays during the architectural development of mango trees. *Journal of Experimental Botany* 64(8), 2467–80. doi:10.1093/jxb/ert105.

Dambreville, A., Lauri, P. É., Normand, F. and Guédon, Y. 2015. Analysing growth and development of plants jointly using developmental growth stages. *Annals of Botany* 115(1), 93–105. doi:10.1093/aob/mcu227.

Davenport, T. L. 2009. Reproductive physiology. In: Litz, R. E. (Ed.), *The Mango, Botany, Production and Uses* (2nd edn.). CAB International, Wallingford, pp. 97–169.

Davenport, T. L. and Núñez Elisea, R. 1997. Stress physiology. In: Litz, R. E. (Ed.), *The Mango, Botany, Production and Uses*. CAB International, Wallingford, pp. 147–74.

De Carvalho, P. J. and de Queiros Pinto, A. C. (Eds). 2002. *A Cultura da Mangueira*. Embrapa Informaçao Tecnológica, Brasilia, 452pp.

Dinesh, M. R., Ravishankar, K. V. and Sangma, D. 2016. Mango breeding in India – past and future. *Journal of Horticultural Science* 11(1), 1–12.

Galán Saúco, V. 2008. *El cultivo del mango* (2nd ed.). MundiPrensa, Madrid, 340pp.

Galán Saúco, V. 2015. Ventajas y desventajas cultivo del mango (*Mangifera indica* L.) en zonas subtropicales y potencial del cultivo bajo invernadero. *Acta Horticulturae* 1075(1075), 167–77. doi:10.17660/ActaHortic.2015.1075.19.

Galán Saúco, V. 2016. Taxonomy, botany, history and distribution of mango around the world. In: *The Mango Tree Encyclopedy* (vol. 2.1). Royal Courts Affairs, Sultanate of Oman, pp. 38–71.

Galán Saúco, V. 2017. Trends in world mango production and marketing. *Acta Horticulturae* 1183(1183), 351–64. doi:10.17660/ActaHortic.2017.1183.51.

Galán Saúco, V. 2018a. Magnesium fertilization in mango. Available at: https://www.mango.org/research-resources.

Galán Saúco, V. 2018b. Mango cultivation practices in the subtropics. In: Galán Saúco, V. and Lu, P. (Eds), *Achieving Sustainable Cultivation of Mangoes*. Burleigh Dodds Science Publishing, Cambridge, UK, pp. 166–83.

Galán Saúco, V. 2019. Mango rootstock. A review. *Acta Horticulturae* (in print).

Galán Saúco, V., Fernández Galván, D., Hernández Conde, J. C. and Morales Navarro, A. 1996. Preliminary studies on fruit-set of mango cultivar Tommy Atkins under greenhouse cultivation in the Canary Islands. *Acta Horticulturae* 455, 530–4.

Galán Saúco, V., Saúco, V. G., Martín, M. J. G., Galván, D. F., Torres, Á. C., Juárez, J. and Navarro, L. 2001. Occurrence of spontaneous tetraploid nucellar mango plants. *HortScience* 36(4), 755–7. doi:10.21273/HORTSCI.36.4.755.

Gazit, S. and Kadman, A. 1980. 13-1 Mango rootstock selection. *HortScience* 57, 81–7.

Goguey, T. 1995. Approche architecturale des mécanismes de la croissance aérienne et de la floraison du manguier. Ph.D. thesis. Université Montpellier II, France.

Goguey, T. 1997. Architectural approach of the mechanisms of canopy growth and flowering of mango trees. *Acta Horticulturae* 455(455), 124–31. doi:10.17660/ActaHortic.1997.455.17.

Hallé, F. and Oldeman, R. A. 1970. *Essai sur l'architecture et la dynamique de croissance des arbres tropicaux*. Masson, Paris, France.

Hermoso Gonzáles, J. M., Guirado Sánchez, E. and Farré Massip, J. M. 2018. *Introducción al cultivo del mango en el Sur peninsular*. Caja Rural Granada, 93pp.

Hernández Delgado, P. M., Aranguren, M., Reig, C., Fernández Galván, D., Mesejo, C., Martínez Fuentes, A., Galán Saúco, V. and Agustí, M. 2011. Phenological growth stages of mango according to the BBCH scale. *Scientia Horticulturae* 130(3), 536–40. doi:10.1016/j.scienta.2011.07.027.

Human, C. F., Rheeder, S. and Sippel, A. D. 2009. New cultivars and hybrid selections from the Mango Breeding Program of the Agricultural Research Council in South Africa. *Acta Horticulturae* 820(820), 119–26. doi:10.17660/ActaHortic.2009.820.10.

Iyer, C. P. A. 2004. Growing mango under organic system. *Acta Horticulturae* 645(645), 71–84. doi:10.17660/ActaHortic.2004.645.4.

Iyer, C. P. A. and Degani, C. 1997. Classical breeding and genetics. In: Litz, R. E. (Ed.), *The Mango, Botany, Production and Uses*. CAB International, Wallingford, pp. 49–68.

Iyer, C. P. A. and Subramanyan, M. D. 1986. Creping, a promising genotype for introduction of dwarfness in mango. *Indian Journal of Horticulture* 43, 221–3.

IPGRI. 2006. *Descriptors for Mango (Mangifera indica L.)*. International Plant Genetic Resources Institute, Rome, Italy.

Issarakraisila, M., Considine, J. A. and Turner, D. W. 1997. Vegetative and reproductive growth aspects of mango growing in a Mediterranean climate in Western Australia. *Acta Horticulturae* 455(455), 56–63. doi:10.17660/ActaHortic.1997.455.8.

Jiron, L. F. and Hedstrom, I. 1985. Pollination ecology of mango (*Mangifera indica* L.) (Anacardiaceae) in the Neotropic Region. *Turrialba* 35(3), 269–77.

Juntamanee, K., Onnon, S., Yingjajaval, S. and Sangchote, S. 2013. Leaf photosynthesis and fruit quality of mango growing under field or plastic roof condition. *Acta Horticulturae* 975(975), 415–20. doi:10.17660/ActaHortic.2013.975.53.

Khan, A. S., Ali, S. and Khan, I. A. 2015. Morphological and molecular characterization and evaluation of mango germplasm. An overview. *Scientia Horticulturae* 194, 353–66. doi:10.1016/j.scienta.2015.08.031.

Kosterman, A. J. G. H. and Bompard, J. M. 1993. *The Mangoes, Their Botany, Nomenclature, Horticulture and Utilization*. Academic Press, Londres, 233pp.

Kuhn, D. N., Dillon, N. L., Innes, D. J., Wu, L.-S. and Mockaitis, K. 2016. Development of single nucleotide polymorphism (SNP) markers from the mango (*Mangifera indica*) transcriptome for mapping and estimation of genetic diversity. *Acta Horticulturae* 1111(1111), 315–22. doi:10.17660/ActaHortic.2016.1111.45.

Kuhn, D. N., Bally, I. S. E., Dillon, N. L., Innes, D., Groh, A. M., Ophir, R., Cohen, Y. and Sherma, A. 2017. Genetic map of mango: a tool for mango breeding. *Frontiers in Plant Science* 8, 577. doi:10.3389/fpls.2017.00577.

Larranaga, N. and Hormaza, J. I. 2016. Advances in genetic diversity analysis in fruit tree crops. In: Lünge, U., Cánovas, F. M. and Matyssek, R. (Eds), *Progress in Botany* (vol. 77). Springer International Publishing, Cham, pp. 245–64. doi:10.1007/978-3-319-25688-7_8.

Ledesma, N. 2016. Distribution and use of *Mangifera caesia* Jack in Bali, Indonesia. *Proceedings of the Florida State Horticultural Society* 129, 115–6.

Ledesma, N. 2017. Feasibility of small scale specially mango production in South Florida. *Proceedings of the Florida State Horticultural Society* 130, 8–10.

Ledesma, N. and Campbell, R. J. 2016. Improving harvest and postharvest practices of Mallika mango for the local industry. *Proceedings of the Florida State Horticultural Society* 129, 130–3.

Ledesma, N., Campbell, R. J. and Wasielewski, J. 2016. Training and pruning a mango orchard to improve blooming and yield in South Florida. *Proceedings of the Florida State Horticultural Society* 129, 192–4.

Ledesma, N., Campbell, R. J., Hass, M. and Campbell, T. B. 2017. Interspecific hybrids between *Mangifera indica* and related species. *Acta Horticulturae* 1183(1183), 83–8. doi:10.17660/ActaHortic.2017.1183.12.

Levin, A. G. 2018. Improving fertilizer and water-use efficiency in mango cultivation. In: Galán Saúco, V. and Lu, P. (Eds), *Achieving Sustainable Cultivation of Mangoes*. Burleigh Dodds Science Publishing, Cambridge, UK, pp. 269–310. doi:10.19103/AS.2017.0026.13.

Levin, A. G., Naor, A., Noy, M., Love, C., Gal, Y. and Peres, M. 2015a. Mango response to deficit irrigation at different phenological periods. *Acta Horticulturae* 1075(1075), 103-13. doi:10.17660/ActaHortic.2015.1075.10.

Levin, A. G., Naor, A., Noy, M., Love, C., Gal, Y., Peres, M. and Bernstein, N. 2015b. The response of field-grown mango (cv Keitt) trees to regulated deficit irrigation at different phenological stages. In: *XI International Mango Symposium*, Darwin, Northern Territory (Australia), 28 September–2 October 2015.

Litz, R. E. 2009. *The Mango, Botany, Production and Uses* (2nd edn.). CAB International, Wallingford, UK.

Lu, P. 2005. Mango water relations and irrigation scheduling. In: Malik, K., et al. (Eds), *International Conference on Mango and Date Palm*, 20–23 June 2005. University of Faisalabad, Pakistan.

Lu, P. 2013. Whole-tree water relations in irrigation scheduling for mango. *Acta Horticulturae* 992, 115–22.

Majunder, P. K., Sharma, D. K. and Singh, R. N. 1982. A study on 'High Density' orcharding in mango (*Mangifera indica* L.) var. Amrapali. *Punjab Horticultural Journal* 22, 123–7.

Majunder, P. K. and Sharma, D. K. 1985. Mango. In: Bose, T. K. and Mitra, S. K. (Eds), *Fruits: Tropical and Subtropical*. Naya Prokash, Calcuta, pp. 1–62.

Majunder, P. K. and Sharma, D. K. 1988. A new concept of orchanding in mango (*Mangifera indica* L.). *Acta Horticulturae* 231, 335–8.

Martínez-García, P. M., Rodríguez-Palenzuela, P., Arrebola, E., Carrión, V. J., Gutiérrez-Barranquero, J. A., Pérez-García, A., Ramos, C., Cazorla, F. M. and de Vicente, A. 2015. Bioinformatics analysis of the complete genome sequence of the mango tree pathogen *Pseudomona syringae* pv. syringae UMAF0158 reveals traits relevant to virulence and epyphytic lifestyle. *PLoS One* 10(8), e0136101.

Medina-Urrutia, V. M., Reyes-Hernández, J. E., Virgen-Calleros, G., Pimienta-Barrios, E. and Robles-González, M. M. 2018. Organic mango production: a review. In: Galán Saúco, V. and Lu, P. (Eds), *Achieving Sustainable Cultivation of Mangoes*. Burleigh Dodds Science Publishing, Cambridge, UK, pp. 229–68. doi:10.19103/AS.2017.0026.12.

Mitra, S. and Bhagwan, A. 2018. Mango cultivation practices in the tropics: good agricultural practices to maximize sustainable yields. In: Galán Saúco, V. and Lu, P. (Eds), *Achieving Sustainable Cultivation of Mangoes*. Burleigh Dodds Science Publishing, Cambridge, UK, pp. 149–63. doi:10.19103/AS.2017.0026.08.

Mukherjee, S. K. 1950. Mango: its allopolyploid nature. *Nature* 166(4213), 196–7. doi:10.1038/166196b0.

Nakagawa, M., Honsho, C., Kanzaki, S., Shimizu, K. and Utsunomiya, N. 2012. Isolation and expression analysis of flowering locus T-like and gibberellin metabolism genes in biennial-bearing mango trees. *Scientia Horticulturae* 139, 108–17. doi:10.1016/j.scienta.2012.03.005.

Nates Parra, G. 2005. Abejas silvestres y polinización. *Manejo Integrado de Plagas y Agroecología (Costa Rica)* 75, 7–20.

Normand, F. and Lauri, P. É. 2018. Advances in understanding mango tree growth and canopy development. In: Galán Saúco, V. and Lu, P. (Eds), *Achieving Sustainable Cultivation of Mangoes*. Burleigh Dodds Science Publishing, Cambridge, UK, pp. 87–119. doi:10.19103/AS.2017.0026.06.

Normand, F., Pambo Bello, A. K., Trottier, C. and Lauri, P. E. 2009. Is axis position within tree architecture a determinant of axis morphology, branching, flowering and fruiting? An essay in mango. *Annals of Botany* 103(8), 1325–36. doi:10.1093/aob/mcp079.

Normand, F., Capelli, M. and Lauri, P. É 2016. The costs of reproduction in plants: a novel approach to study irregular bearing of fruit crops. *Acta Horticulturae* 1130(1130), 207–14. doi:10.17660/ActaHortic.2016.1130.30.

Oosthuyse, S. A. 1993. Mango tree spacing trends and options for yield improvement with special reference to South Africa. *Journal of the Southern African Society for Horticultural-Sciences* 3, 92–6.

Oosthuyse, S. A. 1997. Some principles pertaining to mango pruning and the adopted practices of pruning mango trees in South Africa. *Acta Horticulturae* 455(455), 413–21. doi:10.17660/ActaHortic.1997.455.53.

Oosthuyse, S. A. 2005. Greatly enhanced ease in growing mangoes on small trees. SQM Europe N. V., Belgium. Available at: www.sqm.com. (Printed copy, Author courtesy).

Oosthuyse, S. A. 2006. CropKit. Special plant nutrition guide. Mango. SQM Europe N. V., Belgium. Available at: www.sqm.com. (Printed copy, Author courtesy).

Oosthuyse, S. A. 2018. Management of an ultra-high density mango orchard and benefits of the small-tree system. In: Galán Saúco, V. and Lu, P. (Eds), *Achieving Sustainable Cultivation of Mangoes*. Burleigh Dodds Science Publishing, Cambridge, UK, pp. 205–27. doi:10.19103/AS.2017.0026.11.

Oosthuyse, S. A. and Jacobs, G. 1995. Relationship between branching frecuency, and growth, cropping and structural strength of two-year-old mango trees. *Scientia Horticulturae* 64(1–2), 85–93. doi:10.1016/0304-4238(95)00820-X.

Pavel, E. W., Oosthuyzen, R. G. and Mthembu, G. J. 2000. Effect of shading upon incidense of Bacterial Black Spot and sunburn, flowering, yield and fruit quality in mango. *South Africa Mango Growers Asssociation Yearbook* 20, 100–2.

Pérez, V. and Hormaza, J. I. 2018. Exploiting the mango genome: molecular markers. In: Galán Saúco, V. and Lu, P. (Eds), *Achieving Sustainable Cultivation of Mangoes*. Burleigh Dodds Science Publishing, Cambridge, UK, pp. 3-19. doi:10.19103/AS.2017.0026.01.

Pérez-Barraza, M. H. 2015. Regulación del proceso de floración y desarrollo de frutos de mango 'Ataulfo' mediante biorreguladores de crecimiento. Ph.D. thesis. Colegio de Postgraduados, Montecillo, Texcoco, Estado de México, 148pp.

Pérez-Barraza, M. H. and Osuna-García, J. A. 2018. Advances in understanding flowering, pollination and fruit development in mangoes. In: Galán Saúco, V. and Lu, P. (Eds), *Achieving Sustainable Cultivation of Mangoes*. Burleigh Dodds Science Publishing, Cambridge, UK, pp. 121–46. doi:10.19103/AS.2017.0026.07.

Pérez-Barraza, M. H., Osuna-Enciso, T., Gutiérrez-Espinosa, M. A., Santiago-Cruz, M., de, J., Avitia-García, E. and Cano-Medrano, R. 2015. Thidiazuron y ácido giberélico en amarre y tamaño de frutos de mango 'Ataulfo' polinizados y partenocárpicos. *Interciencia* 40, 677–83.

Pérez-Barraza, M. H., Osuna-Enciso, A.-G., Gutiérrez-Espinosa, T. E., Santiago-Cruz, M. de J., Ramírez, H. and Cano-Medrano, R. 2016. Prohexadiona de calcio reduce crecimiento vegetativo e incrementa brotación floral en mango'Ataulfo'. *Revista Mexicana de Ciencias Agrícolas* 7, 263–76.

Pérez-Méndez, V. 2013. Biología reproductiva del mango (*Mangifera indica* L.). Ph. D. Thesis. Universidad de Málaga. December 2013.

Reyner, K. 2002. The effect of rootstocks on Kensington Pride mango. Lecture (unpublished) at the *Australian Mango Industry Association National Conference and Annual General Meeting*, August 2002, Darwin, Australia.

Rocha, A., Salomao, T. M. F., Siqueira, D. Ld, Cruz, C. D. and Salomao, L. C. C. 2014. Identification of 'Uba' mango tree zygotic and nucellar seedlings using ISSR markers. *Revista Ceres* 61(5), 597–604. doi:10.1590/0034-737X201461040001.

Rossetto, C. J., Ribeiro, I. J. A., Gallo, P. B., Soares, N. B., Sabino, J. C., Martins, A. L. M., Bortoletto, N. and Paulo, E. M. 1997. Mango breeding for resistance to diseases and pests. *Acta Horticulturae* 455(455), 299–304. doi:10.17660/ActaHortic.1997.455.39.

Schaffer, B., Urban, L., Lu, P. and Whiley, A. 2009. Ecophysiology. In: Litz, R. (Ed.), *The Mango*. CAB International, Wallingford, UK.

Sherman, A., Rubinstein, M., Ehed, R., Benita, M., Ish-Shalom, M., Sharabi-Schwager, M., Rozen, A., Saada, D., Cohen, Y. and Ophir, R. 2015. Mango germplasm diversity based on single nucleotide polymorphisms derived from the trancriptome. *BMC Plant Biology* 15(1), 277.10.1186/s12870-015-0663-6. doi:10.1186/s12870-015-0663-6.

Singh, G. 1988. Insect pollinators of mango and their role in fruit setting. *Acta Horticulturae* 231, 629–32.

Singh, S. P. 1999. Wonder mango variety for 21 century. Working Abstracts and Program. In: *The 6th International ISHS Mango Symposium*, Pattaya, Thailand, 6–9 April 1999, p. 72.

Singh, Z. 2009. Gibberellin type and time of application influence fruit set and retention in mango. *Acta Horticulturae* 820(820), 407–12. doi:10.17660/ActaHortic.2009.820.49.

Singh, U. R. and Singh, A. P. 1976. Rootstock studies in mango (*Mangifera indica* L.). *Progressive Horticulture* 8(1), 13–9.

Smith, M. W., Bright, J. D., Hoult, M. D., Renfree, R. A., Maddern, T. and Coombes, N. 2008. Field evaluation of 64 rootstocks for growth and yield of 'Kensington Pride' mango. *HortScience* 43(6), 1720–5. doi:10.21273/HORTSCI.43.6.1720.

Tomer, E., Gazit, S., Lavi, U., Shoker, S., Ripa, M., Zipori, I. and Sáada, D. 1997. Mango breeding in Israel – principles and difficulties. *Acta Horticulturae* 455, 245–51.

Tongumpai, P., Chantakulchan, K., Subhadrabandhu, S. and Ogata, R. 1997. Foliar application of paclobutrazol on flowering of mango. *Acta Horticulturae* 455(455), 175–9. doi:10.17660/ActaHortic.1997.455.24.

Torres, M. D., Guirado, E., Farré, J. M. and Hermoso, J. M. 2009. Prebloom shoot tipping of mango. Influence on yield, fruit size and quality. *Acta Horticulturae* 820, 391–4.

Vargas Ramos, V. H., Pinto, A. C. Q., Junqueira, N. T. V., Gomes,A. C., Andrade, S. R. M. and Cordeiro, M. R. C. 2004. Effect of mono and polyembrionic rootstocks on growth. yield and fruit quality of four mango cultivars in the Central Region of Brazil. *Acta Horticulturae* 645, 201–7.

Yonemoto, J. Y. 2018. Mango cultivation in greenhouses. In: Galán Saúco, V. and Lu, P. (Eds), *Achieving Sustainable Cultivation of Mangoes*. Burleigh Dodds Science Publishing, Cambridge, UK, pp. 185–204. doi:10.19103/AS.2017.0026.10.

Young, T. W. and Sauls, J. W. 1979. The mango industry in Florida. *Florida Cooperative Extension Service Bull* 189, 70p.

Whiley, A. W. and Schaffer, B. 1997. Stress physiology. In: Litz, R. E. (Ed.), *The Mango, Botany, Production and Uses*. CAB International Press, London, pp. 147–73.

Whiley, A. W., Rasmussen, T. S., Saranah, J. B. and Wolstenholme, B. N. 1989. Effect of temperature on growth, dry matter production and starch accumulation in ten mango (*Mangifera indica* L.) cultivars. *Journal of Horticultural Science* 64(6), 753–65. doi:10.1080/14620316.1989.11516018.

# Chapter 19

# Advances in breeding of papaya

*Sunil Kumar Sharma, Indian Agricultural Research Institute, India*



## 1 Introduction

Papaya (*Carica papaya* L.) is a major tropical fruit crop which is widely cultivated in tropical countries, producing 13.02 million tonnes of fresh fruits from 440 629 hectare in 2017 (Anon., 2019). The major share of production comes from Asia, Central and South Americas, and Africa. Asia produces 58% of the world's papaya from 46% of its global area of production. India, Brazil, Nigeria, Indonesia and Mexico are among the major papaya producing countries. India produces about 46% of the world's papaya from 30% of its overall area of production. The global average papaya yield is 29 tonnes/hectare (t/ha). The top five papaya producing countries achieve an average yield of 49 t/ha, one and a half times of the global average (Anon., 2019).

The genus *Carica* has only one species (*C. papaya*) that makes it vulnerable to diseases due to a narrow gene pool (Aradhya et al., 1999; Badillo, 1993, 2000; Kim et al., 2002). Its susceptibility to various biotic and abiotic stresses is the main reason for the variation in the fruit yield. Biotic stresses include fungal, bacterial and viral diseases. While most infections can be managed successfully, viral diseases in particular are difficult to control, particularly papaya ringspot virus (PRSV) which can be devastating. Papaya black spot and

http://dx.doi.org/10.19103/AS.2019.0054.14

papaya die back also pose a serious challenge in certain areas. In addition to these biotic stresses, papaya is also susceptible to abiotic stresses such as high soil moisture and low temperature. Breeding programmes in papaya are now mainly concentrated around introgression of genes which protect against these stresses, though initial breeding objectives were focussed on improvement of fruit yield and quality. This chapter describes the development and current status of papaya breeding.

## 2 Origin and distribution

Papaya belongs to the family Caricaceae. The family originated in Africa and then spread to the New World (Central America) and subsequently, South America (Carvalho, 2015). Out of 34 species and one hybrid (*Vasconcellea* × heilbornii) within six genera (Fig. 1), most genera and species of Caricaceae are distributed in the Neotropical region; however, one species each is found in west (*Cylicomorpha solmsii*) and east Africa (*C. parviflora*). Mexico and Central America contain five genera and ten species while two genera and 25 species are distributed in South America. The genera *Carica* and *Horovitzia* have one species each. Genus *Cylicomorpha* has two species, genus *Jarilla* has three species, genus *Jacaratia* has seven species, and genus *Vasconcellea* has 20 species and one hybrid (Badillo, 1993; Fig. 1). All species of the genus *Vasconcellea* were originally considered to belong to the genus *Carica*. As a result, in literature prior to 2000, *Vasconcellea* species are referred to as species of the genus *Carica* (Badillo, 2000, 2001).

Most of the wild relatives of the *C. papaya* are species of the other five genera of the Caricaceae that is *Cylicomorpha, Horovitzia, Jacaratia, Jarilla* and *Vasconcellea.* The wild forms occur only from southern Mexico to Costa Rica. Some species of the genus *Vasconcellea* have been reported to possess gene(s) with resistance to a number of biotic and abiotic stresses (Table 1).

The genetic relationships among *C. papaya* and other members of Caricaceae are important because of the issue of cross compatibility. Similarities between *Carica* and other genera of the family are limited because of its earlier isolation from other wild relatives. The closest relatives of papaya are four

| Family Caricaceae | | | | | | |
|---|---|---|---|---|---|---|
| Genus (species) | Carica (1) | Cylicomorpha (2) | Jacaratia (7) | Jarilla (3) | Horovitzia (1) | Vasconcellea (20 + 1 hybrid) |

**Figure 1** Distribution of various genera and species in the family Caricaceae.

**Table 1** Sources of gene(s) for important traits in the genus *Vasconcellea*

| Stress | *Species* |
|---|---|
| *Abiotic stress* | |
| Low temperature stress | *V. cundinamarcensis* |
| *Biotic stress* | |
| Viral diseases (Papaya ringspot virus) | *V. cauliflora, V. quercifolia, V. cundinamarcensis (synonym: V. pubescence) and V. stipulata, V. candamarcensis, V. candicans* and *V. x heilbornii nm pentagona* |
| Bacterial diseases | |
| Canker (*Errwinia papaya*) | *V. cauliflora* |
| Black spot (*Asperisporium caricae*) | *V. cundinamarcensis* |
| Bacterial canker | *V. goudotiana* |
| Fungal diseases | |
| Paw paw dieback | *V. parviflora* |
| Phytophthora canker | *V. goudotiana* |
| *Production (High papain production)* | *V. cundinamarcensis, V. x heilbornii cv. 'babaco'* |
| *Fruit quality (High sugar content, TSS)* | *V. quercifolia* |

*Source*: adapted from Sharma and Tripathi (2014).

dioecious species distributed in Mexico, Guatemala and El Salvador: *Horovitzia cnidoscoloides*, *Jarilla heterophylla, J. caudata* and *J. chocola*. Among various species of the genus *Vasconcellea*, the most closely related species to *C. papaya* are *V. stipulata* and *V. pubescens* (Jobin-Décor et al., 1997; Sharon et al., 1992), and *V. pubescens* (Kim et al., 2002). The most distant species are *V. cauliflora* (Jobin-Décor et al., 1997) and *V. goudotiana* (Sharon et al., 1992; Kim et al., 2002). Different *Vasconcellea species* have varying degree of incompatibility with *C. papaya*. According to their cross-compatibility these species were arranged in three groups:

1 *C. monoica, C. cauliflora, C. microcarpa and C. cundinamarcensis*
2 *Carica papaya*
3 *C. goudotiana*

Crosses between group 1 and 2 do not form mature seed, although immature embryos can be cultured artificially. Crosses between group 2 and 3 always give negative results (Singh, 1990).

Around 500 years of selective breeding for fruit size, fruit shape and colour, combined with selfing and inbreeding of preferred bisexual-flowered trees, has resulted in limited genetic diversity in cultivated papaya (Brown et al., 2012). Among various genera and species of Caricaceae, some species have been reported as sources of genes of economic importance (Table 1). Both tubers and fruits of *Jarilla chocola* produce milky latex, and proteinase activity of a papain-like enzyme which may be exploited in breeding programmes. *Horovitzia cnidoscoloides* is likely to carry genes for cold resistance due to habitat adaptation. The introgression of genes of PRSV resistance to *C. papaya* has been a major objective.

Caricaceae is reported to have 2n = 18 chromosomes in the 11 species (Manshardt and Drew, 1998; Storey, 1953, 1976). Carvalho reported chromosomal count 2n = 16 in the genus *Horovitzia* and 14 in *Jarilla*. Differences in the chromosomal number between *C. papaya* and its closest relatives, species of the genera *Jarilla* and *Horovitzia,* makes transfer of genes to *C. papaya* by natural crossings virtually impossible. Molecular methods of gene transfer are needed to achieve the goal.

## 3 Breeding methods

This section covers the following aspects of breeding of papaya:

- Collection of germplasm
- Selection
- Crossing and development of hybrids
- Back crossing
- Mutation breeding
- Marker-assisted selection

### *3.1 Collection of germplasm*

Earlier breeding programmes started with the collection of germplasm with desirable traits, either as primary collections from the areas where papaya is found naturally, or as secondary collections from other centres. Each accession was given a unique name/number and evaluated for desirable traits. Those with attractive horticultural traits were 'introduced' directly for commercial production. The Solo variety in Hawaii was an introduction from Barbados and Sunrise Solo was an introduction to Puerto Rico from Hawaii. Being an open pollinated crop, germplasm is subject to strictly controlled pollination to raise pure progeny for further evaluation. Female flowers of the 'plus plants' of the dioecious lines are crossed with pollens of the male flowers from selected plants of the same line (sib-mating). Hermaphrodite flowers of the plus plants of the gynodioecious lines are covered with paper bags for selfing. Care should be taken to choose only 'elongata' and 'pentandra' types of hermaphrodite flowers for selfing. Further progeny is raised from the seed obtained from the crossed/selfed fruits.

### 3.2 Selection

Selection among collected dioecious germplasm and sib-mating in the further generations has been a commonly-used method of papaya breeding. Selection is made based on desirable phenotypic traits from the population raised by controlled pollination. The procedure is repeated continuously (recurrent selection) for 6–8 generations or until the plants are presumed to be genetically stabilized. Such plants can be released as a variety if they have the required number of positive traits. If they carry only a few desirable traits, they can be maintained as pure lines as a source of genes for those traits in further breeding programme. The variety Coorg Honey Dew was developed by this method from the variety Honey Dew at the ICAR Indian Institute of Horticultural Research (IIHR) in Bengaluru. Honey Gold was developed by selection from Hortus Gold in South Africa. Positive traits in Honey Gold were higher yield, better sugar content and disease resistance. The varieties Pusa Giant, Pusa Delicious and Pusa Majesty were developed by selection from the variety Ranchi at the ICAR Indian Agricultural Research Institute Regional Station at Pusa (in the state of Bihar, India). Tamil Nadu Agricultural University (TNAU) in Coimbatore in India developed the Co.5 by selection from the variety Washington. Bettina and Petersen are examples of inbred pure line cultivars from Australia.

### 3.3 Crossing and development of hybrids

Depending on breeding objectives, crosses are made between plants of two varieties or pure lines. If hybrid vigour (heterosis) is sufficient, $F_1$ plants can be used as a variety (e.g. Red Lady, Sinta etc). If the extent of the desirable traits is not sufficient in the $F_1$ generation, selected plants are taken to the next generation by 'sib-mating and selection' (in dioecious plants) or 'Selfing and Selection' (in gynodioecious plants) for genetic stabilization of traits. Co.3 was developed by crossing Co.2 with Sunrise, followed by 'sib-mating and selection' at TNAU. Bettina was developed as a hybrid of Betty and a Queensland strain. Surya (Sunrise × Pink Flesh Sweet) released from IIHR is an example of stabilized gynodioecious cultivar developed by 'selfing and selection'. A major challenge in the production of $F_1$ hybrids is difficulties in obtaining pure lines in papaya. More than two parents involved in crossing is known as a three way cross ('a x b' x 'c' or 'a' x 'b x c'). Prabhath ('Surya × Tainug-1' × Local Dwarf), released by IIHR, is an example of stabilized gynodioecious cultivars developed by 'selfing and sib-mating'. When the genes of interest are available in plants of another genus (e.g. *Vasconcellea species* carry genes with resistance to PRSV, Bacterial Canker, black spot and Phytophthora canker), the crossing between the donor species and the *C. papaya* are called intergeneric crosses. These are severely restricted by genetic incompatibility between parents.

### 3.4 Back crossing

Back crossing is used in papaya breeding when it is necessary to transfer genes of one trait of the donor parent to the recurrent parent which otherwise has all the favourable traits. Pollens from the selected $F_1$ plant are used to backcross the recurrent parent until the final plant has both the desired trait from the donor parent and all the desirable traits of the recurrent parent. This method is popular in transferring disease resistance to an otherwise good variety with desirable quality characteristics. Dwarf Solo was developed by a backcross of Florida's Betty and Solo.

### 3.5 Mutation breeding

Mutation is created in the population by subjecting seeds to mutagens (chemicals or radiation). The population is observed carefully for mutants with desirable traits. A mutant with desirable traits is rarely developed. Most of the time traits are stabilized in the mutants and they can be used as a variety. Otherwise the 'selection and sib-mating' method may be applied for the stabilization of the desired traits. Pusa Nanha is a successful mutant cultivar developed from Pusa Delicious by gamma radiation at Pusa in India (Fig. 5).

### 3.6 Marker-assisted selection

In this method, selection in the population is based on markers which are closely linked with traits. Traditional breeding used physical or biochemical markers, but now molecular markers closely linked with desirable traits have been developed. Seedlings of progeny are screened in the nursery. Only those seedlings with markers for the desired traits are taken to the field for further evaluation. This cuts down the size of field trials and breeding cycles, thus reducing total breeding time ('foreground selection' and 'background selection').

## 4 Breeding for particular traits: identification of plant sex at seedling stage

Initial breeding programmes were targeted towards enhancing traits contributing to higher plant productivity (e.g. earliness and low height of bearing, reducing summer sterility, narrower fruit cavity). Breeding also focussed on identification of plant sex at nursery stage as a way of increasing the number of productive plants and thus overall productivity. Other breeding targets have included reducing stamen carpelloidy, improving fruit size, shape and skin texture to increase the proportion of marketable fruits. Fruit quality traits, such as pulp colour, sugar content and flavour, as well as shelf life, further enhance marketability. Other objectives include improving disease resistance. Higher latex (papain) production has also been a breeding objective (Giacometti, 1987; Storey, 1953; Fig. 2).

Dioecious papaya varieties are not favoured by farmers given the potential waste of resources on unproductive male plants. Even in gynodioecious varieties, hermaphrodite fruits are preferred over female fruits due to market demand. Elimination of male plants from dioecious varieties, and female plants from gynodioecious plants, is therefore an important breeding objective. Various methods based on morphological, cytological and isozyme-based markers have been used for sex determination at seedling stage, but none of them has proved accurate. Reliable molecular markers (including PCR-based Random Amplified Polymorphic DNA (RAPD), Amplified Fragment Length Polymorphism (AFLP), Sequence Characterized Amplified Region (SCAR) and Inter Simple Sequence

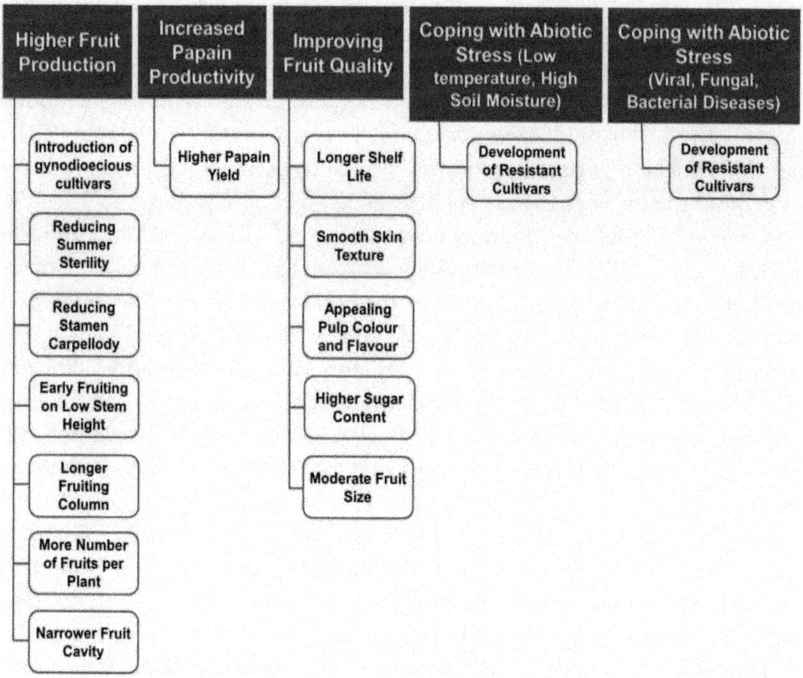


**Figure 2** Breeding objectives of papaya and associated traits.

Repeat (ISSR)) have been developed to identify the sex type at seedling stage (Vashistha et al., 2019). The first sex-linked marker, a microsatellite containing the quadranucleotide repeats (GATA4), was reported by Parasnis et al. (1999). RAPD marker by two primer pairs was 100% successful in predicting the sex type in the varieties Cavite, Cariflora and Sinta. Females produced a single band of 0.8 kb; hermaphrodites produced two distinct bands of 1.3 kb and 0.8 kb, and males had no band (Magdalita and Mercado, 2003). Similarly, OPC09 (1.7 kb) and OPE03 (0.4 kb) markers were identified in male and hermaphrodite plants with OPE19 (2.18 kb) identified in female plants (Niroshini et al., 2008). Being sequence specific, SCAR markers are an improvement over RAPD in terms of reliability, stability and simplicity of use (Paran and Michelmore, 1993). A RAPD Papaya Sex Determination Marker (PSDM) has therefore been developed into a SCAR marker that amplifies fragments from the genomes of male and hermaphrodite plants, but not the female ones (Urasaki et al., 2002). In another study, OPY7, a RAPD marker was developed into a SCAR marker that differentiated hermaphrodite and male plants from female plants (Bedoya and Nuenz, 2006). Among ISSR markers, a 500 bp band (SS72/12) was observed in hermaphrodite plants of the solo group Tainung H and the Formosa group (Da Costa et al., 2011). The microsatellite probes (GACA)4 has generated one band detectable in female and hermaphrodite papaya plants (Gangopdhayay et al., 2007). A microsatellite probe GATA4 generated a 5 kb male-specific band (Parasnis et al., 1999). These markers cannot yet be applied commercially because of cost consideration. Some sequence-based advanced molecular markers like simple sequence repeats (SSR), also termed as simple sequence length polymorphisms (SSLPs)/ microsatellite and short tandem repeats (STRs), and single nucleotide polymorphism (SNP), which have been used in predicting sex in other plants, have the potential to be used in papaya too.

Chan et al. (2008) reported all progeny of a self-pollinated hermaphrodite papaya mutant (SR*) hermaphrodite compared to the normal ratio of 2:1 hermaphrodite to female. Genetic composition of the female (mm) gene complex was considered to be lethal. When SR* was crossed with the cv. Thailand, all genotypes of hermaphrodite plants survived. A 3:1 segregation ratio was obtained from the progeny. The segregation ratio of the $F_2$ generation of randomly selected $F_1$ plants suggested that the mutant had a different form of the M2 allele (M@) which allowed the plants with the M@M2 combination to survive. Genetic studies have proved that the mutant has the M@ml genotype, a new mutant which has enabled it to produce all surviving hermaphroditic progenies. The mechanisms responsible for the all-hermaphrodite phenomenon in the self-pollinated progeny of H*-TSS No.7 (an inbred line of the mutant) has been studied by the application of SCAR markers developed from the *SVP*-like gene that can reliably discriminate among female, male and hermaphrodite papaya (Chen et al., 2019). The study revealed that H*-TSS No.7 contained a recessive lethal allele, *ml*, on the X chromosome. The X*X* homozygous genotype (ml/ml) was detected only in seeds, resulting in loss of germination. This can explain the missing female progeny in the selfed H*-TSS No.7 population. The lethal effect in genotypes of YY, $YY^h$, and $Y^hY^h$ papaya established the need for at least one functional X chromosome necessary for seed development. A lethal allele (*ml*) on X* chromosome H*-TSS No.7 might be the reason for lack of germinating ability in the homozygous female progeny (Chen et al., 2019; Chiu et al., 2015).

## 5 Breeding for disease resistance: papaya ringspot virus (PRSV)

Resistance breeding programmes have been established in papaya to combat bacterial, fungal and viral diseases. Since it is the most devastating and widespread disease, breeding programmes have concentrated particularly on PRSV resistance. Diseases which affect smaller geographical areas, such as papaya dieback and black spot, are also being addressed.

Papaya ringspot virus (PRSV) was first described by Lindner et al. (1945) in Hawaii (USA), and later experiments by Jensen (1949) established the viral nature of infection. The name of the disease, ringspot, is taken from the appearance of rings on the fruit of the infected plants. Other symptoms are: mosaic formation and chlorosis of the leaf lamina, water-soaked oily streaks on the petiole and upper part of the trunk, distortion of young leaves that sometime results in shoe string-like symptoms. Infected plants lose vigour and become stunted with low yield of inferior quality fruits, and no marketable yield in case of severe infection at an early stage (Fig. 3). The virus spreads mainly in the field via several species of aphid vectors in a non-persistent manner. PRSV is a member of the genus *Potyvirus* which is further classified into two types: type P (PRSV-P) which infects cucurbits and papaya, and the type W (PRSV-W) which infects cucurbits but not papaya (Purcifull et al., 1984).

Pathogen diversity, the abundance of aphid vectors and alternate hosts are major challenges to effective PRSV management. Since there is no way of controlling PRSV infection once it becomes established, the main emphasis has been to use cultural practices to minimize yield losses, particularly by avoiding periods of aphid abundance and by avoiding infection at the early stage of plant growth (Sharma et al., 2010). Cultural practices include using healthy (virus-free) seedlings of an appropriate cultivar(s), timing of transplanting when the vector population is naturally low, systematic roguing (removal) of infected plants and controlling the population of aphid vectors to avoid infection during the initial stage of plant growth. Raising a border crop around the main papaya crop can reduce access of aphid vectors to the main plantation (Sharma et al., 2010).

**Figure 3** PRSV infected papaya plants.

Two transgenic varieties (SunUp and UH Rainbow) with PRSV resistance have been released for commercial cultivation in Hawaii, and one (Huanong No.1) in China. However, transgenic resistance may not be effective against different/heterologous strains of infecting PRSV (Gonsalves, 2003). The ideal approach for controlling PRSV is through introgression of resistance genes via conventional breeding, though cross-incompatibility between *C. papaya* and PRSV-resistant *Vasconcellea species* remains a major challenge.

Earlier evidence suggested that the PRSV resistance could be based on a single dominant gene (Micheletti, 1962). A single dominant gene control of the virus resistance in *V. cundinamarcensis* (also known as *V. pubescens*) was suggested by Drew et al. (1995) and O'Brien and Drew (2009). Single gene dominance regulating PRSV resistance in $F_1$ intergeneric hybrids of *C. papaya* and *V. cauliflora* was reported by Magdalita et al. (1997b). Based on the reaction to PRSV in a large population of *C. papaya* × *V. quercifolia* hybrids, Drew et al. (1998) suggested that PRSV resistance in *V. quercifolia* is controlled by a single resistant gene. Genetic mapping studies of *V. cundinamarcensis* × *V. parviflora* hybrids also supported the conclusion that PRSV resistance in *V. cundinamarcensis* is controlled by a single dominant gene (Dillon et al., 2006). A single dominant gene for PRSV resistance in *V. pubescens* was mapped by Alamery and Drew (2014).

*C. papaya* is reported to have limited PRSV resistance (Nishijima, 1994). Cariflora was the first papaya cultivar reported to be tolerant to PRSV infection in Florida, USA (Conover et al., 1986). Similarly, Thapra from Thailand

(Prasartsee et al., 1995); Red Lady and Known You No. 1 from Taiwan and Sinta from the Philippines have also been shown to be PRSV-tolerant cultivars (Siar et al., 2009). These cultivars may become infected with the virus but remain symptomless or develop mild symptoms which mean they are able to produce an economically-viable yield. Tolerance of these cultivars to PRSV varies with the isolates prevalent in different geographical locations.

PRSV resistance has been reported in some species of the related genus, *Vasconcellea*. Horovitz and Jimenez (1967) reported PRSV resistance in some *Vasconcellea species (cauliflora, cundinamarcensis, stipulata* and *candicans*). Conover (1964) reported *V. quercifolia* resistant to PRSV in Florida (USA). PRSV resistance in *V. cauliflora* has also been established by other researchers (Mekako and Nakasone, 1975; Alvizo and Rojkind, 1987; Magdalita et al., 1988). *V. cundinamarcensis* is the only *Vasconcellea species* that has been consistently reported resistant to PRSV under different climatic conditions. In contrast, other resistant species, *V. cauliflora* and *V. quercifolia,* have been reported to be susceptible to PRSV under some climatic condition. The variation in PRSV reaction in the same *Vasconcellea species* in different geographical locations is attributed to the genetic differences in the virus strains and the environmental conditions. Some *Vasconcellea species* (namely, *cauliflora, quercifolia, cundinamarcensis* and *parviflora*) have been used extensively in the introgression of PRSV gene(s) to *C. papaya*.

Breeding programmes for PRSV resistance have focussed on highland papaya (*Vasconcellea species*). Success of crosses is dependent on genetic compatibility of parent species (Aradhya et al., 1999; Costa et al., 2008). PRSV-resistant *Vasconcellea species* used in earlier breeding programme were genetically distant from *C. papaya*. The success of hybridization was restricted by post-zygotic barriers, like abnormal endosperm development or ovule and embryo abortion (Mekako and Nakasone, 1975; Manshardt and Wenslaff, 1989a). Later on, Drew et al. (2006a) proposed that variable chromosome numbers, presence of univalents, lagging chromosomes at anaphase and meiotic irregularities were likely causes for non-functional gametes and infertility among hybrids. In earlier cross-compatibility studies, Horovitz and Jiménez (1967) reported that hybrids between papaya and *V. stipulate* lacked vigour and viability. They often produced non-viable seed or parthenocarpic fruits. The biggest challenge facing breeding programmes was to overcome the cross-incompatibility between *Vasconcellea species* and *C. papaya*. Some hybrids have been developed by crossing *C. papaya* with *V. cauliflora, V. cundinamarcensis, V. quercifolia, V. goudotiana* and *V. stipulata* (Manshardt and Wenslaff, 1989b; Magdalita et al., 1997b; Drew et al., 1998; Siar et al., 2011; Sharma and Tripathi, 2013). However, progressing beyond the $F_1$ generation in intergeneric crosses has been very difficult.

Unlike most of the *Vasconcellea species, V. cauliflora* is a lowland plant and its climatic requirements are similar to those of *C. papaya*. Being resistant to

PRSV infection, it has been a natural choice for breeding programme (Horovitz and Jiménez, 1958; Sawant, 1958a,b; Moore and Litz, 1984; Manshardt and Wenslaff, 1989a,b; Sharma et al., 2015). Use of *V. cauliflora* as a source of PRSV resistance has been severely constrained by its cross-incompatibility with *C. papaya* which is based on their genetic distance (Jobin-Décor et al., 1997). Magdalita et al. (1997a) reported a high level of aneuploidy as one of the causes for the failure of crosses. The failure of hybrids to flower and pollen infertility in hybrids may be due to chromosome imbalance related to incompatibility barriers and lack of homology between parental chromosomes (Magdalita et al., 1997a; Siar et al., 1998). The postzygotic barriers limiting the success of the hybridization were found to be embryo abortion and lack of endosperm development (Manshardt and Wenslaff, 1989a). Some success has been reported by Dinesh et al. (2007) in a PRSV resistance breeding program involving *C. papaya* cv. Surya and *V. cauliflora*. They achieved a maximum number of seed set (13%) when the stigmatic surface of flower was smeared with 5% sucrose solution. In a similar subsequent study involving crosses between *C. papaya* with *V. cauliflora,* true hybrids were produced by Jayavalli et al. (2011). The hybridity of the progenies was established by using an ISSR marker. However, after repeated failures to develop successful crosses with *V. cauliflora,* breeding efforts were expanded to other PRSV-resistant *Vasconcellea species* which were cross-compatible with *C. papaya*.

*Vasconcellea quercifolia* is a better choice as a source for PRSV resistance genes because it is resistant to PRSV and cross-compatible with *C. papaya* (Drew et al., 2005, 2006a; Manshardt and Wenslaff, 1989a). Manshardt and Wenslaff (1989b) reported first successful hybrids between *C. papaya* and *V. quercifolia*. Most of the subsequent PRSV resistance breeding work involving *V. quercifolia* was done by Rod Drew (Griffith University in Australia) and his team in Australia and in the Philippines. They found that a tropical climate was more conducive to breeding than a subtropical climate (Drew et al., 1998, 2005). In a breeding programme between *C. papaya* and *V. quercifolia*, four $F_1$ plants produced embryos when backcrossed to papayas. One plant (clone 54) demonstrated a high level of tolerance against PRSV in both the glasshouse and the field for nine months in Australian conditions (Drew et al., 2005). In a similar study, three quarters of the hybrids showed resistance. Pollens from one male plant were able to backcross successfully with *C. papaya*. $BC_1$ plants showed high mortality and low vigour. Pollen fertility was higher in male $BC_1$ plants, while second backcross generation ($BC_2$) were easier to produce (Drew, 2006b). PRSV-resistant $BC_3$ and $SbC_3$ and $BC_4$ plants were developed subsequently in the Philippines. These plants showed resistance against PRSV both under glasshouse and field conditions. They developed symptoms under field conditions after 18 months of exposure. However, the rate of symptom development was slower in these plants than the control plants raised under glasshouse condition (Davao Solo). A few plants were free from infection for up to 18 months under field conditions. They continued to produce good quality fruits for three years under high disease pressure (Siar et al., 2011). However, the resistance developed by this method was not consistent and there were instances of breakdown.

Subsequent breeding research has focussed on *Vasconcellea cundinamarcensis* with *V. parviflora* as the bridge species. The donor species, *V. cundinamarcensis,* is reported to be consistently resistance against PRSV under different geographical conditions (Conover, 1964; Horovitz and Jiménez, 1967). Crosses between *C. papaya* and *V. cundinamarcensis* resulted in sterile $F_1$ progeny in the tropical lowland conditions of Maracay in Venezuela (Horovitz and Jiménez, 1967). In Hawaii too, several vigorous and PRSV-resistant $F_1$ hybrids between *C. papaya* and *V. cundinamarcensis* were found to be sterile, and backcrossing could produce only infertile sesquidiploids from unreduced megaspores (Manshard et al., 1995). However, the hybrids produced by using embryo rescue and in vitro culture techniques were vigorous and resistant to PRSV, though their growth was slow during summer. The bridge species, *V. parviflora,* is the most closely related to *C. papaya* among all *Vasconcellea species* (Jobin-Décor et al., 1997) with climatic requirements similar to those of *C. papaya* (Scheldeman et al., 2007). However, it is susceptible to PRSV (Horovitz and Jimenez, 1967; Drew et al., 1998). The species is characterized by pink flowers, a dominant characteristic trait, that can be used as a physical marker to identify true crosses. It is cross-compatible both with *C. papaya* and *V. cundinamarcensis* (O'Brien and Drew, 2009).

The concept of bridge crossing was used by Drew and his group in Australia to overcome the incompatibility between *C. papaya* and *V. cundinamarcensis*. Hybrids in all combinations among *C. papaya, V. cundinamarcensis* and *V. parviflora* were produced. PRSV-resistant $F_1$ hybrids from *V. cundinamarcensis* × *V. parviflora* were developed into $F_3$ populations. The disease resistance genes from *V. cundinamarcensis* was successfully backcrossed into *V. parviflora* from $F_2$ hybrids. PRSV-resistant homozygous genotypes (RR) individuals were selected for further crossing (Dillon et al., 2005a,b, 2006; O'Brien and Drew, 2009). The genes were subsequently transferred into the $F_1$ hybrids between *C. papaya* and PRSV-resistant homozygous genotypes (RR) *V. parviflora* [*C. papaya* × (*V. parviflora* backcrossed to *V. parviflora* × *V. pubescens*)]. The successful hybrids retained dominant genetic traits from *V. cundinamarcensis* (seven number of leaf vein), *V. parviflora* (pink flower colour) and *C. papaya* (red-green petiole colour and large plant size). The $F_1$ hybrids carrying the PRSV–P locus were backcrossed to *C. papaya* to generate $BC_1$ generations. They were also sib-crossed with the $F_1$ to generate $F_1$:$F_2$ generations (Kanchana-udomkan et al., 2018).

Other species have also been used to develop resistance to PRSV. Successful hybrids were produced between *V. stipulata* and *V. cundinamarcensis* (Horovitz and Jimenez, 1967). Manshardt and Wenslaff (1989b) crossed *V. stipulata* to *C. papaya* and successfully developed hybrids. *V. stipulata* has been reported as a potential source of PRSV resistance (Scheldeman, 2002). Use of other PRSV-resistant *Vasconcellea species* in PRSV resistance breeding programmes has potential for further investigation.

Breeding research has also focussed on the use of modern techniques such as molecular markers. Molecular markers are being used to determine genetic proximity among parents, in hybrid identification, and to identify particular sequences that are linked with desirable traits (Dillon, 2006; Eustice et al., 2008; Oliveira et al., 2010a). The ideal molecular markers should be closely linked with the target trait and they should exhibit dominant inheritance. The first use of DNA markers was reported for detecting polymorphisms between *C. papaya* genotypes and related species (Sharon et al., 1992). Use of polymerase chain reaction (PCR) techniques in the 1990s resulted in rapid development of molecular markers. The first PCR-based markers for papaya, Randomly Amplified Polymorphic DNA (RAPD), were used to evaluate genetic relationships among papaya cultivars, and between *C. papaya* and *Vasconcellea species* (Jobin-Décor et al., 1997; Stiles et al., 1993). A modification of this technique, Randomly Amplified DNA Fingerprint (RAF), was used to identify markers linked to PRSV resistance in *V. cundinamarcensis* (Dillon et al., 2005a). AFLP markers, which were developed by the integration of AFLP and PCR, were used to understand genetic relationships in the family Caricaceae (Ardhya et al., 1999; Kim et al., 2002; Van Droogenbroeck et al., 2002; Ocammpo Perez et al., 2007; Ratchadaporn et al., 2007). The non-specific and dominant markers (RAPD, RAF and AFLP) can be used easily in papaya crop improvement, for they do not require DNA sequencing analysis. However, these markers are not reliable and stable so have limited uses. SSR/microsatellite markers, an improvement over the dominant markers, have been widely used in studies of papaya genetic diversity (Perez et al., 2006; Oliveira et al., 2010b; Eustice et al., 2008; Wang et al., 2008). Molecular markers linked to the gene *prsv-1* have been used in the marker-assisted breeding programme because of their dominant inheritance and the robust resistance to PRSV (Drew et al., 1998; Magdalita et al., 1997b). Costa et al. (2008) have used fluorescent in-situ hybridization (FISH) to explore genomic and chromosomal relationship among species of Caricaceae (Fig. 4).

Embryo abortion and lack of endosperm development have limited the success of hybridization programmes of *C. papaya* with *Vasconcellea species* (Manshardt and Wenslaff, 1989a). Research has therefore focussed on the rescue of embryos before abortion. A successful protocol to rescue and germinate C. papaya x V. *cauliflora* immature embryos was developed by Magdalita et al. (1996) which resulted in reduced losses of embryos and stronger plantlets in

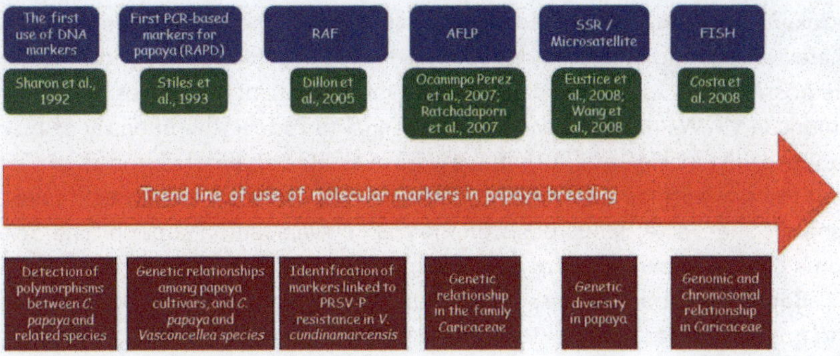


**Figure 4** Development of molecular makers in papaya breeding. (RAPD - Randomly Amplified Polymorphic DNA, RAF - Randomly Amplified DNA Fingerprinting, AFLP - Amplified Fragment Length Polymorphism, SSR - Simple Sequence Repeat, FISH - fluorescent in-situ hybridization)

vitro. Drew et al. (1998) subsequently developed procedures for hybridization of papaya with related PRSV-resistant *Vasconcellea species* (V. *cauliflora*, V. *quercifolia* and V. *cundinamarcensis*). Hybrid plants were produced by forming embryogenic cultures of germinated embryos on a hormone-free agar medium. The protocol was standardized to produce hybrids between C. papaya and V. *quercifolia/V. cundinamarcensis*. Using this technique, intergeneric hybrid plants were produced between C. papaya × *V. goudotiana* and C. papaya × V. *parviflora*. The technique has offered the potential to evaluate both the efficiency of transformation of papaya and the efficacy of the putative resistance genes when transformed from *V. cundinamarcensis* into papaya. The rescue of embryos of crosses between *C. papaya* and *V. cauliflora* was also achieved successfully by Magdalita et al. (2001) with *C. papaya* as the female parent under subtropical conditions.

## 6 Breeding for disease resistance: black spot and bacterial dieback disease

Black spot (causal fungus *Asperisporium caricae*) is an important fungal infection damaging papaya foliage and fruits. Foliage damage leads to reduced photosynthetic area. Infection to the fruits affects their marketability and makes them susceptible to secondary rot (Rezende and Martins, 2005). Breeding for developing genetically resistant cultivars is being undertaken in various places (Dianese et al., 2007; Vivas et al., 2012, 2015). Vivas et al. (2015) screened some genotypes against black spot disease and observed that Cariflora exhibited low severity in the leaves and fruits. Vivas et al. (2016) selected papaya hybrids and verified heterosis expression for black-spot resistance. Disease incidence and severity on leaves were used to estimate genetic distance using Jaccard's

Index. 77 % of hybrids showed heterosis for at least one of the three evaluated characteristics and 58 % of hybrids showed a heterobeltiosis effect, emphasizing the possibility of using hybridization to develop resistant genotypes. The hybrid Americano × Waimanalo has strong potential to reduce the intensity of black spot. In a field trial in Brazil, 20 hybrids were evaluated against resistance for black spot in fruits and leaves. The hybrids were derived from the crosses between superior lines, elite genotypes, commercial cultivars, and a superior line 36/7. Some hybrids showed improved black-spot resistance (Poltronieri et al., 2017).

Bacterial dieback disease (BDB) is caused by the bacterium *Erwinia papaya and Eriwinia mallotivora* (Amin et al., 2010; Gardan et al., 2004). The pathogen enters the host through wounds and then moves into the green stem. The petiole dries up and rotting sets in, showing as 'flag-leaf' symptoms. Stem around the infected petiole develops dark water-soaked streaks or lesions. The tree gradually withers and dies within four to eight weeks (Fullerton et al., 2011; Gardan et al., 2004). All commercial cultivars are susceptible to the infection. The disease in managed by the use of antibacterial compounds (Mohd Khairil and Muhammad Munzir, 2014). A draft genome of *Erwinia mallotivora* BT-Mardi has significant potential to reduce the damage caused by the pathogen (Redzuan et al., 2014). The draft genome would be useful for understanding host-pathogen interaction and the mechanism of infection. Investigation of pathogenicity of genes will decipher their functions and unravel the complex host-pathogen interaction, and their capabilities to enter plant cells, overcome host resistance and induce disease (Bunawan and Baharum, 2015). Wee et al. (2014) identified zinc finger protein, leucine rich protein, aquaporin and peroxidase as the possible defence-related genes. Resistance against papaya dieback has been investigated in Malaysia via systemic acquired resistance (SAR). SAR is characterized by broad spectrum disease control and an associated coordinated expression of a set of pathogenesis-related (PR) genes and proteins (SAR markers). The HrpN recombinant protein was tested for its potential as an elicitor able to increase papaya defences against *E. mallotivora* through the activation of SAR mechanism. Recombinant protein-treated plants showed a slower disease infection rate and severity when compared to control plants. Increased resistance towards dieback was associated with increased expression of selected plant-defined genes after the plants were sprayed with the recombinant HrpN protein. The selected protein induced SAR, increased SAR-associated defence gene expression, and increased resistance against the dieback pathogen (Bakar et al., 2018).

## 7 Case studies: breeding programmes in India and Malaysia

Although India is not a major centre for papaya biodiversity, it contains a large pool of genetic variability in cultivated papaya owing to five centuries of cultivation. Most of the modern *C. papaya* cultivars in India were derived from five horticultural dioecious forms (Bangalore, Ceylon, Honey Dew, Ranchi and Washington) cultivated earlier in India (Singh, 1990). In addition to these, Barwani Red, Barwani Yellow, Santacruz Giant, Singapore Pink, Chianki Local, Ram Nagar Local and Gujarati were also cultivated on a smaller scale in a number of areas. Formal papaya breeding in India was started during the mid-1960s. Pusa, Coimbatore, Pantnagar and Bangalore (now Bengaluru) were the major breeding centres. Earlier papaya cultivars were bred for high yield and gynodioecious character to enhance productivity. The Indian Agricultural Research Institute (IARI) Regional Station at Pusa started working on papaya in the mid-1960s and released a series of popular varieties such as Pusa Giant, Pusa Delicious, Pusa Majesty, Pusa Nanha (Fig. 5) and Pusa Dwarf. TNAU in Coimbatore started papaya-breeding work in the late 1960s. They released another series of papaya cultivars (using Co to denote Coimbatore) such as Co.1, Co.2, Co.3, Co.4, Co.5, Co.6 and Co.7. Govind Bhallabh Pant University of Agriculture and Technology in Pantnagar started a systematic papaya-breeding programme in 1972, releasing Pant Papaya-1, Pant Papaya-2 and Pant Papaya-3. They were suitable for conditions in the Tarai region which is characterized by low winter temperatures and a shallow water table. The IIHR Station at Bengaluru has released three papaya cultivars: Coorg Honey Dew, Surya and Arka Prabhath (Aiyappa and Nanjappa, 1959; Dinesh, 2010; Ram, 1983, 1984; Singh et al., 1984; Sundarrajan and Krishnan, 1984). Many of these cultivars are not in commercial cultivation. The origin of these cultivars and their relationships are shown in Table 2. At present, gynodioecious cultivars from the private sectors are more popular among farmers because of elimination of non-productive

**Figure 5** Pusa Nanha, a dwarf mutant developed by gamma ray exposure.

**Table 2** Origin and relationship among Indian papaya cultivars

| Parentage | Cultivars |
|---|---|
| *1. By selection and sib-mating:* | |
| Ranchi | Co.1, Pusa Giant, Pusa Delicious, Pusa Dwarf, Pusa Majesty |
| Local type | Co.2 |
| Washington | Co.5 |
| Honey Dew | Coorg Honey Dew |
| Pusa Giant | Co.6 |
| *2. By crossing:* | |
| Co.1 × Washington | Co.4 |
| Co.2 × Sunrise | Co.3 |
| Sunrise × Pink Flesh | Arka Surya |
| CP75 × Coorg Honey Dew | Co.7 |
| Surya × Tainug-1 × Local Dwarf | Arka Prabhath |
| *3. By mutation:* | |
| Pusa Delicious | Pusa Nanha |

male plants from the population. Recent efforts are concentrated towards development of PRSV-resistant cultivars involving *Vasconcellea species*, but results have been inconclusive so far.

Breeding programmes in Malaysia initially focussed on yield and fruit quality. Before a formal breeding programme was established in the mid-1960s, Sekaki (meaning 'one foot') was a popular variety. Initial priority was given to selection of baseline cultivars from farmers' fields based on uniformity and yield. Cultivars like Subang 6 and Batu Arang were selected for fresh fruit quality and Sitiawan for canning during the 1970s. The backcross breeding programme, started during the early 1970s, used the high quality Sunrise Solo as the recurrent parent in improving the eating quality of Subang 6. This led to the release of the Eksotika (MARDI backcross Solo) variety in 1987. To reduce quality problems such as freckles on the skin, Eksotika was crossed with freckles-resistant Line 19. The resultant $F_1$ hybrid was Eksotika 2 with attractive, firmer flesh, reduced fruit freckles and significant heterosis in yield. This was claimed to be the first commercial $F_1$ hybrid papaya in the world. Later on, priorities in breeding were shifted to development of $F_1$ MARDI3 pure lines (Line 19 and Line 20) hybrids to exploit heterosis in yield and improve performance under varied environmental conditions. The emphasis was on yield per unit area per unit time with stability in performance, rather than selection based on single tree yield. $F_1$ hybrid varieties have proved superior to self-pollinating pure line varieties. The outbreak of PRSV disease in 1991 resulted into a shift in breeding priority to disease resistance. The donor parents for PRSV resistance genes were Cariflora and Tainung 5. In 1996, biotechnology was brought in to help the industry in solving the PRSV threat as well as improving the shelf life of Eksotika by introducing delayed ripening characteristics (Chan, 2007). Recently, a double-cross method was applied to improve yield, fruit quality and storage life of variety Eksotika. The cross [Eksotika II × (Line 19 × PR217)] included PR217 for longer storage life. The hybrid had higher yield, T.S.S. and longer shelf life compared to Eksotika. Minimal freckles on hybrid improved the appearance (Sarip et al., 2016). Sekeli et al. (2018) has predicted a bright future for the Eksotika variety despite a serious threat from dieback disease and limitation in export marketing due to short fruit shelf life. Current research is focussed on developing resistance against dieback in Eksotika. Future breeding priorities are planned to focus on priorities such as increased nutritional value.

## 8 Conclusion and future trends

Use of intergeneric hybridization seems to be a promising option for the development of PRSV-resistant *C. papaya* with acceptable fruit quality. Selection of PRSV-resistant wild species which are cross-compatible with *C. papaya* will have a better chance of producing viable hybrids. Wild species from the family Caricaceae with resistance genes have varying degrees of cross-compatibility with *C. papaya*. *V. quercifolia* is genetically closest and *V. cauliflora* is the most distant species to *C. papaya*. The chance of successful production of viable hybrids between *C. papaya* and *V. species* has been enhanced by refinement of the embryo rescue technique and the in vitro clonal population. It has reduced dependence on the successful production of viable hybrids, and made it possible to evaluate large number of crosses which, in turn, has facilitated selection of intergeneric hybrids with both resistance and required fertility for backcrossing with *C. papaya*. The inclusion of PRSV-immune *V. cundinamarcensis* in breeding programmes has raised the possibility of obtaining some hybrids with stable resistance. Papaya black spot and die back disease are becoming serious threats in some papaya cultivation areas, with ongoing breeding programmes to improve resistance. Success will depend on combining skills in breeding, biotechnology, acclimatization, cytology and virology.

The discovery by Carvalho and Renner (2012) that *C. papaya* is closest to a clade of herbaceous or thin-stemmed species has added a new dimension to papaya breeding. In addition to woody highland papayas (*Vasconcellea*), other genera of the family closer to the genus *Carica* (i.e. *Cylicomorpha, Horovitzia, Jacarantia* and *Jarilla*) should also be characterized and screened for genes of economic importance as well as their cross-compatibility with *C. papaya* so that they can be used in papaya improvement programmes. This requires thorough investigation of intra-species variation in important traits. Breeding methods should be evolved to include herbaceous or thin-stemmed species. The potential of remaining *Vasconcellea species* in imparting PRSV resistance has not been fully tested. New emerging threats to papaya cultivation should also be taken up in future breeding programme. Considering the threat of losing useful germplasm, there is a need for designated centres for collection, conservation, evaluation and utilization of papaya germplasm.

## 9 Where to look for further information

- The book 'Singh, I.D. (1990), Papaya, New Delhi, Oxford & IBH Publishing Company Private Limited' can be a good starting point to understand the crop, and another book 'Ming, R. and Moore, P. H. 2014. Genetics and Genomics of Papaya. Springer, New York (ISBN 978-1-4614-8086-0, ISBN 978-1-4614-8087-7 eBook, DOI 10.1007/978-1-4614-8087-7)' can give glimpses of modern outlook of papaya breeding.
- The review by 'Barton, N. H. and Hewitt, G. M. 1985. Analysis of hybrid zones. Ann. Rev. Ecol. Syst. 16, 113–48' can be useful considering the issues of intergeneric cross-incompatibility within the family Caricaceae.
- International Society of Horticultural Science is the leading international organization sponsoring academic and research activities in papaya (https://www.ishs.org/papaya). They organized periodic International Symposium on Papaya and publish the peered reviewed research papers in the form of the proceedings.
- The relevant material can also be seen on the websites: http://eprints.utar.edu.my/1969/1/Breeding_the_papaya_-_Carica_papaya.pdf and http://www.itfnet.org/v1/2016/05/papaya-introduction/
- *Some relevant research centres* are:
  - School of Environment and Science, Logan campus, Griffith University, Australia.
  - University of Hawaii, Honolulu, Hawaii, USA.
  - Fengshan Tropical Horticultural Experiment Station Branch, TARI, Taichung, Taiwan.
  - National Chung Hsing University, Taichung, Taiwan.
  - Indian Agricultural Research Institute, Regional Station, Pune, India.
  - Tamil Nadu Agricultural University, Coimbatore, India.
  - Indian Institute of Horticultural Research, Bengaluru, India.
  - EMBRAPA – National Center for Research on Cassava and Tropical Fruit Culture (CNPMF), Capixaba, Brazil.
  - Universidade Estadual do Norte Fluminense (State University of North Fluminense), Darcy Ribeiro, Brazil.

## 10 References

Aiyappa, K. M. and Nanjappa, P. P. 1959. Coorg Honey, a new find in papaya. *Indian Hortic.* 3, 3.

Alamery, S. and Drew, R. A. 2014. Studies on the genetics of PRSV-P resistance genes in intergeneric hybrids between *Carica papaya* and *Vasconcellea quercifolia*. *Acta Hortic.* 1022(1022), 55–61. doi:10.17660/ActaHortic.2014.1022.5.

Alvizo, H. F. and Rojkind, C. 1987. Resistencia al virus mancha anular del papayo en *Carica cauliflora*. *Rev. Mex. Fitopatol.* 5, 61–2.

Amin, N. M., Bunawan, H., Redzuan, R. A. and Jaganath, I. B. 2010. *Erwinia mallotivora* sp., a new pathogen of papaya (*Carica papaya*) in Peninsular Malaysia. *Int. J. Mol. Sci.* 12(1), 39–45.doi:10.3390/ijms12010039.

Anon. 2019. Food and Agricultural Organization of the United Nations. Available at: http://www.apps.fao.org (accessed on 27 April 2019).

Aradhya, M. K., Manshardt, R. M., Zee, F. and Morden, C. W. 1999. A phylogenetic analysis of the genus *Carica* L. (Caricaceae) based on restriction fragment length variation in a cpDNA intergenic spacer region. *Genet. Resour. Crop. Evol.* 46(6), 579–86. doi:10.1023/A:1008786531609.

Badillo, V. M. 1993. *Caricaceae: Segundo esquema*. Issue 43 of Revista de la Facultad de Agronomía de la Universidad Central de Venezuela: Alcance. Universidad Central de Venezuela, Facultad de Agronomía, Maracay.

Badillo, V. M. 2000. *Carica* L. vs. *Vasconcella* St.-Hil. (Caricaceae) con la rehabilitación de este último. *Ernstia* 10, 74–9.

Badillo, V. M. 2001. Nota correctiva *Vasconcellea* St. Hil. y no *Vasconcella* (Caricaceae). *Ernstia* 11, 75–6.

Bakar, N. A., Sohaime, M. Z., Juri, N. M., Badrun, R. and Sarip, J. 2018. Induction of systemic acquired resistance in papaya by foliar application of HrpN recombinant protein for increased resistance against papaya dieback pathogen. *Curr. Invest. Agric. Curr. Res.* 2(3). doi:10.32474/CIACR.2018.02.000136.

Bedoya, G. C. and Nunez, V. 2006. A SCAR marker for the sex types determination in Colombian genotypes of *Carica papaya*. *Euphytica* 153(1–2), 215–20. doi:10.1007/s10681-006-9256-7.

Brown, J. E., Bauman, J. M., Lawrie, J. F., Rocha, O. J. and Moore, R. C. 2012. The structure of morphological and genetic diversity in natural populations of *Carica papaya* (Caricaceae) in Costa Rica. *Biotropica* 44(2), 179–88. doi:10.1111/j.1744-7429.2011.00779.x.

Bunawan, H. and Baharum, S. N. 2015. Papaya dieback in Malaysia: a step towards a new insight of disease resistance. *Iran. J. Biotechnol.* 13(4), 1–2. doi:10.15171/ijb.1139.

Carvalho, F. A. 2015. *Molecular Phylogeny, Biogeography and an E-Monograph of the Papaya Family (Caricaceae) as an Example of Taxonomy in the Electronic Age*. Springer Spektrum, Munich, Germany, 63p.

Carvalho, F. A. and Renner, S. S. 2012. A dated phylogeny of the papaya family (Caricaceae) reveals the crop's closest relatives and the family's biogeographic history. *Mol. Phylogenet. Eval.* 65(1), 46–53.

Chan, Y. K. 2007. Stepwise priorities in papaya breeding. *Acta Hortic.* 740(740), 43–8. doi:10.17660/ActaHortic.2007.740.4.

Chan, T. C., Yen, C. R., Chang, L. S., Hsiao, C. H., Ko, T. S. and Weber, W. E. 2008. All hermaphrodite progeny are derived by self-pollinating the sunrise papaya mutant. *Plant Breed.* 122, 431–4. doi:10.1046/j.1439-0523.2003.00812.x.

Chen, J., Urasaki, N., Matsumura, H., Chen, I., Lee, M., Chang, H., Chung, W. and Ku, H. 2019. Dissecting the all-hermaphrodite phenomenon of a rare X chromosome mutant in papaya (*Carica papaya* L.). *Mol. Breed.* 39(1), 14. doi:10.1007/s11032-018-0918-7.

Chiu, C. T., Wang, C. W., Chen, F. C., Chin, S. W., Liu, C. C., Lee, M. J., Chung, W. C., Chien, Y. W., Chang, H. J. and Lee, C. Y. 2015. Sexual genetic and simple sequence repeat (SSR) analysis for molecular marker development on the all hermaphrodite papaya. *Genet. Mol. Res.* 14(1), 2502–11. doi:10.4238/2015.March.30.8.

Conover, R. A. 1964. Distortion ringspot, a severe virus disease of papaya in Florida. *Proc. Fla. State. Hortic. Soc.* 77, 440–4.

Conover, R. A., Litz, R. E. and Malo, S. E. 1986. "Cariflora" a papaya ringspot virus tolerant papaya for South Florida and the Caribbean. *HortScience* 21, 1072.

Costa, F. R., Pereira, T. N. S., Hodnett, G. L., Pereira, M. G. and Stelly, D. M. 2008. Fluorescent *in situ* hybridization of 18S and 5S rDNA in papaya (*Carica papaya* L.) and wild relatives. *Caryologia* 61(4), 411–6. doi:10.1080/00087114.2008.10589655.

Da Costa, F. R., Pereira, T. N. S., Gabriel, A. P. C. and Pereira, M. G. 2011. ISSR markers for genetic relationships in *Caricaceae* and sex differentiation in papaya. *Crop Breed. Appl. Biotechnol.* 11(4), 352–7. doi:10.1590/S1984-70332011000400009.

Dianese, A. C., Blum, L. E. B., Dutra, J. B., Lopes, L. F., Sena, M. C., Freitas, L. F. and Yamanishi, O. K. 2007. Reação de Genótipos de Mamoeiro à Varíola e à Podridão-dopé. *Fitopatol. Bras.* 32(5), 419–23. doi:10.1590/S0100-41582007000500008.

Dillon, S., Ramage, C., Drew, R. and Ashmore, S. 2005a. Genetic mapping of a PRSV-P resistance gene in 'highland papaya' based on inheritance of RAF markers. *Euphytica* 145(1–2), 11–23. doi:10.1007/s10681-005-8361-3.

Dillon, S. K., Drew, R. A. and Ramage, C. 2005b. Development of a co-dominant SCAR marker linked to a putative PRSV-P resistance locus in 'wild papaya'. *Acta Hortic.* 694(694), 101–4. doi:10.17660/ActaHortic.2005.694.13.

Dillon, S. K., Ramage, C., Ashmore, S. and Drew, R. A. 2006. Development of a codominant CAPS marker linked to PRSV-P resistance in highland papaya. *Theor. Appl. Genet.* 113(6), 1159–69. doi:10.1007/s00122-006-0375-2.

Dinesh, M. R. 2010. Papaya breeding in India. *Acta Hortic.* 851(851), 69–76. doi:10.17660/ActaHortic.2010.851.7.

Dinesh, M. R., Rekha, A., Ravishankar, K. V., Praveen, K. S. and Santosh, L. C. 2007. Breaking the intergeneric crossing barrier in papaya using sucrose treatment. *Sci. Hortic.* 114(1), 33–6. doi:10.1016/j.scienta.2007.05.010.

Drew, R. A., Vogler, J. N., Magdalita, P. M., Mohan, R. E. and Persley, D. M. 1995. Application of biotechnology to *Carica papaya* and related species. In: Terzi, M., Cella, R. and Falavigna, A. (Eds), *Current Issues in Plant Molecular and Cellular Biology*. Kluwer Academic Publishers, Dordrecht, pp. 321–6.

Drew, R. A., O'Brien, C. M. and Magdalita, P. M. 1998. Development of interspecific *Carica* hybrids. *Acta Hortic.* 461, 285–92.

Drew, R. A., Siar, S. V., Villegas, V. N., O'Brien, C. M. and Sajise, A. G. C. 2005. Development of PRSV resistant *Carica papaya* genotypes by introgression of genes from wild *Vasconcellea species*. *Acta Hortic.* 694, 73–4.

Drew, R. A., Siar, S. V., O'Brien, C. M., Magdalita, P. M. and Sajise, A. G. C. 2006a. Breeding for papaya ringspot virus resistance in *Carica papaya* L. via hybridisation with *Vasconcellea quercifolia*. *Aust. J. Exp. Agric.* 46(3), 413–8. doi:10.1071/EA04247.

Drew, R. A., Siar, S. V., O'Brien, C. M. and Sajise, A. G. C. 2006b. Progress in backcrossing between *Carica papaya* x *Vasconcellea quercifolia* intergeneric hybrids and *C. papaya*. *Aust. J. Exp. Agric.* 46(3), 419–24. doi:10.1071/EA04248.

Eustiee, M., Yu, Q., Lai, C. W., Hou, S., Thimmapuram, J., Liu, L., Alam, M., Moore, P. H., Presting, G. G. and Ming, R. 2008. Development and application of microsatellite markers for genomic analysis of papaya. *Tree Genet. Genomes* 4(2), 333–41. doi:10.1007/s11295-007-0112-2.

Fullerton, R. A., Taufa, L., Vanneste, J. L., Yu, J., Cornish, D. A. and Park, D. 2011. First record of bacterial crown rot of papaya (*Carica papaya*) caused by an *Erwinia papayae*-like bacterium in the Kingdom of Tonga. *Plant Dis.* 95(1), 70. doi:10.1094/PDIS-06-10-0455.

Gangopadhyay, G., Roy, S. K., Ghose, K., Poddar, R., Bandyopadhyay, T., Basu, D.. and Mukherjee, K. K. 2007. Sex detection of *Carica papaya* and *Cycas circinalis* in pre-flowering stage by ISSR and RAPD. *Curr. Sci.* 92(4), 524–6.

Gardan, L., Christen, R., Achouak, W. and Prior, P. 2004. *Erwinia papayae* sp. nov., a pathogen of papaya (*Carica papaya*). *Int. J. Syst. Evol. Microbiol.* 54(1), 107–13. doi:10.1099/ijs.0.02718-0.

Giacometti, D. C. 1987. Papaya breeding. *Acta Hortic.* 196(196), 53–60. doi:10.17660/ActaHortic.1987.196.5.

Gonsalves, D. 2003. Transgenic papaya: a case study on the theoretical and practical application of virus resistance. In: Vasil, K. I. (Ed.), *Plant Biotechnology 2002 and Beyond*. Springer, Dordrecht, The Netherlands, pp. 115–8. Chapter 18.

Horovitz, S. and Jiménez, H. 1958. Cruzabilidad entre especies de *Carica*. *Agron. Trop.* 7, 207–15.

Horovitz, S. and Jiménez, H. 1967. Cruzamientos interespecíficos e intergenéricos en Caricáceas y sus implicaciones fitotécnicas. *Agron. Trop. (Maracay)* 17, 323–43.

Jayavalli, R., Balamohan, T. N., Manivannan, M. and Govindaraj, N. 2011. Breaking the intergeneric hybridization barrier *in Carica papaya* and *Vasconcellea cauliflora*. *Sci. Hortic.* 130, 787–94.

Jensen, D. D. 1949. Papaya virus diseases with special reference to *Papaya ringspot*. *Phytopathology* 39, 191–211.

Jobin-Décor, M. P., Graham, G. C., Henr, R. J. and Drew, R. A. 1997. RAPD and isozyme analysis of genetic relationships between *Carica papaya* and wild relatives. *Genet. Resour. Crop. Evol.* 44(5), 471–7. doi:10.1023/A:1008644901727.

Kanchana-udomkan, C., Nantawan, U., Drew, R. and Ford, R. 2018. Progress in introgression of *Papaya ringspot virus* resistance from *Vasconcellea pubescens* to *Carica papaya*. *Acta Hortic.* 1203(1203), 59–66. doi:10.17660/ActaHortic.2018.1203.9.

Kim, M. S., Moore, P. H., Zee, F., Fitch, M. M. M., Steiger, D. L., Manshardt, R. M., Paull, R. E., Drew, R. A., Sekioka, T. and Ming, R. 2002. Genetic diversity of *Carica papaya* as revealed by AFLP markers. *Genome* 45(3), 503–12. doi:10.1139/g02-012.

Lindner, R. C., Jensen, D. D. and Ikeda, W. 1945. Ringspot: new papaya plunderer. *Hawaii Farm Home* 8, 10–4.

Magdalita, P. M. and Mercado, C. P. 2003. *Determining the Sex of Papaya for Improved Production*. Food and Fertilizer Technology Center, Philippines, 11p.

Magdalita, P. M., Villegas, V. N., Pimentel, R. B. and Bayot, R. G. 1988. Reaction of papaya (*Carica papaya* L.) and related species to ringspot virus. *Philipp. J. Crop Sci.* 13, 129–32.

Magdalita, P. M., Adkins, S. W., Godwin, I. D. and Drew, R. A. 1996. An improved embryo-rescue protocol for a *Carica* interspecific hybrid. *Aust. J. Bot.* 44(3), 343–53. doi:10.1071/BT9960343.

Magdalita, P. M., Drew, R. A., Adkins, S. W. and Godwin, I. D. 1997a. Morphological, molecular and cytological analysis of *C. papaya* x *C. cauliflora* interspecific hybrids. *Theor. Appl. Genet.* 95(1-2), 224-9. doi:10.1007/s001220050552.

Magdalita, P. M., Persley, D. M., Godwin, I. D., Drew, R. A. and Adkin, S. W. 1997b. Screening *C. papaya* x *C. cauliflora* hybrids for resistance to *Papaya ringspot virus*-type P. *Plant Pathol.* 46, 837-41.

Magdalita, P. M., Drew, R. A. and Adkins, S. W. 2001. Breeding for resistance to papaya ringspot virus. *J. Trop. Plant Pathol.* 37, 1-15.

Manshardt, R. M. and Drew, R. A. 1998. Biotechnology of papaya. *Acta Hortic.* 461(461), 65-74. doi:10.17660/ActaHortic.1998.461.5.

Manshardt, R. M. and Wenslaff, T. F. 1989a. Interspecific hybridization of papaya with other *Carica* species. *J. Am. Soc. Hortic. Sci.* 114, 689-94.

Manshardt, R. M. and Wenslaff, T. F. 1989b. Zygotic polyembryony in interspecific hybrids of *Carica papaya* and *C. cauliflora*. *J. Am. Soc. Hortic. Sci.* 114, 684-9.

Manshardt, R. M., Lius, S., Sondur, S., Wenslaff, T., Fitch, M., Sanford, J., Zee, F., Gonsalves, D., Stiles, J., Ferreira, S. and Slightom, J. L. 1995. Papaya breeding for PRS resistance. *Acta Hortic.* 370, 27-32.

Mekako, H. U. and Nakasone, H. Y. 1975. Interspecific hybridization among six *Carica* species. *J. Am. Soc. Hortic. Sci.* 100, 237-42.

Micheletti, D. 1962. Descripcion morfologica y cytological de una planta segregante de un cruce entre *Carica cauliflora* Jacq and *C. monoica* Desf. *Agron. Trop. (Maracay)* XI, 193-200.

Mohd Khairil, J. and Muhammad Munzir, M. 2014. Experiences in managing bacterial dieback disease of papaya in Malaysia. *Acta Hortic.* 1022(1022), 125-32. doi:10.17660/ActaHortic.2014.1022.16.

Moore, G. A. and Litz, R. E. 1984. Biochemical markers for *Carica papaya*, *C. cauliflora*, and plants from somatic embryos of their hybrid. *J. Am. Soc. Hortic. Sci.* 109, 213-8.

Niroshini, E., Everard, J. M. D. T., Karunanayake, E. H. and Tirimanne, T. L. S. 2008. Detection of sequence characterized amplified region (SCAR) markers linked to sex expression in *Carica papaya* L. *JNSF* 36(2), 145-50. doi:10.4038/jnsfsr.v36i2.146.

Nishijima, W. 1994. Papaya. In: Ploetz, R. C., Zentmyer, G. A., Nishijima, W. T., Rohrbach, K. G. and Ohr, H. D. (Eds), *Compendium of Tropical Fruit Disease*. US American Phytopathological Society Press, St. Paul, MN, pp. 54-70.

O'Brien, C. M. and Drew, R. A. 2009. Potential for using *Vasconcellea parviflora* as a bridging species in intergeneric hybridisation between *V. pubescens* and *Carica papaya*. *Aust. J. Bot.* 57(7), 592-601. doi:10.1071/BT09111.

Ocammpo Perez, J., d'Eeckenbrugge, C., Risterucci, A. M., Dambier, D. and Ollitrault, P. 2007. Papaya genetic diversity assessed with microsatellite markers in germplasm from Caribbean region. *Acta Hortic.* 740, 93-101.

Oliveira, E. J., Amorim, V. B. O., Matos, E. L. S., Costa, J. L. and Castellen, M. S. 2010a. Polymorphism of microsatellite markers in papaya (*Carica papaya* L.). *Plant. Mol. Biol. Rep.* 28, 519-30.

Oliveira, E. J., dos Santos Silva, A., de Carvalho, F. M., dos Santos, L. F., Costa, J. L., de Oliveira Amorim, V. B. and Dantas, J. L. L. 2010b. Polymorphic microsatellite marker set for *Carica papaya* L. and its use in molecular-assisted selection. *Euphytica* 173(2), 279-87. doi:10.1007/s10681-010-0150-y.

Paran, I. and Michelmore, R. W. 1993. Development of reliable PCR based markers linked to downy mildew resistance genes in lettuce. *Theor. Appl. Genet.* 85(8), 985–93. doi:10.1007/BF00215038.

Parasnis, A. S., Ramakrishna, W., Chowdari, K. V., Gupta, V. S. and Ranjekar, P. K. 1999. Microsatellite (GATA)n reveals sex-specific differences in Papaya. *Theor. Appl. Genet.* 99(6), 1047–52. doi:10.1007/s001220051413.

Perez, J. O., Dambier, D., Ollitraut, P., D'Eeckenbrugge, G. C., Brottier, P., Froelicher, Y. and Risterucci, A. 2006. Microsatellite markers in *Carica papaya* L.: Isolation, characterization and transferability in *Vasconcellea species*. *Mol. Eco. Notes* 6(1), 212–7. doi:10.1111/j.1471-8286.2006.01197.x.

Poltronieri, T. P., Silveira, S. F., Vivas, M., Santa Catarina, R., Cortes, D. F., Azevedo, A. O. and Pereira, M. G. 2017. Selecting black-spot resistant papaya genotypes derived from backcrossing and hybrids. *Genet. Mol. Res.* 16(1). doi:10.4238/gmr16019401.

Prasartsee, V., Kongpolprom, W. and Wichainun, S. 1995. *Thapra 1, Thapra 2, and Thapra 3: Papaya Lines that are Tolerant to Papaya Ringspot Virus*. Brochure, Thailand, p. 8.

Purcifull, D., Edwardson, J., Hiebert, J. E. and Gonsalves, D. 1984. Papaya ringspot virus. CMI/AAB Descriptions of Plant Viruses No. 292 (No. 84 Revised, July 1984). CAB International, Wallingford, UK, p 8.

Ram, M. 1983. Papaya Pusa Nanha. *Int. Agri.* 21(3), 18–9.

Ram, M. 1984. Promising varieties of papaya. *Proceedings of National Seminar on Papaya and Papain Production*, 26–27 March 1984, Coimbatore, India, pp. 37–9.

Ratchadaporn, J., Sureeporn, K. and Khumcha, C. 2007. An analysis on DNA fingerprints of thirty papaya cultivars (*Carica papaya* L.) grown in Thailand with the use of amplified fragment length polymorphic technique. *Pak. J. Biol. Sci.* 10, 3072–8.

Redzuan, R. A., Abu Bakar, N., Rozano, L., Badrun, R., Mat Amin, N. and Mohd Raih, M. F. 2014. Draft genome sequence of *Erwinia mallotivora* BT-MARDI, causative agent of papaya dieback disease. *Genome Announc.* 2(3), 375–14. doi:10.1128/genomeA.00375-14.

Rezende, J. A. M. and Martins, C. 2005. Doenças do mamoeiro. In: Kimati, H., Amorim, L., Rezende, J. A. M., Bergamin Filho, A. and Camargo, L. E. C. (Eds). *Manual de Fitopatologia: Doenças das Plantas Cultivadas* (4ª ed.). Editora Agronômica Ceres, São Paulo, 9p.

Sarip, J., Raveendranathan, P. and Razean Haireen, M. R. 2016. Double-cross: a suitable method to improve Malaysian Eksotikas. *Acta Hortic.* 1111(1111), 67–70. doi:10.17660/ActaHortic.2016.1111.10.

Sawant, A. C. 1958a. A study of the interspecific hybrids, *Carica monoica* x *C. cauliflora*. *Proc. Am. Soc. Hortic. Sci.* 71, 330–33.

Sawant, A. C. 1958b. Crossing relationship in the genus *Carica*. *Evolution* 12(2), 263–6.

Scheldeman, X. 2002. Distribution and potential of cherimoya (*Annona cherimola* Mill.) and highland papayas (*Vasconcellea* spp.) in Ecuador. PhD thesis. Faculty of Agricultural and Applied Biological Sciences, University Ghent, Belgium, p. 165.

Scheldeman, X., Willemen, L., Coppens d'Eeckenbrugge, G., Romeijn-Peeters, E., Restrepo, M. T., Romero Motoche, J., Jiménez, D., Lobo, M., Medina, C. I., Reyes, C., Rodríguez, D., Ocampo, J. A., Van Damme, P. and Goetgebeur, P. 2007. Distribution, diversity, and environmental adaptation of highland papaya (*Vasconcellea species*) in tropical and subtropical America. *Biodivers. Conserv.* 16, 1867–84.

Sekeli, R., Hamid, M. H., Razak, R. A., Wee, C. Y. and Abdulah, J. A. 2018. Malaysian *Carica papaya* L. var. Eksotika: current research strategies fronting challenges. *Front. Plant Sci.* 9, 1380: doi:10.3389/fpls.2018.01380 (accessed 03 December 29).

Sharma, S. K. and Tripathi, S. 2014. Overcoming limitations of resistance breeding in *Carica papaya* L. In: Gaur, R. K., Hohn, T. and Sharma, P. (Eds), *Plant Virus-Host Interaction: Molecular Approaches and Viral Evolution*. Elsevier Academic Press, Amsterdam, The Netherlands, pp. 177–94.

Sharma, S. K., Zote, K. K., Kadam, U. M., Tomar, S. P. S., Dhale, M. G. and Sonawane, A. U. 2010. Integrated management of papaya ringspot virus. *Acta Hortic.* 851(851), 473–80. doi:10.17660/ActaHortic.2010.851.73.

Sharma, S. K., Datar, V. V., Tripathi, S. and Sonawane, A. U. 2015. Morphological traits, growth and yield performance, and *Papaya ringspot virus* strain papaya reaction in *Vasconcellea cauliflora* in Pune, India. *Acta Hortic.* 1111, 55–60.

Sharon, D., Hille, J., Vainstein, A. and Lavi, U. 1992. Application of DNA fingerprints for identification and genetic analysis of *Carica papaya* and other *Carica species*. *Euphytica* 62(2), 119–26. doi:10.1007/BF00037937.

Siar, S. V., Geronimo, S. B., Sierra, Z. and Villegas, V. N. 1998. Cytology of *Carica papaya*, *Carica cauliflora* and their $F_1$ interspecific hybrids. *Philipp. J. Crop. Sci.* 23, 91–6.

Siar, S. V., Beligan, G. A., Drew, R. A. and O'Brien,C. 2009. *Final Report: Development of PRSV-P Resistant Papaya Genotype by Introgression of Genes from Wild Carica species*. Australian Centre for International Agricultural Research, Canberra.

Siar, S. V., Beligan, G. A., Sajise, A. J. C., Villegas, V. N. and Drew, R. A. 2011. Papaya ringspot virus resistance in *Carica papaya* via introgression from *Vasconcellea quercifolia*. *Euphytica* 181(2), 159–68. doi:10.1007/s10681-011-0388-z.

Singh, I. D. 1990. *Papaya*. Oxford & IBH Publishing Company Private Limited, New Delhi.

Singh, I. D., Hari Har Ram, Ranjeet and Sirohi, S. C. 1984. Germplasm evaluation and selection in papaya at Pantnagar. *Proceedings of the National Seminar on Papaya and Papain Production*, 26–27 March 1984, Coimbatore, India, pp. 29–34.

Stiles, J. I., Lemme, C., Sondur, S., Morshidi, M. B. and Manshardt, R. 1993. Using randomly amplified polymorphic DNA for evaluating genetic relationship among papaya cultivars. *Theor. Appl. Genet.* 85(6–7), 697–701. doi:10.1007/BF00225007.

Storey, W. B. 1953. Genetics of papaya. *J. Hered.* 44(2), 70–8. doi:10.1093/oxfordjournals.jhered.a106358.

Storey, W. B. 1976. Papaya, *Carica papaya*. In: Simmonds, N. W. (Ed.), *Evaluation of Crop Plants*. Longman, London, UK, pp. 21–4.

Sundarrajan, S. and Krishnan, B. M. 1984. Improved varieties of papaya. *Proceedings of National Seminar on Papaya and Papain Production*, 26–27 March 1984, Coimbatore, India, pp. 40–1.

Urasaki, N., Tokumoto, M., Tarora, K., Ban, Y., Kayano, T., Tanaka, H., Oku, H., Chinen, I. and Terauchi, R. 2002. A male and hermaphrodite specific RAPD marker for papaya (*Carica papaya* L.). *Theor. Appl. Genet.* 104(2), 281–5. doi:10.1007/s001220100693.

Van Droogenbroeck, B., Breyne, P., Goetghebeur, P., Romeiin-Peeters, E., Kyndt, T. and Gheysen, G. 2002. AFLP analysis of genetic relationships among papaya and its wild relatives (Caricaceae) from Ecuador. *Theor. Appl. Genet.* 105(2–3), 289–97. doi:10.1007/s00122-002-0983-4.

Vashistha, P., Yadav,A., Dwivedi, U. N. and Yadav, K. 2019. Genetics of sex chromosomes and sex linked molecular markers in papaya (*Carica papaya* L.). *Mol. Plant Breed.* 7(28), 1–18.

Vivas, M., Silveira, S. Fd, Cardoso, D. L., Pereira, M. G., Santos, P. H. Dd and Ferreguetti, G. A. 2012. Capacidade combinatória e heterose para resistência a pinta preta em mamoeiro por meio de análise dialélica. *Trop. Plant Pathol.* 37(5), 326–32. doi:10.1590/S1982-56762012000500004.

Vivas, M., Silveira, S. F., Pio-Viana, A., Amaral-Júnior, A. T., Ferreguetti, G. A. and Pereira, M. G. 2015. Resistance to multiple foliar diseases in papaya genotypes in Brazil. *Crop Prot.* 71, 138–43.doi:10.1016/j.cropro.2015.02.007.

Vivas, M., Ramos, H. C. C., Santos, P. H. D., Silveira, S. F., Pereira, T. N. S., Amaral Jr., A. T. and Pereira, M. G. 2016. Heterosis and genetic diversity for selection of papaya hybrids for resistance to black spot and phoma spot. *Trop. Plant Pathol.* 41(6), 380–9. doi:10.1007/s40858-016-0109-1.

Wang, J., Chen, C., Na, J., Yu, Q., Hou, S., Paull, R. E., Moore, P. H., Alam, M. and Ming, R. 2008. Genome-wide comparative analysis of microsatellites in papaya. *Trop. Plant. Biol.* 1(3–4), 278–92. doi:10.1007/s12042-008-9024-z.

Wee, C. Y., Muhammad Hanam, H., Mohd Waznul Adly, M. Z. and Khairun, H. N. 2014. Expression of defense-related genes in papaya seedling infected with *Erwinia mallotivora* using real-time PCR. *J. Trop. Agric. Fd. Sc.* 42(1), 73–82.

# Chapter 20

# Advances in papaya cultivation

*Parmeshwar Lal Saran, ICAR – Directorate of Medicinal and Aromatic Plants Research, India; and Ravish Choudhary, Indian Agricultural Research Institute, India*



## 1 Introduction

Papaya (*Carica papaya* L.) is one of the most important fruits belonging to the family Caricaceae. It originated from the Caribbean coast of Central America and is now cultivated throughout the tropical and subtropical regions of the world (Teixeira da Silva et al., 2007). World production of papaya in 2017 was 13016281 tonnes. India was the largest producer (5940000 tonnes), followed by Brazil (1057101 tonnes) (FAO, 2017). In India, Tamil Nadu, Gujarat, Karnataka, Maharashtra, Uttarakhand, Bihar, Jharkhand, West Bengal and Andhra Pradesh are major papaya growing states (Saran et al., 2015). Given characteristics such as fruiting ability throughout the year, early fruiting, high productivity per unit area and varied uses, the crop is widely grown both domestically and commercially, and is seen as having significant potential for expansion (Saran and Choudhary, 2013).

http://dx.doi.org/10.19103/AS.2019.0054.15

Papaya is regarded as an excellent source of vitamin C (ascorbic acid), is a good source of carotene and riboflavin, as well as containing iron, calcium, thiamin, niacin, pantothenic acid and vitamin B-6 (Bari et al., 2006; Adetuyi et al., 2008; Saxholt et al., 2008). The edible portion of the fruit contains several minerals such as sodium, potassium, calcium, magnesium and phosphorus. Micro-minerals in papaya include iron, copper, zinc, manganese and selenium (USDA, 2009).

Ripe papaya fruit and papaya products are consumed for both their flavour and nutritional value (Saran, 2010). The fruit is also used to make fruit salads, drinks, jam, jelly, candies, marmalade, soup and other processed products like puree (Matsuura et al., 2004; Nwofia et al., 2012; Saran et al., 2015), juice, frozen slices or chunks, mixed beverages, papaya powder, baby food, concentrated and candied items (OECD, 2005; OECD, 2008; OGTR, 2008). The unripe papaya fruit, which has high sugar content, is used as an additive in ice creams. Green papaya fruit are cooked (often boiled) prior to consumption to denature the enzyme papain in the latex (Odu et al., 2006). Unripe fruit are also used as a source of papain, which has industrial and medicinal applications (Mendoza, 2007; Mano et al., 2009). Papaya seeds are an excellent source of 18 different amino acids, edible oil and also used to adulterate whole black pepper (Morton, 1987). Papaya leaves contain carpaine and pseudocarpaine (bitter alkaloids), which have nutraceutical properties. Young stems are cooked and older stems are grated after peeling, the bitter juice squeezed out, and the mash mixed with sugar and salt (Saran and Choudhary, 2013).

## 2 Plant morphology

The papaya is a perennial herb, semi-woody, latex-producing, usually single-stemmed, short-lived, shrubby or arborescent plant with a well-developed system of articulated laticifers. The usually sparsely branched plants are mostly dioecious and bear berry-like fruit (Badillo, 2000). High photosynthetic rates of short-lived leaves, fast growth and high reproductive output characterize plant growth. Under favourable environmental conditions, papaya seeds undertake epigeal germination and seed emergence is completed in 2–3 weeks (Fisher, 1980). Papaya plant development is very fast, taking 3–8 months for the juvenile phase from seed germination to flowering and 9–15 months for harvesting (Paterson et al., 2008). Because of its flat root system, papaya is vulnerable to wind damage, especially in a monoculture. Healthy roots are whitish cream in colour without laticifers (Carneiro and Cruz, 2009).

Papaya trees are dioecious or hermaphroditic, with cultivars producing only female or bisexual (hermaphroditic) flowers. Papaya is sometimes described as 'trioecious' where separate plants bear either male, female or bisexual flowers. The bisexual (cylindrical) and female flowers (pear shaped) are waxy, ivory white in colour and borne on short peduncles in leaf axils along the main stem with a superior ovary. Bisexual plants are preferred over female or male plants since they produce the most desirable fruit. Identification of desirable males is very important in breeding programmes. The male flower colour correlates with the flesh colour of ripe fruit. The intensity of the whiteness of male flower colour is responsible for the flesh colour (yellow to pink) of the fruit (Fig. 1).

Plants start fruit bearing 6–12 months after transplanting. Individual fruit mature in 5–9 months which depends on cultivar and temperature. Fruit have a central seed cavity and are large, oval to round in shape with a waxy skin (Morton, 1987). Immature green fruit are rich in white latex. As the ripening process progresses, the skin becomes light or deep-yellow-orange, while the thick wall of juicy flesh becomes aromatic, yellow orange or various shades of salmon or red (Morton, 1987). Approximately, 400–600 seeds per fruit can be obtained from well-pollinated fruits. The seeds are covered by a pluristratified epidermis of the outer integument, which is mucilaginous in nature. At physiological maturity, the embryo is enclosed by sarcotesta (a gelatinous layer) (Ming and Moore, 2014). Seeds of wild papayas are photosensitive and dormant in nature at maturity so germination may be triggered using light (Paz and Vazquez-Yañes, 1998). In general, papaya seeds have intermediate seed storage behaviour so that seed storage is very difficult for more than 5 years (Ellis et al., 1991; Wood et al., 2000). Chin et al. (1984) and Hofmann and Steiner (1989) showed that papaya seeds are recalcitrant in nature, making papaya propagation using seed difficult due to rapid seed deterioration after harvesting. The seed viability of papaya is reduced by microbial deterioration of the sarcotesta. Dormancy induced by desiccation can be removed by heat-shocking rehydrated seeds.

**Figure 1** Relationship between flower and flesh colour of the fruit in papaya. Yellowish white (a), bright yellow (b), pinkish (c) and reddish (d) for flesh colour.

## 3 Floral morphology, biology and pollination

The male papaya plant has a long inflorescence with a pendulous panicle bearing many flowers. Male flowers are small in size (about 2.0–3.0 cm long), have a minute calyx with five small combined sepals, and a long corolla. The corolla is pale-yellow in colour and is often odorous (Fig. 2) (Saran et al., 2016).

Female flowers are much larger than male flowers (around 4.0–5.0 cm long), have a short calyx-tube with five short lobes, and five waxy, yellow petals united at the base. The ovary is large, superior, sessile, globular and green with five stigmas. All the female flowers are pistillate which means that visible stamen vestiges are not present and remain free from physiological influences that cause sex reversals in certain hermaphrodite and male trees. Sex reversal or modification is found mainly in hermaphrodite plants. The sex reversal takes place only from male to female but not at all from female to male.

Terminal flowers are selected and tagged for breeding and emasculated before the initiation of anthesis (Sharma and Bajpai, 1969). Under artificial conditions, highest pollen germination has been observed with 5% sucrose solution. Papaya pollens retain 50% viability up to 10 days when stored in desiccators containing 64.8% $H_2SO_4$ and 10% relative humidity at room temperature. The stigma turns receptive from before 48 hours of anthesis to

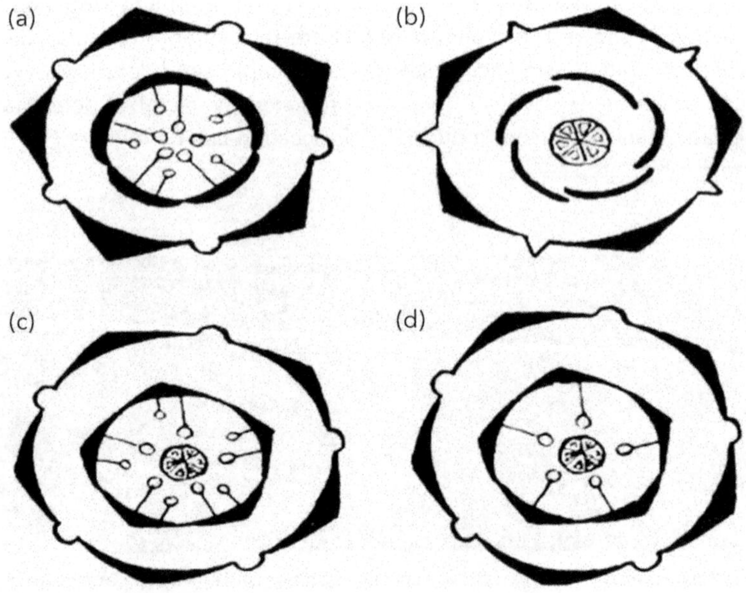


**Figure 2** Floral morphology of Papaya staminate (a), pistillate (b), normal hermaphrodite (c) and elongata hermaphrodite (d).

**Figure 3** Different pollinators of Papaya.

72 hours after anthesis. The maximum number of fruits are set by those female trees which are pollinated from the flowers of bisexual trees (Saran et al., 2015). Nocturnal moths such as hawkmoths or sphinx moths are the best pollinators (Martins and Johnson, 2009). Other pollinators include beetles, skipper butterflies, bees, flies and hummingbirds (Fig. 3).

## 4 Genetics and plant breeding

*C. papaya* is the sole species of the genus *Carica* with a somatic chromosome number, $2n = 2x = 18$, that includes six genera with at least 35 species (Storey, 1976; Carvalho and Renner, 2013). Caricaceae consists of four genera:

1 *Carica* (22 species)
2 *Jacaratia* (6 species)
3 *Jarilla* (1 species)
4 *Cylicomorpha* (2 species).

The first three (*Carica, Jacaratia* and *Jarilla*) are indigenous to tropical America and *Cylicomorpha* to equatorial Africa. Edible fruit are found only in species

**Figure 4** *Carica cauliflora*.

of *Carica*. A virus resistant papaya species is *C. cauliflora* (Fig. 4), and *C. candamarcensis* and *C. pentagonia* are known to be frost resistant (Singh, 1964). These *Carica* species can be arranged in three different groups according to their crossability and these groups provide a foundation for breeding:

- Group A (*C. monoica, C. microcarpa, C. cauliflora* and *C. candamarcensis*)
- Group B (*C. papaya*)
- Group C (*C. goudotiana*).

Papaya is a polygamous species (with many forms of inflorescence). Papaya has three types of flowers namely, staminate, pistillate and hermaphrodite, generally. It has been classified into six categories with many variations within each type (Fig. 5) (Ray, 2002; Saran et al., 2016):

1 Typical female flower
2 Pentandria

**Figure 5** Different type of flowers in papaya.

3 Hermaphrodite intermediate flower
4 Hermaphrodite elongated flower
5 Hermaphrodite sterile flower
6 Typical male flower.

Papaya plants in orchards sometimes fail to develop the right shaped fruit due to fusion of stamen tissue to the gynoecium to complete carpeloidy of antepetalous stamens. The hermaphrodite plants are sexually ambivalent, producing staminate, perfect and pistillate flowers. The pistillate plant is stable.

Through these alterations, carpeloidy/pistillody is restricted to the process of feminization of stamen (Saran et al., 2015).

Papaya has nine chromosome pairs [metacentric (7) and submetacentric (2)] (Ming et al., 2008). Papaya chromosomes are small and uniform in morphology but there have been difficulties in identifying sex chromosomes in early studies of papaya sex determination due to arm ratios or banding patterns (Wai et al., 2010). Sex in papaya is controlled by three sex-determining complexes in the sixth chromosome, which have five pairs of genes due to tight linkage between the genes. Hofmeyr's (1967) hypothesis involves genic balance and the symbols $M_1$ and $M_2$ represent inert or inactivated regions of slightly different lengths on sex chromosomes from which vital genes are missing. This zygotic lethality of $M_1M_1$, $M_2M_2$ and $M_1M_2$ genotypes are given below (Table 1).

Papaya germplasm ranges from very primitive types to major commercial varieties. On account of high degree of variability in production characteristics, several selections from the existing populations have been made in different papaya growing areas of the world (Saran et al., 2016). In India alone, over 10 such selections are available for commercial cultivation. To widen the genetic base, a large number of superior cultivars have been introduced in India from other countries (Table 2). Some hybrids are also available for use in further variety improvement programmes. In India, papaya breeding and selection work has been carried out for over 30 years beginning with 100 introduced strains and 16 local variations. The different varieties developed in different parts of India are shown in Table 3. A papaya breeding programme started at the ICAR

**Table 1** Crossing among the basic sex forms in papaya and ratio of segregating progenies

| | Progenies | | | |
|---|---|---|---|---|
| Cross | Female (mm) | Hermaphrodite ($M_2$m) | Male ($M_1$m) | Non-viable (Lethal) |
| F × S (mm × $M_1$m) | 1 | - | 1 | - |
| F × H (mm × $M_2$m) | 1 | 1 | - | - |
| H × H ($M_2$m × $M_2$m) | 1 | 2 | - | 1 ($M_2M_2$) |
| S × S ($M_1$m × $M_1$m) | 1 | - | 2 | 1 ($M_1M_1$) |
| H × S ($M_2$m × $M_1$m) | 1 | 1 | 1 | 1 ($M_2M_1$) |
| S × H ($M_1$m × $M_2$m) | 1 | 1 | 1 | 1 ($M_1M_2$) |

F = Female; H = Hermaphrodite; S = Staminate.
*Source*: adapted from Hofmeyr (1967).

**Table 2** Improved papaya varieties in different countries

| Country | Variety |
|---|---|
| South Africa | Hortus Gold, Honey Gold |
| Sri Lanka | Peradeniya |
| Hawaii (USA) | Solo, Solo-10, Higgins, Wilder, Waimanalo, Homestead |
| Trinidad | Santa Cruz Giant (hermaphrodite), Cedros (dioecious), Singapore Pink (hermaphrodite) |
| Indonesia | Semangka (Red-Fleshed Fruits); Thailand |
| Venezuela | Maradol Roja, Cubana, Paraguanera |
| Florida (USA) | Betty, Washington, Honeydew |
| Australia | Improved Peterson, Peterson, Bettina |
| Surinam | Santo 3, Santo 4, Santo 7 |

*Source*: adapted from Saran et al. (2016).

**Table 3** Important papaya varieties cultivated in different states of India

| State | Varieties |
|---|---|
| Andhra Pradesh | Honeydew, Coorg Honeydew, Washington, Solo, CO-1, CO-2, CO-3, Sunrise Solo, Taiwan |
| Jharkhand and Bihar | Ranchi selection, Honeydew, Pusa Delicious, Pusa Dwarf and Pusa Nanha, Pusa Majesty, Pusa Nanha, Pusa Giant |
| Karnataka and Kerala | Coorg Honeydew, Coorg Green, Pusa Delicious and Pusa Nanha |
| West Bengal | Ranchi selection, Honeydew, Washington, Coorg Green |
| Orissa | Pusa Delicious, Pusa Nanha, Ranchi selection, Honeydew, Washington, Coorg Green |

*Source*: adapted from Ray (2002).

(Indian Agricultural Research Institute – Regional Station in Pusa, Bihar state) in 1966. Breeding programmes also started at Tamil Nadu Agricultural University (Coimbatore, India) in 1970 and at the G.B. Pant University of Agriculture and Technology (Pantnagar, Uttarakhand state, India) in 1972. These programmes have led to the release of five, seven and three varieties, respectively.

## 5 Climate and cultivation requirements

Papaya is a tropical plant and very sensitive to frost, with cultivation limited to the region between 32° N and 32° S of the Equator. Flooding for 48 hours is fatal. Brief exposure to 0°C is damaging; prolonged cold without overhead sprinkling will kill the plants (DAIS, 2009). Although a mature papaya tree can withstand a temperature of –2°C, production is only recommended in areas where the average daily minimum temperature during mid-winter never drops below 5°C. Ideally, night temperatures should not drop below 12°C. The optimum temperature range for papayas is between 25°C and 30°C. Temperatures higher than 36°C and lower than 16°C for extended periods of time will adversely affect the growth of the trees. Strong, cool, hot and dry winds are not preferred. In open and high-lying areas, plants can be damaged by strong winds (Saran et al., 2016). To ensure successful establishment of a papaya plantation, it is better to use a shelterbelt with full sunshine (Saran et al., 2015). Staking, earthing up (to support the trunk) and windbreaks can decrease the damage to plants from strong winds.

Papayas grow and produce well on a wide variety of soils. The trees often develop a strong taproot shortly after planting. Well-drained sandy-loam soil with adequate organic matter is most important for papaya cultivation. In high rainfall areas, if drainage is poor and roots are continuously drenched for 24–48 hours, plants may die. In such cases, raised beds and drainage ditches are recommended (Saran et al., 2015). Application of mulch and organic material. Planting a cover crop (legume) provides a good source of organic matter. In such cases, the crop must be planted about six months before transplanting (DAIS, 2009). Papayas grow best in soils with pH values of 6.0–6.5. Calcium carbonate or lime can be used to overcome deficiencies of nitrogen, phosphorus, potassium, calcium and magnesium under acidic soils (pH values below 5.5), and sulphur is very effective for deficiencies of magnesium, manganese, iron, zinc, copper and boron under alkali soil (pH values above 8.0) (Medina et al., 2003).

A papaya orchard should be free from unwanted plants/weeds for better growth. This requires techniques such as herbicide application, hand weeding, mulching and deep hoeing, either in single or in combination. Applications of Fluchloralin or Alachlor or Butachlor (2.0 g/ha) two months after transplanting can effectively control weeds for a period of four months. Inorganic materials such as plastic mulches are also used for weed control. Among different mulching materials and methods (such as organic mulch, plastic mulch or mounding in bare soil), the best treatment combination has been found to be plastic mulch with mounding which is also good in preventing phytophthora (root rot) disease (Saran et al., 2015).

Earthing up is also useful to avoid overcrowding or water-logging after the 4th, 6th and 8th months of transplanting before onset of the monsoon season. Earthing up in 30 cm radius around the plants also helps the plants to stand erect in sandy-loam soils (Saran et al., 2014). Staking is done with bamboo and other types of sticks during the fruit development stage to protect the plant against winds and handle fruit load, especially after heavy rainfall (Fig. 6) (Saran et al., 2014).

Thinning prevents damage to fruits due to overcrowding, avoids pressure injury and competition for space and nutrients in some dwarf cultivars like Pusa Dwarf and Pusa Nanha (Saran et al., 2014, 2015). As soon as the plants start flowering, about 5–10% of the male plants are kept in the orchards for

**Figure 6** Papaya plant damage after heavy rainfall.

adequate pollination. The hermaphrodite plants produce complete flowers and should not be confused with male flowers while removing them from the orchard (Saran et al., 2014). Bagging with cheesecloth/glassine paper/wax paper bags can be an effective control measure for fruit fly in small plantings. Cold weather may interfere with pollination and cause shedding of unfertilized female flowers. Removing the side shoots of the stem during the first crop and cutting the old, dry, diseased leaves and petioles are standard practices to ensure a healthy harvest. Pollarding is required for ratoon crops (one-year-old) in order to reduce the plant height (and vulnerability to wind damage) and enhance yield and productivity (Prakash et al., 2014).

## 6 Papaya propagation

Even though scion grafting and rooting of cuttings are possible (Allan and MacMillan, 1991), these methods are not routinely used for commercial papaya propagation. Farmers generally collect fruits of good quality from their orchards and extract seeds for subsequent plantings. Commercial papaya seeds are costly as they are produced under controlled conditions to maintain their genetic purity. There are many problems associated with farmer-collected seeds such as poor seed germination. Seed quality deterioration sets in early after extraction, resulting in low germination and seedling vigour. The subsequent reduction in seed germination after storage is also very rapid due to accumulation of inhibitors in the sarcotesta (Rayes et al., 1980). The higher the moisture content of the seeds, and the higher the temperature of the storage environment, the quicker is the loss of viability (Saran et al., 2015). Priming seeds with 1000 ppm $GA_3$ and 1M $KNO_3$ showed significant improvement in seed quality (Dhinesh Babu et al., 2010). The time to germination after soaking in KNO3 solution treatment is reduced, and maximum germination is achieved sooner when untreated seeds are sown (Yogananda et al., 2004).

As it is a cross-pollinated crop, plants raised from seeds have a mixed inheritance, which make them highly variable in performance. Clonal propagation is necessary for improvement of papaya given the high cost of the seeds. Maximum shoot multiplication was observed in alternate subculturing of basal medium and MS (Murashige and Skoog) medium + 0.5 mg/L BAP (6-benzylaminopurine) + 0.1 mg/L NAA (Naphthaleneacetic acid) and gave highest number of shoots per explants. Sucrose 30 g/L in medium was found to be more favourable for maximum number of shoots and the length of longest shoots. When using a proliferation medium, the maximum length of shoots, number of shoots and growth rate was observed in MS medium fortified with 160 mg/L adenine sulphate. Rooting treatment consisting of half MS medium supplemented with 1.0 mg/L IBA (Indole-3-butyric acid) was found to be the best for early induction of roots (28 days), maximum number of roots/shoot (5.00) and length of root (6.00 cm). Among the potting mixtures tested, soil: sand: farmyard manure (1/1/1: V/V/V) was found to be suitable for hardening *in vitro* raised papaya plantlets (Patel et al., 2013).

## 7 Raised bed sowing

Smallholder farmers may not have infrastructure such as a polytunnel or greenhouse to raise seedlings under protected conditions. Seedlings can be raised in nursery beds (300 cm × 100 cm × 10–15 cm) as well as in root trainers, pots or polythene bags. The seed rate is 250–300 g/ha in gynodioecious cultivars and 300–500 g/ha in dioecious cultivars (Ram, 1986). The seeds are sown 1–2 cm deep in rows 10 cm apart and covered with fine compost, leaf mould, r *khar* (dried leaves of *Saccharum*), polythene sheets or dry paddy straw to protect the seeds, provide shade and maintain moisture. Germination may take 10–12 days during the summer and rainy season, while during the winter season it takes 25–30 days due to low temperatures. The average maximum germination and seedling height was observed at temperatures of 29.8°C and 29.5°C during July and August, respectively, whereas the minimum was found at temperatures of 14°C and 15.9°C during January and December, respectively. The best time for nursery sowing was July to August for an autumn season crop under subtropical conditions in India (Saran et al., 2015).

Papaya seedlings are transplanted 40–45 days after seed sowing when they are 15–20 cm in height. Tall and vigorous varieties are planted with greater spacing, while medium and dwarf ones are planted with closer spacing. Spacing of 1.4 m × 1.4 m or 1.4 m × 1.6 m is best suited for papaya cv. Pusa Delicious under subtropical condition. A closer spacing of 1.2 m × 1.2 m for Pusa Nanha has been adopted for high-density planting, accommodating 6400 plants yielding 103 tonnes of fruits per hectare (Singh et al., 2010).

## 8 Fertilization

Papaya is a heavy feeder crop. If a sufficient amount of balanced fertilizer is not applied to the plants, they fail to produce fruit or give meagre fruiting. Adequate and efficient manuring of young and mature trees is essential to maintain the health of papaya and to obtain a profitable yield. Organic manures like farmyard manure, sheep manure, neem cake, wood ash and bone meal improve yield and quality of papaya fruits. Manuring in papaya should be half from organic manures and half from chemical fertilizers (Saran et al., 2016).

The doses listed in Table 4 should be applied in six split applications, once in two months commencing from the second month of planting. The fertilizers should be applied 20–30 cm away from the main stem and covered through earthing up or mixed in soil around the stem. The orchard should be irrigated lightly after fertilizer application (Saran et al., 2015). Inappropriate application of nitrogenous fertilizers may induce boron deficiency through crop growth dilution factors. Boron deficiency leads to bunchy top and deformed fruit (Fig. 7). Boron deficiency is a major production constraint of papaya in sand to sandy-loam soils.

It is recommended to apply only one spray of boron in the initial fruit development stage. If both foliar and soil applications are combined, it may prove toxic to the plant. The basal application of borax of 5g/plant at the onset of the reproductive stage is very effective in controlling bumpy fruit disorder in papaya (Saran et al., 2017). Spraying of boric acid at 0.1% at a 3-month interval from the sixth month after transplanting is effective and economic (Oliveira et al., 2018). Soil type, mulching, irrigation and fertilization influence the water and nutrient supply to the plant, which in turn, affect the nutritional quality of the fruit. High calcium uptake by fruit has been reported to reduce respiration rates and ethylene production which delays ripening, increases firmness, reduces incidence of physiological disorders and decay, and ultimately increases postharvest life. High nitrogen content is often associated with

**Table 4** Recommended doses of manures and fertilizers (tree$^{-1}$) for papaya cultivation

| Organic manures (basal dose) | Amount (kg) | Inorganic fertilizers (top dressing) | Amount (g) |
|---|---|---|---|
| Compost | 20 | Nitrogen | 200–250 |
| Cake | 1 | Phosphorus | 200–250 |
| Bone meal | 1 | Potassium | 250–500 |

**Figure 7** Deformed (left) and healthy (right) papaya fruit.

reduced postharvest life and decay (Kader and Rolle, 2004). Over application of nitrogen fertilizers results in fruit softening (Desai and Wagh, 1995).

## 9 Pests and diseases affecting papaya

Papaya is damaged by a wide range of insect and mammalian pests, including birds, squirrels, monkeys and elephants as well as snails, earthworms, ants, mites and webworms. Birds such as crows can become a great nuisance during winter months. Crows mostly damage ripe fruit.

Mealy bug infestation appears as clusters of cotton-like masses on the above-ground portion of plants with long waxy filaments. Immature and adult stages of *P. marginatus* suck the sap of the plant and weaken it. Heavy infestations make fruit inedible by covering it with a thick white waxy coating, especially in warm and dry weather. Use of healthy planting material, orchard sanitation, sticky bands, alkathene sheet or a band of insecticide on the main stem to prevent movement of insects from the ground up the plant, are effective practices in pest control.

Papaya fruit fly (*Toxotrypana curvicauda*) is the principal insect pest of papaya throughout tropical and subtropical areas, depositing its eggs in the fruit. After about 12 days, the larvae emerge and feed on the developing seeds and internal portions of the fruit. Infested fruit subsequently turn yellow and eventually fall from trees prematurely. Fruit from infected areas are restricted from export to regions that do not have these pests. Use of paper bags just after the flower parts have fallen, traps and trap crops like *Ocimum sanctum* are effective management practices.

Whitefly (*Trialeuroides variabilis*) is a sucking insect, which coats the leaves with honeydew that forms the basis for sooty mould development. Aphids are vectors for papaya ringspot virus (PRSV). Trees that are infected at a young stage remain stunted and will not produce an economical crop. Fruit from infected trees may have bumps similar to that observed on fruit of plants with boron deficiency and often have 'ring spots', which is the basis for the disease's common name. A severe PRSV isolate from Taiwan is also known to induce systemic necrosis and wilting and chlorosis (Saran et al., 2016).

Nehra et al. (2019) reported that viruses, mainly begomoviruses and potyviruses, cause significant losses in papaya. They have also identified a new species of begomovirus and betasatellite causing infection in India. Viral diseases have been recognized as the greatest threat to the papaya industry. It is thought that symptoms like ring spots, leaf curling and yellowish-green lamina/leaf size reduction are due to viral disease. It has been suggested that diseases are spread in part by tapping of green fruit for their latex. Papaya is not a preferred host of aphids, the main vectors for PRSV. Disease management by controlling the vector within the papaya orchard is very difficult as aphids visit but do not colonize papaya orchards. Papaya plants infected with PRSV must be destroyed in order to minimize the spread of the virus. Virus problems can be avoided by planting genetically resistant cultivars. PRSV is not transmitted via seeds, but it can be spread to areas where it is not present by transporting infected seedlings. In areas where PRSV is already present, it is advisable to raise seedlings in nurseries close to the planting site to limit the possibility of spreading the virus. Early detection of infected plants and prompt removal can also reduce the further spread of the disease. The main options for managing viral diseases are vector control, planting in areas where there is no virus, rouging, netting, cross protection and resistance (Tripathi et al., 2008; Gonsalves et al., 2010).

PRSV-resistant genetically engineered papaya lines have been designed (such as 55-1 and 63-2) using virus coat protein, based on identification of the papaya ring spot potyvirus gene (CaMV 35S uidA) leader sequence. These transgenic varieties have been produced by biolistic (particle bombardment) transformation of embryogenic cultures of the papaya cultivar 'Sunset' with DNA-coated tungsten particles. Papaya lines 55-1 and 63-1, were used to create the cultivars 'SunUp' and 'Rainbow'. 'SunUp' is a transgenic version of its red-fleshed, progenitor cultivar, 'Sunset' and is homozygous for the coat protein gene. 'Rainbow' is an $F_1$ hybrid created by crossing transgenic 'SunUp' with non-transgenic cultivar, 'Kapoho'.

Cross protection is the phenomenon whereby plants that are systemically infected with a mild strain of a virus are protected against the effects of infection by a more virulent-related strain. Cross protection was practiced quite extensively in Taiwan and to a very limited extent in Hawaii in the 1980s. At both locations, the same mild strain was used, which was developed by selecting a mild mutant from nitrous acid treatment of tissue extracts of leaves infected with a severe strain of PRSV in Hawaii. This mutant strain caused mild symptoms on both papaya and cucurbits and gave proper cross protection against severe strains of PRSV in Hawaii. The first attempts to use this mild strain for large-scale protection were initiated in Taiwan, where PRSV was widespread and causing severe damage. The practice was not fully commercialized because the mild strain caused quite severe symptoms on Sunrise papaya. In addition, the logistics for raising the inoculum, uniformly infecting plants with the mild strain, and delivery of the mild strain-infected plants to the farmers, did not work out economically (Tripathi et al., 2008; Gonsalves et al., 2010). Field sanitation such as removal and destruction of affected plants reduce the spread of the disease (Saran et al., 2015).

Physiological disorders of papaya include skin freckles, pulp gelling or pulp flesh translucency, pulp softening, hard lumps in pulp and chilling injury. Skin freckles is a disorder of abiotic origin that appears as dark green to brown spots on the fruit's skin. The disorder known as 'pulp flesh gelling', 'pulp gelling', or 'pulp flesh translucency' causes serious damage to papaya fruit quality, presenting a problem for papaya growers in some Brazilian production areas where incidence of the disorder is high. Standard softening of papaya pulp occurs 6–12 days after the harvesting of green, ripe fruit (known as colour break). Internally, the fruit pulp shows changes in the mesocarp colour that extend from the region close to the seed cavity toward the skin. The occurrence of hard lumps in the pulp is characterized by the formation of highly distinct areas of the mesocarp in which tissue is firmer than in the surrounding areas. Chilling injury is a disorder generally observed in fruit of tropical and subtropical origin in which plant tissues exposed to low temperatures show physiological injury. In general, the most common symptoms on fruit are skin scald, hard lumps in the pulp around the vascular bundles, uneven ripening, inferior fruit flavour and carotenoids content, water soaking of flesh, and increased susceptibility to postharvest pathogens. All of these disorders appear seasonally, with peak periods of occurrence followed by periods with no such incidence, a fact that suggests the influence of weather factors on their occurrence (Oliveira et al., 2018).

The major postharvest diseases are anthracnose and stem-end rot. Postharvest diseases, especially anthracnose become apparent when fruit have 25% or more of skin yellowing. The anthracnose fungus can infect immature green fruit attached to the plant and remain until the fruit begins to ripen, with small well-defined dried pink spots on the surface of ripening fruit. Stem-end rot usually induces a wider and softer water-soaked margin and greater internal discoloration, which is more rapid on ripe and half ripe fruit. The affected portion becomes shrivelled and turns dark brown. Postharvest diseases are effectively controlled by three preharvest consecutive sprays of copper-based or mancozeb (1 g $l^{-1}$ of water) fungicide at intervals of 12-15 days and/or postharvest hot water at 49°C for 20 minutes (Saran et al., 2017).

## 10 Ripening and harvesting

It is particularly important to pick fruit at the right time. Picking too early or too late increases the risk of postharvest physiological disorders. The best quality of fruit for eating is when it is allowed to ripen on the plant. Papaya fruit are harvested 125–140 days after flowering. The period from the fruit colour-break stage to full skin yellowing takes 10–15 days. Fruit ripening in papaya cultivars varies widely in terms of softening, skin colour changes and shelf life. Yellow colour in the fruit skin has been used as a harvest index criterion to assure adequate ripening and maximum postharvest life. For instance, in Solo type papayas yellow colour must cover 6% of the fruit skin surface to attain maximum total soluble solids. Papaya fruits begin ripening with the appearance of light longitudinal stripes that turn progressively yellow (Santamaría et al., 2009). In addition to fruit colour as a good maturity index, total soluble solids (TSS) and flesh firmness values can also be used as quality standards (Basulto et al., 2009). Papaya fruit experiences changes in firmness, which is correlated with activity from two of the main enzymes: polygalacturonase (PG) and pectin methyl esterase (PME) and with the increase of respiration and production of ethylene (Sancho et al., 2010).

The best time of harvesting of papaya fruit is when the skin colour changes from dark green to light green (Fig. 8) and sometimes one yellow streak begins from the base upwards. Papaya skin, approximately 80%, when changed to yellow colour represents its highest peak of flavour, while immature or less mature fruit are lower in sugar (Kader, 2006). At fruit maturity stage, fruit latex becomes watery and TSS should be minimum ($8^0$ brix) (DAIS, 2009).

Fruit are harvested by hand using sharp knives to trim the stem close to the base of the fruit. When harvesting tall trees, ladders or platforms rigged to tractors are used to lift workers. For shipping long-distance, fruits should be picked when mature but still unripe to cope with postharvest handling (Kader and Płocharski, 1999). The cartons used for packaging vary in size from 4.5 to 10 kg for easy transportation. Foam mesh sleeves, foam padding on the bottom of cartons, or paper wrapping of fruit prevent abrasion injury. Fruit are stored and transported at 8–10°C with 90–95% relative humidity.

## 11 Conclusion and future trends

Papaya is one of the most important fruit crops in tropical and subtropical areas, given factors such as high photosynthetic rates of leaves, reproductive

**Figure 8** Ripening symptoms (color change) of papaya fruit.

precocity, high reproductive output and low production costs. A continuing constraint is that small and medium-scale farmers are not able to purchase seeds or tissue-cultured plants of papaya due to high costs. Production of gynodioecious varieties for good yield, quality, good shelf life and tolerance against PRSV and frost are future research priorities.

## 12 Where to look for further information

Papaya ring spot virus and frost are major problems in papaya cultivation. Availability of quality seeds and disease-free planting material is also the need of ours. Therefore, further research on Good Agricultural Practices (GAP), post-harvest management, value addition and crop improvement specially for viral diseases and frost resistant should be carried out. All the agricultural institutes/universities can provide region specific information for papaya.

## 13 References

Adetuyi, F. O., Akinadewo, L. T., Omosuli, S. V. and Lola, A. 2008. Antinutrient and antioxidant quality of waxed and unwaxed pawpaw *Carica papaya* fruit stored at different temperatures. *African Journal of Biotechnology* 7, 2920–4.

Allan, P. and MacMillan, C. N. 1991. Advances in propagation of *Carica papaya* L. cv. honey gold cuttings. *Journal of the South African Society for Horticultural Science* 1, 69–72.

Badillo, V. M. 2000. *Carica* L. vs. *Vasconcella* St.-Hil. (Caricaceae) con la Rehabilitación deeste ultimo. *Ernstia* 10(2), 74–9.

Bari, L., Hassen, P., Absar, N., Haque, M. E., Khuda, M. I. I. E., Pervin, M. M., Khatun, S. and Hossain, M. I. 2006. Nutritional analysis of two local varieties of papaya (*Carica papaya*) at different maturation stages. *Pakistan Journal of Biological Sciences* 9(1), 137–40. doi:10.3923/pjbs.2006.137.140.

Basulto, F. S., Duch, E. S., Gil, F. E., Plaza, R. D., Saavedra, A. L. and Santamaría, J. M. 2009. Postharvest ripening and maturity indices for maradol papaya. *Interciencia* 34(8), 583–8.

Carneiro, C. E. and Cruz, J. L. 2009. Caracterização anatômica de órgãos vegetativos do mamoeiro. *Ciência Rural* 39(3), 918–21. doi:10.1590/S0103-84782009005000011.

Carvalho, F. A. and Renner, S. A. 2013. The phylogeny of Caricaceae. In: Ming, R. and Moore, P. H. (Eds), *Genetics and Genomics of Papaya*. Springer Science + Business Media, New York, pp. 81–92.

Chin, H. F., Hor, Y. L. and Lassim, M. B. 1984. Identification of recalcitrant seeds. *Seed Science and Technology* 12, 429–36.

DAIS. 2009. Cultivating papayas. Department of Agriculture, Forestry and Fisheries, Directorate Agricultural Information Services, Pretoria, South Africa. Available at: www.nda.agric.za/docs/Infopaks/papayas.pdf (accessed on 19 March 2019).

Desai, U. T. and Wagh, A. N. 1995. Papaya. In: Salunke, D. J. and Kadam, S. S. (Eds), *Handbook of Fruit Science and Technology: Production, Composition, Storage, and Processing*. Marcel Dekker, New York, pp. 4–314.

Dhinesh Babu, K., Patel, R. K., (Akath) Singh, A., Yadav, D. S., (Akath) Singh, A., Yadav, D. S., De, L. C. and Deka, B. C. 2010. Seed germination, seedling growth and vigour of papaya under North East Indian condition. *Acta Horticulturae* 851(851), 299–306. doi:10.17660/ActaHortic.2010.851.46.

Ellis, R. H., Hong, T. D. and Roberts, E. H. 1991. Effect of storage temperature and moisture on the germination of papaya seeds. *Seed Science Research* 1(1), 69–72. doi:10.1017/S0960258500000659.

FAO. 2017. FAOSTAT. Food and Agriculture Organization of the United Nations, Rome, Italy. Available at: http://www.fao.org/faostat/en/#data/QC (accessed on 20 May 2019).

Fisher, J. R. 1980. The vegetative and reproductive structures of papaya (*Carica papaya* L.). *Lyonia* 1, 191–208.

Gonsalves, D., Tripathi, S., Carr, J. B. and Suzuki, J. Y. 2010. Papaya ringspot virus. *The Plant Health Instructor*. doi:10.1094/PHI-I-2010-1004-01.

Hofmann, P. and Steiner, A. M. 1989. An updated list of recalcitrant seeds. *Landwirtschaftliche Forschung* 42, 310–23.

Hofmeyr, J. D. J. 1967. Some genetic and breeding aspects of *Carica papaya*. *Agronomia Tropical* 17, 345–51.

Kader, A. A. 2006. Papaya: recommendations for maintaining postharvest quality. Postharvest Technology Research Information Center. Department of Plant Sciences, University of California, p. 3.

Kader, A. A. and Płocharski, W. 1999. Fruit maturity, ripening, and quality relationships. *Acta Horticulturae* 485(485), 203–8. doi:10.17660/ActaHortic.1999.485.27.

Kader, A. A. and Rolle, R. S. 2004. *The Role of Postharvest Management in Assuring the Quality and Safety of Horticultural Produce*. FAO Agricultural Services Bulletin 152, Publishing and Multimedia Service, Information Division, FAO, pp. 1010–365.

Mano, R., Ishida, A., Ohya, Y., Todoriki, H. and Takishita, S. 2009. Dietary intervention with Okinawan vegetables increased circulating endothelial progenitor cells in healthy young woman. *Atherosclerosis* 204(2), 544–8. doi:10.1016/j.atherosclerosis.2008.09.035.

Martins, D. J. and Johnson, S. D. 2009. Distance and quality of natural habitat influence hawkmoth pollination of cultivated papaya. *International Journal of Tropical Insect Science* 29(3), 114–23. doi:10.1017/S1742758409990208.

Matsuura, F. C. A. U., Folegatti, M. I. D. S., Cardoso, R. L. and Ferreira, D. C. 2004. Sensory acceptance of mixed nectar of papaya, passion fruit and acerola. *Scientia Agricola (Piracicaba, Braz.)* 61(6), 604–8. doi:10.1590/S0103-90162004000600007.

Medina, J. De La C., Gutierrez, G. V. and Garcia, H. S. 2003. Pawpaw: post-harvest operation. Edited by AGSI/FAO: Danilo Mejía, PhD, AGST, FAO (Technical), p. 70.

Mendoza, E. M. T. 2007. Development of functional foods in the Philippines. *Food Science and Technology Research* 13(3), 179–86. doi:10.3136/fstr.13.179.

Ming, R. and Moore, P. H. 2014. *Genetics and Genomics of Papaya*. Plant Genetics and Genomics: Crops and Models (vol. 10). Springer Science + Business Media, New York, 438p. doi:10.1007/978-1-4614-8087-7_2.

Ming, R., Yu, Q., Blas, A., Chen, C., Na, J. K. and Moore, P. H. 2008. Genomics of papaya, a common source of vitamins in the tropics. In: Moore, P. H. and Ming, R. (Eds), *Genomics of Tropical Crop Plants* (vol. 1). Springer Science+Business Media, New York, pp. 405–20.

Morton, J. 1987. Papaya. In: Morton, J. F. and Miami, F. L. (Eds), *Fruits of Warm Climates*. New Crop Resource Online Programme, Purdue University, pp. 336–46. Available at: http://www.hort.purdue.edu/newcrop/morton/papaya_ars.html (19 March 2019).

Nehra, C., Marwal, A., Verma, R. K., Mishra, M., Sharma, P. and Gaur, R. K. 2019. Papaya yellow leaf curl virus: a newly identified begomovirus infecting *Carica papaya* L. from the Indian Subcontinent. *The Journal of Horticultural Science and Biotechnology* 94(4), 475–80. doi:10.1080/14620316.2019.1570827.

Nwofia, G. E., Ogimelukwe, P. and Eji, C. 2012. Chemical composition of leaves, fruit pulp and seed in some morphotypes of *C. papaya* L. morphotypes. *International Journal of Medicinal and Aromatic Plants* 2, 200–6.

Odu, E. A., Adedeji, O. and Adebowale, A. 2006. Occurrence of hermaphroditic plants of *Carica papaya* L. (Caricaceae) in Southwestern Nigeria. *The Journal of Plant Sciences* 1(3), 254–63. doi:10.3923/jps.2006.254.263.

OECD (Organisation for the Economic Cooperation and Development). 2005. Consensus document on the biology of papaya (*Carica papaya*). Series on Harmonisation of Regulatory Oversight in Biotechnology, OECD Environment Directorate, Paris, 33.

OECD (Organisation for the Economic Cooperation and Development). 2008. Consensus document on compositional considerations for new varieties of tomato: key food and feed nutrients, toxicants and allergens. Series on the Safety of Novel Foods and Feeds, OECD Environment Directorate, Paris, 17.

OGTR (Office of the Gene Technology Regulator, Australia). 2008. The biology of *Carica papaya* L. (papaya, papaw, paw paw), Version 2: February 2008, Australian Government, Department of Health and Ageing, OGTR. Available at: http://www

.ogtr.gov.au/internet/ogtr/publishing.nsf/content/papaya-3/$FILE/biologypapaya08.pdf (19 March 2019).

Oliveira, J. G., Morales, L. M. M., Silva, W. B., Filho, A. G. and Paull, R. E. 2018. Papaya. In: de Freitas, S. T. and Pareek, S. (Eds), *Postharvest Physiological Disorders in Fruits and Vegetables* (1st edn.). CRC Press, Taylor & Francis Group, Boca Raton, London and New York, pp. 399-421. Chapter 22.

Patel, J. R., Patel, R. M., Patel, S. R. and Singh, V. 2013. *In vitro* micro propagation of papaya var. Red Lady. *International Journal of Plant Protection* 6(2), 449–56.

Paterson, A. H., Felker, P., Hubbell, S. P. and Ming, R. 2008. The fruits of tropical plant genomics. *Tropical Plant Biology* 1(1), 3-19. doi:10.1007/s12042-007-9004-8.

Paz, L. and Vazquez-Yanes, C. 1998. Comparative seed ecophysiology of wild and cultivated *Carica papaya* trees from a tropical rain forest region in Mexico. *Tree Physiology* 18(4), 277–80. doi:10.1093/treephys/18.4.277.

Prakash, J., Das, S. P., Bhattacharjee, T. and Singh, N. P. 2014. Studies on effect of pollarding in papaya. *Indian Journal of Horticulture* 71(3), 419–20.

Ram, M. 1986. Seed production in papaya. *Punjab Horticultural Journal* 25, 95–101.

Ray, P. K. 2002. *Breeding Tropical and Subtropical Fruits*. Narosa Publishing House, New Delhi (India), pp 1-337.

Rayes, M. N., Perez, A. and Cueveas, J. 1980. Detecting endogenous growth regulators of the sarcotesta, sclerotesta, endosperm and embryo by paper chromatography in fresh and aged seed of two varieties of papaya. *Journal of the Agricultural University of Puerto Rico* 15, 164–72.

Sancho, L. E., Yahia, E. M., Martínez-Téllez, M. A. and González-Aguilar, G. A. 2010. Effect of maturity stage of papaya maradol on physiological and biochemical parameters. *American Journal of Agricultural and Biological Sciences* 5(2), 194-203. doi:10.3844/ajabssp.2010.194.203.

Santamaría, B. F., Sauri, D. E., Espadas y Gil, F. L., Díaz, P. R., Larqué, S. A. and Santamaría, J. 2009. Postharvest ripening and maturity indices for *Maradol papaya*. *Interciencia* 34, 583–8.

Saran, P. L. 2010. Screening of papaya cultivars under Doon Valley conditions. *Pantnagar Journal of Research* 8(2), 246–47.

Saran, P. L. and Choudhary, R. 2013. Drug bioavailability and traditional medicaments of commercially available Papaya – a review. *African Journal of Agricultural Research* 8(25), 3216–23.

Saran, P. L., Choudhary, R. and Solanki, I. S. 2014. Papaya: Agri-hort system. *Agriculture Today* 17(2), 40–1.

Saran, P. L., Choudhary, R., Solanki, I. S. and Kumar, P. R. 2015. Stamen feminization in hermaphrodite papaya: a new report. *Progressive Horticulture* 47(1), 173–4. doi:10.5958/2249-5258.2015.00032.9.

Saran, P. L., Solanki, I. S. and Choudhary, R. 2016. *Papaya: Biology, Cultivation, Production and Uses*. CRC Press, Taylor and Francis Group, Boca Raton, London and New York, p. 266. ISBN: 978-1-4987-3560-5.

Saran, P. L., Choudhary, R. and Solanki, I. S. 2017. Fruit deformity in papaya: field screening, nutrient composition and amelioration by boron application. *Indian Journal of Experimental Biology* 55(12), 870–5.

Saxholt, E., Christensen, A. T., Moller, A., Hartkopp, H. B., Hess Ygil, K. and Hels, O. H. 2008. Danish food composition databank. revision 7, Department of Nutrition, National

Food Institute, Technical University of Denmark. Available at: http://www.foodcomp.dk/ (accessed on 19 March 2019).

Sharma, H. C. and Bajpai, P. N. 1969. Studies on floral biology of papaya. *Indian Journal of Science and Technology, Section A* 3, 9–18.

Singh, R. N. 1964. Papaya breeding – a review. *Indian Journal of Horticulture* 21(2), 148–54.

Singh, K., Ram, M. and Kumar, A. 2010. Forty years of papaya research at Pusa, Bihar, India. *Acta Horticulturae* 851(851), 81–8. doi:10.17660/ActaHortic.2010.851.9.

Storey, W. B. 1976. Papaya. In: Simmonds, N. W. (Ed.), *Evolution of Crop Plants*. Longman, London, pp. 21–4.

Teixeira da Silva, J. A., Rashid, Z., Nhut, D. T., Sivakumar, D., Gera, A., Souza Jr., M. T. and Tennant, P. F. 2007. Papaya (*Carica papaya* L.) biology and biotechnology. *Tree and Forestry Science and Biotechnology* 1(1), 47–73.

Tripathi, S., Suzuki, J. Y., Ferreira, S. A. and Gonsalves, D. 2008. Papaya ringspot virus-P: characteristics, pathogenicity, sequence variability and control. *Molecular Plant Pathology* 9(3), 269–80. doi:10.1111/j.1364-3703.2008.00467.x.

USDA (United States Department of Agriculture). 2009. National nutrient database for standard reference, Release 22. Nutrient Data Laboratory home page. Agricultural Research Service. Available at: ndb.nal.usda.gov/ndb/foods (accessed on 19 March 2019).

Wai, C. M., Ming, R., Moore, P. H., Paull, R. E. and Yu, Q. 2010. Development of chromosome-specific cytogenetic markers and merging of linkage fragments in papaya. *Tropical Plant Biology* 3(3), 171–81. doi:10.1007/s12042-010-9054-1.

Wood, C. B., Pritchard, H. W. and Amritphale, D. 2000. Desiccation-induced dormancy in papaya (*Carica papaya* L.) seeds is alleviated by heat shock. *Seed Science Research* 10(2), 135–45. doi:10.1017/S0960258500000143.

Yogananda, D. K., Vyakaranahal, B. S. and Sekhargouda, M. 2004. Effect of seed invigoration with growth regulators and micronutrients on germination and seedling vigour of bell pepper cv. California wonder. *Karnataka Journal of Agricultural Sciences* 17(4), 811–3.

# Chapter 21

# Advances in breeding and cultivation of pomegranate

*Ibrahim Kahramanoglu, Serhat Usanmaz and Turgut Alas, European University of Lefke, Cyprus*



## 1 Introduction

The pomegranate is one of the earliest domesticated fruit crops, first planted between 4000 and 3000 BC, and is even mentioned in the Quran and the Bible. The scientific name (*Punica granatum* L.) of the pomegranate is derived from the name 'seeded apple', which means *Pomum* (apple) *granatus* (grainy). The pomegranate is native to Central Asia, especially to Iran, and it is believed to have spread to the rest of the world from Iran (Levin, 1994; Verma et al., 2010; Kahramanoğlu and Usanmaz, 2016). The pomegranate family has a single genus *Punica* with two species, namely, *P. granatum* and *P. protopunica*. *P. protopunica* is the only relative of the cultivated pomegranate and is endemic to the Socotra islands (Moriguchi et al., 1987). The cultivated species of pomegranate is known to contain 2n = 16 or 18 chromosomes. Pomegranate varieties are often cross-pollinated and are genetically heterozygous. Despite the intense selection of pomegranates during domestication, there is little difference

http://dx.doi.org/10.19103/AS.2019.0054.18

between the natural and domesticated forms (Harlan, 1992). Jalikop and Kumar (1990) reported that approximately 13% crossing occurs in pomegranates and that seedlings are not the 'true to type', which means that there are high levels of variation among the resultant plant characteristics. The domestication of pomegranates is reported by Lye (2008) to have begun in the north of Iran and the Himalayas. However, pomegranate selections were and continue to be made on the basis of flower and fruit characteristics, resistance to biotic and abiotic stresses, yield and so forth (Hancock, 2004; Levin, 2006; Holland et al., 2009). Today, more than 500 globally distributed varieties of pomegranate are being grown in subtropical and tropical regions throughout the world, including Mediterranean and Asian countries as well as the United States, Brazil, Chile, South Africa, Australia and Israel (Kahramanoğlu and Usanmaz, 2016). The most important cultivar, in terms of total consumption in the world is the cv. Wonderful (Fig. 1). Very little breeding work has been carried out for pomegranate trees with specific objectives using modern breeding techniques. The most commonly used methods in these works are hybridization, mutation and polyploidy. The main aim of breeding efforts is to suit the needs of growers (in terms of cultivation), exporters (in terms of postharvest storage) and consumers (in terms of marketing quality).

According to Langley (2000), '[The] pomegranate represents regeneration, life and marriage in the Greek mythology'. The pomegranate has been used by human beings since the dawn of civilization not only as a raw fruit for eating but

**Figure 1** Trees and the fruits of a 'Wonderful' cultivar pomegranate.

also as an herbal healing agent. The pomegranate has found a place even in ancient literature by Dioscorides, Hippocrates, Pliny and Soranus. The immense potential of health benefits induced by the pomegranate have enabled it to achieve the title of 'super fruit'. Recent scientific studies have confirmed the traditional knowledge about the health benefits of pomegranate fruit. Jurenka (2008) reported that pomegranate fruit, flowers and other tree parts contain beneficial phytochemicals that are antioxidant and antimicrobial and act against serious diseases such as cancer and diabetes. Numerous studies of the pomegranate (Gil et al., 2000; Aviram and Dornfeld, 2001; Lansky et al., 2005; Türk et al., 2008; Haidari et al., 2009) have confirmed its beneficial effects on human health. The valuable results of these studies have increased the public awareness of pomegranates and induced an increase in the consumption of pomegranate fruit throughout the world. However, consumption is still limited mainly due to the difficulty in extracting arils, and production is limited mainly due to physiological disorders, pest and disease problems and postharvest problems.

## 2 Advances in pomegranate breeding

For many years, the aims of genetic improvement of the pomegranate were to improve juice content, seed mellowness and appearance of the fruits as well as to increase the fruit yield (Jalikop, 2010). The most preferred characteristics are fruits with bold, juicy and red arils with small, soft seeds. On the other hand, fruits with relatively high juice content are more desirable than large fruit size. However, changes in consumers' preferences and increases in demand caused a change in the aims of breeding, which moved in accordance with the demand in international markets; these changes included fruit size, shape, skin colour, easily separable arils, size, sugar/acid ratio, nutrient content, ripening time and postharvest quality, as well as resistance to pests, diseases and physiological disorders, including fruit cracking, aril browning, chilling injury and so forth. The pomegranate is a cross-pollinated crop and is genetically heterozygous. The out-crossing that occurs in the pomegranate is not 'true to type', resulting in a large variation from plant to plant (Jalikop and Kumar, 1990). Most of the cultivars grown today are the result of human selection from naturally occurring intraspecific variation. Breeding in pomegranates might achieve seedling selection and hybridization. Mutation breeding has some limitations because it is only a 'hit-and-miss' method, and it requires a large number of plants to obtain a desirable genotype (Jalikop, 2010).

Shao et al. (2003) reported the application of 10.0 mg $L^{-1}$ colchicines along with 1.0 mg $L^{-1}$ BA (6-benzyladenine) and 0.1 mg $L^{-1}$naphthalene acetic acid (NAA) to pomegranate shoots of the cv. Nana propagated *in vitro* for 30 days produced 20% tetraploid plants. They also noted that treating the shoots with 5000 mg $L^{-1}$ colchicines for 96 h produced mixoploid mutants with narrow leaves. The researchers noted that the *in vitro* tetraploid plants had shorter leaves and roots as well as wider leaves and larger flowers than the diploid plants. In summary, the researchers concluded that tetraploidy can lead to the evolution of genotypes with larger fruits of improved quality. Nafees et al. (2015) conducted a study in 42 pomegranate accessions using 13 morphological traits of fruits and 29 simple sequence repeat (SSR) markers with the objective to breed pomegranates to estimate their morphological and molecular diversity. The researchers reported that the fruit weight was highly diverse and correlated with other characteristics. They indicated that there was a very high morpho-genetic diversity in the selected pomegranate accessions that could be efficiently used in pomegranate breeding programmes.

Rouholamin et al. (2015) used real-time PCR analysis to study the expression levels of some transcription factors/genes and their role in the skin colour of pomegranates. The researchers noted that the bright red and black skin genotypes accumulated the lowest AN1 (bHLH) level, whereas the white and green skin genotypes had the highest AN1 (bHLH) levels. With regard to the expression of the three transcription factors in all the genotypes, the researchers reported that the complex of MYB–bHLH–WD40 confers skin colour in the pomegranate.

Turkey, after India and Iran, is the third largest producer of pomegranates, and 'Hicaznar' is the most highly produced and exported cultivar that was originally developed through selective breeding. Yazıcı and Şahin (2016) conducted a study to isolate the lines with improved productivity and quality. To do so, Hicaznar was self- or open-pollinated with the sweet pomegranate cultivars Ernar and Fellahyemez. The researchers reported that 3118 lines were developed in total, and 67 were considered for important traits. Luo et al. (2018) conducted a study to analyse the genetic structure of 136 samples of pomegranate using fluorescent SSR markers. These researchers are the first to report 13 SSRs using a fluorescent SSR panel according to the reference genome information, and they detected the genetic diversity and population structure of the studied pomegranate varieties. According to the researchers' notes, the average number of alleles per locus was 6.31, where the gene diversity ranged from 0.16 to 0.37. Researchers used the population structure results and divided the varieties into three subpopulations (Q1, Q2 and Q3). They noted that the varieties originating from the same geographical region are more closely related, while extensive gene flow had occurred between the varieties from different regions. A similar study was carried out in Turkey by Çalışkan et al. (2017), who studied the genetic diversity of 78 pomegranate accessions and provided essential information to manage pomegranate breeding programmes and to establish conservation strategies for preserving local pomegranate genetic resources. In their study, 41 alleles were characterized in total with an average of 4.6 alleles per locus and a mean probability of identity (PI) value of 0.366.

Currently, many plant breeding projects focus on the determination of genetic resources. Until the twenty-first century, classical techniques were generally used, but biotechnological techniques have recently been used in pomegranate breeding. Plant tissue culture and molecular techniques can be used to shorten the breeding duration.

## 3 Advances in propagation methods

Pomegranate seeds have out-crossing, and seedlings are not true to type. Therefore, pomegranate plants grown from seed vary widely and are undesirable. Therefore, trees must be propagated vegetatively. The cutting method is universally used to raise pomegranate plants on a commercial basis while air-layering and tissue culture are the other methods. However, some studies report that air-layering requires 60–75 days, while stem cuttings require 100–120 days for rooting. Both air-layering and cuttings bear fruit 2 years after planting and reach maturity in 5–7 years.

The cuttings should preferably be 10–15 cm long and be collected from 1-year-old, fully mature wood. The best time for cutting is the time when the plants shed leaves before opening in later fall or early winter. The cuttings should be approximately 6–12 mm in diameter. The suckers from the base of the plant or from the interior of the canopy often make some of the best wood for cuttings. The cuttings are preferentially buried in soil or submerged in modified growing medium, leaving only 50–80 mm of the top of the cutting exposed. It is preferable to have at least three nodes under the surface. The root suckers can also be used to produce new plants. Table 1 summarizes the recent studies on the propagation of pomegranate trees.

Alikhani et al. (2011) studied the effects of different growing media and cuttings on the growth and rooting of pomegranate cuttings. They reported that cuttings with more than three buds had the highest number of leaves and branches. According to the authors, a growing medium of sand had the highest rooting potential; however, the greatest root length was observed in sand + peat media. Indole butyric acid (IBA) and NAA are the most widely used auxins for rooting stem cuttings and are required for the initiation of adventitious roots. Damar et al. (2014) investigated the effects of growth regulators and biofertilizers on the survival of pomegranate (*Punica granatum* L.) stem cuttings and reported that biofertilizers and growth regulators have significant effects on most of the shoot and root characteristics of cuttings. They noted that the combination of phosphorus solubilizing bacteria (PSB) and 2000 ppm IBA was the best application for maximum rooting, growth and success of pomegranate cuttings followed by the combination of *Azotobacter* and 2000 ppm IBA.

**Table 1** Recent studies on the propagation of pomegranate trees

| Propagation method | Tested-growing media | Tested treatment | Reference(s) |
|---|---|---|---|
| Cuttings (3+ buds) | • Sand<br>• Sand and peat moss | • IBA<br>• NAA | Alikhani et al. (2011) |
| Stem cuttings | • Rooting media (N/A) | • PSB: phosphorus solubilizing bacteria;<br>• IBA;<br>• *Azotobacter* and IBA | Damar et al. (2014) |
| Semi-hardwood cuttings (15–20 cm) | • River bed sand | • Borax;<br>• IBA;<br>• NAA | Sharma et al. (2009) |
| Hardwood stem cuttings | • Sandy loam and farmyard manure (1:1) | • IBA | Singh (2014) |
| Semi-hardwood shoots (15 cm) | • Moistened perlite | • Melatonin + IBA | Sarrou et al. (2014) |
| Hardwood cuttings (20 cm) | • Sand, soil and farmyard manure (FYM) | • *p*-Hydroxy benzoic acid (PHB) + IBA | Kaur et al. (2016) |
| Air-layering | • Rooting media: sand, leaf mould, FYM and sandy clay soil in 1:1:1:1 ratio | • IBA;<br>• NAA | Tomar and Tomar (2011) |
| Air-layering | • Rooting media: sand, soil and FYM in 1:1:1 ratio | • IBA | Munde et al. (2016) |

Previously, Sharma et al. (2009) reported that the application of 1% borax with 500 ppm IBA or 2% borax with 300 ppm IBA improves the rooting and survival of cuttings. Moreover, Singh (2014) reported that 5 g $L^{-1}$ IBA provides the highest efficacy of rooting in pomegranate cuttings. The combination of the human hormone melatonin with IBA has also been reported to produce positive effects on the rooting of pomegranate cv. Wonderful (Sarrou et al., 2014). Kaur et al. (2016) noted that the application of *p*-hydroxy benzoic acid (PHB) improves the efficiency of IBA on the rooting of pomegranate cuttings.

Air-layering is also a successful method that results in rooting within shorter periods than cutting but is less preferred. Air-layering is similar to the cutting method, except that rooting is performed while the branch is still connected to the mother tree. Plant stems are wounded by girdling approximately 20 mm rings into the bark and are wrapped in plastic-covered media filled with suitable growing medium to allow the growth of adventitious roots. The roots that form around the cut area are allowed to grow until they are well developed.

Tomar and Tomar (2012) carried out a study to determine the effects of different dosages of growth regulators on rooting air-layered pomegranate. They conducted studies of different concentrations of IBA and NAA on the rooting and survival percentage of pomegranate and reported that IBA has a better effect than NAA. They also reported that among the different concentrations of NAA and IBA, 2000 ppm is the best. Munde et al. (2016) also noted that IBA is an effective method for rooting cv. Bhagwa using the air-layering method. The duration of air-layering is also of utmost importance for the efficacy of the method. For example, in the Maharashtra region of India, air-layering in July provides higher efficacy in root initiation, rooting percentage, length of root, fresh and dry weight of root and number of leaves, while air-layering in August provides efficacy in the number of roots and root volume (Tayade et al., 2017).

Tissue culture is the growth and development of cells or tissues separate from the organism. A cultured tissue might be a single cell, a population of cells or a whole or part of an organ. Fragments of tissue from a plant are transferred to an artificial environment, and they continue to survive and function.

Singh et al. (2013) conducted an *in vitro* mass-multiplication study from cotyledonary nodal explants of cv. Ganesh. They reported that treatment with $HgCl_2$ (0.1%) for 3 min gave good sterilization of the cotyledonary nodal explants. The researchers also noted that the maximum percentage of multiple shoots (86.33%) in cotyledonary explants was observed following treatment with MS, 1.0 mg $L^{-1}$ BAP, 1.0 mg $L^{-1}$ kinetin and 200 mg $L^{-1}$ activated charcoal. Previously, Naik et al. (1999) reported that thidiazuron was not effective for all the cytokinins tested, while 1.0 mg $L^{-1}$ IBA provided 86% of the regenerated shoots. Kaji et al. (2013) reported that the application of NAA to the pomegranate tissues in an *in vitro* culture provides high efficiency for rooting shoots and improves the acclimatization when the rooted plantlets are transferred into the soil. Singh and Patel (2014) studied the factors influencing *in vitro* growth and shoot multiplication of pomegranate. Maximum shoot proliferation (78.25%), maximum number of shoots (3.75) per explant and maximum shoot length (3.06 cm) were reported by the researchers to be obtained with MS media supplemented with 1.0 mg $L^{-1}$ BAP, 1.0 mg $L^{-1}$ kinetin and 40 mg $L^{-1}$ adenine sulphate. The researchers noted that the multiplication and growth of shoots were significantly influenced by the different levels of the adenine sulphate, sucrose, light intensity and media pH.

Grafting is rare as a propagation method for pomegranates. Grafting could result in success, but due to the bark characteristics of the plant, trees can be broken when bearing fruit. However, there are some studies that report success for pomegranate grafting, mainly published after 2010. Karimi and Farahmand (2011) reported that grafting is practical and could be optimized for pomegranate propagation, but further studies are required. Karimi (2011) also first reported that stenting grafting might be used for the propagation of pomegranate with 500 ppm IBA treatments. Paiva et al. (2015) mentioned that budding patch grafting is not viable for the propagation of the pomegranate tree, whereas side slit and cleft grafting provide greater growth and dry-matter accumulation in pomegranate. According to Ranpise and Ahire (2016), the type of rootstock also affects the success of grafting, whereas wedge grafting or the patch budding method gives better stionic growth of the pomegranate grafts. Similarly, Karimi and Nowrozy (2017) noted that rootstock type significantly affects the success of grafting and that the 'Gorj-e-Dadashi' rootstock is effective on the scions of 'Rabab-e-Neyriz' and 'Khafr-e-Jahroom'.

## 4 Advances in planting systems, training and pruning

Plants require adequate energy to grow and develop, and they obtain this energy from the sun. Therefore, the planting system and orchard orientation must be properly designed to ensure better sun penetration. The potential light interception by trees in north/south rows is always higher than the equivalent of trees grown in east/west rows. Planting density is another important contributing factor to yield. Optimum spacing is important for the maximum utilization of land and a good income over a long period. The preferred planting dimensions for the pomegranate are 5 × 3 m, but the dimensions are dependent upon the variety of pomegranates. For the dwarf varieties, 1 m between plants and 3 m between rows may also be used (Kahramanoğlu and Usanmaz, 2016).

A number of studies on planting systems focussed on the research of high-density systems after 2010. Haneef et al. (2014) carried out a study with cv. Bhagwa growing in a high-density planting system (2 m × 2 m) and reported that obtaining a high yield from 4-year-old trees is possible with the recommended dose of fertilizer and irrigation. Shanmugasundaram and Balakrishnamurthy (2017) also noted that high-density growth of pomegranate is possible and that the application of three sprayings of NAA (10 ppm) with GA3 (50 ppm) resulted in the highest yields obtained from the pomegranate cv. Mridula. Marathe et al. (2017a) reported a unique plantation method for pomegranate in arid and semi-arid regions named the 'bedding system'. They reported that the bedding system in plantations led to the appropriate aeration of the active root zone, increased the nutrient uptake and resulted in a higher yield than did the traditional methods.

Pomegranate trees may be trained to a single stem or multiple stems. Pomegranate trees naturally tend to produce many suckers from the base. The suckers arising from the stem or root at ground level should be removed regularly. Due to the sensitivity of pomegranate trees to frost injury, farmers may choose the multiple-stem system in regions with the possibility of frost injury (Fig. 2). Pruning is common for pomegranates to encourage flowering and fruiting and to minimize the infestation of pests and diseases.

There is little research about the effects of training systems and pruning on pomegranate yields. Hiremath et al. (2018) reported that pruning branches to 30 cm results in the best growth and yield performance in the pomegranate cv. Super Bhagwa. Similar results were reported by Sharma and Singh (2018), who noted that the 15–30 cm length of fruiting shoots gives better fruit size, fruit weight and marketable yield. They also indicated that pruning proves to be beneficial in controlling bacterial blight on fruit. Pomegranate trees are shrub-like plants with many trunks, and some researchers have observed that the trees collapse and break easily (Atzmon, 2015). Moreover, trellis systems might be used to shape and support the branches and trunks of the tree (Fig. 3).

**Figure 2** Single- (left) and multiple- (right) stem pomegranate trees.

**Figure 3** Pomegranate orchard with a trellis system.

## 5 Advances in irrigation and fertilization

Regular irrigation is recommended for pomegranate trees with a period of 1–7 days depending on the variety, climate and soil conditions. Shorter intervals are recommended for sandy soils, while up to 7-day watering intervals may be used on clay-loam soils. Irrigation is very important and, if not regularly performed, is known to induce fruit cracking. The fertilizer requirements of pomegranate trees depend on the variety and age of the trees. The pure nitrogen, phosphorus and potassium requirements of a well-developed pomegranate tree are approximately 300, 180 and 250 g tree$^{-1}$ year$^{-1}$, respectively (Kahramanoğlu and Usanmaz, 2016) (Table 2). Water is one of the key elements of agricultural production throughout the world, and water scarcity can cut production and badly impact food security worldwide. Due to the depletion of clean water resources in the world, scientists are focussing on new technologies for improving water-use efficiency.

**Table 2** General daily water (L tree$^{-1}$) and monthly pure fertilizer (g tree$^{-1}$) requirements of pomegranate trees

| Months | Items | 1 year | 2 year | 3 year | 4 year | 5+ year |
|---|---|---|---|---|---|---|
| Month 1[a] | Water (L tree$^{-1}$ day$^{-1}$) | 4 | 7 | 10 | 11 | 13 |
| | N (g tree$^{-1}$ month$^{-1}$) | 10 | 15 | 22.5 | 37.5 | 45 |
| | P (g tree$^{-1}$ month$^{-1}$) | 10 | 12.5 | 20 | 30 | 45 |
| | K (g tree$^{-1}$ month$^{-1}$) | 0 | 0 | 0 | 0 | 0 |
| Month 2 | Water (L tree$^{-1}$ day$^{-1}$) | 6 | 12 | 19 | 20 | 21 |
| | N (g tree$^{-1}$ month$^{-1}$) | 15 | 20 | 30 | 50 | 60 |
| | P (g tree$^{-1}$ month$^{-1}$) | 15 | 20 | 32 | 48 | 72 |
| | K (g tree$^{-1}$ month$^{-1}$) | 0 | 0 | 0 | 0 | 0 |
| Month 3 | Water (L tree$^{-1}$ day$^{-1}$) | 8 | 14 | 19 | 22 | 27 |
| | N (g tree$^{-1}$ month$^{-1}$) | 15 | 25 | 37.5 | 62.5 | 75 |
| | P (g tree$^{-1}$ month$^{-1}$) | 15 | 15 | 24 | 36 | 54 |
| | K (g tree$^{-1}$ month$^{-1}$) | 0 | 8 | 13 | 20 | 25 |
| Month 4 | Water (L tree$^{-1}$ day$^{-1}$) | 10 | 17 | 21 | 28 | 40 |
| | N (g tree$^{-1}$ month$^{-1}$) | 10 | 15 | 22.5 | 37.5 | 45 |
| | P (g tree$^{-1}$ month$^{-1}$) | 5 | 2.5 | 4 | 6 | 9 |
| | K (g tree$^{-1}$ month$^{-1}$) | 0 | 8 | 13 | 20 | 25 |
| Month 5 | Water (L tree$^{-1}$ day$^{-1}$) | 12 | 20 | 26 | 32 | 50 |
| | N (g tree$^{-1}$ month$^{-1}$) | 10 | 10 | 15 | 25 | 30 |
| | P (g tree$^{-1}$ month$^{-1}$) | 0 | 0 | 0 | 0 | 0 |
| | K (g tree$^{-1}$ month$^{-1}$) | 0 | 20 | 32.5 | 50 | 62.5 |
| Month 6 | Water (L tree$^{-1}$ day$^{-1}$) | 12 | 20 | 26 | 32 | 45 |
| | N (g tree$^{-1}$ month$^{-1}$) | 5 | 5 | 7.5 | 12.5 | 15 |
| | P (g tree$^{-1}$ month$^{-1}$) | 0 | 0 | 0 | 0 | 0 |
| | K (g tree$^{-1}$ month$^{-1}$) | 0 | 20 | 32.5 | 50 | 62.5 |
| Month 7 | Water (L tree$^{-1}$ day$^{-1}$) | 10 | 18 | 24 | 28 | 40 |
| | N (g tree$^{-1}$ month$^{-1}$) | 5 | 5 | 7.5 | 12.5 | 15 |
| | P (g tree$^{-1}$ month$^{-1}$) | 0 | 0 | 0 | 0 | 0 |
| | K (g tree$^{-1}$ month$^{-1}$) | 0 | 16 | 26 | 40 | 50 |
| Month 8 | Water (L tree$^{-1}$ day$^{-1}$) | 7 | 15 | 20 | 24 | 30 |
| | N (g tree$^{-1}$ month$^{-1}$) | 5 | 5 | 7.5 | 12.5 | 15 |
| | P (g tree$^{-1}$ month$^{-1}$) | 0 | 0 | 0 | 0 | 0 |
| | K (g tree$^{-1}$ month$^{-1}$) | 0 | 16 | 26 | 40 | 50 |

[a] Month 1 is the month of waking up (bud breaking), which corresponds to February or March in the northern hemisphere and September or October in the southern hemisphere.

It is well known that regular irrigation improves tree growth and total yield when compared with basin irrigation. According to Prasad et al. (2003), the application of 8 L h$^{-1}$ day$^{-1}$ for 3 h increased the yield from 17.7 kg plant$^{-1}$

(basin irrigation) to 28.2 kg $plant^{-1}$ in arid regions. Increasing the irrigation water amount (from 7 to 9–11 $m^3$) significantly decreased the fruit-drop percentage. On the other hand, increasing the humic acid doses (from 32 to 48 g) and amino acids (from 8 to 16 g $tree^{-1}$) were reported to enhance the vegetative growth and fruiting on the cv. Manfalouty (Khattab et al., 2012). The comparison of the trunk diameter with the midday stem-water potential ($\Psi_{stem}$) and midday leaf conductance ($g_l$) in the cv. Mollar de Elche pomegranate trees might be used in the preparation of the irrigation schedules (Galindo et al., 2013).

The drip irrigation (DI) method is known to have the greatest effect on the yield while increasing the water-use efficiency. Irrigation is also important for the postharvest fruit quality. It was reported that the fruits from the trees that sustained deficit irrigation (SDI) (with 50% of the water from $ET_C$ irrigation and 50% from regulated deficit irrigation during ripening) showed lower levels of weight loss during storage compared to those of the control treatment (Laribi et al., 2013). Peña et al. (2013) also noted that SDI has similar effects on the postharvest quality of fruits belonging to the cv. Mollar de Elche. The success of DI has been confirmed by many studies (Ghosh et al., 2015). However, the amount of water and fertilizer applied also play an important role in the yield. On the other hand, a new method with the name 'partial root drying (PRD)' is becoming important to reduce water use while maintaining plant growth and yield. The method involves irrigating approximately half of the root system of a given crop while the other half is left to dry. Following a certain period, the dry half of the root system is then irrigated, while the wet half is left to dry. Parvizi et al. (2014) investigated the effect of different irrigation and fertilizer regimes on the yield of 9-year-old pomegranate cv. Rabab trees. They tested and compared five different irrigation practices. The test treatments of the researchers were irrigating one side of trees with 50% and 75% $ET_C$ (DI50, DI75); irrigating alternate sides of trees with 50% and 75% $ET_C$ (PRD50, PRD75); and full irrigation (FI) with received 100% $ET_C$. Researchers also tested three types of fertilizers, including sheep manure, M (50 mg $ha^{-1}$); chemical, CF (urea: 120, diammonium phosphate: 75 and potassium sulphate: 60 kg $ha^{-1}$); and foliar, FF (Ecoquel micromix: 1 kg $ha^{-1}$dissolved in 1000 L water, Defender calcium and Futop: 2 L mixed in 1000 L water) fertilizer. Researchers indicated that in comparison with the FI treatment, the PRD75 treatment increased the yield by 5.6–8.3%, but PRD50 decreased it by 15.8–17%. The PRD strategies were found to have superior results in increasing the yield and fruit weight and decreasing the fruit cracking compared to those of the DI strategies.

Apart from macronutrients, some micronutrients, that is zinc (Zn), calcium (Ca) and boron (B), are known to be very important for fruit quality and for the prevention of fruit cracking. For example, zinc deficiency is commonly observed in pomegranates, and foliar application of zinc was reported by Khorsandi et al. (2009) to improve the marketable fruit quality. The application of animal manure is common not only in pomegranate orchards but also in almost all types of crops. However, the composition of animal manure is important in terms of both nutrients and disease problems. Marathe et al. (2017b) carried out a study on the cv. Bhagwa to investigate the effects of organic treatment with farmyard manure (FYM), vermicompost, poultry manure (PM), *in situ* green manure application and neem (*Azadirachta indica*). The application of FYM was reported to result in the greatest availability of most of the nutrients in the soil. The maximum fruit yield was attained from the application of PM (3.96 kg tree$^{-1}$) followed by FYM (3.86 kg tree$^{-1}$). As in many other crops, the nutrient needs of pomegranate trees change depending on the stage of fruit development, which are mainly distinguished as the flowering, fruit development and first harvesting stages (Gosavi et al., 2017).

## 6 Advances in pest and disease management

Among all the pests and diseases, aphids (*Aphis punicae*), Mediterranean fruit flies (*Ceratitis capitata*) and *Alternaria* black spot pomegranate butterflies (*Deudorix 'Virachola' livia*), as well as black heart and bacterial blight (*Xanthomonas axonopodis* pv. *punicae*), are the most devastating, and these agents cause the greatest losses for growers. Table 3 summarizes the recent studies on the control of pests and diseases of pomegranates.

A detailed experiment was conducted by Kahramanoğlu and Usanmaz (2013) investigating the management strategies for some of the important pests and diseases. The results of this study suggest that contact insecticides are less effective than systematic insecticides against *P. citri,* while fruit thinning in combination with Spirotetramat were reported to be effective. The researchers also noted that attract-and-kill traps (30 traps ha$^{-1}$) are effective in controlling *C. capitata*. On the other hand, two applications with indoxacarb as the active ingredient were reported to be the most effective in controlling *D. livia* by reducing the damage from 15% to <2%. For the control of aphids, imidacloprid (1 µL mL$^{-1}$), thiacloprid (1 µL mL$^{-1}$) and thiamethoxam (0.35 µL mL$^{-1}$) are effective (Rouhani et al., 2013). Temperature is an important factor in the development of aphids, and it is reported that the damage caused by aphids on pomegranate increases at 27°C (Ahmadi and Poorjavad, (2016). The same insecticides were reported to be effective by Elango et al. (2017), who also reported that spinosad might be effective in increasing the mortality of aphids by 54.4% and 75.5%, respectively, at 24 and 48 h after treatment in laboratory conditions. Apart from studies with synthetic chemicals, there are some biological control studies on the control of aphids. One of them was carried out by Farrokhzadeh et al. (2017), who reported that *Binodoxys angelicae* (Haliday; Hymenoptera: Braconidae) and *Ephedrus persicae* (Frogatt; Hymenoptera: Braconidae) have the potential to be used as biological control agents of *A. punicae*.

**Table 3** Recent studies on pest management

| Pest/disease | Recommendation | Reference(s) |
|---|---|---|
| Mealybug (*Planococcus citri*) | • Spirotetramat | Kahramanoğlu and Usanmaz (2013) |
| Aphids (*Aphis punicae*) | • Imidacloprid;<br>• Thiacloprid;<br>• Thiamethoxam | Rouhani et al. (2013)<br>Elango et al. (2017) |
| | • *Binodoxys angelicae* (Haliday; Hymenoptera: Braconidae)<br>• *Ephedrus persicae* (Frogatt; Hymenoptera: Braconidae) | Farrokhzadeh et al. (2017) |
| Pomegranate butterfly (*Deudorix livia* = *Deudorix isocrates*) | • Indoxacarb | Kahramanoğlu and Usanmaz (2013) |
| | • Emamectin benzoate | Kambrekar et al. (2015) |
| | • Chlorantraniliprole | Nadaf (2017) |
| Mediterranean fruit fly (*Ceratitis capitata*) | • Cera-trap | Pezhman (2016) |
| Black heart (*Alternaria* spp.) | • Tebuconazole<br>• Propiconazole | Kumar et al. (2017a) |
| | • Boscalid + Pyraclostrobin<br>• Propiconazole + Difenoconazole | Kahramanoğlu et al. (2018b) |
| Bacterial blight disease (*Xanthomonas axonopodis*) | • Sadaphuli leaf extract | Raut et al. (2015) |

The pomegranate butterfly is one of the most harmful pests of pomegranate crops, causing approximately 65–70% yield loss worldwide. It has been reported that *D. isocrates* takes 63.92 ± 2.87 days to complete its life cycle from oviposition to adult emergence on pomegranate (Kumar et al., 2017b). Treatment with emamectin benzoate (90.83 g $ha^{-1}$) and spinosad (88.91 g $ha^{-1}$) were reported to result in the highest quantity of marketable fruits in terms of controlling pomegranate butterfly (Kambrekar et al., 2015). Chlorantraniliprole (0.15 mL $L^{-1}$) is a newly developed insecticide that has been reported to have a significant effect on *D. isocrates* and improves the marketable yield of pomegranate (Nadaf, 2017).

The Mediterranean fruit fly (Medfly: *Ceratitis capitata*) is a serious pest on pomegranate in many European countries. The population density of the Medfly varied from 44.2% to 53.68% from October to November in pomegranate orchards in Turkey (Demirel, 2016). A similar study in the same region in Turkey showed that the damage to the fruits of the cv. Acco caused by the Medfly is 5.2%, while the damage to persimmon is only 1.35% (Kasap and Arslan, 2016). Cera-trap is recommended for mass trapping of Medfly in pomegranate orchards (Pezhman, 2016).

One of the most important causal agents of black heart is *Alternaria alternata*. The black heart (or heart rot) of pomegranate fruit was previously considered to be a minor disease. However, the listing of the pathogen on the restricted pests list restricted the trade of the crop with some of the major export markets where quarantines were put in place. Thus, studies have focussed on the identification and management of black heart. The first appearances of *Alternata alternata* in pomegranate orchards were observed after 2010 in some of the European countries (i.e. Cyprus (Kahramanoğlu et al., 2014), Spain (Berbegal et al., 2014) and Italy (Faedda et al., 2016)). The symptoms included black spots on leaves and fruits as well as chlorosis and premature abscission of leaves. *Alternaria alternata* also causes black spots on fruits known as *Alternaria* black spots, and these also cause significant economic losses. The azole group of fungicides (Tebuconazole and Propiconazole) were reported to be effective in inhibiting pathogen growth (Kumar et al., 2017a).

Bacterial blight disease (*Xanthomonas axonopodis* pv. *punicae*) is one of the most serious threats to pomegranate production. Raut et al. (2015) performed a study on the efficacy of different plant extracts for the management of bacterial blight disease on pomegranate fruits. The researchers reported that 10% sadaphuli leaf extract is effective for controlling *X. axonopodis*, resulting in 8.22% inhibition, and 20% sadaphuli leaf extract results in 9.06% inhibition. Maity et al. (2018) reported that salicylic acid (SA) application after N application to plants exhibited the greatest efficacy in reducing disease infestation.

## 7 Advances in fruit thinning

Many trees, including the pomegranate, have a tendency to produce high numbers of flowers (Sheikh, 2015). However, when growing conditions are suitable, most of these flowers set, and the trees cannot adequately feed all the fruits to meet market standards in terms of size (Southwick and Glozer, 2000). Large-sized fruits are of significant importance for the premium markets (Wetzstein et al., 2011). When excessive fruit set occurs, removal of fruit is necessary to improve fruit size and quality of the remaining fruits. Correct fruit thinning also allows each fruit to receive more light and improves fruit flavour and colour. Fruit thinning also prevents the occurrence of some pests such as the mealy bug. Glozer and Ferguson (2011) suggested allowing closer spacing near the base of the branch and a wider spacing near the tip of the branch to avoid branch bending. Gibberellic acid and ethephon can be used to perform fruit thinning and are reported to improve fruit size (El-Mahdy et al., 2010). Mohsen and Osman (2015) studied the effects of four thinning levels (1, 2, 3 or 4 fruits cluster$^{-1}$). They reported that one fruit per cluster returned the best physical and chemical properties but reduced the yield per tree.

## 8 Advances in physiological disorder management

Some physiological disorders, such as sun scald, fruit cracking and aril browning, are common in pomegranate fruits. Sun-scald damage becomes severe when fruits are exposed to direct sunlight. Correct pruning (proper canopy management), application of kaolin and use of paper covers are known to be the best ways to protect fruits from sun burning (Kahramanoğlu and Usanmaz, 2016). Ghorbani et al. (2015) reported that covering fruits with 18 or 22 holes cm$^{-2}$ covering materials provides an approximately 50% reduction in sunburn. It was also noted that ammonium nitrate fertilizer (600, 900, 1200 and 1500 g tree$^{-1}$) and/or calcium chloride (0%, 1% and 2%) alone or in combination enhanced the vegetative growth parameters and reduced the fruit cracking and sunburn damage on pomegranate fruits (Bakeer, 2016). Using red nets over mature pomegranate trees at intermediate shade levels (35% PAR and 50% PAR) has also been reported to provide better quality and to prevent sunburn on fruits (Meena et al., 2016).

Fruit cracking is a serious problem of pomegranate fruits. The main causes of fruit cracking are irregular irrigation and boron and calcium deficiency. Regular irrigation to maintain soil moisture at the desired level and spraying of calcium compounds or 0.2% borax or 250 ppm GA3 on young fruits are reported to minimize fruit cracking. Fruit cracking was reported to decrease significantly with the application of 300 ppm paclobutrazol, while grain %, fruit juice %, TSS % and acidity % were also improved (Khalil and Aly, 2013). Foliar calcium fertilization in the form of a nano-Ca formulation (0.50 g Ca $L^{-1}$) or in the form of $CaCl_2$ (2.73 and 5.45 g Ca $L^{-1}$) reduces fruit cracking in pomegranates (Davarpanah et al., 2018).

Aril browning is the major internal defect of pomegranate fruits. Aril browning, similar to heart rot, does not have any external symptoms and is more difficult than heart rot to manually identify, thus posing serious problems for export; the mature fruit looks good and intact, but after opening some arils in a particular segment or the whole of the fruit are found to be brown or black, and they give off a foul smell in the advanced stage of the disorder. Aril browning is a complex caused by genotypes, season of fruiting, soil fertility and an imbalance of nutrients at the stage of maturity. Aril browning onset occurs 90 days after anthesis, and the intensity increases if fruits are retained for more than 140 days. The exact cause for this malady is unknown even now (Kahramanoğlu and Usanmaz, 2016). Sing et al. (2016) reported that inter-fruit competition (unequal distribution of nutrients) might be the causal agent of this disorder. It was reported that the foliar application of SA and sodium nitroprusside significantly reduced the percentage of aril browning (Khodaei et al., 2015).

## 9 Advances in postharvest handling

The main challenge for pomegranate fruit producers is maintaining the postharvest quality. Pomegranates are very susceptible to storage conditions and lose weight and quality during long/inappropriate storage periods/conditions. Other important problems associated with the storage of pomegranate fruits are chilling injury, decay and changes in total soluble solid content and antioxidant activity as well as changes in total anthocyanin, titratable acidity and ascorbic acid. Weight loss causes huge losses not only in the fruit quality but also in the income for marketers (Caleb et al., 2012). Many scientific studies have been conducted to investigate the favourable conditions for pomegranate storage in terms of the prevention of weight loss, and they suggest maintaining stores at 5–7°C and >90% relative humidity (RH) for the reduction of weight loss (Elyatem and Kader, 1984; Küpper et al., 1995; Al-Mughrabi et al., 1995; Opara et al., 2008; Kazemi et al., 2013). Recent studies on the postharvest storage of pomegranate fruits are summarized in Table 4.

Modified atmosphere packaging (MAP) is a dynamic process of altering the gaseous composition inside a package. The interaction between the respiration rate (RR) of the crop and the permeability of the packaging material are the two important elements of MAP, with no further control exerted over the initial gas composition (Farber et al., 2003; Mahajan et al., 2007; Caleb et al., 2012). It has also been reported to be successful for prolonging the storage duration of pomegranate fruits for up to 3 months while reducing

**Table 4** Recent studies on postharvest storage of pomegranate fruits

| Storage condition | Temperature (°C) | Relative humidity (%) | Additional treatment | Successful storage duration | Reference(s) |
|---|---|---|---|---|---|
| Modified atmosphere packaging (MAP) | 5–7 | 90–95 | - | 90 days | Kader (1995), Palou et al. (2007), Mirdehghan and Ghotbi (2014) |
| Storage room | 2 | 90 | Methyl jasmonate | 84 days | Sayyari et al. (2017) |
| Storage room | N/A | N/A | Salicylic acid, chitosan and salicyloyl chitosan | 150 days | Sayyari et al. (2016) |
| Storage room | 4.5 | N/A | Commercial resin, carnauba waxes and chitosan | 120 days | Meighani et al. (2015) |
| MAP | 7 ± 1 | 95 | - | 120 days | Shaarawi and Nagy (2017) |
| MAP | 6.5 ± 1 | 95 | Propolis and black seed oil | 150 days | Kahramanoğlu et al. (2018a) |

N/A: not applicable.

the weight loss and chilling injury and preventing decay (Kader, 1995; Palou et al., 2007; Mirdehghan and Ghotbi, 2014). Temperature and RH are the two main elements of a successful postharvest storage life for pomegranate fruits. However, weight loss, chilling injury and decay, the three main causes of postharvest deterioration in pomegranates, are differentially influenced by temperature and RH. Temperatures below 5°C increase susceptibility to chilling injury, but temperatures above 5°C enhance fruit decay and increase weight losses. Moreover, RH above 90% enhances fruit decay while RH below 90% increases weight loss.

Intermittent heating and methyl jasmonate are known to prevent chilling injury in pomegranate fruits. Sayyari et al. (2011) reported that methyl jasmonate with vapour treatment alleviated chilling injury and enhanced the antioxidant potential during the postharvest storage of pomegranates. Methyl jasmonate alone is also noted to reduce chilling injury and improve the quality characteristics of pomegranates during postharvest storage (Sayyari et al., 2017). Salicyloyl chitosan has also been reported to reduce chilling injury in pomegranate fruits (Sayyari et al., 2016).

Commercial resin and carnauba waxes can also significantly lower RR and weight loss, and carnauba wax can maintain considerably higher fruit quality and bioactive compounds and has a very good potential to extend the storage life of pomegranate fruits (Meighani et al., 2015). Storing pomegranate fruits with MAP at 7 ± 1°C and 95% RH remarkably reduced the weight loss and decay incidence and maintained the fruit quality compared to the control outcomes (Shaarawi and Nagy, 2017). MAP can also be used for the bulk storage of pomegranate fruits, which allows the direct transfer of large volumes of fruit from the field to cold storage rooms and saves precious time and extra packaging and labour costs (Porat et al., 2016).

Fludioxonil is a well-known fungicide against the pomegranate postharvest pathogens *Botrytis* sp., *Penicillium* sp. and *Pilidiella* (Palou et al., 2007). Fludioxonil application prevents fruit decay, prevents weight loss, increases phenolic contents and improves sensory attributes, but it also negatively affects the aril colour and lowers the ascorbic acid content (Atukuri et al., 2017). Thus, olive-mill wastewater has the potential to be used as an alternative for the control of fungal pathogens of pomegranate fruit (Quaglia et al., 2016). On the other hand, some biological agents, that is, crab shell chitosan, have been reported to have positive effects on the control of these pathogens (Munhuweyi et al., 2016,2017). The combined application of chitosan with MAP is reported to prevent weight loss and control chilling injury (Candir et al., 2018). Recent studies by Kahramanoğlu et al. (2018a) indicated that the combination of MAP bags with propolis (0.1%) and/or black seed oil (0.5%) treatments help to maintain fruit weight and quality for up to 150 days (Fig. 4). The application of black seed oil or propolis with MAP was also reported to be effective in

**Figure 4** Effects of propolis, black seed oil and MAP on the fruit quality of the cv. Wonderful pomegranate after 150 days of storage.

controlling grey mould development and slowing the occurrence of chilling injury.

## 10 Conclusion and future trends

The verification of the health benefits of pomegranate fruit on human health have caused tremendous increases in consumption and production. However, pomegranate cultivation still faces several problems, such as fruit cracking, bacterial blight, black heart, yield reduction due to water scarcity, quality loss during storage and so forth, which further studies have the capacity to resolve. Since 2010, many of the studies have focussed on the effects of natural extracts as well as biochemical and nanotechnological applications (i.e. black seed oil, chitosan, MAP, nano-Ca etc.) on physiological disorders, pests and diseases and postharvest quality. These applications are environmentally friendly and have no negative effects on consumers' health, thus helping to ensure sustainable production. The successful results of those studies offer hope for the future of the pomegranate sector. However, it is also important that some of the limitations imposed by tree and fruit characteristics are explored by pomegranate breeding. The breeding or selection of varieties that are resistant to the above-mentioned problems and/or varieties producing fruits with a higher demand in the market are the key issues for the future of the pomegranate sector.

## 11 Where to look for further information

The following websites provide further information about advances in the breeding and cultivation of pomegranate fruits:

- The International Society for Horticultural Science (www.ishs.org) began to carry out 'international symposiums on pomegranate' in 2008. The first one was held in Turkey and the fourth symposium took place in Spain in

2017. The proceedings are being published in *Acta Horticulturae*, which is a good source for further reading.

- Options Méditerranéennes is another good source for further reading and is a collection of scientific papers intended to aid the development of Mediterranean agriculture and has been published by CIHEAM since 1970. To date, CIHEAM has organized two symposiums about pomegranates, and the full texts of the papers are available in:
  - http://om.ciheam.org/option.php?IDOM=343#prettyPhoto
  - http://om.ciheam.org/option.php?IDOM=1005

## 12 References


Ahmadi, M. and Poorjavad, N. 2016. The impact of temperature and pomegranate cultivars on biological characters of pomegranate aphid, *Aphis punicae* (Hem.: Aphididae). *Plant Pest Research* 6(1), 37–47.

Alikhani, L., Ansari, K., Jamnezhad, M. and Tabatabaie, Z. 2011. The effect of different mediums and cuttings on growth and rooting of pomegranate cuttings. *Iranian Journal of Plant Physiology* 1(3), 199–203.

Al-Mughrabi, M. A., Bacha, M. A. and Abdelrahman, A. O. 1995. Effects of storage temperature and duration on fruit quality of three pomegranate cultivars. *Journal of King Saud University* 7(2), 239–48.

Atukuri, J., Fawole, O. A. and Opara, U. L. 2017. Effect of exogenous fludioxonil postharvest treatment on physiological response, physico-chemical, textural, phytochemical and sensory characteristics of pomegranate fruit. *Journal of Food Measurement and Characterization* 11(3), 1081–93. doi:10.1007/s11694-017-9485-6.

Atzmon, I. 2015. Growing pomegranates using a trellising system. *Acta Horticulturae* 1089(1089), 427–30. doi:10.17660/ActaHortic.2015.1089.59.

Aviram, M. and Dornfeld, L. 2001. Pomegranate juice consumption inhibits serum angiotensin converting enzyme activity and reduces systolic blood pressure. *Atherosclerosis* 158(1), 195–8. doi:10.1016/S0021-9150(01)00412-9.

Bakeer, S. M. 2016. Effect of ammonium nitrate fertilizer and calcium chloride foliar spray on fruit cracking and sunburn of Manfalouty pomegranate trees. *Scientia Horticulturae* 209, 300–8. doi:10.1016/j.scienta.2016.06.043.

Berbegal, M., Lopez_Cortes, I., Salazar, D., Gramaje, D., Perez-Sierra, A., Garcia-Jimenez, J. and Armengol, J. 2014. First report of *Alternaria* black spot of pomegranate caused by *Alternaria alternata* in Spain. *Disease Notes* 98(5), 689.

Caleb, O. J., Opara, U. L. and Witthuhn, C. R. 2012. Modified atmosphere packaging of pomegranate fruit and arils: a review. *Food and Bioprocess Technology* 5(1), 15–30. doi:10.1007/s11947-011-0525-7.

Çalışkan, O., Bayazıt, S., Öktem, M. and Ergül, A. 2017. Evaluation of the genetic diversity of pomegranate accessions from Turkey using new microsatellite markers. *Turkish Journal of Agriculture and Forestry* 41, 142–53. doi:10.3906/tar-1606-124.

Candir, E., Ozdemir, A. E. and Aksoy, M. C. 2018. Effects of chitosan coating and modified atmosphere packaging on postharvest quality and bioactive compounds of pomegranate fruit cv. 'Hicaznar'. *Scientia Horticulturae* 235, 235–43. doi:10.1016/j.scienta.2018.03.017.

Damar, D., Barholia, A. K., Lekhi, R. and Haldar, A.2014. Effect of growth regulators and biofertilizers on survival of pomegranate (*Punica granatum* L.) stem cuttings. *Plant Archives* 14(1), 347–50.

Davarpanah, S., Tehranifar, A., Abadía, J., Val, J., Davarynejad, G., Aran, M. and Khorassani, R. 2018. Foliar calcium fertilization reduces fruit cracking in pomegranate (*Punica granatum* cv. Ardestani). *Scientia Horticulturae* 230, 86–91. doi:10.1016/j.scienta.2017.11.023.

Demirel, N. 2016. Population density and damage ratios of Mediterranean fruit fly, *Ceratitis capitata* (Wiedemann) (Diptera: Tephritidae) on pomegranate orchards in Turkey. *Entomology and Applied Science Letters* 3(5), 1–7.

Elango, P. K., Sridharan, S., Saravanan, P. A. and Balakrishnan, S. 2017. Laboratory evaluation of insecticides and biopesticides against pomegranate aphid *Aphis punicae*. *International Journal of Chemical Studies* 5(5), 1810–2.

El-Mahdy, T. K. R., Mohamed, A. K. A. and Mohamed, N. I. A. 2010. Effect of flower thinning and spraying with gibberellic acid and ethephon on yield and fruit quality of Manfalouty pomegranate cultivar. *ASSIUT Journal of Agricultural Sciences* 40(4), 69–91.

Elyatem, S. M. and Kader, A. A. 1984. Post-harvest physiology and storage behavior of pomegranate fruits. *Scientia Horticulturae* 24(3–4), 287–98. doi:10.1016/0304-4238(84)90113-4.

Faedda, R., Granata, G., Pane, A., Evoli, M., Giudice, V. L., Lio, G. M. S. and Cacciola, S. O. 2016. Heart rot and soft rot of pomegranate fruit in southern Italy. *Acta Horticulturae* 1144, 195–8. doi: 10.17660/ActaHortic.2016.1144.28.

Farber, J. N., Harris, L. J., Parish, M. E., Beuchat, L. R., Suslow, T. V., Gorney, J. R., Garrett, E. H. and Busta, F. F. 2003. Microbiological safety of controlled and modified atmosphere packaging of fresh and fresh-cut produce. *Comprehensive Reviews in Food Science and Food Safety* 2(s1), 142–60. doi:10.1111/j.1541-4337.2003.tb00032.x.

Farrokhzadeh, H., Moravvej, G., Awal, M. M., Karimi, J. and Rashed, A. 2017. Comparison of molecular and conventional methods for estimating parasitism level in the pomegranate aphid *Aphis punicae* (Hemiptera: Aphididae). *Journal of Insect Science* 17(6), 1–7. doi:10.1093/jisesa/iex087.

Galindo, A., Rodríguez, P., Mellisho, C. D., Torrecillas, E., Moriana, A., Cruz, Z. N., Conejero, W., Moreno, F. and Torrecillas, A. 2013. Assessment of discretely measured indicators and maximum daily trunk shrinkage for detecting water stress in pomegranate trees. *Agricultural and Forest Meteorology* 180, 58–65. doi:10.1016/j.agrformet.2013.05.006.

Ghorbani, M., Dabbagh, G. R., Yousefi, D., Khademi, S. and Taki, M. 2015. The effect of application of different kinds of covers on the sunburn and internal qualities of pomegranate in Iran. *Biological Forum – An International Journal* 7(1), 64–8.

Ghosh, S. N., Bera, B. and Wathugala, D. L. 2015. Yield and fruit quality of pomegranate (*Punica granatum* L.) grown in laterite soil under drip and basin irrigation during summer. *International Journal of Minor Fruits, Medicinal and Aromatic Plants* 1(1), 25–30.

Gil, M. I., Tomas-Barberan, F. A., Hess-Pierce, B., Holcroft, D. M. and Kader, A. A. 2000. Antioxidant activity of pomegranate juice and its relationship with phenolic composition and processing. *Journal of Agricultural and Food Chemistry* 48(10), 4581–9. doi:10.1021/jf000404a.

Glozer, K. and Ferguson, L. 2011. Pomegranate Production in Afghanistan. UC Regents Davis Campus, 32p. Available at: http://ucanr.edu/sites/Pomegranates/files/164500.pdf (accessed on 08 January 2018).

Gosavi, A. B., Deshpande, A. N. and Maity, A. 2017. Identifying nutrient imbalances in pomegranate (cv. Bhagwa) at different phonological stages by the diagnosis and recommendation integrated system. *Journal of Plant Nutrition* 40(13), 1868–76. doi: 10.1080/01904167.2016.1267209.

Haidari, M., Ali, M., Casscells, S. W. and Madjid, M. 2009. Pomegranate (*Punica granatum*) purified polyphenol extract inhibits influenza virus and has a synergistic effect with oseltamivir. *Phytomedicine: International Journal of Phytotherapy and Phytopharmacology* 16(12), 1127–36. doi:10.1016/j.phymed.2009.06.002.

Hancock, J. F. 2004. *Plant Evolution and the Origin of Crop Species* (2nd edn.). CABI Publishing, Cambridge, MA, p. 313.

Haneef, M., Kaushik, R. A., Sarolia, D. K., Mordia, A. and Dhakar, M. 2014. Irrigation scheduling and fertigation in pomegranate cv. Bhagwa under high density planting system. *Indian Journal of Horticulture* 71(1), 45–8.

Harlan, J. R. 1992. *Crops and Man* (2nd edn.). American Society of Agronomy and Crop Science Society of America, Madison, WI, p. 295.

Hiremath, A., Patil, S. N., Hipparagi, K., Gandolkar, K. and Gollagi, S. G. 2018. Influence of pruning intensity on growth and yield of pomegranate (*Punica granatum* L.) cv. Super Bhagwa under organic conditions. *Journal of Pharmacognosy and Phytochemistry* 7(2), 1027–31.

Holland, D., Hatip, K. and Bar-Ya'akov, I. 2009. Pomegranate: botany, horticulture and breeding. In: Janick, J. (Ed.), *Horticultural Reviews* (Vol. 35). John Wiley & Sons Inc., Hoboken, NJ, pp. 127–91.

Jalikop, S. H. 2010. Pomegranate breeding. *Fruit, Vegetable and Cereal Science and Biotechnology* 4(2), 26–34.

Jalikop, S. H. and Kumar, P. S. 1990. Use of a gene marker to study the mode of pollination in pomegranate (*Punica granatum* L.). *Journal of Horticultural Science* 65(2), 221–3. doi:10.1080/00221589.1990.11516050.

Jurenka, J. S. 2008. Therapeutic applications of pomegranate: a review. *Alternative Medicine Review* 13(2), 128–44.

Kader, A. A. 1995. Regulation of fruits physiology by controlled and modified atmosphere. *Acta Horticulturae* 398(398), 59–70. doi:10.17660/ActaHortic.1995.398.6.

Kahramanoğlu, İ. and Usanmaz, S. 2013. Management strategies of fruit damaging pests of pomegranates: *Planococcus citri*, *Ceratitis capitata* and *Deudorix* (*Virachola*) *livia*. *African Journal of Agricultural Research* 8(49), 6563–8.

Kahramanoğlu, İ. and Usanmaz, S. 2016. *Pomegranate Production and Marketing*. CRC Press, Boca Raton, FL, 148pp.

Kahramanoğlu, İ., Usanmaz, S. and Nizam, İ. 2014. Incidence of heart rot at pomegranate fruits caused by *Alternaria* spp. in Cyprus. *African Journal of Agricultural Research* 9(10), 905–7.

Kahramanoğlu, İ., Aktaş, M. and Günsüz, Ş. 2018a. Effects of fludioxonil, propolis and black seed oil application on the postharvest quality of 'Wonderful' pomegranate. *PLoS ONE* 13(5), e0198411. doi:10.1371/journal.pone.0198411.

Kahramanoğlu, İ., Usanmaz, U., Alas, T., Helvacı, M. and Aşkın, M. A. 2018b. Fungicides effects on heart rot infestations at pomegranate fruit. *International Journal of Agriculture, Forestry and Life Sciences* 2(2), 1–5.

Kaji, B. V., Ershadi, A. and Tohidfar, M. 2013. In vitro propagation of pomegranate (*Punica granatum* L.) cv. 'Males Yazdi'. *Albanian Journal of Agricultural Science* 12(1), 1-5.

Kambrekar, D. N., Biradar, A. P. and Karabhantanal, S. S. 2015. New insecticides for the management of pomegranate fruit borer, *Deudorix isocrates* (f.). *Indian Journal of Entomology* 77(3), 240-4. doi:10.5958/0974-8172.2015.00050.4.

Karimi, H. R. 2011. Stenting (cutting and grafting) - a technique for propagating pomegranate (*Punica granatum* L.). *Journal of Fruit and Ornamental Plant Research* 19(2), 73-9.

Karimi, H. R. and Farahmand, H. 2011. Study of pomegranate (*Punica granatum* L.) propagation using bench grafting. *Journal of Fruit and Ornamental Plant Research* 19(2), 67-72.

Karimi, H. R. and Nowrozy, M. 2017. Effects of rootstock and scion on graft success and vegetative parameters of pomegranate. *Scientia Horticulturae* 214, 280-7. doi:10.1016/j.scienta.2016.11.047.

Kasap, A. and Aslan, M. M. 2016. Akdeniz meyve Sineğinin feromon Tuzaklarla (Diptera: Tephritidae)'nin Nar ve Hurmadaki Populasyon Takibi ve Zarar Oranının Tespiti (*Ceratitis capitata* Wied. (Diptera: Tephritidae)'nin Nar ve Hurmadaki Populasyon Takibi ve Zarar Oranının Tespiti). *KSU Journal of Natural Science* 19(1), 43-50.

Kaur, S., Kaur, A. and Kaur, G. 2016. Effect of IBA, PHB and time of planting on rooting of pomegranate (*Punica granatum* L.) cuttings cv. Ganesh. *Asian Journal of Science and Technology* 7(11), 3757-61.

Kazemi, F., Jafararpoor, M. and Golparvar, A. 2013. Effects of sodium and calcium treatments on the shelf life and quality of pomegranate. *International Journal of Farming and Allied Sciences* 2, 1375-8.

Khalil, H. A. and Aly, H. S. H. 2013. Cracking and fruit quality of pomegranate (*Punica granatum* L.) as affected by pre-harvest sprays of some growth regulators and mineral nutrients. *Journal of Horticultural Science and Ornamental Plants* 5(2), 71-6.

Khattab, M. M., Shaban, A. E., El-Shrief, A. H. and Mohamed, A. S. E. 2012. Effect of humic acid and amino acids on pomegranate trees under deficit irrigation. I: growth, flowering and fruiting. *Journal of Horticultural Science and Ornamental Plants* 4(3), 253-9.

Khodaei, M., Nahandi, F. Z., Motallebi-Azar, A. and Dadpour, M. 2015. Effect of salicylic acid and sodium nitro proside on the pomegranate aril browning disorder. *Biological Forum - An International Journal* 7(2), 1014-20.

Khorsandi, F., Yazdi, F. A. and Vazifehshenas, M. R. 2009. Foliar zinc fertilization improves marketable fruit yield and quality attributes of pomegranate. *International Journal of Agricultural Biology* 11, 766-70.

Kumar, A., Chahal, T. S., Hunjan, M. H. S., Kaur, H. and Rawal, R. 2017a. Studies of *Alternaria* black spot disease of pomegranate caused by *Alternaria alternata* in Punjab. *Journal of Applied and Natural Science* 9(1), 156-61. doi:10.31018/jans.v9i1.1164.

Kumar, K. P., Jayanthi, P. D. K., Naik, S. O., Verghese, A. and Chakravarthy, A. K. 2017b. Biology of Anar Butterfly, *Deudorix isocrates* (Fab.) (Lycaenidae: Lepidoptera) on pomegranate, *Punica granatum* L. *International Journal of Pure & Applied Bioscience* 5(1), 498-503. doi:10.18782/2320-7051.2564.

Küpper, W., Pekmezci, M. and Henze, J. 1995. Studies on CA-storage of pomegranate fruit (*Punica granatum* L., cv. Hicaz). *Acta Horticulturae*398, 101-8.

Langley, P. 2000. Why a pomegranate? *BMJ* 321(7269), 1153-4. doi:10.1136/bmj.321.7269.1153.

Lansky, E. P., Jiang, W., Mo, H., Bravo, L., Froom, P., Yu, W., Harris, N. M., Neeman, I. and Campbell, M. J. 2005. Possible synergistic prostate cancer suppression by anatomically discrete pomegranate fractions. *Investigational New Drugs* 23(1), 11-20. doi:10.1023/B:DRUG.0000047101.02178.07.

Laribi, A. I., Palou, L., Intrigliolo, D. S., Nortes, P. A., Rojas-Argudo, C., Taberner, V., Bartual, J. and Pérez-Gago, M. B. 2013. Effect of sustained and regulated deficit irrigation on fruit quality of pomegranate cv. 'Mollar de Elche' at harvest and during cold storage. *Agricultural Water Management* 125, 61–70. doi:10.1016/j.agwat.2013.04.009.

Levin, G. M. 1994. Pomegranate (*Punica granatum*) plant genetic resources in Turkmenistan. *Plant Genetic Resources Newsletter* 97, 31–6.

Levin, G. M. 2006. *Pomegranate Roads: A Soviet Botanist's Exile from Eden* (1st edn.). Floreant Press, Forestville, CA, pp. 15–183.

Luo, X., Cao, Shangyin, Hao, Z., Hou, Lefeng, Cao, D., Zhang, J., Li, Haoxian, Niu, J., Xue, H. and Chen, L. 2018. Analysis of genetic structure in a large sample of pomegranate (*Punica granatum* L.) using fluorescent SSR markers. *Journal of Horticultural Science and Biotechnology* 93(6), 659–65. doi:10.1080/14620316.2018.1432994.

Lye, C. 2008. Pomegranate: preliminary assessment of the potential for an Australian industry. Rural Industries Research and Development Corporation of Australian Government offices Publication No. 08/153, p. 17.

Mahajan, P. V., Oliveira, F. A. R., Montanez, J. C. and Frias, J. 2007. Development of user-friendly software for design of modified atmosphere packaging for fresh and fresh-cut produce. *Innovative Food Science and Emerging Technologies* 8(1), 84–92. doi:10.1016/j.ifset.2006.07.005.

Maity, A., Sharma, J., Sarkar, A., More, A. K., Pal, R. K., Nagane, V. P. and Maity, A. 2018. Salicylic acid mediated multi-pronged strategy to combat bacterial blight disease (*Xanthomonas axonopodis* pv. *punicae*) in pomegranate. *European Journal of Plant Pathology* 150(4), 923–37. doi:10.1007/s10658-017-1333-3.

Marathe, R. A., Sharma, J., Babu, K. D. and Murkute, A. A. 2017a. Bedding system: a unique plantation method of pomegranate in arid and semi-arid region. *National Academy Science Letters* 40(4), 249–51. doi:10.1007/s40009-017-0567-0.

Marathe, R. A., Sharma, J., Murkute, A. A. and Babu, K. D. 2017b. Response of nutrient supplementation through organics on growth, yield and quality of pomegranate. *Scientia Horticulturae* 214, 114–21. doi:10.1016/j.scienta.2016.11.024.

Meena, V. S., Kashyap, P., Nangare, D. D. and Singh, J. 2016. Effect of coloured shade nets on yield and quality of pomegranate (*Punica granatum*) cv. Mridula in semi-arid region of Punjab. *Indian Journal of Agricultural Sciences* 86(4), 500–5.

Meighani, H., Ghasemnezhad, M. and Bakhshi, D. 2015. Effect of different coatings on post-harvest quality and bioactive compounds of pomegranate (*Punica granatum* L.) fruits. *Journal of Food Science and Technology* 52(7), 4507–14. doi:10.1007/s13197-014-1484-6.

Mirdehghan, S. H. and Ghotbi, F. 2014. Effects of salicylic acid, jasmonic acid and calcium chloride on reducing chilling injury of pomegranate (*Punica granatum* L.) fruit. *Journal of Agricultural Sciences and Technologies* 16, 163–73.

Mohsen, F. S. and Osman, A. 2015. Improving physico-chemical aspects of Wonderful pomegranate fruits through fruit thinning. *International Journal of Current Research in Biosciences and Plant Biology* 2(9), 95–103.

Moriguchi, T., Omura, M., Matsuta, N. and Kazaki, I. 1987. In vitro adventitious shoot formation from anthers of pomegranate. *Horticultural Science* 22, 947–8.

Munde, G. R., Nainwad, R. V., Maske, S. N. and Pawar, J. V. 2016. Effect of IBA and other chemicals on air layering in pomegranate cv. Bhagwa. *BIOINFOLET - A Quarterly Journal of Life Sciences* 13(2a), 291–3.

Munhuweyi, K., Lennox, C. L., Meitz-Hopkins, J. C., Caleb, O. J., Sigge, G. O. and Opara, U. L. 2016. In vitro effects of crab shell chitosan against mycelial growth of *Botrytis* sp., *Penicillium* sp. and *Pilidiella granati*. *Acta Horticulturae* 1144. doi:10.17660/ActaHortic.2016.1144.60.

Munhuweyi, K., Lennox, C. L., Meitz-Hopkins, J. C., Caleb, O. J., Sigge, G. O. and Opara, U. L. 2017. Investigating the effects of crab shell chitosan on fungal mycelial growth and postharvest quality attributes of pomegranate whole fruit and arils. *Scientia Horticulturae* 220, 78–89. doi:10.1016/j.scienta.2017.03.038.

Nadaf, A. M. 2017. Bioefficacy of newer insecticides against anar butterfly, *Deodorex isocrates* Fab., on pomegranate. *Journal of Entomology and Zoology Studies* 5(3), 1655–7.

Nafees, M., Jaskani, M. J., Ahmed, S. and Awan, F. S. 2015. Morpho-molecular characterization and phylogenetic relationship in pomegranate germplasm of Pakistan. *Pakistan Journal of Agricultural Science* 52(1), 97–106.

Naik, S. K., Pradeep, S. and Chand, P. K. 1999. In vitro propagation of pomegranate (*Punica granatum* L. cv. Ganesh) through axillary shoot proliferation from nodal segments of mature tree. *Scientia Horticulturae* 79(3–4), 175–83. doi:10.1016/S0304-4238(98)00218-0.

Opara, U. L., Mahdoury, A. A., Al-Ani, M., Al-Khanjari, S. A., Al-Yahyai, R., Al-Kindi, H. and Al-Khanjari, S. A. 2008. Physiological responses and changes in postharvest quality attributes of 'Helow' pomegranate variety (*Punica granatum* L.) during refrigerated storage. In: *International Conference of Agricultural Engineering*, 31 August–4 September 2008, Iguassu Falls City, Brazil.

Paiva, E. P., Rocha, R. H. C., Sousa, F. A., Nobre, R. G., Guedes, W., Moreira, I. S. and Sá, F. V. S. 2015. Growth and physiology of pomegranate nursery tree cv. Wonderful propagated by grafting. *Revista Brasileira de Ciências Agrárias* 10(1), 117–22.

Palou, L., Crisosto, C. H. and Garner, D. 2007. Combination of postharvest antifungal chemical treatments and controlled atmosphere storage to control gray mold and improve storability of 'Wonderful' pomegranates. *Postharvest Biology and Technology* 43(1), 133–42. doi:10.1016/j.postharvbio.2006.08.013.

Parvizi, H., Sepaskhah, A. R. and Ahmadi, S. H. 2014. Effect of drip irrigation and fertilizer regimes on fruit yields and water productivity of a pomegranate (*Punica granatum* (L.) cv. Rabab) orchard. *Agricultural Water Management* 146, 45–56. doi:10.1016/j.agwat.2014.07.005.

Peña, M. E., Artés-Hernández, F., Aguayo, E., Martínez-Hernández, G. B., Galindo, A., Artés, F. and Gómez, P. A. 2013. Effect of sustained deficit irrigation on physicochemical properties, bioactive compounds and postharvest life of pomegranate fruit (cv. 'Mollar de Elche'). *Postharvest Biology and Technology* 86, 171–80. doi:10.1016/j.postharvbio.2013.06.034.

Pezhman, H. 2016. Comparison of various protein hydrolesates for mass trapping of *Ceratitis capitata* (Wiedman) (Dip.: Tephritidae) in a pomegranate orchard in Shiraz region. *Plant Pest Research* 6(1), 61–70.

Porat, R., Kosto, I. and Daus, A. 2016. Bulk storage of 'Wonderful' pomegranate fruit using modified atmosphere bags. *Israel Journal of Plant Sciences* 63(1), 45–50. doi:10.1080/07929978.2016.1152839.

Prasad, R. N., Bankar, G. J. and Vashishtha, B. B. 2003. Effect of drip irrigation on growth, yield and quality of pomegranate in arid region. *Indian Journal of Horticulture* 60(2), 140–2.

Quaglia, M., Moretti, C., Cerri, M., Linoci, G., Cappelletti, G., Urbani, S. and Taticchi, A. 2016. Effect of extracts of wastewater from olive milling in postharvest treatments of pomegranate fruit decay caused by *Penicillium adametzioides*. *Postharvest Biology and Technology* 118, 26–34. doi:10.1016/j.postharvbio.2016.03.012.

Ranpise, S. A. and Ahire, D. B. 2016. Effect of different grafting methods and rootstocks on growth and survival of pomegranate (*Punica granatum* L.). *International Journal of Minor Fruits, Medicinal and Aromatic Plants* 2(2), 22–8.

Raut, G. B., Ambadkar, C. V. and Dhawan, A. S. 2015. Efficacy of different phytoextracts for management of bacterial blight disease of pomegranate caused by *Xanthomonas axonopodis* pv. *punicae*. *International Journal of Plant Sciences* 10(2), 118–21. doi:10.15740/HAS/IJPS/10.2/118-121.

Rouhani, M., Samih, M. A., Izadi, H. and Mohammadi, E. 2013. Toxicity of new insecticides against pomegranate aphid, *Aphis punicae*. *International Research Journal of Applied and Basic Sciences* 4(3), 496–501.

Rouholamin, S., Zahedi, B., Nazarian-Firouzabadi, F. and Saei, A. 2015. Expression analysis of anthocyanin biosynthesis key regulatory genes involved in pomegranate (*Punica granatum* L.). *Scientia Horticulturae* 186, 84–8. doi:10.1016/j.scienta.2015.02.017.

Sarrou, E., Therios, I. and Dimassi-Theriou, K. 2014. Melatonin and other factors that promote rooting and sprouting of shoot cuttings in *Punica granatum* cv. Wonderful. *Turkish Journal of Agriculture and Forestry* 38, 293–301.

Sayyari, M., Babalar, M., Kalantari, S., Martínez-Romero, D., Guillén, F., Serrano, M. and Valero, D. 2011. Vapour treatments with methyl salicylate or methyl jasmonate alleviated chilling injury and enhanced antioxidant potential during postharvest storage of pomegranates. *Food Chemistry* 124(3), 964–70. doi:10.1016/j.foodchem.2010.07.036.

Sayyari, M., Aghdam, M. S., Salehi, F. and Ghanbari, F. 2016. Salicyloyl chitosan alleviates chilling injury and maintains antioxidant capacity of pomegranate fruits during cold storage. *Scientia Horticulturae* 211, 110–7. doi:10.1016/j.scienta.2016.08.015.

Sayyari, M., Salehi, F. and Valero, D. 2017. New approaches to modelling methyl jasmonate effects on pomegranate quality during postharvest storage. *International Journal of Fruit Science* 17(4), 374–90. doi:10.1080/15538362.2017.1329051.

Shaarawi, S. A. M. A. and Nagy, K. S. 2017. Effect of modified atmosphere packaging on fruit quality of 'Wonderful' pomegranate under cold storage conditions. *Middle East Journal of Agriculture Research* 6(2), 495–505.

Shanmugasundaram, T. and Balakrishnamurthy, G. 2017. Exploitation of plant growth substances for improving the yield and quality of pomegranate under ultra high density planting. *International Journal of Current Microbiology and Applied Sciences* 6(3), 102–9. doi:10.20546/ijcmas.2017.603.011.

Shao, J. Z., Chen, C. L. and Deng, X. X. 2003. In vitro induction of tetraploid in pomegranate (*Punica granatum*). *Plant Cell, Tissue and Organ Culture* 75(3), 241–6. doi:10.1023/A:1025871810813.

Sharma, D. P. and Singh, N. 2018. Effect of rejuvenation pruning on the growth, productivity and disease incidence in declining trees of pomegranate (*Punica granatum* L.) cv. Kandhari Kabuli. *Journal of Applied and Natural Science* 10(1), 358–62.

Sharma, N., Anand, R. and Kumar, D. 2009. Standardization of pomegranate (*Punica granatum* L.) propagation through cuttings. *Biological Forum - An International Journal* 1(1), 75–80.

Sheikh, M. K. 2015. Effect of growth regulators and hand thinning of flowers/fruits on size of pomegranate (*Punica granatum* L. 'Ganesh'). *Acta Horticulturae* 1089(1089), 407–10. doi:10.17660/ActaHortic.2015.1089.55.

Singh, K. K. 2014. Effect of IBA concentrations on the rooting of pomegranate (*Punica granatum* L.) cv. Ganesh hardwood cuttings under mist house condition. *Plant Archives* 14(2), 1111–4.

Singh, P. and Patel, R. M. 2014. Factors influencing *in vitro* growth and shoot multiplication of pomegranate. *The Bioscan* 9(3), pp. 1031–5.

Singh, P., Patel, R. M. and Kadam, S. 2013. *In vitro* mass multiplication of pomegranate from cotyledonary nodal explants cv. Ganesh. *African Journal of Biotechnology* 12(20), 2863–8.

Singh, H., Singh, N. and Dolu, S. 2016. Effect of inter-fruit competition on development of physiological disorder 'Aril browning' in pomegranate (*Punica granatum* L.). *Journal of Applied and Natural Science* 8(4), 1835–8. doi:10.31018/jans.v8i4.1050.

Southwick, S. M. and Glozer, K. 2000. Reducing flowering with gibberellins to increase fruit size in stone fruit trees: applications and implications in fruit production. *Hort Technology* 10(4), 744–51.

Tayade, S. A., Joshi, P. S., Raut, H. S. and Shete, M. B. 2017. Effect of time and air layer per shoot on rooting and survival of air layers in pomegranate cv. Bhagwa. *International Journal of Minor Fruits, Medicinal and Aromatic Plants* 3(1), 20–4.

Tomar, K. S. and Tomar, S. S. 2012. Effect of different dosage of growth regulators on rooting & survival percentage of pomegranate (*Punica granatum* L.) air layers. *Green Farming* 3(6), 701–3.

Türk, G., Sönmez, M., Aydın, M., Yüce, A., Gür, S., Yüksel, M., Aksu, E. H. and Aksoy, H. 2008. Effects of pomegranate juice consumption on sperm quality, spermatogenic cell density, antioxidant activity and testosterone level in male rats. *Clinical Nutrition* 27(2), 289–96. doi:10.1016/j.clnu.2007.12.006.

Verma, N., Mohanty, A. and Lal, A. 2010. Pomegranate genetic resources and germplasm conservation: a review. *Fruit, Vegetable and Cereal Science and Biotechnology* 4(2), 120–5.

Wetzstein, H. Y., Zhang, Z., Ravid, N. and Wetzstein, M. E. 2011. Characterization of attributes related to fruit size in pomegranate. *HortScience* 46(6), 908–12. doi:10.21273/HORTSCI.46.6.908.

Yazıcı, K. and Şahin, A. 2016. Characterization of pomegranate (*Punica granatum* L.) hybrids and their potential use in further breeding. *Turkish Journal of Agriculture and Forestry* 40, 813–24.

# Index